Fuel Cycle to Nowhere

Fuel Cycle to Nowhere

U.S. Law and Policy on Nuclear Waste

Richard Burleson Stewart
Jane Bloom Stewart

Vanderbilt University Press

Nashville

© 2011 by Richard Burleson Stewart and Jane Bloom Stewart
Published by Vanderbilt University Press
Nashville, Tennessee 37235
All rights reserved
First printing 2011

This book is printed on acid-free paper made
from 30% post-consumer recycled content.
Manufactured in the United States of America

Cover design: Gary Gore
Text design: Dariel Mayer

Library of Congress Cataloging-in-Publication Data

Stewart, Richard B.
 Fuel cycle to nowhere : U.S. law and policy on nuclear waste /
Richard B. Stewart and Jane B. Stewart.
   p. cm.
 Includes bibliographical references and index.
 ISBN 978-0-8265-1774-6 (cloth edition : alk. paper)
 1. Hazardous wastes—Law and legislation—United States.
2. Radioactive waste disposal—Law and legislation—United
States. 3. Radioactive wastes—Transportation—Law and legis-
lation—United States. 4. Radioactive wastes—United States.
I. Stewart, Jane B. II. Title.
 KF3946.S78 2011
 344.7304′622—dc22
                          2011003004

For our children

Contents

Acronyms and Abbreviations   xi
Acknowledgments   xv
Introduction   1

1   The Evolution of U.S. Nuclear Waste Law and Policy   15
    Nuclear Weapons Buildup and the
       Rise of Nuclear Power, 1946–1970   17
    The Rise of the Environmental Movement
       and the End of Reprocessing   30
    Federal Nuclear Waste Disposal Initiatives and Legislation   56
    Yucca Abandoned: Repository Limbo, Orphan Waste Challenges   73

2   Radioactive Waste Classification and Regulation   84
    Nuclear Wastes and Their Regulatory Classification   85
    Nuclear Waste Regulatory Classification and Requirements   95
    HLW Reclassification Initiatives   102
    Mixed Waste   111
    Toward a More Risk-Based System of Radioactive
       Waste Classification and Regulation   113
    Conclusion   122

3   Nuclear Waste Transport   123
    The Evolution of Nuclear Waste Transport Regulation   124
    The Current Nuclear Waste Transport Regulatory Structure   129
    Experience with Nuclear Waste Transport   133
    Transportation of TRU for Disposal at WIPP   135
    Transportation of SNF and HLW to Yucca Mountain   140
    Assessing the Nuclear Waste Transportation Regime   142
    Conclusion   143

4 Low-Level Waste Disposal — 145
Early Developments — 146
The 1980 Low-Level Radioactive Waste Policy Act — 147
The 1985 Low-Level Radioactive Waste Policy Act Amendments — 149
The Current LLW Disposal Situation — 154
Options for Addressing LLW Disposal Problems — 157
Conclusion — 160

5 WIPP: The Rocky Road to Success — 162
The Origins of WIPP — 162
Restriction of WIPP to Defense TRU — 163
New Mexico's Successful Efforts to Gain a Role in Decision Making Regarding WIPP — 168
New Mexico: Federal Interactions and Resolution of Conflicts, 1980–1992 — 171
Controversy over Land Withdrawal for WIPP, 1989–1992 — 175
Restoring WIPP's Credibility, 1992–1996 — 179
WIPP Moves into Full-Scale Operation as a TRU Repository — 180
Conclusion — 184

6 Yucca Mountain: Blueprint for Failure — 186
The 1982 Nuclear Waste Policy Act and Its Implementation by DOE — 187
DOE's Implementation of the NWPA Siting Process — 195
Crisis in the NWPA Siting Scheme — 201
The 1987 Nuclear Waste Policy Act Amendments and Resistance to a Yucca Repository — 207
Nevada's Legal Actions to Thwart a Repository at Yucca Mountain — 210
The Battle for Public Opinion — 214
Responses to Repository Delay — 216
Twists and Turns in the Technical Debate over Yucca — 217
Federal Designation of Yucca for a Repository, and Nevada's Continuing Resistance — 222
DOE's Yucca License Application to NRC — 225
Obama's Election as President: Political Victory for Nevada — 226
Conclusion — 230

7	**Options for Orphan Wastes**	231
	Continued SNF Storage at Reactors	232
	Consolidated SNF Storage	235
	Evaluating Interim Storage Options	239
	SNF Reprocessing	242
	Conclusion	253
8	**Nuclear Waste in the United States: Lessons Learned and Future Choices**	254
	Evolution of U.S. Nuclear Waste Policy: Recapitulation	254
	Current Nuclear Waste Dilemmas and Options	258
	Lessons Learned and Future Strategies for U.S. Nuclear Waste Policy	272

Appendix A
Operating U.S. Nuclear Power Units by Year   307

Appendix B
Uranium Oxide Spot Prices   309

Appendix C
The Hanford Waste Cleanup Agreement and Program   311

Notes   315
Bibliography   397
Index   413

Acronyms and Abbreviations

AEA	Atomic Energy Act
AEC	Atomic Energy Commission
AIF	assured isolation facility
ANDRA	Agence nationale pour la gestion des déchets radioactifs (French National Radioactive Waste Management Agency)
APA	Administrative Procedure Act
ASLB	Atomic Safety and Licensing Board
BIA	Bureau of Indian Affairs
BLM	Bureau of Land Management
BPA	Bonneville Power Administration
BRC	below regulatory concern
C&C	Consultation and Cooperation (Agreement)
CERCLA	Comprehensive Environmental Response, Compensation, and Liability Act
C.F.R.	Code of Federal Regulation
CH-TRU	contact-handled transuranic waste
CPI	Consumer Price Index
CRBR	Clinch River Breeder Reactor
CRESP	Consortium for Risk Evaluation with Stakeholder Participation
DEIS	Draft Environmental Impact Statement
DHS	Department of Homeland Security
DOD	Department of Defense
DOE	Department of Energy
DOI	Department of the Interior
DOJ	Department of Justice
DOT	Department of Transportation
DSEIS	Draft Supplemental EIS
DSEIS II	Draft Supplemental EIS II
DU	depleted uranium
EEG	Environmental Evaluation Group
EHS	environmental, health, and safety
EIS	Environmental Impact Statement
EM	Office of Environmental Management, DOE
EnPA	Energy Policy Act
EPA	Environmental Protection Agency

EPC	Executive Planning Council
ERA	Energy Reorganization Act
ERDA	Energy Research and Development Administration
EW	exempt waste
FEIS	Final Environmental Impact Statement
FBI	Federal Bureau of Investigation
FCC	Federal Communications Commission
FFCA	Federal Facilities Compliance Act
FLPMA	Federal Land Policy and Management Act
FOIA	Freedom of Information Act
FRSA	Federal Railroad Safety Act of 1970
FSEIS II	Final Supplemental EIS II
GAO	Government Accountability Office (previously General Accounting Office)
GNEP	Global Nuclear Energy Partnership
GTCC	greater-than-class-C waste
HASC	House Armed Services Committee
HEW	Department of Health, Education, and Welfare
HLW	high-level waste
HMTA	Hazardous Materials Transportation Act
HMTUSA	Hazardous Materials Transportation Uniform Safety Act
HSWA	Hazardous and Solid Waste Amendments (RCRA)
IAEA	International Atomic Energy Agency
ICC	Interstate Commerce Commission
INL	Idaho National Laboratory
IFNEC	International Framework for Nuclear Energy Cooperation
IRG	Interagency Review Group on Nuclear Waste Management
ISF	intermediate scale facility
ITF	intermodal transport facility
LAMW	low-activity mixed waste
LCF	latent cancer fatalities
LILW	low- to intermediate-level waste
LILW-LL	low- to intermediate-level waste–long-lived
LILW-SL	low- to intermediate-level waste–short-lived
LLRWPA	Low-Level Radioactive Waste Policy Act
LLRWPAA	Low-Level Radioactive Waste Policy Amendments Act
LLW	low-level waste
LLRW	low-level radioactive waste
LMFBR	liquid metal fast breeder reactor
LULU	locally undesirable land use
MOU	memorandum of understanding
MOX	mixed oxide
MRS	monitored retrievable storage (facility)
MTHM	metric ton of heavy metal
MWe	megawatt electrical
NAS	National Academy of Sciences
NASA	National Aeronautic and Space Administration

NARM	naturally occurring and accelerator-produced radioactive material
nCi	nanocurie
NCNP	Nevada Commission on Nuclear Projects
NCRPM	National Council of Radiation Protection and Measurement
NEA	Nuclear Energy Agency (within OECD)
NEPA	National Environmental Policy Act
NFMDA	Nuclear Fuel Management and Disposal Act
NFS	Nuclear Fuel Services
NGA	National Governors Association
NIMBY	Not in My Backyard
NORM	naturally occurring radioactive material
NPL	National Priorities List (CERCLA)
NRC	Nuclear Regulatory Commission
NRDC	Natural Resources Defense Council
NRTS	Nuclear Reactor Test Site
NTS	Nevada Test Site
NWF	Nuclear Waste Fund
NWMO	Nuclear Waste Management Organization (Canada)
NWPA	Nuclear Waste Policy Act
NWPAA	Nuclear Waste Policy Act Amendments
NWPO	Nuclear Waste Project Office (Nevada)
NWTRB	Nuclear Waste Technical Review Board
OCRWM	Office of Civilian Radioactive Waste Management, DOE
OECD	Organization for Economic Cooperation and Development
OMB	Office of Management and Budget
ONWN	Office of Nuclear Waste Negotiator
ORNL	Oak Ridge National Laboratory
OTA	Office of Technology Assessment (Congress)
PA	Performance Assessment
PFS	Private Fuel Storage, LLC
PUREX	plutonium uranium extraction
R&D	research and development
RCRA	Resource Conservation and Recovery Act
rem	roentgen equivalent in man
RH	remote handled
RH-TRU	remote-handled transuranic waste
ROD	Record of Decision
RWCA	Radioactive Waste Consultation Act
SARA	Superfund Amendments and Reauthorization Act
SDWA	Safe Drinking Water Act
SNF	spent nuclear fuel
SRS	Savannah River Site
SSNM	surplus special nuclear material
SSEB	Southern States Energy Board
STB	Surface Transportation Board (within DOT)
Sv	sievert

T	time it takes for 95 percent of a radioactive substance to decay
TPA	Tri-Party Agreement
TRANSCOM	Transportation Command (United States)
TRU	transuranic waste
TRUPACT	transuranic packaging container
TSCA	Toxic Substances Control Act
TSPA	total system performance assessment
TVA	Tennessee Valley Authority
UCS	Union of Concerned Scientists
USGS	United States Geological Survey
VA	viability assessment
WIPP	Waste Isolation Pilot Plant
WIPPLWA	Waste Isolation Pilot Plant Land Withdrawal Act
WIR	waste incidental to reprocessing
WVDPA	West Valley Demonstration Project Act

Acknowledgments

This book emerged out of research that we undertook as part of our work with the Consortium for Risk Evaluation with Stakeholder Participation (CRESP), a multidisciplinary consortium of academics at eight universities and a medical school. CRESP conducts and publishes research on the technical, scientific, economic, legal, and policy elements of U.S. nuclear waste policies.[1] We found that parts of the policies and their history had been told, some in great depth and in a number of cases informed by the authors' participation or close familiarity with the events. But we could not find a work that synthesized the various elements into a single comprehensive and objective account of the history and current situation of U.S. nuclear waste law and policy. This book seeks to meet this need. The authors are an academic and an environmental lawyer who have both long worked in the environmental law and policy realm but were not directly involved in the events recounted.

In researching and writing this book, we sought to consult original sources but have also necessarily drawn widely on the important contributions of many earlier authors who have written on specific aspects of the history, and on those of experts in government, the academy, and nongovernmental organizations who have examined particular regulatory and legal topics. We are deeply indebted to all of them. Their works are found in the Bibliography.

We are deeply grateful for the steadfast support and assistance of the co-directors of CRESP, David Kosson and Charles Powers, and other CRESP colleagues, including especially Lisa Bliss, Jim Clarke, Michael Greenberg, Henry Meyer, and Frank Parker. We are much indebted to Tom Isaacs for reading through an earlier draft of the entire book and offering many valuable comments and suggestions. This book could not have been written without the extensive work of a raft of New York University law students, who undertook substantial portions of the underlying research and cheerfully and diligently checked citations. They include Jacob Berman, Scott Blair, Bridget Burns, Eli Corin, Kirti Datla, Justin Gundlach, Ryan Hooper, Carolyn Kelly, Daniel Kesack, Isaac MacDonald, Kenneth C. Michaels, George Mustes, Derek Scadden, Brandon Schwartz, Philip Smithback, and Chris Suh. We owe special thanks to Alice Byowitz, who worked full time for six months as our principal research assistant, coordinating and integrating all the parts of the book without dropping a stitch, and to James Chapman, who preceded Alice in this role; to George Minot, who read and edited several drafts of the book to the great benefit of the final product; and to our assistant Basilio Valdehuesa for his indefatigable energy, skill, and good cheer in securing research materials, generating the bibliography, handling document formatting and production, and assisting us in countless other ways.

We also gratefully acknowledge the financial and other support of CRESP, Vanderbilt University, and the U.S. Department of Energy, and of the Filomen D'Agostino and Max E. Greenberg Research Fund at New York University School of Law.[2] We also want to thank our patient and highly supportive team at Vanderbilt University Press: the director, Michael Ames; the managing editor, Ed Huddleston; and our exceptional copy editor, Bobbe Needham.

Finally, we thank our children, Ian and Emily, for accepting—in most cases with understanding and patience—all the hours that we spent working on this book on weekends and during summer vacation in Maine, and for insistently reminding us of the many joys in life beyond research and writing. This book is, in many ways, truly for them and their brothers and sister, Will, Paul, and Elizabeth.

Fuel Cycle to Nowhere

Introduction

This book presents the first comprehensive account of the history and current status of U.S. nuclear waste regulatory law and policy. The history, extending over sixty years, is extraordinarily rich, with interacting technological, scientific, economic, political, social, and international security dimensions. The U.S. legal and regulatory regime for nuclear waste is also highly complex, even labyrinthine. It is a palimpsest composed of federal and state statutes, presidential executive orders, administrative regulation and guidance documents, reports of expert bodies and government agencies, and court decisions contributed by a variety of actors over many decades.

One principal aim of this book is to unravel this tangle of events, reports, laws, and policies and to present the elements in an ordered and accessible narrative while remaining faithful to the complexities and contingencies in the record. The opening chapter presents a historical overview. Subsequent chapters examine specific topics, including nuclear waste classification and regulation; waste transportation; the contrasting stories of two deep geologic repositories for permanent waste disposal, the Waste Isolation Pilot Plant (WIPP) and Yucca Mountain; and the current dilemmas posed by so-called orphan wastes that have no disposal pathway. This history has produced the nuclear waste dilemma that now confronts the administration of President Barack Obama and the nation, which leads to the second goal of this book: to analyze our nation's present legal and policy conundrum and examine the potential means for resolving it. The United States has a huge legacy of highly radioactive wastes accumulated over many decades from weapons production and nuclear power generation. Yet the Obama administration has abandoned the waste disposal repository at Yucca Mountain in Nevada that Congress designated to receive these wastes. In doing so, the administration repudiated what has been for more than two decades the country's only plan—albeit a flawed one—to deal with these wastes. The Blue Ribbon Commission on America's Nuclear Future has been appointed by the administration to study and present solutions. In order to help policy makers and the public solve our nuclear waste challenge, the final chapter of this book draws lessons from the history examined in the preceding chapters, analyzes the available options, and offers recommendations for moving forward.

This Introduction first presents an overview of the nuclear waste legal and policy questions that we confront and the means for addressing them. It then summarizes the principal events, broader developments, and institutional structures—elaborated in detail in the book—that have generated that history.

The Current Nuclear Waste Policy Dilemma

The nuclear waste dilemma arises at the same time the nation faces momentous energy policy choices posed by the need for decisive actions to mitigate climate change and to reduce dependence on foreign oil. As a major part of an initiative to promote development of low-carbon and renewable energy resources, President Obama, with broad support in Congress but significant dissent from the public, has strongly supported big government subsidies and other initiatives to stimulate construction of large numbers of new nuclear power plants after twenty-five years in which not a single such plant has been built. George W. Bush's administration invoked both climate and energy security goals in proposing the Global Nuclear Energy Partnership (GNEP), a major initiative to make new fuel from uranium and plutonium extracted from spent nuclear fuel (SNF), after an almost forty-year period in which there was a national moratorium on SNF reprocessing. GNEP was roundly criticized on technical, security, and cost grounds by a National Academy of Sciences (NAS) committee and numerous other independent experts; it was cancelled by the Obama administration. The brief U.S. experiment with SNF reprocessing ended in the 1970s after a dismal record of operational, financial, and environmental failures. Presidents Gerald Ford and Jimmy Carter applied the coup de grace by halting federal aid for reprocessing because of the proliferation risks posed by the plutonium it produces. GNEP, however, triggered active discussion of reprocessing options, notwithstanding a barrage of powerful criticisms.

Even if the public were otherwise prepared to go along with a major expansion of nuclear power, much less reprocessing, it is unlikely to do so without a new, credible regime for disposing of our existing and future nuclear power wastes. SNF continues to accumulate at seventy-seven nuclear power plant sites across the country without a disposal destination or even a plan for one. Meanwhile, the federal government is subject to mounting liabilities, running to many billions of dollars, to nuclear utilities for breach of its commitments to take charge of SNF and start disposing of it beginning in 1998.

In addition to the massive SNF waste problem, there still reposes at Department of Energy (DOE) facilities a huge Cold War legacy of highly hazardous reprocessing wastes—high-level wastes (HLW)—from weapons production; these are being addressed by a massive, ongoing DOE cleanup effort expected to cost several hundreds of billions of dollars. DOE is obligated by agreements with the states where these facilities are located to remove these HLW by specified deadlines, but it will be unable to meet them unless the Yucca Mountain facility is built or another repository is developed soon.

The Obama administration's dilemma is this: it needs to solve the nuclear waste problem in order to advance its nuclear power agenda, but it has repudiated Yucca, the only waste solution available under current law. The nation's dilemma is somewhat different. There is an imperative need to deal responsibly with the large quantity of nuclear wastes we already have and those that will continue to be generated at existing power plants, whatever the fate of the "nuclear renaissance" advocated by the administration and many in Congress. Obama's abandonment of Yucca is currently being challenged in litigation. The wastes are left in legal and policy limbo in or near a host of communities large and small throughout the country. At this point, there is neither a plan to develop a repository nor a considered policy for how and where to store them in the interim.

This does not mean that Yucca must at all costs be built. The technical merits of

Yucca—at a location chosen by Congress pursuant to a raw political power play—have been sharply challenged. Even if after long legal battles, a repository at Yucca is eventually licensed, and even if Congress funds its construction, the repository would not open for many years. Developing a brand-new repository elsewhere could take even longer. Another option could be to open WIPP to HLW and SNF, which is prohibited under current law due to New Mexico's opposition to hosting these wastes in the past. Practical politics as well as sound principle dictate that to proceed with this option would require first winning New Mexico's assent.

To help it solve the nation's dilemma, the administration has constituted a distinguished blue-ribbon commission charged to "conduct a comprehensive review of policies for managing the back end of the nuclear fuel cycle, including all alternatives for the storage, processing, and disposal of civilian and defense used nuclear fuel and nuclear waste."[1] The commission has been established at a watershed moment for U.S. nuclear waste policy, much like the Interagency Review Group (IRG) created by President Carter just over thirty years ago; IRG stimulated wide-ranging debate and made proposals for fundamental changes that significantly influenced future policy. The commission is tasked to provide to the president an interim report of its findings by August 2011, and a final report by January 2012. It should interpret its charge broadly in order to examine nuclear waste issues in relation to the technological and social systems of which they are but a part, including terrorism, proliferation, and energy security; the wisdom or folly of reviving reprocessing of spent fuel; and the institutional capacity of federal, state, and local actors effectively to safeguard nuclear wastes awaiting permanent disposal. Most of all, we hope that the commission will study, and learn from, the history of our nation's law and policy on nuclear waste to date—including mistakes, missed opportunities, and successes.

Lessons from Prior Government Programs for Nuclear Waste Disposal

This book is in important respects a tale of two repositories. As we researched the history of U.S. nuclear waste policy, which really began in the 1950s with a key report by an NAS committee recommending deep geologic repository disposal for the nation's most radioactive waste, we became fascinated by the question of why the WIPP repository for intermediate-level defense transuranic wastes succeeded in becoming the world's only operating nuclear waste repository, while Yucca Mountain encountered seemingly unalterable opposition and has altogether foundered. The starkly opposed fates of the twin repositories raised a host of questions with critical implications for the future development of HLW and SNF repositories and interim storage facilities.

Over the past sixty years, the federal government has undertaken three distinct programs for disposal of nuclear wastes. Congress designed two of these programs, one for HLW and SNF embodied in the 1982 Nuclear Waste Policy Act (NWPA), and a second for low-level waste (LLW), the 1980 Low-Level Radioactive Waste Policy Act (LLRWPA), and subsequent amendments to both laws. It is now clear that these statutes have failed to achieve their goals and that the programs fashioned by Congress are bankrupt.

NWPA imposed a blueprint on DOE to develop two repositories for burial of HLW and SNF, one in the eastern and the other in the western part of the United States. In 1987, congressional power brokers from other states decreed that only one repository should be built—at Yucca Mountain. Nevada steadfastly opposed the facility and delayed its development. Nevada's resistance and the fortuities of the presidential campaign process led candidate Obama to oppose Yucca. Now that he is president, his administration has dropped Yucca, and with it the longstanding statutory design for developing a repository. Today we seem to be no closer to solving our HLW-SNF waste disposal challenge than we were over fifty year ago, when the NAS issued its call to action. A repository to dispose of these wastes is a mirage on the policy horizon, steadily receding even as we attempt to claim it.

The LLRWPA has also failed. In thirty years, the regional compact system that Congress sought to establish has not developed any new disposal facilities to handle the growing stocks of the more hazardous class B and C low-level radioactive wastes (LLW) stored at generator sites in many states that lack access to disposal facilities licensed to receive these wastes. In addition, DOE, which is responsible for the most hazardous category of LLW wastes—those greater-than-class C (GTCC)—has yet to identify a method for their disposal. Dangerous wastes from medical, industrial, and commercial radioactive devices are not adequately regulated.

The third and only successful federal nuclear waste disposal program is WIPP. It was not designed in Washington but emerged gradually over twenty years through a step-by-step process of contestation, litigation, and negotiation between DOE and the State of New Mexico, which finally accepted the facility after its concerns and interests had substantially been accommodated. Congress played a largely reactive role through intermittent legislation that determined and cemented some important elements of the evolving WIPP project. Since the facility opened in 1999, WIPP has been disposing of long-lived transuranic (TRU) wastes from nuclear weapons production, without major incident. The tale of the two repositories—failed Yucca and successful WIPP—has important lessons for future policy.

Nuclear Waste Disposal: Meeting the Political and Institutional Challenge

There are significant technical challenges in identifying the most suitable sites and designing and constructing secure nuclear waste repositories and storage facilities, but—as WIPP demonstrates—we have the means to overcome them. The history shows that the most important and difficult challenges are not technical but political, institutional, and social. Siting and developing repositories and other nuclear waste facilities is a politically fraught enterprise. Many nuclear wastes are indeed highly hazardous if not properly managed. The knowledge and technical means are available to safely manage, store, and dispose of these wastes. But jurisdictional fragmentation, misguided federal policies, and political challenges have impeded our progress in doing so, most notably in the failure to develop and open a repository for disposition of the most highly radioactive defense wastes and spent fuel from civilian nuclear power reactors. The policy challenges of waste disposal are magnified by the public perception of the risks of nuclear waste storage, transportation, and disposal as far greater than most experts estimate. This public perception also has its basis in an important nontechnical reality, however: the government has in the past failed to safely manage the vast amounts of nuclear waste generated at its weapons production facilities. Moreover, there continue to be significant concerns raised about the adequacy of federal government safety regulation of the tens of thousands of tons of SNF stored at nuclear power plants near populated localities and tribes across the nation, most of it packed in cooling ponds that are potential targets for terrorist attacks. Seen in this light, it is not too difficult to understand public skepticism of the notion that a repository will be safely sited by the government and that the nuclear wastes disposed there will be safely isolated from people and environmental resources in perpetuity.

Understanding these and other key factors that shape public perception of nuclear waste risks can help point to sound solutions. Public views of risks are significantly influenced by equity concerns. Localities and states tend to oppose hosting nuclear or other wastes generated elsewhere and want assurances that the burden of hosting nuclear facilities is being fairly shared. Familiarity with nuclear risks and economic factors is also important. There are indications that people who live and work near existing nuclear facilities regard the risks that they present as appreciably lower than do those who live farther away. This difference in perceived risk directly informs political support or opposition. Some localities have expressed interest in hosting nuclear waste facilities, only to be overridden by political opposition at the statewide level. Finally, beyond dramatic accidents, as in Japan, public perceptions of and attitudes toward risks are strongly influenced by the processes for making decisions regarding nuclear wastes and the public's trust or lack thereof in the institutions that make these decisions.

Accordingly, a government or a private developer of a nuclear waste disposal or storage facility must win public confidence in its safety and integrity and thereby secure the voluntary assent of jurisdictions to host them. It must provide full and open information about the facility and its risks; afford meaningful opportunity for public participation; ensure that host jurisdictions have access to and are given adequate funding for independent technical experts; and follow a step-by-step approach to planning and developing the facility that ensures a genuine opportunity for host jurisdictions to be engaged in project decisions and have their concerns addressed and met.[2] Economic and other inducements—on a far more substantial scale than offered by the federal government in the past—to both localities and states to host facilities are also very likely to be needed. Further, the environmental safety of the facility's design, construction, and operation must be secured by trusted independent regulators. Regulators can also help win back the public trust by vigorously enforcing existing safety rules for nuclear power plants and by adopting new, more rigorous rules for their wastes, including requiring beefed up security at SNF storage facilities, not allowing spent fuel assemblies to be overcrowded into cooling ponds, and removing SNF after it has sufficiently cooled into dry-cask storage hardened against terrorist attack.

The federal government eventually adhered to many of these prescriptions at WIPP, but only after being forced to do so as the result of litigation by New Mexico and pressure from its congressional delegation. In other instances it has failed to do so, most notably in the case of the decision process for the Yucca Mountain site, which failed utterly to meet its goal of developing a repository. The Nuclear Waste Policy Act, both as enacted in 1982 and amended in 1987, disregarded the need for host-state engagement and assent by imposing top-down systems for the siting of a repository for HLW and SNF with federal override of host-state objections. Furthermore, the federal agencies charged with responsibility for nuclear waste facilities—originally the Atomic Energy Commission (AEC), later the Energy Research and Development Administration (ERDA), and now DOE—have often acted in a secretive and arrogant manner, pursuing a largely unilateral approach to decision making. The federal government must embrace a radically different approach to decision making to have a chance of success in solving the nation's nuclear waste dilemma.

Nuclear Waste Policy Options and Choices

Despite much controversy, in recent years there has emerged significant consensus across the political and policy spectrum on several nuclear waste issues. It is accepted that the only long-run solution for dealing with highly radioactive wastes, including HLW and unreprocessed SNF, is permanent disposal in a geologic repository. There is further agreement that the federal government should proceed forthwith to develop at least one repository, although views are divided on the best means of achieving this goal. Since the fate of Yucca is currently in litigation and will remain uncertain for a considerable time, the government should not wait before actively pursuing other alternatives. It should explore with New Mexico and affected localities the potential for locating a new repository on federal lands near WIPP or opening the WIPP facility itself to HLW, SNF, or both; Congress would have to authorize a new repository or approve any such change in WIPP's mission. The government should also initiate a search for at least one new repository site, preferably in another part of the country so as to restore a measure of geographic equity to the siting program; this will require Congress to amend the Nuclear Waste Policy Act, which currently bars any government efforts to develop a site other than Yucca. The search should develop candidates based on the technical suitability of sites and potential receptivity by localities and states, and should explore sites in various geologic media and various parts of the country.

There is also broad consensus that government must adopt a considered policy for managing HLW and SNF while one or more repositories are developed, although there is disagreement on what that policy should be. HLW and defense SNF could continue to be stored at the DOE sites where they were generated while continuing to treat HLW destined for repository disposal, or could consolidate treated HLW and SNF for storage at one or more DOE sites, in both cases subject to renegotiation of agreements with host states. The options for power plant SNF include continuing but significantly upgrading storage at reactor sites, moving the wastes to new government or private consolidated storage facilities, and reprocessing—or some combination of these.

In choosing between at-reactor and consolidated storage of SNF, there are competing cost, transport, security, and other considerations. Government development of at least one consolidated storage facility—at the least for SNF from decommissioned nuclear power plants—sited in accordance with the same principles for siting repositories, combined with the simultaneous upgrading of SNF storage at reactor sites could hedge bets on the choice of storage strategies, signal to the public that the government is taking concrete steps to address the waste problem, and provide valuable experience in developing new approaches to siting The government could also support opportunities for private sector development of consolidated SNF storage facilities, as long as high levels of public involvement in decision making and public access to information and public oversight are provided, financing to properly operate and decommission the facility is assured, and states and localities are fairly compensated for additional burdens imposed by the facility. Finally, wherever located, SNF should be moved as promptly as practicable from pool storage, which is vulnerable to terrorist attack, into more secure dry-cask storage containers hardened against terrorist attack.

SNF reprocessing has never been able to produce fuel at a cost competitive with fuel from virgin uranium. Moreover, the plutonium produced by reprocessing raises serious proliferation and security concerns. Nor do the energy security benefits from reprocessing seem compelling, given the availability of uranium supplies from friendly countries, including Canada and Australia. Past federal subsidies for commercial reprocessing produced failed facilities and environmental contamination, including a costly and lengthy cleanup at West Valley, New York, which after thirty years is still not complete. Nonetheless, the fuel resources contained in SNF represent a potential energy source that could be further explored, including through measured government investments in reprocessing research and development (R&D).

In the case of LLW, the federal government must assume a leadership role, as was recommended in the late 1970s by President Carter's IRG but rejected by Congress, in order to overcome the endemic collective action problems in siting new disposal facilities. Siting should follow the principles and procedures outlined here earlier. The failed LLRWPA compact system should either be abandoned or supplemented by federal initiatives and incentives for development of new facilities. A regulatory and disposal program must be adopted for GTCC wastes and other currently unregulated but dangerously radioactive wastes that fall into the broad LLW category.

In addition to waste disposal and storage facilities, federal law and policy must also move forward to grapple with a host of other neglected questions with the help of the blue-ribbon commission. Congress needs to establish new and more assured arrangements for financing repositories and storage facilities, simultaneously resolving the government's escalating liabilities for failure to take SNF from utilities. Serious consideration should be given to removing the nuclear waste facility siting and waste management functions from DOE and creating new entities specifically charged with the respective responsibilities, as recommended by earlier government task forces. Siting requires a body that is politically attuned, open and flexible, and capable of winning local and state assent to new facilities. Waste management requires more of a business model, with a clearly defined mission, results-based management, high-quality, technically adept personnel, and assured funding; such a body might take the form of a government corporation with utility participation, on the model successfully followed in a number of European countries. Waste transport and storage policies must more fully address terrorism risks and find ways to provide sufficient information about those risks and the government's response to them to instill and justify public confidence.

Moreover, the existing regulatory system for classifying nuclear wastes is a patchwork that in some significant respects fails to match regulatory requirements with the hazards posed by different wastes. Specific areas that should be reexamined include disposal requirements for extremely low-activity LLW; inadequate regulation of dangerously radioactive wastes from medical and industrial devices; and the classification and disposal of GTCC wastes. A more consistently risk-based approach to classifying and regulating these and similar wastes, potentially including the low-activity portions of some reprocessing wastes, should be developed in line with approaches followed in other countries and recommended by expert bodies in the United States. Finally, the current fragmented arrangements for nuclear waste regulation, which evolved without any overall plan and now include three federal agencies as well as state and local authorities, needs careful examination to determine whether some form of streamlining or reconfiguration is justified.

Evolutionary Dynamics in U.S. Nuclear Waste Law and Policy

We intend the history presented in this book not only to provide lessons to help improve future nuclear waste law and policy, but also to be of intrinsic interest for what it reveals about the processes of historical change, the interplay of contingencies and systemic factors, and the nature of U.S. political and legal institutions.

The history falls into two distinct eras. During the first three decades—from the beginning of the Cold War to the late 1970s—the problem of disposing of the most highly radioactive wastes was largely ignored. During the second three decades, the nation abruptly changed course and mobilized what was intended to be a crash effort to develop repositories for SNF and HLW. That effort has now ended in apparent failure. We stand on the threshold of a third era.

The first three decades were dominated by Cold War weapons production and the push to develop civilian nuclear power. The 1946 Atomic Energy Act gave sweeping powers to the AEC. Environmental concerns took a backseat to the arms race. Many nuclear wastes were stored or buried under unsafe conditions, leading to significant releases of radioactivity to the air, soil, and groundwater at AEC facilities. At the same time, AEC successfully pushed and provided significant subsidies for reprocessing SNF from commercial power plants, postponing any consideration of what to do with the waste that would be generated by reprocessing. The 1957 NAS committee report, calling for development of deep geologic repositories for the most dangerously radioactive wastes, was largely ignored.

The policy ground began to shift in the late 1960s, in part due to accident and contingency. The role of chance is reflected in the events that produced WIPP. In 1969, a major fire broke out at AEC's Rocky Flats, Colorado, weapons production facility, forcing AEC to scramble to find a site for disposal of the wastes that had been generated and were being stored there. After several misadventures, the government responded to an overture from city officials of Carlsbad, New Mexico, who expressed interest in hosting a repository. After many further twists and turns, WIPP eventually opened more than twenty years later.

More systemic factors—most notably the rise of the environmental movement along with a growing distrust of government, fueled by its callous nuclear bomb testing, the Vietnam War, and other factors—drove a more far-reaching shift in U.S. nuclear waste law and policy. Congress created a raft of ambitious new environmental statutes, while the courts provided expanded access to and remedies for environmental plaintiffs.[3] During the 1970s, nuclear power opponents, using new legal tools created by Congress and the courts, mounted extensive litigation that delayed the construction of new nuclear plants and increased their costs. Opponents also succeeded in securing the adoption in California and a dozen other states of moratoriums on construction of new nuclear power plants until disposal of their wastes was assured. These steps, along with widespread public opposition to nuclear power, contributed to a halt in development of new nuclear power plants in the United States. A contingency—or perhaps an accident waiting to happen—also played a role: the Three Mile Island incident supplied the final nail in the coffin of expanding nuclear power. Meanwhile, the sole facility for reprocessing power plant SNF, in West Valley, New York, closed. President Carter, driven by rising international nuclear proliferation concerns, effectively blocked further reprocessing in the United States by withdrawing its federal financial lifeline, thus removing that option for dealing with SNF. India's detonation of a nuclear bomb fueled by plutonium from SNF reprocessing was a key factor behind President Carter's initiative.

These developments alarmed the nuclear industry and its government supporters and spurred a striking turnaround in U.S. nuclear waste policy beginning in the late 1970s. President Carter's IRG called for federal development of at least two HLW/SNF repositories to open by the end of the century and other initiatives to address nuclear wastes. In the 1982 NWPA, Congress abruptly reversed its previous approach of leaving it up to AEC, ERDA, and DOE to design and execute a repository development program, and instead prescribed its own program, including highly detailed tasks, targets, and forced-march timetables for DOE to site and builds two repositories for SNF and HLW. The rigid structure of the NWPA, however, soon proved unworkable. Congress in 1987 short-circuited the technocratic-meritocratic administrative process for siting repositories that it had called for in 1982. Steamrolling politically weak Nevada, powerful members of Congress from other states made a deal, designating Yucca Mountain as the sole candidate site. Nevada responded with a sustained campaign of bitter opposition to the facility, including extensive litigation against DOE that eventually delayed the project. At that point, fortune again intervened. The 2008 Nevada Democratic presidential primary was the third such primary in the nation. All the Democratic candidates, including Barack Obama, lined up in opposition to Yucca. Also, Nevada senator Harry Reid had been elected U.S. Senate majority leader. As president, Obama would need Reid's support to carry his legislative agenda in Congress. Not long after taking office, the Obama administration began shutting down the Yucca repository project.

Federalism and the Federal Courts

The history of nuclear waste policy underscores the importance of two fundamentals in our country's governance structure: federalism and the federal judiciary. Both of these have critical roles as legal checks and policy balances against the political branches of the federal government. State and local governments are an important locus of political power and legal authority that can contest and blunt assertions of federal power. The federal courts operate as a check on Congress and the executive branch and play a critical role in ensuring that federal agencies adhere to the rule of law in making decisions and are accountable to citizens affected by those decisions.

During the first several decades of the nuclear era, the states and the federal courts largely failed to provide a significant check on federal nuclear policies, which were largely determined by the AEC with the backing of the exceedingly powerful Joint Committee on Atomic Energy in Congress. The Atomic Energy Act granted vast powers to AEC in sweeping discretionary terms that provided few legal handholds for litigation challenges. Nor were courts inclined to closely securitize AEC decisions in the few cases that were brought. Federalism checks were also weak during World War II and the early Cold War period as states supported the federal government's push to develop nuclear weapons. During this period AEC, in large measure due to secrecy about the government's weapons production effort, was able to site and develop all its major weapons-building facilities—including Hanford, the Savannah River Site, Oak Ridge, Idaho National Laboratory, Rocky Flats, and others—without major difficulty. Further, many states and localities followed the lead of AEC and their local electric utilities in embracing nuclear power as a technological and economic bonanza.

The easing of public Cold War fears, the discredited Vietnam War effort, and the rise of environmentalism all served to undermine the AEC's hegemony in nuclear policy. Concerns over conflicts between AEC's environmental regulatory and nuclear development roles led Congress in 1974 to break up AEC, giving its regulatory authority to the Nuclear Regulatory Commission (NRC) and its management and R&D tasks to ERDA, later DOE. The Environmental Protection Agency (EPA) also acquired nuclear regulatory authority, further fragmenting the AEC's nuclear policy monopoly within the federal government, and thereby creating openings for opponents of new nuclear facilities.

States and localities also began to resist the federal government's development of new nuclear facilities, as exemplified by the moratoriums on new nuclear plants that many states adopted. DOE was eventually able to develop WIPP, but only because it was operating within a loosely defined—indeed, in key respects, undefined—legal framework that was sufficiently flexible and mutable to allow New Mexico to play a role in key decisions and allow DOE and Congress to meet the states' concerns. The highly prescriptive NWPA system, however, broke against determined state resistance, both when DOE initially tried to site waste disposal and storage facilities under the 1982 act and later when it attempted to develop Yucca pursuant to the 1987 NWPA amendments. The courts have consistently held that the federal government has plenary constitutional authority to build nuclear waste facilities on federal lands, notwithstanding objections by the state in which the lands are located. But the federal government's authority is counterbalanced by deep political and institutional safeguards of federalism that make it very difficult for the federal government to impose such facilities by fiat against the determined opposition of host jurisdictions. The states are represented in Congress through their delegations. New Mexico, Tennessee, and many other states successfully called on their delegations to block federal nuclear waste facilities. Nevada was eventually able to exploit the intimate connection between local and national politics to block Yucca.

Starting in the late 1960s, the federal courts began extending rights to participate in administrative proceedings and to obtain judicial review of agency decisions—formerly limited to regulated entities—to a greater segment of, as well as new, nonprofit environmental, civil rights, consumer, and other reform-oriented advocacy organizations as well as to individuals concerned with such issues. These new plaintiffs increasingly spoke for beneficiaries of regulatory programs and others indirectly affected by agency decisions. The judicial creation of "interest representation" in administrative law responded to the perception that government agencies had been captured by the regulated industry or had systematically disregarded environmental, consumer, and other societal considerations outside their central mission. This expansion of procedural rights and judicial remedies, which was extended further by Congress in "citizen suit" provisions in many federal environmental regulatory statutes, enabled critics of nuclear power and those seeking stricter regulation to participate and present evidence and analysis in support of their positions in AEC, NRC, and EPA licensing and environmental standard-setting proceedings and to obtain judicial review of adverse or inadequate agency decisions.

States opposed to DOE's efforts to site nuclear facilities also successfully resorted to the courts to protect their interests. New Mexico won court rulings that DOE was obliged to conclude a cooperation and consultation agreement with the state to enable the state to take part effectively in decisions regarding WIPP. Further, it was judicially decided that DOE could not proceed to construct and operate the WIPP repository without specific authorization from Congress. Nevada obtained court rulings that DOE was obliged to grant the state funding to conduct its own technical studies on Yucca, and that EPA was required to revise its Yucca environmental standards to assure protection from radioactive releases over a far longer time than the ten thousand year period originally adopted by EPA. The courts ruled that "mixed waste" contained both radioactive and chemically hazardous constituents and thus was subject to federal hazardous substances regulatory laws that enabled states and environmental groups to sue DOE to force cleanup of its weapons facilities. This decision also enabled New Mexico to gain Resource Conservation and Recovery Act licensing authority over WIPP, a site disposing of mixed waste, thus giving the state significant oversight power and leverage in decision making about the repository.

These and other judicial rulings required greater accountability from DOE and other federal agencies for their decisions, and restored checks and balances within the federal system. Rather than imposing specific substantive outcomes, the courts promoted more open administrative and political decision-making processes in which environmental and social interests and states and localities could play a greater role. The courts' rulings also strengthened the political safeguards of federalism by enabling states to be more effective players in federal nuclear waste decision making by, for example, requiring DOE to come to the table to negotiate with New Mexico or ensuring that Nevada could tap experts to evaluate DOE's plans for Yucca. When the courts required DOE to go to Congress for specific authorization to build WIPP, the New Mexico delegation was able to ensure that the authorization later enacted contained important protections for New Mexico. In these and other ways, the courts' and states' growing roles in federal nuclear policy decision making reinforced each other.

The open and fluid character of the WIPP regime, evolving step-by-step, enabled New Mexico to mount successful legal challenges, exercise influence in Congress, and assert regulatory authority to check DOE and gain leverage to influence key decisions about the project. This in turn enabled the state to meet its concerns and led it eventually to accept the project. The closed, top-down Yucca regime afforded few access points for Nevada to assert real leverage or influence key project decisions; as a result, the state was forced to keep fighting a guerilla war and to drag out the NWPA script as long as possible.

The Coming Third Era

As we stand on the threshold of a new era, the ethical foundations of our nuclear waste law and policy bear reflection. A striking and pervasive quality of the federal government's stance in both earlier eras is arrogance: first, the arrogance of AEC and the rest of the nuclear establishment in pushing forward with weapons production and nuclear power heedless of the huge debts they were accumulating by ignoring waste disposal and other environmental problems; second, Congress's arrogance when it finally came to terms with the waste legacy by setting irresponsible deadlines and mandates in NWPA and imposing waste facilities on states against their will. In doing so, Congress was driven by the proposition, previously embraced by IRG, that the then-current generation, having reaped the benefits of nuclear production, was correspondingly obliged to dispose of the resulting wastes as quickly as possible so as not to impose them on future generations. This motivating proposition blinked reality in positing that disposal could be achieved within fifteen to twenty years. But its normative premises were also unduly simplistic.

It is now clear that the work of safely disposing of nuclear wastes cannot be accomplished by a single generation alone and will require a partnership across generations. To the extent that benefits such as nuclear-based security and low-carbon energy have accrued to current and past generations from the sources of nuclear waste, future generations will also share in some of those benefits. Our ability to evaluate repository sites and to advance the effectiveness of technologies for containing wastes, moreover, is likely to improve. A more considered ethical formulation therefore might be that our obligation is to give succeeding generations a real choice and the opportunity to make their own decisions, while not imposing a burden that those future generations may not be able to manage.[4] This obligation requires, among other matters, that we move forward with all deliberate speed to build at least one repository for waste disposal and, we believe, at least one consolidated SNF storage facility. In doing so, the federal government must win the assent of host states and localities. This requirement has ethical as well as pragmatic foundations. Congress ignored both in NWPA. As the history shows, the political safeguards of federalism give states considerable ability to block new facilities. But it is also deeply unfair to impose the long-run burden of other jurisdictions' nuclear wastes on a polity without rewarding it for the burdens it is shouldering on behalf of the common good, and affording it the opportunity to engage in and influence key decisions about a waste facility and ultimately to dissent or assent to the decisions.

As the history recounted in this book abundantly confirms, dealing successfully with nuclear wastes is far from a purely technocratic enterprise. Although a sound technical base for decisions is essential, it is also, as the tales of WIPP and Yucca illustrate, a highly political and social and also a deeply normative enterprise. Given the significant obstacles, success also requires patience, openness, flexibility, and fairness, as well as the maneuverability to capitalize on opportunities that fortune may afford. Rebuilding trust in DOE and the federal government is also critical. The events described herein remain very much in flux. As we prepared to send our final manuscript to the printer, respective decisions by a court and by the NRC Atomic Safety and Licensing Board have resuscitated, at least for the time being, the potential construction of a private SNF storage facility on Skull Valley Goshute tribal land in Utah and the Yucca repository, both of which had been given up for dead. And New Mexico officials have recently signaled their openness to disposal of HLW and SNF at WIPP.

Also, as this book goes to press, a deep nuclear power plant crisis is still unfolding in Japan. Nuclear reactors, struck by a severe earthquake and resulting tsunami that disabled cooling systems at the plants, are suffering core meltdowns, exposure of spent fuel in cooling ponds, and releases of radioactivity. The extent of contamination and threats to human health are as yet unknown, but as of this writing the situation at the Fukushima Daiichi reactors appears grave and difficult to bring under control. Whatever the outcome, these events could chill further development of nuclear power in the United States and justifiably stimulate greater attention to the safety of existing nuclear plants, including (but by no means limited to) redressing the practice of often crowded long-term storage of SNF in cooling ponds at reactor sites. The Japanese crisis vividly reminds us of the need for the utmost precaution and vigilance in regulating nuclear power and the dire consequences that may result when cooling water is lost, whether from earthquake, tsunami, terrorist attack, poor plant maintenance, or other causes.

The story told here is also a cautionary tale of the limitations of prescriptive law and the importance of process. In dealing with HLW, SNF, and LLW, Congress proceeded as though legislating a result would make it happen. It didn't. The reasons why are complex but instructive in thinking about how to do things better in the future. Process really does matter—in fact, it's the key. Our legal and political system makes it very difficult for the federal government to impose a fait accompli on a state that is unwilling to accept it. But a law that, first, provides substantial compensation and inducements to individual localities and states for bearing the nation's collective burden and, second, establishes and supports an equitable and fair process that engages the public and potential host jurisdictions and gives them a meaningful role in decisions, can make a great difference in the outcome.

We trust this book provides helpful background and insights to those who are now studying the way forward, including the blue-ribbon commission, DOE, and others in the administration, as well as Congress and the broader nuclear policy community, by enabling them to look at the paths taken—and forgone—on this controversial subject. But we equally hope that it will help inform public debate on the important issues involved—debate which has not taken place since the Carter era and is long overdue. The blue-ribbon commission should actively seek to stimulate this debate by holding public meetings around the country on the key issues involved, following in the footsteps of the IRG panel, which did so as it formulated its seminal report in the late 1970s.[5] We also hope that journalists will widely report on these issues and on the commission's work. The nation has not grappled openly with these highly important issues in many years. Today's public needs to be informed about and engaged in the issues so that it can participate effectively in deciding how we as a nation will address them. Public engagement at all levels will be needed to transform U.S. policy for the back end of the nuclear fuel cycle—presently, a policy that leads nowhere—into a well-considered and fair program that will ensure safe waste management and disposal and thereby resolve our current dilemma.

Chapter 1

The Evolution of U.S. Nuclear Waste Law and Policy

For nearly thirty years after the inception of the nuclear era during World War II, the federal government paid little serious attention to disposing of the most highly radioactive wastes generated by nuclear weapons manufacture and the rise of nuclear power. Disposal of defense high-level wastes took a backseat to the Cold War buildup of the nation's nuclear arsenal. Further, it was assumed that the spent nuclear fuel at power plants would be reprocessed to make new fuel; the problem of disposing of the highly radioactive wastes produced by reprocessing was deferred to the indefinite future. Beginning in the late 1960s, the rise of the environmental movement and of concerns over nuclear proliferation, as well as a number of complementary contingencies, produced fundamental changes in U.S. nuclear policy.

Sharp opposition to nuclear power, which sprang from nuclear power plant safety issues, the unresolved waste problem at reactor sites, and growing public concern over poor government management of wastes from weapons production, prompted the Carter administration to form an interagency task force that recommended federal government initiatives to dispose permanently of power plant and weapons wastes. In response, Congress enacted the 1982 Nuclear Waste Policy Act (NWPA), which mandated the Department of Energy (DOE) to implement an ambitious blueprint for rapid siting and construction of two repositories for disposal of these wastes, with a 1998 target for federal takeover of nuclear power wastes. Due to opposition from potential host states and delays and cost overruns in the siting program, Congress in 1987 short-circuited the siting process: it designated Yucca Mountain in Nevada as the sole candidate site for a repository and cancelled the second repository. Congress also responded to the lack of disposal facilities for low-level radioactive waste—a problem also identified by the Carter task force—by enacting legislation in 1980, modified in 1985, aimed at fostering a system of regional state compacts to develop such facilities.

Both sets of statutory programs have failed completely, leaving the nation's nuclear waste policies in disarray. Under the NWPA, Nevada strenuously resisted federal imposition on the state, over its objections, of a repository to dispose of all of the nation's most highly radioactive wastes. It succeeded in delaying Yucca's development until the election as president of Barack Obama, who opposed Yucca during the 2008 Nevada Democratic presidential primary. The Obama administration has since terminated the Yucca repository, although DOE's authority to withdraw its Yucca license application to the Nuclear Regulatory Commission (NRC) has been challenged and is currently being litigated. Although Yucca may yet be built, it faces many obstacles; these will be overcome, if at all, only after many years—probably decades. This leaves the nation today with no disposal pathway for massive amounts of the most highly radioactive wastes. In addition, the regional compact scheme envisaged by Congress to develop new disposal sites for growing volumes of civilian low-level radioactive wastes has failed to establish a single new facility.

There is, however, a brighter aspect to the story of U.S. nuclear waste policy. DOE eventually succeeded in developing, over a twenty-year period, the Waste Isolation Pilot Plant (WIPP), a repository near Carlsbad, New Mexico, for burial of highly radioactive transuranic (TRU) wastes from weapons fabrication. WIPP is now the only operating deep geologic repository in the world; it has successfully been receiving and disposing of TRU wastes since 1998. This was accomplished not according to any blueprint made in Washington, but through a long process of contestation, negotiation and compromise between DOE and the State of New Mexico, which eventually concluded that its concerns and interests had largely been met and accepted the facility.

Recently, President Obama has strongly supported expansion of nuclear power in the United States as part of a strategy to address climate change and to serve economic and other objectives. History strongly suggests, however, that the public and many states may not support—indeed, may effectively block—any major expansion of nuclear power so long as there is no credible means for dealing with nuclear power wastes. By canceling Yucca, Obama has abandoned the only location for disposing of that waste that is permitted under current law. The courts may eventually hold that NWPA requires the administration to move forward with licensing Yucca. But Nevada remains implacably opposed. Even if Yucca is eventually licensed after court review, Congress, which has recently refused to appropriate monies for Yucca, would still have to fund construction of the repository. The government has no plan or legal authority to develop an alternative to the Yucca repository.

Energy Secretary Steven Chu has appointed the Blue Ribbon Commission on America's Nuclear Future to address and develop solutions for the nation's nuclear waste dilemmas. The commission must carefully review and analyze the failures and successes of the nation's efforts to date for dealing with nuclear wastes and provide recommendations in order to help set a new and more successful course for the future. This book provides a comprehensive overview of that history and lessons to be learned from it. We hope it will inform the commission's work and future policy.

This chapter lays out the general history. The remaining chapters address specific issues and important aspects of the history in greater detail. Chapter 2 addresses nuclear waste classification and regulation. Chapter 3 covers nuclear waste transportation. Chapter 4 deals with low-level radioactive waste. Chapters 5 and 6 examine in detail the WIPP and Yucca projects, respectively. Chapter 7 addresses the options for dealing with the highly radioactive wastes that currently lack a disposal pathway as a result of Yucca's cancellation. And Chapter 8 offers insights culled from the history recounted and analyzed in the preceding chapters and provides recommendations for the path forward.

Nuclear Weapons Buildup and the Rise of Nuclear Power, 1946–1970

Shortly after the rapid development of the atomic bomb during World War II and the wartime demonstration in 1945 of the terrible power of nuclear technology, Congress adopted the Atomic Energy Act of 1946 (AEA), which created the Atomic Energy Commission (AEC) to run a tightly centralized federal monopoly on all applications of nuclear technologies.[1] Established as an extremely powerful agency with sweeping authority over all nuclear activities, AEC was overseen and supported within Congress by the powerful Joint Committee on Atomic Energy. The Cold War had begun, and AEC's primary mission was to rapidly build up the nation's nuclear arsenal. Involving various national laboratories, AEC carried out research and development on military and other applications of nuclear technologies, and manufactured nuclear weapons for the military. AEC built and operated a huge nuclear weapons production complex consisting of facilities, located in a number of different states, to enrich uranium, irradiate fuel rods through controlled fission in reactors, reprocess the irradiated fuel to extract plutonium, produce other weapons materials, and fabricate bombs. Military imperatives controlled AEC's project mission and activities, and relatively little attention was paid to the large volumes of radioactive and chemically toxic wastes generated in the course of the push to produce weapons.

The main facilities in the AEC weapons production complex were the Oak Ridge Reservation in Tennessee (Oak Ridge), the Los Alamos Scientific Laboratory in New Mexico (Los Alamos), the Hanford Site in the state of Washington (Hanford), the Savannah River Site in South Carolina (SRS), and the Idaho National Laboratory site in Idaho (INL).[2] Oak Ridge housed two plants that separated uranium isotopes to produce enriched uranium that would sustain fission, as well as the first reactor dedicated to nuclear materials production, which produced small amounts of plutonium. Los Alamos was the research center of the Manhattan Project, where the first atomic bombs were designed and manufactured. Hanford Site eventually housed nine nuclear reactors, five reprocessing plants, and a plutonium-finishing plant to produce plutonium for weapons. SRS housed five heavy-water reactors to produce plutonium and tritium. INL was reserved for the design and testing of nuclear reactors and reprocessing plants. The most highly radioactive wastes were generally stored at the sites where they were produced, without any plan for their disposal. Less radioactive wastes were generally disposed of in on-site landfills. These various wastes resulted in significant contamination of the soil and groundwater at many sites.[3]

Although the AEC initially focused most of its attention on nuclear weapons development and testing, the agency was also in charge of developing nuclear reactors to generate electricity for the public. The federal government decided to adapt the basic process it had developed for controlled fission of enriched uranium in reactors to produce plutonium and use it to heat water in reactors to generate steam in order to propel nuclear submarines and to generate electricity for civilian use. The thorium fuel cycle for electricity production, which has significant potential advantages over the uranium fuel cycle in terms of ore abundance, wastes generated, and proliferation resistance while posing distinctive technical challenges, was considered at various times but never fully developed.[4] The 1946 AEA granted AEC sweeping authority to regulate the use, possession, and transfer of nuclear technology and materials, including nuclear wastes.[5] The commission was empowered to grant licenses for the use or possession of nuclear materials, on the condition that the recipient adhere to AEC "safety standards to protect health." AEC standards for licensing possession or use of fissionable material required licensees, in the most general terms, to "minimize danger from explosion or other hazard to life or property."[6]

In 1953 President Dwight Eisenhower announced the Atoms for Peace initiative, under which the private sector, with assistance from AEC and subject to its regulatory oversight, would develop, construct, own, and operate nuclear electric generating plants. In 1954, Congress amended the AEA to authorize and regulate civilian uses of nuclear materials and spur expansion of civilian nuclear power.[7] A central purpose of the amended statute was to "encourage widespread participation in the development and utilization of atomic energy for peaceful purposes."[8] Designed to promote development of a private nuclear industry, it included provisions authorizing private ownership of commercial production facilities. It required AEC to distribute "special nuclear material" such as enriched uranium, capable of sustaining fission, to private parties under license.[9] Congress at the same time confirmed that "the processing and utilization of . . . nuclear material must be regulated in the national interest and in order to provide for the common defense and security and to protect the health and safety of the public." AEC's authority to regulate and license nuclear fuel cycle materials and related activities remained extremely broad.[10] The commission's regulatory authority extended to private possession and use of special nuclear material; "source material" such as uranium ore, for making special nuclear material; and "byproduct material," including nuclear wastes.[11]

With substantial financial and technical help from AEC, private industry soon launched into research and development of commercial-scale nuclear power plants.[12] AEC oversaw construction at INL in 1951 of the first reactor—an experimental breeder-type reactor—to produce electricity from nuclear energy.[13] AEC also sponsored the first large-scale commercial light-water reactor nuclear power plant, which began operating in 1957 at Shippingport, Pennsylvania.[14] By the early 1960s, commercial nuclear power plants had begun operation. Most plants were owned and operated by privately owned electric utilities. By 1971, 22 commercial nuclear power plants were operating around the country, and 41 more plants had been ordered.[15] In addition, private firms and institutions were authorized and began to use nuclear materials for medical, research, and other purposes. AEC regulation of private sector nuclear-related activities under AEA extended to all aspects of the nuclear reactor fuel cycle, including ore processing; fuel assembly; fission to generate heat used to produce electricity; and reprocessing of spent fuel rods.

Because AEA does not grant AEC/NRC authority to regulate non–fuel cycle materials and wastes, including naturally occurring radioactive materials (such as radium) and accelerator-produced radioactive materials, the states are responsible for regulating civilian uses of these materials in medical, research, industrial, commercial, and other applications, as well as for the wastes that result from these activities. (However, Congress gave NRC authority over certain naturally occurring radioactive materials in the Energy Policy Act of 2005.)[16]

The development of civilian nuclear power and the desire of the states for a role in that development led Congress in 1959 to amend the AEA in order "to clarify the respective responsibilities . . . of the States and [AEC] with respect to the regulation of [fuel cycle] nuclear materials."[17] The legislation, codified as Section 274 of AEA, authorized AEC to enter into agreements with states, allowing states with qualifying regulatory programs to regulate with respect to one or more categories of fuel cycle nuclear materials, including waste byproduct materials. This provision has allowed AEC and its regulatory successor, NRC, to determine which states are capable of taking over regulatory authority (known as Agreement States), while retaining to itself authority over activities in other states.[18] The amendments also provided for training and assistance to states to ensure that public health and safety are protected adequately under state regulatory regimes.[19]

Under the 1959 amendments, however, the commission retained exclusive authority over the construction or operation of any "production or utilization" facility, which would include nuclear reactors and reprocessing facilities, and also over disposal of nuclear wastes that the commission determined to be so hazardous that they should not be disposed of without a license from the commission. AEC implemented this provision by allowing Agreement States to license and regulate disposal of commercial low-level radioactive wastes pursuant to agreement, while reserving to itself direct control over the disposition of more hazardous forms of wastes, including spent nuclear fuel and high-level wastes from reprocessing.[20] A savings clause in the statute provided that nothing in the statute "shall be construed to affect the authority of any State or local agency to regulate activities for purposes other than protection against radiation hazards."[21]

The AEA has long been understood as preempting state regulation of fuel cycle materials, except where Congress has withdrawn regulatory authority over such materials from AEC/NRC or has specifically provided for state regulation, as for example in the 1959 AEA amendments establishing the Agreement State arrangement and saving state regulatory authority except with respect to radioactive hazards. While there is no explicit preemption provision in the act, courts have consistently held that the comprehensive sweep of the powers granted to AEC manifested an intent by Congress to preempt state regulation of any matters within the commission's authority, unless Congress has specifically otherwise provided.[22]

For example, *Northern States Power Company v. Minnesota*, a 1971 decision by the Eighth Circuit Court of Appeals that was affirmed without opinion by the Supreme Court, held that any regulation of radiation hazards by a state that was not an Agreement State was preempted by AEA, and that accordingly Minnesota could not impose restrictions on radioactive waste releases from nuclear power plants that were more stringent than AEC regulations. The court asserted that the federal government "had occupied the entire field of nuclear safety concerns" except in cases of regulation by Agreement States, or in other limited circumstances where the federal government had expressly ceded authority to the states.[23] This conclusion was endorsed by the Supreme Court in its 1983 *Pacific Gas & Electric* decision addressing the validity of a California law that banned new nuclear power plants until a means had been demonstrated to deal with the nuclear wastes they would produce. The Court, however, sustained the California measure, finding that it was based not on environmental but on economic regulatory concerns, a long-recognized field of state regulation that had not been preempted by AEA.[24] In support of this conclusion, the Court pointed to Section 271 of AEA, which provides: "Nothing in this chapter shall be construed to affect the authority or regulations of any Federal, State or local agency with respect to the generation, sale, or transmission of electric power produced through the use of nuclear facilities licensed by the Commission."[25]

Regulation and Disposal of Nuclear Wastes

Neither the AEA of 1946 nor the 1954 amendments to it specifically addressed nuclear waste or its disposal. The act classified three basic types of nuclear material. It defined "source material" as the raw or processed ore that could be refined to produce useful nuclear materials.[26] "Special nuclear material" (called "fissionable material" in the 1946 version of the act) referred to materials that could be used to produce or release atomic energy, whether for power generation or for military purposes: "plutonium, uranium enriched in the isotope 233 or in the isotope 235," and similar material so designated by the AEC.[27] "Byproduct material" was defined as any radioactive material except special nuclear material that resulted from the production of special nuclear material.[28] Nuclear wastes, including spent nuclear fuel and reprocessing wastes, were not specifically designated as a separate category; rather, they fell within the definition of "byproduct material." The act gave AEC sweeping authority over possession, use and disposition of all of these materials: "In the performance of its functions the [AEC] is authorized to ... establish by rule, regulation, or order, such standards and instructions to govern the possession and use of special nuclear material, source material and byproduct material as [the AEC] may deem necessary or desirable to promote the common defense and security or to protect health or to minimize danger to life or property."[29]

The regulatory authority granted by AEA to AEC and its successor agencies over fuel cycle nuclear materials continues to form the basis of most of the corpus of U.S. nuclear environmental, health, and safety regulation, including regulation of nuclear wastes generated by both private and governmental activities. Through an accretion over time of federal statutes and regulations, fuel cycle wastes came to be classified, for purposes of management and disposal, into four basic categories:

- **Spent nuclear fuel (SNF)** consists of fuel rods that have been irradiated in fission reactors to produce electricity, or in connection with weapons production or research, and that contain a number of different, highly radioactive elements. Most SNF has been, and continues to be, produced in connection with commercial generation of electricity at nuclear power plants.
- **High-level wastes (HLWs)** are highly radioactive wastes produced by reprocessing SNF to extract plutonium for weapons production, or to extract fissionable uranium and plutonium for reuse as fuel. Almost all HLWs have been produced by the government in connection with nuclear weapons production, although a small amount was generated from reprocessing civilian SNF.
- **Transuranic (TRU) wastes,** almost all of which have been generated in weapons production, contain high proportions of elements with atomic numbers greater than that of uranium; these elements have moderate to moderately high radioactivity levels and half-lives greater than twenty years.
- **Low-level wastes (LLWs)** are a catchall residual category that includes all wastes that are not SNF, HLW, or TRU wastes. LLWs are generated from a great variety of weapons production, nuclear power, research, medical, and other activities, and have levels of radioactivity that are generally (although not always) lower than those in the first three categories. The radioactive characteristics of these different wastes vary widely. They are classified into subcategories based on their relative degree of radioactivity: **A** (lowest), **B, C,** and greater-than-class C (**GTCC**) (high-

est). Class A wastes constitute the great bulk of LLW by volume but contain a very small proportion of the total radioactivity of LLW.

As explained more fully in Chapter 2, nuclear wastes vary according to the types of radiation emitted, which pose different health and environmental risks; the levels of radiation emitted at any given time; the rates at which radioactivity decays over time; and the concentrations of radioactive elements. These variations, in turn, determine the appropriate regulatory requirements and methods for treating, storing, and eventually disposing of these wastes to best limit radioactive exposure to people and other life forms, in order to protect health, safety, and the environment. Nuclear wastes can be chemically toxic as well as radioactive; those with both sets of properties eventually came to be defined as "mixed wastes." Such wastes are currently subject both to NRC regulation of their radioactive components and to regulation by the Environmental Protection Agency and delegated states of their chemically hazardous constituents pursuant to the Resource Conservation and Recovery Act.

Other radioactive wastes include uranium mill tailings; certain other byproducts from the uranium fuel cycles used for electricity generation and weapons production; stocks of waste plutonium and uranium originally intended for weapons production that are currently surplus because the United States has been reducing its stock of nuclear weapons pursuant to international agreements on nuclear disarmament; and naturally occurring and accelerator-produced radioactive materials. These are treated only briefly in this book.

Pursuant to AEA, AEC in 1957 promulgated rules for management and disposal of nuclear waste from private sector nuclear power generation and other activities subject to AEC licensing.[30] These rules did not apply to AEC's management of the wastes that it had produced unless it disposed of such wastes in private facilities, at which point the waste became subject to the AEC regulations for those facilities. The AEC regulations pertaining to private sector facilities provided for disposal methods that included transfer to another licensed recipient, release into the sewer system, shallow burial in soil, or other methods specifically approved by the AEC on a case-by-case basis. The sewer and soil methods were available only for isotopes with relatively low levels of radioactivity. Ocean disposal was authorized in some cases.

AEC took the lead in the development of commercial LLW nuclear waste disposal facilities. These were sited on federal or state land and operated by commercial lessees. Six commercial facilities that disposed of LLW by shallow land burial were operating by 1971. They experienced a variety of economic, operating, and environmental regulatory compliance difficulties, including problems stemming from the lack of "systematic site selection criteria or design requirements that could be used to establish the best mix of features necessary to contain and isolate the wastes."[31] A number of these LLW disposal facilities were eventually closed.

Most of the high-level, TRU, and SNF wastes and some of the LLW generated by AEC and its successors in connection with weapons production and other activities were stored at the sites where they were produced, on the assumption that methods to dispose of them would become available within the lifetime of the storage facilities.[32] Much LLW and some TRU was disposed of by shallow land burial and other means on-site or off-site. The largest volume of AEC wastes consisted of liquid HLW from reprocessing for weapons production stored in large in-ground tanks. Incidents of tank leakage at Hanford and other sites caused environmental releases of wastes and radiation, fueled public concern, and ultimately prompted greater attention to studying and improving methods for disposing of these wastes.[33]

In the case of civilian SNF, the prevailing assumption during the late 1950s and the 1960s was that spent fuel rods from nuclear power plants would in due course be reprocessed to extract the substantial amounts of fissionable uranium and plutonium they contained, and that these elements would be used for additional power generation.[34] Reprocessing had become a familiar technology in the context of weapons production, and hence it was not difficult to envisage its application in the nuclear power context. While reprocessing would generate substantial residuals of HLW that would have to be disposed of, it was further assumed that, since nuclear plants were just beginning operation, large volumes of such wastes would not be produced for some time. As a result of these assumptions, which prevailed until the late 1970s, the problem of disposing of highly radioactive wastes from civilian power plants was largely ignored.

Studies of Options for Permanent HLW Disposal

In 1955, AEC approached the National Academy of Sciences (NAS) about the creation of a body, called the Steering Committee, to study and recommend ways of dealing with the issue of nuclear waste disposal.[35] AEC had come to recognize that shallow burial and disposal into the sewage system were inappropriate for the most highly radioactive wastes being produced by AEC weapons facilities, and for those expected to be produced by reprocessing of SNF from civilian nuclear power plants. AEC was concerned about the threat of nuclear wastes to public health, and more particularly about the scientific and technical uncertainties regarding means for their disposal.[36] AEC staff also believed that AEC needed a public relations strategy, because as "the atomic industry expanded and moved into populated areas, public concern about nuclear safety generally and waste disposal specifically seemed likely to increase."[37] Future AEC chair Glenn Seaborg, among others, saw public concerns about nuclear risks, including those posed by wastes, as the biggest threat to the country's nuclear power future.[38] The 1954 act had allowed for the commercialization (and thus potential boom) of nuclear power, and AEC was eager to resolve any issues that might stand in its way.[39]

The 1957 NAS Advisory Committee on Nuclear Waste Report

In 1957, the same year that electricity produced by a civilian nuclear reactor was first connected to the electrical grid, the Steering Committee, renamed the Committee on Waste Disposal of the Division of Earth Science, released its report.[40] It identified and addressed many of the issues regarding nuclear waste that have been discussed and debated ever since. The committee's report stated that, "unlike the disposal of any other type of waste, the hazard related to radioactive waste is so great that no element of doubt should be allowed to exist regarding safety. . . . *Safe* disposal means that the waste shall not come in contact with any living thing."[41]

The report stated that, in addition to the substantial amounts of HLW from AEC defense facilities, the country would have to deal with large additional amounts of HLW as a byproduct of reprocessing commercial nuclear power SNF to extract new fuel for reuse at nuclear power plants.[42] The committee estimated that by the turn of the millennium, there could be as many as 350 nuclear power plants operating in the United States, producing up to 50 percent of the country's electricity. It calculated that the total annual volume of liquid HLW produced by reprocessing the SNF from these plants would be about fifty-two million gallons, requiring seven million cubic feet of space, if underground disposal were chosen[43]—enough to fill almost a million fifty-five-gallon drums.[44]

The committee considered various options for dealing with these wastes.[45] Its report found that geologic burial of HLW in "cavities in salt" and underground salt domes represented "the most promising method of disposal of high level wastes at this time."[46] The committee favored disposal of HLW in abandoned salt mines or other specially excavated cavities in salt formations over other types of disposal because "salt domes would be self-healing around fractures," and salt formations might provide "long-enduring" containment of wastes. Other advantages of salt formations included substantial heat conductivity and a high melting point.[47] A potential danger that the committee identified and recommended for further research, however, was that cavities mined out of salt formations to hold the radioactive wastes might be subject to cave-ins unless the openings to the cavities were the right shape and size.[48] An appendix to the report identified salt formations in a number of states and regions of the country that might ultimately prove suitable for disposal of radioactive wastes.[49] Cautioning that no serious research had been done on disposal of high-level radioactive waste by these methods, the committee noted that research was needed to verify the soundness of its conclusions, which had been made on the basis of limited existing data. It recommended storage of reprocessing wastes in on-site tanks until such research could be conducted.[50]

The committee determined that the "second most promising method" was to stabilize the waste in an insoluble form by forming silicate bricks out of the waste, and to store these in surface sheds in arid areas, dry mines, or in large cavities in salt.[51] Deep geologic disposal by underground injection of liquid wastes into porous rock (a method used by the oil and gas industry for brine wastes) was also identified as a potential option for the future; but significant problems, such as likely clogging of the pore spaces of receiving formations, and difficulty predicting and controlling the migration of injected fluids, were thought to require further study. Other alternatives the committee considered, but found less attractive, included disposal in the ocean; burial under the ocean seabed; disposal in natural caverns, granite quarries, and abandoned mines; and excavations in non-porous rock such as clay or shale.[52]

Further, the committee suggested that research should begin on how to separate out the different waste streams contained in reprocessing wastes, so that elements with different radioactivity levels, half-lives, and other characteristics could be dealt with separately. This approach could allow for use of waste management and disposal methods matched to the characteristics of different types of waste, rather than allowing the constituents posing the greatest risks or technical difficulty to define the management and disposal strategy for all of them.[53] In this connection, the committee emphasized that removing the cesium and strontium isotopes, which generate high levels of radioactivity and heat for several hundred years, from reprocessing or other fission product wastes would "greatly simplify the general problem of waste disposal."[54]

Subsequent Evaluations of HLW Disposal and Treatment Methods

The options reviewed by the NAS committee in its 1957 report were further examined in the following years by scientists and policy makers. Deep geologic disposal on land has remained the favored means for permanent disposal of long-lived, highly radioactive wastes, although serious attention has been given to several other options. In 1966, a successor NAS committee that had reviewed AEC's radioactive waste management program issued a report that reaffirmed the conclusion that its predecessor had reached nine years earlier, namely that salt beds were the most promising sites for permanent disposal of HLW.[55]

Some scientists have at various times thought that disposal beneath the ocean seabed would offer effective containment, as well as the potential to minimize risks of exposure in the event of leakage, due to the high dilution factor of oceans.[56] Indeed, from 1946 to 1969, some low-level defense waste was sealed into large concrete blocks and disposed of at the bottom of the Pacific and Atlantic Oceans. Until 1959, this disposal method was employed by the U.S. Navy and thereafter by companies licensed by AEC.[57] There was a strong public outcry over ocean dumping in the 1960s that continued despite AEC's efforts to publicize scientific studies establishing the safety of the technique. In response, AEC shifted toward land burial of LLW, which was cheaper and less controversial.[58] AEC ceased ocean disposal in 1970.[59] The practice had already aroused enough controversy to come to international attention, and it was thought to be very difficult to achieve the international political consensus necessary to dispose of such waste in the global commons. International regulation of ocean disposal of wastes was regulated by the London Convention beginning in 1975; amendments to the Convention banned the disposal of radioactive waste at sea in 1983.[60]

Another option that later received attention was sending waste into outer space, including shooting it into the sun.[61] A primary concern with this method was the provision of adequate protection in the event of a failure at launch—an event euphemistically referred to as "unanticipated re-entry."[62] In the case of SNF, the high cost of space transport would likely require that the waste be reprocessed to reduce the amount being sent up.

Disposal of waste underneath ice sheets also received significant consideration.[63] Technical obstacles to developing this method included the lack of understanding of long-term ice flow and mass, as well as the irretrievability of the waste.[64] Additional problems arose from legal restrictions or uncertainties regarding disposal in ice. In 1959, use of Antarctic ice for this purpose was specifically prohibited by treaty.[65] The absence of any international agreement to date on the use of Arctic resources makes use of Arctic ice for nuclear waste disposal legally and diplomatically problematic. The current melting of Arctic ice due to climate change provides another, most probably decisive, reason to discard this option.

An NAS Commission on Geosciences, Environment, and Resources issued a report in 1996 on one option earlier recommended by the 1957 NAS committee: separating out radioactive elements in reprocessing wastes in order to apply the management and disposal methods appropriate for the different waste components. The report neither favored nor opposed separation of wastes, but it did note that the technology to separate nuclear wastes efficiently had not yet been demonstrated. The committee also examined the potential for transmutation of long-lived fission products into less radioactive elements and, in the same 1996 report, found "no evidence that applications of advanced [separations technology and transmutation] have sufficient benefit for the U.S. HLW program to delay the development of the first permanent repository for commercial spent fuel."[66] Among other problems, use of transmutation technology for SNF would require reprocessing it, which would create additional wastes to be dealt with, as well as the expenses and associated challenges entailed.[67] These conclusions have been echoed in subsequent studies and reports, including a 1999 report by the Nuclear Energy Agency (NEA) of the Organization for Economic Cooperation and Development, which stated that "the waste management community does not [regard waste separation] as an alternative [to ultimate geologic disposal]; at best it reduces the volume, or changes the isotope distribution, of wastes requiring deep disposal."[68]

The NAS Board on Radioactive Waste Management conducted a comprehensive review of the options for dealing with the most highly radioactive wastes, and concluded in a 1990 report: "There is a strong worldwide consensus that the best, safest long-term option for dealing with HLW is geological isolation.... Although the scientific community has high confidence that the general strategy of geological isolation is the best one to pursue, the challenges are formidable."[69] In 1999, the NAS Commission on Geosciences, Environment, and Resources revisited the subject.[70] It endorsed the international scientific consensus in favor of disposal in a geological repository reflected in the Collective Opinions published by NEA, the International Atomic Energy Agency, and the European Union.[71] Other alternatives (space, seabed burial, transmutation, and continued storage) were reconsidered but rejected.

In sum, although various NAS and other expert groups have revisited and reexamined, for more than fifty years, the options for disposal of the most highly radioactive wastes, including SNF, HLW, and TRU, burial in a deep geologic repository has remained the consensus choice.

AEC Investigations of Potential HLW Repository Sites

The 1957 NAS committee report recommended that "the necessary geologic investigation of [potential disposal sites] must be completed and the decision as to a safe disposal means established before authorization" for construction of nuclear power plants is given. Noting that such an investigation could "take several years and cause embarrassing delays" in reactor licensing, the committee urged "starting investigation now of a large number of potential future sites as well as the complementary laboratory investigations of disposal methods."[72]

AEC nonetheless proceeded to license new civilian nuclear power reactors without first investigating or establishing disposal sites for the resulting wastes. The commission, however, took some steps to investigate disposal. Oak Ridge National Laboratory (ORNL) and other facilities conducted extensive studies of various options for burial of highly radioactive wastes in deep geologic formations. AEC's Office of Reactor Development had commissioned a report in 1958 to identify salt deposits potentially suitable for a repository. The report identified the Permian Basin (which includes the Delaware Basin in eastern New Mexico and much of Kansas, West Texas, and Oklahoma) as a potential location for a radioactive waste repository.[73] From 1957 to 1963, AEC commissioned research by ORNL on disposal of radioactive waste in salt formations.[74] In 1960, Harry Hess (who had chaired the 1957 NAS committee) warned AEC chair John A. McCone in a letter that the search was not advancing fast enough. He counseled AEC to begin to develop a comprehensive geologic disposal plan, and advised that none of the sites being actively considered by AEC (mostly AEC facilities where defense wastes were being produced and stored) was suitable.[75]

In 1962, ORNL initiated Project Salt Vault, the first field experiment designed to assess the thermal and radiation effects of HLW on salt, to determine whether salt media could safely contain such wastes. The project was located at an abandoned salt mine in Lyons, Kansas. From 1965 to 1967, canisters of experimental test reactor irradiated fuel were emplaced in the mine as the basis for scientific experiments.

In its 1966 report, the NAS Advisory Committee on Geologic Aspects of Radioactive Waste Disposal emphasized the need for AEC to proceed promptly with developing an HLW repository.[76] Leaks and other releases of defense reprocessing wastes stored in tanks at Hanford and other AEC sites continued to occur. Also, by the late 1960s, efforts to develop and operate facilities for reprocessing civilian nuclear wastes were beginning to encounter increasingly evident technological, environmental regulatory, and economic problems.[77] Notwithstanding these problems, AEC's studies of disposal options proceeded slowly and without priority attention from the commission.

Finally, in 1970, AEC announced a policy for disposal of civilian HLW reprocessing wastes in geologic formations, recommending salt beds in preference to other types of geologic formations. AEC's "Policy Relating to the Siting of Fuel Reprocessing Plants and Related Waste Management Facilities" was promulgated as Appendix F to its regulations at 10 C.F.R. Section 50, the regulations governing civilian but not defense wastes. The AEC's policy statement in the *Federal Register* notice on issuance of Appendix F envisioned disposal of civilian HLW from SNF reprocessing plants in bedded salt repositories developed and owned by the federal government. The plan was that these wastes would be stored at the reprocessing facilities in liquid form for five years; then, when sufficiently cooled, they would be rendered into solid form and buried in the repositories within ten years after they had been generated.[78] Consistent with the NAS committee's 1957 and 1966 recommendations, AEC favored salt formations because "bedded salt . . . is widespread and abundant; it has good structural properties; it is relatively inexpensive to mine; its thermal properties are better than those of most other rock types; and it occurs generally in areas of low seismicity. Most important, salt deposits are free of circulating groundwaters and completely isolated from underground aquifers by essentially impermeable rock formations. Furthermore, this situation tends to be preserved because any fractures which might develop are readily healed by plastic deformation of the salt."[79] Appendix F also set out the first formal definition of HLW—specifically, of "high-level liquid radioactive waste"—as "those aqueous wastes resulting from the operation of the first cycle solvent extraction system, or equivalent, and the concentrated wastes from subsequent extraction cycles, or equivalent, in a facility for reprocessing irradiated reactor fuels."[80]

The expected problem of dealing with large volumes of civilian HLW from reprocessing power plant SNF, however, failed to materialize. Only one civilian SNF reprocessing plant, in West Valley, New York, ever went into operation. It closed in 1972 due to environmental regulatory, operating, and financial problems. The facility produced a limited amount of HLW, along with TRU and other radioactive wastes, left a pool full of unreprocessed SNF, and caused radioactive and chemical contamination throughout the site. Two other SNF reprocessing plants under development at the time never began operation. In fact, no further reprocessing of commercial SNF has taken place in the United States since West Valley closed. As described later, President Carter halted federal support for commercial reprocessing of SNF because of nuclear proliferation concerns. Also, commercial reprocessing of SNF to extract and recycle uranium and plutonium for electricity generation has proven to be far more costly than using virgin uranium to make new fuel rods. The end of SNF reprocessing has left large and increasing amounts of SNF accumulating in storage at nuclear power plant sites throughout the country without a disposal pathway.

AEC's plans for disposing of commercial SNF reprocessing wastes by solidifying them for burial in salt beds did not extend to defense HLW, which at the time constituted 95 percent of all HLW in the country. AEC argued that the radioactive characteristics of defense HLW were significantly different from those of civilian HLW. In order to maximize the production of plutonium for weapons production, fuel rods are generally irradiated for a shorter period than in the reactor process used for producing electricity. AEC cited the higher costs of solidification and substantially lower heat output and radioactivity of the reprocessing wastes produced from these fuel rods, as well as their much higher current volume, as justifications for allowing defense HLW to remain in liquid form in tank storage. AEC stated that it was seriously considering the feasibility of disposal of defense HLW in underground caverns beneath the SRS and Hanford facilities.[81] Since 1958, AEC had also intermittently studied the option of disposing of defense HLW in bedrock vaults under SRS. It persevered in considering this option even though the Radioactive Waste Management Committee of the OECD Nuclear Energy Agency declared in 1966 that the project was "dangerous" and should not be pursued.[82] Eventually, AEC abandoned the idea of burying waste in Hanford caverns or in SRS bedrock vaults, leaving defense HLW also without a disposal pathway.[83]

The Defense TRU Waste Problem and the Search for a Repository

In addition to HLW from reprocessing, substantial quantities of TRU wastes containing plutonium and other transuranic elements had been generated by AEC in the course of fabricating nuclear weapons. Before 1970, AEC had managed TRU and LLW as a single waste category, and disposed of them by burial in shallow trenches at Hanford, INL, Los Alamos, ORNL, SRS, and Sandia National Laboratories in New Mexico.[84] This practice was eventually halted, and most defense TRU was simply stored at AEC facilities around the country where the TRU was produced.[85]

AEC's nuclear weapons production facility at Rocky Flats, Colorado, which was regarded as vital to national defense efforts, was one of the main sites at which defense TRU was being generated. This waste was sent to INL for burial. On May 11, 1969, a serious fire broke out at Rocky Flats.[86] The fire caused damage and losses amounting to $70.7 million, making it the most expensive industrial accident in U.S. history at the time.[87] Had it not been contained, the fire could have spread airborne plutonium throughout the Denver metropolitan area.[88] While the fire did not immediately attract sustained public attention, it eventually led environmental organizations, independent scientists, and other concerned citizens to begin investigating Rocky Flats, generating a movement in Colorado of environmentalists, peace activists, and health professionals that sought to permanently close and clean up Rocky Flats.[89] The fire, along with numerous other safety violations at Rocky Flats, provoked wider public concerns about the safety of nuclear waste management activities at the government's weapons facilities generally.[90]

Cleanup at Rocky Flats after the fire filled hundreds of railroad cars with plutonium-contaminated debris, TRU, and other wastes. About 330,000 cubic feet of waste was sent for disposal to INL.[91] The waste's destination was public information but at first generated little interest. In June 1969, the *New York Times* ran a story about the Rocky Flats fire, noting the waste's destination in Idaho.[92] Bob Erkins, an Idaho trout farmer, was trying to sell one of his trout farms near the plant. A prospective buyer of his trout farm mailed him a clipping of the article and said he didn't want to buy a trout farm next to a nuclear waste burial site.[93] Erkins worried that the plutonium buried at INL might leak into the aquifer underlying INL that supplied water to his trout farms. He sent letters protesting the waste burial to the governor and to newspapers all over the state, many of which published the letters or quoted from them. The resulting publicity created quite a stir. Others in the same agricultural community began to share Erkins's fears about the aquifer, which supplied water to a large portion of southern Idaho. Don Samuelson, the governor of Idaho, created a state task force to investigate the issue, and Idaho senator Frank Church commissioned a study at the federal level.[94]

As a result of the controversy, AEC in March 1970 announced a new policy that all TRU waste nationwide would be segregated from other types of nuclear waste and that it would be placed in retrievable storage; it also directed that all TRU buried at INL be dug up.[95] The AEC also informally promised Idaho that all TRU buried at INL would be recovered and removed from Idaho by 1980.[96] Commissioner Seaborg indicated that the waste would be sent to a federal repository in a salt mine.[97] These events would help spur AEC and its organizational successors responsible for weapons plants, the Energy Research and Development Administration (ERDA) and DOE, to develop a geologic repository near Carlsbad, New Mexico, for disposal of all defense TRU wastes. This repository, named the Waste Isolation Pilot Plant (WIPP), would eventually be constructed.

The public and political furor that resulted from the Rocky Flats fire finally forced AEC to stop its endless series of studies of disposal options and begin to develop a federal repository.[98] AEC focused on salt bed formations for potential repository sites. At various stages in the process, AEC contemplated disposing of both civilian SNF and defense wastes, including the TRU at INL and the reprocessing HLW being stored at Hanford and SRS, at the repository it was planning.[99]

On June 17, 1970, AEC announced its tentative selection of the Permian Basin as the site for a "demonstration repository." AEC specifically focused on Project Salt Vault, the Lyons, Kansas, salt mine in which ORNL had been conducting field experiments on the interactions of HLW and salt formations. AEC stated its expectation that this would be the site of the first permanent repository for highly radioactive nuclear waste.[100] This announcement was rushed, surprising even the scientists who had been conducting field tests at the site; they had not envisioned the site being used as a geologic repository for nuclear waste and did not consider it suitable for this purpose.[101]

Not long after this announcement, Project Salt Vault began to unravel. In rushing forward, AEC had discounted serious technical concerns that had been identified by the Kansas State Geological Survey and Kansas political leaders.[102] The discovery of boreholes penetrating the salt formation led to serious concern about containment at the site. The Kansas State Geological Survey staff disputed AEC's suggestion that oil and gas boreholes located nearby could be plugged effectively; the dispute was reported nationally.[103] In July 1971, roughly 175,000 gallons of water used for hydraulic fracturing in a nearby salt mine inexplicably disappeared.[104] This event, first uncovered by state officials, contradicted key findings of AEC's eight-page report on the site's suitability and solidified opposition to the site by Kansas political officials and the State Geological Survey.[105]

AEC abandoned Project Salt Vault and the Lyons site in 1972. The commission's botched effort in Kansas had long-term repercussions. It spawned public mistrust of the federal government's ability to select a safe means of disposing of radioactive waste, diminished states' receptivity to hosting a federal waste repository, and helped fuel the anti-nuclear political groundswell of the early 1970s.[106]

From 1972 to 1976, AEC/ERDA reviewed salt formations around the nation to identify potential repository sites, and expanded the search for a geologic repository to sites in nonsalt media. In 1975, ERDA informed the governors of thirty-six states that it would be developing several repositories by 2000, and asked for their cooperation. It received in response negative reactions and denials of permission to explore the sites from virtually all the states, followed by strangulating budget cuts from the Office of Management and Budget. Efforts by AEC/ERDA to site repositories were often clumsy. For example, ERDA failed to consult political leaders in Michigan and New York before sending in teams to examine local salt formations; once word got out, political rejection was swift and vocal.[107] ERDA was forced to cut back on its effort to identify a range of potential repository sites nationwide, and began to focus its investigations only on sites in Texas, Louisiana, Mississippi, Washington, and Nevada.[108]

In searching for alternatives to the Lyons site, AEC/ERDA eventually identified southeastern New Mexico as the most promising geologic area for siting a deep nuclear waste repository. At the same time, local leaders in Carlsbad, New Mexico, having heard about the demise of the planned Lyons repository, proposed that their community host such a repository in order to revitalize the declining economy of the city.[109] In 1972, the Los Medanos site, on federal land thirty miles east of Carlsbad, was proposed by AEC as a potential site for a pilot federal repository project.[110] Field tests began the following year and culminated in the selection, in 1975, of the site that eventually became the WIPP repository.[111] The repository project began to move forward under the auspices of ERDA.[112] The State of New Mexico did not play an active role in the process at this point. Under contract with ERDA, Sandia National Laboratory undertook a geologic survey, generated a facility design, and prepared a draft environmental impact statement for the repository.

In 1976, ERDA applied to the Department of the Interior to administratively withdraw 17,200 acres of federal lands at the Los Medanos site from use for other purposes and designate them for detailed site characterization and use by ERDA for what was referred to as a "waste isolation pilot plant." The *Federal Register* notice of the application provided no further details.[113] The filing of the application triggered a two-year withdrawal, which allowed ERDA to conduct studies at the site, designated in the application as the Waste Isolation Pilot Plant (WIPP). Although various federal officials envisaged that WIPP might potentially be used for both civilian and defense wastes, no explicit decision at the time was made regarding the type or amount of wastes, safety standards or criteria, supporting transportation infrastructure, the roles and authority of the various state and federal actors, or the procedures for making final project and regulatory decisions.[114]

The Rise of the Environmental Movement and the End of Reprocessing

Beginning in the late 1950s and gathering steam through the 1960s, the modern environmental movement and the rise of organized opposition to nuclear power in the United States emerged on parallel tracks, as postwar horror at the human consequences of the atomic bomb deepened and awareness of the environmental consequences of industrial development grew. These often intertwined social and political movements also displayed certain affinities with the civil rights movement (including use of litigation to promote political and societal change) and the movement against the Vietnam War. Both were fueled by increasing public distrust of government and the military-industrial complex. Environmental advocates and opponents of nuclear power, with the increasing support of the American public, eventually succeeded in checking the expansion of nuclear power and in imposing legal obligations on DOE to clean up the legacy of toxic and radioactive contamination left by AEC's ill-conceived and unsafe nuclear waste storage and disposal practices at its weapons sites. Concerns over nuclear proliferation and environmental contamination, combined with the dismal economic performance of commercial SNF reprocessing, also led to the end of civilian SNF reprocessing in the United States, which helped trigger a sense of crisis regarding nuclear waste disposal.

The modern environmental movement in the United States is a phenomenon of the past sixty years, though built on a firm foundation of American conservationism that stretches far back into the nineteenth century.[115] Modern environmentalism was sparked by compelling and influential accounts of the adverse health effects and ecological destruction caused by new technologies that were deployed on a wide scale during and after World War II. Concerned about unchecked industrial pollution and the environmental consequences of new industries and development projects, citizens around the country protested and organized to demand control of pollution and oppose a wide variety of new industrial facilities and development projects whose environmental impacts had not been understood or considered by the government bodies responsible for undertaking, funding, or licensing them.

The movement gathered remarkable political steam during the 1960s, culminating in passage by Congress of the landmark 1969 National Environmental Protection Act (NEPA). Newly emerging national environmental organizations, staffed with scientific experts and well-trained lawyers, worked to document and publicize environmental problems and propose the means to address them; working with Congress and the media, they helped stimulate and mobilize public concern over environmental problems and widespread public support for addressing them. These efforts led Congress during the 1970s to enact a set of powerful federal environmental laws, including the Clean Air Act, the Clean Water Act, the Toxic Substances Control Act, and laws dealing with hazardous wastes. The parallel rise of numerous, more grassroots-oriented environmental organizations throughout the country helped spur the adoption of environmental legislation at the state and local levels, and rallied citizens to take concerted action against environmentally damaging facilities and development projects.[116]

Opposition to nuclear weapons and nuclear power had originated in the aftermath of World War II. It began with protests against nuclear weapons production and testing by the government and later included opposition to civilian nuclear power plants on environmental, health, and safety grounds. The civilian deaths and suffering inflicted by the atomic bombings of Hiroshima and Nagasaki provoked feelings of intense horror and anguish in many Americans. Prominent scientists who had helped develop the atomic bomb through their work on the Manhattan Project publicly expressed remorse and regret for their role in the development of nuclear weapons, and a small but growing postwar antinuclear peace movement began actively to oppose the production and deployment of atomic weapons.

The U.S. government in 1945 initiated an intensive program of nuclear weapons testing that continued throughout the Cold War, lasting until 1992.[117] Much of the weapons testing took place at the Nevada Test Site (NTS), although tests were also conducted in New Mexico, Alaska, Colorado, Mississippi, and at the Pacific Proving Grounds in the Marshall Islands. From 1945 until 1962, the tests occurred at or above ground, causing radioactive fallout.[118] A 1954 weapons test in the Marshall Islands released very large amounts of radioactive fallout, forced evacuation of islanders, and caused severe health effects and long-term contamination of the area.[119] Government studies released decades later documented that "downwinders" in proximity to NTS and other 1950s test sites were at far greater risk than the general population of contracting thyroid and other cancers.[120] At the time of the testing, however, the government denied or downplayed any adverse impacts resulting from its nuclear weapons tests.

In a storied public relations campaign in 1953, AEC invited local residents to sit in the open air and observe the mushroom cloud from a test as it rose from NTS and afterward invited observers onto the test site itself.[121] Farmers whose sheep had died or were born with malformations after nuclear weapons testing sued the government, which denied any liability, in federal district court. Based on the testimony of government witnesses, the court in its 1956 decision in *Bulloch v. United States* determined that there was no evidence that the sheep had died of radiation.[122] In well-publicized congressional hearings held in 1979, decades later, the government's cover-ups were finally revealed.[123] After the hearings, the judge in *Bulloch* vacated his original decision and ordered the cases processed for retrial for "fraud practiced upon the court by representatives of the United States Government."[124] The government's dissemblings, denials, and apparent disregard of the public health and other consequences of its weapons testing program ultimately sowed widespread distrust of the government's nuclear programs, and later helped discredit the government's efforts to site one or more repositories and interim storage facilities for the most dangerous radioactive wastes in the nation.

In the early 1960s, however, public sentiment toward nuclear power was generally positive, and protests against new plants were localized.[125] Attitudes toward nuclear power changed more broadly in the 1970s, part of a sharp rise in public concern over national environmental quality generally. By the mid-1970s, substantial public concern had arisen about reactor safety and the absence of disposal sites for safe long-term disposition of SNF, together with growing skepticism over the government's competence to regulate and oversee reactor operations at nuclear power plants. Large, well-organized, and highly visible antinuclear protests in the 1970s at a number of planned and operating nuclear power plants, including the Diablo Canyon plant in California and Seabrook Station in New Hampshire, commanded widespread public attention. Throughout the 1970s, opponents of nuclear power, including environmental, consumer, and labor organizations as well as a number of states, launched significant legal and political initiatives, many based on nuclear waste issues, to halt the further development of nuclear power. Concerns about the already-high and ever-increasing costs of constructing nuclear power plants—and the resulting regulatory pass-through of these costs to ratepayers, as well as the economic risks to bondholders who financed the plants—also contributed significantly to halting construction of new nuclear plants.

The near-disaster at the Three Mile Island nuclear plant in 1979 posed by partial meltdown of a reactor core, and then the complete reactor core meltdown and release of large amounts of radiation from a reactor at the Chernobyl, Ukraine, power plant in 1986, cemented public opposition to further expansion of nuclear power.[126] These accidents were hugely important events in the antinuclear movement. Although they occurred after the most significant initiatives against nuclear power in the 1970s, they served to seal broad public opposition and, together with economic factors, led to the end of new nuclear plants in the United States. The Chernobyl accident also provoked public concern in the United States about the safety of DOE nuclear facilities, leading to a comprehensive DOE review of their safety and environmental security. This review in turn helped launch, in 1990, a huge, extraordinarily costly, and still ongoing DOE program to clean up its facilities pursuant to congressional legislation imposing waste remediation obligations on federal agencies. This cleanup effort will not be completed for decades more.[127]

The Surge of Environmental Legislation and Regulation

The National Environmental Policy Act

The National Environmental Policy Act of 1969 required the preparation by the responsible federal agency of an environmental impact statement (EIS) for any "major Federal actions" undertaken, funded, or granted regulatory approval by the federal government "significantly affecting the quality of the human environment."[128] In litigation brought by states, individual citizens, environmental and consumer organizations, and others, the courts gave flesh to NEPA's broad generalities, holding that an EIS must discuss all material environmental consequences of a proposed action and consider alternatives for mitigating them, including abandonment of the proposed action.[129] Further, the courts held that the responsible federal agency must prepare and make publicly available a draft EIS for public comments, and thereafter prepare a final EIS that discusses and responds in detail to the issues and contentions presented in comments.[130] The courts also held that the general remedy for an agency's failure to meet these requirements was to enjoin further action on a project until the failure was remedied.[131] These requirements were held applicable to AEC/NRC licensing of commercial construction and operation of power plants and nuclear waste storage and disposal actions, as well as to actions by AEC/ERDA/DOE to develop new nuclear facilities or undertake significant nuclear waste management or disposal initiatives.[132] Because a host of environmental issues could be invoked during the EIS process, this new legal regime created a powerful weapon for opponents of individual power plants or of nuclear power generally to slow down or stop the development of new nuclear facilities and other initiatives.

The Freedom of Information Act

Another vital new legal tool for the burgeoning environmental movement was the federal Freedom of Information Act (FOIA).[133] Enacted in 1966 and expanded and strengthened in 1974, FOIA empowered private citizens and advocacy groups to obtain, with limited exceptions (including those based on national security and personal privacy), all records in the government's possession without the requirement to show any particularized need for them. If the government unlawfully failed to produce a requested record, the requester could obtain a court order requiring its production.[134]

While FOIA applies to records of all kinds held by the government, it was widely used by a host of individuals, by national, state and local organizations, and by others concerned about nuclear weapons and nuclear power plants to obtain information about and expose problems with nuclear power plants, weapons production facilities, nuclear weapons testing, and nuclear waste management—and government failures to adequately address these problems. Use of FOIA also enabled organizations and individual citizens to obtain information needed to participate more effectively in government regulatory and policy-making proceedings; to support "impact" litigation designed to stimulate law and policy reforms; and to better mobilize public awareness of, and hence opposition to, unsafe or otherwise unjustified government practices, policies, and projects relating to nuclear facilities, wastes, and weapons systems.

For example, the Southwest Research Information Center in 1978 used FOIA to obtain the nonpublic draft EIS prepared by Sandia National Laboratory on the WIPP project. That information enabled the center to more effectively participate in the administrative proceedings relating to the project.[135] The Hanford Education Action League, the Environmental Policy Institute, and other groups used FOIA in 1986 to force DOE to release many thousands of pages of records documenting both planned and accidental releases of radioactive wastes from the Hanford facility that had contaminated the air, soil, and water.[136] The documents revealed Hanford's dismal safety record and allowed thousands of citizens exposed to radiation to file suit.[137] FOIA was also instrumental in exposing the impact of nuclear testing at NTS.[138]

Creation of the Environmental Protection Agency

Facing pressure from the environmental movement, President Richard Nixon determined to show environmental initiative by reorganizing the administration of federal environmental regulatory programs and assigning them to a single new agency.[139] In 1970, Nixon issued an executive order creating the Environmental Protection Agency and transferring environmental regulatory authorities exercised by other federal agencies to the new body.[140] The powers transferred to EPA included AEC's authority under AEA to adopt "generally applicable environmental standards" limiting radioactivity exposures from commercial and defense waste in order to protect human health and the environment.[141] The order conferred on EPA the authority to establish "limits on radiation exposures or levels, or concentrations or quantities of radioactive material, in the general environment outside the boundaries of locations under the control of persons possessing or using radioactive material."[142] The AEA regulatory powers retained by

AEC, however, gave it broad powers over commercial nuclear power plants and operations as well as its own operations, provoking a series of conflicts between the two agencies over their respective regulatory authority.[143]

When EPA proposed to set separate exposure and release limits for each of three types of nuclear facilities—nuclear fuel supply operations, light-water reactor plants, and fuel reprocessing plants—AEC protested, claiming that these were not "generally applicable" standards.[144] The dispute was referred to the Office of Management and Budget, which agreed with AEC. It advised EPA that it should set only generic standards for the total amount of radiation in the environment from nuclear installations. It could not establish specific radiation standards for the different types of facilities AEC was responsible for regulating, as that would infringe on AEC's regulatory responsibilities.[145] However, EPA could review proposed AEC standards (and AEC was advised to take EPA's comments into account). EPA acquiesced in this determination, and after working with AEC/NRC, in 1977 adopted a basic annual limit of twenty-five millirems of radiation exposure to the whole body and seventy-five millirems to the thyroid for persons in the general environment from all components of the uranium fuel cycle.[146] The annual radiation at the "fencepost" boundary of any nuclear facility could not exceed this level.[147] This standard is still in effect.[148] AEC (later NRC) was given responsibility for setting facility-specific requirements to ensure compliance with EPA's generic exposure standards and for enforcing compliance with EPA radiation standards.[149]

1974 Energy Reorganization Act: Division of AEC into NRC and ERDA/DOE

The 1974 Energy Reorganization Act (ERA) split AEC into two new agencies.[150] The Nuclear Regulatory Commission, an independent regulatory agency headed by five commissioners, assumed AEC's responsibility for environmental, health, and safety, and other regulation of civilian nuclear power and other private sector nuclear activities and the wastes they generated. The Energy Research and Development Administration (ERDA), headed by an administrator answerable to the president, took over AEC's responsibilities for weapons production and nuclear research and development, including new nuclear fuel cycle technologies. ERDA was responsible for ensuring the environmental, health, and safety of its facilities and operations, including the nuclear wastes they generated. NRC was given authority to regulate ERDA facilities for the receipt and storage of high-level radioactive wastes resulting from commercial reprocessing activities licensed by the commission.[151] In 1977, ERDA became the Department of Energy.[152] DOE took over ERDA's responsibilities for defense sites and wastes as well as its research and development activities in support of nuclear power. The earlier transfer to EPA of responsibility for setting general radiation standards for the environment was not affected by these reorganizations.

The 1974 ERA was enacted primarily in response to the widespread view that there was a serious and inherent conflict of interest in allowing one agency to exercise both the power to develop and promote nuclear technologies, and the power to regulate their use so as to ensure environmental, health, and safety protection.[153] AEC, which had struggled to reconcile the various and sometimes conflicting roles and goals that it had been assigned, came under intense scrutiny in the 1970s as public concerns over radioactive releases from nuclear power plants mounted.[154] The adequacy of AEC's safety regulations for nuclear power plants was the subject of sharp controversy.[155] ERA expressed and implemented Congress's conclusion that it was "in the public interest that the licensing and related regulatory functions of the Atomic Energy Commission be separated from the performance of the other functions of the Commission."[156]

The net effect of the creation of the 1970 EPA and enactment of the ERA was to fragment regulation of nuclear facilities and wastes among three agencies—NRC, ERDA (later, DOE), and EPA—thereby creating a regulatory divide between management, storage, and disposal of civilian radioactive wastes on the one hand, and of defense nuclear wastes on the other. Further complicating matters was the enactment of separate federal statutes for cleanup and regulation of chemically hazardous and radioactive mixed waste at both civilian and DOE sites, as described next.

RCRA and CERCLA

In response to dramatic and widely publicized examples of soil, water, and air pollution from chemically toxic wastes that had been dumped or had leaked into the ground, Congress enacted two far-reaching statutes to clean up the accumulated hazardous waste problems created by decades of neglect and to prevent their recurrence. These statutes were the 1976 Resource Conservation and Recovery Act (RCRA) and the 1980 Comprehensive Environmental Response, Compensation, and Liability Act (CERCLA), popularly known as Superfund.[157] These statutes, and subsequent legislation extending them to federal government agencies and facilities, provided the foundation for the massive and still ongoing DOE program to clean up the vast quantities of radioactive and chemically hazardous wastes at its facilities.

RCRA established a comprehensive system of hazardous waste regulation to protect human health and the environment.[158] The stated purposes of RCRA are to "(1) minimize the generation of hazardous waste by encouraging process substitution and recycling, and (2) comprehensively regulate the generation, transportation, treatment, storage, and disposal of hazardous waste."[159] RCRA accomplishes these goals through a comprehensive and detailed strategy of cradle-to-grave waste management. Hazardous waste generators and transporters and facilities for treatment, storage, and disposal of hazardous waste are subject to elaborate systems of permits and requirements to track such waste and ensure that it is safely managed, stored, transported, and disposed of. Amendments to RCRA enacted in 1984 established corrective action (cleanup) requirements, imposed through RCRA permits, for hazardous waste facilities that are subject to RCRA regulation.[160] EPA is responsible for implementation and enforcement of the RCRA program; EPA can, however, delegate its regulatory and enforcement authority for various parts of or for the entire RCRA program to states that apply for such authorization and can demonstrate that they have substantially equivalent hazardous waste management programs.[161]

In 1980, Congress added a second legal weapon to deal with the nation's legacy of toxic waste pollution by enacting CERCLA, which imposes a far-reaching system of strict joint and several retroactive and ongoing liability for the costs of cleanup on owners and operators of sites at which a release or threatened release of a hazardous substance has taken place, as well as on transporters and generators of such wastes. EPA itself was authorized to undertake cleanup of abandoned toxic dump sites on a national priorities list; to recover its cleanup costs from responsible parties; and to use court injunctions and administrative orders to force responsible parties to undertake cleanup of contaminated sites.

These statutes raised two sets of questions regarding DOE wastes. First, were the laws legally binding on DOE, which, as an agency of the federal government, enjoys sovereign immunity against suit, unless waived by Congress? And if so, what legal remedies or other arrangements might be available to force DOE to comply with such obligations? These questions were answered by Congress's enactment of the 1986 Superfund amendments and Reauthorization Act (SARA) and the 1992 Federal Facility Compliance Act (FFCA), which extended the requirements of CERCLA and RCRA, respectively, to DOE and other federal government facilities and provided mechanisms to achieve compliance with these obligations by federal agencies.

Second, did the new requirements applied to mixed wastes, namely, wastes that are both chemically toxic and radioactive and are therefore already subject to regulation under AEA? The notion that CERCLA cleanup applied to mixed wastes was accepted without strenuous controversy. But ERDA/DOE strongly and repeatedly maintained that RCRA did not apply to mixed wastes. Pointing out that the radioactive and chemically toxic components of mixed wastes often cannot feasibly be separated, ERDA/DOE emphasized the potential for conflicting regulation of such wastes under RCRA and the AEA. (As described in Chapter 2, DOE eventually lost this battle as a result of court decisions establishing that RCRA applies to the chemically hazardous portion of mixed wastes.)

SARA and the Federal Facility Compliance Act

The SARA provisions dealing with federal facility cleanup waive sovereign immunity and extend CERCLA obligations and liabilities to federal facilities.[162] As required by SARA, EPA has established a Hazardous Waste Compliance Docket for DOE and other federal facilities, which collects information on contamination by and releases of hazardous chemicals at the facilities. EPA regularly conducts assessments of the contamination problems at federal facilities and determines which facilities should be placed on the National Priorities List (NPL). While EPA does not have authority to sue another federal agency in court because of the unitary executive theory, SARA requires federal agencies in charge of facilities listed on the NPL to conclude an interagency agreement (sometimes also referred to as a federal facilities agreement), with EPA specifying cleanup actions and a schedule for achieving them.[163] The public has access to information and opportunity for participation in the development of the cleanup program. Once an agreement has been signed, EPA monitors the cleanup schedule and milestones and oversees its requirements to ensure proper implementation of each cleanup. The agreements typically provide for EPA assessment of stipulated penalties against the responsible federal agency for noncompliance with certain terms of an agreement, including binding milestones for cleanup progress.[164] The responsible agency must make annual reports to Congress on its progress. DOE currently has nineteen facilities on the NPL, all covered by interagency agreements specifying cleanup actions, schedules, and milestones.[165]

In 1992 Congress, impatient with lack of cleanup progress, adopted FFCA, which amended RCRA to further waive the sovereign immunity of federal agencies, including DOE, making them subject to federal, state, and local hazardous waste regulatory laws, including laws requiring cleanup; and also making them subject to suits by states or private litigants, including individuals, local community groups, and environmental organizations, to enforce compliance with those laws.[166] As a result of this statute, DOE became subject to waste cleanup obligations at its 114 facilities for the large amounts of mixed waste—including HLW, TRU, and LLW—that had been generated in the course of nuclear weapons production and other activities throughout the entire Cold War period.

Adoption of FFCA was prompted by widespread dissatisfaction with the failure of DOE and the Department of Defense to clean up their hazardous waste legacies, coupled with continuing legal disputes and uncertainties regarding the federal government's sovereign immunity to waste cleanup litigation and liability.[167] In the House floor debates, Senator David Durenberger expressed a widely held view: "I sense that most of the departments and agencies are now eager to get on with the job of cleanup. The one holdout is [DOE]. They have historically been the worst polluter and it is clear that they will be the last to change their attitude about public health and safety."[168] Representative Charles Luken similarly asserted, "DOE has absolutely failed in every respect to do anything about meaningful cleanup."[169] Members also decried the operation of a legal double standard, under which private and state facilities were subject to cleanup obligations, but federal facilities were not.[170] Senator Howard Metzenbaum asserted that FFCA was necessary in order to "tell DOE once and for all that it is not above Federal law. It cannot be allowed to ignore the harmful pollution problems it helped create."[171] Further, there had been confusion in the courts over whether and to what extent RCRA had waived federal sovereign immunity.[172] By enacting the FFCA, Congress sought to explicitly waive sovereign immunity for federal facilities and subject them to regulatory requirements for waste management and cleanup obligations under RCRA, and also under state and local hazardous waste laws.[173]

In order to accomplish these objectives, Section 103 of FFCA amends the RCRA definition of a "person" subject to RCRA to include "each department, agency and instrumentality of the United States."[174] Section 102 provides:

> Each department, agency, and instrumentality of the executive, legislative, and judicial branches of the Federal Government (1) having jurisdiction over any solid waste management facility or disposal site . . . shall be subject to, and comply with, all Federal, State, interstate, and local requirements, both substantive and procedural (including any requirement for permits or reporting or any provisions for injunctive relief and such sanctions as may be imposed by a court to enforce such relief), respecting control and abatement of solid waste or hazardous waste disposal and management in the same manner, and to the same extent, as any person is subject to such requirements, including the payment of reasonable service charges. The Federal, State, interstate, and local substantive and procedural requirements referred to in this subsection include, but are not limited to, all administrative orders and all civil and administrative penalties and fines, regardless of whether such penalties or fines are punitive or coercive in nature or are imposed for isolated, intermittent, or continuing violations. The United States hereby expressly waives any immunity otherwise applicable to the United States with respect to any such substantive or procedural requirement (including, but not limited to, any injunctive relief, administrative order or civil or administrative penalty or fine referred to in the preceding sentence, or reasonable service charge).[175]

As a result of this provision, DOE became subject not only to RCRA but also to state and local hazardous waste requirements, even if those requirements are more stringent than federal requirements under RCRA. It is subject to injunctive relief for noncompliance and also to civil and administrative fines and penalties established by state and local law. In addition to enforcement by state authorities, private individuals and environmental groups who can establish a risk of harm that satisfies constitutional requirements for standing to sue can use the citizen suit provisions of RCRA Section 7002(a) to enforce RCRA requirements against federal facilities.[176]

Section 105 of FFCA amended RCRA to require DOE to create and submit "a plan for developing treatment capacities and technologies to treat all of the facility's mixed wastes" for all facilities where DOE generates or stores mixed wastes; and to submit the plan for approval to EPA or state regulatory agencies exercising delegated RCRA authority.[177] It also required DOE to submit inventory reports on all mixed waste; plans for the development of treatment capacities and technologies; and schedules and progress reports. When a cleanup plan is approved, EPA, the responsible state agency, and DOE enter into an agreement providing for compliance. These RCRA agreements, developed with opportunity for public participation, are known as federal facility compliance agreements but may have other names. They often include cleanup undertakings pursuant to CERCLA as well as RCRA, in which case they may be known as tri-party agreements. Pursuant to RCRA, state agencies can enforce provisions of these several different types of agreements through litigation against DOE;[178] as previously noted, EPA cannot.[179] Many of the agreements also include dispute settlement procedures and stipulated damages for DOE's breach of their terms.[180]

DOE's Cleanup Initiatives

The Chernobyl disaster in 1986 and subsequent investigations of nuclear installations in the United States focused public and political attention on the serious contamination and environmental hazards created by the accumulated wastes and poor waste management conditions at the nation's nuclear weapons complex. Media interest skyrocketed. In a little over three months at the end of 1988, the *New York Times* alone carried 108 stories on the problems at the DOE's facilities, 37 of which were front-page stories.[181] Numerous congressional hearings were held. Retired Admiral James Watkins, who became secretary of energy in 1989, was determined to clean up the legacy wastes at DOE facilities. His resolve was heightened by the widely publicized raid by FBI agents and EPA officials at DOE's Rocky Flats facility to investigate and address suspected environmental violations. Watkins recruited a management team dedicated to cleanup and did much to install a new culture of environmental performance at DOE. He established a DOE Office of Environmental Management (EM) to coordinate and prioritize the nuclear- and non-nuclear-related cleanup at sites and facilities across the nation that had previously been carried out through separate offices within DOE. In 1989, DOE issued its first comprehensive cleanup plan.[182] EM was charged with cleaning up 134 sites associated with the mining and processing of nuclear weapons–related materials or the research, development, testing, and production of nuclear weapons.[183] The extent of contamination, the means for remediating it, and the costs of doing so were largely unknown. The scope and complexity of the cleanup problem presented an unprecedented challenge (see Map 1.1).

SARA and FFCA created very powerful legal drivers for waste cleanup by DOE.

Map 1.1. DOE environmental management sites

Sources: Office of Environmental Management, DOE, *www.em.doe.gov/Pages/SitesLocations.aspx?PAGEID=MAIN*; Office of Legacy Management, DOE, *www.lm.doe.gov/Sites_Map.aspx*; Mayer et al., *Challenges of Nuclear Weapons Waste* (2010).

The latter act was especially important because it empowered states hosting DOE facilities to sue or threaten to sue DOE under RCRA; this effectively forced DOE to sign agreements with cleanup commitments enforceable through court actions and penalties. The agreements, often including CERCLA cleanup obligations, set out in varying detail remediation goals and timetables that DOE agreed to follow. In retrospect, it is evident that many of these agreements were signed hurriedly under political pressure before sites were adequately characterized and without full appreciation of the complexities of remediation. Political expediency—and not the magnitude and nature of risk—was often the basis for the terms of agreement.[184]

There are currently thirty-seven legally enforceable cleanup agreements covering twelve DOE sites.[185] The six major DOE sites subject to such agreements are Hanford, SRS, INL, Rocky Flats, Oak Ridge, and the Fernald site in Ohio; these sites have received approximately two-thirds of DOE's cleanup funding.[186] Besides establishing what cleanup activities will be undertaken and setting milestones for achieving them, the agreements establish methods for resolving disputes arising out of the agreements and guarantee facility access for state and federal inspectors.

In addition to obligations established by these agreements, DOE facilities are subject to RCRA corrective action requirements in RCRA permits, consent orders, compliance orders, and site treatment plans; requirements established by CERCLA records of decision; obligations under miscellaneous other laws and permits, including those imposed under the federal Clean Air Act, Clean Water Act, and Toxic Substances Control Act; and obligations under state and local laws.[187] Independent oversight of DOE's cleanup activities is provided by the Defense Nuclear Facilities Safety Board, an independent federal agency created by Congress in 1988 specifically to oversee DOE's nuclear weapons activities. EPA and state agencies also provide oversight through their implementation of RCRA and CERCLA.[188]

DOE and the other parties to these agreements periodically renegotiate their terms and those of other legal instruments requiring cleanup, including orders, permits, and decrees, in order to deal with new information or developments, including problems encountered in meeting milestones. Some milestones are legally enforceable, while a larger number of other milestones are "planned," "rolling," or "target" milestones that typically become enforceable milestones at a later date. Large DOE facilities are subject to many milestones, each covering different components of cleanup.[189] DOE also maintains a list of "enforceable milestones at risk," obligations that it may not be able to fulfill by the date currently set.[190] Table 1.1 summarizes the major sources of cleanup obligations and the number of enforceable milestones for major DOE facilities.

The system of milestones has also been criticized: while DOE has met many of the milestones, these often concern administrative steps and are not necessarily indicative of actual cleanup progress.[191] There have also been complaints that state regulators are too tolerant of missed deadlines, fail to take enforcement actions, and are too ready to agree to amendments.[192] Others see the root of the problem in the piecemeal character of the governing environmental regulatory system, the result of a multiplicity of statutes, regulations, and decisions adopted by various jurisdictions and agencies.[193] They criticize the array of cleanup obligations at various sites under various laws as preventing DOE from adopting a coordinated national approach to cleanup that would allocate resources in accordance with overall risk management priorities. DOE's efforts to develop such a strategy have not proven successful, in part because the management of sites is highly decentralized, generally accomplished through a series of agreements with contractors responsible for doing the cleanup work at each site.

Overview of DOE Waste Cleanup Progress

In 1997, EM published *Linking Legacies: Connecting the Cold War Nuclear Weapons Production Processes to Their Environmental Consequences*, a comprehensive report on the nation's accumulated 36 million cubic meters of nuclear weapons wastes. A 2009 DOE report to Congress, *Status of Environmental Management Initiatives to Accelerate the Reduction of Environmental Risks and Challenges Posed by the Legacy of the Cold War*, updated with revised data in 2010, provides a review of progress made with cleanup since 1997.[194]

From its inception in 1989, EM has been responsible for a massive legacy of radioactive wastes, many of which had been mismanaged. Large amounts of HLW and lesser but still significant amounts of TRU and SNF remain at DOE sites, along with very large quantities of LLW. EM has made significant progress in securing and reducing the risks posed by these materials, including removing high-risk waste from inadequate storage conditions and isolating it pending treatment and ultimate disposal. Still, final disposition of much of DOE's legacy HLW requires construction of facilities for treating it,

Table 1.1. Agreements, consent orders, and enforceable milestones at Office of Environmental Management sites

Site	Agreements	Enforceable Milestones
Brookhaven National Laboratory	Federal Facility Agreement, 1992	1
Energy Technology Engineering Center	FFC Act Site Treatment Plan/Compliance Order, 1995 RCRA Consent Order, 2007	14
Idaho	Federal Facility Agreement, 1991 RCRA Consent Orders, 1992, 1999, 2000 and 2001 Batt Settlement Agreement, 1995 FFC Act Site Treatment Plan/Consent Order, 1995	21
Los Alamos National Laboratory	RCRA Consent Agreement, 1993 FFC Act Site Treatment Plan/Compliance Order, 1995 Federal Facility Compliance Agreement, 2005 RCRA Consent Order, 2005	42
Nevada Test Site	RCRA Consent/Settlement Agreements, 1992, 1994 FFC Act Site Treatment Plan/Consent Order, 1996 Federal Facility Agreement, 1996	25
Oak Ridge	Federal Facility Agreement, 1992 FFC Act Site Treatment Plan/Compliance Order, 1995 TSCA Compliance Agreement, 1996	65
Paducah	Federal Facility Agreement on Toxic Substances Control Act (TSCA), 1992 FFC Act Site Treatment Plan/Compliance Order, 1997 Federal Facility Agreement, 1998	10
Portsmouth	RCRA/TSCA Consent Decree, 1989 FFC Act Site Treatment Plan/Compliance Agreement, 1992, 1995 RCRA/CERCLA Consent Order, 1997 RCRA Compliance Order, 1998	7
Richland and River Protection (Hanford)	Federal Facility Agreement (Tri-Party Agreement), 1989 Clean Air Act Compliance Agreement, 1994 TSCA Compliance Agreement, 1996 RCRA Consent Decree on Tank Interim Stabilization, 2000 RCRA Consent Order on Tank Integrity, 2000 Settlement Agreement on National Environmental Policy Act and Waste Disposal, 2006	116
Sandia National Laboratories	RCRA Consent Order, 2004	2
Savannah River	RCRA Consent Orders, 1985, 1999 (2) RCRA Settlement Agreements, 1987 (2), 1988, 1989, 1991 Federal Facility Agreement, 1993 FFC Act Site Treatment Plan/Consent Order, 1995 Clean Water Act Consent Order, 1999	29
Stanford Linear Accelerator Center	Regional Water Quality Control Board/Compliance Order, 2005	13

Source: Office of Environmental Management, DOE, *Status of Environmental Management Initiatives*, 69–70 tbl.2.2 (2009).

and disposal of the treated waste as well as of the SNF at DOE sites remains contingent on the construction of a geologic repository.[195] Until 1970, AEC had buried TRU and LLW together in shallow trenches in the soil at generation-sites, discharged contaminated water into soils or surrounding bodies of water, and stored wastes in tanks operating beyond their expected performance lifetimes.[196] Many wastes that previously had been thought permanently disposed of have been found not to meet contemporary waste management and disposal standards and to require further remediation and proper disposal.[197]

High-Level Waste

In 1997, EM confronted the need to deal with 100 million gallons of HLW in underground storage tanks at SRS, INL, West Valley, and Hanford.[198] At that time, much of that waste was in danger of leaking or breaking through the tanks. At Hanford, DOE has had to drain nearly three million gallons of reprocessing wastes from single-hulled tanks, many of which were operating beyond their designed lifetimes, and transfer them to new double-hulled tanks to await vitrification and ultimate disposal.[199] The West Valley HLW has already been vitrified, and other means to treat and solidify reprocessing HLW for ultimate disposal are in progress at SRS and INL. DOE has also emptied and sealed two older HLW tanks at SRS and eleven HLW tanks at INL.[200] Currently, EM manages eighty-eight million gallons of liquid HLW; the cost of monitoring and maintaining the tanks exceeds $500 million per year.[201] The vitrified HLW at West Valley is being stored on-site awaiting development of and shipment to a repository for its disposal.[202]

EM's strategy for dealing with certain reprocessing and other wastes in tanks is to separate out the high-activity and low-activity fractions, vitrify or otherwise solidify the high-activity fractions in preparation for geologic disposal, and treat and manage the low-activity fractions for disposal as LLW. If such treatment eliminates its chemically hazardous constituents, the low-activity waste could be disposed of on-site as LLW, rather than having to be disposed of as mixed LLW in RCRA landfills. At SRS, vitrification plant test runs have produced vitrified wastes that contain no detectable levels of chemically hazardous constituents.[203] The treated high-activity fractions must be stored until a geologic repository is constructed.[204] At SRS, the high-activity fraction is being processed by SRS's existing Defense Waste Processing Facility, while several intermediate facilities for processing the low-activity fraction are already complete.[205] A salt waste–processing facility to deal with the low-activity fraction of SRS reprocessing wastes is under construction and scheduled to begin operations in 2014 at an estimated cost of $1.3 billion. At Hanford, an enormous waste treatment and immobilization plant vitrification facility is being constructed to handle both the high- and low-activity fractions from tank wastes. Currently, its cost is projected at $12.3 billion; it is scheduled to come online in 2019.[206] A sodium-bearing waste facility is currently under construction at INL to treat the remaining 0.9 million gallons of liquid HLW at the site. The facility is expected to cost $551 million and begin operations in 2011.[207]

EM is concerned that the amount of HLW after it is treated for repository disposal, combined with the amount of SNF under DOE's authority, will exceed DOE's NWPA allotment of 10 percent of the storage space, or seven thousand metric tons of heavy metal (MTHM), in the initial geologic repository, currently Yucca. It is also concerned that delay in opening a repository will prevent its removing HLW from DOE facilities in accordance with facility cleanup agreement deadlines. Under either of these scenarios, EM would have to construct additional storage facilities to store excess waste.[208]

Spent Nuclear Fuel

EM manages 2,400 MTHM of SNF at various DOE sites. In 1992 nearly all SNF managed by DOE was stored in wet pools. Since then EM has moved nearly all SNF from pool storage to dry storage, which reduces the risk of the waste contaminating groundwater and nearby surface water.[209] EM is evaluating options for treatment and disposal of the radioactive sludge and debris that has collected at the bottom of the SNF pools at Hanford. While awaiting the opening of a geologic repository to dispose of the SNF, EM is also considering reprocessing SNF at SRS to remove the uranium, which could be converted into commercial nuclear fuel to save space at the geologic repository. Otherwise, the dry casks will be moved to an interim storage facility on-site to await the opening of the geologic repository. An additional concern is DOE's commitment to remove SNF from Idaho by 2035.[210]

Transuranic Waste

EM is responsible for 150,000 cubic meters of contact-handled TRU, which can be handled by workers, and 7,000 cubic meters of remote-handled TRU, which must be enclosed in lead-shielded containers. WIPP, which began receiving shipments in 1999, will be the final disposal point for all TRU under DOE management. Thus far, 36 percent of DOE's total inventory of TRU waste has been disposed of at WIPP. The TRU wastes that have not yet been disposed of are spread among fourteen DOE sites.[211] EM is working on making the shipment of wastes to WIPP more efficient and designing improved containers for remote-handled TRU. DOE expects to continue generating reduced levels of TRU through 2050; EM is studying how to efficiently dispose of these wastes after most TRU has been shipped to WIPP.[212] There is also a small amount of civilian TRU at West Valley awaiting a disposal pathway; current law precludes disposal of any civilian waste at WIPP.

Low-Level Waste

All told, DOE is currently managing or has already disposed of more than 10 million cubic meters of LLW. About 1.4 million cubic meters of LLW was generated by defense activities during the Cold War. These wastes are being disposed of on-site at INL, SRS, Oak Ridge, Los Alamos, Hanford, and the Nevada Test Site and at commercial facilities. EM has disposed of about 75 percent of this waste. DOE site cleanups have produced a further 9 million cubic meters of LLW waste, mostly resulting from soil remediation, that have already been disposed of at Hanford, INL, the Nevada Test Site, and Oak Ridge. EM anticipates that further cleanup at Hanford, INL, Oak Ridge, SRS, Paducah, and West Valley will yield at least 400,000 additional cubic meters of LLW requiring disposal.[213]

Surplus Special Nuclear Material (SSNM)

EM is responsible for 13 MT (metric tons) of plutonium, which was produced for use in nuclear weapons but never fabricated into weapons and which is currently in excess of defense needs. EM is in the process of consolidating SSNM at SRS. All SSNM from Rocky Flats has been shipped to SRS, and shipments have begun from INL and

Hanford.[214] EM plans to process the 13 MT of plutonium at a mixed oxide fuel fabrication facility at SRS that is currently under construction in order to produce mixed oxide reactor fuel. The role played by MOX and SNF in the Fukushima Daiichi crisis may cause this plan to be scuttled.

Facility Cleanup and Closure

EM is responsible for the cleanup of 107 sites. It has completed closure of 12 and cleanup at 86.[215] The sites range in size from small, privately owned areas where DOE activities left contamination, to entire DOE sites with no future mission. Significant progress has been made in cleaning up the smaller DOE sites, but because of tank waste cleanup at the three largest DOE sites—Hanford, SRS, and INL—the smaller sites have recently fallen on EM's priority list.[216] Three major facilities—Rocky Flats, Mound, and Fernald—have been completely remediated.[217] At Hanford, SRS, and INL, significant challenges remain. For instance, at Hanford, 450 billion gallons of radioactive water were discharged into the ground, resulting in eighty square miles of contaminated groundwater that require remediation. Though EM expects to finish groundwater cleanup sooner than originally anticipated, seventy-five square miles of the original complex's land will still require remediation.[218] (Appendix C describes in greater detail DOE's cleanup efforts at Hanford.) Cleanup at SRS has been targeted on one of fourteen major cleanup areas; the area was completed two years ahead of schedule and for $37 million less than originally budgeted.[219] Significant progress has been made at INL, where cleanup of four reactors has been completed and cleanup of another is expected to be complete by the end of 2010.

The Costs of Cleaning Up the Cold War Defense Legacy

DOE's cleanup of federal facilities is a vast enterprise that has made progress, but that progress has come at a very high cost. In its 2009 report on its cleanup program, EM reported that it has spent $96 billion to date.[220] The cleanup program still has a long way to go; DOE projected that it will not be complete until 2050.[221] DOE estimated the total life-cycle costs of cleanup as between $273 and $327 billion.[222] The Consortium for Risk Evaluation with Stakeholder Participation (CRESP), after a review of EM's report, issued a report in 2010 concluding that cleanup will not be complete until 2070 and that the total life-cycle cost will exceed $560 billion.[223]

The End of Spent Nuclear Fuel Reprocessing

As previously noted, from the mid-1950s to the mid-1970s AEC proceeded on the assumption that power plant SNF would be reprocessed in commercial facilities.[224] In fact, however, only a small amount of such reprocessing took place from 1966 to 1972, and that at only one site: West Valley, New York. Since that time, there has been no reprocessing of civilian SNF in the United States.

AEC's initial push for reprocessing was based on the view that a uranium shortage was imminent and that reprocessing was essential to provide fuel for the nuclear power industry.[225] AEC announced in 1956 that it would provide private firms with the technology to construct reprocessing plants, building on its successful experience with reprocessing at its weapons production facilities.[226] It also promised to provide firms with a guaranteed supply of defense SNF until the industry could financially self-sustain.[227] At least two-thirds of the SNF reprocessed at West Valley was supplied by AEC.[228] In addition, AEC signed generous contracts with Nuclear Fuel Services to reprocess AEC SNF, with advance payments in order to help the firm raise the large amounts of capital needed to build the plant.[229] Additional federal support came in the form of exemptions from certain AEC regulatory requirements, including a requirement that reprocessing facilities solidify final waste products.[230] These various government inducements stimulated development of three private reprocessing plants.

Midwest Fuel Recovery Plant

General Electric undertook to build the Midwest Fuel Recovery Plant at Morris, Illinois, adjacent to the site of the Commonwealth Edison Company Dresden reactors. Capable of reprocessing three hundred tons of SNF per year, it was completed in 1974 at a cost of $64 million.[231] Equipment failures and technical problems prevented the plant from ever achieving full-scale operation. Its longest period of operation was twenty-six hours. The facility was declared inoperable and the project was terminated in 1974.[232]

West Valley Plant

Nuclear Fuel Services began developing a reprocessing plant at West Valley, New York, in 1962 with significant support from the state and AEC. Governor Nelson Rockefeller had created the New York State Atomic and Space Development Authority to fund the promotion and development of new nuclear facilities. Nuclear Fuel Services leased the plant site from the Development Authority.[233] When breaking ground at West Valley, Rockefeller called the plant the "focal point for an atomic industry."[234] The West Valley facility (acquired in 1969 by Getty Oil Company) began operation in 1966. It used a plutonium uranium extraction technology capable of reprocessing three hundred tons of SNF per year; it operated from 1966 to 1972, generating approximately 600,000 gallons of liquid HLW. The facility encountered significant operational and environmental regulatory problems and in 1972 suspended operations, with a plan to reopen after plant reconstruction.

The plant, however, never reopened. One reason was that newly adopted AEC environmental regulations required substantial modifications to the plant, including adding the capacity to solidify reprocessing wastes. Getty estimated that the cost of meeting this and other requirements would be $600 million.[235] Exercising rights Nuclear Fuel Services had negotiated in its lease, Getty turned the plant over to the development authority, which was left with both SNF and reprocessing wastes and a highly contaminated site that would require a major and extremely expensive cleanup effort.

In 1980, Congress enacted legislation governing disposition of the HLW, SNF, and other nuclear waste at West Valley; DOE was assigned the task of cleaning up the site and treating and disposing of the waste that had been generated. The West Valley Demonstration Project Act specified that the State of New York would pay 10 percent of the cleanup project costs and the federal government would provide the remaining 90 percent.[236] The West Valley Nuclear Services Company has operated the West Valley site for DOE since 1982. Processing of nearly 600,000 gallons of highly radioactive liquid waste was finally completed in 2002. The waste was vitrified into 275 ten-foot-tall cylinders of hardened glass sealed in ten-foot-tall steel canisters.[237] With the cancellation of the Yucca repository, this waste lacks a disposition pathway and remains stored on-site.

LLW was formerly buried at the site; it is now being shipped to the EnergySolutions facility in Utah and to NTS for disposal.[238]

Barnwell Plant

In 1970, Allied General Nuclear Services began construction of a reprocessing plant capable of processing 1,500 tons a year of SNF at Barnwell, South Carolina, adjacent to DOE's SRS site, on federal land donated by the federal government to Allied as part of AEC's efforts to support reprocessing.[239] The plant was opposed by environmental groups but had popular support among the residents of nearby towns. A press account reported that "residents of Barnwell County, who have lived with the federal Savannah River Plant in their backyard since the 1950s, have grown used to such warnings, and appear to believe the economic advantages are worth the physical risks."[240] The plant was due to begin operation in 1974, but after delays in construction and licensing, it still had not been completed or licensed when President Carter ended federal support for reprocessing in 1977.[241]

Halting Federal Support for Reprocessing

In October 1976, President Ford issued a comprehensive statement on nuclear policy that included a temporary moratorium on federal support for SNF reprocessing, motivated in large part by concerns about nuclear weapons proliferation associated with plutonium produced by reprocessing SNF.[242] Part of the background for Ford's action was that in 1974 India had shocked the world and embarrassed the United States by successfully testing a nuclear weapon made with plutonium produced by reprocessing SNF from a reactor bought from Canada that used uranium. To make matters worse, the reprocessing facility was built using technical knowledge gained through training provided by the United States.[243] Ford's speech announcing the move made no mention of environmental concerns; the focus was on nuclear weapons proliferation: "The reprocessing and recycling of plutonium should not proceed unless there is sound reason to conclude that the world community can effectively overcome the associated risks of proliferation . . . ; the United States should no longer regard reprocessing of used nuclear fuel to produce plutonium as a necessary and inevitable step in the nuclear fuel cycle, and . . . we should pursue reprocessing and recycling in the future only if they are found to be consistent with our international objectives."[244] President Ford directed federal agencies to delay further support for commercialization of SNF reprocessing until uncertainties were resolved. Ford's statement also called for construction of "the first demonstration repository for high-level wastes" to be licensed "by the independent NRC to assure its safety and acceptability to the public" and completed by 1985.[245]

Following President Ford's lead, President Carter announced on April 7, 1977: "We will defer indefinitely the commercial reprocessing and recycling of plutonium produced in the U.S. nuclear power programs." Further, "the plant at Barnwell, South Carolina, will receive neither federal encouragement nor funding for its completion as a reprocessing facility."[246] Subsequently, Carter vetoed S. 1811, the ERDA Authorization Act of 1978, thereby halting government support for reprocessing.[247] The removal of federal support proved to be the coup de grace for reprocessing. Allied General Nuclear Services eventually terminated the Barnwell project in 1981, citing its belief that reprocessing was commercially impractical.[248] Unsuccessful efforts, led by then–secretary of energy James B. Edwards, to resuscitate Barnwell continued into the 1980s, though they never generated interest from industry.[249]

The demise of reprocessing dramatically changed the expectations of the technical community about the types of nuclear power wastes that would need to be managed, and left no clear path for disposal of the large amounts of SNF that were beginning to accumulate at power plants around the country.[250] Champions of reprocessing asserted that it would reduce by a factor of four the overall volume of SNF requiring disposal.[251] But it would also produce HLW and TRU, as well as other wastes, including iodine-129, krypton-85 gas, incinerated solvents, and contaminated equipment, which would present significant technical challenges for secure disposal.[252] If cesium and strontium were separated out from the reprocessing HLW waste streams and stored for several centuries for cooling and decay, they could be buried as LLW, but this prospect would pose the formidable and untested challenge of assuring safe and secure storage for several centuries.

DOE responded to the end of SNF reprocessing by hastily announcing a new search for six potential repository sites to dispose of commercial SNF. The Los Medanos site in New Mexico, as a defense installation, was initially excluded from consideration as a possible site for disposal of these civilian wastes.[253]

The Demise of the Clinch River Breeder Reactor

In the 1960s, after commercializing light-water electric power reactors, AEC began to focus on developing a successful commercial-scale electric power breeder reactor program. AEC's concern for potential uranium supply shortages led it to push development of the breeder reactor, which uses a fuel mix and fission regime similar to that used to produce plutonium for weapons and is designed to produce more fuel, in the form of plutonium obtained through reprocessing, than it consumes.[254] AEC settled on the liquid metal fast breeder design in preference to alternatives, including the molten salt breeder, light-water breeder, and gas-cooled fast breeder.[255]

Congress authorized the Clinch River Breeder Reactor (CRBR) project in 1970.[256] This was a demonstration project designed to test whether breeder reactors were commercially feasible, in hopes of encouraging future private and public investment.[257] AEC/ERDA/DOE had first developed the Fast Flux Test Facility, designed to test fuels and materials but incapable of generating electricity.[258] It then developed, in partnership with the private sector, the 375-megawatt electrical CRBR Demonstration Project near Oak Ridge, Tennessee. The CRBR was supposed to be followed by a larger, commercial-scale breeder reactor facility in the 1990s.[259]

The liquid metal fast breeder program was plagued with problems. Breeder reactors were much more costly to build and operate than the established light-water reactors.[260] Like reprocessing of SNF from light-water reactors, the breeder reactor made economic sense only in the event of sharply higher uranium prices. While uranium prices in the late 1970s had shot up, by the 1980s, they were falling swiftly rather than continuing to rise as predicted.[261] Beyond these problems, the fast breeder program did not have the support of the utilities. They believed that even if a large demonstration plant could be built, it would be built on government land, would be run by government staff, would operate without an NRC license, and accordingly would not provide a realistic or useful demonstration of the breeder technology for civilian applications.[262]

The Carter administration halted funding for the CRBR project in 1977, calling it "technically obsolete and economically unsound" and also invoking proliferation concerns.[263] But Congress continued to fund the project.[264] The Reagan administration revived it. The Natural Resources Defense Council (NRDC) filed suit, challenging DOE's actions in beginning CRBR site development without first issuing a final supplemental EIS. In September 1982, the district court issued a preliminary injunction, temporarily halting the project. The government then appealed to the Eleventh Circuit, which granted expedited review, dismissed the action on the grounds that there was a legally binding agreement between DOE and EPA permitting site preparation prior to issuance of the final supplemental EIS, and dissolved the injunction.[265] When construction recommenced in 1983, however, the project proved a fiasco. There were long delays. Construction costs mushroomed.[266] Much of the money for the project was consumed in remedying safety failures or swallowed up by fraud (including overcharges due to fake bids submitted by purchasing officials) and misappropriation of resources.[267]

Congress finally voted to completely end funding for the project in late 1983. At that point, Clinch River needed $1.5 billion in federal funds and $1 billion in private funds to complete construction. The private investors demanded significant federal guarantees in order to continue with the project.[268] In a report and testimony to Congress, the director of the Congressional Budget Office, Rudolph Penner, concluded that the proposed financing scheme was "highly advantageous to investors" and would cost the government $250 million more than if Congress funded the project with federal dollars alone.[269] Appalled by the price tag, congressional opponents of the project succeeded in withdrawing all funding for the facility, ending the fast breeder reactor program. From 1966 through 1981, the government had spent between $5 billion and $6 billion on the stillborn liquid metal fast breeder program.[270]

Legal Strategies of Nuclear Power Opponents

Litigation and legislation were used during the 1970s and early 1980s by various opponents of new nuclear power plants—including states; labor unions; environmental, consumer, antinuclear, and other organizations; local community groups; and others—to block or retard construction and operation of new nuclear power plants. Some opponents wanted to scuttle nuclear power altogether as inherently unsafe. Others were not irrevocably opposed to nuclear power as such but believed that current NRC regulation was seriously inadequate, that the plants were uneconomic, and/or that there were better ways of meeting energy needs, including energy conservation.[271] The major initiatives included:

- Challenges to licensing of new nuclear power plants based on reactor safety and other general environmental risks
- Challenges to AEC/NRC licensing of plants based on environmental risks posed by unresolved waste disposal problems
- Challenges to NRC regulatory approvals for utilities to expand on-site storage of SNF

- State legislative moratoriums on construction of new nuclear power plants, justified on the basis of economic risks to ratepayers associated with the lack of assured disposal for the SNF that such plants would generate

Three of these four initiatives were based specifically on nuclear waste disposal problems.

Litigation Challenging Reactor Safety

In the 1960s and 1970s, opponents of nuclear power used the AEC/NRC licensing process to delay and frustrate the building of new nuclear power plants. AEC followed a two-step process for licensing new plants: a construction license, followed by an operating license.[272] The first litigation was brought in 1960 by labor unions, challenging AEC's grant of a construction license to the Fermi experimental fast breeder reactor in Detroit; the litigation was encouraged by the House-Senate Joint Committee on Atomic Energy.[273] The unions asserted that the safety of the facility had not been adequately demonstrated, specifically with respect to the risk of an accidental core reactor meltdown. AEC's Advisory Committee on Reactor Safeguards in a 1956 report had advised AEC against licensing construction of the plant without investigating and resolving this risk more thoroughly.[274]

AEC asserted that it would address these issues at the second, operating license stage. The unions, however, contended that this bifurcated approach was unsound, because the grant of a construction license would lead the commission to slight safety issues at the subsequent operating license stage, when the consequence of adverse findings could be to prohibit an already constructed plant from operating. The unions prevailed in the D.C. Circuit Court of Appeals, but the Supreme Court, in its 1961 decision in *Power Reactor Development Co. v. International Union of Electric, Radio and Machine Workers*, upheld AEC's two-step procedure.[275] This was the first in a series of Supreme Court victories for AEC/NRC in cases challenging its regulation of nuclear power plants.

In the 1970s, nuclear power opponents began to challenge AEC/NRC licensing and other regulatory decisions on a regular basis. In 1973, during the first oil shock of the 1970s, plaintiffs in *Nader v. Ray*—the consumer advocate Ralph Nader and Friends of the Earth—sued AEC, seeking a preliminary injunction against continued operation of twenty nuclear power plants for failure to comply with AEC's interim acceptance criteria for emergency core-cooling systems, which incorporated new requirements designed to prevent a nuclear meltdown accident. AEC had required compliance with interim acceptance criteria for new but not for existing plants.[276] After discussing potential adverse impacts on the nation's energy supply, the district court denied a preliminary injunction on the ground that plaintiffs had failed to exhaust all available administrative remedies.[277] The emergency core-cooling systems issue was litigated again in *Union of Concerned Scientists v. AEC* in 1974 and *Nader v. NRC* in 1975.[278] In both cases, the D.C. Circuit upheld AEC's position.

In *York Committee for a Safe Environment v. NRC*, the petitioners—local community groups and the national Environmental Coalition on Nuclear Power—challenged NRC's grant of a license for the Peach Bottom Atomic Power Station in Pennsylvania, arguing that NRC could not rely on uniform numerical radiation standards in licensing individual plants, but rather had to consider whether a lower limit was feasible and appropriate in the circumstances of each particular plant.[279] The D.C. Circuit Court agreed and remanded the matter to NRC to consider whether a lower limit was appropriate but declined to suspend the plant's operating permit.[280] Antinuclear groups also attempted to require EPA to adopt radiation standards under the Clean Water Act for power plant discharges to water bodies, on the ground that radiation is a "pollutant" within the meaning of the act. The Supreme Court, however, read the act as retaining the exclusive authority of AEC over radiation releases from fuel cycle nuclear materials.[281]

Maryland public interest groups seeking to stop the licensing of a nuclear power plant under construction on the Chesapeake Bay won a landmark victory in the 1971 D.C. Circuit decision in *Calvert Cliffs' Coordinating Committee v. AEC*, which overturned AEC's granting a construction license for a new plant without first having issued a comprehensive EIS. The court interpreted NEPA as establishing "environmental protection as an integral part of the Atomic Energy Commission's basic mandate."[282] It held that environmental groups could bring litigation to challenge AEC's failure to issue adequate EISs in licensing new plants and that the presumptive remedy for AEC's failure to do so was to enjoin the grant of a license until an adequate EIS had been prepared. After *Calvert Cliffs*, all nuclear power plant licenses were held up for at least twelve months while the commission prepared EIS documents for them.[283]

Nuclear power plant opponents used NEPA and other legal grounds to challenge the licensing of most new plants. "Between 1970 and 1972, for instance, 73 percent of all new nuclear license applications were legally contested."[284] Environmental groups and interested individuals formed an association based in Washington, D.C. called the Consolidated National Interveners to coordinate their interventions in AEC/NRC licensing hearings and to pursue legal challenges in court.[285] The AEC/NRC licensing proceedings involved formal trial-type adjudicatory hearings. Nuclear power plant opponents used the hearing procedures and the EIS development process to raise a host of issues, a strategy that led to protracted proceedings. They often challenged unfavorable licensing decisions of AEC/NRC in court. Most of their court victories, however, were based on rulings that the commission had made procedural errors; courts were consistently reluctant to overrule the agency's grant of a license on substantive issues of power plant safety. Eventually the Supreme Court's 1979 *Vermont Yankee* decision, discussed in the next subsection, curtailed the procedural claims available to nuclear power plant opponents.[286]

While opponents failed to win a decision to entirely halt AEC licensing of new power plants, they were able to overturn a number of NRC decisions and to force NRC to address additional issues and follow additional procedures. By these means, they succeeded in increasing the complexity of licensing proceedings and prolonged the licensing process. The resulting procedures for licensing a new plant featured "six review stages, four appeals stages, and six separate federal agencies (as well as state and local agencies) involved in making the permit decision."[287] It was not uncommon for the litigation challenging a license grant to last two to three years, followed in a number of cases by further AEC/NRC administrative proceedings.[288] Delays and uncertainties increased costs for utilities, as interest accrued on loans, returns on invested capital were postponed, and later start dates resulted in higher labor and equipment costs due to inflation. During the lengthy period that a license application was pending, the commission would often adopt new safety and design regulations, and the applicant would have to redesign the plant to meet the new requirements.[289] The average time between initial license application and operational start-up in the first half of the 1980s was 11.2 years in the United States, compared, for example, to 5.3 years in Japan.[290]

Interventions in licensing proceedings and lawsuits made the agencies more responsive to environmental concerns and forced negotiation and cooperation between the utilities and nuclear power plant critics and opponents. The ability of opponents to delay the licensing process often led utilities to make substantive concessions in order to avoid protracted proceedings on a contested issue. For example, a utility estimating the costs of delay at $1 million per month agreed to modify and improve its reactor's waste and radiation treatment systems. Another agreed to enhance its water-cooling system in return for a promise from the Friends of the Earth to waive its right to intervene in the administrative process.[291]

Litigation by opponents of nuclear power did not end nuclear power in the United States, but it imposed significant costs, delays, and uncertainties on utilities and proved far more effective than anyone had anticipated in the early 1970s in slowing and contributing to the ultimate cessation of new nuclear power plants construction in the United States.[292] Delays and regulatory changes increased the costs of plants and created sig-

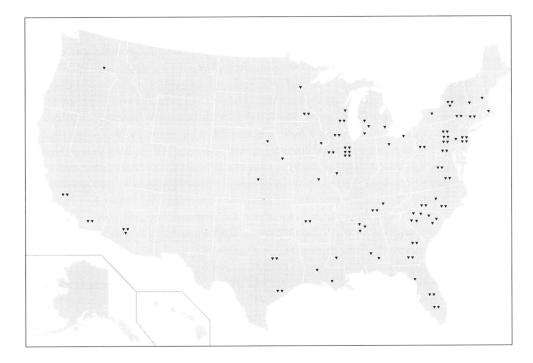

Map 1.2. Operating U.S. nuclear power reactors

Source: NRC, "Operating Nuclear Power Reactors by Location or Name," *www.nrc.gov/info-finder/reactor.*

nificant uncertainty that likely helped discourage potential investors in new plants. These developments contributed significantly to the virtual end of new nuclear power plant licensing in the United States by the end of the 1970s. No new nuclear power plant license application was submitted to NRC after 1980 until recently. The last new nuclear power plant to be constructed in the United States went on line in 1996. (See Appendix A for the number of operating nuclear power plants in the United States over the years, and Map 1.2 for currently operating plants.)

Litigation Challenging New Reactor Licensing Based on Waste Disposal Issues

In addition to challenges based on reactor safety, nuclear power opponents brought NEPA claims demanding that AEC, in decisions on whether to license new reactors, consider the environmental impacts of storage and disposal of the nuclear wastes that they would produce. Eventually AEC decided not to address this issue on a case-by-case basis in adjudicatory proceedings for licensing individual plants, fearing that opponents could use formal hearing procedures to further delay licensing. Instead, AEC used more flexible notice-and-comment rulemaking procedures to adopt a uniform rule on the issue. It then sought to use the determinations incorporated in the rule to resolve the waste issues in individual licensing proceedings without affording opponents the ability to reopen the generic findings made by the commission in the rulemaking. Over a number of years, NRC adopted several versions of a rule, with a table (termed Table S-3) evaluating the environmental impacts associated with the nuclear waste that would be generated by a new power plant, consistently concluding that the impacts should be treated as zero—meaning that nuclear waste carried no environmental or safety risk whatsoever—for licensing purposes.[293] By adopting Table S-3 through more streamlined generic rulemaking procedures and then applying the rule in all individual licensing decisions, the commission would, it asserted, comply with the law but avoid further licensing delays.[294]

AEC initiated rulemaking on the waste issue in 1972 and issued an initial rule in 1974 that incorporated Table S-3 from AEC's 1972 Environmental Effects of the Nuclear Fuel Cycle. Table S-3 addressed the environmental impact of the entire fuel cycle: uranium mining and milling, production of uranium hexafluoride, uranium enrichment, fuel fabrication, spent-fuel transport and disposal, reprocessing, transportation of radioactive material, and management and disposal of LLW and HLW.[295] Based on Table S-3, the AEC's 1974 rule determined that nuclear waste from new plants would most likely be properly managed, stored, and disposed of with no release of radioactivity.[296] It accordingly determined that the environmental impacts of the waste generated by new plants would be zero.

Following the S-3 rulemaking, the AEC/NRC applied its zero-release determination in resolving the waste disposal issues raised by environmental intervenors in an operating license proceeding for the Vermont Yankee power plant. NRDC and other litigants challenged both the S-3 rule and AEC's granting of Vermont Yankee's operating license. In a 1976 decision, the D.C. Circuit found that in adopting the S-3 rule, the commission had complied with the notice-and-comment rulemaking procedures specified in the federal Administrative Procedure Act, but that, in the context of the complex and contested scientific and technical issues presented by nuclear waste, the commission should have adopted unspecified additional procedures, such as the opportunity for a hearing on key technical issues or other steps to ensure fuller development of, and public input on, the questions presented.[297] Because the S-3 rule was procedurally invalid, NRC could not rely on it in the Vermont Yankee proceeding. Accordingly, NRC had been required to conduct an adjudicatory hearing to address the environmental impacts of the waste that the plant would produce; the failure to conduct such a hearing rendered its grant of a license to Vermont Yankee invalid.

NRC responded to the D.C. Circuit decision by publishing a revised survey of environmental impacts and a new interim rule supplementing Table S-3 but reaffirming the zero-release value.[298] It also sought review of the D.C. Circuit's decision by the Supreme Court. Subsequently, NRC reopened proceedings on the S-3 rule to determine whether the interim rule should be adopted or modified.

In its 1978 decision in *Vermont Yankee*, the Supreme Court sharply reversed the D.C. Circuit's decision, finding that the commission's rulemaking procedures had fully complied with APA and were therefore valid. However, the Court remanded the case to the Court of Appeals for a determination as to whether the substantive reasons given by the commission for adopting the zero-release determination in the 1974 rule were sound and had adequate factual support in the rulemaking record.[299]

NRC, after further rulemaking, adopted a further, final Table S-3 rule in 1979.[300] It retained the zero-release assumption of the original rule, but its decision was accompanied by a much more extensive record and rationale.[301] NRC noted the existence of uncertainties as to whether a repository would be sited and whether it would work as anticipated, but declined to include those uncertainties in Table S-3 on the ground that the result of doing so would be to require examination of the uncertainties in each individual licensing decision, which would not be sound or workable. It stated that it was reasonable to use what the commission believed to be the most probable result as the generic assumption for licensing decisions. It found highly probable the successful development of a repository in a salt dome to safely contain the SNF generated by nuclear power plants, and based the zero-release value in the final S-3 on that determination.[302]

The State of New York and NRDC challenged NRC's final rule. In 1982 the D.C. Circuit Court of Appeals again invalidated the S-3 rule and the commission's Vermont Yankee license grant.[303] The court held that NRC had violated NEPA by failing to take account of evidence in the rulemaking record (much of it provided during the rulemaking comment process by environmental organizations, states, and other opponents of the plant) of technical, institutional, and political uncertainties surrounding the permanent disposal of nuclear power plant waste. It further held that NRC's finding that there was no significant risk of a radioactivity release from a future repository for SNF was a "clear error of judgment" because of the risk of theft of nuclear material prior to disposal.[304] NRC again sought review by the Supreme Court, which again reversed the D.C. Circuit in its *Baltimore Gas & Electric* decision.[305] The Court found that NRC had indeed complied with NEPA and had provided adequate reasons for its conclusions regarding waste disposal; it strongly asserted that judges should not second-guess the commission's expert judgments regarding long-term waste management. The Court pointed to the NRC Waste Confidence Proceeding (WCP) then in progress (discussed later) as evidence that NRC was continuing to examine the environmental issues posed by waste disposal that had been addressed by the S-3 rulemaking.[306]

Litigation Challenging Expansion of At-Reactor SNF Storage

A third initiative by some nuclear power opponents consisted of litigation challenging NRC's authorization of expanded SNF pool storage at nuclear power plants on the ground that this measure created undue environmental risks from leaks, theft, or terrorism. As NRC wrangled with NRDC and other litigants over the S-3 rule, nuclear waste continued to accumulate in reactor pools around the country. In the absence of a disposal repository or away-from-reactor storage facilities, pool storage capacity limitations threatened to shut down nuclear power plant operations. Two power plants, Vermont Yankee and Prairie Island, applied to NRC in 1976 for amendment of their licenses to authorize expansion of their storage pools; NRC granted the requested amendments. Rejecting claims by opponents, the commission concluded that it was not required by NEPA to consider the environmental consequences of extended SNF storage at reactor sites, because there was reasonable confidence that a federal repository would be developed to receive and dispose of the SNF.[307] On review, the D.C. Circuit held that NRC could only make such a determination on the basis of a fact-finding proceeding.[308] NRC responded by initiating a waste confidence proceeding rulemaking in 1979.[309] The proceeding addressed not only the issue of whether a federal repository would be built, but also the broader question of whether nuclear waste could be managed safely through on-site or off-site storage and in ultimate repository disposal. This broad issue had been raised in an earlier NRDC petition requesting that NRC determine "whether radioactive wastes can be generated in nuclear power reactors and subsequently disposed without undue risk to the public health and safety," and asking that NRC refrain from granting further operating licenses until such a "definite finding of safety" was made. NRC had denied that petition in 1977.[310]

NRC issued its final waste confidence decision in 1984, reaffirming its conclusion of "reasonable confidence" that SNF would be safely stored pending development of a federal repository to dispose of it. NRC believed a repository would be available by 2007 or 2009.[311] To support this finding, NRC cited a relative lack of technical barriers to developing a repository, and the congressionally mandated scheme for development of a federal repository in the 1982 Nuclear Waste Policy Act. The waste confidence decision and its subsequent invocation by the NRC in individual licensing decisions to expand on-site storage capacity were not challenged in court. The S-3 Rule and the waste confidence decision were adopted by the commission to serve similar purposes: to block use of adjudicatory licensing proceedings to delay or preclude construction or operation of nuclear power plants. Both were based on the assumption that SNF would be disposed of in a federal repository that would be developed in a timely fashion.

Dry-cask storage, in which SNF is encased in a metal container, which is then placed in a metal or cement outer-cask shell, was developed in the late 1970s and early 1980s as a response to the problem of limited pool storage capacity at reactor sites. The first at-reactor dry-cask storage facility was licensed by NRC in 1986.[312] Since then, an increasing percentage of the accumulating SNF at reactors has been stored in dry casks. As of June 2009, forty-five sites with operating or shut-down reactors in thirty states had dry-cask storage facilities.[313] However, only about 15 percent of accumulated SNF was being housed in dry casks at reactor sites; the balance remained in storage pools.[314]

State Nuclear Power Plant Moratoriums

The fourth set of antinuclear initiatives consisted of state legislative moratoriums on the development of new power plants. Beginning in the late 1960s, many opponents of nuclear power saw little prospect of convincing Congress to reign in the expansion of nuclear power and decided to focus on promoting state-level legislation. In 1975, bills were introduced in twenty-four states that would prohibit or restrict the development or operation of nuclear power plants.[315] As discussed here earlier, the AEA vested exclusive authority to regulate the safety of nuclear power plants in AEC/NRC. Because these state bills were generally based on the environmental risks of nuclear power, had they been enacted, they would almost certainly have been held preempted by federal law if challenged in court. As previously noted, California took a different tack in an effort to avoid the problem of federal preemption. In 1976, the state adopted legislation, as the result of a voter initiative, that banned the certification of new nuclear power plants until a California state authority "finds that there has been developed and that the United States through its authorized agency has approved and there exists a demonstrated technology or means for the disposal of high-level nuclear waste."[316] Although a central concern—if not the main concern in terms of the background politics—underlying the state's opposition to nuclear power plant development was environmental risk, the legislation was styled as an economic rather than an environmental measure. The state justified the moratorium as a means to protect ratepayers against the high costs of dealing with nuclear wastes for which there was no method of disposal.[317]

Antinuclear sentiment was reinforced by the 1979 Three Mile Island accident. Demonstrations against nuclear power spiked, and public support for new nuclear plants plummeted.[318] In 1977, 69 percent of the American public had approved of building new nuclear power plants; on April 5, 1979, a week after the Three Mile Island accident, only 46 percent approved. In 1979, the governor of South Carolina, the location of DOE's SRS facility, was elected on an anti–nuclear waste platform.[319]

Utilities challenged the California measure in court, contending that it was preempted by the AEA.[320] The Supreme Court upheld the moratorium in its 1983 decision in *Pacific Gas & Electric Co. v. State Energy Resources Conservation and Development Commission*.[321] The Court stated that, although the AEA had "occupied the field" of nuclear safety regulation, and therefore federal law preempted any state safety-based regulation of nuclear plants, the AEA preserved the states' traditional regulatory authority over the economic aspects of power production. The Court upheld the California measure as economic-based regulation that accordingly was not preempted.

Other states followed California's lead. Currently, Oregon, Maine, Montana, Connecticut, Massachusetts, West Virginia, Wisconsin, Kentucky, Kansas, Illinois, New Jersey, Minnesota, and Pennsylvania also have laws on the books that restrict or prohibit the development of new nuclear power plants.[322] The New Jersey and West Virginia measures tie approval of new nuclear power plants to adequate waste disposal capacity.[323] Montana's referendum gives voters the power to approve or reject any proposed nuclear power plant; the state later adopted a referendum prohibiting nuclear waste storage.[324] Oregon's statute institutes comprehensive state regulation of nuclear power, including the creation of its own Department of Energy and a requirement that voters approve all nuclear projects. California, Maine, Massachusetts, Connecticut, and Wisconsin condition approval of new nuclear power plants on identification of a permanent federal nuclear waste solution. These measures, however, have been under reconsideration in some states in light of emerging concerns over climate change and energy security.[325] It is unclear which of these measures would survive a preemption challenge under the logic of the Supreme Court's *Pacific Gas & Electric* decision.

Federal Nuclear Waste Disposal Initiatives and Legislation

The Carter Administration Interagency Review Group

In response to the end of reprocessing, the state moratoriums on new power plants, and growing public concern over the failure of the federal government to develop a coherent policy for disposal of the more highly radioactive forms of both civilian and defense wastes, President Carter sought to engage experts and the broader public in an effort to build consensus on a new comprehensive national nuclear waste policy. In March 1978, he convened the Interagency Review Group on Nuclear Waste Management (IRG), a commission comprising representatives of fourteen federal agencies and chaired by Secretary of Energy John M. Deutch, to report to him within seven months.[326] IRG evaluated technical strategies for disposal of a broad range of defense and civilian nuclear wastes, including HLW, TRU, SNF, LLW, uranium mill tailings, and wastes from nuclear power plant decommissioning.[327] It also comprehensively examined the institutional and management issues facing nuclear waste disposal.

IRG's work was initiated through a DOE task force paper and was later informed by the reports of six staff subgroups of IRG and a technical advisory committee, as well as by reviews of the earlier work of NAS and other expert bodies. IRG obtained a very broad range of viewpoints and strove to maintain strict political neutrality in its work and deliberations.[328] Draft reports from the six IRG subgroups were issued for public comment. An extensive public involvement process was conducted by IRG to solicit nationwide input, including public hearings and meetings throughout the country. Seven of these meetings, held in Texas and New Mexico, were specifically dedicated to WIPP.[329] Small group meetings were held with members of Congress and scientists, and with representatives from state and local governments, Native American tribes, industry, environmental organizations, and other stakeholders. In 1979, IRG issued a draft report and recommendations. More than 3,300 public comments, most of them disagreeing with parts of the report or opposing further action without more information, were received. At public meetings, almost a quarter of the comments were opposed to further nuclear power development and waste generation.[330] Notwithstanding this expressed opinion, IRG issued its final report in 1980 without major changes.

HLW/SNF

IRG's recommendations for HLW and SNF were based on the premise that the then-current generation, having enjoyed the benefits of nuclear technologies, could not rightfully leave to future generations the burden of the highly radioactive wastes it had generated and therefore must secure, as soon as possible, their permanent and perpetual burial in one or more deep geologic repositories. In short, "the responsibility for establishing a waste management program shall not be deferred to future generations."[331] In accordance with this principle, IRG proposed that several repositories be developed to begin receiving wastes within twenty years.

The 1979 draft IRG report recommended that detailed studies of specific potential repository sites "in different geologic environments," including granite, salt, shale, and tuff (volcanic rock, such as that found at Yucca), should begin "immediately" in order to identify two, possibly three, repositories for HLW and SNF that could become operational by the end of the twentieth century. These repositories should be located "ideally in different regions of the country."[332] The report recommended that in the near term there be a single repository for both TRU and HLW. It further recommended siting one or more "intermediate scale facilities" for potentially retrievable emplacement of SNF; these were defined as facilities that could hold up to a thousand spent-fuel assemblies.[333] IRG also suggested that, if necessary, DOE develop facilities for away-from-reactor storage of SNF until a repository for its disposal opened. IRG believed that such storage would be part of "conservative and prudent planning" and would provide flexibility in the development of a permanent repository.[334]

IRG saw deep seabed and deep borehole disposal as potential options in the future, suggesting ten to fifteen years for development of such options before implementation was realistic. Transmutation, space disposal, and placement in a mined cavity that led to rock melting were all seen as more distant, due to the "scientific, engineering or institutional problems that must be overcome."[335]

TRU

IRG recommended proceeding with construction of a TRU repository and suggested that "the implementation of this recommendation could conceivably be by means of the Waste Isolation Pilot Plant."[336] The IRG report criticized and proposed modification of what it called DOE's "fast track" approach to certifying WIPP for TRU storage; its alternative approach called for the site to be certified by NRC and EPA.[337] It stated that the TRU repository could be co-located with an intermediate-scale facility (ISF) for HLW but that the TRU repository could and should go forward without an ISF in the event that construction of an ISF was held up by NEPA review.[338] IRG recommended separate tracks to develop TRU and HLW disposal facilities in accordance with a "conservative and stepwise approach" to address the waste problem as quickly as possible by not tying projects to each other.[339] IRG's findings and recommendations were central to DOE's efforts to take WIPP from the experimental and research phase to construction of a geological repository.

LLW

IRG recommended changes to the nation's LLW disposal system that would have increased the federal government's role. LLW disposal had been plagued by closure of some disposal facilities, temporary shutdowns of others due to problems with environmental performance, and failure to develop new facilities. IRG proposed that a new federal Executive Planning Council reporting directly to the president have lead responsibility for developing new LLW facility sites.[340] NRC would set new more detailed and stringent LLW disposal standards, which would be enforced by Agreement States or directly by NRC in non–Agreement States.

Institutional Issues

The 1979 IRG report recommended that a new Executive Planning Council (EPC) be created, which would have primary responsibility for setting nuclear waste policy. EPC would be composed of federal representatives designated by the president, as well as state and local representatives designated by the National Governors Association, the U.S. Conference of Mayors, and similar organizations representing state and local governments. DOE would exercise planning responsibilities pursuant to the policy direction of EPC. Operational regulatory authority would be retained by existing federal and states agencies.

The report further advocated the creation of a uniform national plan for LLW classification and disposal to be developed by DOE and NRC in place of the existing unplanned and uncoordinated arrangements. In the past, disposal sites were developed by private firms (sometimes with AEC backing), which would either lease land from federal or state government or acquire private land and deed it to a government. IRG envisaged a system of national planning that would ensure an adequate number of sites located near waste sources.[341] To facilitate a national plan, IRG recommended that states have the option of transferring control of LLW sites to the federal government.

IRG's LLW reforms were unpopular with state governors and legislatures and never came to fruition. The National Governors Association instead proposed a radical reworking of the nation's LLW disposal system. The Association's proposal made the states responsible for commercial LLW generated within their borders and created a set of federal incentives for states to enter into compacts to develop regional LLW disposal facilities that could be used by the compact member states but could be closed to wastes from nonmember states.

IRG also recommended the creation of uniform waste transportation standards by the Department of Transportation. IRG's recommendations on institutional issues are summarized in Table 1.2.

Congressional Response to IRG's Recommendations

The recommendations in IRG's reports provided the baseline for Congress's efforts throughout 1980s to develop and enact a comprehensive omnibus bill to overhaul the nation's radioactive waste disposal laws and policies.[342] Title I of the omnibus bill dealt with HLW and SNF. It generally followed IRG's recommendations for a highly centralized federal program to develop permanent repositories for these wastes on an expedited basis. Title II dealt with LLW and provided for a system of interstate compacts to develop new LLW sites, rejecting IRG's proposal in favor of the approach recommended by the National Governors Association. WIPP and the question of a TRU repository were not addressed in the omnibus bill because WIPP was already proceeding to development by DOE.[343] In late 1980, when it proved impossible to reach agreement on the details of Title I of the omnibus bill dealing with SNF and HLW sections, they were deleted; Title II dealing with LLW was enacted separately, becoming the Low-Level Radioactive Waste Policy Act of 1980.[344]

DOE's FEIS for Civilian Nuclear Waste Disposal

Shortly after IRG issued its report, DOE issued its 1980 Final Environmental Impact Statement (FEIS) for Management of Commercially Generated Radioactive Waste, which was a generic EIS on disposal of civilian nuclear waste, including HLW, SNF, and

Table 1.2. Recommendations by the Interagency Review Group on Institutional Structures for Nuclear Waste Management (1979)

Issue	Recommendation
Overall	Establish an executive planning council composed of federal, state, and local officials to oversee the federal government's approach to waste disposal.
	Assign overall planning responsibility to DOE; other agencies handle implementing regulation.
	Involve state, local, and tribal authorities in the overall planning process.
SNF/HLW	Existing governance structure is adequate.
LLW	DOE/NRC develop and help implement a uniform national plan for LLW disposal.
	NRC develop strengthened LLW disposal regulations. Agreement States retain the ability to regulate pursuant to federal direction and oversight.
Transport	DOT establish uniform regulations for waste transportation.

TRU.[345] The FEIS examined disposal options at the programmatic level, anticipating that future EISs would be prepared for any specific site and facility selected for development. The FEIS recommended using mined geological repositories for HLW, SNF, and TRU, with continuing R&D on other options. It also endorsed consideration of developing a geologic repository on a remote island.[346]

Although it saw some potential for SNF and HLW disposal by what it termed "Very Deep Hole Waste Disposal," the FEIS found that it would be economically infeasible to dispose of substantial volumes of wastes, such as TRU, in that manner and that the technical feasibility of the whole concept was not fully supported.[347] The rock melt concept was dismissed as unsuitable for certain wastes, such as reactor hulls, end fittings, and TRU wastes.[348] The FEIS found that both shallow and deep well injection would be appropriate only for reprocessing wastes or SNF that had been treated to form a liquid or slurry, and noted significant concerns about the feasibility of both methods.[349] The technical feasibility of seabed disposal was noted, but the "international and domestic legal problems to its implementation" were found to be too problematic.[350] Ice sheet disposal was not seen as having a sufficiently solid scientific basis.[351] Space disposal was rejected on two grounds: "the risk of launch pad accidents and low earth orbit failures" and a lack of practicality and economically feasibility for high-volume wastes.[352] Space disposal was seen as potentially useful, however, for certain isotopes, such as radioactive iodine.[353]

Debates over WIPP's Mission and Its Development as a TRU Repository

A fire at the Rocky Flats weapons facility and Idaho's opposition to disposal of Rocky Flats TRU within the state persuaded AEC/ERDA to undertake serious efforts to site a TRU repository, leading to development of the WIPP facility, as noted earlier. While it was accepted that WIPP would dispose of TRU, serious consideration was given to expanding its mission to include SNF and possibly also HLW.

The need to find a site for disposal of SNF mounted as a result of the decisions of Presidents Ford and Carter, respectively, to suspend and terminate federal support for SNF reprocessing and state legislative initiatives to ban new nuclear power plants unless and until a method for SNF disposal was demonstrated. Notwithstanding the rosy assurances of NRC in fending off legal challenges based on the SNF disposal issue, there was no credible federal program to deal with accumulating volumes of SNF piling up at power plants across the country.[354] *Congressional Quarterly* noted at the time (1977): "The swimming pools in the backyards of the nation's nuclear reactors are filling up with used fuel, and there's no firm plan to dispose of the deadly radioactive waste."[355] ERDA approached a number of states about the possibility of hosting an SNF repository but made no headway.[356] The only feasible near-term option for permanent disposal seemed to be burial at WIPP.

ERDA raised the possibility of disposing of commercial SNF at WIPP beginning in 1977. In January of that year, WIPP's project manager announced that WIPP's design would make the site suitable for licensing by NRC, implying SNF disposal.[357] A month later, the project manager told the New Mexico governor's advisory committee that "consideration would obviously be given to making [WIPP] a commercial site."[358] James Schlesinger, the entering secretary of the newly minted Department of Energy, openly favored disposing of commercial HLW at WIPP.[359] A General Accounting Office (GAO) report, issued in September 1977, also endorsed this solution for HLW, TRU, and SNF, reasoning that DOE should seize the opportunity made available by favorable "public and official sentiment in New Mexico."[360] Studies of the WIPP site through the 1970s identified no insurmountable technical barriers to the co-location there of SNF, HLW, and TRU wastes.[361]

In the 1980 DOE National Security and Military Applications of Nuclear Energy Authorization Act, however, Congress restricted WIPP to disposal of defense wastes.[362] Subsequently, the 1992 Waste Isolation Pilot Plant Land Withdrawal Act limited WIPP to disposal only of TRU.[363] The 1980 legislation excluding civilian SNF from WIPP was the outcome of political skirmishing that involved, among others, Representative Mel Price, chair of the House Armed Services Committee; the governor of New Mexico; the state's congressional delegation; DOE; President Carter; and others. Chairman Price opposed commercial SNF disposal at WIPP because it would entail NRC regulatory jurisdiction over the facility, defeating exclusive control by the executive and his congressional committee.[364] New Mexico, which did not have a nuclear power plant, was opposed to disposing of SNF from other states.

As the development of WIPP as a defense TRU repository moved forward, New Mexico began to assert its interests in the facility's design and operation through a combination of state legislation, initiatives by a state executive task force, pressures through its congressional delegation, and litigation against DOE. In 1978, the state had successfully pushed DOE to enter into a memorandum of understanding that established and funded an Environmental Evaluation Group, which evolved into an independent scientific body that provided technical expertise to the state on repository issues.[365] In the 1980 legislation restricting WIPP to defense wastes, Congress rejected giving New Mexico a veto over the project but mandated a process of "consultation and cooperation" between DOE and New Mexico.[366] In 1981, Jeff Bingaman, the attorney general of New Mexico, brought suit against DOE to halt construction at WIPP.[367] New Mexico and DOE agreed to a settlement, providing in part that the state and DOE would execute a legally binding consultation and cooperation (C&C) agreement.[368] The C&C agreement and several subsequent amendments to it gave New Mexico a significant voice in decision making on WIPP and the right to enforce the terms of the C&C agreement in court. These arrangements forced DOE to come to the bargaining table to meet the state's concerns about the facility, including issues relating to repository location, design, and safety; state regulatory authority over the wastes sent to WIPP; and waste transportation, including upgrading New Mexico highways over which waste would be transported and construction of bypasses around several cities.[369] These concerns were extensively debated, negotiated, and in some instances further litigated, but they were eventually accommodated—although not without considerable and often sharp contention and disagreement.

Delays in opening WIPP because of the need to deal with New Mexico and other factors created political problems for DOE with other western states impatient to see removal of TRU from DOE facilities within their borders. Despite AEC's promises to Idaho, AEC continued to ship TRU to Idaho for storage well into the 1980s. In 1988, Idaho governor Cecil Andrus dramatically stopped a train car full of Rocky Flats TRU waste from entering Idaho.[370] In the resulting standoff, waste began piling up at Rocky Flats; within a few months, the amount in storage at Rocky Flats would exceed the storage limit agreed upon by Colorado and DOE.[371] Governor Roy Romer of Colorado pushed for the Rocky Flats waste to be sent to WIPP, while Governor Garrey Carruthers of New Mexico refused to accept it. DOE met with the three governors to resolve the issue.[372] By the end of the meeting, DOE had agreed to look for a new interim site for Rocky Flats waste outside Idaho, to continue to work toward opening WIPP, and to stop shipments to INL by August 1989, so long as it could resume them in the interim.[373]

Shipment to INL stopped, as agreed, in August 1989, and DOE began searching for somewhere to ship the TRU from Rocky Flats. The search was soon halted, however, when a November 1989 raid of Rocky Flats by EPA and FBI agents aimed at suspected safety violations resulted in a temporary shutdown of plutonium operations at the plant.[374] The Cold War ended soon thereafter; the impetus for making new nuclear weapons dissipated, and production at Rocky Flats never resumed.[375] Since no more TRU was being generated at the plant, DOE stopped looking for an interim site to store Rocky Flats TRU pending opening of WIPP.[376] However, TRU waste continued to be generated as a result of the decontamination and decommissioning of Rocky Flats. This waste—as well as waste from plutonium production that had not yet been sent to Idaho—was kept on-site for eventual shipment to WIPP.[377]

As DOE moved to develop WIPP into a full-scale TRU repository, New Mexico in 1991 again brought litigation against DOE, challenging the facility on the grounds that the withdrawal of federal lands for the facility in 1976 and again in 1979 had been only for a pilot project and that federal legislation was necessary to authorize use of the lands for a full-fledged operating repository. New Mexico prevailed in the litigation.[378] As a result, DOE was forced to obtain congressional authorization for the facility. In 1992, Congress enacted the Waste Isolation Pilot Plant Land Withdrawal Act authorizing use of federal land for construction of the repository.[379] The act also imposed various requirements and restrictions on the WIPP facility and its operations, all designed to meet the state's concerns. The act directed EPA to issue revised radiation release standards for repositories; EPA's prior standards had been partially invalidated as a result of a litigation challenge. The act also directed EPA to certify the facility's compliance with EPA's standards and other requirements and recertify it every five years thereafter.[380] It also gave EPA broad oversight over testing, disposal, retrieval, and decommissioning at the facility.[381] Congress enacted these provisions to ensure that a regulatory body independent of DOE determined WIPP's safety.[382] NRC had no regulatory authority over WIPP, a consequence of Melvin Price's careful steering of the 1979 WIPP Authorization Act to exclude civilian waste from WIPP, thereby avoiding NRC jurisdiction over the facility.[383] Although Congress gave NRC licensing authority over Yucca in the NWPA of 1982, in the case of WIPP it did not reverse its prior decision to exclude NRC jurisdiction over WIPP and instead gave licensing authority over the facility to EPA.

As required by the Waste Isolation Pilot Plant Land Withdrawal Act, EPA in 1993 issued revised generic radiation standards for repositories and monitored retrievable storage facilities, which applied to WIPP.[384] EPA reinstated without change the portions of the earlier standards that had not been invalidated in litigation, including the standards governing predisposal activities and most of the disposal standards.[385] EPA tightened the individual protection requirements in the disposal standards to a limit of fifteen millirems of exposure to any individual (instead of the previous twenty-five millirems) for a period of ten thousand years (instead of the previous one thousand years).[386] EPA also adopted groundwater protection standards for final disposal, requiring "reasonable assurance" that, for ten thousand years, groundwater outside the "controlled area" would meet EPA's Safe Drinking Water Act groundwater standards.[387] The controlled area was defined as an area no larger than a hundred square kilometers, extending no more than five kilometers in any direction from the repository.[388]

Later, following further litigation brought by the state, DOE was also obliged to obtain an RCRA facility permit for WIPP, as a facility disposing of mixed waste due to the chemically hazardous components of the TRU, from the State of New Mexico, which had been delegated RCRA implementing and enforcement authority by EPA. WIPP's RCRA permit, issued by New Mexico, specifies a variety of conditions that the facility must meet, overseen and enforced by the state.[389]

WIPP received its first shipment of TRU waste in 1999 and has been disposing of TRU wastes since that time.[390] It was certified by EPA in 1998 and recertified in 2006 and 2010.[391] The facility has been operating pursuant to its New Mexico RCRA permit since 1999; the state recently issued a ten-year renewal of the permit.[392] WIPP is the world's only operating full-scale deep geologic repository for highly radioactive wastes.

Low-Level Radioactive Waste Disposal Legislation

As of the late 1970s, civilian LLW was being disposed of in commercial landfills licensed and regulated by AEC/NRC or Agreement States. DOE disposed of some defense LLW in commercial landfills, but most of it was buried at DOE sites. Six commercial landfills had been developed under the initiative of AEC/NRC, sited on federal or state land leased to the facility operators. These facilities, however, suffered from chronic operating, environmental regulatory, financial, and other problems. By 1979, three of the six facilities had closed permanently, and two of the remaining three were shut down temporarily. States were extremely reluctant to host new facilities, especially following a 1978 Supreme Court ruling, based on the Commerce Clause of the Constitution, that prohibited states from excluding out-of-state wastes from disposal facilities within their borders.[393] In 1980, Congress enacted the Low-Level Radioactive Waste Policy Act (LLRWPA) "to stimulate development of more LLW disposal capacity on a regional basis" through a system of interstate compacts.[394]

LLRWPA makes each state responsible for the disposal of LLW generated within its borders, with the exception of waste produced by DOE, for which DOE is responsible.[395] LLRWPA authorizes, but does not require, states to enter into regional compacts that share a single "regional disposal facility" for LLW.[396] The 1985 LLRWPA amendments provide that the states in a congressionally approved compact can refuse to accept waste generated outside their compact, and institutes other incentives to encourage states to develop disposal facilities whose availability could be limited to other states within their compact.[397]

Since 1985, forty-four states have entered into compact arrangements. But only one new disposal facility—in Andrews, Texas—has been developed; it is not yet in operation.[398] This commercial facility is being developed under the auspices of the Texas Compact, which cannot be regarded as a true compact of the sort envisaged by Congress. The Texas Compact has only two members (Texas and Vermont) and is totally controlled by Texas. (See Chapter 4, Table 4.2, which lists current LLW compacts and disposal facilities.)

Because of the lack of disposal capacity for class B and C wastes and the high fees charged by the few operating disposal facilities, LLW generators have implemented strategies to reduce the total volume of waste requiring disposal, including supercompaction, incineration, and decontamination. These approaches have reduced the volume of LLW but have not solved the disposal problem.[399] Since the Barnwell, South Carolina, facility restricted itself to in-compact waste in 2008, 36 states have had no disposal site that will accept their class B and class C LLW.[400] The commercial Clive, Utah, facility, which currently disposes of almost all of the nation's class A LLW, might be opened to class B and class C wastes, but this would require approval by the Utah legislature, which has thus far been unwilling to grant it.[401] Other options for LLW disposal include opening DOE disposal sites to commercial wastes and legislation that makes class B and class C wastes a federal responsibility.[402] Meanwhile, substantial amounts of class B and class C wastes are being stored by generators of the wastes at the generator sites. NRC has recently authorized at-generator storage, but it is expensive and burdensome for waste generators.[403]

A new LLW disposal facility designed to accept class A, B, and C wastes is under construction near Andrews, Texas, under the auspices of the Texas Compact. The Texas Low-Level Radioactive Waste Commission recently voted to authorize the Andrews facility to accept class A, B, and C LLW from as many as thirty-six states.[404] If this facility is opened to out-of-compact class A, B, and C wastes, it would relieve existing class B and C disposal problems and provide competition with the Clive, Utah, facility for disposing of class A LLW; as discussed in Chapter 4, however, additional sites need to be developed to assure a stable and equitable system of LLW disposal for the long term.

The LLW legislation makes DOE responsible for disposal of civilian and defense GTCC wastes, some of which are relatively highly radioactive. DOE is currently studying this issue and has not yet developed a plan for disposing of these wastes.[405]

Disposal of HLW and SNF: The 1982 Nuclear Waste Policy Act

The new Congress that convened in 1981 took over the task, which its predecessor failed to accomplish, of passing legislation to deal with SNF and HLW. It succeeded in enacting the Nuclear Waste Policy Act of 1982 (NWPA).[406] This law and later amendments to it have defined U.S. policy on disposal and interim storage of SNF and HLW to the present day. In accordance with IRG's imperative to dispose of the nation's waste legacy as soon a possible, NWPA mandated rapid development of two repositories, with the first to begin receiving wastes by 1998.

In the deliberations preceding passage of NWPA, Congress debated the respective merits of extended storage of SNF and HLW, either at reactor sites or in consolidated storage facilities; early development of one or more repositories for permanent disposal; or some combination of these approaches. Different interests favored different proposals. States that were potential candidates to host a repository typically favored extended storage, while states that were candidates to host consolidated storage facilities tended to favor a repository. Some communities located near nuclear power plants were anxious simply to have SNF taken off the plant sites and moved elsewhere. DOE, a number of environmental organizations, and the nuclear power utilities were strongly in favor of establishing a permanent federal repository and feared that allowing prolonged interim storage would diminish the impetus for developing one. The utilities believed that without a repository "solution" for disposing of nuclear waste, the public would never accept the expansion of nuclear power. The utilities also sought to eliminate a point of serious friction with local communities by having the federal government assume responsibility for removing the accumulated SNF at reactor sites for storage or disposal.[407]

NWPA struck a complex balance between the various competing interests and objectives.[408] First, NWPA mandated the development of deep geologic repositories as the centerpiece of federal waste policy. It also made the federal government responsible for the disposal of commercial as well as defense HLW and SNF and for siting, constructing, and operating geological repositories for this purpose.[409] Further, it required that HLW produced through defense activities be deposited in a repository for commercial wastes, unless the president determined that this waste should be disposed of in a separate repository.[410] President Reagan eventually determined that defense HLW would be co-disposed with commercial HLW and SNF.[411]

The 1982 NWPA envisaged siting and construction of two federal repositories on a tight timetable, with the second repository in the East sited shortly after the first in the West in order to ensure regional equity.[412] The first repository was to be operational by 1998.[413] It was limited to disposal of seventy thousand metric tons of waste.[414] This cap, which meant that the first repository would not be able to accommodate all accumulating SNF, was designed to force development of a second repository.[415] Although two repositories were initially envisioned, the statute's authorization for siting and construction would allow additional repositories to be built.[416]

In the first round of siting, DOE was required to identify very quickly an unspecified number of states containing "potentially acceptable sites."[417] In the next phase of the process, it was to nominate five sites from those states as suitable for "characterization" (i.e., detailed investigation) and, finally, to recommend, by January 1, 1985, three of these five candidates to the president for characterization by DOE.[418] The president would approve or disapprove characterization of each candidate site.[419] Following characterization, DOE would recommend one or more of the final three candidate sites to the president, who would choose one to recommend to Congress for the first repository.[420] The state or tribe where the recommended site was located could issue a notice of disapproval, which in turn could be overridden by a joint resolution of Congress.[421]

NWPA also established a system of contracts between DOE and utilities under which the utilities would underwrite the costs of the projected repositories by paying a fee on nuclear energy produced.[422] In return, the federal government undertook to "accept title to all SNF and/or HLW, . . . provide subsequent transportation for such material to the DOE facility, and dispose of such material" no later than January 31, 1998.[423] While it was widely understood that it probably would not be feasible to site, construct, and open a repository for full-scale operation within the statutorily allotted sixteen years, the 1998 deadline was inserted to keep pressure on DOE to move forward and on Congress to fund the program.[424]

NWPA specified key roles for three federal agencies in the development of federal repositories. First, the statute created within DOE the Office of Civilian Radioactive Waste Management and charged it with locating, evaluating, nominating, constructing, and operating repository sites for permanent geologic disposal of SNF and HLW.[425] Second, NRC was given the task of establishing the design and operational requirements for the repositories, including requirements for how the waste should be packaged and stored and how a repository should be sealed or decommissioned.[426] Once a repository site was selected, the statute mandated that DOE apply to NRC for authorization to construct a repository. DOE was to obtain from NRC an authorization to receive and possess HLW/SNF for disposal and, at the end of the repository's operating life, a further authorization to close and decommission the repository.[427] Finally, NWPA charged EPA with issuing environmental standards to protect humans and the environment from the release of radiation from repositories and gave it the responsibility of assessing the security and protectiveness of the site over time.[428]

DOE immediately undertook the process for selecting the first repository. The process was framed as meritocratic, aimed at selecting the best-qualified site by taking into account variables such as the suitability of the hydrogeology for a repository, number of people located nearby, transportation, and cost. In 1984, DOE published repository siting and construction guidelines for comparative evaluation of candidate sites.[429] DOE sought to evaluate sites with diverse rock types and geohydrological settings.[430] In 1983, DOE identified nine potential sites for further evaluation.[431] In the course of the further site selection process, states, Native American tribes, environmental groups, and private associations initiated dozens of lawsuits to challenge DOE's actions and resist selection of various sites.[432] Three years later, in 1986, DOE narrowed the nine potential sites to five sites: Richton Dome in Mississippi, Deaf Smith County in Texas, Davis Canyon in Utah, Yucca Mountain in Nevada, and Hanford in Washington.[433] The Yucca Mountain site in Nevada was ranked first under DOE's scoring criteria, although the scores of all five sites were relatively close.[434] Shortly thereafter, DOE selected three western sites with different geologies—Yucca Mountain (volcanic rock), Hanford (basalt), and Deaf Smith (salt dome)—to recommend for full characterization.[435]

DOE also identified a number of potential sites for a second repository in the East but abandoned the second repository siting process in 1986 because of the political furor caused by its designation of potential sites for further study.[436] Jettisoning the second repository angered western states that had been targeted for the first repository, who came to view the repository-siting effort as politically motivated and unfair.

While implementing the repository-siting process, DOE also took steps toward developing a consolidated storage facility for commercial SNF/HLW. Section 141 of NWPA authorized government development of a monitored retrievable storage (MRS) facility to permit "continuous monitoring, management and maintenance of [civilian SNF/HLW] for the foreseeable future" and to provide for "ready retrieval" of such wastes "for further processing or disposal."[437] The MRS was intended to operate as an insurance policy in case the repositories were significantly delayed.[438] But no MRS was ever developed, because of state opposition to hosting such a facility. Between 1982 and 1987, the Office of Civilian Radioactive Waste Management evaluated and developed a plan for an "integrated MRS" co-located with a permanent disposal repository, in which the MRS would serve as a staging facility for SNF on its way from reactors to a repository.[439] This concept also failed to bear fruit.[440] In addition to the MRS provisions, the NWPA included a provision to deal with exhaustion of SNF storage capacity at reactor sites. Section 135 required DOE to take possession of a quite limited amount—up to 1,900 metric tons—of commercial SNF if requested by the utilities.[441] The utilities never requested such storage, and DOE's authority to enter into contracts to provide this interim storage expired in 1990.[442]

NWPA also left open the possibility of developing private consolidated storage facilities not located on federal lands. As discussed later in this chapter, efforts to develop such facilities have also not succeeded.

The 1987 NWPA Amendments and Congressional Selection of Yucca for a HLW/SNF Repository

In 1982, Congress had established a timetable for selection of a first repository site by 1989 (see Chapter 6). As of 1987, however, none of the three finalist sites had been characterized. Completion of the characterization process was now expected to take far longer than had originally been predicted, and the price tag for site characterization had jumped from $60–$100 million to almost $1 billion per site.[443] In 1987, Congress aborted the ongoing DOE repository site selection process by amending NWPA to require that only the Yucca Mountain site could be characterized and considered as a candidate site for a repository: "The Secretary shall terminate all site specific activities (other than reclamation activities) at all candidate sites, other than the Yucca Mountain site, within 90 days after the date of enactment."[444]

The 1987 NWPA amendments also postponed indefinitely plans for siting a second repository, providing that DOE "may not conduct site-specific activities with respect to a second repository unless Congress has specifically authorized and appropriated funds for such activities."[445] This decision reflected the intense opposition by eastern states to hosting a repository, as noted earlier (see Chapter 6). The abandonment of the effort to locate a repository in the East flagrantly disregarded and destroyed the promise of geographical equity that had enabled Congress to pass NWPA in 1982.[446]

Congress's action in short-circuiting the 1982 NWPA site selection process and its designation of Yucca as the sole potential repository was due in substantial measure to the influence and actions of three powerful members of Congress: Senator J. Bennett Johnston of Louisiana, chair of the Senate Committee on Energy and Natural Resources; Representative Tom Foley of Washington, House whip; and Representative Jim Wright of Texas, House majority leader (see Chapter 6). Johnston expressed concern that delays, the high costs of characterizing multiple sites, and a building political backlash would jeopardize any drawn-out siting process; he pushed hard for Congress to designate a site.[447] Johnston was also concerned about preventing the siting of a repository in the salt domes of Louisiana.[448] Foley and Wright wanted to ensure that the candidate sites in their states (Hanford in Washington and Deaf Smith County in Texas) were not chosen.[449] Nevada was politically weak and unable to resist steamrolling, although members from Nevada tried but failed to block the measure.[450] The naked political process by which Yucca was chosen, overriding the technocratic administrative system of site selection established under the NWPA, provoked outrage and relentless opposition from Nevada.[451]

Unless very serious technical problems with the Yucca site emerged, the new law made approval of Yucca Mountain as the repository location a near fait accompli. The 1987 amendments enforced Congress's determination to develop a repository at Yucca through a variety of means. The secretary of energy could determine only whether Yucca met applicable regulatory standards and whether to recommend to the president to proceed with construction and operation.[452] The president could then only approve or disapprove the Yucca site.[453] Although Nevada could object to the president's approval of Yucca through a notice of disapproval, Congress could override the state's objection by passing a joint resolution.[454] It was highly likely that Congress would override Nevada's inevitable notice of disapproval; members from other potential host states would want to avoid reopening the siting process, while states currently hosting stored SNF and HLW would want to see a federal repository to receive these wastes developed—and developed soon.

As a further means of assuring that the Yucca Mountain site would be approved, the amendments provided that no MRS facility for interim storage of SNF could be sited until after DOE had recommended the Yucca site to the president.[455] Also, as noted earlier, to prevent the prospective MRS facility from becoming a de facto repository, the 1987 amendments set a storage cap on a government MRS of ten thousand metric tons of heavy metal, a fraction of the total amount of accumulating SNF.[456] Further, in response to DOE's efforts to develop an MRS site in Tennessee, the Tennessee congressional delegation obtained a provision in the 1987 amendments that precluded the siting of any MRS facility in that state.[457]

The 1992 Energy Policy Act and Environmental Standards for Yucca

The 1992 Energy Policy Act (EnPA) directed EPA to adopt environmental standards specifically for Yucca in lieu of its generic repository/MRS standards. The background of the generic standards must be understood in order to appreciate the significance of these provisions and EPA's implementation of them. Pursuant to NWPA, EPA in 1985 had promulgated generic standards, found at 40 C.F.R. Part 191, limiting the amount of general environmental radiation from SNF, HLW, and TRU repositories and MRS facilities.[458] Subpart A placed limits on radiation exposure from the management or storage of waste prior to its disposal.[459] Subpart B provided four types of standards for final disposal, requirements for general containment, assurance, individual protection, and groundwater protection.[460] Several environmental groups and states challenged the standards in court. In 1987, the First Circuit found that the groundwater protection requirements were both inconsistent with the Safe Drinking Water Act and arbitrary and capricious.[461] The court also determined that the individual protection requirements—which limited annual radiation exposure to any individual to twenty-five millirems to the whole body and seventy-five millirems to any critical organ, over one thousand years—were arbitrary and capricious because the agency did not adequately explain the thousand-year compliance period, which petitioners challenged as too short.[462] The court vacated and remanded the disposal regulations (Subpart B) in their entirety.

By 1992, EPA had still not reissued the disposal regulations, and Congress was becoming impatient. In 1992, it enacted two statutes that directed EPA to issue new standards. The first, the WIPP Land Withdrawal Act, included directives to EPA to issue new generic standards, which would apply to WIPP. Their application to Yucca was specifically excluded.[463] The other 1992 statute, EnPA, focused on Yucca.

The House version of the EnPA legislation required EPA to reinstate all the provisions of its 1985 generic standards not found unlawful by the circuit court and to issue final revisions of the remanded provisions within nine months.[464] In the conference committee, the bill was amended to require EPA to promulgate new environmental radiation standards specific to Yucca, which would be the only radiation standards applicable to the facility.[465] The bill was also amended to provide that EPA's Yucca-specific standards were to be based on and consistent with a study to be conducted by NAS.[466] These changes were pushed by Senator Bennett Johnston, chair of the Senate Energy Committee.[467]

The Nevada delegation opposed adoption of the conference bill due to these changes. One member of the delegation criticized the NAS committee provision as an attempt to undermine EPA's control over its own regulations by legislatively dictating what the committee could consider in its study.[468] Representative Barbara Vucanovich characterized it as a secret backroom deal orchestrated by Johnston, a strong supporter of nuclear power. She also argued that the provision for Yucca-specific standards was designed to make it easier to license the facility, based on concern that it would be unable to meet EPA's original standards.[469] Natural Resources Defense Council scientist Thomas Cochran shares this view, noting that NRC and DOE had reportedly complained that the EPA generic repository standards were too stringent; he has also asserted that Johnston pushed the NAS committee provision in the expectation that it would prevent EPA from setting excessively stringent standards.[470]

The NAS committee released its report in 1995.[471] EPA promulgated its Yucca-specific radiation standards in 2001.[472] Subpart A of the regulations limited exposure from storage and management of waste at Yucca (as distinct from final disposal) to members of the public to fifteen millirems annually.[473] Subpart B provided three different types of standards for disposal. First, the individual protection standard required that DOE must demonstrate a reasonable expectation that for ten thousand years the "reasonably maximally exposed individual" will not receive more than fifteen millirems of exposure annually.[474] The "reasonably maximally exposed individual" is defined as a hypothetical person that is representative of a person who lives above the area of highest contamination.[475] Second, there were separate standards to deal with human intrusions into the repository.[476] Third, groundwater protection standards required DOE to show a "reasonable expectation" that releases from the repository within ten thousand years will not cause radioactivity levels in groundwater to increase beyond certain levels outside the "controlled area."[477] EPA's Yucca-specific standards define the "controlled area" differently than do its generic repository standards. The Yucca controlled area can total no more than three hundred square kilometers, whereas the generic standards limit the area to a hundred square kilometers. Both sets of standards stipulate that the boundaries of the controlled area cannot extend beyond five kilometers from the repository, but the Yucca-specific standards allow the controlled area to extend to a point eighteen kilometers from the site in the direction of groundwater flow.[478] Finally, EPA provided that in making a performance assessment to demonstrate to NRC its compliance with EPA's standards, DOE need not consider events that a have less than a one in ten thousand chance of occurring within ten thousand years.[479] Parts of these standards were eventually vacated by the D.C. Circuit in *Nuclear Energy Institute v. EPA* in litigation brought by Nevada and environmental groups challenging the EPA Yucca regulations.[480]

The Yucca site designation process moved forward as scripted by Congress in the 1987 amendments. In February 2002, DOE secretary Spencer Abraham determined that the site met applicable regulatory standards and recommended the Yucca Mountain site to President Bush, who approved it.[481] Nevada then issued a notice of disapproval.[482] In July 2002, Congress adopted a joint resolution overriding Nevada's opposition.[483] A few weeks later, the president signed the resolution approving Yucca. Construction of the repository, however, could not begin until NRC issued a construction license to the facility. Despite Nevada's attempts to stop characterization of the Yucca Mountain site by DOE and the U.S. Geological Survey, DOE finally submitted its Yucca construction license application to NRC on June 3, 2008.[484] On September 8, 2008, NRC docketed the application for decision.[485]

Before NRC could decide DOE's application, however, the commission had to await issuance by EPA of revised radiation standards for Yucca following a successful court challenge to EPA's 2001 Yucca standards by Nevada and environmental groups.[486] Key provisions of those standards had required compliance with exposure limits for ten thousand years. The report of the NAS Committee on the Technical Bases for Yucca Mountain Standards had concluded that limiting compliance to ten thousand years had "no technical basis." It found that it was feasible to assess the physical and geologic aspects of Yucca repository performance over a period on the order of one million years. Explaining that the period of greatest exposure risk is a function of the rate of decay in the radioactivity of the buried wastes and of the rate at which the repository's engineered and geologic barriers to waste migration deteriorate, the committee recommended that EPA determine a compliance period that corresponded to the period of greatest risk of repository failure resulting in significant radiation releases, and concluded that peak radiation exposures from Yucca might not occur until tens to hundreds of thousands of years after disposal.[487] The D.C. Circuit held, in *Nuclear Energy Institute v. EPA*, that the ten-thousand-year compliance period in the EPA standards was contrary to the statutory requirement that the standards be "based upon and consistent with" the NAS committee report, and that the standards were arbitrary.[488] As a result, the court vacated the EPA standards for Yucca to the extent that they incorporated the ten-thousand-year compliance period.[489]

In 2008, EPA issued amendments to the standards to replace the vacated sections.[490] The new standards created two relevant compliance periods: the first ten thousand years, and from ten thousand years until the end of the estimated period of Yucca geologic stability, or one million years, after disposal.[491] The second time period was established to correspond to the estimated time of greatest risk as recommended by the NAS committee. The new individual-protection standard limited exposure to the "reasonably maximally exposed individual" to an annual dose of fifteen millirems for the first ten thousand years, and one hundred millirems after that.[492] The "human-intrusion standard" was altered similarly—a limit of fifteen millirems annually for the first ten thousand years, and one hundred millirems thereafter.[493] The regulations also changed what must be considered in DOE's performance assessments of the repository's ability to comply with the EPA standards. DOE is allowed to disregard only events that have less than a one in one million chance of occurring in any given year. However, DOE does not have to consider events that will not change the results of the performance assessment for the first ten thousand years.[494]

Upon issuance of the revised standards, NRC started considering the Yucca license application. NWPA mandates that NRC decide on the Yucca Mountain license application within three years of its submission, subject to NRC granting a one-year extension.[495] Tables 1.3 and 1.4 summarize the history of EPA radiation standards and NRC and EPA licensing requirements for repositories.

By 2008, the political fortunes of Nevada and the prospects for establishing a repository at Yucca Mountain had undergone a complete transformation. In January 2007, Nevada senator Harry Reid, a longtime vehement opponent of Yucca, became U.S. Senate majority leader. In November 2008, Barack Obama was elected president. Obama had indicated his opposition to Yucca as early as 2007 and reiterated that opposition in the 2008 Nevada Democratic primary (see Chapter 6). To accomplish his legislative agenda, the new president would need Reid's support.

In February 2009, the new secretary of energy, Stephen Chu, made clear that the administration would not proceed with development of a Yucca repository.[496] The FY 2010 DOE budget reduced the funds for development of Yucca from $288 million to $197 million, leaving just enough money for DOE to continue its NRC license application and begin a preliminary examination of alternatives to Yucca.[497] Chu stated that DOE hoped to apply the knowledge gained through the Yucca project and license application process to inform siting and development of a repository elsewhere.[498] On February 1, 2010, President Obama released the administration's budget for FY 2011, which withdrew all funds for development of Yucca, on the grounds that "developing a repository at Yucca Mountain, Nevada, is not a workable option and that the Nation needs a different solution for nuclear waste disposal."[499] The same day, DOE moved to stay the Yucca licensing proceedings before the NRC's Atomic Safety and Licensing Board (ASLB).[500] On March 3, DOE filed a motion with NRC requesting withdrawal of the application "with prejudice."[501] DOE expressly did not claim that the Yucca site was technically unsuitable but rather justified withdrawal of the application on grounds of "policy."[502]

On June 29, 2010, the Atomic Safety and Licensing Board denied DOE's withdrawal

Table 1.3. History and provisions of EPA radiation standards

Standards	History	Citation	Apply to	Key provisions
Generic radiation protection standards	Original 1985 Vacated 1987 Revised 1993	40 C.F.R. 191	WIPP, potential non-NWPA repositories, MRS facilities, private interim storage	1993 standards: exposure limits to any individual, 0–10,000 years: 15 mrems/yr*
Yucca-specific standards	Original 2001 Vacated 2004 Revised 2008	40 C.F.R. 197	Yucca Mountain	2008 standards: exposure limits to "reasonably maximally exposed individual," 0–10,000 years: 15 mrems/yr; 10,000–1,000,000 years: 100 mrems/yr**

*The exposure limits apply to all individuals outside the controlled area, defined as an area no more than 100 km2 extending no more than 5 km from the site (40 C.F.R. § 191.12 (2009)). Annual exposure to any individual is limited to 25 mrems. Id. § 191.03.

**In the event of an intrusion during the first 10,000 years, exposure caused by the intrusion is limited to 15 mrems. Between 10,000 and 1,000,000 years, exposure is limited to 100 mrems (40 C.F.R. § 197.25 (2009)). The Yucca-specific disposal standards do not limit annual exposure to all individuals but to the "reasonably maximally exposed individual," defined as a hypothetical person who lives above the area of highest contamination outside the controlled area (Id. § 197.21). The controlled area is defined as an area no more than 300 km^2 extending no more than 5 km from the site, except that it can extend 18 km in the direction of groundwater flow (Id. § 197.12).

Table 1.4. History and provisions of EPA and NRC compliance/licensing standards

Agency	Standards	Citation	History	Applies to	Key provisions
EPA	WIPP compliance criteria	40 C.F.R. 194	Original 1996 Amended 1998, 2004	WIPP	Compliance application must include system design and performance assessments for compliance with applicable radiation standards
NRC	Generic licensing standards for repositories	10 C.F.R. 60	Original 1981 Amended frequently, most recently 2009	Potential NWPA repositories other than Yucca Mountain	License application must include system design, schedules, security measures, site characterization information, and a safety analysis report.*
NRC	Yucca-specific licensing standards	10 C.F.R. 63	Original 2001 Amended frequently, most recently 2009	Yucca Mountain	Incorporates Yucca-specific radiation standards. License application must include system design, schedules, security measures, site characterization information, and a safety analysis report.**

*NRC's generic standards also include a limit on exposure of 5 rems to any individual as a result of natural or human-induced events considered "unlikely" (10 C.F.R. § 60.136 (2010)).

**NRC's Yucca-specific standards also include a limit on exposure of 5 rems to any individual as a result of natural or human-induced events considered "unlikely" (10 C.F.R. § 63.111 (2010)).

motion. The board invoked Section 114(d) of NWPA, which provides that NRC "shall consider" the application submitted to it and "issue a final decision."[503] It stated that NWPA "does not give the Secretary [of Energy] the discretion to substitute his policy for the one established by Congress in the NWPA that, at this point, mandates progress toward a merits decision by the Nuclear Regulatory Commission on the construction permit."[504] It found that Congress had determined that a repository should be built at Yucca if NRC determined that it would meet applicable regulatory standards, and that DOE's effort to derail the NWPA decision process by asserting a policy judgment that Yucca is not a workable option contravened Congress's intent.[505] DOE appealed the board's decision to the commission, which heard oral argument on July 9. However, because of divisions among the commissioners, the commission has been unwilling to bring the matter to a vote. At the same time, a group of cases brought by South Carolina, Washington, and others seeking review of DOE's effort to withdraw the Yucca license application and presenting essentially the same legal issues as the commission's proceeding have been pending in the D.C. Circuit Court of Appeals.[506] The court had originally stayed proceedings on the cases pending a commission decision, but on January 10, 2011, it scheduled oral arguments for March 22.[507] DOE's attempted license withdrawal has also prompted fresh litigation over DOE's failure to take responsibility for the SNF as mandated under NWPA.

Yucca Abandoned:
Repository Limbo, Orphan Waste Challenges

As a consequence of the abandonment of Yucca by the Obama administration, the nation's accumulated SNF and HLW now lack a disposal pathway. The NWPA strategy for dealing with the nation's most highly radioactive wastes—a strategy that has been the fundamental underpinning of government nuclear waste regulatory and policy decisions for more than twenty-five years—lies in ruins.

At the same time, President Obama has embraced the expansion of nuclear power. In his 2010 State of the Union address, the president proposed "building a new generation of safe, clean nuclear power plants."[508] On February 16, 2010, he announced more than $8 billion in federal loan guarantees to the Southern Company for construction of two new nuclear power plants, which would the first built in the United States since the 1980s, and promised support for more plants. The total amount of new federal loan guarantees in Obama's budget proposal for 2011 is $36 billion. There is $18.5 billion already authorized for this purpose, for a total of $54.5 billion in federal loan guarantees.[509] This generous subsidy has been sharply criticized by a number of groups, including the Union of Concerned Scientists, Physicians for Social Responsibility, the National Taxpayers Union, Taxpayers for Common Sense, the George Marshall Institute, the Heritage Foundation, and the Non-Proliferation Policy Education Center.[510]

Broad public support for the president's nuclear initiative may well depend on there being a credible plan for dealing with the country's growing accumulation of nuclear power wastes. There is no single centralized source of information on how much commercial SNF is currently in storage at reactor sites, but recent estimates place it between 62,500 and 65,000 MTHM.[511] The waste is stored in pools or dry casks at 104 nuclear power plants and several other facilities located at seventy-seven sites in thirty-five states.[512] New SNF is generated at a rate of approximately 2,000 MTHM per year.[513] Assuming a steady rate of generation, accumulated stocks of commercial SNF would reach an estimated 165,000 MTHM by 2060.[514] Commercial SNF inventories either already have or shortly will exceed the statutory cap of 63,000 MTHM on commercial SNF that may be disposed of at Yucca.

DOE is responsible for another 2,500 MTHM of SNF, which includes defense and other government SNF as well as some commercial, research, and foreign SNF.[515] This includes 2,129 MTHM at Hanford, 280 MTHM at INL, 28 MTHM at SRS, 15 MTHM at Fort St. Vrain, and 3 MTHM at other sites, including Argonne, Brookhaven, and Sandia National Laboratories and some university research reactors.[516] DOE is also responsible for approximately ninety megagallons of liquid HLW, primarily at Hanford and SRS with a quite small amount at INL, and 4,400 cubic meters of dry HLW at INL.[517] DOE's Office of Environmental Management estimates this SNF/HLW to be the equivalent of between 8,000 and 17,000 MTHM when rendered in a form suitable for repository disposal, while its Office of Civilian Radioactive Waste Management estimates it at 10,300 MTHM.[518] DOE's allocation of Yucca repository space for HLW and SNF combined is 7,000 MTHM. By any calculation, then, DOE inventory has exceeded that number.

Devising a solution to the waste problem will be the task of the Blue Ribbon Commission on America's Nuclear Future appointed in January 2010 by Energy Secretary Chu. Obama directed that the commission "conduct a comprehensive review of policies for managing the back end of the nuclear fuel cycle, including all alternatives for the storage, processing, and disposal of civilian and defense used nuclear fuel and nuclear waste."[519] Among other matters, it will have to deal with where to store accumulating SNF, including whether to prolong indefinite storage at reactor sites or to develop new consolidated storage facilities; whether to invest in developing new reprocessing and other fuel cycle technologies; how to develop one or more new repositories; and what to do about the government's mounting liabilities for failure to take SNF from nuclear utilities.

NWPA provides no current authority for development of any repository other than Yucca. It provides that a second repository cannot be built "unless Congress has specifically authorized and appropriated funds for such activities" and orders DOE to report to Congress on the need for a second repository.[520] In December 2008, only a month before President Obama took office, DOE issued its report, stating that : "The Secretary recommends that the preferred course of action is legislative removal of the statutory capacity limit of 70,000 MTHM on disposal at Yucca Mountain."[521] Removal of this statutory limit would defer the urgency in evaluating the issues associated with a second repository."[522] With the current administration's termination of Yucca, this option is now off the table. There will almost certainly be no new repository for a long time. Unless Yucca is revived as a result of further political changes, or WIPP is opened to HLW and SNF, Congress will have to legislate a new arrangement for developing one or more HLW/SNF repositories and addressing interim storage. Thus, the process begun decades ago to establish a safe and secure repository for orphan SNF and HLW has come full circle, putting the country back to where it was in 1982 or, indeed, in 1957.

Nuclear Renaissance?

President Obama's embrace of nuclear power in his first State of the Union Address and his push for loan guarantees to support new nuclear plants have been celebrated as heralding a "renaissance" of nuclear power in the United States. There is a rising tide of political support for the technology, not only in the United States but also in many other countries.[523] After three decades in which not a single new U.S. nuclear power plant had been started, eighteen companies have applied to NRC for new combined operating licenses to build and operate twenty-eight reactors; currently, thirteen applications for twenty-two units are under active review.[524] Even if all these plants are built, given the lead times needed to bring new plants on line, expected retirements of existing plants (taking into account NRC extensions of their operating licenses), and projected increase in total U.S. electricity production, these new plants would not significantly increase the share of electricity production contributed by nuclear plants because of the concomitant growth in other forms of generation.[525]

The push for new nuclear power plants began under the George W. Bush administration, encouraged and supported by subsidies and other incentives provided under the Energy Policy Act of 2005.[526] NRC is rebuilding its resources and capacity to process license applications and regulate the anticipated increase in operating plants. Rising concern with climate change has helped build new support for nuclear power, even among some states and environmentalists who formerly opposed it.[527] No carbon dioxide or other forms of conventional air pollution are emitted directly by nuclear plants, and the amount of indirect carbon dioxide emissions attributable to such plants is quite low. Nuclear power plants are thus highly climate friendly, especially compared to coal combustion, the leading method used for electricity generation in the United States.[528] There is, however, considerable debate about how much the share of nuclear power in U.S. electricity generation could realistically be increased, and how much it could contribute to climate change mitigation, over the next fifty to one hundred years. Skeptics invoke a well-known Massachusetts Institute of Technology (MIT) report, finding that replacing coal-fired plants with nuclear plants on a large scale would result in only a relatively small (though not insignificant) reduction in carbon dioxide emissions: 1.5 million tons of carbon per year per plant.[529]

Nonetheless, many thoughtful observers have concluded that expanded reliance on nuclear power is a necessary and significant element in the broad suite of measures that will be needed to reduce greenhouse gas emissions, along with very significant investments in energy efficiency in all sectors, renewable energy generation, and carbon capture and storage.[530] Some states that adopted moratoriums on new nuclear plants are reconsidering their position.[531]

The operating efficiency of nuclear plants has improved substantially over the past decades, and NRC has extended operating licenses for existing plants by up to 20 years.[532] R&D is moving forward on new advanced fuel cycle reactor designs, including several different very high temperature designs that promise greater production efficiency and would use passive safety designs that are less complex and more reliable than those used in current light-water reactor (LWR) designs.[533] Some of the new designs would allow construction of smaller reactors (two hundred maximum gross weight, compared to one thousand maximum gross weight or more for typical LWR) of standardized design that are cheaper and quicker to build. Some models could be built in manufacturing plants and transported to sites for installation. Also in the early planning stages are new advanced fuel cycle reprocessing methods and reactor designs, discussed here later.

Significant concerns remain, however, about the safety of nuclear power plants, the effectiveness of their security arrangements, the adequacy of NRC regulation, and the still-unresolved waste problem.[534] Vermont recently decided not to extend the operating life of the Vermont Yankee nuclear plant, despite local support for it, based on radioactive leaks and misstatements in testimony by plant officials before the state legislature.[535] Additional concerns relate to the economics of nuclear power, which entail huge up-front costs and long periods for licensing and construction, at least for large models constructed on-site. Unless the United States adopts climate regulatory legislation that imposes a substantial price on carbon dioxide emissions, electricity generated by new nuclear plants would be substantially more expensive than that generated by either coal or natural gas.[536] MIT Professors Joskow and Parsons calculated and compared the cost of electricity from new coal, natural gas, and nuclear power. Using an estimate of moderate coal and natural gas prices, they calculated that nuclear energy is 2.2 cents per kilowatt-hour more expensive than coal, and 1.9 cents per kilowatt-hour more expensive than natural gas.[537] Nuclear power may thus be economically competitive only if it receives favorable treatment from state utility regulators or additional, very substantial government subsidies (many other energy sources receive some subsidies).[538] According to opinion polls, a significant segment of the public remains skeptical or opposed to new nuclear power construction—although overall there is now greater public support, prompted by climate change concerns, for increasing nuclear power production, than there was in past years.[539]

As this book goes to press, a deep nuclear power plant crisis is unfolding in Japan. A severe earthquake and a resulting tsunami seriously damaged a number of nuclear reactors and knocked out regional power, causing cooling systems at the reactors to fail and resulting in at least partial core reactor meltdowns and potential releases at SNF cooling ponds. Substantial radioactive material has been released into the atmosphere, and over a hundred thousand people have been evacuated. As of this writing, the situation at the stricken plants has not been brought under control; the extent of human health risk posed by radioactive releases from the plants will not be known for some time. Whatever the outcome, these events have already caused many Europeans and Americans to question the wisdom of expanding nuclear power. These developments will also focus long-overdue attention on the safety of existing nuclear plants, on the wisdom of steps contemplated by NRC to further extend the operating licenses of older U.S. plants, and on the risks posed by current practices for long-term storage of SNF in crowded cooling ponds. At the very least, the loss of coolant in the stricken Japanese plants underlines the need to move SNF into dry-cask storage as promptly as is feasible. The blue-ribbon commission will have to confront these and other issues resulting from the Japanese power plant crisis and their implications for America's nuclear future.

The biggest boost to nuclear power would be adoption of climate legislation that would impose a significant price on carbon dioxide emissions and thereby substantially raise the price of coal-fired electricity relative to that of nuclear.[540] The Obama administration's effort to enact climate regulatory legislation that includes a cap-and-trade system has failed, at least for now; U.S. climate and energy legislation is moving in the direction of granting greater subsidies for nuclear power and renewable energy sources rather than imposing a significant price on greenhouse gas emissions.

Federal Liability for Failure to Dispose of Civilian SNF

NWPA provided that DOE would take ownership of civilian SNF for disposal beginning in 1998 and required private utility companies in turn to pay a fee based on their nuclear-powered electricity generation.[541] The proceeds of this fee—an average charge of one mil (one tenth of one cent) per kilowatt-hour—are designated for a federal Nuclear Waste Fund (NWF). NWPA also imposed a retroactive tax of one mil per kilowatt-hour on nuclear electricity generated before its enactment. According to the Congressional Budget Office, total NWF revenues plus accumulated interest less expenditures through FY09 amounted to $23.8 billion.[542]

In developing SNF repositories and storage facilities, DOE may only expend such funds as Congress authorizes every three years and appropriates annually.[543] Through FY 2008, a total of $7.1 billion had been appropriated by Congress and spent by DOE for dealing with SNF, much of it for activities relating to development of Yucca Mountain. Because total NWF revenues amounted to over $31 billion, approximately $24 billion in fees paid by utilities (and ultimately by electricity consumers) has not been spent for dealing with SNF. These revenues, however, are not set aside in a pot of money earmarked for SNF storage and disposal. Instead, Congress has already used the revenues for general government spending. In seeking congressional authorizations and appropriations for its SNF activities, DOE must compete with all other discretionary (nonentitlement) federal programs for the funds available under overall federal spending caps.[544] Thus the NWF is simply an accounting convention without real financial or operational significance. Effectively, excess NWF revenues over expenses have been used by Congress over the years for purposes other than dealing with SNF.[545]

The contracts executed by DOE and the utilities pursuant to NWPA provide that DOE will take title to and begin to dispose of the utilities' SNF not later than January 31, 1998. Because no federal repository or interim storage facility has been built, the government has not taken charge of the utilities' SNF. Asserting that the government has been in default of its obligations since 1998, utilities have brought litigation against it seeking judgments requiring the government to take the SNF waste and pay monetary damages for its failure thus far to do so. The Standard Contract between DOE and utilities provides that neither party will be liable for damages caused by DOE's failure to perform its obligations under the contract "if such failure arises out of causes beyond the control and without the fault or negligence of the party failing to perform."[546] Anticipating that it would be unable to take SNF by the 1998 deadline, DOE in 1994 took the position that unless a federal repository or other SNF facility becomes available, "it has no statutory obligation to accept spent nuclear fuel beginning in 1998."[547] Acknowledging that it had "created an expectation" contrary to this view, DOE promised to "explore" different forms of cost sharing with the utilities.[548] Subsequently, the government argued that the failure to develop a repository relieved it of liability pursuant to the contract provision regarding unavoidable delays; the courts rejected this claim and held the government in breach of its obligations to take the utilities' SNF.[549]

As regards remedies for the government's breach, the federal courts have not granted the utilities specific performance of the contracts, ruling that the government does not have to take title to the SNF but must pay damages based on the economic losses that the utilities have incurred by reason of the government's breach.[550] For example, in *Sacramento Municipal Utilities District v. United States*, the U.S. Court of Federal Claims ordered damages to be paid for the materials and labor required for the construction and maintenance of at-reactor dry-cask storage. It did not, however, require the government to pay for the additional cost of dual-use dry storage casks capable of being used in the transport of waste to a repository or MRS.[551] Other utility lawsuits have reached similar results.[552] A still unresolved legal issue in determining damages is how much SNF DOE is contractually obligated to take, at what rate, and the priority accorded SNF at different reactor sites.

Utilities have filed more than seventy suits against the government on this matter, of which fifty-one remain pending. The rest were settled, litigated to final judgment, or withdrawn.[553] While there was discussion under the George W. Bush administration of seeking Supreme Court review of a group of adverse court decisions, the Obama Department of Justice has been silent on this possibility. The courts have ordered that hundreds of millions of dollars in damages be paid to the utilities.[554] The cost to date to the government solely for litigating the claims filed has been estimated at $154 million.[555] The courts have also ruled that DOE may not use the NWF to pay damages, and that it may not offset damages through compensating adjustments to a utility's future NWF fee payments.[556] Damages are paid out of the U.S. Treasury's Judgment Fund. The nuclear industry has recently claimed that the government's liabilities could mount to $50 billion.[557] DOE, by contrast, has estimated that its liability will be around $11 billion if the government begins accepting nuclear waste in 2020, plus an additional $500 million for each additional year thereafter that it fails to take waste.[558]

The administration's recent cancellation of Yucca raises the potential liability stakes significantly. In the cases decided thus far, DOE has been held to be in partial breach, on the premise that Yucca would eventually open.[559] If, however, Yucca is conclusively terminated, utilities could argue that the government had committed a complete breach, because there is no prospect that a repository will be available in the foreseeable future. Currently, the utilities sue for the extra costs they have incurred to date for storage of SNF on-site. In the event of a complete breach, the utilities could seek permanent damages based on the net present cost to them of either building or operating a consolidated storage facility or indefinite at-reactor storage.

The anticipated cancellation of Yucca has triggered proposals to modify or eliminate the current NWF financial arrangements. A group of Senate Republicans introduced a bill to liquidate the NWF if Yucca is cancelled and distribute an amount equal to 75 percent of the accrued unspent revenues to utility consumers.[560] Maine adopted a resolution urging President Obama and Congress to reduce fees that utilities pay and pass on to consumers, while Michigan adopted a resolution to put fees paid by the utilities into an escrow account.[561] The Senate Appropriations Committee report accompanying the FY 2010 appropriations bill recommended suspending the collection of payments from the utilities.[562] The National Association of Regulatory Utility Commissioners wrote to Secretary Chu requesting that he suspend payments into the fund.[563] DOE, however, has rejected requests that it stop collecting NWF fees.[564]

Options for Dealing with Orphan Waste

Large amounts of defense HLW and smaller amounts of SNF for which DOE is responsible were earmarked for disposal at Yucca Mountain. They will presumably remain at the DOE sites where they are currently stored, although there may be some transfers among DOE sites. Putting some or all of these wastes into consolidated storage is also an option. The uncertainties and delays involved in siting, licensing, and constructing a new repository for disposal of DOE's nuclear waste will most likely make it impossible for DOE to comply with the deadlines in some tripartite agreements requiring HLW and SNF at its facilities to be shipped off-site.[565] Delays and uncertainties in the availability of a repository also raise questions about the resources that should currently be devoted to treating HLW for repository disposal.

The disposition of SNF from nuclear power plants presents more options and more difficult policy and political issues. As discussed earlier, over 62,500 metric tons of SNF are currently stored at seventy-seven sites in thirty-five states.[566] It is possible that under a future Republican president and Congress (with Harry Reid no longer Senate majority leader), Yucca might be revived in some form. It is also possible, if New Mexico were given sufficient inducements to agree, that WIPP could be opened to SNF. But absent those eventualities, and given the time (decades, at a minimum) until a repository at another location could be sited, licensed, built, and opened, existing stocks of SNF have no destination pathway. SNF could continue to be stored at reactor sites, although pressure from local communities to move the waste elsewhere is likely to build, especially after the reactors that created the SNF are decommissioned. Other options include SNF storage at existing DOE sites or at new government or private consolidated storage facilities, or reprocessing of SNF.

Interim Storage of SNF

As of 2009, 85 percent of accumulated SNF was being held in concrete-lined pools of water at reactors; the remaining SNF was housed in dry storage casks at reactor sites.[567] Dry-cask storage is more expensive than pool storage but is also regarded as safer. NRC has regarded the risk that an earthquake, accident, or terrorist attack could drain a storage pool's water, possibly sparking a fuel assembly fire and releasing radioactivity, as quite small.[568] NRC, however, did not consider the possibility of a terrorist attack on the storage pools.[569] It is reexamining this assessment in light of the Fukushima crisis. A 2007 report by the Union of Concerned Scientists (UCS) found that the risk and potential consequences of terrorist attacks on SNF stored at reactor sites have been significantly underestimated. UCS contends that it would be relatively easy for terrorists to overwhelm guards at a power plant and drain water from the cooling pool to start a fire in fuel assemblies, threatening widespread releases of radioactivity. It concluded that both pool storage and dry-cask storage need to be "hardened" against terrorism and reactor site security upgraded.[570] The NAS Board on Radioactive Waste Management recommended in a 2001 report that most or all SNF at reactor pools be moved to dry-cask storage as soon as possible due to vulnerability to terrorism.[571] Nevertheless, NRC has regularly granted utility requests to increase the amount of SNF stored in pools rather than use more costly dry casks. Dry-cask storage is, however, expected to increase as reactors run out of pool storage capacity.[572]

NRC has determined that SNF storage in dry casks, even if not entirely free from risk, is "safe and environmentally sound."[573] NRC currently issues licenses for dry-cask interim storage for twenty years, with the option of renewal.[574] DOE secretary Chu has stated that dry-cask storage can be relied upon for half a century, perhaps longer, while alternatives to Yucca are explored and developed.[575] Temporary pool storage will always be needed, however, because SNF requires, on average, five years of cooling in a pool before it can be transferred to dry-cask storage, where it is air cooled.[576]

There are various possibilities for storing SNF away from reactor sites for a substantial period pending development of a repository. One option, floated by DOE, is storage at existing DOE sites in conjunction with co-developed energy parks, which could include advanced fuel cycle reprocessing and reactor facilities and other installations at the same sites that could provide additional economic and other benefits to host jurisdictions. But potential hosts may be quite leery at the prospect of bringing more highly radioactive wastes onto DOE sites that may well be unable to comply with commitments to remove the wastes already stored there. Also, DOE has concluded that NWPA precludes it from storing SNF at its sites without further legislation by Congress.[577]

Another option is federal government development of consolidated storage facilities at new sites. As discussed, NWPA provided for development of one or more federal MRS facilities for civilian SNF, but DOE's efforts to develop such a facility were blocked by potential host states and greatly restricted by the 1987 NWPA amendments. DOE has not undertaken any significant steps to site an MRS facility since that time.[578] The Office of the Nuclear Waste Negotiator, authorized by the amendments and since lapsed, also failed to develop an MRS.

Finally, NWPA does not preclude a private SNF storage facility. The Private Fuel Storage (PFS) consortium of nuclear utilities came close to developing such a facility on lands of the Skull Valley Band of Goshute Indians in Utah. The facility, which would have had the capacity to store forty thousand metric tons of SNF, was granted an NRC license in 2006 following a nine-year licensing process.[579] Construction of the PFS facility, however, was strongly opposed by the State of Utah and by some dissenting tribe members. It was blocked by the Interior Department's Bureau of Indian Affairs (BIA), which refused to approve the tribe's land lease to PFS, and by its Bureau of Land Management (BLM), which denied the consortium a right-of-way for transporting spent fuel.[580] In 2007, PFS and the Skull Valley Band of Goshute Indians filed suit in federal district court in Utah challenging the BIA and BLM permit denials.[581] The court overturned the denial of the permits in July 2010, and remanded the decisions back to the agency.[582] It remains to be seen whether the facility will eventually receive the authorizations and be built.

Proposals to Revive Reprocessing: GNEP

After a thirty-year moratorium on federal support for SNF reprocessing, the Bush administration in 2006 proposed to revive reprocessing as part of the Global Nuclear Energy Partnership (GNEP).[583] GNEP proposed development and widespread adoption of advanced reprocessing technologies significantly different from the plutonium uranium extraction (PUREX) process for making mixed oxide fuel used in the past in the United States and still being used in France and Russia, the only countries that currently operate full-scale SNF reprocessing plants.[584] GNEP proposed that the private sector would use the new technologies to reprocess SNF to produce a new form of plutonium-based fuel that would be burned in new advanced fuel cycle reactors.[585] GNEP was strongly criticized by many experts, was unpopular with the Democratic Congress, and has been dropped by the Obama administration.[586] On April 15, 2009, DOE declared that "the long-term fuel cycle research and development program will continue but not the near-term deployment of recycling facilities or fast reactors."[587] In June 2009, DOE cancelled an environmental impact statement on GNEP, effectively terminating the project.[588] Nonetheless, the GNEP proposal has served to put reprocessing back in play as a potential component of any nuclear power expansion in the United States (see Chapter 7 for additional discussion of GNEP, including the technical issues involved).

Although currently abandoned, GNEP exemplifies the types of advanced fuel cycle technologies that might be pursued in the future. Energy Secretary Chu has expressed interest in R&D on such technologies.[589] GNEP proposed that SNF would be reprocessed and its constituents separated into four basic groups: (1) uranium would be used to generate electricity in the United States in a new type of light-water reactor or provided to developing countries for electricity production; (2) Sr-90 and Cs-137, two relatively shorter-lived but highly radioactive fission products, would be placed in surface storage, most likely at the reprocessing plants, for several hundred years until the waste had decayed to the point that it could be disposed of as LLW;[590] (3) other fission products, including long-lived elements such as I-129, Cs-135, Tc-99, Sn-126, and Se-79, would be disposed of as HLW in a repository; and (4) plutonium and other transuranic elements would be fabricated into a new form of fuel that would be burned in new "fast" reactors and run through the reprocessing/reactor fuel cycle several times, substantially reducing the amount of transuranic waste that would remain to be disposed of at the end of these cycles.[591] DOE claimed that, under GNEP, the volume and heat of the wastes requiring deep geologic disposal would be appreciably reduced relative to SNF or the wastes produced by existing reprocessing technologies, such that there might be no need for a second repository or some of the technical challenges for siting a repository would be alleviated.[592] Proponents also claimed that the GNEP technologies would produce fuel forms that would resist proliferation, and that reprocessing would enhance U.S. energy security.

The GNEP proposal included an international component in the form of a multilateral nonproliferation regime that would include the United States, other fuel cycle developed countries engaged in reprocessing, and developing countries wishing to expand nuclear power.[593] The developed countries would lease low-enriched uranium extracted during reprocessing to developing countries for use in light-water reactors to generate electricity, in exchange for developing countries' agreement to refrain from using uranium enrichment or reprocessing technologies that could produce weapons-grade uranium or plutonium.[594] A number of developed and developing countries signed on to the prototype for such a regime. To obtain participation by developing countries, however, the fuel cycle countries supplying the uranium might have to take back and assume responsibility for the SNF produced, aggravating the domestic politics of nuclear waste in those countries.[595]

The complex, ambitious, and costly GNEP proposal encountered strong opposition from many quarters, including independent expert bodies, environmental groups, and experts on proliferation.[596] A 2008 report by an NAS committee charged with examining DOE's R&D program concluded that GNEP would be enormously expensive and unlikely to produce meaningful benefits anytime soon.[597] GAO also expressed strong skepticism about GNEP.[598] The proposal was unpopular with Democrats in Congress.[599]

The critics underscored the very high costs and major technical challenges presented by developing to scale advanced fuel cycle reprocessing technologies such as those proposed in the GNEP program. They also asserted that the plutonium produced by reprocessing poses grave proliferation risks and sharply challenged claims that the proposed GNEP technologies would significantly reduce those risks.[600] As regards energy security, critics noted that the biggest U.S. problem is imported oil. Nuclear power is used for base-load electricity generation. It can accordingly do little to reduce oil imports for transportation, at least until the use of electric cars becomes widespread or hydrogen-based transportation systems that use electricity to produce hydrogen fuel are developed. SNF does represent a secure potential source of future nuclear fuel. The United States imports 92 percent of the uranium currently used in nuclear fuel (a large percentage from Russia, from dismantled nuclear warheads).[601] Supplies of uranium worldwide, however, are currently plentiful, and major supplier countries such as Canada and Australia are reliable U.S. allies.[602] Yet as surplus defense inventories are depleted through conversion to fuel and as worldwide use of nuclear power expands, significant increases in uranium prices may well occur.

Although proponents claim that use of advanced reprocessing technologies would significantly reduce the waste burden, some critics of reprocessing charge that it would actually generate more radioactive waste that requires storage and disposal than does the current, once-through cycle.[603] The waste forms produced by reprocessing include large amounts of TRU as well as other wastes, including iodine-129, krypton-85 gas, incinerated solvents, and contaminated equipment, which would present significant technical challenges for secure disposal.[604] If the cesium and strontium components of SNF were separated out from the reprocessing HLW waste streams and stored for several centuries for cooling and decay, they could be buried as LLW, but this prospect would pose the formidable and untested challenge of assuring safe and secure storage for hundreds of years.[605]

DOE prepared a preliminary EIS analyzing the anticipated wastes generated by GNEP reprocessing over fifty years. The EIS predicted that this reprocessing would increase LLW by a factor of seven. This would not be just low-activity waste. Reprocessing was estimated to increase the volume of GTCC waste generated over the next fifty years from 2,500 to 416,500 MTHM.[606] The estimated total volume of SNF that would be generated by electric power reactors over that period is between 100,000 and 158,000 MTHM.[607] If the increase in GTCC waste volume, which may well require geological isolation in repositories, exceeds the waste volume reduction from recycling SNF, then reprocessing may not save space in repositories, although initial heat loadings would be greatly reduced.[608] DOE's EIS also found that dealing with reprocessing wastes would result in significant increases in transportation and handling exposures.[609]

Critics also challenge the economics of reprocessing. Use of mixed oxide (MOX) fuel for electricity generation is currently substantially more expensive than using fuel from virgin uranium.[610] A 2003 Harvard study found that even with relatively conservative assumptions, using MOX fuel in the United States would not be economic until uranium prices rise to more than $360 per kilogram.[611] Historical prices for processing uranium have been far below this level (see Appendix B). As of July 2010, the price of uranium was around $90 per kilogram.[612] In France, where about 70 percent of SNF is currently reprocessed, the government estimates that reprocessing increases the annual cost of nuclear power generation by 12 percent.[613] Changes in future uranium processing, reductions in the cost of making MOX fuel, and rising concerns about energy security may well change the relative economics of mixed oxide fuel reprocessing and use of virgin uranium fuel.[614]

Notwithstanding all the criticism that it merited, GNEP succeeded in reviving serious interest in reprocessing as an option to consider in future nuclear waste and nuclear power policies. On the international front, GNEP still meets as an international network but without the full backing of the United States. Responding to concerns of nuclear proliferation like those that that led to the international component of the GNEP proposal, the Obama administration has proposed support for a global uranium fuel bank to serve the same anti-proliferation ends. The uranium fuel bank would provide a supply of low-enriched uranium to be used in power plants around the world and would provide nuclear rogue countries like Iran with uranium for power plants, with the goal of dissuading them from developing enrichment facilities that could produce uranium for use in nuclear weapons.[615]

The idea of a uranium fuel bank has earned broad multilateral support, including that of the International Atomic Energy Agency (IAEA). IAEA first proposed the fuel bank back in 2006 as part of its Nuclear Threat Initiative, and as of March 2009, the nations of the world have pledged $100 million to develop one.[616] However, there has been opposition to a uranium fuel bank from countries interested in developing nuclear power. In 2008, Iran indicated it would join a fuel bank only as a supplier, while countries as diverse as South Africa, Egypt, and Italy have balked at abstaining from uranium enrichment while the United States and other developed nations maintain facilities to do so.[617]

Even if the United States were to move toward some form of SNF reprocessing, any advanced fuel cycle reprocessing and reactor technologies would take at least several decades to be developed and brought on line at scale.[618] Accordingly, even under the most optimistic projections—and assuming the as-yet-undemonstrated proposition that reprocessing could alleviate the need to build geological repository facilities—reprocessing cannot realistically be viewed as a near-term means for addressing the SNF disposal problem. If, however, an entirely new repository, other than Yucca or WIPP, must be developed for SNF, the extended period needed for its development (with SNF in storage) might recommend further consideration of reprocessing options.

Chapter 2

Radioactive Waste Classification and Regulation

The classification in legally defined categories of radioactive material produced by nuclear facilities has important regulatory and liability consequences, including requirements as to management, storage, treatment, packaging, transportation, and disposal. Most radioactive wastes are produced by the various stages in the uranium fuel cycle for producing electric power or nuclear weapons material. Smaller amounts of wastes are produced in connection with the use of radioactive materials that exist naturally (for example, radium) or that are produced by particle accelerators.

This chapter first presents an overview of the radioactivity characteristics of different types of wastes in relation to the wastes' origins, the associated health and ecological risks, the existing regulatory waste classifications for civilian and defense wastes, and the ways in which these classifications have evolved and been implemented by the Atomic Energy Commission (AEC), the Environmental Protection Agency (EPA), the Nuclear Regulatory Commission (NRC), and the Department of Energy (DOE). Occupational risk regulation is not included. The chapter also examines controversial efforts by NRC and DOE to classify and treat certain reprocessing wastes as other than high-level waste (HLW), and by NRC and EPA to classify certain very low activity wastes as exempt from nuclear waste disposal requirements. It further examines the treatment of mixed wastes that have both radioactive and chemically toxic components and therefore are potentially subject to regulation by EPA and states under the Resource Conservation and Recovery Act (RCRA), as well as by DOE and NRC under the Atomic Energy Act (AEA).

The current U.S. system of waste classification is an amalgam of various provisions in statutes and regulations that has evolved in piecemeal, patchwork fashion over many years. The result is a mixed constellation of categories, definitions, precedents, and regulations whose legal consequences do not always reflect relative hazards or sensible waste management policies and priorities. The implications are significant in terms of environmental health and safety protection, management and disposal options, cost, feasibility (fiscal, social and political), and the design of waste regulatory management and regulatory institutions. The concluding section of the chapter outlines a number of steps toward a more risk/hazard-informed and performance-based approach.

Nuclear Wastes and Their Regulatory Classification

Environmental, health, and safety (EHS) regulation of nuclear wastes must consider the radioactive characteristics of various types of wastes, the human health and ecological risks they pose, and the appropriate means of managing those risks at an acceptable level—including, potentially, precluding their production. The hazards posed by the various radioactive elements contained in nuclear wastes depend on the type, level, and longevity of the radioactivity they emit, as well as on their concentrations. Risks to living things depend on the extent of exposures to radiation, as well as its character. EHS risks from radioactive wastes can be reduced by engineered barriers, including containment materials and shielding; ground disposal in suitable soils or geologic formations; measures to close exposure pathways; siting of storage and disposal facilities away from humans and other life forms; and use of institutional controls to prevent humans and other living things from coming into proximity to wastes. The solubility of various waste elements is also a salient factor in evaluating and mitigating the risk of waste releases and diffusion and resulting exposure.

Waste management and regulation must consider not only these different risk factors and means for reducing exposures but also administrative feasibility and the ability to monitor and enforce compliance. In light of these considerations, regulators seek to devise the most workable and appropriate way of classifying wastes and matching regulatory requirements with those classifications. In doing so, they must weigh the tradeoffs between uniform requirements and context-specific measures.

Hazards Presented by Radioactive Elements

Radioactivity occurs when the nucleus of a radioactive element decays and releases energy particles. Three main types of radiation exist, each with its own decay product that poses type-specific health and environmental hazards and calls for type-specific measures to prevent harmful exposures.

In the first type of decay, alpha decay, the nucleus releases an alpha (α) particle, a large, positively charged, relatively heavy mass of nucleons. This form of radioactivity is the least able to penetrate substances: a sheet of paper—or human skin—can stop most alpha radiation. But if ingested or inhaled by a human or an animal, alpha particles can cause serious or fatal injuries.

In the second type of decay, beta decay, the nucleus releases a beta (β) particle that consists of a free electron or positron (an electron's positively charged counterpart) traveling at speeds near that of light.[1] More significant barriers, like a metal plate, are required to block beta radiation.

In the third type of decay, gamma (γ) decay, the nucleus emits a photon, a massless, chargeless quantum of electromagnetic radiation. Because gamma particles have no mass and high energy relative to other electromagnetic radiation, they penetrate matter readily and deeply, and require large amounts of dense shielding material to prevent exposures.

The amount of radiation emitted is also a key factor in determining the extent and types of measures required to prevent harmful exposures. Radioactivity—the number of radioactive transformations that occur in a given amount of a given element over a given period of time—is the measure of this emission rate. This rate is equal to the density of the radioactive element multiplied by its decay rate. The most common unit for measuring radioactivity is the becquerel (Bq), defined as one radioactive transformation per second; the curie (Ci) is also used in the United States.[2] A related quantity, called the equivalent dose rate, is used to relate the radioactivity of a substance to its biological effects by measuring the amount of radioactive energy per unit of organic mass exposed. This rate is then multiplied by a factor to adjust for the relative damage caused by the different types of radiation.[3] The international unit of equivalent dose rate is the sievert (Sv).[4] Also used in the United States is the roentgen equivalent in humans (rem).[5]

The radioactivity of a given element is a function of the density of the radionuclide and its stability. Any particular radioactive event occurs at random, but together a large number of radioactive nuclei follow an exponential decay pattern. The rate of radioactive decay for a given element is usually characterized by a quantity called the half-life (λ), which represents the average time it will take for half the nuclei to decay, i.e., to emit half of their radioactivity. As decay occurs, the level of radioactivity emitted by the undecayed portion declines correspondingly. Because significant amounts of the radioactive mass may be left after multiple half-lives have elapsed, a measure that may be more useful for purposes of regulatory policy is the time it will take for 95 percent of a radioactive element to decay (T_{95}).[6] The half-lives of radioactive elements vary widely: cesium-137 (Cs-137), among the more important fuel cycle byproducts, has a half-life of only 30 years and a T_{95} of 130 years; while plutonium-239 (Pu-239), the principal ingredient in nuclear weapons, has a half-life of 24,000 years and a T_{95} of about 100,000 years. Uranium-238 (U-238)—sometimes called "depleted uranium" or "the isotope remaining after enrichment"—is even longer-lived, with a half-life of 4.5 billion years. Its radioactivity level is, however, low. Radioactivity levels and lives are inversely correlated. An element with high radioactivity decays more quickly than the equivalent amount of an element with low radioactivity.

In addition to potential exposures from nuclear wastes, humans are regularly exposed to low-level radiation from radioactive elements in the earth's crust, from extraterrestrial radiation that penetrates the earth's atmosphere, and from anthropogenic sources such as x-rays and televisions (see Table 2.1). The degree of risk to living creatures from radiation depends on its type, the amount of exposure, and the type of tissue exposed. The equivalent dose rate is a tool for comparing differences in risk due to exposures to different types and levels of radiation.

The purpose of regulating nuclear waste storage and disposal is to reduce, as far as possible, exposure to humans and the environment. The most basic and often least costly approach is to package waste in protective material—such as dry casks for spent nuclear fuel—that blocks the escape of radiation, and then to store it on or just under the ground surface. Fences and buildings to help keep interlopers away can be supplemented by measures such as signage and guards. While this approach may be suitable for certain short-lived wastes, barriers and other more robust and long-lasting means of isolation must be used for waste that contains elements with longer half-lives. For materials with a limited degree of radioactive contamination or radioactive elements with relatively short half-lives and low levels of radioactivity, disposal in landfills is often appropriate, perhaps supplemented by engineered barriers and institutional controls.

Table 2.1. Average annual whole-body radiation dose from common sources

Radiation source	Average annual whole-body dose (millirems/year)
Natural: cosmic	29
Terrestrial	29
Radon	200
Internal (K-40, C-14, etc.)	40
Synthetic: diagnostic x-ray	39
Nuclear medicine	14
Consumer products	11
All others: fallout, air travel, occupational, etc.	2
Average annual total	360*

*Tobacco smoking adds approximately 280 millirems/year for a person who smokes.

Source: Office of Environmental Health and Safety, Princeton University, "Open Source Radiation Training," *web.princeton.edu/sites/ehs/osradtraining/backgroundradiation/background.htm*.

For concentrated wastes containing relatively long-lived elements, deeper burial in a site isolated from groundwater and erosion may be necessary. For very long-lived and highly radioactive wastes, burial in deep geological repositories is necessary to prevent future exposure. Such waste may be vitrified to keep it from seeping into groundwater and migrating.[7] In the very long run, over hundreds of thousands of years, some degree of release and migration of the most long-lived wastes is inevitable, because the effectiveness of all containment methods will eventually deteriorate. Consequently, confidence in the use of engineered barriers and institutional controls to prevent release, migration, and exposure to radionuclides is greater and more justified in the near term than in the long term.

If waste is sufficiently radioactive, storage and disposal are complicated by significant heat generation. When initially withdrawn from the reactor, SNF can emit energy at rates around two to twenty kilowatts per cubic meter, which requires cooling measures to be taken lest the fuel melt surrounding metal containment cladding. As an illustration of the heat loads involved, an air conditioner capable of cooling a midsized room would be needed to adequately cool and keep at a stable temperature a single cubic meter of SNF. In sum, the types of radiation and activity levels and corresponding decay rates are key in determining what methods of disposal are appropriate for different types of radioactive wastes.

The Nuclear Fuel Cycle and Other Sources of Nuclear Wastes

To better understand the appropriate methods for managing and regulating various nuclear wastes, one must consider their source and character. The most radioactively significant nuclear wastes are produced in the process of generating electricity or making nuclear weapons through use of a nuclear fuel cycle that involves controlled fission of uranium in a reactor. Nuclear wastes are also generated by use of sealed radioactive sources in medical, research, and commercial applications; these sources become waste when no longer used.

The Once-Through Fuel Cycle for Electricity Generation

The typical fuel cycle used in the United States for generating electricity is the once-through thermal fuel cycle. The term "once-through" refers to the practice of using the uranium in fuel rods in the reactor fission process only once, and not reprocessing the spent fuel to extract fissionable elements (uranium and plutonium) remaining in the rods for future use as fuel.[8]

Uranium deposits extracted as ore from the earth's crust are concentrated and processed into a powder known as yellowcake. The uranium ore contains two main uranium isotopes, U-235 and U-238. U-235 constitutes 0.72 percent of the total ore and U-238 the balance. In order to sustain fission in a reactor or a bomb, a higher percentage of U-235 is required; the fuel used in a power plant consists of 3–5 percent U-235. To increase the U-235 concentration, centrifuges or (less frequently) gaseous diffusion technologies are used to separate out U-235 from the naturally occurring U-235/U-238 mixture by taking advantage of the different atomic weights of the two isotopes.[9] The enriched uranium mixture is fabricated into uranium-hexafluoride pellets and then packed into metal alloy cladding in the form of a rod, which generally has a radius of one centimeter and a length of four meters. Fuel rods are bundled together with metal frames in arrays of 14×14 or 17×17 rods; such arrays are called fuel assemblies.

The fuel assemblies are inserted into a reactor core to feed the fission process. The fission process accelerates radioactive decay and generates substantial heat. In the light-water reactor that is generally used to generate electricity in the United States, this heat is used to heat water to drive steam electric generating turbines.[10] Fuel rods are withdrawn from the reactor after they are no longer capable of sustaining a fission reaction because of changes in their elements as a result of the fission process. At this point, fuel rods become spent nuclear fuel (SNF). A typical thousand-megawatt light-water reactor will produce about twenty-seven metric tons of SNF per year, which corresponds to seventy-five cubic meters of waste.[11]

About 98 percent of the spent fuel by mass from commercial reactors consists of "unburned" uranium. About 0.8 percent of this total is U-235; the rest is U-238.[12] The remainder of SNF consists of fission byproducts. Uranium fission produces a very wide variety of elements with different lives and different patterns and levels of radioactivity. Some of the byproducts, known as daughter products, are produced when fission breaks up the uranium atoms in the fuel; they consist of elements with lower atomic weights than uranium's. Others are formed when some of the uranium, instead of fissioning, absorbs neutrons, thus creating transuranics—elements with a larger atomic number than uranium's. Plutonium is one of these transuranic byproducts of fission; 1 percent of the total mass of SNF consists of plutonium. Other transuranics and daughter products constitute another 1 percent of the total mass of SNF. The U-235 and plutonium in SNF, as discussed later, can be extracted through reprocessing and used to create fresh fuel, but reprocessing has not been practiced in the United States for thirty-five years.

Only two transuranics (plutonium and neptunium) occur on earth naturally; they are found, in very small amounts, in uranium ore. The other transuranics are solely the product of human nuclear activities. The transuranics tend to be very long-lived alpha emitters. For example, plutonium-240 (Pu-240) is an alpha emitter with a half-life of 6.5 thousand years and decays into uranium-236 (U-236), itself an alpha emitter with a half-life of 23 million years.[13] Such elements require very long-term isolation of the kind only deep geological repositories can currently afford. Although they are easily shielded, they must remain so for an extremely long time to make sure they are not inhaled or ingested by humans or other living beings while they still emit dangerous amounts of alpha radiation.

The daughter fission products include elements with varying types and levels of radioactivity and lives. The most important of these are certain highly active, short-lived, and extremely hot varieties, such as Cs-137 and strontium-90 (Sr-90). Both are high-level beta emitters; Cs-137 also emits gamma radiation during its decay. These elements have half-lives of about thirty years and also emit high levels of heat, which complicates their disposal, as well as the disposal of any SNF that contains them. Cesium is very toxic to humans, and its water solubility puts it at high risk for groundwater contamination. Strontium, if it enters the human body, tends to be incorporated into the bones, where it can irradiate marrow and cause leukemia. Nevertheless, Sr-90 and Cs-137 remain dangerous for relatively short periods of time: heat generation is not a significant problem after about eighty years,[14] and radiation falls to very low levels after a few hundred years. If there is reasonable assurance that effective institutional controls and other regulatory and response capabilities can be maintained over such a period (an issue that is sharply disputed), strontium and cesium wastes could be separated out from other fission products with much longer lives and stored while they decay to levels that would permit their disposal as LLW.[15] Their removal would reduce the technical demands on repositories posed by SNF, which contains substantial amounts of these "hot" elements.[16]

Other daughter fission products are not as highly radioactive and thermally hot as cesium and strontium, but many of them have longer lives. Zirconium 93 (Zr-93) is an example: it is a very long-lived fission product (half-life 1.53 million years), but its long life corresponds with low activity. Zr-93 emits "soft" beta and gamma radiation with lower energies than that of Cs-137 or Sr-90, and presents a lesser environmental threat than those shorter-lived counterparts. It is not likely to leach out of buried waste quickly, but because of its very long life, it requires very long-term isolation. The composition of the fission products (including transuranics and daughter products) in SNF is summarized in Table 2.2.

The complex mix of nuclides in SNF makes disposal far from straightforward. Short-lived fission daughter products initially release too much heat for dry storage. When retired from use, spent fuel rods require storage in cooling ponds for at least one year, although five to ten years is more common, to allow some of the short-lived fission products (notably cesium and strontium) to decay.[17] Then the fuel rods can be kept in

Table 2.2. Characteristics of key radioactive isotopes in SNF

Isotope	Product type	Half-life (years)	Years to reach 1% of original activity	Other characteristics
Strontium-90	Fission product	29	191	Highly radioactive and thermally hot
Cesium-137	Fission product	30	200	Highly radioactive and thermally hot; water soluble
Americium-241	Transuranic	432	2,870	Fissile; proliferation threat
Plutonium-240	Transuranic	6,563	43,604	Not fissile; limited proliferation risk*
Plutonium-239	Transuranic	24,200	160,781	Key weapons isotope; proliferation threat
Technetium-99	Fission product	211,000	1,401,854	Low-activity fission product; highly soluble
Zirconium-93	Fission product	1,530,000	10,165,100	Used in reactor rods
Neptunium-237	Transuranic	2,140,000	14,217,852	Fissile; highly mobile
Uranium-235	Uranium	703,800,000	4,675,945,986	Principal uranium isotope for reactor fuel
Uranium-238	Uranium	4,468,000,000	29,684,749,456	Most common uranium isotope; cannot support fission on its own

*Pu-240 is one of the fission products for nuclear reactors but does not undergo fission itself and thus builds up the longer the fuel spends in the reactor. Nuclear weapons can be created only out of reactor fuel with less than 7 percent Pu-240. This is one reason light-water reactors are considered "proliferation resistant"; a reactor will exceed the 7 percent threshold within four months of continuous operation, and the entire reactor must be shut down to retrieve the fuel to use for weapons. Such a shutdown can easily be detected and investigated by an organization like the United Nations or IAEA. In heavy-water reactors and some breeder reactors, fuel can be removed without shutting down the reactor, rendering it effectively impossible to control proliferation risks.

the cooling ponds or put in dry casks and stored, either at the power plant site (the practice in the United States today) or at other facilities—including, potentially, consolidated SNF storage facilities—until a deep geological repository opens to receive them.[18]

The Once-Through Fuel Cycle with Reprocessing for Weapons Production

When fuel rods are irradiated to produce plutonium for weapons rather than to generate electricity, they are left in the reactor fission process for a shorter time. This process is designed to produce a "younger" form of plutonium, Pu-239, which is generated earlier in the fission cycle. When fuel rods are left in a reactor for a longer time to generate electricity, a higher proportion of other transuranics (such as Pu-240) and daughter products (including highly radioactive and thermally hot cesium and strontium) is produced; these elements are not suitable for weapons use. Thus, fuel rods irradiated for weapons production contain a different—and in some respects less radioactive and thermally hot—mixture of elements than SNF from power plants. These fuel rods are then reprocessed with chemical solvents that break down the various elements in SNF in order to extract the plutonium and uranium components for use in weapons production.[19] The various other SNF elements that remain in the soupy reprocessing liquids are categorized as HLW. The general plan for dealing with HLW is to solidify and vitrify them for deep geologic burial in a repository.

Mixed Oxide Fuel Cycle with Reprocessing

Reprocessing can also be carried out with SNF produced by electricity generation. Rather than storing and disposing of SNF, as in the once-through fuel cycle, the spent fuel rods can be broken down by chemical means, allowing plutonium and uranium to be extracted for use in new fuel rods, thereby utilizing 25 percent more of the useful energy in the fuel rods than the once-through process.[20] Although no reprocessing occurs in the United States, a few countries have reprocessed SNF, generally through the PUREX process, to produce mixed oxide fuel, or MOX fuel. In this process, plutonium extracted from the SNF is mixed with depleted uranium—uranium with a substantially reduced concentration of U-235 left over from the enrichment process—then heated to high temperatures and pressed into fuel pellets that are placed into fuel rods. In Europe, more than thirty-five reactors currently use MOX fuel, derived mostly from SNF reprocessed in France and the United Kingdom.[21]

According to the private company Areva, which has been making MOX fuel with plutonium from reprocessed SNF for forty years in France, depleted uranium mixed with 5–10 percent plutonium has the same power generation potential as 4.5 percent enriched U-235.[22] Areva claims that reprocessing as few as seven SNF fuel assemblies can create one new fuel assembly. It also asserts that, because most of the actinide wastes in SNF are reused in its MOX fuel, a single Areva PUREX reprocessing and MOX fuel production step results in a fivefold reduction in the volume of the most highly radioactive wastes requiring disposal, relative to the once-through fuel cycle.[23] However, critics are quick to point out that reprocessing increases the total volume of other types of radioactive wastes, some of which can be as radioactive as SNF.[24] Reprocessing is also costly; MOX fuel is substantially more expensive than fuel made from virgin uranium. Accordingly, MOX fuel must be subsidized to enable it to compete with natural uranium fuel and avoid large surpluses of reprocessed plutonium. Reprocessing also creates risks of terrorism and proliferation associated with the plutonium produced.

MOX fuel can also be made using weapons plutonium. Modernization and reduction of nuclear warhead inventories as a result of international agreements and weapons modernization has yielded a significant surplus stock of plutonium that can be used to make MOX fuel. In 2000, the United States and Russia each agreed to dispose of 34 metric tons of weapons-grade plutonium as reactor fuel. Each country promised to build MOX fuel fabrication facilities, but funding committed to Russia by the United States and western European countries failed to materialize, stalling the program. The two countries recently renegotiated the deal and now stand committed to disposing of the excess plutonium by making MOX fuel.[25] These programs are similar to an earlier agreement, the Megatons for Megawatts program, which has since 1994 converted 382 metric tons of weapons-grade uranium into nuclear fuel.[26] Additionally, in 2007, the Bush administration designated nine metric tons of plutonium as surplus to defense needs and directed it for disposal partly as MOX reactor fuel and partly as HLW.[27] It remains to be seen, however, whether the surplus weapons plutonium will be successfully used as commercial fuel in the United States. Duke Energy's Catawba plant in South Carolina used MOX fuel fabricated from surplus U.S. weapons plutonium by Areva in France on an experimental basis for four years. Duke, however, declined to renew its contract with Areva, citing delays in the construction of the Savannah River reprocessing plant, which is slated to convert surplus plutonium into MOX fuel for the United States, and a tightening uranium market that makes Duke reluctant to be locked in with an unreliable supplier.[28]

Advanced Fuel Cycles with Reprocessing: GNEP
A number of advanced fuel cycle processes beyond the current PUREX process have been explored. One example is the UREX+ reprocessing system that was part of the GNEP program proposed by the George W. Bush administration (GNEP is discussed more fully in Chapters 1 and 7).[29] Under this approach, uranium isotopes would first be separated from SNF reprocessing liquids and either reused in light-water reactors in the United States or foreign countries, or disposed of as low-level waste (LLW) in landfills. Sr-90 and Cs-137 would be separated and placed in surface storage for several centuries until the waste has decayed long enough that it could be disposed of as LLW.[30] Next, other fission products would be separated out and disposed of as HLW in a repository. The transuranics remaining in the reprocessing liquids, including plutonium, would be used to make a fuel mixture to be burned in new fast reactors.[31] Unlike the PUREX process, UREX+ does not separate out pure plutonium but makes a fuel that contains all the transuranics in SNF.[32] The plan was to burn and reconstitute the transuranics fuel mixture through several cycles, with the goal of burning most of the transuranics without the need for separate disposal.[33] The process would, however, produce substantial increases in other types of wastes. Thus, while DOE projected that the GNEP reprocessing system would greatly reduce the volume and heat load of the most highly radioactive wastes requiring deep geologic disposal relative to SNF, it would increase the volume of LLW, including the more hazardous class B and C wastes, by a factor of seven, and the volume of moderately to highly radioactive GTCC waste may increase more than a hundredfold.[34]

Other Fuel Cycle Wastes

Various other radioactive wastes are produced over the course of the fuel cycle. The processing of uranium ore leaves some residual radioactive material that cannot be extracted from the ore waste. This waste is classified as mill tailings, and its disposal is regulated separately from other fuel cycle wastes; mill tailings wastes are not covered in this book.[35] Reactor components become contaminated gradually over time, and at the terminal end of the cycle, repair, maintenance, and decommissioning of nuclear power plants create large amounts of LLW, including class B and C wastes. NRC maintains a regulatory program for decommissioned reactors to make sure that these materials are properly classified and disposed of.[36]

As SNF is reprocessed and the recovered plutonium is converted into weapons, tools and protective clothing become contaminated with transuranic nuclides. DOE manages such material under the category of transuranic waste (TRU). A quite small amount of TRU has also been produced in connection with civilian nuclear power production and with SNF reprocessing at West Valley.

Non–Fuel Cycle Radioactive Wastes

LLW is also generated by civilian activities that use radioactive materials other than fuel cycle materials, including naturally occurring radioactive elements such as radium and elements produced by particle accelerators. Radioactive sources are used for radiation therapy, commercial applications, and scientific research. Very small amounts of americium are used in smoke detectors, and glow-in-the-dark watch hands contain very small amounts of radium or tritium. When these sources are retired from use or the products that contain them are discarded, nuclear wastes are created.

Basic Nuclear Waste Regulatory Categories

The basic regulatory categories for nuclear wastes, and the corresponding requirements for their storage and disposal, are examined in detail in the remainder of this chapter. First, an overview of these categories:

- **SNF (spent nuclear fuel)** includes spent nuclear fuel from both civilian and weapons fuel cycles. Burial in a deep geologic repository is required unless SNF is reprocessed.
- **HLW (high-level waste)** includes wastes generated as byproducts of the reprocessing of spent fuel, either to extract plutonium to produce nuclear weapons or to produce fuel for electricity production. Burial in a deep geologic repository is required for disposal of many components of reprocessing wastes, although some less radioactive components may appropriately be classified as LLW and disposed of in landfills.
- **TRU (transuranic waste)** consists of wastes contaminated by plutonium and other elements with atomic numbers greater than that of uranium; most TRU wastes are produced as byproducts in nuclear weapons production. These wastes require permanent isolation in repositories due to their long half-lives.
- **LLW (low-level waste)** is a residual category that encompasses all other appreciably radioactive wastes not included in the first three categories listed. It has four

subcategories: **greater-than-class C (GTCC)**, **class C**, **class B**, and **class A**. These subcategories are based on radioactivity concentration limits, which vary depending on the elements contained in and the characteristics of the waste involved. The GTCC category has no radioactivity ceiling; it includes some wastes that are more hazardous than many TRU wastes and that approach the radioactivity of SNF and HLW. Below GTCC, in descending order of radioactivity, are categories C, B, and A. Any materials with concentrations of nuclides with half-lives shorter than five years are considered class B waste. Class A LLW consists of wastes that have very low concentrations of radioactivity. (NRC's definitions of the A, B, and C classes of LLW are represented in Table 2.4.) The greatest volume of LLW consists of components of decommissioned power plants that have been exposed only to attenuated radiation. Current NRC regulations and DOE policies require land burial of these wastes at various depths but do not require deep geological isolation. Wastes that exceed the class C limits are GTCC. DOE is currently studying disposal options for GTCC wastes.

- **NARM (naturally occurring and accelerator produced radioactive material)** is a category of radioactive materials that either occur naturally or have been created by particle accelerators and are not associated with the nuclear fuel cycle. Radioactive materials that occur naturally are subclassified as NORM (naturally occurring radioactive materials), while those created in particle accelerators are subclassified as ARM (accelerator-produced radioactive materials). NARM includes material contaminated with radium gas, and many radioactive sources used in medicine and industry.[37] NARM was not originally regulated under AEA; it was regulated by the states. However, the Energy Policy Act of 2005 brought some NORM under NRC authority.[38]
- **BRC (below the level of regulatory concern)** wastes include those nuclear wastes with very low radioactivity that would otherwise be classified and regulated as class A LLW but have been given individual exemptions by NRC. One such example is the americium-241 (Am-241) used in smoke detectors. Am-241 is medium-lived (with a half-life of 432.7 years), and its decay chain involves all three forms of radiation. But concentration of this element in used smoke detectors is low enough that NRC allows for its disposal in ordinary solid waste landfills. States may regulate any waste so exempted by NRC.
- **UMT (uranium mill tailings)** are subject to a distinct regulatory scheme and are not considered in this book.[39]
- **DU (depleted uranium)** refers to the uranium mixture remaining after enrichment has removed most of the U-235 from uranium-bearing ores; DU consists almost entirely of U-238.[40] Because enriched fuel requires a U-235 content so much higher than the naturally occurring rate, a large amount of DU is produced during fuel production. Some of this mildly radioactive, alpha-emitting mix can be used for civilian purposes, such as ballast weights in aircraft and boats or as a gamma radiation shield in radiography cameras. The U.S. military has also exploited DU's high density by incorporating it into weapon munitions: a DU projectile penetrates armor easily, making DU rounds desirable as antitank weapons. Significant controversy surrounds use of DU for weapons, however, because of possible long-term health effects caused by possible exposure.[41] NRC has determined that

DU generated by civilian activities should be regulated as LLW but has not yet determined how to do so.[42]
- **Special nuclear material** is defined by AEA as plutonium, U-233, any mix of uranium isotopes enriched with concentrations of U-235 or U-233 above naturally occurring levels, or other material designated as such by NRC. The United States has substantial stocks of plutonium and U-235 enriched uranium that have no military use, since the country has ceased to expand its nuclear arsenal.[43] However, because these materials are statutorily defined as special nuclear material, they are not classified or regulated as byproduct waste. The federal government has designated portions of its plutonium and enriched uranium stockpiles as surplus, which DOE is consolidating and storing.[44] The government plans to convert most of this surplus special material into fuel, either by down-blending U-235 enriched uranium to concentrations suitable for use as fuel, or by using plutonium to make MOX fuel. Some of the surplus material will, however, be packaged and disposed of in the same manner as HLW.[45]

Table 2.3 provides a summary of the most important components of the U.S. nuclear waste regulatory classification system.

Nuclear Waste Regulatory Classification and Requirements

This subsection provides a more detailed account of current U.S. regulatory classifications for the most important radioactive wastes and associated regulatory requirements and disposal practices.

Low-Level Waste

LLW is defined by the Low-Level Radioactive Waste Policy Amendments Act of 1985 (LLRWPAA) to consist of all radioactive material other than HLW, SNF, TRU, and certain nonreactor wastes, including radium-226 sources and material irradiated during the production of special nuclear material.[46] LLW is thus a residual category. Most nonreactor wastes, including accelerator-produced and naturally occurring materials (NARM/NORM), are excluded and are not subject to federal regulation.

NRC regulates civilian LLW, which it classifies by regulation. It subdivides LLW into four classes, from lowest to greatest hazard: class A, class B, class C, and greater-than-class C (GTCC).[47] LLRWPAA makes disposal of class A, B, and C wastes the responsibility of the states, except for material owned or generated by DOE and U.S. Navy decommissioning programs. Disposal of civilian GTCC waste, along with DOE and U.S. Navy waste, is the responsibility of the federal government.[48] DOE self-manages its own LLW, which is not subject to NRC's jurisdiction, but DOE's published internal regulations echo NRC's regulations for LLW waste.[49] Although the term "low level" connotes little danger, the nomenclature is misleading. Some GTCC LLW waste can be as radioactive as some HLW, and both class B and class C wastes can also pose quite substantial radiation hazards if not properly managed. The sub-classifications of LLW are delineated according to tables of radioactivity concentration, as illustrated by Table 2.4.[50]

Table 2.3. U.S. nuclear waste regulatory classification system

Source of waste	Waste classification	Definition	Civilian generated	Government generated	Regulatory authority*	Tools for management
Fuel cycle waste	Spent nuclear fuel (SNF)	Irradiated reactor fuel rods declared as waste	Yes	Yes	NRC, DOE	Geiger counter, NaI detector, heavy protective clothing
	High-level waste (HLW)	Waste derived from SNF reprocessing	Minimal**	Yes	Principally DOE	Geiger counter, NaI detector, heavy protective clothing
	Transuranic waste (TRU)	Isotopes of uranium and heavier elements derived mainly from government weapons manufacture and irradiated protective gear	Negligible	Yes	DOE	Geiger counter, NaI detector, shielding
	Low-level waste (LLW)	Everything not falling into other categories	Yes	Yes	NRC, DOE	Varies by subcategories
	Uranium/thorium mill tailings	Leftover radioactive ore materials from nuclear fuel production	Yes	Yes	NRC, DOE	Geiger counter, NaI detector, heavy protective clothing
Other reactor waste	Irradiated reactor parts and shielding (classified as LLW)	Waste from decommissioned reactors and protective gear	Yes	Yes	NRC, DOE	Geiger counter, NaI detector, lighter protective clothing
Naturally occurring and accelerator-produced waste (NARM)	Accelerator-produced waste (ARM)	Nuclear material produced in accelerators, often for medical or research purposes	Yes	Yes	NRC, states	Geiger counter, NaI detector, lighter protective clothing
	Naturally occurring radioactive material (NORM)	Radioactive materials that occur naturally	N/A	N/A	States***	Geiger counter, NaI detector, lighter protective clothing

*Any type of waste mixed with other toxic chemicals (mixed waste) is subject also to EPA's regulatory authority.

**A small amount of civilian HLW remains; it is onsite at West Valley, the only facility that actively reprocessed civilian SNF. Reprocessing was halted by the Carter administration.

***The Energy Policy Act of 2005 expanded AEA's definition of byproduct material to include some NORM, specifically that which "would pose a threat similar to the threat posed by a discrete source of radium-226" (Energy Policy Act of 2005 § 651, 42 U.S.C. § 2014(e) (2006)). NRC published regulations exerting authority over radium-226 and some NARM materials but has not included any other NORM in the byproduct material category (72 Fed. Reg. 55,864, at 55,867–55,869 (Oct. 1, 2007)). NORM now defined as byproduct material is typically regulated by NRC as LLW.

Class A, B, and C wastes are disposed of by burial in near-surface vaults or trenches at landfill facilities. Generally, class A wastes are those that will decay within one hundred years or less to levels deemed "safe," meaning that buried materials are no more radioactive than other common, naturally occurring material in the soil.[51] Class B wastes are more radioactive; they may take up to three hundred years to decay to background soil radiation levels and are subject to correspondingly stricter disposal and other regulatory requirements than class A wastes. Class C wastes may take up to five hundred years to decay to safe levels and must be isolated from the environment by still stricter packaging and disposal requirements.[52] Very dilute concentrations of more highly radioactive elements like Cs-137 are classified as B and C waste.

GTCC wastes are those that exceed the class C radioactivity elements and do not fall into any other waste category. Such wastes have much longer lives or much higher radioactivity levels than other LLW and must be kept isolated for longer periods of time, either in a deep geologic repository or other facility that will ensure long-term isolation. GTCC wastes include portions of the reactor core and the metal in fuel assemblies that are exposed to the fission process, as well as materials from certain medical and industrial uses of radioactive elements. Many of the more highly radioactive GTCC wastes—such as Cs-137 and actinides that are used for medical treatment and research, respectively—are produced from civilian applications.

NRC has statutory authority to exempt specific class A wastes from otherwise applicable packaging and disposal requirements if it deems them to be below regulatory concern.[53] Class A wastes that are designated BRC may be disposed of in ordinary waste landfills or, if chemically hazardous, in RCRA landfills. NRC, however, defines and applies the BRC exemption in a very narrow way. Those seeking to dispose of low-activity material as BRC must apply to NRC on a case-by-case basis for an exemption, and those materials exempted are subject to state regulatory authority.[54]

NRC must license disposal facilities for any waste produced by an NRC-licensed activity, including civilian waste such as GTCC or SNF reprocessing waste that is managed by DOE.[55] The licensing requirements for class A, B, and C LLW disposal facilities limit the allowable release of radiation into the air and groundwater and dictate the allowable forms of waste. The NRC performance objectives for civilian LLW facilities are set out in 10 C.F.R. Sections 61.40 to 61.44. They specify that the radiation dose will not exceed twenty-five millirems for a "reasonable maximally exposed individual," such as an inadvertent intruder into a radioactive waste disposal site.[56] The NRC regulations develop LLW disposal site certification criteria based on this target dose rate. The regulations further delineate specific LLW sub-classification thresholds for various radionuclides in Table 1 of 10 C.F.R. Section 61.55. This table and the target dose rates were determined by NRC, although they conform to recommendations submitted by EPA during the regulation's notice period.[57] In order to ensure that disposal facilities comply with those standards, certification requirements for individual facilities are established in 10 C.F.R. Sections 61.50 to 61.59, and testing for compliance is provided for in 10 C.F.R. Section 61.81. In licensing a LLW disposal facility, NRC determines that the waste packaging, burial methods, engineered barriers (if any), and institutional controls will ensure conformance with the applicable radiation standards.[58] The disposal requirements are based on the assumption that no active institutional controls will be available after one hundred years, meaning that a LLW disposal site must be secure enough after one hundred years that the combination of physical barriers (e.g., concrete vaults), public

Table 2.4. Low-level waste category definitions

Radionuclide	Class A limit (Ci/m3)	Class B limit (Ci/m3)	Class C limit (Ci/m3)
Carbon 14	0.8	N/a	8
Carbon 14 in activated metal	8	N/a	80
Nickel 59 in activated metal	22	N/a	220
Niobium 94 in activated metal	0.02	N/a	0.2
Technetium 99	0.3	N/a	3
Iodine 129	0.008	N/a	0.08
α-emitting transuranic nuclides (λ > 5 yrs)	110 nCi/g	N/a	1,100 nCi/g
Plutonium 241	1,350 nCi/g	N/a	13,500 nCi/g
Curium 242	12,000 nCi/g	N/a	120,000 nCi/g
All nuclides with λ < 5 yrs	700	No limit	No limit
Hydrogen 3 (tritium)	40	No limit	No limit
Cobalt 60	700	No limit	No limit
Nickel 63	3.5	70	700
Nickel 63 in activated metal	35	700	7000
Strontium 90	0.04	150	7000
Cesium 137	1	44	4600

land-use records, and passive warning devices will suffice to protect even an inadvertent intruder from hazardous exposures. DOE follows an approach similar to that of NRC for its own LLW disposal facilities.

The NRC requirements for civilian GTCC disposal have not yet been specified because no facility for these wastes has yet been proposed. NRC regulations presume that GTCC wastes require disposal in a geological repository unless an alternative disposal method is developed and approved by NRC.[59] Disposal of both civilian and defense GTCC wastes is the responsibility of DOE, which is studying disposal options.

Transuranic Waste

TRU consists of wastes contaminated by plutonium and other elements with atomic numbers greater than that of uranium. It includes such items as gloves, protective clothing, metal tools and pipes, rubber boots, brushes, tape, and other miscellaneous items that come into contact with plutonium or other transuranics in the course of weapons production or SNF reprocessing.[60] Only plutonium and neptunium are naturally occurring transuranic elements; the remainder are created in the process of irradiating fuel rods in nuclear reactors. Most TRU consists of byproducts from weapons production. A limited amount of TRU was also produced in reprocessing civilian SNF.

EPA and DOE define TRU as wastes other than SNF and HLW that contain more than one hundred nanocuries per gram of alpha-emitting, transuranic isotopes with half-lives greater than twenty years.[61] TRU wastes are subdivided into two regulatory categories: the less radioactive contact-handled TRU (CH-TRU), which has sufficiently low activity that the waste packages can be handled by workers without protective gear, and the more radioactive remote-handled TRU (RH-TRU), which requires protective shielding and remote handling.[62] There is no NRC TRU category for the relatively small amounts of civilian transuranic wastes similar to defense TRU. Because they do not fall into any other category, under the 1985 LLRWPAA such wastes are classified as LLW, which is a residual category. Those with activity of over 1,100 nanocuries per gram are classified by NRC as GTCC waste.[63] In this book, these wastes will be referred to as civilian TRU.

Although there are exceptions, transuranics tend to be alpha emitters with very long half-lives that are most dangerous when inhaled or ingested; their half-lives range from many thousands of years to millions of years.[64] Both NRC and DOE consider TRU unsuitable for near-surface disposal.[65] It must be kept securely isolated for geologic time periods, and it is particularly important to prevent TRU from dissolving in water that might be taken up in the biosphere. Before 1970, however, AEC did not differentiate between TRU and LLW and buried 138,000 cubic meters of TRU in near-surface landfills.[66] In 1974, AEC declared TRU unsuitable for near-surface disposal and put a moratorium on civilian TRU disposal.[67] Only in the early 1980s did various regulations begin to categorize TRU as a separate category of wastes meriting its own regulatory framework.[68]

DOE divides TRU into two subcategories according to the level of radioactivity.[69] CH-TRU is less hazardous, radiating less than two hundred millirems per hour at the surface of the container. RH-TRU, defined as TRU that radiates more than two hundred millirems per hour, must be managed with greater precautions, including remote handling by workers of its packaging. TRU resulting from defense activities is disposed of at WIPP. TRU from civilian SNF reprocessing, currently limited to TRU produced at West Valley, is not eligible for disposal at WIPP. Pursuant to the LLRWPA (and, in the case of West Valley TRU, the WVDPA), it is managed by DOE subject to NRC regulation and currently lacks a disposal plan.

Spent Nuclear Fuel

The Nuclear Waste Policy Act of 1982 (NWPA) defines SNF as "fuel that has been withdrawn from a nuclear reactor following irradiation, the constituent elements of which have not been separated by reprocessing."[70] The vast majority of SNF comes from commercial power reactors and has been highly irradiated.[71] A relatively small amount of SNF consists of unreprocessed spent fuel from DOE weapons reactors; there is also some SNF from navy nuclear power reactors used to power nuclear submarines and some large surface vessels. Defense SNF has not undergone the same amount of fission as SNF from civilian nuclear power reactors, and therefore defense SNF has a lower proportion of highly radioactive elements, such as Sr-90 and Cs-137. DOE manages approximately 2,150 MTHM of SNF from defense and other operations, compared to 62,500 MTHM of commercial SNF that has been generated by nuclear power plants.[72] Under NWPA, all SNF, both commercial and defense related, must be disposed of in a deep geologic repository.[73]

High-Level Waste

HLW consists of wastes from reprocessing SNF. Today, some reprocessing wastes may be classified and managed as other than HLW; unlike HLW, they do not need to be disposed of in a geologic repository. But just what reprocessing wastes are not HLW is a matter of great uncertainty and controversy, fostered by the various definitions of HLW adopted in statutes and federal agency regulations and statements over the years. There is no standard, unambiguous definition of HLW. Accordingly, in determining whether specific wastes are HLW in a given context, it is essential to identify the relevant legal authorities and carefully examine their language, context, and history.

AEC issued the first definition of HLW in 1970 when the commission codified the licensing requirements for civilian spent-fuel reprocessing facilities. Appendix F of 10 C.F.R. Part 50 defines HLW as "those aqueous wastes resulting from the operation of the first cycle solvent extraction system, or equivalent, and the concentrated wastes from subsequent extraction cycles, or equivalent, in a facility for reprocessing irradiated reactor fuels." This inclusive definition seems to designate all reprocessing wastes as HLW, without distinguishing materials that are more hazardous from those that are less hazardous. As discussed later, the aim of this definition was not to establish a category of wastes according to radiation levels and risks posed. Its rationale was jurisdictional: to identify those civilian wastes generated pursuant to AEC/NRC licenses for civilian reprocessing plants under 10 C.F.R. Part 50 that would be transferred to AEC/ERDA/DOE regulatory jurisdiction. Congress in 1972 used AEC's Appendix F definition of HLW, adding SNF, to define "high level nuclear waste" in the Marine Sanctuaries Act, which forbids the dumping at sea of hazardous materials, including high-level nuclear waste, and expressly precludes any waiver of this prohibition with respect to "high level nuclear waste" in particular.[74]

When AEC was split by the Energy Reorganization Act of 1974 into NRC and what is now DOE, NRC was assigned the responsibility of licensing both civilian and defense HLW disposal facilities.[75] HLW was not defined by the act or its legislative history. In 1981 NRC incorporated Congress's Marine Sanctuaries Act HLW definition to interpret HLW in the context of licensing repositories, although the definition originated as a criterion of regulatory jurisdiction rather than of relative hazard and was subsequently adopted by Congress in the different context of ocean dumping.[76]

The West Valley Demonstration Project Act (WVDPA), however, enacted in 1980, adopted a still different definition of HLW, namely, "high level radioactive waste which was produced by the reprocessing [center at West Valley]. Such term includes both liquid wastes which are produced directly in reprocessing, dry solid material derived from such liquid waste and such other material as the [NRC] designates as high level radioactive waste for purposes of protecting the public health and safety."[77] The WVDPA definition of HLW might be read as including only reprocessing waste deemed "high level" in terms of its danger to "public health and safety," although it could also be read as including all reprocessing wastes, liquid and solid, plus other materials designated by NRC.

Section 2(12) of the 1982 NWPA, which requires all HLW to be buried in a geologic repository, adopted still a different definition of HLW: The term 'high-level radioactive waste' means—(A) the highly radioactive material resulting from the reprocessing of spent nuclear fuel, including liquid waste produced directly in reprocessing and any solid material derived from such liquid waste that contains fission products in sufficient concentrations; and (B) other highly radioactive material that the Commission (NRC), consistent with existing law, determines by rule requires permanent isolation."[78] The NWPA definition of HLW contains several ambiguities. The modifier "highly radioactive" might be understood to indicate that only a subset of reprocessing wastes should be regarded as HLW. But is all "liquid waste produced directly in reprocessing" deemed by the statute to be "highly radioactive" and therefore HLW, or may regulators classify some portion of such wastes as not highly radioactive and thereby treat that as other than HLW? Also, what does "produced directly in reprocessing" mean? Further, what does "sufficient concentrations" mean? And does the codicil "contains fission products in sufficient concentrations" modify solely "solid material" or the entire phrase "liquid waste . . . and solid material"?

The legislative history of NWPA answers some of these questions. The HLW definition in the NWPA originated in the House Armed Services Committee, which explained in its report that the definition permits the responsible regulatory agency (which it identified as EPA, the agency setting radiation standards) "to determine the concentration of fission products and transuranic elements that require permanent isolation." This makes clear that some low-activity reprocessing wastes can be classified and disposed of as non-HLW if the hazards they present are sufficiently low as not to require repository disposal.[79]

Finally, to the extent regulatory agencies have some discretion under any or all of these terms to classify some reprocessing wastes as non-HLW, including the determination of "sufficient concentrations," which agency has that authority: NRC, DOE, or EPA? The doctrine of agency deference laid out in the Supreme Court case *Chevron v. NRDC*, applied by courts to interpretations of ambiguous statutory provisions like this, would ordinarily grant substantial deference to the interpretation of the agency administering NWPA.[80] But in the case of this statute, there are multiple agencies involved.[81] DOE is in charge of most HLW, including its management, storage, repository siting and construction, and disposal. But NRC has regulatory authority over civilian HLW generally and civilian and defense HLW transport to a HLW repository; NRC also is responsible for licensing the Yucca Mountain repository, while EPA is in charge of setting radiation standards for the repository. While the House Armed Services Committee regarded EPA as the responsible agency, its passing remark, not linked to any statutory language, is not determinative. DOE and NRC have far more direct roles in managing and regulating HLW and its disposal, and have a stronger claim to deference to their interpretations.

In 1983 NRC adopted a different definition of HLW for NWPA purposes. Both 10

C.F.R. Section 60.2 (for waste destined for geologic repositories) and 10 C.F.R. Section 63.2 (for Yucca Mountain specifically) include SNF, liquid, and solid reprocessing wastes in their definitions of HLW. Section 60.2 offers this: "HLW means: (1) Irradiated reactor fuel, (2) liquid wastes resulting from the operation of the first cycle solvent extraction system, or equivalent, and the concentrated wastes from subsequent extraction cycles, or equivalent, in a facility for reprocessing irradiated reactor fuel, and (3) solids into which such liquid wastes have been converted."[82] Section 63.2 is slightly more flexible, as it also includes "other highly radioactive material that the [NRC], consistent with existing law, determines by rule requires permanent isolation," but otherwise the two definitions are the same.[83] The NRC definition, which echoes AEC's original 10 C.F.R. Part 50 Appendix F definition, removes some of the ambiguities in the NWPA definition, but in doing so it seems to include all liquid and solid reprocessing wastes as HLW. This would seemingly remove the flexibility potentially afforded by NWPA to classify as non-HLW some solid reprocessing waste components with relatively low radioactivity on the ground that they do not have "sufficient concentrations" of radioactivity to require their disposal in a repository.

Because DOE manages nearly all existing HLW, its own interpretation is also highly significant. The DOE Order 435.1 *Radioactive Waste Management Manual* definition of HLW repeats verbatim Subsection (A) of the NWPA definition of HLW but substitutes for Subsection (B) "other highly radioactive material that is determined, consistent with existing law, to require permanent isolation," thereby removing any reference to NRC and implying that DOE is to make such determinations itself.[84]

The interrelation between the HLW definitions in the DOE manual, NRC regulations, and NWPA has been a source of controversy relating to the classification or reclassification of HLW. NWPA does not explicitly apply to defense waste managed by DOE at its facilities, although it requires that all wastes that are classified as HLW be disposed of at a geologic repository licensed by NRC pursuant to the act, and NRC regulations apply to the packaging and handling of the waste transported to the repository.[85] Further, after enactment of NWPA, AEA was amended by Congress to adopt the NWPA definition; the amended AEA might accordingly be construed to apply the NWPA definition to waste managed by DOE.[86] If so, NRC might assert the authority to interpret the ambiguous provisions in subsection (A) of the NWPA HLW definition and make the determinations provided in subsection (B) with respect to DOE wastes.

An important issue for future policy is whether certain reprocessing wastes can be separated out and appropriately disposed of as other than HLW, which requires extensive and costly predisposal treatment, such as vitrification, and then disposal in a deep geologic repository. After the demise of Yucca, no such repository is in sight.

HLW Reclassification Initiatives

This section reviews the legal and policy issues presented by DOE plans for classifying and disposing of certain reprocessing wastes as LLW, rather than disposing of them as HLW in a repository. Given the high costs of treating and vitrifying wastes for repository disposal and the uncertainty as to whether or when a repository to receive such wastes may be available, the issues involved have broad policy significance. The NAS

Committee on Risk-Based Approaches for Disposition of Transuranic and High-Level Radioactive Waste analyzed the HLW stored at Hanford, SRS, and INL, and found that many components of these wastes posed hazards similar to class C or lower LLW and could be safely disposed of in LLW facilities for such wastes (whether on or off DOE sites), thereby achieving significant resource savings and, in many cases, earlier disposal.[87]

The first DOE reclassification initiative concerned tank waste residues. In the late 1980s, DOE concluded that it would not be practicable or prudent in some cases to remove all the reprocessing wastes held in tanks at Hanford and SRS for treatment and disposal as HLW. Their unique chemistry and characteristics made complete retrieval of certain wastes quite difficult, even with advanced technology.[88] Citing technical difficulties, high costs, and worker safety risks in trying to extract the entirety of the wastes in some tanks, DOE planned to leave a waste "tank heel"—a relatively small amount of waste sludge—in the tanks after diluting it with cement grout to immobilize the waste and lower radioactivity to LLW levels. The tanks would then be topped off with concrete and left in place, in a fashion similar to common near-surface LLW disposal practices. DOE determined that this method of treatment and disposal would limit the radiation from the wastes to concentrations that would comply with requirements for LLW disposal.

As a second initiative, DOE began in 1999 to consider the possibility of separating the reprocessing liquids extracted from waste tanks at Hanford, INL, and SRS into high- and low-activity fractions. The latter would include reprocessing waste components such as salts—or, specifically at Hanford, aluminum—which would interfere with waste vitrification unless they were removed.[89] DOE intended to use washes or other methods to separate out these components, which would have LLW radioactivity levels. They would be solidified and disposed of in LLW land burial facilities either on-site or off-site.[90] The remaining fractions in the reprocessing wastes would be vitrified for repository disposal as HLW.

These several DOE plans, however, confronted the question of whether it is lawful to dispose of reprocessing wastes traditionally regarded as HLW through such methods, taking into account, most importantly, the definition of HLW in NWPA. NRC initiated a rulemaking to clarify the NWPA definition of HLW but, as described later, ultimately abandoned the effort as too difficult and controversial. Thereafter, DOE adopted a new approach to classifying HLW in connection with it management of reprocessing wastes, which was challenged in litigation. Ultimately, Congress addressed and clarified some of the legal issues presented by DOE's plans and left others unresolved.

NRC Proposed Rulemaking

Against the background of post-NWPA legal uncertainties regarding DOE's management of reprocessing wastes, NRC in 1987 proposed rulemaking to modify the definition of HLW in its regulations, originally adopted by AEC in 1970 under AEA, which classified as HLW all reprocessing wastes, including any liquid or solidified reprocessing byproducts.[91] Its stated goal was to "follow more closely the statutory definition in [NWPA]."[92] NRC framed the basic issue as follows:

Clause (A) of the NWPA definition of HLW refers to wastes produced by reprocessing spent nuclear fuel and thus is essentially identical to the Commission's current HLW definition in 10 C.F.R. Part 60. Clause (A) is, however, different in one respect. The NWPA wording would classify solidified reprocessing waste as HLW only if such waste "contains fission products in sufficient concentrations"—a phrase that may reflect the possibility that liquid reprocessing wastes may be partitioned or otherwise treated so that some of the solidified products will contain substantially reduced concentrations of radionuclides.

The question, then, is whether Commission should (1) numerically specify the concentrations of fission products which it would consider "sufficient" to distinguish HLW from non-HLW under Clause (A); or (2) define HLW so as to equate the Clause (A) wastes with those which have traditionally been regarded as HLW.[93]

NRC proposed to adopt regulations setting numerical criteria by which waste could be classified as "highly radioactive," suggesting as a model the table of radioactivity classifications for LLW contained in 10 C.F.R. Section 61.55. Only waste both sufficiently radioactive and long-lived would need deep geologic isolation and be considered HLW. Waste that could be safely stored in "intermediate" facilities would not.[94]

NRC reported that virtually all commenters agreed with NRC on one point: "use of the term 'high-level waste,' at least under Clause (B) of the NWPA definition, serves to identify those wastes which require the degree of isolation afforded by a deep geologic repository."[95] Commenters disagreed with each other, however, over which wastes fit that definition. Some felt virtually all radioactive waste required geologic disposal, while others wanted "to reclassify as low-level large quantities of defense reprocessing wastes long regarded as HLW."[96] There was also criticism that classifying wastes according to a numerical radioactivity table was an invitation to dilute waste until it fell below the HLW threshold.

Further complicating the reclassification proposal, NRC proposed to compare the containment performance of a hypothetical intermediate facility to that of a geologic repository as a reference point in drawing a line separating HLW from wastes that would not be categorized as HLW and could therefore be disposed of in an intermediate facility.[97] However, a study published by Oak Ridge National Laboratory concluded that analysis of an intermediate disposal facility would be inherently so site specific that it was pointless to codify a waste classification framework based on what could safely be disposed of at a particular hypothetical site.[98]

Given the wide range of sharply opposed views in the public comments, including significant opposition to classifying any reprocessing wastes as other than HLW, and the evident difficulty of developing numeric rules of general applicability for classifying wastes as HLW/non-HLW, NRC in 1988 abandoned the quest. It simply retained its existing definition of HLW, notwithstanding the apparent inconsistency between it and the NWPA definition that it had previously noted. Yet in the statement terminating the rulemaking, NRC reserved future flexibility in the determination of HLW: "[HLW] would include the primary reprocessing waste streams at DOE facilities, though not the incidental wastes produced in reprocessing."[99] It did not elaborate as to how it might determine which reprocessing wastes are non-HLW "incidental wastes." At the same time, NRC adopted a rule requiring that, in the absence of a facility for GTCC waste disposal licensed by NRC, all GTCC wastes require disposal in a deep geologic repository.[100]

Following the abandonment of the rulemaking, DOE would from time to time consult with NRC on a case-by-case basis when it proposed to treat certain defense reprocessing wastes as non-HLW.[101] NRC would raise no objection to DOE's disposal of certain wastes as non-HLW "incidental wastes" if DOE could show that they could be safely disposed of without recourse to a deep geologic repository. For instance, DOE consulted with NRC over plans to dispose of low-activity fractions of some Hanford and Idaho reprocessing wastes as LLW, and to close tanks at SRS and INL containing some tank heel waste grouted in place.[102] Since 2002, DOE has closed eleven tanks at INL and two tanks at SRS by this method.[103]

A later NRC staff guidance document described the DOE-NRC consultation process: "DOE would periodically request that NRC provide technical advice for specific waste determinations. The NRC staff reviewed DOE's waste determinations to assess whether there were sound technical assumptions, analyses, and conclusions with regard to meeting the applicable incidental waste criteria. . . . NRC's advice was provided in an advisory manner and did not constitute regulatory approval [since] NRC does not have regulatory authority over DOE. These types of reviews were completed for waste intended to be removed from tanks at Hanford, tank closure at [SRS], waste intended to be removed from tanks at [INL], and tank closure at INL."[104] Through this process, DOE and NRC would cooperate in decision making without squarely addressing or attempting to resolve the allocation of legal authority over waste classification. A 1993 NRC staff memo had concluded that DOE consultation with NRC staff regarding waste incidental to reprocessing (WIR) determinations would be appropriate if DOE concluded that the jurisdiction of NRC in the matter was unclear.[105] Evidently DOE either agreed with this conclusion or for other reasons concluded that consultation with NRC was prudent.

In 1993, the States of Washington and Oregon and the Yakima Indian Nation petitioned NRC to conduct a rulemaking to establish standards by which NRC would make case-by-case determinations on whether reprocessing wastes are HLW.[106] The petitioners were concerned over DOE's plans to grout and dispose of in place as LLW certain reprocessing waste sludges in tanks at Hanford. The petition asked NRC to establish standards by which it would determine "whether wastes are 'highly radioactive material' or 'solids derived from [liquid reprocessing wastes] that contain fission products in sufficient concentrations'" within the meaning of the NWPA definition of HLW requiring repository disposal.[107] The petitioners also asked NRC to apply such standards to the sludges that DOE proposed to dispose of in place and set limits on the residual radioactivity that could remain in such wastes. They argued that if the wastes were HLW, then NRC would have authority over them under Section 202 of the 1974 Energy Reorganization Act, which "defines Commission authority over retrievable surface storage facilities and other facilities authorized for the express purpose of subsequent long-term storage of high-level radioactive waste generated by DOE which are not used for, or are part of, research and development activities."[108]

NRC denied the petition, blandly stating: "The petition is being denied because the NRC concludes that the principles for waste classification are well established and can be applied on a case-by-case basis without revision to the regulations."[109] Further:

The basis for the Commission's conclusion [that the tank grout wastes at issue were not HLW] is that the reprocessing wastes disposed of in the grout facility would be "incidental" wastes because of DOE's assurance that they: (1) have been processed (or will be further processed) to remove key radionuclides to the maximum extent that is technically and economically practical; (2) will be incorporated in a solid physical form at a concentration that does not exceed the applicable concentration limits for Class C LLW as set out in 10 C.F.R. Part 61; and (3) are to be managed, pursuant to the *Atomic Energy Act of 1954*, as amended, so that safety requirements comparable to the performance objectives set out in 10 C.F.R. Part 61 are satisfied.[110]

Although NRC did not dispute that it had the authority to license repositories for receipt of DOE HLW, it declined to assert regulatory authority over wastes still being managed or stored by DOE and deemed by DOE to be non-HLW.[111] It further stated: "[NRC] staff concluded that the expected residual waste would not be high-level waste and would thus not be subject to NRC licensing authority. The staff thereupon advised DOE that NRC agreed that the criteria used by DOE for classification of the grout feed [were] appropriate and that the grout facility for the disposal of the double-shell tank waste would not be subject to NRC licensing authority."[112]

NRC's statements in denying the petition set two important precedents.[113] First, they squarely adopted the concept of "waste incidental to reprocessing" as a subcategory of reprocessing wastes that, under certain conditions, would not be treated as HLW requiring repository disposal. Second, they established a norm of informal consultation and cooperation between DOE and NRC regarding applications for specific wastes managed by DOE to be treated as WIR. NRC would give an informal sign-off when it agreed with DOE's position that certain wastes could be disposed of as other than HLW without resolving the question of which agency might have final legal authority on the question.

DOE Manual 435.1 WIR Provisions

In 1999, DOE promulgated a new *Radioactive Waste Management Manual* (DOE M 435.1) and *Implementation Guide* to govern its management of nuclear wastes.[114] DOE M 435.1 in effect asserts that DOE is authorized to identify some reprocessing waste as WIR and to manage it "in accordance with the requirements for transuranic waste or low-level waste, as appropriate."[115] In justifying this authority, DOE cited prior NRC and DOE practices for case-by-case disposition of certain reprocessing wastes as WIR rather than HLW.[116]

DOE M 435.1 provides two procedures that DOE may follow for reclassifying reprocessing wastes as WIR and, in turn, as either LLW or TRU: citation and evaluation. The citation method allows DOE field managers to classify as WIR waste regularly regarded by NRC in the past as "incidental" to reprocessing, such as contaminated tools, gloves, and other material, without the need to obtain approval from DOE headquarters.[117] The evaluation method, by contrast, can apply to reprocessing wastes in the form of liquids or sludges and requires case-by-case testing and documentation to prove that reclassification as either LLW or TRU is justified. The manual provides, in the case of reclassification as LLW:

> In order to be reclassified as LLW, the DOE must determine that the wastes:
> 1: Have been processed, or will be processed, to remove key radionuclides to the maximum extent that is technically and economically practical; and
> 2: Will be managed to meet safety requirements comparable to the performance objectives set out in [NRC regulations at]10 C.F.R. Part 61, Subpart C, *Performance Objectives*; and
> 3: Are to be managed, pursuant to DOE's authority under the *Atomic Energy Act of 1954*, as amended, and in accordance with the provisions of Chapter IV of this Manual provided the waste will be incorporated in a solid physical form at a concentration that does not exceed the applicable concentration limits for Class C low-level waste as set out in 10 C.F.R. 61.55, *Waste Classification*; or will meet alternative requirements for waste classification and characterization as DOE may authorize.[118]

Similar requirements apply for reclassifying reprocessing wastes as TRU. The three criteria that DOE adopted for reclassification through the evaluation method build on the WIR criteria articulated by NRC in its 1993 decision to deny the rulemaking petitions of the states and tribe, discussed here earlier. In that decision, NRC had identified the first two criteria as appropriate for determining whether reprocessing wastes could be safely managed as non-HLW. DOE's third criterion, compliance with class C LLW limitations, had also been articulated by NRC in its 1993 decision.[119] However, in DOE M 435.1, DOE added to the third criterion the phrase "or will meet alternative requirements for waste classification and characterization as DOE may authorize," thereby giving itself authority and discretion to depart from class C limitations in individual cases.

DOE stated that it regarded its action as broadly consistent with the views expressed by NRC staff in the context of uncertainty regarding the precise allocation of regulatory authority between the two agencies.[120] NRC later issued a staff guidance document to standardize the handling of DOE's anticipated requests for WIR classification.[121]

NRDC v. Abraham

Natural Resources Defense Council (NRDC) brought litigation challenging the DOE M 435.1 evaluation procedures for reclassification as contrary to the NWPA definition of HLW.[122] NRDC argued that the "technically and economically practical" criterion and the "alternative requirements" clause added by DOE to the third criterion departed from the NWPA definition of HLW, which, NRDC contended, defined all reprocessing wastes as HLW. NRDC further asserted that the new WIR rules allowed DOE too much discretion and gave it the ability to circumvent any oversight, thus "creat[ing] an open-ended process for exempting high-level radioactive waste from the stringent requirements of the NWPA."[123] DOE responded that NWPA did not apply to defense wastes, and that DOE's evaluation method was in any event consistent with NWPA because the NWPA definition provides that only wastes with "fission products in sufficient concentrations" are defined as HLW.[124]

In *NRDC v. Abraham*, the Idaho federal district court ruled that, because NWPA required DOE to dispose of HLW only at Yucca, it governed DOE's efforts to classify certain reprocessing wastes as non-HLW.[125] The court further ruled that because the *Manual* allowed DOE to consider technical and economic factors when deciding what would be considered non-HLW, while NWPA's language referred only to the level of radioactivity, DOE M 435.1 conflicted with NWPA, which makes "sufficient concentrations" of radioactivity the only criterion, and therefore was invalid. "The first [factor DOE considers when making a WIR determination] is that key radionuclides are removed to the extent technically and economically practical. This means that if DOE determines that it is too expensive or too difficult to treat HLW, DOE is free to reclassify it as incidental waste."[126] Additionally, the court found problematic DOE's claim of authority to decide independently what waste would be considered non-HLW without regard to existing EPA or NRC guidelines for LLW. "The second [factor DOE considers when making a WIR determination] is that HLW incorporated into a solid form must either meet the concentration levels for class C low-level waste or meet such alternative requirements for waste classification and characterization as DOE may authorize. These 'alternative requirements' are not defined, and thus are subject to the whim of DOE."[127] The court, however, did not endorse the claim in the states' amicus brief that DOE lacked any authority unilaterally to reclassify wastes.[128]

On appeal, the Ninth Circuit reversed the district court's decision and dismissed NRDC's challenge, holding that, because DOE had not actually reclassified any wastes under the new manual, there was no concrete action for the court to review, and the matter was therefore not ripe for judicial review.[129] The court expressed doubt that DOE was trying to undermine the legislative scheme and indicated that NWPA did not apply to DOE's management of HLW:

> The NWPA decidedly does not purport to control DOE's management of nuclear waste. Indeed, it recognizes DOE's managerial authority. When it comes to high-level waste, however, NRDC contends that permanent disposal is quite another matter....
>
> NRDC complains that closely parsing the language describing high-level waste requirements can lead to a conclusion that DOE will, or might, simply dub high-level waste as something else, and then actually dispose of it improperly. Perhaps DOE *could* do that, but it denies any intention of so doing, and the Manual does not require that interpretation. In fact, DOE assures us that what it does do will be documented and will be publicly available. It does not plan a camisado. [Emphasis in original.][130]

While the court thus expressed a rather sympathetic view of DOE's actions, its dismissal of the case as unripe left the extent of DOE's WIR classification authority unclear. The district court's judgment was vacated; its opinion remains the only (and adverse) court pronouncement on the legality of DOE's M 435.1 reclassification restrictions, thereby casting something of a cloud on their validity.[131]

2005 National Defense Authorization Act

In 2000, during the ongoing NRDC litigation, DOE, with approval from South Carolina, grouted in place sludges in two HLW tanks at the Savannah River Site. Because the Idaho District Court's decision cast doubt on the legality of this and any similar steps by DOE to classify reprocessing wastes as non-HLW, DOE sought congressional legislation to validate the authority it had asserted in Order 435.1. Congress responded by enacting Section 3116 of the 2005 National Defense Authorization Act.[132] The act authorizes DOE, in consultation with NRC, to classify reprocessing waste located in South Carolina (the location of SRS) and Idaho (the location of INL) as non-HLW upon its determination that such waste does not require permanent isolation in a deep geologic repository for SNF and HLW, has had highly radioactive radionuclides removed to the maximum extent practical, does not exceed the NRC radioactivity concentration limits for class C wastes, and will be disposed of pursuant to a state-issued permit.[133] Further, NRC is obligated to monitor DOE's compliance with these requirements and notify DOE and the host state of any noncompliance.[134]

In the Senate debate on the legislation, senators from various states with DOE facilities alternatively supported and attacked the amendment authorizing DOE to reclassify certain wastes as WIR. Although South Carolina had filed an amicus brief for the plaintiffs in *NRDC v. Abraham*, Senator Lindsey Graham of South Carolina lobbied for passage of the legislation, commenting on the need for flexibility in the regulatory scheme.[135] Graham also emphasized the risks to workers in not allowing classificatory flexibility, stating that "the last 1.5 inches of waste that is in the bottom of these rather large tanks cannot be environmentally remediated in a manner safe for South Carolina that would prevent people from unnecessarily risking their lives to go get that last inch and a half."[136]

Senators from Colorado (where DOE had closed the Rocky Flats facility) and Idaho became cosponsors.[137] Both senators from the state of Washington, however, criticized DOE for not being willing to work with the state over disposition of wastes at Hanford, accusing DOE of blackmail and bad faith.[138] Senator Hillary Clinton of New York, the state that hosts the West Valley facility's reprocessing wastes, worked to narrow the legislation before it came out of committee so as to restrict DOE's authority to reclassify HLW to a few sites which did not include West Valley.[139] The most controversial issue in the congressional process was DOE's insistence that it had unilateral decision-making authority to reclassify waste and that it was not required to work with state and other agencies.[140] The legislation passed by Congress addressed this concern by requiring state licensing for disposal of reclassified HLW and compliance monitoring by NRC.[141]

The act also directed NAS to investigate and report on the suitability of DOE's plans.[142] In 2006, the NAS Committee on the Management of Certain Radioactive Waste Streams Stored in Tanks at Three Department of Energy Sites issued a report that summarizes the authority granted by Congress and the differences between it and the authority asserted in DOE's Manual 435.1:

1. Section 3116 [of the 2005 National Defense Authorization Act] addresses only wastes that are to be disposed of on-site and which are subject to a state compliance agreement whereas the provisions of DOE Manual 435.1 could encompass any waste and its planned destination.
2. Section 3116 sets out roles for the host states and [NRC], which are absent from Order 435.1.
3. Section 3116 and Order 435.1 differ in their description of the degree of removal of the highly radioactive fraction:

- Section 3116(a)(2): "has had highly radioactive radionuclides removed to the maximum extent practical"
- Manual 435.1-1 (p. II-1): "has been processed, or will be processed, to remove key radionuclides to the maximum extent that is technically and economically practical"[143]

The report also pinpointed the crux of the tank waste controversy: "Ideally, all wastes would be removed from the tanks [used to store HLW at INL SRS, and Hanford]. However, it is widely recognized that it is prohibitive in terms of worker risk and economic cost to exhume the tanks or remove all of the wastes from all of the tanks. The debate now is over how much removal of the wastes from the tanks is enough, how much removal of radionuclides from the retrieved waste is enough, and whether grout is an adequate form of residual waste immobilization onsite."[144]

Current Status of DOE's WIR Legal Authority

Section 3116 of the 2005 National Defense Authorization Act clarified to a considerable degree the legal rules governing DOE's tank heel disposal plans at SRS and INL. But the effect of the new legislation on DOE's authority at Hanford and any other site is unclear. DOE has taken no significant further action to reclassify reprocessing wastes at Hanford as non-HLW. One could interpret the legislation as leaving the matter of DOE's authority to classify waste incidental to reprocessing (WIR) at Hanford where it was before, in the uncertain state left by the NRDC litigation. Alternatively, Congress's grant of WIR authority for DOE at SRS and INL, but not at Hanford, might be interpreted as implicitly negating any such authority at Hanford. However, Congress apparently intended to leave undisturbed existing law and agreements, including those that apply to Hanford.[145] Thus, the failure to include Hanford in section 3116 does not appear to have any implications one way or the other regarding DOE's legal authority to classify low-activity reprocessing wastes at Hanford as non-HLW.

One set of ongoing regulatory issues relates to reprocessing waste tank heels. The NAS committee report agreed with DOE that the reclassification was appropriate for some tank waste heels but disagreed with its plan to permanently seal tanks at SRS once they were emptied of the rest of the reprocessing wastes.[146] It recommended that DOE instead wait to seal residual sludges and close the tanks until governing deadlines established by federal facilities compliance agreements draw nearer, in the hope that technology developments will enable DOE to safely remove or more effectively seal some portions of the heel that are currently impractical to extract.[147]

A second set of reclassification issues concerns segregating liquid reprocessing wastes into high-activity and low-activity portions. The NAS committee approved of DOE's plan to separate, immobilize, and dispose of certain low-activity portions as LLW at SRS, INL, and Hanford.[148] As discussed here earlier, reprocessing liquids often have components, including salts and aluminum, that interfere with vitrification but that are not themselves strongly radioactive. When separated out, they may be too large in volume to be treated and disposed of as HLW at a reasonable cost. Accordingly, DOE plans to treat and dispose of them as LLW.[149]

Removal of Cs-137 and Sr-90 from reprocessing wastes leaves large amounts of salts in reprocessing wastes. At SRS, INL, and Hanford, DOE treats such wastes to remove much of the salt content, producing a large volume of weakly radioactive salt wash. DOE plans to solidify these wastes and dispose of them as LLW.[150] The NAS committee endorsed this plan, finding that DOE's measurement and characterization of the radioactivity and chemical makeup of the salt washes justified their disposal as LLW.[151]

In addition to salts, Hanford has additional elements in some of its reprocessing liquids, most notably aluminum, that also interfere with the process used for vitrification for HLW disposal unless they are removed. DOE proposes to separate out these materials, which have low activity, vitrify them in bulk, and dispose of them as LLW.[152] Because of these variations in WIR, each facility is pursuing a different immobilization method for its low-activity fractions: saltstone conversion at SRS, steam reforming at INL, and vitrification at Hanford.[153] Implementation of these pre-disposal waste treatment strategies awaits the completion of processing facilities at all three sites.[154]

Mixed Waste

The 1987 Superfund Amendments and Reauthorization Act (SARA) Amendments to the Comprehensive Environmental Response, Compensation, and Liability Act (CERCLA, also known as Superfund) applied hazardous substance cleanup liabilities and requirements to DOE and other federal facilities, while the 1992 Federal Facilities Compliance Act (FFCA) amended RCRA to subject DOE and other federal facilities to the requirements of RCRA and to state and local hazardous waste regulations, enforcement actions, and sanctions (see Chapter 1 for details).[155] An issue of great practical importance is whether these laws apply to mixed wastes—that is, nuclear wastes that are both chemically hazardous and radioactive. They are of three basic types, depending on their radioactive characteristics: low-level mixed waste, high-level mixed waste, and transuranic mixed waste.[156] According to EPA, almost all mixed waste in the United States has been generated by DOE. Only about 2 percent of the total volume of mixed waste is the product of civilian activities.[157]

While it had been accepted without much controversy that CERCLA's cleanup obligations extend to mixed wastes, there was hot dispute over whether EPA and states exercising delegated RCRA regulatory authority could regulate mixed wastes. RCRA specifically excludes "source, special nuclear, [and] byproduct materials" from regulation as solid waste.[158] It is, however, often impracticable or impossible to separate radionuclides from the non-radioactive chemical media in which they are "contained, dissolved, or suspended."[159] Did Congress intend to exempt from RCRA regulation all mixed wastes, including their chemically hazardous components, or only the radionuclides in them? This issue was the subject of much litigation and regulatory maneuvering. The stakes are significant because of the high costs and technical difficulties in treating radioactive materials to achieve compliance with RCRA requirements for storage, treatment, transportation, and disposal and at the same time meet parallel but quite different—and sometimes incompatible—regulatory standards set and enforced by DOE and NRC for the radioactive components. Furthermore, subjecting DOE's huge legacy of mixed wastes to RCRA would mean that states could regulate them and enforce cleanup requirements against DOE.

For years, DOE took the position that mixed waste was not subject to RCRA requirements, and EPA did not actively challenge DOE's position. However, environmental groups seeking cleanup of mixed wastes at DOE weapons facilities initiated litigation under RCRA's citizen suit provisions, contending that DOE's mixed waste is subject to regulation under RCRA.[160] In 1984, a federal district court in Tennessee held that "the most reasonable reconciliation of the RCRA and the AEA is that AEA facilities are subject to the RCRA except as to those wastes which are expressly regulated by the AEA: nuclear and radioactive materials."[161] In a subsequent case, a federal district court in Colorado noted in 1990 that "Sierra Club and DOE agree[d] that the AEA 'appears directed only to the radioactive component of nuclear waste,' and that the hazardous component of 'mixed waste' must be managed as hazardous waste whether or not the radioactive component of the mixed waste is subject to regulation."[162] This illustrates DOE's acquiescence in the position embraced by the Tennessee federal district court.

Thereafter, EPA, NRC, and DOE issued guidance documents to clarify the status of mixed waste regulation. They provide that NRC or DOE regulates the radioactive components and EPA the hazardous components, an arrangement which often results in two agencies regulating the same waste.[163] In cases where it is not possible to comply with both sets of requirements, pursuant to RCRA Section 106(a), NRC/DOE regulation under the Atomic Energy Act takes precedence over regulation under RCRA.[164]

The 1984 Hazardous and Solid Waste Amendments (HSWA) to RCRA significantly strengthened the RCRA regulatory scheme by, among other things, imposing stringent restrictions on the storage and land disposal of hazardous waste.[165] Congress passed HSWA to "force the creation of new technology and to create a disincentive to generate waste that is currently regarded as difficult to treat."[166] To qualify for land disposal, hazardous waste must either be pretreated to meet specified hazardous constituent levels in order to reduce the threat that these constituents will migrate, or such waste must receive an exemption based on a demonstration that there will be "no migration" of hazardous constituents out of the land disposal unit for as long as the waste remains hazardous.[167] In order to prevent hazardous waste storage from becoming a de facto disposal method, the land disposal regulations prohibit the storage of hazardous waste for more than ninety days unless "such storage is solely for the purpose of the accumulation of such quantities of hazardous waste as are necessary to facilitate proper recovery, treatment or disposal."[168]

To accommodate the difficulties facing mixed waste generators in complying with these requirements, EPA delayed application of the land disposal regulations to mixed waste and made enforcement of the storage restrictions a low priority.[169] In addition, Congress enacted legislation allowing government agencies to store mixed waste, and subsequently both Congress and EPA took action to provide for the land disposal of mixed waste.[170]

The Federal Facilities Compliance Act, while waiving the sovereign immunity of federal facilities and subjecting them to state and local waste regulations and enforcement, also contains provisions to accommodate DOE's massive mixed waste inventory. It effectively exempts DOE facilities from RCRA's prohibition of generator storage of hazardous wastes for more than ninety days by waiving liability for federal facility storage of mixed wastes in accordance with an approved mixed waste management plan.[171] Commercial generators still risked being penalized for storing their mixed waste beyond the RCRA ninety-day limit. In 2001, EPA issued 40 C.F.R. Part 266, the Mixed Waste Rule.[172] Under Subpart N of this rule, low-level mixed waste may be stored "as [if it were] solely radioactive" thereby permitting the indefinite storage of low-level mixed waste.[173] Subpart N of the Mixed Waste Rule also exempts eligible low-level mixed waste from RCRA's transportation and disposal requirements.[174] EPA promulgated this exemption to reduce the administrative hurdles facing waste generators and to increase disposal options for low-level mixed waste by allowing this waste to be put into low-level radioactive waste disposal facilities.[175] Implicit in the Mixed Waste Rule is EPA's belief that NRC regulation is sufficient to protect "human health [and] the environment" from the dangers of low-level mixed waste.[176] States with delegated RCRA authority are not obligated to follow EPA's Mixed Waste Rule and can continue to regulate low-level mixed wastes in accordance with their RCRA authority.[177] Some of these states may be unwilling to surrender such regulatory control.[178]

The National Defense Authorization Act for Fiscal Year 1997 exempted TRU mixed waste designated by DOE for disposal at WIPP from the provisions of RCRA prohibiting land disposal of hazardous wastes without extensive pretreatment to reduce their hazard.[179] According to DOE, complying with the pretreatment requirements could cost an additional $500 million in operating costs over the life of WIPP.[180] Both EPA and DOE agreed that the RCRA land disposal requirements were not needed to protect either human health or the environment in the context of TRU disposal at WIPP.[181] WIPP is still subject to the rest of the RCRA waste management standards, including the EPA facility standards in 40 C.F.R. Part 264.[182] Because New Mexico exercises delegated RCRA authority, the New Mexico Environment Department issued WIPP's RCRA permit and enforces compliance with it, as well as with RCRA requirements generally.

In 2006, three companion bills were introduced in Congress in response to the Bush administration's legislative proposal for a Nuclear Fuel Management and Disposal Act (NFMDA).[183] Under these bills, any material owned by DOE and stored or transported in a NRC-certified container would be exempt from the RCRA provisions codified at 42 U.S.C. Section 6961(a), requiring federal government compliance with "all Federal, State, interstate, and local requirements, both substantive and procedural . . . respecting control and abatement of solid waste or hazardous waste disposal."[184] These bills, if enacted, could thus have eliminated all state authority under RCRA to regulate mixed waste stored or transported in NRC licensed containers.[185]

These bills did not make it out of committee. In 2007, the Bush administration proposed a modified version of NFMDA that limited the RCRA exemption "to materials being transported to or stored at the Yucca Mountain repository." To qualify, waste must either be in transit to Yucca and contained in an NRC-licensed container or be located at Yucca and managed under an NRC license.[186] The administration took the position that "the NRC licensing process is complex and comprehensive," and that "largely duplicative reviews under a different regulatory scheme" were not needed to protect human health or the environment.[187] All three bills fizzled after being referred to committee.

Toward a More Risk-Based System of Radioactive Waste Classification and Regulation

As this chapter abundantly demonstrates, the current U.S. nuclear waste regulatory classification system has evolved in patchwork fashion over decades, through various congressional and state statutes; federal and state and agency regulations, guidance documents, and regulatory practices; and court decisions. The current system is a mix between classification and regulation based on:

- the process generating the waste (e.g., HLW defined as wastes produced by reprocessing);
- the identity of the entity generating the wastes (e.g., civilian versus defense wastes subject to different regulatory systems);
- the characteristics of the facility accepting the waste for storage or disposal (e.g., repositories versus shallow land burial);
- the characteristics of the wastes themselves (e.g., TRU); and
- categorization based on a waste not being any other type of waste (LLW).[188]

Finally, the identity of the regulatory authority—NRC, DOE, EPA, and state authorities—varies depending in some cases on waste origins and in others on waste or disposal site characteristics.[189]

While patchwork in character, this system nonetheless provides for appropriate regulation of many types of wastes, including requirements for disposal of highly radioactive HLW, SNF, and TRU in repositories. But in other instances the existing classifications are not matched with hazard and risk. Examples include low-activity fractions of reprocessing wastes that are, at least presumptively, classified as HLW; wastes that have radioactivity levels near background but are classified as LLW and required to be disposed of in NRC regulated facilities; and used commercial and research instruments that contain hazardous radioactive sources which lack an effective regulatory system assuring safe disposal. In addition, GTCC wastes are currently in a regulatory/disposal limbo. These problems point to the need to modify the U.S. waste classification/regulation system to achieve a better match between risk and regulation. The remainder of this section examines the considerations involved in making modifications to the system and proposals to address key areas of weakness, as well as the implications of such changes for current U.S. arrangements.

Challenges in Moving toward a Sounder and More Practical Nuclear Waste Classification and Regulatory System

Ideally, a waste classification and regulation scheme should combine three elements. It should be informed by the hazard presented by different types of wastes, in terms of the type of radiation that they emit, their activity level, and their half-lives. It should also be informed by risk—the degree of harm to health and the environment that wastes with various hazards pose. Finally, it should be performance based, in that the regulatory classification should generate the level of regulatory control commensurate with the risk posed, while allowing those managing and disposing of wastes a degree of flexibility to choose the most appropriate and least costly means of achieving such regulatory control, with the result that the system will promote consistency, predictability, and cost effectiveness. In practice, it is not easy (and probably impossible) to devise a classification/regulation scheme that fully satisfies all these goals in a system that can be workably administered, monitored, and enforced.

As an illustration of the complexities that a waste classification and regulation system must face, different wastes and their components are composed of different elements that present different characteristics and hazards. For example, long-lived radioactive waste elements pose risks over an extended time period, but such wastes are long-lived because their atoms are more stable and less radioactive. In this respect, they pose a lower level of risk at any given time than do less stable, more active elements that are correspondingly shorter-lived. Some highly active elements, such as Cs-137 and Sr-90, could be handled by storing them for a cool-down period of several hundred years—assuming it becomes possible to assure institutional controls throughout such a long period—and then disposing of them as LLW. Also, some elements, such as cesium, are water-soluble; if they were to leach out of storage or disposal containers over time, they could enter groundwater and readily migrate. Also, different elements are present in different concentrations in different wastes. It would be quite challenging to develop a system for classification and regulation of wastes that took into account and tailored regulatory requirements to each and all such differences.

Variations in the locations where wastes are stored, transported, and disposed of, and in the soils, geologic formations, and hydrologic conditions where they are buried, also substantially affect the risk posed to humans and the environment, as do the proximity to and number of living organisms potentially exposed. Further, regulatory classifications should in principle consider the risks to workers and others involved in implementing the requirements to isolate, remove, treat, or transport radioactive material, as well as the risk of unforeseen accidents during these activities. The performance, especially over long time horizons, of different means of managing the risks posed by different wastes, including packaging and other engineered barriers, soils and geologic media, and institutional controls, introduces additional variables and uncertainties.

These various determinants of hazard and risk cannot be captured with a single "rating." Experience shows that, to successfully craft effective new regulations, it is imperative to consider administrative feasibility and accountability in implementing risk classifications and correlative regulatory requirements. Should we have a few crude but clear waste/regulatory categories and associated requirements, or should the system be highly context specific, taking into account the many variables just outlined? The former may be relatively simple to administer and enforce, but these advantages are offset by inflexibility, higher cost, and barriers to innovation. Without flexibility in waste classification and regulatory requirements, there may be little incentive for developing waste treatment, management, and storage options that are not compatible with the regulatory status quo. A much more flexible scheme has the opposite virtues and vices of a highly prescriptive approach. Also, regardless of which approach is taken, decisions must be made about how precautionary the system should be in the face of uncertainties regarding risks and the efficacy of risk reduction measures. The political and institutional complications and values in a federal form of government present additional challenges. These and other trade-offs in waste classification and regulation involve complex technical issues but simultaneously present basic conflicts of values and policy trade-off considerations that should be addressed and resolved through open, accessible, and accountable decision-making processes.

Obviously, underclassifying and underregulating dangerous wastes can pose health and safety risks. For example, under the current system, some radiation sources contained in industrial and commercial devices that are no longer in service are lumped in the LLW residual category because they do not fit any other definition. The LLW waste regulatory system, however, does not adequately address the particular types of hazards and risks that these sources pose.[190] On the other hand, classifying wastes in ways that result in overregulation may not only drive up the costs to society of dealing with wastes and preclude development of smarter ways of doing so, but can also in some cases create health risks. For example, if all reprocessing waste without qualification is classified as HLW and must be disposed of in a repository, the scale-like residue left when liquid reprocessing wastes are removed from storage tanks must be scraped out for repository burial. This procedure potentially exposes workers to more radiation than if the waste were covered with cement and left in place. On the other hand, not removing such wastes from tanks might in the longer term result in harmful exposures to more people if, for example, grouting in place does not effectively immobilize radioactive constituents and they migrate into drinking-water sources. Cost and feasibility inevitably have to be considered at some point when drawing the lines to determine how much risk reduction is enough. Yet an absolutist law may provide no lawful scope for regulators to make exceptions or provide guidance on where to draw the line. In the cases of reprocessing wastes stored in tanks, for example, existing classifications and regulations provide no well-defined regulatory stopping point in removal and treatment at which a waste storage tank can be declared "empty" and sufficiently clean to be sealed with cement and left in place.[191]

Institutional Issues in Waste Classification and Regulation

The federal government's record in dealing with its own wastes has improved substantially in the past two decades, due in major part to litigation and political action by environmental and local citizens' groups that helped produce state and federal legislation to force DOE to clean up its act. Janus-faced, AEC/ERDA/DOE carried both the role of waste generator, operating under powerful incentives to minimize the costs and burdens of dealing with wastes, and the role of regulator of its own waste practices. For much of its history, the first role dominated the agency's decision making. More recently, DOE's Office of Environmental Management has become a specialized entity pushing forward with cleanup. Congress has devoted substantial resources to this effort, to the point where spending on cleanup has become a sometimes dysfunctional political football for DOE contractors and communities near its sites.

This history shows that confidence in the government institutions that manage any waste classification and regulation system is a fundamental consideration in designing such a system. Clearly defined classifications with little scope for agency discretion may seem warranted where government regulators (especially self-regulators) have lost the public's trust. A lack of trust in DOE's willingness or ability to responsibly exercise its discretion, for example, was evident in NRDC's initiation of the *Abraham* litigation, and the district court's decision invalidating the discretion that DOE asserted for itself in classifying certain reprocessing wastes as other than HLW. It may also have influenced Congress's decision to give both NRC and EPA a prominent role in the regulation of deep geologic repositories developed by DOE.

Moving from the existing U.S. waste classification and regulatory system to one that is more hazard/risk informed and performance based could lead to adoption of less stringent regulation of some wastes than is required under current law, and more stringent regulation of others. Applied as a practical tool, waste classification necessarily operates as a divisive force among opposing interests: waste owners have economic and operational incentives to prefer less stringent regulation and to oppose changes that would increase regulatory costs and burdens. Environmental and community groups generally favor more stringent regulation and tend to oppose as regulatory backsliding changes that result in less stringent requirements. Their stance is quite understandable against the historical background of lax regulation of LLW disposal and the abysmal past record of AEC/ERDA/DOE in dealing with TRU and HLW. The government's record includes not only shoddy storage and disposal practices but also cover-ups of releases to the environment and exposures of the public; in some cases, such as at Hanford in the 1950s, the releases and exposures were deliberate.[192] The conflicts provoked by efforts to modify existing waste classifications are illustrated by DOE's WIR initiatives, the *NRDC v. Abraham* litigations, and the 2005 National Defense Authorization Act.

These institutional issues assume particular importance because Congress may, as a practical matter, be unable to fashion and adopt a new and comprehensive waste classification and regulation system in statutory form. Administrative agencies must in any event play a major role in implementing any new scheme. Agencies such as DOE that regulate themselves encounter serious conflicts of role and interest, compounded in the case of DOE by its lingering culture of secrecy and by public distrust of the agency. On the other hand, because of its operational experience with waste management, DOE has unique and valuable insights regarding risk management practicalities and trade-offs, and is in the best position to understand context-specific risks and to develop and implement practical and innovative means for managing them. By the same token, independent regulators without DOE's level of operational responsibility or experience may fail sufficiently to understand and appreciate the practical realities and opportunities presented by nuclear waste management. To the extent that regulatory agencies, especially self-regulatory agencies such as DOE, are given flexibility to adjust the classification of wastes and their corresponding regulatory treatment, it is essential that such discretion be exercised through open, public processes with full information disclosure and opportunity for public comment. The agency must give reasoned justifications for its decisions, adequately supported by an evidentiary record, with opportunity for judicial review. An opportunity for review and concurrence by an independent regulatory authority such as NRC should also be provided. Congress appropriately incorporated many of these safeguards in Section 3116 of the 2005 National Defense Authorization Act, along with a requirement for concurrence by host-state regulators. They should be included in any future efforts to change the current waste classification and regulation scheme. Furthermore, any reform of existing U.S. nuclear waste classifications should most likely be implemented incrementally and prospectively to avoid disrupting agreements and practices, such as the measures for cleanup of DOE facilities contained in tripartite agreements negotiated under the existing classifications.

The International Atomic Energy Agency Framework for Waste Classification and Regulation

International leadership in developing a risk-based approach to nuclear waste classification and regulation has been exercised by the International Atomic Energy Agency

(IAEA). Its work was favorably cited in a 2006 NAS report by the Committee on Improving Practices for Regulating and Managing Low-Activity Radioactive Waste, which proposed a more risk-based approach for U.S. nuclear waste classification, with specific reference to LLW.

In 1994, IAEA released its second recommended radioactive waste classification framework.[193] The purpose was to synthesize the various systems developed by individual nuclear countries into a single model framework for countries to reference in creating or revising their nuclear waste standards. Many countries have adopted the IAEA approach to varying degrees.[194] Austria, Brazil, Spain, and Sweden have adopted the IAEA framework explicitly, while others, such as France, use approaches that are substantially similar.[195]

The IAEA framework provides three broad, risk-informed categories into which waste would be assigned: high-level waste (HLW); low- to intermediate-level waste (LILW); and exempt waste (EW). LILW was further subdivided into two categories based on the length of time over which such wastes would be radioactive: long-lived LILW (LILW-LL), and short-lived LILW (LILW-SL). LILW waste is considered LILW-LL if it includes alpha-emitting isotopes with half-lives longer than thirty years, and either a single component isotope emits 4,000 Bq/g of radioactivity, or the emissions from all long-lived alpha emitting isotopes totals more than 400 Bq/g, averaged across the total mass of the waste. The agency also provided specific standards for classification based on the radioactivity of different radionuclides, and detailed the considerations necessary for determining an appropriate disposal method. The IAEA standards differ from the current U.S. approach in several key respects. The most significant is the IAEA provision for an exempt waste (EW) category, which comprises wastes that are predicted—on the basis of computer modeling incorporating a set of assumptions about human exposure and other relevant factors—to pose very small health and environmental risks.[196] The framework provides that these wastes may be disposed of in ordinary landfills. IAEA limits the EW category to material which could, after taking into account likely exposure pathways, expose people near the disposal site to a dose of .01 mSv/yr. In comparison, the average dose rate from background radiation is 2.4 mSv/yr. A substantial number of countries, including France, Japan, the United Kingdom, Sweden, and Spain, have adopted a general policy of classifying very low activity wastes as exempt from specialized nuclear waste disposal requirements, permitting their disposal in landfills.[197] U.S. law, however, does not recognize any such general category of wastes.

The second IAEA waste category, low- to intermediate-level waste (LILW), encompasses a wide variety of wastes with different hazard levels. It ranges from wastes with radioactivity just high enough to require some form of modest regulatory controls to those that generate enough heat and radiation to require cooling and repository disposal.[198] To help distinguish these wastes, LILW is subdivided into short-lived (SL) and long-lived (LL) species: LILW-LL is typically destined for repositories; LILW-SL is disposed of as its radioactivity requires, typically in facilities similar to U.S. LLW disposal sites. The aim of the classification approach, however, is not to provide "a general boundary between near surface and geological disposal of radioactive waste ... as activity limitations will differ between individual radionuclides or radionuclide groups and will be dependent on the actual planning for a near surface disposal facility."[199] The precise line at which the waste's lifetime and radioactivity concentrations preclude it from near-surface burial is dependent on the properties of the nuclides that make up the waste, the specific near-surface facility, and a country's policy preferences and regulatory approaches. Each country will have to make its own determinations as to classifying specific wastes and their precise disposition, but the LILW-LL and LILW-SL subcategories serve as a rough guide based on waste characteristics and two basic types of disposal pathways.

The third category, HLW, includes the highly radioactive portion of reprocessing wastes, spent nuclear fuel, and other wastes with comparable nuclide concentrations, such as Cs-137 and Sr-90 fractions that have not aged to the point where their radioactivity and heat have been significantly reduced.[200] These materials require geological isolation. Additional design and engineering features are needed for repository disposal of elements such as Cs-137 and Sr-90 in order to manage their heat.

The IAEA definitions for HLW and LILW-LL eliminate the source-based U.S. category distinctions of SNF, HLW, TRU, and GTCC LLW, under which some wastes posing equivalent risks to the environment and populace fall into different regulatory categories with different disposal and handling consequences.[201] The IAEA approach is to establish broad categories that include wastes with similar ranges of hazard, and then account for the special characteristics and requirements of different waste streams that fall within a given category through appropriate differences in management, packaging, and disposal, including through the development of subcategories based on relative hazard.[202] The latter approach is much like the class A, B, and C subcategories in the U.S. LLW system.[203]

New U.S. Approaches to Waste Classification and Regulation

The IAEA framework provides one possible approach for adjusting the U.S. waste classification and regulatory system to achieve a better fit between regulatory requirements and the hazards posed by different types of wastes. A number of expert U.S. groups have recommended similar approaches. Such an approach, if endorsed by Congress, could provide an overall legal and policy structure and set of procedures to foster gradual development of a more rational and consistent overall system, in contrast to the ad hoc, improvised, and often controversial efforts of DOE, EPA, and NRC to introduce some flexibility in the current legal and regulatory patchwork.

Revising the U.S. system in accordance with the IAEA model would be relatively straightforward in concept. For example, SNF and the long-lived hazardous components of reprocessing wastes would be in the category equivalent of the IAEA HLW category. TRU and many GTCC wastes (which display a range of hazardous characteristics, with some wastes as radioactive as TRU) could be placed in a category similar to the IAEA LILW-LL, requiring repository disposal. Some reprocessing wastes, such as the low-activity reprocessing waste fractions at Hanford and SRS or Cs-137 and Sr-90, after being aged for several centuries to sufficiently reduce their heat and radioactivity, could be placed in a category similar to the LILW-SL category, with further subcategorization to match management, packaging, and disposal requirements with hazard. The existing U.S. class C and B LLW subcategories, along with class A LLW, other than very low-activity exempt wastes, could also be classified as LILW-SL; the C, B, and A subcategories could be retained to determine the appropriate disposal methods.

The most conspicuous change required to implement the IAEA approach in the

United States would be the establishment of an exempt waste class. Section 81 of AEA authorizes NRC to "establish classes of byproduct material and to exempt certain classes or quantities of material or kinds of uses or users from the requirements for a license set forth in this section when it makes a finding that the exemption of such classes or quantities of such material or such kinds of uses or users will not constitute an unreasonable risk to the common defense and security and to the health and safety of the public."[204] NRC has exercised this authority to exempt ordinary smoke detectors and some radiological drugs from disposal regulations as below regulatory concern (BRC).[205] However, these exemptions are ad hoc and limited. In 1986 and 1990, NRC issued two policy statements that set general standards for exempting very low activity waste as BRC under Section 81 of AEA and Section 10 of LLRWPA.[206] This initiative met strong resistance from the states and environmental groups, and NRC was challenged in court over it.[207] Congress responded with Section 276 of the Energy Policy Act of 1992 (EnPA), which effectively revoked the two NRC policy statements and gave the states power to regulate any materials that NRC exempts as BRC.[208] More recently, EPA solicited comments on a potential rule to allow, with NRC approval, limited amounts of very low activity mixed waste to be disposed of in RCRA landfills, but a deluge of adverse comments has stalled rulemaking on the issue since 2003.[209] Against the background of this experience, NRC has taken the view that it lacks authority to establish a new general category of very low activity waste regulated other than as LLW.[210] Even if NRC were to classify very low activity class A LLW as BRC, states would, under EnPA, be free to regulate such wastes as they saw fit.

The NAS Committee on Improving Practices for Regulating and Managing Low-Activity Radioactive Waste issued a report in 2006 proposing approaches to developing a more hazard/risk-informed approach to what it characterized as a "regulatory patchwork that has evolved over almost 60 years" in the United States, based in large part of the origin of wastes rather than their characteristics.[211] Acknowledging that there is "no easy way to reform the existing system," the committee recommended that the federal government proceed incrementally, within the parameters for waste classification developed by IAEA, toward a "risk-informed approach . . . based on information provided by science-based risk assessment but includes stakeholders as a central component in decision making."[212] The committee noted approvingly that "international organizations, especially the European Commission (EC) and the International Atomic Energy Agency (IAEA), are making significant progress in developing consistent, risk-based standards for managing [low-activity waste]. Their approaches include a number of important elements of a risk-informed system."[213] The committee also recommended that stakeholder social and economic concerns should play a role in determining acceptable risk, and that incorporating such concerns in the design and implementation of a new waste classification framework in the United States would promote public confidence in it, especially as regards the determination of exempt wastes.[214]

One of the specific opportunities for reform emphasized in the committee report concerns the "large volumes of wastes from decommissioning and site cleanup [that] often contain practically no radioactive material, but [that] cannot exit the regulatory system because Class A has no lower boundary," unlike the IAEA approach, which provides for an exempt waste category.[215] The committee estimated costs of $4.5 billion to $11.7 billion for disposing of ten million tons of very low activity wastes in NRC-regulated landfills, and concluded that other, less costly alternatives including disposal in RCRA hazardous waste landfills or exemption from regulation would be appropriate, would be sufficiently protective of these wastes, and should be actively explored.[216] These materials consist mostly of lightly contaminated building materials or other shielding from decommissioned reactors.[217] They must be transported to and disposed of in NRC-licensed LLW disposal sites, the most important of which is in Clive, Utah, distant from most waste-generating facilities. By contrast, as the NAS committee noted, class B and C wastes account for only 900 cubic meters out of around 65,000 total cubic meters of LLW generated annually but contain over 90 percent of total LLW radiation.

The NAS committee also addressed problems of underregulation posed by existing federal regulatory programs for radiation sources (including those using radium) used for medical, industrial, and research applications that are no longer in service. Although classified as LLW, these contain highly concentrated radiation sources, are often "orphaned" with no one clearly responsible for them, and "can pose acute risks to the public and the environment." The report did note some recent progress in the area, but it still emphasized the potential hazard from these sources.[218]

In 2002, the National Council on Radiation Protection and Measurements (NCRPM), a congressionally chartered technical body tasked with researching and communicating information about radiation, put forth a reclassification proposal entitled Risk-Based Classification of Radioactive and Hazardous Chemical Wastes.[219] The proposed system is almost exactly the same as the IAEA framework. It outlines a three-category classification system (exempt waste, low-hazard waste, and high-hazard waste) and gives examples of how some extant waste categories would fit into the proposed system. The NCRP proposal varies from the IAEA proposal in that it seeks to bring hazardous chemical waste into the same regulatory framework as radioactive waste. Regulators would model the hazards to the public and the environment associated with any given hazardous chemical, radioactive, or mixed waste, and wastes would be would be classified into categories defined by specific hazard thresholds. The high-hazard wastes would be disposed of in geologic isolation, while the low-hazard category would be disposed of in regulated, near-surface facilities. Exempt waste would have no disposal requirements.[220] These categories could be further subdivided to provide more targeted regulatory requirements.[221] NCRP also outlined the legal and regulatory changes necessary for its proposal to be implemented. The most important would be creating an exempt waste class and eliminating the source-based classifications in current law.[222]

None of these proposals, however, are meant to provide complete systems for waste regulation. Rather, they offer conceptual frameworks for grouping wastes in broad categories based on hazard levels. The next step in developing a revised, more consistently risk-based U.S. classification/regulation system would be to adopt subclassifications within the categories, based on radiation types and levels like those in the U.S. LLW classifications (class A, B, and C). The source-based distinctions that determine many other U.S. waste categories would be eliminated. Once established, this system would be substantially transparent and relatively straightforward to administer. By contrast, the many different frameworks and distinctions embedded in the existing U.S. system makes it very difficult to adopt changes to better align the classification and regulation of wastes with their hazards and the risks that they present. Such efforts are likely to seem ad hoc and opportunistic, provoking controversy, as illustrated by DOE's efforts to classify and dispose of certain reprocessing wastes as other than HLW and the efforts of NRC and EPA to develop a BRC category for very low activity wastes. It would, however, be a major challenge to develop the degree of political consensus needed to adopt a new general waste classification framework and process.

Conclusion

Any nuclear waste regulatory system must classify wastes according to the hazards they present and seek to match regulatory requirements with the classifications in order to appropriately protect the public health and the environment. Because of institutional and other practical considerations, no regulatory system can achieve a perfect fit. The current U.S. system of classification and regulation was not developed according to a single plan or consistent set of principles. It is patchwork in character. Nonetheless, it does a fairly good job of matching regulation and risk in the cases of HLW, SNF, TRU, and much LLW. However, it fails to provide clear direction or, in some cases, adequate legal authority for dealing with certain low-activity reprocessing wastes that do not require repository disposal; GTCC wastes that present significant hazards; very low activity wastes that do not require disposal in LLW facilities; and highly hazardous radioactive sources that have been used in civilian applications. The Blue Ribbon Commission on America's Nuclear Future should address and recommend measures to meet these problems. Congress should step forward to solve them through appropriate legislation that promotes a more rational and consistent risk-based nuclear waste classification and regulation system.

Chapter 3

Nuclear Waste Transport

The primary federal agencies responsible for regulating the transportation (including packaging and handling) of nuclear wastes are DOE, NRC, and the Department of Transportation (DOT). They work in conjunction with state and local authorities and regional transportation planning boards. The current distribution of regulatory authorities is summarized here first, then discussed in greater detail.

DOE is responsible for undertaking and managing the transportation of defense wastes, including TRU destined for disposal at WIPP, and LLW disposed of at commercial facilities. Under NWPA, DOE is also responsible for transporting commercial and defense HLW and SNF to a federal repository for disposal, for transporting commercial HLW and SNF to a federal MRS facility for storage, and for transfers of nuclear wastes and other nuclear materials between DOE facilities. In addition, in siting federal repositories and MRS facilities, DOE must take transportation issues into account, select the mode of transportation of wastes to such facilities, and determine rail transportation routes. In carrying out these tasks, DOE is subject to the nuclear waste container and transportation regulations issued by NRC and DOT, respectively.

Transport of commercial LLW for disposal and of civilian SNF to consolidated storage facilities (other than a federal MRS) or to reprocessing facilities is a private sector responsibility, subject to regulation by NRC and DOT in coordination with state and regional authorities. NRC is responsible for establishing standards for the containers that will be used in the transportation of civilian nuclear wastes other than class A LLW, for approving their design, and for regulating the proper handling of the containers and related matters. DOT sets standards for containers used for civilian class A LLW, and for determining highway routing for nuclear waste transport by truck. It regulates the actual transportation of the waste by truck or rail and the carriers that undertake it. With respect to transport by truck, DOT regulates such matters as the placarding of nuclear waste transport convoys, maintenance and use of the vehicles that will be used in transportation, truck-driver training and responsibilities, and positioning of railcars carrying nuclear waste. State, tribal, and regional bodies and carriers have leeway to determine the routes for truck shipments, but their selections are subject to the standards set by DOT. As for rail transport of nuclear wastes, DOT exercises broad authority to regulate all aspects of rail safety and establishes standards for rail carriers' determinations of routes for shipments and the positioning of the railcars. States and local and tribal governments have no authority to regulate rail routing, but rail carriers are required to consult with state, local, and tribal governments when determining routing.

DOE is responsible for managing and regulating shipments of its own wastes, as well as of SNF from naval reactors and from foreign research reactors. It generally uses private truck and rail carriers to make these shipments, which are subject to NRC and DOT regulation. In carrying out these activities, DOE has, over the years, as a result of federal legislation and its own contractual commitments, become obliged to adhere to NRC packaging and associated regulations and to DOT transportation safety regulations. DOE also complies with NRC and DOT regulations as a matter of policy, even when not legally obligated to do so, except when it determines that, for national security reasons, following such regulations would be inappropriate.

The most comprehensive study of nuclear waste transportation is that of the NAS Committee on Transportation of Radioactive Waste, whose report, *Going the Distance?*, was published in 2006.[1] The committee found that, overall, the current regime for nuclear waste transport is regarded as working well. More than ten thousand rail and truck shipments of various nuclear wastes, including regular shipments of TRU wastes to WIPP over the past eleven years, have been made without an accident involving any release of radioactivity.[2] The earlier safety record, however, has been found wanting by at least one commentator, who reported a significant number of LLW transport accidents involving some radioactive releases.[3] Moreover, there are significant unknowns regarding the degree and magnitude of terrorism risks that are difficult to assess because much of the information gathered by the government on the issue remains classified; the NAS committee reported that it did not have access to key information on terrorism in preparing *Going the Distance?* and was therefore unable to evaluate the terrorism risks entailed in nuclear waste transportation.[4] Moreover, shipment of the nation's large accumulated stocks of HLW and SNF to a repository or (in the case of SNF) to federal or private consolidated storage facilities would present logistical and risk-management challenges on a much larger scale than faced to date. Nonetheless, the NAS committee found that there are no significant technical barriers to the safe large-scale transportation of SNF and HLW, and this conclusion has been endorsed by other experts.[5] The committee, however, emphasized that to succeed with such a project, the federal government will need to take steps to win the confidence and trust of the public, which continues to express significant concerns about the safety of nuclear waste transport,[6] and gain the cooperation of states and localities.

The Evolution of Nuclear Waste Transport Regulation

AEC was originally assigned by Congress the regulatory responsibilities over nuclear waste transportation that are now charged to NRC, DOT, and DOE. Under the broad authorities granted to AEC by the 1954 Atomic Energy Act, which includes authority over "the transfer, delivery, receipt, acquisition, possession, and use of nuclear materials,"[7] AEC regulated and controlled all aspects of the transportation of nuclear materials, including nuclear wastes, at every point in their life cycle. Later, in the Energy Reorganization Act of 1974, Congress divided AEC's broad management and regulatory powers, which included general authority to regulate transportation of nuclear wastes, between NRC (civilian activities and wastes) and ERDA, which later became DOE (defense activities and wastes).[8]

Early Federal Radioactive Waste Transportation Legislation

Congress first specifically addressed transportation of radioactive materials in 1960, when it passed legislation authorizing the Interstate Commerce Commission (ICC) to require "the best-known practicable means for securing safety in transit" by regulating packaging, marking, loading, and handling of shipments of radioactive materials by land, including by truck and by rail, and to "prescribe the route or routes over which . . . radioactive materials . . . shall be transported."[9] Congress required ICC to consult with AEC before issuing regulations relating to radioactive materials.[10] ICC regulation applied only to commercial shipments of radioactive materials; the legislation specifically exempted shipments of radioactive materials by the Department of Defense or AEC for "national security purposes."[11] The statute also did not address the role of state and local authorities in regulating hazardous materials transportation.

When Congress created the Department of Transportation in 1966, it did not transfer jurisdiction over transportation of radioactive materials to the new department; ICC retained that jurisdiction.[12] ICC authority over road and rail transportation of hazardous materials increasingly overlapped with other federal efforts, such as DOT's regulation on railroad safety and its advisory role on hazardous materials shipping, and state and local regulation of shipping.[13] The division of responsibility between ICC, with authority over radioactive waste transport, and AEC, with broad authority over radioactive materials, was not clearly defined in law, nor was the role of state and local governments.[14] The result was a confusing and uncoordinated web of regulations and enforcement at a time when public awareness of the risks of hazardous materials was growing.[15]

1970 Federal Railroad Safety Act

The Federal Railroad Safety Act of 1970 (FRSA), granted DOT authority over rail transportation in an effort to make the "laws, rules, regulations, orders, and standards relating to railroad safety . . . nationally uniform to the extent practical" and to "reduce deaths and injuries . . . [and] damage to property caused by accidents involving any carrier of hazardous materials."[16] FRSA granted DOT very broad regulatory authority to "prescribe, as necessary, appropriate rules, regulations, orders, and standards for all areas of railroad safety."[17] State law and regulation of railroad safety were permitted on matters DOT had not addressed, but once DOT issued regulations on a subject, state law that imposed additional or more stringent railroad safety requirements than DOT regulations were generally preempted. The only state laws exempted from preemption were those "necessary to eliminate or reduce an essentially local safety hazard," provided that they did not conflict with DOT regulations or create an undue burden on interstate commerce.[18]

1975 Hazardous Materials Transportation Act

In 1975, Congress sought to address continuing confusion over transportation regulation, especially regarding road transportation, by enacting the Hazardous Materials

Transportation Act (HMTA), which consolidated all federal regulatory authority over transportation of hazardous materials within DOT. Under HMTA, DOT was authorized to designate materials as hazardous upon determination that they may pose an "unreasonable risk to health and safety or property"; radioactive materials were specifically identified as an example of materials the DOT was to designate as hazardous.[19] DOT was empowered to issue regulations "govern[ing] any safety aspect of the transportation of hazardous materials . . . , including, but not limited to, the packing, repacking, handling, labeling, marking, placarding, and routing . . . of hazardous materials, and the manufacture, fabrication, marking, maintenance, reconditioning, repairing, or testing of" containers used for transporting hazardous materials.[20] DOT was required to consult with ICC before issuing regulations under the statute.[21] HMTA, in contrast to earlier federal statutes, expressly preempted any state or local requirement that was inconsistent with HMTA or DOT's implementing regulations.[22] States could petition DOT for a waiver of preemption, however, if the state or local requirement "(1) afford[ed] an equal or greater level of protection to the public than is afforded by [HMTA or DOT implementing regulations] and (2) [did] not unreasonably burden commerce."[23]

Despite Congress's purpose and efforts to promote national uniformity by passing HMTA, a maze of conflicting state, local, and federal regulations continued to proliferate, especially with regard to highway transportation.[24] In 1990, the House Committee on Energy and Commerce concluded that HMTA had "failed to provide either a timely or a final resolution of [preemption] issues. . . . The current process for administrative determinations of whether a state law is preempted has failed to work effectively, causing delay, litigation, and confusion."[25] The committee noted that attempts to amend HMTA had stalled for several years because of disagreement on how to fix the statute's problems. States sought to maximize their ability to regulate hazardous waste transportation within their borders and particularly opposed any federal restrictions on state enforcement of state and local laws and regulations on transport of hazardous materials, including nuclear wastes. Industry, on the other hand, argued that the lack of uniform national regulation was economically wasteful and even threatened safety.[26]

HMTA authorized but did not require DOT to issue rules on routing of trucks carrying hazardous materials. DOT initially declined to regulate routing, allowing states and localities to enact their own routing rules.[27] This proved to be highly problematic. After New York City and other localities blocked or restricted shipments of radioactive materials, DOT became concerned about piecemeal regulation of truck routing and the resulting effects on interstate commerce.[28] DOT issued a proposed rule in 1980 and promulgated a final rule in 1981 on highway routing of trucks carrying radioactive material: the final rule established that radioactive materials could generally be shipped on interstate highways.[29] States were permitted to require shippers to use alternative routes, but state-designated routes had to comply with DOT guidelines. Despite the greater clarity, litigation continued.[30]

1980 Low-Level Radioactive Waste Act

Congress in 1980 enacted the Low-Level Radioactive Waste Policy Act (LLRWPA), amended in 1985, to establish a system of interstate compacts to create incentives for states to cooperate in developing new facilities for LLW disposal (see Chapter 4). The act acknowledged the role of states in regulating LLW transport but contained a provision that nothing in the LLRWPA "may be construed to confer any new authority on any compact commission or State . . . to regulate the packaging, generation, treatment, storage, disposal, or transportation of low-level radioactive waste in a manner incompatible with the regulations of the Nuclear Regulatory Commission or inconsistent with the regulations of the Department of Transportation."[31] This did little to clarify uncertainty about the role of state and local regulation.

1990 Hazardous Materials Transportation Uniform Safety Act

Congress sought to resolve the continuing problems by enacting the Hazardous Materials Transportation Uniform Safety Act (HMTUSA) of 1990. In the statute's findings, Congress stated: "Many States and localities have enacted laws and regulations which vary from Federal laws and regulations pertaining to the transportation of hazardous materials, thereby creating the potential for unreasonable hazards in other jurisdictions and confounding shippers and carriers which attempt to comply with multiple and conflicting registration, permitting, routing, notification, and other regulatory requirements."[32] HMTUSA vested DOT with authority to regulate the transportation of all nuclear materials, including nuclear wastes.[33] It required DOT to issue regulations for the safe transportation of SNF and HLW by rail; previously, DOT had been permitted but not required under FRSA to address nuclear waste transportation.[34] HMTUSA also required DOT to prepare a study identifying factors that carriers and transporters of SNF and HLW should consider when selecting rail or highway routes.[35]

Congress also provided for a rather elaborate preemption scheme in HMTUSA. First, Congress expressly preempted state and local regulations on certain topics related to truck and rail transportation, including the designation of hazardous materials, documentation for shipments, reporting procedures for accidents, and design of containers, unless the state requirements were substantially the same as DOT regulation on those matters.[36] Second, state or local regulations on other hazardous materials transport matters were preempted under HMTUSA if compliance with both the state or local law and DOT regulations was impossible, or if the state or local law "create[d] an obstacle to the accomplishment and execution" of HMTUSA or regulations established under the act.[37] States could, however, petition DOT for waivers of preemption. The act established specific procedures for such waiver petitions, including procedures for administrative and judicial review of DOT's decisions on such petitions.[38] Third, HMTUSA generally left intact FRSA's provisions regarding preemption of state or local regulation of rail transport.

HMTUSA also extensively addressed highway routing of shipments of hazardous materials, including radioactive waste. It gave states and tribes authority to designate and enforce specific highway routings for transporting hazardous materials, but subjected them to federal conditions and standards and DOT oversight.[39] In designating highway routings, states and tribes were required to comply with standards that would be issued by DOT, which would obligate them to consider public safety, to allow public participation, to ensure that shipments were able to cross their territories on designated routes, and to ensure that designated routes gave hazardous waste carriers reasonable access to fuel, repairs, rest stops, and the origins and destinations of the shipments. States were also required to consider other factors, including population density along routes, types of highways, types of hazardous materials being shipped, emergency-response capabilities, "exposure and other risk factors," terrain, and effects on commerce.[40] Before issuing standards, DOT was required to consult with the states.[41] Another important requirement was for states and tribes to consult with other state, local, and tribal governments affected by routing decisions; this was intended to encourage the states to address hazardous waste transportation in regional bodies such as the Western Governors' Association and the Southern States Energy Board, to ensure that the routings selected by different states were coordinated.[42] HMTUSA also recognized the roles of shippers and carriers in planning shipments of hazardous materials to the extent that federal and state regulations allowed them a choice of routes; it required DOT to develop guidelines to ensure that shippers and carriers selected routes and modes of transportation that enhanced public safety.[43]

2007 9/11 Commission Act

HMTUSA contained a comprehensive scheme for highway routing of hazardous materials but did not provide a similar scheme for rail routing. Congress addressed rail routing of radioactive materials in the Implementing Recommendations of the 9/11 Commission Act of 2007, which implemented recommendations of the 9/11 Commission for dealing with the risk of terrorist attacks.[44] Congress instructed the Department of Homeland Security (DHS), in consultation with DOT, to designate certain materials as "security-sensitive" if their potential use by terrorists poses a national security threat; DHS was specifically directed to consider designating radioactive materials as "security sensitive."[45] DHS issued a final rule in November 2008 designating radioactive materials as security sensitive.[46] The act also instructed DOT to publish a final rule in 2008 requiring rail carriers transporting security-sensitive materials to produce annual safety and security risk analyses of the routes used for shipping these materials and for alternative routes;[47] DOT issued this final rule, also in November 2008, implementing rail-routing requirements for transportation of nuclear waste and other hazardous materials.[48] Finally, the 9/11 Commission Act also instructed DOT to require rail carriers to use the safest and most secure routes, based on the annual analyses, for transporting security-sensitive materials.[49] The act did not provide a role for states, tribes, or local governments in the rail route analysis and selection process, but it did require rail carriers to seek information from state, tribal, and local governments regarding security risks along rail lines.[50]

HMTUSA did not address the relationship between DOT's power to regulate transportation of radioactive materials and the AEA-based authority of AEC/NRC to regulate the "transfer" of the materials.[51] This issue is addressed by an interagency agreement signed in 1979 by DOT and NRC.[52] DOT develops safety standards for packaging of civilian class A LLW radioactive materials, for marking of packages and vehicles, for qualifications of carriers, and for loading, unloading, and handling procedures; NRC develops safety standards for packaging of civilian radioactive materials exceeding class A limits. Each agency consults with the other when formulating standards, and each agency issues regulations requiring transporters to follow the standards of both agencies.[53]

Federal Shipments of Radioactive Materials

The statutes providing for regulation of transportation of radioactive materials have not applied to certain government shipments. The 1960 legislation authorizing ICC regulation exempted AEC or Department of Defense (DOD) shipments made for the "purpose of national security," including shipments by contractors on behalf of the agencies.[54] When Congress enacted HMTA in 1975, it provided that the act only applied to "transportation in commerce" of hazardous materials.[55] Shipments by DOE or DOD designated by them as related to national security remained exempt from DOT regulation.[56] The original House version of the legislation that became HMTUSA included a provision requiring "shipments of high-level radioactive waste or spent nuclear fuel made by or under the direction of the Federal Government [to] conform with" DOT regulations; this language, however, was not included in the version of the bill eventually enacted.[57]

Other legislation, however, imposed a variety of regulatory requirements on DOE's transport of nuclear wastes. The Nuclear Waste Policy Act of 1982 made DOE's transportation of SNF and HLW wastes to a federal repository or MRS facility subject to NRC packaging requirements and also required DOE to notify state authorities of shipments.[58] The WIPP Land Withdrawal Act, passed in 1992, required DOE to study transportation options, to use NRC-certified packaging, to give states and tribes advance notice of TRU shipments, and to coordinate with states and tribes on emergency preparedness; it did not require DOE shipments of TRU to WIPP to adhere to DOT regulations.[59] However, in 1987, DOE signed an agreement with New Mexico containing provisions committing DOE to adhere to DOT and NRC regulations for transportation of nuclear waste to the WIPP facility and to package all waste shipped to WIPP in NRC-certified packages.[60] In addition, DOE has declined to characterize its shipments of naval SNF to Idaho as national security shipments, thereby agreeing to subject them to DOT regulation.[61] DOE has also coordinated with regional organizations of states on transportation issues, and has entered into agreements with regional organizations on its radioactive materials transportation practices.[62]

Currently, "as a matter of policy," DOE complies with NRC and DOT transportation regulations for all shipments of radioactive materials, even when not required to by legislation or agreement, except when DOE determines that national security requires that it deviate from NRC and DOT standards.[63] Even for national security shipments, DOE states that it coordinates with other federal agencies and with state, tribal, and local governments "to the extent practicable" and undertakes to meet or exceed safety and security requirements for comparable commercial shipments.[64]

The Current Nuclear Waste Transport Regulatory Structure

Nuclear Regulatory Commission (NRC)

NRC regulates nuclear waste packaging and related transportation matters, such as labeling packages, lifting and tie-down procedures, protecting workers and the public from radiation during transport, and receiving and opening packages upon delivery.[65] Found at 10 C.F.R. Parts 20, 71, and 73, these standards specify acceptable doses of radiation for both occupational and general public exposure; establish the tests that packages must undergo; provide standards for package design certification; and establish various procedures, such as measures for detecting and preventing unauthorized access to packaging while loaded on the truck or railcar, and coordination with law enforcement authorities before shipping nuclear waste, for the physical security and protection of the waste and its packaging while in transit.[66] Rather than specify a predetermined list of materials or designs, 10 C.F.R. Part 71, Subpart E, sets out the basic performance standards for all packages. The most important performance standard is that the use of the package will not result in loss or dispersal of radioactive contents or in significant increase in external radiation levels. Subpart E also regulates the lifting and tie-down standards for all packages; external radiation standard for all packages; additional requirements for Type B packages; and additional requirements for Type B packages that contain more than 10^5 A_2 (a measure of maximum radioactivity).[67] Subpart F of Part 71 specifies the tests to ensure proper function of the package under normal conditions of transport, and proper function after a series of hypothetical accident conditions.[68] Subpart G of Part 71 sets out the operating controls and procedures for shippers or carriers, which include preliminary inspections before the first use of a package, routine inspections of the physical condition of packages, and opening instructions; Subpart G also requires shippers and carriers to allow the NRC to conduct inspections and tests of their material, packages, and facilities.[69] Finally, 10 C.F.R. Part 73 includes requirements for protection of wastes while in transit, including requirements for threat assessments, trained security escorts, and notification of NRC of shipments; Part 73 also requires shippers to obtain advance approval from NRC of road and rail routes to be used for shipment.[70]

The 2006 report by the NAS Committee on Transportation of Radioactive Waste, *Going the Distance?*, contains an extensive review of the NRC regulations and of NRC's testing of packaging for SNF. The report noted the critical role of packaging in preventing release of radiation in transportation accidents, which will inevitably occur, and the importance of effective response capability. Its principal recommendations were that NRC should continue to conduct full-scale stress tests and should conduct more testing of package performance in long-duration, fully engulfing fires, and to promote development and coordination of response capacity by federal, state, and local authorities.[71]

Department of Transportation (DOT)

DOT has issued nuclear waste transportation regulations, which address factors state and tribal authorities are required to consider in designating preferred routes for highway shipping of radioactive materials; guidelines for shippers and carriers for choosing among available highway or rail routes; the inspection of motor vehicles and railcars used to transport nuclear waste; nuclear waste handling; the licensing of transporters of nuclear materials; provision of grants to local authorities to aid in the planning and training of personnel in the case of accidents and incidents; and inspection and monitoring of the transportation process.[72]

DOT regulations at 49 C.F.R. Part 107 set forth hazardous materials regulations and procedures, including requirements for posting warning/informational placards on transport vehicles, conducting inspections, and confirming certification of transporting parties.[73] Procedures for states, tribes, and local communities to obtain DOT grants for training of hazardous material emergency personnel are contained in 49 C.F.R. Part 110.[74] DOT rules requiring rail carriers to identify and use the safest and most secure routing practicable, and to develop security plans for nuclear waste shipments, are contained in 49 C.F.R. Part 172, Subpart I and Appendix D.[75] Safety regulations for rail transport of hazardous materials, including nuclear waste, appear in 49 C.F.R. Part 174.[76] The regulations for actual transportation of hazardous materials, including parking, smoking, fueling, tires, documentation, and highway routing, are set forth in 49 C.F.R. Part 397.[77]

DOT regulations in 49 C.F.R. Part 397, Subpart D, require highway carriers of radioactive material to follow "preferred routes"; deviation from preferred routes is permitted only to access pick-up and delivery points; for necessary food, fuel, and repair stops; and in emergencies. Where several possible preferred routes exist, the transporter must choose the route that minimizes radiological risk, considering factors such as accident rates along the routes, transit times, population densities, and time of day and day of the week. Deviations from a preferred route for such reasons must follow the shortest-distance route, or a route that minimizes radiological risk and is not twenty-five miles longer, or five times longer, than the shortest route.[78]

DOT's Subpart D regulations recognize "any Interstate System highway" as the preferred route for truck transportation of radioactive materials, unless states have specified an alternative route as preferred, in accordance with DOT's Guidelines for Selecting Preferred Highway Routes for Highway Route Controlled Quantity Shipments of Radioactive Materials.[79] The DOT guidelines, which are based on the routing requirements in HMTUSA, specify three primary and four secondary factors that states must consider in designating alternative routes. The primary factors are potential radiation exposure from normal transport, public health risk from the accidental release of radioactive materials, and risk of economic harms from such a release. The secondary factors are emergency response capability, locations of special facilities (such as schools and hospitals), ability to evacuate in case of accidental release of radioactive materials, and expected traffic fatalities and injuries related to increase in transportation activity but not related to the nature of the cargo.[80] Further, states must notify DOT of a selected alternative route before it comes into effect.[81] State authorities are responsible for maintaining emergency-response capabilities and for responding in the event of a transportation incident.[82]

Regional boards comprising representatives of states facilitate transportation by coordinating planning and interstate shipment of radioactive materials.[83] Among other matters, they ensure that the routes selected by different states meet at contiguous borders. Regional boards include the Southern States Energy Board (SSEB) Radioactive Materials Transportation Committee, which comprises eight southern states;[84] the Technical Advisory Group of the Western Governors' Association (WGA), which includes ten states;[85] the Midwestern Radioactive Materials Transportation Committee of the Council of State Governments, Midwest, which has representatives from twelve states;[86] and the Northeast High-Level Radioactive Waste Transportation Project of the Council of State Governments, Eastern Regional Conference, which consists of ten states.[87] These regional bodies maintain cooperative agreements with DOE that allow them to provide input to the DOE decision-making process and jointly resolve transportation issues.[88]

Since 2005, DOT has required that highway shipments of SNF, HLW, or TRU pass inspection under North American Standard Level VI Inspection guidelines, which the Commercial Vehicle Safety Alliance (CVSA) developed in cooperation with DOE. CVSA is a quasi-governmental organization of commercial vehicle safety enforcement officials throughout the United States, Canada, and Mexico. CVSA receives DOE funding to train and certify Level VI inspectors.[89]

The provision in HMTUSA for state routing authority over radioactive waste is limited to transport by trucks and highways and does not extend to carriage by trains and rail. The 9/11 Commission Act provides that rail carriers themselves are responsible for determining routes, but carriers are required to follow DOT rules when analyzing and selecting routes for radioactive materials.[90] DOT announced in 2008 its final rule on procedures for determining rail routing, based on the requirements in the 9/11 Commission Act; these procedures are contained in 49 C.F.R. Part 172.[91] Section 172.820, contained in Subpart I of Part 172, includes requirements for rail carriers transporting radioactive materials to collect data on routing of these materials; conduct annual analyses of the safety and security of shipping nuclear waste along the currently used and practicable alternative routes; select routing for nuclear waste based on the results of their analyses; and, for security reasons, restrict access to route analyses.[92] Subpart I also requires rail carriers to produce a security plan for rail shipments of hazardous materials, which must include plans for providing security personnel, restricting access to the materials, and ensuring origin-to-destination security for the shipment.[93] Appendix D to Part 172 lists factors carriers must consider when analyzing rail routes, including volume of hazardous materials transported, traffic density, route length, track types, potential terrorist targets along the route, and population density.[94]

DOT rail-routing rules promulgated under the 9/11 Commission Act apply only to "railroad carriers transporting security-sensitive materials in commerce"; the act does not directly address whether it applies to shipments by DOE or by private carriers on behalf of DOE.[95] DOE claims the authority "to regulate all aspects of activities involving radioactive materials that are undertaken by DOE or on its behalf, including . . . transportation,"[96] and it asserted this authority when it selected a rail corridor and alignment for shipments of nuclear wastes to the Yucca repository.[97] Nevada sued DOE, claiming that DOE could not designate a rail line without the approval of the Surface Transportation Board, an agency located within DOT. In 2006, the D.C. Circuit dismissed Nevada's case as not ripe; it also noted that STB's authority extends only to rail lines operated as common carriers, and there was no indication that DOE would establish a Yucca branch line operating as a common carrier.[98] The court noted approvingly that DOE intended to "obtain all necessary regulatory approvals" before beginning construction. Nevertheless, neither the court opinion nor DOE's ROD designating the route identified which federal regulations might apply.[99]

Department of Energy (DOE)

As the agency with operational authority for disposal of TRU, SNF, and HLW, DOE determines the method of transportation of wastes to repositories and federal MRS facilities, and, when it chooses rail transport, selects the rail corridors to be used.[100] DOE examines transportation options in the environmental impact statement (EIS) process for development of such facilities. If rail is selected in the EIS ROD, DOE prepares another EIS to select the rail corridor to be used. If truck transport is chosen, DOE identifies routes likely to be selected according to the preferred highway routes determined by DOT.[101] Ultimate selection of the highway routes, however, as described earlier, is the responsibility of the states and tribes and carriers, subject to compliance with DOT regulations.[102]

DOE consults and cooperates with state and regional transportation authorities on truck routings,[103] and it selects rail routings for shipments of SNF and HLW in consultation with carriers, states, and tribes. DOE allows carriers transporting LLW to determine routing, in accordance with applicable routing regulations.[104] It also provides state and local authorities funds and assistance for various initiatives related to the transportation of radioactive waste, including road improvements in anticipation of increased road usage from the transportation of waste by trucks, accident preparedness, emergency response, and the monitoring of transportation incidents.[105]

Experience with Nuclear Waste Transport

Nuclear wastes, including LLW, SNF, and TRU waste, have been shipped regularly and extensively throughout the country for decades. No current, systematically gathered data exist, however, on all these shipments.

We nonetheless know a great deal. For instance, large volumes of class A LLW have been shipped from reactor and other generator sites to disposal facilities, including by rail to the Clive, Utah, facility. Shipping from commercial facilities is done either by utilities and other generators themselves, or by brokers who provide collection and transportation.[106] The major portion of transported LLW comes from the decommissioning of nuclear power plants and cleanup of DOE sites. GAO reported that shipments of class A waste to commercial disposal facilities increased from four million cubic feet (113,000 cubic meters) in 1999 to twelve million cubic feet (340,000 cubic meters) in 2003; the 300 percent increase was attributable to accelerated cleanup of DOE sites.[107]

Substantial shipments of both government and commercial SNF have been made over the years, subject to a complex allocation of responsibility among regulatory authorities.[108] Most commercial shipments have been small-quantity shipments, either transfers of SNF between reactors owned by the same utility, to economize on storage space, or transfers of SNF to research facilities for testing.[109] Other shipments of commercial SNF have included the transfer of the damaged SNF from the Three Mile Island Unit to INL.[110] Some shipments in the 1960s and 1970s included transport of SNF to commercial reprocessing facilities. For example, the reprocessing facility at Morris, Illinois, received about 700 MTHM of SNF from commercial power plants, although the plant never became operational; this SNF remains in pool storage at Morris.[111] In addition to commercial SNF shipments, considerable amounts of naval and research SNF have been shipped domestically. From 1996 to 2002, twenty to sixty casks of SNF were transported annually from domestic research reactors or from foreign research reactors sent to the United States pursuant to treaties signed in the 1950s as part of the Atoms for Peace program.[112] Because foreign SNF is not subject to NRC licensing, DOE asserts authority over its shipment; it pledges to meet or exceed NRC and DOT safety and security standards for such shipments.[113] Most foreign SNF is received by ship at

Charleston, South Carolina, and then shipped to SRS by truck or rail, where most of it is stored, although some has been transported to INL by truck.[114] SNF from university or other domestic research reactors is shipped by truck to INL by commercial shippers in accordance with DOT and NRC requirements.[115]

Naval SNF for nuclear-powered submarines and ships is sent from navy facilities to INL by commercial train on government-owned railcars; 19.5 MTHM of naval SNF had been shipped to Idaho as of 2005.[116]

Overall, the NAS Committee on Transportation of Radioactive Waste estimated that about 3,400 commercial SNF shipments, totaling more than 3,000 MTHM, took place in the United States between 1964 and 2004.[117] Overall, 85 percent of shipments have taken place by highway and 15 percent by rail; however, because trains can carry more SNF per shipment, more than 70 percent of SNF by weight has been shipped by rail. The proportion of SNF shipped by rail has increased significantly; between 1998 and 2004, 766 MTHM of commercial SNF were transported on 261 rail shipments, versus 16 MTHM on 102 highway shipments.[118]

About 90 megagallons and 4,400 cubic meters of defense HLW are currently stored at three government sites: Hanford, INL, and SRS. Liquid sludge is stored at Hanford (53 million gallons), SRS (37 million gallons), and INL (less than 1 megagallon), while dry HLW is stored at INL (4,400 cubic meters).[119] DOE is vitrifying and packaging this HLW into HLW canisters and will need approximately 22,000 containers to contain all the waste.[120] The West Valley site has an additional 275 canisters of HLW derived from commercial reprocessing; both the site and the HLW belong to the State of New York.[121] DOE plans to transport the HLW to a federal repository in the same containers used for commercial SNF.[122]

Regarding TRU wastes, as of December 2010, WIPP had received 9,207 truck shipments of TRU; these shipments, from twenty-one DOE sites across the country, traversed more than 10.4 million miles (see Map 5.1). Over 72,000 cubic meters of contact-handled TRU (CH-TRU), and 220 cubic meters of remote-handled TRU (RH-TRU), had been disposed of at WIPP as of that date.[123] Because DOE lacks the infrastructure or capacity at some of its sites to package TRU as required for disposal at WIPP,[124] it also transports some TRU to the Advanced Mixed Waste Treatment Project in Idaho for consolidation and repackaging before shipment to WIPP. Under an agreement between DOE and Idaho, TRU shipped to Idaho for packaging must be transported out of the state within six months of receipt.[125] DOE has also shipped by rail to SRS about 300 cubic meters of TRU from the Mound Plant in Ohio; much of the TRU at the Mound Plant would not fit into NRC-certified CH-TRU transuranic packaging containers (TRUPACT-II), as required for shipment to WIPP. SRS processes the waste for packaging in TRUPACT-II containers for transport to WIPP.[126]

The NAS Committee on Transportation of Radioactive Waste found that existing DOE routing arrangements, including arrangements for consultation and cooperation with states and regional bodies, are generally working well.[127] The committee did, however, call for greater efforts to develop emergency-response and preparedness programs and capabilities in states and localities.[128]

While data on all shipments of nuclear waste have not been systematically kept, the available data on transport accidents show, overall, a favorable safety record. Four accidents involving highway shipments of SNF and five accidents involving train shipments of SNF occurred between 1971 and 2003, none of which involved the release of radiation.[129] Trucks transporting TRU to WIPP were involved in three accidents between 1999 and 2005, again without any release of radioactivity.[130] Several incidents of minor radioactive releases from SNF transportation accidents were reported prior to 1984, but changes in waste packaging designs since then have reduced the risk of similar incidents occurring.[131] A significant number of releases from transportation of LLW prior to 1985 have also been reported.[132]

Notwithstanding the overall positive record, the NAS Committee on Transportation of Radioactive Waste cautioned that any large-scale transport of SNF to a repository or of SNF to a consolidated storage facility would represent a major increase in shipments and a commensurate challenge. The committee reported that estimated shipments of SNF to Yucca would be twenty times the total number of SNF shipments made in the United States before 2006. It stated that these shipments would be primarily by rail; the number of planned rail shipments to Yucca would be eighteen times the number of rail shipments that occurred up to 2006. The committee estimated that HLW shipments to a repository, which would be in addition to the estimated SNF shipments, would comprise about a fifth of the total shipments to the facility.[133]

The NAS committee noted that transport of SNF to the planned interim private fuel storage (PFS) facility in Utah would also be a major undertaking, involving an estimated thirteen times as many SNF shipments as had taken place up to 2006 in the United States.[134] The site of the planned facility does not have access to rail. PFS transport plans provided for shipment of SNF in NRC-approved containers by dedicated rail from reactor sites or other points of origin to an "intermodal transport facility" (ITF) twenty-four miles from the PFS facility, where the containers would be transferred to heavy-haul trucks for transportation to PFS. Two dedicated trains would arrive at the ITF weekly, and two to four truck shipments of SNF would be needed weekly between the ITF and PFS.[135] The federal government has thwarted efforts by PFS to obtain transportation rights-of-way over federal lands in order to reach its facility. A planned rail access route across public lands was precluded when Congress included the relevant lands in a wilderness area designation. The Bureau of Land Management then denied a right-of-way for construction of the ITF and truck access to the PFS site on the ground, among others, that its own EIS on the proposal was inadequate. DOI's Bureau of Indian Affairs also refused to approve the Goshute tribe's lease of its land for the facility to PFS. However, the U.S. District Court for the District of Utah vacated and reversed DOI's decisions in July 2010, finding DOI's actions arbitrary and capricious;[136] only time will tell whether the facility will ever come to fruition.

Transportation of TRU for Disposal at WIPP

The shipment of TRU wastes to WIPP represents the largest-scale U.S. program thus far for shipment of highly radioactive wastes. Given that there have been few accidents and no transportation-related releases of radiation documented since shipments of TRU began in 1999, the NAS Committee on Transportation of Radioactive Waste regards the WIPP transportation program as successful.[137] The program may thus provide some lessons for meeting the challenges involved in future shipments of SNF and HLW to repositories or consolidated storage facilities. However, the NAS committee identified a number of important distinctions between issues raised by transportation of TRU to

WIPP and issues that would be encountered in transport of SNF and HLW to one or more repositories, and cautioned that these distinctions may limit the relevance of the WIPP transportation program experience. These include differences in support by the host and transit states; number of routes used for shipments; mode of transportation (TRU is being trucked to WIPP, whereas DOE has planned to ship SNF and HLW to a repository by rail); and the volume of material to be transported.[138] As noted earlier, WIPP received 9,207 truck shipments of TRU from 1999 to December 2010.[139] In comparison, more than 50,000 truck shipments of SNF and HLW would be required to fill the Yucca repository to capacity over its twenty-four-year operational period.[140] Also, SNF and HLW shipments generally consist of more highly radioactive material than do shipments of TRU, which tends to consist of miscellaneous used items contaminated with some amount of plutonium or other long-lived radioactive material; in many but not all instances, the amounts of contamination are small and the activity level low. Relative to TRU, SNF and HLW generate more heat and are more radioactive; they require more shielding to protect workers and the public during transit; and accidents involving breaches of SNF or HLW containers could have more severe radiological consequences.[141]

Transportation Issues under DOE–New Mexico Agreements

In the early 1980s, New Mexico was able—through a strategy combining legal, public relations, and political action—to gain sufficient leverage to induce DOE to enter into a series of agreements with the state regarding the WIPP repository.[142] New Mexico effectively used these agreements, and the procedures established by them, to voice the state's concerns and advance its interests regarding a number of key issues, including transportation (see Chapter 5). For instance, the 1981 Stipulated Agreement provided that DOE would assist New Mexico in gaining new funding to upgrade the roads that would be used to transport TRU waste, as well as funding for training and emergency response and preparedness.[143]

The Supplemental Stipulated Agreement of 1982 addressed in more detail a broader range of transportation issues. The agreement provided that DOE was responsible for paying the costs of cleanup related to WIPP transportation accidents; gave New Mexico the right independently to monitor the transportation of nuclear waste to and from the WIPP site; allowed the state to review transportation records; gave state scientists access to the WIPP site to monitor transportation activities; and permitted the state to inspect waste shipments at the points of entry into New Mexico, as well as to conduct inspections of TRU shipments at their points of origin—including those located outside the state. The agreement further obliged DOE to pay for the equipment and the salary of a state scientist to monitor transportation on and off the WIPP site, and also to notify the state before shipping the waste.[144] New Mexico in turn agreed to allow the use of highways designated in DOE's 1980 WIPP FEIS for transporting wastes, and DOE and New Mexico both agreed that neither party would designate alternative routes without the written consent of the other party.[145] The supplemental agreement provided that DOE would aid New Mexico in obtaining funding from the federal government to upgrade the designated highways and explicitly absolved the state from the responsibility for upgrading existing roads or constructing new ones that might be needed for future operations at WIPP.[146]

Further, in the 1987 Second Modification to the Agreement for Consultation and Cooperation, DOE agreed to comply with the relevant DOT and NRC regulations in connection with the transportation of nuclear waste to WIPP and to package all waste shipped to WIPP in NRC-certified packages.[147] Until this modification, DOE had not been under any legal obligation to comply with NRC regulations regarding packaging and carriage of packages, or to comply with DOT transport safety regulations. Accordingly, these contractual agreements by DOE represented a significant new regulatory commitment on its part.

Section 16 of the 1992 WIPP Land Withdrawal Act (WIPPLWA) contains transportation-related provisions that impose on DOE many of the regulatory requirements to which DOE had previously committed by contract in its agreements with New Mexico, including requirements that NRC certify TRU containers shipped to WIPP; that DOE provide advance notification of waste shipments to the state; and that DOE assist the state with accident prevention, transportation safety, and emergency preparedness.[148] Section 16(e) specifically bars the transportation of TRU from Los Alamos National Laboratory in New Mexico to WIPP until the federal government has provided the state with funding for a highway bypass around Santa Fe and the bypass has been constructed.[149] Finally, Subsections (c) and (d) of Section 16 of the act contain detailed provisions relating to transportation safety and emergency response, including federal funding for related equipment.[150]

In various agreements over the years, DOE had undertaken to assist New Mexico in securing federal funding for WIPP transportation-related matters, such as highway upgrades and personnel expenses. Funding was eventually provided through Section 15 of WIPPLWA, which authorized federal grants to New Mexico of $20 million per year for fifteen years without any restriction on their use. State officials reportedly regarded such funds as "the cost of doing business in New Mexico."[151]

Rail versus Truck Transportation

In the WIPP FEIS of October 1980, DOE anticipated using a mixed transportation regime of 75 percent train and 25 percent truck. Both methods would be required to use a Type A TRUPACT-I container for CH-TRU waste.[152] In the draft supplemental EIS (DSEIS) of 1989, however, DOE modified its position and proposed two alternatives: (1) 100 percent transport by truck, and (2) transport by train for all TRU wastes from facilities with rail access; under this alternative, wastes from the two DOE facilities lacking railheads (Los Alamos National Laboratory and the Nevada Test Site) would be trucked.[153]

A second round of supplemental EISs (DSEIS-II in November 1996 and FSEIS-II in September 1997) indicated that regular use of rail for transporting TRU to WIPP would overall have the lowest cost and lowest human health risk.[154] DOE estimated that using trucks for thirty-five years of waste transportation would result in fifty-six accidents, thirty-nine injuries, and five fatalities; maximum use of regular rail would result in half the number of accidents and casualties of truck use. Although it found that fatalities per mile are about the same for rail and truck transport, half as many shipments would be needed if waste were transported by rail rather than truck. Radiation from truck shipments would have an estimated non-occupational impact of 3.0 latent cancer fatalities (LCFs), and an occupational impact of 0.3 LCF. Rail use would decrease these impacts by a factor of ten for non-occupational impact and a factor of one hundred for occupational impact, because of the availability of additional shielding during stops, greater distance between the casks and the crew during transport, and fewer shipments. The estimated total LCF impact from a transportation accident without a breach of containment was estimated to be less than 1.0 for either option. In the event of a breach, assuming conditions that would maximize exposure to the population, DOE estimated the impact of a cask breach to be 3 to 16 LCFs for TRUPACT-II containers for CH-TRU, and 0.4 to 16 LCFs for casks for RH-TRU, depending on the concentration of radionuclides in the waste. The projected "worst case" risk would double for rail transport, because railcars would carry twice as many containers as trucks, and so a severe accident during rail transport could result in twice as many breaches in containment.[155]

Despite rail's overall lower risk, DOE in FSEIS-II identified trucks as the preferred transport mode, leaving open the possibility of rail use in the future.[156] DOE considered rail transport "less reasonable at this time" than truck shipping because of limited interest by rail carriers in shipping TRU; the high cost of dedicated rail; additional initial costs for three times as many containers for TRU packaging under the rail option; and DOE's inability to obtain assurance from potential/prospective rail carriers that TRU containers could be transported and unsealed within sixty days, as required by NRC regulations.[157] The State of New Mexico did not appear to have any objections to the selection of highway over rail transportation. Indeed, the state was given an opportunity to oppose the proposal in DOE's DSEIS-II to use trucks over railcars but did not do so.[158] In its 1998 record of decision (ROD) based upon the FSEIS-II, DOE confirmed that it would use trucks for transportation of waste, while continuing to leave open the option of using rails in the future.[159]

Following DOE's decision, the New Mexico State Highway and Transportation Department selected trucking routes based on DOT guidelines.[160] Interstate routes to WIPP from all DOE facilities nationwide where TRU is stored were also identified.[161] The routes to be used were then initially proposed by DOE in its FEIS, based on a variety of factors, including accident rates, traffic counts, highway segment length, vehicle speeds, population distribution, and nearby land use.[162] Affected states could then conduct their own studies and propose alternative routes. California, Colorado, and New Mexico elected to do so, and the routes they selected ultimately were established.[163]

TRU Waste Packaging

The initial package design proposed in 1980 by DOE for transport of CH-TRU to WIPP was known as the transuranic packaging container (TRUPACT).[164] The U.S. Senate Surface Transportation Subcommittee expressed concern that the packaging was not subject to review and approval by any agency independent of DOE.[165] TRUPACT was also criticized because of lingering technical concerns, including the potential for explosions due to accumulations of hydrogen gas within the containers. When scientists at Sandia National Labs proposed adding a ventilation system to TRUPACT to address this concern, New Mexico strongly objected to any continued use of the design; its position was supported by the Environmental Evaluation Group (EEG), the independent expert advisory group established by New Mexico with DOE funding.[166] DOE responded by agreeing, in 1987, to submit its TRU packaging design to NRC for approval, and by upgrading the packaging to the TRUPACT-II design.[167] The TRUPACT-II package, using a double hull, was submitted to intense stress tests, including extended exposure to fires and a series of drop tests. NRC approved the TRUPACT-II design in 1999.[168]

Transport of TRU Wastes to WIPP

In March 1999, the first shipment of CH-TRU waste arrived at WIPP from Los Alamos. This was followed by shipments of CH-TRU from INL and Rocky Flats in April 1999, Hanford in 2000, SRS in 2001, and Argonne National Laboratory-East in 2003.[169] Furthermore, DOE developed and received NRC certification for a RH-TRU cask, enabling RH-TRU to be shipped to WIPP beginning in 2007.[170] WIPP officials reported fourteen traffic incidents involving TRU shipments to WIPP during the first ten years of its operation; none involve a release of radioactivity to the environment.[171] As detailed earlier, through December 2010, 9,207 truckloads of TRU had been successfully shipped to WIPP for emplacement. Assuming that the average amount of TRU per shipment remains the same, about 13,000 more shipments to WIPP will take place before the repository reaches its capacity.[172] However, DOE announced in 2004 its plan to use TRUPACT-III packages for shipping large containers of CH-TRU to WIPP; the TRUPACT-III would reduce the need to repackage TRU too large to fit into TRUPACT-II containers, thereby reducing the number of shipments to WIPP by nearly 3,000, and eliminating intermediate shipments of TRU for repackaging.[173] NRC certified TRUPACT-III in June 2010.[174]

Cooperation on Transportation between the Western Governors' Association and DOE

The transportation of radioactive waste to WIPP was initially addressed as a bilateral issue between New Mexico and DOE. This changed in 1989 with the involvement of the Western Governors' Association (WGA), which addresses issues of common concern to the western states. In 1989, Idaho governor Cecil Andrus attempted to halt DOE shipments of TRU wastes to INL. This action and the resulting controversy over the disposition of TRU wastes roused interest by western states in TRU transport issues.

WGA's newly formed Transportation Safety Technical Advisory Group met with DOE and DOT and submitted a report to Congress in 1989 explaining its vision of "safe and uneventful" nuclear waste transport arrangements.[175] A follow-up report in 1991 recommended procedures for implementation.[176] Acting on these reports, the secretary of energy initiated a cooperative agreement between WGA and DOE to develop a coordinated program for WIPP shipments, and to work with local governments on the issue.[177] Congress codified this joint approach in the WIPP Land Withdrawal Act of 1992, which requires that DOE cooperate with the states in developing a transportation program for WIPP. The act also provided funding to the states for this purpose.[178]

In 1995, DOE and WGA executed a memorandum of agreement committing both parties to adhere to the WGA *WIPP Transportation Safety Program Implementation Guide* (*WIPP Guide*), which the parties have since repeatedly revised and updated.[179] The *WIPP Guide*'s protocols and regulations addressed such topics as security, route designation, public information, and especially accident prevention and emergency preparedness. These standards, which exceeded those of the federal government, were deemed necessary by WGA for the safe transportation of nuclear waste. DOE produced a WIPP Transportation Plan in 1998 (and revised it in 2002), setting forth its procedures and requirements for transporting TRU to WIPP. It incorporated protocols from the *WIPP Guide*.[180]

WGA and DOE agree that the WIPP transportation program has been highly successful, with more than 8,700 shipments made without incident since 1999.[181] The WGA reported that, although there have been some "minor accidents and incidents over the past decade, none have resulted in the release of radioactive materials."[182] The *WIPP Guide* and WIPP Transportation Plan have been reaffirmed by WGA and DOE every three years since 2000. The success of this approach prompted DOE to adapt the guide and plan protocols and regulations to transportation programs for non-nuclear hazardous materials, such as nitric acid.[183] In response to DOE's decision to keep the rail option open for the future, WGA in 2004 published the *WIPP Rail Transportation Safety Program Implementation Guide*, which reflected the western states' determination that the standards and procedures applied to potential rail transportation be comparable to those developed for truck shipments.[184] However, there is no indication currently that DOE is actively considering rail transport of TRU to WIPP.

Recently, in the wake of the defunding of the Yucca Mountain repository, WGA expressed concerns to DOE that it might from now on have a reduced role in the nuclear waste transportation process.[185] In response to these concerns, WGA and DOE released another memorandum of agreement in June 2009 affirming their commitment to the *WIPP Guide* and to the ongoing partnership.[186]

DOE has, however, been criticized by at least one critic for its reliance on guidelines produced by WGA and other "quasi-governmental" entities. This criticism emphasizes that regulatory guidelines authored by such organizations are not subject to the rulemaking requirements, public participation, or other procedural protections applicable to decision making by federal agencies. Nor are these organizations subject to direct congressional or presidential oversight.[187]

Transportation of SNF and HLW to Yucca Mountain

The NAS Committee on Transportation of Radioactive Waste found in 2006 that transporting SNF and HLW to Yucca Mountain "will be a daunting task" because of the amount of material that must be shipped, the large number of stakeholders involved, the extent of the transportation network, and the long time frame during which the transportation network would operate.[188] It also found that DOE had made only limited progress in developing a comprehensive Yucca transportation plan, let alone in engaging the public in its development.[189] The issues presented by transportation of HLW and SNF to Yucca offer a preview of the transportation issues that will be presented if it or another repository is eventually opened; the same issues would be presented by the prospect of large-scale shipments of SNF to a consolidated SNF storage facility.

In February 2002, DOE issued a FEIS for the Yucca repository identifying rail as the preferred mode for nuclear waste transport to the site.[190] In its April 2004 ROD, DOE selected rail transport but stated that trucks or barges would be used when the originating site was either not capable of loading rail packages or lacked immediate access to rail lines.[191] DOE estimated that nine to ten thousand railcar shipments to Yucca, and about one thousand truckload shipments from originating sites to railroads, would be required to transport SNF and HLW to Yucca over the twenty-four-year period planned for receiving wastes at the facility. DOE planned to use a satellite-based tracking and communication system such as TRANSCOM to maintain communication with the train or vehicle operators and to track the progress of the shipments.[192]

DOE justified its selection of rail on the grounds that fewer individual shipments would be necessary by rail than by truck, which would in turn lower the number of predicted traffic fatalities, as well as potential radiation exposures to both the general population and the workers involved in waste loading and handling. Nevada had indicated that, unlike New Mexico, it preferred rail transportation.[193] Nevada asserted, however, that DOE was overestimating its ability to rely on trains for shipping and that DOE's mostly rail plan would still require significant shipments of nuclear waste by truck.[194]

In 2005, DOE further determined that dedicated trains would be most suitable for transportation to Yucca, since the FEIS found that regular rail would require a number of transfers between trains, while dedicated rail involving only railcars destined for the repository would eliminate the need for transfers. DOE chose dedicated rail transport because the consequent decrease in transit time would result in lower risk of radioactive exposure, less residence time in railyards, increased command and control capabilities, reduced operating costs due to lower fleet size and the fleet's lower operating and maintenance costs, and greater flexibility and predictability in routing and scheduling for shipments.[195] The problem of rail carriers being uninterested in transporting SNF, which had been cited by DOE as an important factor in its selection of trucks for TRU transport to WIPP, was not cited by DOE as a significant factor in its decision to opt for rail transport of SNF to Yucca.

The NAS Committee on Transportation of Radioactive Waste endorsed DOE's selection of dedicated rail transport to ship wastes to Yucca. The committee noted that transport by rail rather than by truck would reduce the number of shipments needed by a factor of five and thus would reduce significantly the potential for serious accidents that would result in radiation exposure. The committee also noted that rail shipments are more physically isolated from the public than truck shipments are, and that a rail-based transportation system would be logistically easier for DOE to manage than one based on truck shipments.[196] Further, the committee approved DOE's selection of dedicated trains over general rail freight, noting that GAO had estimated that transporting nuclear waste in dedicated trains would be at least twice as fast as shipping on general trains, and found that derailments and catastrophic fires would be less likely on dedicated trains because railcars carrying SNF or HLW would not be commingled with other, lighter freight cars, including cars carrying flammable materials.[197]

As DOT and the states do not have explicit authority to regulate rail transport of SNF and HLW by DOE or by private carriers on its behalf, DOE asserted sole authority to select rail routes for waste transport to Yucca. It selected the Caliente Corridor, a 319-mile route that would run from a Union Pacific Railroad interchange at Caliente, Nevada, around the northern boundary of the Nellis Air Force Range, and southwest toward the Yucca site. DOE determined that this corridor had the fewest existing land-use or other conflicts in relation to the other corridors under consideration, and that this advantage would offset the fact that the Caliente corridor was one of the longest corridors and would therefore have among the highest initial construction and operating costs.[198]

In 2005, Nevada filed a petition for review in the U.S. Court of Appeals for the D.C. Circuit, challenging DOE's treatment of waste transportation issues in the 2002 FEIS on the Yucca Mountain repository and the 2004 ROD selecting rail transport for shipments of waste to Yucca.[199] The state argued that approval by DOT's Surface Transportation Board would be required for the construction of a rail line; that the state had not been adequately consulted during DOE's decision-making process; and that DOE's multistep approach of first identifying a rail corridor and later selecting the rail alignment within that corridor was improper.[200] In 2006, the court dismissed most of the claims as not ripe for review. It also held that DOE had in fact adequately satisfied its obligation, imposed by environmental regulations promulgated under NEPA, to obtain comments from state and local officials by consulting with the Nevada state engineer during the FEIS process.[201]

In 2008, after issuing another FEIS on the selection of a rail route to Yucca Mountain,[202] DOE published a ROD announcing its choice of the Caliente corridor.[203] Nevada responded in 2009 by filing a motion with DOT's Surface Transportation Board, asking it to suspend DOE's pending application to construct a rail line to the proposed repository on the ground that broad opposition to the Yucca mountain repository makes it unlikely that it will ever open, obviating the need for the proposed rail line.[204] The board has not yet addressed the state's motion. Yucca transportation issues are presently on hold pending resolution of DOE's efforts to withdraw its Yucca license application to NRC.

Assessing the Nuclear Waste Transportation Regime

The overall performance of the current nuclear waste transportation system for TRU, SNF, and HLW—to date based primarily on intermittent shipments of SNF and sustained shipments of substantial amounts of TRU from a relatively small number of sites to WIPP—has been positive and appears to have satisfactorily addressed states' concerns about the risks involved. The greatest challenge going forward will be the need at some future point for transport of large volumes of SNF and HLW for permanent disposal or consolidated interim storage from many points of origin. Such a program will be far more extensive and complex than the nuclear waste transport experience to date. It is likely to present greater risks, including risks of human error, and will no doubt stir significantly heightened public concerns over the safety of nuclear waste transportation.

The NAS Committee on Transportation of Radioactive Waste found that the health and safety risks of shipping nuclear waste are generally well understood, with the major and troubling exception of terrorism risks. In addition to flagging its inability to assess terrorism risks because relevant information is classified, the NAS committee recommended additional study of the risks presented by large sustained fires resulting from nuclear waste transportation accidents.[205] Members of Congress have echoed these concerns.[206]

The NAS committee found that there are no significant technical barriers to transporting large amounts of SNF and HLW safely, but emphasized the need to develop stronger emergency-preparedness programs at all levels.[207] It also found that the social risks of waste transport need to be better understood and systematically addressed, in particular by understanding and meeting the concerns of the public and of localities and states regarding the risks presented, and by instilling confidence that they are being properly evaluated and managed. The report, however, is short on concrete recommendations for successfully managing these social risks, beyond calls for enlisting experts and the need for federal agencies to engage more closely with states, local government, and citizens.[208] The Western Governors' Association issued a policy resolution calling upon NRC to implement the NAS committee's recommendations to improve information sharing with state and local governments regarding security of SNF and HLW shipments.[209] DOE's statement that it plans to work with carriers, state and local governments, and other stakeholders to develop routings for SNF and HLW shipments is consistent with the NAS committee's recommendations.[210] The NAS committee also noted the conflict between the need for open information and discussion regarding terrorism risks, on the one hand, and, on the other, the government's concerns about threats to national security if materials assessing terrorism risks are disclosed. Failure to resolve this conflict could stand in the way of public acceptance and confidence in the safety of transporting SNF and HLW and could hamper the development of effective state and local emergency preparedness and response programs.[211]

The WIPP transportation program, with more than 8,700 shipments to date, confirms that openness, consultation, and engagement by the federal government with states and local communities regarding transportation arrangements is essential to the development of a successful nuclear waste transportation program. But the WIPP experience also indicates that steps beyond information disclosure and consultation may be needed. New Mexico, through hard-won agreements with DOE and WIPP-specific legislation, eventually gained a substantial role in regulating the acceptance of waste at WIPP and in ensuring that the arrangements for transporting that waste through the state met safety concerns. Similar arrangements may well be necessary to obtain acceptance from other host states for large-scale shipments of SNF and HLW. Close engagement with transit states through mechanisms like the DOE-WGA cooperative regime are also indicated.

Conclusion

The U.S. regime for regulation of nuclear waste transport has evolved into a mature and rather complex system involving three federal regulatory agencies, states, and private carriers. In recent years, the regime appears to have performed well in assuring safe transportation of the relatively small number of shipments of SNF, HLW, and TRU shipped to date. Large-scale transport of HLW and SNF, whether to repositories for disposal or to consolidated interim storage facilities, will represent a new order of magnitude in nuclear waste transportation efforts, make much greater demands on the regulatory regime, and consequently present major challenges—logistical, technical, national security and terrorism related, and in terms of addressing public concerns about the risks of nuclear waste transport. Federal authorities must cooperate and work closely with states and localities in order to meet those challenges and to provide the public with justified assurances of safety. More explicit attention should be given to assessing terrorism risks and developing effective programs to address them, as well as to ways for providing the public with sufficient information to evaluate both terrorism risks and the adequacy of regulatory programs for addressing such risks.

Chapter 4

Low-Level Waste Disposal

Low-level waste (LLW) is a residual regulatory category encompassing all radioactive fuel-cycle waste that is not spent nuclear fuel (SNF), high-level waste (HLW), or transuranic (TRU) waste. LLW comes from a variety of sources, including nuclear power plants, government defense programs, and commercial and research activities. LLW from defense sources, including DOE nuclear weapons–related waste and navy waste, is managed and regulated by DOE and is not subject to NRC regulation. All other LLW is regulated by NRC.

Since 1959, the Atomic Energy Act has authorized AEC and its successor, NRC, to delegate primary responsibility for regulating commercial LLW to qualified state authorities, subject to AEC/NRC requirements.[1] States that have been delegated such authority are called Agreement States. NRC grants Agreement State status on a determination that a state has statutes, regulations, personnel, and inspection and enforcement programs consistent with NRC requirements.[2] There are currently thirty-seven Agreement States, with one more in process.[3]

NRC by regulation classifies LLW (which it terms "low-level radioactive waste," or LLRW) in order of increasing radioactivity, as class A, class B, class C, and greater-than-class-C (GTCC) waste.[4] NRC regulates civilian LLW according to these categories. As a consequence of the 1974 Energy Reorganization Act, NRC lacks regulatory authority over DOE's management of federal LLW wastes. In practice, DOE uses a LLW classification scheme similar to NRC's, but uses the term "LLW-like waste"—class A-like waste, class B-like waste, class C-like waste, and GTCC-like waste—presumably in order to avoid any suggestion that such wastes are subject to NRC regulation.[5]

The states are responsible for disposal of civilian LLW, with the exception of GTCC waste, which is DOE's responsibility. Civilian LLW is disposed of in commercial landfills regulated by NRC. Such disposal has long suffered from chronic and serious capacity limitations because of the difficulties in siting new facilities.[6] DOE generally has ample LLW disposal capacity at several of its facilities; in some cases it disposes of LLW at commercial facilities. DOE has yet to decide on a disposal policy for defense and civilian GTCC wastes. LLW is usually transported by rail but may also be transported by truck. Transportation of LLW is subject to NRC and Department of Transportation (DOT) regulation, in coordination with states and regional bodies.

Although systematic historical data on volumes of LLW and of the different categories of LLW are lacking, the available information shows that far more LLW (class A, B, and C) by volume is generated than any other category of nuclear waste. Yet these

LLW represent only between 0.1 percent and 1 percent of the total radioactivity (measured in curies) of all nuclear wastes.[7] Almost all LLW is class A waste. For instance, class B and C waste represented only 0.5 percent of the total volume of LLW disposed of between 1999 and 2003.[8] While class B and C waste volumes are relatively small, these wastes contain far more radioactivity than the much larger volume of class A waste. In each of the last ten years, for example, class C wastes represented between 69 percent and 97 percent of the total radioactivity (measured in curies) present in LLW.[9]

Between 1999 and 2003, annual commercial class A and DOE A-like waste volumes tripled, from 4 million cubic feet to 12 million cubic feet. DOE generated 78 percent of this waste, or 9.3 million cubic feet.[10] Class B waste volumes declined over this period from 23,500 cubic feet to 12,400 cubic feet, due largely to adoption by generators of waste reduction strategies stimulated by the lack of disposal capacity.[11] Volumes of class C wastes fluctuated greatly over the same period, with amounts varying from 11,000 cubic feet to 23,000 cubic feet per year.[12] This variation tracks the irregular pattern of nuclear power plant decommissioning, which yields large amounts of LLW in the form of dismantled reactor structures and components; most of this waste has been relatively lightly irradiated, but a small portion is more radioactive. The vast majority of B and C wastes are currently stored at generator sites because of limited available disposal capacity.

Early Developments

Between 1962 and 1971, six commercial disposal sites for civilian LLW were built by private parties. They were located at Beatty, Nevada; Sheffield, Illinois; Richland, Washington; Maxey Flats, Kentucky; West Valley, New York; and Barnwell, South Carolina.[13] The Nevada, Kentucky, New York, and South Carolina facilities were all licensed by their host Agreement States. The Richland and Sheffield facilities were directly licensed by AEC, as neither Washington nor Illinois was an Agreement State.[14] These facilities buried LLW in shallow reinforced concrete trenches, the same basic approach still employed today. Once filled, these trenches were sealed with impermeable synthetic barriers to prevent the waste from leaching out while radioactivity decreased to safe levels.[15] When these six facilities were being sited, there were no systematic site selection criteria or requirements.

Several of the early LLW sites, including those at Barnwell, Richland, and Beatty, were chosen as a result of AEC initiative, largely due to their proximity to existing AEC nuclear weapons facilities. Beatty is adjacent to the Nevada Nuclear Test Site, Richland is located on the Hanford Reservation, and Barnwell is located near DOE's Savannah River Site. Richland was located on federal lands, the other five sites on state lands, with the sites leased to the commercial operators of the facilities.[16] These facilities were licensed solely on the basis of AEC's 10 C.F.R. Part 20 generic radiation protection standards; no specific waste-handling, storage, or disposal regulations were promulgated until the 1980s.[17] Environmental considerations were slighted, most severely at Maxey Flats and West Valley, as AEC and host states rushed to promote the nuclear industry during the early Atomic Age.[18] Widespread waste leakage from the trenches at Maxey Flats ultimately led EPA in 1991 to designate it a Superfund site.[19] The West Valley LLW disposal facility, as well as the rest of the West Valley site of which it forms a part, has been undergoing extensive remediation off and on since its formal closure in the 1970s.[20]

In 1975, soon after it was created, NRC began work on developing a comprehensive system of LLW regulation. This effort culminated in 1982 in NRC's adoption, pursuant to AEA, of 10 C.F.R. Part 61 to establish "procedures, criteria, and terms and conditions" for the use of any near-surface LLW disposal techniques.[21] NRC also created the current system of classifying LLW in subcategories of ascending radioactivity and hazard. These new regulations, combined with improved management oversight and more sophisticated forms of shallow land burial, soon bore fruit: by the late 1980s, the problem of radionuclide leakage from LLW disposal sites had largely been solved.[22]

The 1980 Low-Level Radioactive Waste Policy Act

While NRC was drafting its LLW regulations, a LLW disposal crisis was looming. By 1978, the Sheffield, West Valley, and Maxey Flats facilities had been closed because radionuclides were leaking from the facilities' protective trenches.[23] In 1979, Beatty and Richland were temporarily shut down for several weeks when arriving waste containers at both facilities were discovered to be leaking.[24] For a time, only one LLW facility remained operational: the Barnwell facility in South Carolina. South Carolina's governor, concerned that South Carolina might become the nation's LLW dumping ground, acted to limit the volume of waste that the site would accept.[25]

The disposal crisis was exacerbated by the Supreme Court's ruling in *Philadelphia v. New Jersey*, in which the Court struck down a New Jersey law prohibiting the importation of "solid or liquid waste which originated . . . outside the territorial limits of the State."[26] The Court held that the Commerce Clause barred the states from imposing restrictions on waste based on its source unless Congress specifically provided otherwise.[27] Although the *Philadelphia* decision did not deal directly with LLW, the implications were clear: a state opening a LLW disposal facility would have to open it to all comers, absent explicit congressional authorization to the contrary.

These ongoing problems, combined with a mounting SNF waste disposal crisis, prompted the Carter Administration Interagency Review Group on Nuclear Waste Management (IRG) to devise a national strategy to deal with ever-increasing levels of nuclear waste. Among other recommendations, it proposed that the federal government assume a greater role in LLW policy. Under IRG's plan, NRC would adopt uniform, detailed national LLW disposal regulations; Agreement States could maintain more stringent requirements above the federal baseline. Previously, Agreement States had wide discretion in regulatory programs, subject only to conformance with AEC/NRC generic radiation standards that did not include specific requirements for design and operation of disposal facilities.[28] IRG also proposed that an eighteen-member presidential commission exercise siting authority for LLW facilities, with "constructive state participation" playing a role. The commission would be made up of representatives from federal agencies with some form of jurisdiction over LLW, along with state and local representatives designated by organizations such as the National Governors Association (NGA) and the National Conference of Mayors.[29]

IRG's proposals regarding LLW were unpopular with state governors and legislatures, which demanded an absolute veto over any proposed disposal site.[30] NGA issued a report that made a counterproposal, under which siting and operational regulatory responsibility for commercial LLW would rest with the states.[31] NGA argued that siting was fundamentally a land-use decision, one best left to the knowledge and expertise of state and local governments, and that the Agreement State framework had proved that states had the requisite expertise to oversee the operation of LLW disposal sites.[32] NGA justified its recommendations by asserting that the federal government had a "poor waste management track record."[33] It also argued for a cooperative approach under which groups of states would develop disposal facilities that the members of the group would share. The states lobbied Congress heavily, hoping to avoid the burden of hosting a nuclear waste facility imposed by a federal authority.[34] NGA's proposal proved influential with Congress, which had become dissatisfied with federal LLW regulation in light of its prior failings.

In response to IRG's various recommendations (described in Chapter 1)—for HLW and SNF, as well as for LLW—the House and Senate developed versions of an omnibus bill that would totally revamp waste disposal law and policy.[35] The omnibus bill's approach to LLW, in contrast to its approach to SNF/HLW, did away with the idea of a "centralized, federal LLRW disposal siting policy."[36] Congress debated IRG's proposals for LLW but ultimately rejected them, opting for the NGA approach, which called for a set of federal incentives to induce states to enter into voluntary, cooperative arrangements for LLW disposal.[37] NGA envisaged a series of regional interstate compacts, each with its own LLW disposal facility, shown in Table 4.1.

The LLW provisions of the omnibus bill, which were directly adopted from the NGA proposal, made each state responsible for the disposal of LLW generated within its borders.[38] Each state would be required either to have its own disposal facility or to contract for the use of another disposal site.[39] States would be required to use NRC's waste disposal standards as minimum standards. The provisions authorized, but did not require, states to enter into compacts to arrange for joint LLW disposal facilities.[40] Before becoming effective, each compact would have to be approved by Congress. Approved compacts would be able to site and build LLW regional disposal facilities that would be shared among the compact members. After January 1, 1986, each compact would be allowed to close its borders to LLW generated outside the compact.[41] States and compacts were given broad discretion when it came to siting; the only limit was that state-regulated facilities could not violate minimum federal standards.[42]

The primary benefit of joining a compact was the ability to exclude outside waste; if a state built a disposal facility without joining a compact, that state could not exclude waste coming from other states.[43] Congress envisaged that this incentive would be sufficient to encourage development of new LLW disposal facilities and thereby reduce the burden on Washington, Nevada, and South Carolina, where the only existing sites were located.[44]

Congress had settled on the details of its LLW approach by the 1980 election but was deadlocked over the details of the omnibus bill's approach to SNF and HLW. In the waning days of the Carter administration, Congress decided to split the bill in two, enacting the LLW provisions just described as the 1980 Low-Level Radioactive Waste Policy Act (LLRWPA) and deferring SNF/HLW legislation.[45]

**Table 4.1. Proposed LLW compacts,
from 1980 National Governors Association report**

Compact	States	Regional disposal facility
Southeast	SC, NC, GA, FL, AL, TN	Barnwell, SC
Southwest	NV, CA, AZ, NM, CO, UT	Beatty, NV
South Central	TX, LA, MS, AR, MO	
Midwest	IL, IN, OH, MI, WI, MN, IA	
Northwest	WA, AK, ID, MT, OR, WY	Richland, WA
Northeast	ME, NH, VT, MA, RI, CT, NY, NJ, PA	
Multiple possibilities*	HI, ND, NE, KS, OK, DC, KY, VA, WV, MD, DE, SD	

*The NGA report envisioned that the states listed could be logically included in two or more compacts.

The 1985 Low-Level Radioactive Waste Policy Act Amendments

The compact system adopted by Congress in 1980 failed to produce any new LLW facilities. By 1985, thirty-seven states had entered into compacts, but not a single one of these compacts had been approved by Congress, because of opposition from states without sites and states from other compacts.[46] Many states without sites were loath to have Congress approve the three compacts that included the three disposal facilities still operating—Beatty, Richland, and Barnwell—for fear that their LLW would be excluded from disposal at those facilities.[47] Efforts by other compacts to develop new facilities faltered. In most instances, no compact state volunteered to host a facility. To deal with that eventuality, compacts provided for selection of a site by a compact commission, a body established by the participating states to implement the compact. However, the host states designated by the commissions strongly resisted. No state wanted to open one of the first new regional disposal facilities. If a state opened a LLW disposal site, but states outside the compact without access to disposal sites blocked congressional approval of the host state's compact, the host state would have to open the new facility to all other states.[48] It became apparent that it would be impossible to open new disposal facilities by 1986.[49] At the same time, states with existing disposal facilities reiterated that they would not continue to serve as dumping grounds for the nation's LLW and threatened to limit access to their facilities unless Congress took more effective steps to force other states to deal with their own wastes.[50]

Congress amended LLRWPA in 1985 to allow states and compacts more time to develop waste facilities and also to impose penalties against states that failed to provide, through compact arrangements or otherwise, for the disposal of their own LLW.[51] Under the Low-Level Radioactive Waste Policy Amendments Act (LLRWPAA), each of the three LLW disposal facilities still operating would be required to continue to accept out-of-compact LLW for the next seven years, subject to a statutory cap on the total waste volume set for each individual facility. Based on 1983 waste levels at the existing facilities, the statutory caps were imposed to encourage the development of new facilities and to promote waste volume reduction strategies; prior to 1983, waste reduction strategies were virtually nonexistent.[52] Beginning in 1992, compacts with LLW disposal facilities could stop accepting out-of-compact waste. NRC retained the ability, in emergencies, to order regional disposal facilities to accept out-of-compact waste.[53]

The 1985 amendments required states without disposal facilities to meet a series of milestones in order to demonstrate progress toward developing new regional disposal facilities.[54] If these milestones were not met, a graduated penalty would be charged for the use of existing disposal facilities, and the offending states ran the risk of completely losing access to these facilities.[55] States that failed to provide for the disposal of their LLW by January 1, 1996, were required to take title to LLW generated within their borders.[56] Once title to LLW passed to the state, the state became liable for any harm caused by the waste.[57] This mix of stronger incentives, including penalties and liability, was adopted against the background of threats from the governors of South Carolina, Nevada, and Washington to limit access to their disposal facilities if no new LLW facilities were built.[58]

In *New York v. United States*, the State of New York and two upstate New York counties faced with the prospect of a new LLW disposal facility being sited in their region, challenged the graduated penalty, denial of access, and take-title provisions of LLRW-PAA.[59] The Supreme Court upheld the first two provisions but struck down the take-title requirement on the grounds that take-title constituted an unconstitutional compulsion of the states, in violation of the Tenth Amendment. The Court found that the federal government could neither order a state to build a LLW disposal site nor require that state to take title to private waste. It then concluded that imposing on states a Hobson's choice between these two independently invalid requirements also offends the Constitution.[60] The Court's decision in *New York* eliminated a key LLRWPAA incentive for states to develop new disposal facilities. Before *New York*, the threat of take-title liability was sufficient incentive to prod the states into planning LLW disposal facilities; California, for instance, had taken the lead in forming a compact and volunteered to host a regional disposal facility in order to avoid take-title liability.[61] The Court's decision in *New York* took the steam out of the push to develop new facilities, and designated host states delayed or killed proposed facilities outright.

Implementation of the 1985 Amendments

By the time of the 1992 *New York* decision, all but six states had joined compacts. Of the ten compacts then existing, seven had generated license applications for new regional disposal facilities.[62] But in the end, the compact system envisaged by Congress has failed to secure the construction and operation of new LLW disposal facilities. Only the Texas Compact has commenced construction on a regional disposal facility, and that only after twenty years of contentious legal and political battles.[63] Moreover, the Texas Compact is the alter ego of the State of Texas, which outvotes Vermont, the only other member; it is not the sort of compact that NGA or Congress envisaged.

The NGA report originally envisioned six geographically contiguous regional compacts, each with one LLW disposal facility. This did not happen. Instead, LLRWPA and the 1985 amendments ended up creating incentives for large waste-producing states and states with disposal facilities to enter into gerrymandered compacts, often with noncontiguous states, and sometimes with states at opposite ends of the country. Of the nine LLW compacts that currently exist, four have a noncontiguous configuration: the Southwest, Atlantic, Central Midwest, and Texas Compacts. Seven states, and the District of Columbia and Puerto Rico, are not members of any compact.

The unusual geographic configuration of the Southwest Compact, composed of California, Arizona, and the Dakotas, illustrates some of the incentives generated by the congressional scheme. California, seeking to avoid take-title liability for its substantial volumes of wastes, decided to build a disposal facility. It was able to secure modest construction subsidies from the sparsely populated Dakotas (which produced little waste) in exchange for the right to dispose of their waste at the proposed facility and their participation in creation of a compact that would allow California to exclude waste from other states.[64] The California facility, however, was never built, due to strident opposition.

The Atlantic Compact is another illustration of a gerrymandered compact. Originally composed of Connecticut and New Jersey, it admitted South Carolina in 2000, after the South Carolina legislature sought to control the volume of waste entering the Barnwell facility.[65]

The Texas Compact, consisting of Texas and Vermont, was initiated by Texas to allow it to control access by other states to a new privately operated LLW waste disposal facility that it proposed to develop. Texas required a $25 million buy-in from states wishing access to its proposed facility. Also, it excluded from eligibility states with substantial populations, or those that produced relatively large amounts of LLW.[66] Texas ultimately succeeded in finding a compact partner in faraway Vermont, the forty-ninth-largest state by population. Vermont's membership in the Texas Compact is all but a formality. The compact's governing board is composed of six Texas appointees and two from Vermont, meaning that Texas effectively makes all compact decisions, including admission of additional states and decisions on whether to open the facility to out-of-compact wastes.[67] Table 4.2 lists LLW compacts and disposal facilities as of July 2010.

Problems in Siting Regional Disposal Facilities

LLW compact agreements developed under the 1985 amendments to LLRWPA specify procedures for siting a regional disposal facility if no state comes forward to host one. In that event, the compact commission chooses a site based on specified technical criteria of suitability and policy and political factors.[68] In theory, the compact regime was supposed to give incentives sufficient for at least one compact state to volunteer to host a site; in practice, after the decision in *New York*, no state has volunteered to do so except Texas, which in effect formed its own compact.[69] Efforts to find a compact member state that would volunteer to host a regional site or, failing that, to develop a compact facility at a location designated by the compact commission, encountered stiff opposition. Opponents of LLW disposal sites, often a mix of local citizens' groups and national nonprofits, raised numerous technical and legal grounds in challenging proposed facilities, including claims of lax governmental oversight and weak security.[70]

Compact commissions are composed of representatives of each member state and

Table 4.2. Current LLW compacts and disposal facilities

Compact	States	Disposal facilities	Wastes accepted	Noncompact waste accepted?
Northwest	WA, OR, ID, MT, WY, UT, AK, HI	Clive, UT Richland, WA	A A, B, C	Yes Rocky Mountain Compact states only
Rocky Mountain	NV, CO, NM	None, but agreement with Northwest Compact facility		Disposal at Richland available
Southwestern	CA, ND, SD, AZ	None		
Texas	TX, VT	Andrews, TX (under construction)	A, B, C	DOE waste only
Midwest	MN, WI, IA, MO, IN, OH	None		
Appalachian	PA, WV, DE, MD	None		
Central Midwest	IL, KY	None		
Southeast	AL, FL, GA, MS, TN, VA	None		
Atlantic	NJ, SC, CT	None		
Central	NE, KS, OK, LA, AR	None		
Unaffiliated	MA, MI, NH, NY, RI, NC, DC, PR	None		

Source: Data drawn from NRC, "Low-Level Waste Disposal," www.nrc.gov/waste/llw-disposal.html.

generally lack a staff or a political base of their own, requiring them to rely heavily on often-reluctant state agencies in developing new facilities. Licensing a LLW disposal facility can require expertise from twenty-two disciplines—expertise that the compacts lack.[71] The compact-state agencies involved in siting usually answer to the state governors, who can orchestrate administrative processes to delay and obstruct the development of LLW disposal facilities in order to sidestep the political battles involved in siting them.[72]

At least two states designated as hosts by their compacts, Michigan and North Carolina, refused outright to serve as host states and withdrew from their respective compacts under acrimonious circumstances. In Michigan, the state legislature passed a siting authorization bill so restrictive that no site could ever pass statutory muster; as a result, the Midwest Compact ejected Michigan for bad faith.[73] North Carolina, the Southeast Compact's designated host state, failed to site a facility and subsequently left the Southeast Compact to join the Atlantic Compact.[74] The other Southeast Compact states sued North Carolina under the Supreme Court's original jurisdiction.[75] The special master appointed by the court has recommended that North Carolina be held liable for its failure to site and construct a facility; the Court has heard argument from the parties on the recommendation and rejected North Carolina's attempts to dismiss the case, but has not yet issued a decision.[76]

A third state, Nebraska, failed to site a facility designated by the Central Compact commission and was, as a result of litigation, subjected to heavy liabilities. Local Nebraska politicians and community activists turned the proposed Boyd County LLW facility into a major issue in Nebraska's 1994 gubernatorial election.[77] Then-governor (now U.S. senator) Ben Nelson was instrumental in derailing the siting process, leading to two license denials. The other states in the Central Compact and the site developer sued Nebraska for failure to abide by the compact terms.[78] In 2002, a federal district court ruled in the plaintiffs' favor, holding that Nebraska spuriously denied the proposed LLW facility an operating license based on extraneous political reasons rather than on the merits of the site.[79] Nebraska was ordered to pay damages of $151 million—$88 million in compensatory damages plus $46 million in interest to the Central Compact; $3 million in community development funds plus $1 million interest to the Central Compact; and $6 million in reliance damages plus $6 million in interest to U.S. Ecology, the site developer. Nebraska appealed to the Eighth Circuit, which affirmed the district court's ruling.[80]

Siting Strategies of Potential Host States

States that have sought to site LLW facilities within their borders, either on their own initiative or as a result of designation as hosts by their compacts, have faced significant challenges. Finding a site with the requisite combination of conditions—such as cheap land, limited rainfall, isolation, good transport access, and favorable geology—is difficult in many areas. In addition, proposed facilities have drawn opposition at both local and state levels. States have used two strategies for siting: top-down and bottom-up.

The top-down approach, used by Texas and California, relies on studies that use technical criteria (including geology, transportation links, rainfall, and other factors), as well as on the existence of local support, to identify one or more candidate sites, which are then announced to the public.[81] Texas has successfully used this strategy to site and develop a LLW facility in the far western part of the state near Andrews, Texas, that will take class A, B and C waste. But this process required nearly twenty years of continued support from Texas's political establishment to overcome opposition, notwithstanding the fact that the site had excellent geohydrology (salt formation), little rainfall, sparse population, and strong local support for nuclear-related industries.[82] California came close to successfully siting a facility on federal land in the Mojave Desert but ultimately failed for a diverse set of reasons that included environmental challenges (including potential impacts on endangered tortoises), earthquake risks, and problems in arranging necessary intergovernmental land swaps.[83] After selecting a developer, finding a dry, isolated site with stable geology and railway access, and issuing a license for the Ward Valley Regional Disposal Facility, California ultimately gave up on the facility in 2002 after a decade of litigation.[84]

In contrast to the top-down planning model, Illinois and Nebraska used a bottom-up approach that allowed individual communities to decide whether to be considered as hosts.[85] Both states were able to find interested communities, but neither managed to site a facility. In Illinois, the public lost faith in the siting process because of mismanagement.[86] In Nebraska, the low-level waste disposal site became the victim of local rivalries as well as state politics. The villages in Boyd County, Nebraska, had long been subject to internecine squabbling over the county's few precious nonagricultural jobs.[87] Butte, the proposed host community for the Central Compact's proposed LLW disposal site, had recently lost its village school to neighboring Naper and was in danger of disappearing as a community as teaching jobs were eliminated.[88] Desperate for jobs that would ensure the town's continued viability, Butte, population 366, offered itself for LLW site consideration; it also stood to gain $3 million per year in community development funds from the compact commission.[89] The surrounding villages, stoked by a mix of jealousy and NIMBYism, embarked on a campaign against the proposed facility, enlisting state-level officials, including then-governor Ben Nelson.[90] The facility was never built.

The Current LLW Disposal Situation

Civilian LLW

Under LLRWPAA, states with existing disposal facilities were no longer required, after December 31, 1992, to provide access to out-of-state waste generators.[91] A flurry of activity sprang up soon after this deadline passed. Bowing to political pressures, the governor of Nevada closed the Beatty site to LLW by executive order.[92] Washington limited the Richland site to LLW generated by the Northwest and Rocky Mountain Compacts. South Carolina limited the Barnwell site to Southeast Compact waste.[93] Then, in 1995, South Carolina left the Southeast Compact because North Carolina had failed to open a regional disposal facility as required by the agreement that established the compact.[94] As a result, the Barnwell facility became open to LLW from every state except North Carolina.[95] Between 1992 and 2003, the cap on waste surcharges at the facility increased from $40 to more than $300 per cubic foot.[96] After joining the Atlantic Compact, South Carolina closed the Barnwell facility to out-of-compact waste effective July 1, 2008, citing capacity problems. Up to that point, Barnwell had been the only disposal site in the country accepting all classes of LLW from all states.

In 1991, a commercially developed disposal facility at Clive, Utah, began receiving class A LLW. It is open to wastes from all states and has become the site for disposal of almost all the nation's class A LLW. Since Clive opened, Barnwell has had difficulty attracting the Atlantic Compact's class A waste, due to its high disposal fees.[97] Volumes of class B and class C wastes generated by the states in the Atlantic Compact have been low, due to the small size of the compact.[98] It is possible that the Barnwell facility could tap sufficient reserve capacity to reopen to out-of-compact class B and C LLW, if demand pushes prices high enough. Alternatively, the Atlantic Compact could add additional members in order to provide more compact waste for the site. Unaffiliated states such as New York, Michigan, and Massachusetts all produce substantial quantities of class B or C LLW and lack an available disposal facility.

The Texas Compact's regional disposal facility in Andrews, Texas, has acquired a permit and is under construction. Thanks to its control of the Texas Compact Commission, Texas can control licensing and operation of the facility and determine whether to accept or exclude out-of-compact waste and what fees to charge. Under the current Texas Compact arrangements, the Andrews facility is licensed to accept class A, B and C waste from the Texas Compact, as well as federal government waste.[99] Recently, the Texas Low-Level Radioactive Waste Commission voted to authorize the Andrews facility to accept class A, B, and C LLW from as many as thirty-six states.[100] If this plan goes forward, Texas will have the sole disposal site for class B and C wastes from thirty-six states, thereby gaining significant market power. It will have succeeded in using the compact system to its own distinct advantage, squarely contrary to what Congress envisaged in its LLW legislation.

It is by no means assured, however, that Andrews will be opened to all of these thirty-six states. Two members of the Texas Low-Level Radioactive Waste Commission voted against the measure. One of the dissenters stated that the facility and the commission were not prepared to deal with a huge influx of waste from other states. Local groups may well challenge the commission decision in litigation. In response to criticism that the decision would open Texas to a flood of waste from other states, a commission spokesman stated: "It's actually just the opposite of opening up the floodgates. It's putting a gate in place . . . [t]o determine what comes in and what doesn't."[101] The statement reflects the purpose to use the majority that Texas has in the Texas Compact Commission and the provisions of LLRWPAA to choose which out-of-compact states' wastes to receive and thereby exploit its market power in order to dictate terms. Meanwhile, the new governor of Vermont opposes the commission decision and has stated that he would replace the two Vermont members of the Texas Compact Commission with members more in line with his views.[102] While the Vermont members can be outvoted by the Texas members, there is the potential that Vermont could withdraw from the compact and effectively dissolve it, thereby depriving Texas of the leverage conferred by the LLRWPAA to compact states, forcing Texas to open the facility to waste from all other states and undermining its market power.

The only disposal facility currently accepting LLW from all states is EnergySolutions' Clive, Utah, facility, which takes only class A wastes.[103] About 99 percent of the total volume of the nation's class A LLW disposed of in 2005 was disposed of at the Clive facility, much of it transported long distances from generator sites.[104] EnergySolutions, formerly known as Envirocare, is a private firm that developed the Clive facility as a private venture outside the compact system. The Northwest Compact, which includes Utah, already had a LLW facility at Richland, Washington, which it has designated as a regional disposal facility, and accordingly the compact has not had to develop such a facility pursuant to the LLRWA.

The Clive facility's status as an independent commercial venture operating outside the compact system is unique among current LLW facilities. The underlying land is owned by EnergySolutions, not by the State of Utah. Funds to cover estimated future cleanup costs were placed in escrow, in the event of EnergySolutions' financial insolvency.[105] All other LLW disposal sites—Richland, Barnwell, and Andrews—although originally commercially developed, have become compact-designated regional disposal facilities. Barnwell and Andrews lie on state-owned land, while Richland lies on federal land at DOE's Hanford facility; the lands are leased to the operating companies.

Clive is regulated by the State of Utah, an Agreement State. The site was first licensed by the state in 1987 to receive uranium mill tailings. The license was amended by Utah in 1991 to allow Clive to accept some LLW as well, on the condition that the Northwest Interstate Compact (of which Utah is a member) consent to the importation to the facility of LLW from outside the compact's borders.[106] In 2001, Utah further amended EnergySolutions' license to allow for the disposal of class B and C wastes, but in 2005 EnergySolutions dropped its plans for disposal of such wastes, citing mounting political opposition in the Utah legislature. Notwithstanding EnergySolutions' efforts to prevent it, the Utah legislature in 2005 passed a ban on storage of class B and C waste in the state.[107] EnergySolutions' license was reissued by Utah in 2008 with the stipulation that state approval is required if the facility is to receive waste generated within the Northwest Compact's borders; the compact's approval is not required for waste generated outside the compact's borders.

The Northwest Compact responded by placing limitations on the origin of the wastes that the Clive facility could receive, specifically seeking to bar the class A wastes from Europe. EnergySolutions had stirred the controversy by proposing to import LLW from Italy for disposal at the Clive facility and offering Utah 50 percent of the profits.[108] EnergySolutions sued, challenging the Northwest Compact's jurisdiction over the Clive facility.[109] A federal district court found that EnergySolutions was a purely private facility, not subject to the Northwest Compact's authority.[110] The court went on to find that the purpose of LLWRPA—to ensure the availability of LLW disposal sites—would be contradicted if compacts had the authority effectively to shut down private facilities by denying them the right to import wastes from outside or within a compact's boundaries.[111] The State of Utah, the Rocky Mountain Compact, and the Northwest Compact have appealed the ruling.[112] Legislation has been proposed in Congress to forbid the importation of LLW from outside the U.S. unless the president grants approval.[113] EnergySolutions has dropped its proposal but plans to consult with Italy and other nations on how best to handle their nuclear wastes.[114]

Federal LLW

DOE is responsible for managing, regulating, and disposing of DOE, Navy, and other nuclear weapons-related LLW, as well as both civilian and defense GTCC waste. Wastes managed by DOE are not subject to NRC regulation, but, as noted earlier, DOE follows the NRC's basic LLW subclassifications. Currently, LLW disposal facilities at DOE's Nuclear Test Site and Hanford facility serve as the primary disposal sites for LLW generated at other DOE sites.[115] LLW facilities at Los Alamos National Laboratory, Idaho National Laboratory, Oak Ridge National Laboratory, and the Savannah River Site serve as secondary disposal sites. In 1993, DOE adopted a policy of disposing of its LLW in commercial facilities when the use of DOE facilities is "not practical or when a situation-specific cost-benefit-analysis favors non-DOE site disposal."[116] The department used commercial facilities for disposing of some of its LLW between 1999 and 2003.[117]

GTCC Waste

Greater-than-class-C LLW is excluded from the LLW compact system. Whether generated by civilian or defense activities, GTCC is exclusively the responsibility of DOE, which is studying disposal options.

The two major sources of GTCC waste are "commercial sealed sources"—devices containing significant amounts of radioactive materials used in medical, research, and commercial applications—and activated metal wastes from decommissioned civilian and defense nuclear reactors; government fuel and weapons research has also produced some GTCC.[118] Commercial GTCC waste makes up less than 0.01 percent of all LLW by volume generated annually. GTCC LLW remains radioactive for significantly longer than do other classes of LLW (more than 500 years) but does not necessarily require the kind of ultra-long-term (100,000 years or longer) precautions necessary to dispose of SNF or HLW; in this respect it is somewhat similar to TRU.[119] Still, like TRU, the more highly radioactive GTCC wastes may call for disposal in a repository. At present, 4,600 cubic feet of commercial GTCC waste and 31,000 cubic feet of DOE GTCC-like waste are stored at the sites where they were generated. After current civilian nuclear power plant decommissioning projects and DOE nuclear waste cleanup projects are completed, DOE estimates that storage and disposal capacity for another 200,000 cubic feet of GTCC and GTCC-like waste will be required.[120]

GTCC waste is statutorily excluded from disposal at WIPP.[121] Congress would have to amend existing law for GTCC waste to be disposed of there. NWPA restricts Yucca to HLW and SNF but grants NRC the power to designate any "highly radioactive material that the Commission, consistent with existing law, determines by rule requires permanent isolation" as HLW.[122] In 1989, NRC promulgated a regulation requiring that GTCC waste be disposed of in a geologic repository, absent another NRC-approved disposal option.[123] It is unclear whether this regulation by itself is sufficient to fulfill the NWPA condition for disposing of GTCC at Yucca; neither DOE nor NRC has ever claimed that NRC's regulation does so. Accordingly, GTCC wastes, which represent a range of hazard levels (some relatively high that probably call for repository disposal of the wastes in question), are currently consigned to regulatory and disposal limbo.

In February 2011, DOE issued a draft GTCC waste disposal EIS to explore disposal options.[124] It sets forth five alternatives without indicating a preference. Alternative 1, the no-action baseline, appears in the EIS because a no-action baseline is required in every EIS by NEPA. Alternative 2 would dispose of GTCC waste with TRU at WIPP. Alternative 3 explores use of hundred-foot-deep boreholes. Alternative 4 would dispose of GTCC waste in near-surface trenches similar to those used for LLW but with more sophisticated barriers to prevent radionuclides from migrating. Alternative 5 considers enhanced above-grade vaults covered with soil to discourage intruders. For the last three options, DOE is analyzing the possibility of disposal at seven federal sites, as well as at a generic commercial LLW site.[125] All seven federal sites being examined under alternatives 3–5 currently host facilities for LLW disposal.[126]

Options for Addressing LLW Disposal Problems

The Clive facility, which handles 99 percent of the nation's commercial class A waste, has ample capacity, without further expansion, to accommodate projected volumes of class A waste for the next fifteen years.[127] Barnwell's 2008 closure to out-of-compact waste, however, poses a major challenge for disposal of class B and class C LLW. This gap may well be met by the Andrews, Texas, facility, but at the cost of ceding extensive power to Texas, which will be able to effectively dictate the terms of class B and C LLW disposal to most of the nation. Further, relying entirely on just two disposal facilities to handle all of the nation's civilian LLW is not an equitable solution, nor is it necessarily a stable one because of the politics of NIMBYism, already apparent in local reaction to the Andrews vote of the Texas Low-Level Radioactive Waste Commission.

Waste storage capacity limits—and consequent rising disposal costs—have created economic incentives that have led waste generators to take steps to reduce LLW waste volumes through a variety of means, including super-compaction, shredding, incineration, and decontamination. Super-compaction has achieved up to a 5,000 percent reduction in the volume of some class A wastes. Techniques to reduce waste volume alone, however, do not remove radionuclides, so the processed waste materials are still radioactive and must still be disposed of as LLW.[128] Furthermore, an increase in radionuclide concentrations as a result of compaction can trigger more stringent regulation if the waste categorization radioactivity thresholds in 10 C.F.R. Section 61.55 (class A to class B; or class B to class C)[129] are exceeded. Industry has also succeeded in replacing some radioactive materials with non-radioactive ones and limiting the amounts of class B and C wastes produced. For example, nuclear power plant operators now change filters frequently to prevent a buildup of radioactivity in them.[130]

NRC has reacted to LLW disposal capacity problems by authorizing indefinite storage of LLW at the site where they are generated. Because it is generally not economical to store large volumes of class A waste on-site, and the Clive, Utah, facility has been available to dispose of such waste, in practice only class B and C wastes, which together represent less than 5 percent of the LLW generated by volume, are stored on-site.[131] Yet class B and C wastes collectively contain about 90 percent of the total radioactivity present in all current LLW, excluding GTCC wastes.[132]

NRC used to limit storage of LLW at the site of generation to five years but has since dropped this limitation, in part because it failed to spur construction of new disposal facilities.[133] NRC now grants licenses permitting on-site storage of LLW for "as long as it is safe," without a specified time limit.[134] Communities, environmental organizations, and opponents of nuclear facilities have objected to indefinite on-site storage of class B and C LLW. Generators complain of the regulatory compliance burdens and expense of such storage. NRC prefers disposal of these wastes to storage.[135] If the Andrews, Texas, facility is opened to out-of-compact class B and C wastes, it can provide for disposal of these wastes, but it may not prove to be a stable or equitable long-term solution.

One potential option for providing additional and assured long-term disposal capacity, suggested by GAO, would be to open DOE disposal facilities to commercial class B and C wastes. GAO reported in 2004 that, as of 1999, DOE's LLW disposal facilities at the Nuclear Test Site and Hanford could accommodate a total of 171 million cubic feet of LLW; DOE had estimated that it would need less than 30 million cubic feet of this space to handle projected LLW from its own sites.[136] By comparison, Barnwell accepted 142,000 cubic feet of class B and C wastes in 2005.[137] LLRWPAA gives DOE responsibility for federal wastes and GTCC wastes and gives states responsibility for other LLW, but nothing in the statute explicitly prohibits DOE from executing a contract with the various state compacts or noncompact states for DOE to accept their commercial LLW.[138]

Another possible option, promoted by the Central Compact in the 1990s, would be to build assured isolation facilities (AIFs) that function "as engineered, above-ground monitored storage systems" for LLW.[139] AIF technology and its leak-resistance performance were touted by U.S. Ecology, the operator of Richland, and by the Central Compact in the 1990s as a safer and surer alternative to shallow land burial, but this has failed to break the siting logjam within the compact.[140] Ohio, alone among the states, has promulgated AIF regulations; no regional disposal facility based on the AIF standard has been developed. NRC is not currently proceeding with AIF regulations.[141]

NRC has proposed down-blending class B and C wastes with class A waste to reduce aggregate radioactivity to class A levels. A study on this approach, requested by the NRC chair appointed by President Obama, Gregory Jaczko, is currently ongoing. A preliminary report supports in principle the idea of blending homogenous class B and C wastes down into class A waste, subject to further examination and development of specific regulations governing the various categories of LLW.[142] This proposal has been strongly supported by EnergySolutions, which operates the Clive and Barnwell facilities, as such blending would permit class B and C wastes to be sent as class A LLW to the Clive facility for disposal.[143] U.S. Ecology and Valhi, the operator of Andrews, have opposed this proposal because it would diminish their current and prospective class B and C waste disposal business.[144] The State of Utah has also stridently opposed NRC's proposal, viewing it as an end run around a 2004 state law that prohibits the importation of class B and C wastes to Clive.[145] If Andrews becomes the disposal site for class B and C wastes for the thirty-six states that currently lack access to such a site, that would take the steam out of any down-blending initiative.

Another set of issues concerns wastes with very low levels of radioactivity. In 2003, EPA proposed that certain low-activity mixed waste (LAMW)—a category of LLW waste that qualifies as both class A LLW (subject to regulation by NRC) and hazardous waste (subject to regulation by EPA under RCRA)[146]—be disposed of in RCRA hazardous waste landfills. Currently, such waste must be disposed of in more-costly class A LLW facilities unless granted a special dispensation from NRC and EPA.[147] EPA's proposal was for the agency to relinquish to NRC its authority to regulate LAMW as RCRA hazardous waste; NRC would license RCRA landfills to accept LAMW under LLRWPAA.[148] EPA reportedly revived this plan in 2008 but has not moved forward on it under the Obama administration.[149] In 2006, the NAS Committee on Improving Practices for Regulating and Managing Low-Activity Radioactive Waste issued a report recommending a more risk-informed approach to nuclear waste classification and regulation.[150] Among other matters, the report found unjustified the current practice of regulating as LLW large volumes of very low activity debris, contrary to the policy followed in some European countries, which permits disposal of such wastes in ordinary landfills. The report estimated a cost of $4.5 billion to $11.7 billion for disposing of ten million tons of concrete and metal debris as class A LLW, versus as RCRA hazardous waste.[151] The committee recommended developing alternatives, including disposal in RCRA hazardous waste landfills.[152] In addition, twenty years ago NRC proposed to establish a general category of very low activity wastes as below regulatory concern, allowing for their disposal in ordinary landfills (see Chapter 2). This proposal, which embodied a policy recommended by both U.S. and international expert groups, was strongly resisted by many states and environmental groups and was torpedoed by Congress.

In addition to instances of what it viewed as overregulation, such as with respect to LAMW, the committee also found instances where the current classification system results in underregulation of serious nuclear waste risks. For example, some wastes now classified as LLW are in fact highly radioactive, but neither these wastes nor their radioactive sources are being adequately regulated. Specifically, the committee found that there are currently no federal regulatory programs for radiation sources (including those using radium) contained in devices used for medical, industrial, and resource applications that are no longer in service. Although these devices, when no longer used, are classified as LLW, the committee found that in many cases they contain highly concentrated radiation sources, are often "orphaned" with no one clearly responsible for them, and "can pose acute risks to the public and the environment," and that accordingly federal regulation of these sources needs to be strengthened.[153]

Conclusion

After thirty years, the LLRWPA system for disposal of civilian LLW through the state compact system has proven a failure. The act, stripped by the *New York* decision of a key goad, has failed to generate sufficient incentives for states to develop a national system of regional disposal facilities. The compact system has also been riddled by gerrymandering and political and financial gamesmanship that have sabotaged achievement of the statute's goals.

The 1980 NGA report envisioned six compacts, each with its own regional disposal facility, in which the member states, acting in good faith, would develop the compact's disposal facility in collaboration with each other. As of this writing, there are only four licensed LLW disposal facilities, none of which was developed under the compact system as envisioned. Barnwell and Richland date back to the AEC era. Clive is a private facility, and Andrews, now under construction, exists only by virtue of a highly artificial two-state Texas Compact that is entirely at odds with the compact scheme envisioned by Congress. The recent decision by Texas authorities to authorize the opening of Andrews to out-of-compact class B and C wastes from thirty-six states will, if implemented, provide an immediate solution to the problem of class B and C waste disposal capacity. But it is not necessarily a stable long-term solution, nor an equitable one.

In light of the surge of class B and C LLW that will sooner or later result from nuclear power plant decommissioning, new approaches at the national level may well be necessary. The federal government could seek to fix the LLW disposal system without altering the LLRWPA structure by subsidizing new LLW disposal facilities that accept out-of-compact wastes; host jurisdictions could share in the subsidies by taxing the disposal facilities that receive them. As a further step, Congress could impose fees on out-of-state or out-of-compact wastes shipped to these facilities to cover the federal subsidies. In the event of a temporary unavailability of disposal facilities, NRC might seek to exercise its LLRWPA emergency authority to require the sites at Barnwell and Richland to accept the out-of-compact class B and C wastes in the interim,[154] although it would be prudent and desirable to obtain host jurisdiction assent to such a move. Alternatively, DOE sites could be opened to commercial LLW disposal.

A more fundamental and transformative change would be to discard the broken compact system and install a system, as recommended by IRG in 1979, of centralized federal leadership in LLW facility siting. Notwithstanding regulatory failures, AEC initiatives succeeded in producing six disposal sites distributed relatively evenly geographically with sufficient projected capacity to meet the nation's anticipated LLW disposal needs indefinitely. In rejecting IRG's proposed model, Congress seems to have supposed that the serious environmental problems encountered in the 1970s at many LLW disposal facilities—which were attributable to AEC's primitive regulatory tools and its failures in determining and implementing proper disposal methods—pointed to the need to abandon federal leadership over LLW disposal in favor of the NGA approach. If so, the LLRWPA devolution cure was the result of serious misdiagnosis. LLW disposal presents a powerful collective action problem among the states that the IRG had anticipated and that LLRWPA has failed to solve. A strategy like that envisaged by IRG that relies on a much stronger federal government role is warranted.

Under a revived strategy of federal leadership, NRC (or preferably a new specialized entity) would take the lead in siting and developing new facilities, located with a view to regional balance. NRC would set general regulatory requirements and directly regulate civilian LLW disposal practices at facilities not subject to Agreement State jurisdiction. In siting new facilities, the federal government would need to follow the steps discussed in Chapter 8 to secure acceptance of such facilities by host communities and states, including openness and full information disclosure, step-by-step decision making with genuine opportunity for local and state involvement, a respected independent source of technical information, and assured mechanisms for proper regulation and compliance. This strategy could also employ subsidies and fees, as discussed earlier. Arrangements for transition from the current compact system (which might include retention of successful compacts) would also be needed.

Chapter 5

WIPP: The Rocky Road to Success

The Waste Isolation Pilot Plant (WIPP) is the only operating deep geologic repository for disposal of long-lived nuclear wastes in the world. It has been receiving and emplacing defense TRU wastes for more than a decade. This chapter recounts the checkered, uncertain, but ultimately successful story of its development over more than twenty-five years. In doing so, the chapter identifies key features of the WIPP development process that help explain how siting, construction and operation of WIPP were able to be achieved, despite obstacles, and what lessons this process holds for future development of nuclear waste disposal and storage facilities.

The Origins of WIPP

A 1969 fire at the Rocky Flats nuclear weapons facility set off a train of events that eventually spurred the federal government to site and develop a geologic repository for transuranic wastes—and also possibly for defense HLW and civilian and defense SNF. After an initial AEC attempt to develop a repository at Lyons, Kansas (Project Salt Vault), ended in failure,[1] AEC's search for another salt-bed location eventually, in 1975, targeted the Los Medanos site within the Salado Formation, located on federal lands near the town of Carlsbad, New Mexico.

Carlsbad's political leaders actively courted a federal nuclear waste repository for the economic benefits such a project might provide for their beleaguered local economy. The project began to move forward in 1975 under the auspices of the Energy Research and Development Administration (ERDA).[2] There was no blueprint for the project that eventually became known as the Waste Isolation Pilot Plant (WIPP), nor for the process or path to be taken in creating it. In the early stages, no explicit decisions were made regarding the wastes that would go into the repository, the applicable safety standards or criteria, the supporting transportation infrastructure, the roles and authority of the various state and federal actors, or the procedures for making final project and regulatory decisions.[3] These issues and their resolution evolved over time through interactions among ERDA (and its successor, DOE), the State of New Mexico, Congress, the federal courts, environmental and citizens' groups, independent experts, and other actors.[4] In 1979, Congress limited WIPP's mission to serving as a "research and development facility to demonstrate the safe disposal" of defense wastes, ruling out civilian SNF.[5] It was not until 1992 that Congress restricted WIPP to disposal of defense TRU, thus excluding both civilian and military SNF and HLW from the repository.[6]

In April 1975, shortly after it decided to focus on Los Medanos, ERDA turned over management of the project to Sandia National Laboratory. Sandia eventually undertook a comprehensive geologic survey of the site, designed a pilot disposal plant, prepared a draft environmental impact statement for the plant, and mounted a public information campaign.[7] New Mexico's governor, Bruce King, established a technical committee, chaired by the head of Sandia, to monitor and report to him on project activities, although the state at this stage did not play an active role in the project.[8]

ERDA considered a number of specific locations in Los Medanos for the WIPP repository, all of which were located on federal lands. Several proposed sites had to be abandoned at the outset when exploratory boreholes encountered high-pressure brine pockets, a feature of salt formations that must be avoided in constructing a repository.[9] In 1976, after a suitable site was found, ERDA submitted to the Department of Interior an application pursuant to provisions in the Federal Land Policy and Management Act to administratively withdraw 17,200 acres of federal lands at the Los Medanos site and limit their use to detailed site characterization by ERDA for a "waste isolation pilot plant," thereby giving the project the name by which it has become known; the notice provided no further details.[10] The notice provided no further details or explanation. The filing of this application effected a temporary segregation of the land from other uses for two years.[11] Two years later, in 1978, DOE filed for a second administrative withdrawal of the WIPP site that extended temporary segregation of the land for another two years. Again, the notice provided no details or explanation.[12] The type of waste to be tested at the plant—i.e., SNF, HLW, or TRU—was not specified in either application. At this early point, public statements by state officials were "moderately supportive" of the project.[13]

Restriction of WIPP to Defense TRU

From the earliest stages, the exact type of nuclear waste to be handled at WIPP was unresolved; this issue remained in flux for many years. There is evidence that ERDA initially envisioned that WIPP would serve as a pilot facility for disposal of HLW from SNF reprocessing (which would require NRC licensing of the facility), but in 1975, at the time when federal support for SNF reprocessing was ending, it decided not to proceed with this concept and redefined WIPP as a TRU disposal facility (which would not require NRC licensing).[14] Yet the plant's project manager announced in January 1977 that WIPP's planned design would make the site suitable for licensing by NRC. This was widely interpreted as a signal that the site was now being considered for storage or disposal of commercial SNF.[15] A month later, the project manager told the New Mexico governor's advisory committee that "consideration would obviously be given to making [WIPP] a commercial site."[16] A Sandia draft environmental impact statement (DEIS), completed in April 1977, in fact described WIPP as a repository for defense TRU waste, as well as for a quantity of commercial high-level waste.[17] The DEIS, however, was not released until October 1978, after the Southwest Research and Information Center of New Mexico filed a Freedom of Information Act request for the document.[18]

Few in Washington, D.C., disputed that disposing of commercial SNF at WIPP made good sense. As described earlier, a series of actions by President Carter in the mid-1970s had made it clear that SNF would no longer be reprocessed. Carter decided in April 1977 to maintain indefinitely President Ford's temporary moratorium on federal support for SNF reprocessing.[19] Carter also announced that "the plant at Barnwell, South Carolina . . . will receive neither federal encouragement nor funding for its completion as a reprocessing facility."[20] The president also subsequently vetoed S. 1811, the ERDA Authorization Act of 1978, which would have authorized construction of a breeder reactor and a reprocessing facility.[21] By essentially precluding reprocessing as an option for SNF, these executive actions signaled that development of a repository to dispose of SNF would have to be given priority attention. In 1976, ERDA sent letters to the governors of the thirty-six states with existing nuclear facilities, including the three states that were home to mothballed SNF reprocessing facilities, proposing to site SNF repositories in their states. When no state indicated interest, ERDA/DOE was compelled to start exploring other options.[22]

The only feasible near-term option for permanent disposal of the commercial SNF that was piling up at commercial sites around the country seemed to be WIPP.[23] James Schlesinger, the entering secretary of the newly minted Department of Energy, was openly in favor of disposing of commercial waste at WIPP.[24] A GAO report issued in September 1977 also endorsed SNF disposal at WIPP, reasoning that DOE should seize the opportunity made available by favorable "public and official sentiment in New Mexico."[25] DOE notified NRC in November 1977 that it intended to change WIPP's scope so that it could also handle commercial SNF, and that it would therefore seek an NRC operating license for the facility.[26]

The federal government's maneuvering to include SNF in plans for WIPP soured many in New Mexico on the WIPP project, inspiring the state legislature to come very close to passing a constitutional amendment that would ban storage of any radioactive waste brought into the state.[27] New Mexico had no nuclear power plants, and the state was strongly opposed to hosting other states' nuclear power wastes. Because of its role in the development of the atomic bomb and the ongoing presence of government nuclear facilities, including Sandia, in New Mexico, the state was less averse to hosting defense wastes. The Sandia DEIS, which had explicitly contemplated disposal of SNF at WIPP, prompted New Mexico's congressional delegation to send a letter to Secretary Schlesinger in November 1977 warning against "major restructuring of the functions of WIPP."[28] The delegation explained that they were upset not only because SNF was more hazardous than TRU waste, but also because the decision had been made "without any prior consultation with any of the members of this delegation" and "without the informed concurrence of the people of New Mexico."[29] Expanding WIPP's mission to include SNF, the delegation charged, would represent a major change in the project. This incident helped give impetus to the state's demands that it be given an expanded role in decision making on the project.[30]

Secretary Schlesinger responded by publicly offering New Mexico's governor, Jerry Apodaca, the opportunity to veto WIPP altogether.[31] Schlesinger pressed, despite Apodaca's professed indifference to the offer, insisting that the state could reject WIPP at any point.[32] In April 1978, when Schlesinger addressed the question of a state veto or right of concurrence before the House Committee on Interior and Insular Affairs, he said: "As a practical matter, the states have a right of concurrence. . . . Whether under the law that is the case, I think would best be left unresolved. It is a gray area of the law and I think it is more convenient to leave it there rather than to define it too precisely."[33] But the offer was later revealed to be both legally groundless and disingenuous in an opinion of the U.S. Justice Department Office of Legal Counsel made public in 1979. The opinion concluded that, under existing law, New Mexico had no authority to veto a federal WIPP facility on federal lands, nor could its concurrence with the facility be required; further, DOE had no legal authority to confer veto or concurrence authority upon the state. The opinion also revealed that DOE's Office of General Counsel had arrived at the same conclusion in an internal memorandum in March 1978.[34] Secretary Schlesinger's misleading offer to New Mexico and his misrepresentation of the state's power to halt the facility further cemented mistrust of the pilot project by New Mexico's state and congressional representatives, as well as by many of its citizens.[35]

Notwithstanding growing concern in New Mexico, further impetus for SNF disposal at WIPP came from the public release in March 1978 of the draft report of the DOE-appointed Task Force for Review of Nuclear Waste Management, headed by ERDA's director of energy research, John Deutch; the task force was formed as part of President Carter's efforts to create a national nuclear waste management plan.[36] The draft, commonly known as the Deutch report, stressed the importance of demonstrating that commercial nuclear waste could be disposed of safely and permanently in a repository and therefore recommended that WIPP be used to test storage of one thousand commercial spent fuel assemblies.[37] The revelations about DOE's attempts to expand WIPP's function, however, colored reception of the Deutch report.[38]

Salt formations had long been considered as a promising geologic medium that should be considered for a repository, and mining out a bedded salt site specifically for a repository, as DOE planned to do at WIPP, was more technically defensible than using an existing abandoned salt mine, as AEC/ERDA had tried to do at Lyons, Kansas.[39] Yet substantial concerns were raised in connection with the development of WIPP about the long-term suitability of using bedded salt as a disposal medium. In a January 1978 memorandum commissioned by the task force, scientist William Luth criticized DOE for failing to examine seriously other geologic media.[40] A panel of earth scientists noted their concern over the potential for small amounts of water located in the salt to decrepitate and possibly even explode when roasted by the heat of decaying SNF or HLW, impairing the waste containment security, while other scientists raised the possibility of brine migration toward the disposed waste, which could degrade the protective waste canisters and result in release of radioactivity to the environment.[41] A series of 1978 reports by a Sandia geologist, the White House Office of Science and Technology Assessment, and the U.S. Geological Survey (USGS), respectively, raised technical concerns about the WIPP site because of the potential presence of high-pressure brine pockets below the repository.[42] The concern was that at some point in the future, after institutional control over the site might have been lost, drilling for oil or natural gas would cause the brine to rise up through the repository to the surface, bringing radionuclides with it.[43] As previously noted, discovery of this very hazard during the site investigation process had several times led DOE to relocate the site for WIPP within Los Medanos.

Concerns about DOE's approach to technical and scientific issues in developing WIPP were voiced by various parties. In his 1978 memorandum, William Luth complained that ERDA was overly focused on evaluating alternative policies instead of on building technical consensus. He criticized ERDA's, and later DOE's, "unseemly emphasis . . . on demonstrating the technical feasibility of geologic disposal instead of on arriving at a scientifically objective assessment."[44] New Mexicans outside Carlsbad were becoming wary of DOE and WIPP.[45] DOE's tactics had created a more adversarial relation with the state.[46]

In late 1978 and early 1979, DOE crossed swords with Melvin Price, chair of the House Armed Services Committee, which had jurisdiction for authorizing funding for siting and characterization of WIPP. Price was against giving New Mexico a say in WIPP decision making.[47] He was strongly invested in making sure that the facility was limited to taking defense wastes; while Price supported nuclear power, he staunchly opposed the disposal of commercial SNF at WIPP, which would trigger NRC regulation of the repository and thereby undermine his committee's exclusive jurisdiction over the site.[48]

DOE's DEIS on WIPP, published in April 1979, nonetheless made clear that DOE intended to develop WIPP as a repository for both military and commercial waste.[49] Price countered by refusing to authorize funds for DOE to seek a commercial waste disposal license from NRC.[50] He also blocked efforts by DOE to obtain funds through the House Interior Committee.[51]

Notwithstanding Price's obstruction of funding, DOE refused to rule out that it might ultimately seek NRC licensing for commercial waste disposal at WIPP.[52] In December 1978, Secretary Schlesinger wrote directly to Price to suggest that WIPP be converted from a military facility to a commercial one.[53] Price replied that, if the secretary wanted to do that, he should seek funding from some other committee; Price then got to work preventing this very possibility.[54]

The Interagency Review Group on Nuclear Waste Management (IRG) issued its final report in March 1979. The report supported DOE's position by proposing the co-location at WIPP of a TRU repository and an SNF demonstration repository.[55] Encouraged by the IRG report,[56] DOE's April 1979 DEIS for WIPP (a separate document from the earlier Sandia DEIS for WIPP) followed the IRG recommendation and contemplated a facility for disposing of up to one thousand SNF fuel assemblies, with the capability, but without the expectation, of retrieval.[57]

There were, however, several decidedly negative responses from the state and Congress to DOE's maneuvers to dispose of SNF at WIPP. New Mexico's legislature came close to submitting authorization of WIPP to a statewide referendum.[58] Instead it passed the New Mexico Radioactive Waste Consultation Act,[59] which provided that "no person shall store or dispose of radioactive waste in a disposal facility until the state has concurred in the creation of the disposal facility." In May 1979, the House Armed Services Committee reported out a bill that deleted all further WIPP funding for fiscal year 1980.[60] The Senate Armed Services Committee did not cut off all funding, but it restricted the funds authorized for WIPP to "the pursuit of a demonstration project for the long-term storage of nuclear wastes generated by the nuclear weapons research, development, and production complex."[61] Unable to find another way to fund WIPP for development as a commercial facility, DOE capitulated in July 1979 by abandoning its plans for co-locating SNF at WIPP.[62]

The debate over whether WIPP would handle commercial or defense waste ended in December 1979, when Congress enacted the DOE National Security and Military Applications of Nuclear Energy Authorization Act of 1980, colloquially known as the WIPP Authorization Act. Section 213 of the act authorized funding for development of WIPP, while restricting its mission to defense wastes. Senator Henry M. Jackson of Washington, a powerful figure on the Senate Armed Services Committee, pushed for this restriction. He understood that authorizing WIPP as a defense waste disposal facility did not require NRC licensing—a clear requirement if WIPP were to take civilian waste—and was convinced that this would provide the best chance for a repository to open in the near future.[63] The House-Senate conferees on the act noted that the "constant attempt to change the purpose of the WIPP has resulted in delay and confusion" and stated that Congress intended the act to resolve any remaining questions about WIPP's purpose.[64] The WIPP Authorization Act thus limited WIPP to the "express purpose of providing a research and development facility to demonstrate the safe disposal of radioactive wastes resulting from the defense activities and programs of the United States exempted from regulation by the Nuclear Regulatory Commission."[65] While the WIPP Authorization Act thus limited WIPP's purpose as "research and development facility" for "demonstration" to defense waste, it did not explicitly address the disposal of nondefense waste in the event that a repository were to be constructed on the site. Also, it left open the question of just what types of defense wastes WIPP could accommodate. DOE's 1980 FEIS for WIPP described, as an "authorized alternative," use of WIPP as a repository for 6.2 million cubic feet of contact-handled TRU waste, up to 250,000 cubic feet of remote-handled TRU waste, and 150 cubic feet of "high-level waste for experiments." But, despite the provisions of the recently enacted WIPP Authorization Act, the FEIS gave as DOE's "preferred alternative" the potential co-location of commercial HLW with defense TRU at WIPP.[66]

In February 1980, President Carter announced that he had "decided that the Waste Isolation Pilot Plant project should be cancelled" in favor of a new, comprehensive radioactive waste disposal program; the new program would "expand and diversify DOE's program of geologic investigation before selecting a specific site for development."[67] Carter sought development of a repository that would be open to civilian wastes and required to obtain an NRC license.[68] The president, whose administration was extremely weak at that point, was ignored in all quarters. WIPP retained committed backers in Congress, including Melvin Price of the House Armed Services Committee.[69] Notably, New Mexico's congressional delegation, having been handed the opportunity to acquiesce in Carter's attempt to cancel WIPP, backed the project.

In the late 1970s and early 1980s, Congress began to move forward on a plan for disposing of commercial SNF and both defense and commercial HLW in a repository independent of WIPP. The 1982 Nuclear Waste Policy Act (NWPA) directed DOE to select and build two repositories, one in the East and one in the West, on an accelerated schedule, after sites had been selected in accordance with a procedure specified in the statute. NWPA expressly excluded WIPP from consideration as a potential site.[70] Congress's later decision to target Yucca as the sole NWPA repository candidate site caused DOE to cancel many of the experiments it had been performing at WIPP to determine the suitability of bedded salt as an appropriate geologic setting for a commercial nuclear waste repository.[71] On October 4, 1991, during the middle of negotiations for land withdrawal to construct the WIPP repository, Secretary of Energy James Watkins wrote a letter to Governor Bruce King of New Mexico promising that, no matter how the negotiations turned out, WIPP would not be used for emplacement of SNF or HLW.[72] The 1992 Waste Isolation Pilot Plant Land Withdrawal Act (WIPPLWA) finally banned SNF and high-level waste from being shipped to or disposed of at WIPP, confirming the repository's restriction to disposal of defense TRU alone.[73]

New Mexico's Successful Efforts to Gain a Role in Decision Making Regarding WIPP

Beginning in the late 1970s, the State of New Mexico began to play a far more assertive role in the development of WIPP. The New Mexico Radioactive Waste Consultation Act created an executive task force, which led and coordinated the state's activities and policies regarding WIPP and its interaction with DOE and other stakeholders.[74] The state began to use an effective combination of state legislation, initiatives by the task force, political pressures through the New Mexico delegation in Congress, and federal court litigation against DOE to exert influence over WIPP's development. The state's hopes of gaining legal authority to block the facility were unsuccessful; the Justice Department's opinion effectively foreclosed that option, unless Congress was willing to enact a statute to give the state such authority, which it declined to do. And, as evidenced by its congressional delegation's support of WIPP when it was in danger of being cancelled by President Carter, the state over time came to support the project, provided that key conditions could be assured. The state developed a number of ways to play a significant role in establishing those conditions and to influence evolving decisions about the facility.

The 1978 Memorandum of Understanding and the Environmental Evaluation Group

In 1978, New Mexico successfully pressured DOE to sign a memorandum of understanding (MOU) with the state concerning WIPP. Among other matters, the MOU established an Environmental Evaluation Group (EEG), which evolved into an independent scientific body that provided technical expertise on WIPP issues. DOE's charter for EEG established it as a scientific body that would "conduct an independent technical evaluation of the [WIPP] Project to ensure the protection of the public health and safety and the environment."[75] Pursuant to a DOE contract with the state, EEG was funded by DOE but housed in the New Mexico Environmental Improvement Division, part of the New Mexico Health and Environment Department.[76] Subsequently, Congress specifically authorized EEG in the National Defense Authorization Act, fiscal year 1989. The act defined EEG as "conduct[ing] independent reviews and evaluations of the design, construction, and operations" of WIPP.[77] It provided that EEG would be run by scientists who were well known within the field and were "free from any biases related to the activities of the WIPP, and ... widely known for their integrity and scientific expertise."[78] DOE continued to pay for EEG's work throughout its entire existence with the understanding, later observed in practice, that neither federal nor state actors would attempt to bias the group's conclusions.[79] In order to formalize EEG's independence from both state and federal influence, the act required that the secretary of energy contract with a state research university, the New Mexico Institute of Mining and Technology, to house EEG, thus removing it from the New Mexico Health and Environment Department.[80]

Deputy Secretary of Energy John O'Leary, who had served as New Mexico's secretary of natural resources before taking his position at DOE, was instrumental in establishing EEG. He realized from direct experience that, even if the federal government had constitutional authority to build a repository on federal land at Los Medanos, the state had political, legal, and practical capacities to impede or even thwart federal development of the repository. Accordingly, he promoted the formation of EEG in order to accommodate the state.[81]

EEG became a trusted and independent source of technical expertise.[82] Its willingness to critique flaws in the WIPP project ultimately worked to reassure the state and other stakeholders of the repository's technical soundness and to bolster the credibility of the project. The group, as stated in its reports, was "to conduct an independent technical evaluation of the [WIPP] Project to ensure the protection of the public health and safety and the environment. . . . EEG performs independent technical analyses of the suitability of the proposed site; the design of the repository, its planned operation, and its long-term integrity; suitability and safety of the transportation systems; suitability of the Waste Acceptance Criteria and the compliance of the generator sites with them; and related subjects."[83] EEG was respected for its technical advice and its "independent scientific oversight" for the duration of its existence (1978–2004).[84] The U.S. Office of Technology Assessment noted in 1991 that "EEG's full-time, long-term presence, permanent staff and consistent resources are unique elements that contribute to its effectiveness. Also of importance is EEG's ability to remain independent of DOE, even though its funding comes from the Department."[85]

Over the course of its existence, EEG issued eighty reports and gave testimony before Congress. It identified a number of significant problems in the project, including DOE's faulty assessments of geology at Los Medanos[86] and deficiencies in the design of containers used for transporting TRU to WIPP.[87]

By ensuring the state access to impartial scientific and technical data regarding the site and repository design, EEG was crucial to New Mexico's ability to participate effectively in decision making regarding WIPP. EEG, which never recommended halting the project, also gave DOE an opportunity to correct problems as they were identified, and thus to keep the project on course. For example, EEG made important contributions in the development of Sandia National Laboratory's Performance Assessment model, which was used to ensure the repository's compliance with EPA radiation standards. EEG regularly critiqued new iterations of Sandia's calculations and made detailed recommendations for improvement.[88]

In addition to raising questions about DOE's plans and assessments, EEG addressed issues and contentions raised by critics of WIPP.[89] For example, a geophysicist from Sandia raised concerns in 1982 about whether the geology of the WIPP site was potentially "karst" geology—porous underground features created by the dissolution of soluble bedrock.[90] Such conditions could allow excessive moisture into the repository and provide a path for radionuclides to escape into the environment. EEG, working with USGS, conducted a thorough investigation of the issue. In 1985 it released a report evaluating the implications of karst and recommended specific studies for further work.[91] Subsequent evaluations by EPA and NAS's Committee on the Waste Isolation Pilot Plant relied and expanded upon the EEG analysis.[92] Subsequently, DOE conducted a number of studies over a twenty-year period, concluding that karst was not a problem at WIPP.[93]

Implementing New Mexico–DOE Consultation and Cooperation

In the congressional proceedings leading up to enactment of the 1980 WIPP Authorization Act, there was wide debate over what role the state should play in the siting and characterization process. The range of options for empowering the state vis-à-vis DOE included veto, "concurrence" authority, "consultation and cooperation," and "consultation and review."[94] Negotiations continued through the summer and into the fall of 1979 among the House Armed Services Committee, various senators, DOE, and New Mexico's congressional delegation. No one openly sought the veto option—not even New Mexico Senator Pete Domenici, who had earlier advocated giving the state such authority.[95] The House Armed Services Committee put WIPP's funding back on the table but would not accept state "concurrence," whose practical and legal meanings were difficult to distinguish from a state veto.[96] The Senate Armed Services Committee and New Mexico's governor were dissatisfied with the preference of DOE and the House Armed Services Committee for "consultation and review," which they viewed as an insufficiently strong mechanism for state involvement in the decision-making process.[97] A compromise was achieved in the conference committee, which adopted a requirement that DOE "consult and cooperate" with the state, leaving it to DOE and New Mexico to work out exactly what the phrase meant:

> § 213(b)(1) The Secretary shall consult and cooperate with the appropriate officials of the State of New Mexico, with respect to the public health and safety concerns of such State in regard to such project and shall . . . give consideration to such concerns and cooperate with such officials in resolving such concerns.
>
> (2) The Secretary shall seek to enter into a written agreement with the appropriate officials of the State of New Mexico, as provided by the laws of the State of New Mexico, not later than September 30, 1980, setting forth the procedures under which the consultation and cooperation required by paragraph (1) shall be carried out.[98]

The conference committee explained this provision as follows:

> The conferees believe that the language [in Section 213(b)(1)] directing the Secretary of Energy to consult and cooperate with officials responsible under New Mexico laws . . . , along with existing applicable Federal laws, Executive orders, and regulations, are sufficient to permit the State to protect its legitimate interests. At the same time, the language will permit the Secretary of Energy to carry out his legal responsibilities and to balance the national interest with that of the state. . . .
>
> The conferees recognized the possibility that the Secretary of Energy and the State might not be able to negotiate a mutually satisfactory agreement. However, in view of the long history of cooperation by the State with the Federal Government in atomic energy matters and in view of the national significance of the WIPP project, the conferees fully expect such an agreement can be successfully negotiated. In this spirit of cooperation, no specific legislative remedy has been included to resolve a situation where there is no agreement.
>
> The consultation and cooperation agreement (or lack of such an agreement) and the proper execution of the process specified by such an agreement should not

delay the progress of the WIPP project. Work should proceed on the WIPP project as planned. . . .

The specifics of the agreement . . . are left to the parties. However, the conferees were adamant that the Secretary of Energy may not enter into any agreement or make any commitment under which the State of New Mexico . . . could in effect veto such project.[99]

This statutory provision led DOE and New Mexico to enter into a legally binding agreement for consultation and cooperation (C&C) that gave the state significant voice in decisions regarding WIPP and forced DOE to come to the bargaining table to meet the state's concerns about the facility.[100] The agreement was renegotiated by the state and DOE and modified three times.[101] The state's concerns, including waste transportation issues, were extensively negotiated and eventually accommodated.

New Mexico: Federal Interactions and Resolution of Conflicts, 1980–1992

This section describes how disputes between state and federal officials during WIPP's development were resolved under the structure established by the WIPP Authorization Act. Resolution of the first of these conflicts, which yielded a formal agreement between DOE and the state in July 1981,[102] framed and furthered resolution of subsequent disputes.

The 1981 Stipulated Agreement and Agreement for Consultation and Cooperation

The 1980 WIPP Authorization Act required DOE to "consult and cooperate" with New Mexico officials "with respect to the public health and safety concerns" regarding WIPP and to "seek to enter into a written agreement" with the state establishing "procedures" for such consultation and cooperation, as noted earlier.[103]

At the outset, New Mexico, now armed with DOE-funded scientific expertise in the form of EEG,[104] sought a consultation and cooperation agreement that was legally enforceable.[105] Monthly negotiation meetings began early in 1980 and progressed apace to meet the September 30, 1980, statutory deadline. But on September 28, New Mexico's attorney general, Jeff Bingaman, announced that New Mexico would not sign unless DOE stipulated that the agreement would be legally binding and its implementation subject to judicial review.[106] New Mexico's governor backed Bingaman, DOE stood firm, and no agreement was signed.[107] Moreover, just what "consultation and cooperation" amounted to remained undefined.

On May 14, 1981, Bingaman brought suit against DOE and the Department of Interior, asserting four claims: (1) under New Mexico state law, DOE was obliged to seek state concurrence before commencing construction; (2) the WIPP Authorization Act either should be interpreted to require DOE to resolve disputes with New Mexico through a negotiated C&C agreement or was unconstitutional under the Tenth Amendment; (3) DOE's EIS failed to comply with the National Environmental Policy Act (NEPA) and with EPA regulations; and (4) DOE's land withdrawal was invalid under FLPMA.[108] On May 25, after DOE had not relented, Bingaman sought a preliminary injunction to halt construction.

By July 1981, however, New Mexico and DOE had reached a settlement. Both sides felt that the state had a good case, and the parties were eager to resolve the matter. It was also known that Judge Guerrero Burciaga of the federal district court of New Mexico sympathized with states in conflicts with the federal government.[109] From the outset, Bingaman's office made clear that the suit did not reflect opposition to WIPP, but rather a need to ensure reasonable authority, assurances, and compensation for New Mexico.[110] U.S. attorney Ray Hamilton was aware of the situation's multiple political sensitivities. The Reagan administration—with its vocal commitment to states' rights—could not be seen as bullying New Mexico into accepting a radioactive waste repository. But it would be even worse to fail to move WIPP forward.[111] In addition, WIPP was setting a precedent; many political actors believed that any success in dealing with the politically sensitive issues of nuclear waste in other settings would depend on states' willingness to accept storage or disposal facilities within their borders.[112] For all these reasons, Hamilton made the case to his superiors at the Department of Justice that New Mexico's demands presented the federal government with an opportunity, rather than an obstacle. New Mexico was seeking voice and compensation, not a means to block WIPP; and the negotiation-contract-compensation-repository template was one that DOE could hold out to other states as evidence that its siting efforts were responsible and trustworthy.[113]

Thus, within six weeks of New Mexico's bringing suit, the parties signed the agreements they had spent more than a year negotiating. These included a stipulated agreement that guaranteed New Mexico access to DOE technical and planning information and, as an appendix to the stipulated agreement, an agreement for consultation and cooperation (C&C) that spelled out New Mexico's various demands.[114] Important provisions in the C&C agreement included a list of key events and milestones that triggered a federal obligation to give advance notice to the state.[115] The C&C agreement also prohibited permanent disposal of HLW at WIPP.[116]

Both the C&C agreement and the stipulated agreement provided for judicial review through a combination of provisions within the stipulated agreement itself and court orders. The stipulated agreement provided that DOE "shall comply with all of the terms and conditions contained in the provisions of this Stipulated Agreement" and that the C&C agreement "shall be a binding, enforceable agreement between the Department of Energy and the State of New Mexico."[117] On July 1, 1981, the court ordered that both parties "shall in good faith comply with the provisions of the Stipulated Agreement."[118] On December 28, 1982, a supplemental stipulated agreement was entered into the court record, and on January 10, 1985, the court ordered that *New Mexico ex rel. Bingaman v. Department of Energy* could be reopened by a party "upon a claim of noncompliance with this court's order of July 1, 1981, the Stipulated Agreement of July 1, 1981, or the Supplemental Stipulated Agreement of December 28, 1982."[119] The stipulated agreement had already made clear that in agreeing to the C&C agreement and the stipulated agreement, the state did not waive its right to challenge as unlawful final agency actions taken by DOE in conjunction with the WIPP project.[120]

The C&C agreement provided for a "conflict resolution" process that required DOE to address state concerns promptly and in a form that the state could then challenge in court.[121] The process involved mediation between the State of New Mexico and the DOE management office in Albuquerque by a conflict resolution hearing officer (a "mutually agreed upon recognized, independent expert in the particular field"), who would hear the case, take detailed notes and prepare transcripts, and pass along his or her recommendations to the secretary of energy.[122] The secretary would then "issue a written decision . . . which shall . . . constitute final agency action on and resolution of the matter."[123] DOE's decision could then be challenged in court by New Mexico.

The 1981 and 1982 agreements set out the rules of the game but did not immediately produce cooperation. For example, in July 1983, DOE surprised New Mexico officials when it announced that construction would begin on WIPP; New Mexico governor Toney Anaya responded by arguing that the state's concerns must be fully addressed before construction could begin.[124] In addition, DOE consistently refused EEG's requests for access to technical documents during the first months of WIPP's construction.[125]

Implementing and Supplementing the Stipulated Agreement

In addition to creating the agreement for consultation and cooperation, the stipulated agreement provided for conclusion by the parties of a Supplemental Stipulated Agreement Resolving Certain State Off-site Concerns over WIPP.[126] These "off-site concerns" included state liability for WIPP-related transportation and other incidents, emergency-response preparedness, independent monitoring by the state, and upgrades/maintenance to state highways.[127]

In negotiating the supplemental agreement, DOE's position had been weakened by a significant technical problem contemporaneously encountered at the repository site.[128] In late November 1981, at EEG's urging, DOE's contractor deepened shaft WIPP-12—a mile north of WIPP's central shaft—to determine the existence of and extent of geologic changes in the salt formation below WIPP. At a depth of about three thousand feet, the contractor penetrated a pressurized brine pocket, and almost 2.5 million gallons of brine rapidly flowed to the surface.[129] After discovery of the brine pocket, DOE decided, upon EEG's recommendation, to build the repository in a location extending south from the central shaft instead of north as originally planned.[130]

The parties reached agreement on a supplemental stipulated agreement in December 1982. The supplemental agreement required DOE to support New Mexico on several fronts. First, DOE undertook to assist the state in obtaining favorable terms in the upcoming congressional reauthorization of the Price-Anderson Act of 1957, which governs federal and state liability in the event of nuclear accidents.[131] Second, DOE committed to help the state on a variety of matters relating to waste transportation to WIPP. These included assisting the state to obtain federal funds to upgrade state highways to be used for transporting waste to WIPP, allowing the state to monitor transport activities and inspect waste shipments, and providing financial support to the state for emergency-response preparedness.[132] Further, DOE agreed to provide funding in the amount of $600,000 annually to enable EEG to assess the repository's environmental effects.[133] Of critical importance to New Mexico, the supplemental agreement provided for judicial review of DOE's noncompliance with any of the agreed terms.[134] DOE, however, in keeping with a pattern that eventually became familiar, later tried to reduce the amount of state funding for transportation improvements that it had promised in the supplemental agreement.[135]

DOE proceeded with construction of surface facilities at the site, and in its July 1983 Defense Waste Management Plan announced that WIPP would begin operation in October 1988. The process was not so simple, however. At each stage of the WIPP project, DOE was required under FLPMA to obtain approval from the Department of Interior for a "withdrawal" of the federal land in question from Bureau of Land Management control and general public uses.

Interior did not act on the first two federal land withdrawals that DOE had requested, in 1976 and 1978. As a result of Interior's inaction, pursuant to FLPMA the lands requested in each application were automatically withdrawn, but only temporarily, for a period of two years per application. DOE had in 1980 made a third land withdrawal application to Interior that stated that DOE planned to use the land for a Site and Preliminary Design Validation Program for WIPP. Although the application provided no details, DOE's plan for the program was to dig two shafts to repository level and construct a series of tunnels in order to verify the integrity of the site.[136] In 1982, Interior acted upon the application, granting DOE an eight-year withdrawal for this purpose.[137] Since the land withdrawal limited the use of the site to the Site and Preliminary Design Validation Program, DOE applied for a fourth land withdrawal in January 1983, which would allow it to construct the WIPP repository but only on the condition that no waste would be stored or disposed of at WIPP without further applications and approval.[138] This fourth land withdrawal application was also acted upon by Interior and was approved in July 1983 for a period of eight years.[139]

An additional hurdle to constructing WIPP was provided in the stipulated agreement, which required that the "State of New Mexico be given the opportunity to have a final resolution of all essential and integral off-site state government concerns involving health, safety, and public welfare issues prior to a final decision to commence construction of permanent WIPP facilities."[140] After 1982, nearly all disputes between New Mexico and DOE focused on how to define and address off-site environmental, health, and safety risks presented by the facility and on transportation arrangements.

DOE–New Mexico Disputes, Concessions, and Delays: 1983–1987

Soon after assuming his second term as governor in January 1983, New Mexico's governor, Toney Anaya, expressed concern that, notwithstanding enactment of NWPA, the problems that would be encountered by the federal government in siting repositories for SNF and HLW would revive DOE efforts to conduct "experiments" with defense HLW at WIPP. As a result, Anaya feared, WIPP would become a de facto repository for these wastes.[141] Also, residents of Santa Fe and Albuquerque reacted with alarm to the news that DOE planned to transport radioactive waste destined for WIPP by truck rather than rail, using highways that passed through these cities.[142]

DOE failed to notify New Mexico stakeholders before announcing in July 1983 that construction of WIPP's underground structures would begin.[143] According to New Mexico, this was a "key event" that, according to the C&C agreement, required notification of the state.[144] As a result, the state threatened to invoke the conflict resolution provisions contained in the C&C agreement.[145] The resulting political backlash led DOE to agree to a first modification to the C&C agreement, signed in November 1984.[146] Key provisions of the first modification addressed the following matters:

- limits on the volume and waste characteristics of RH-TRU that could be accepted at the repository[147]
- temporary experimental storage of defense HLW at WIPP[148]
- a demonstration of retrievability of both defense HLW and TRU waste prior to permanent disposal[149]
- post-closure control and responsibility by DOE[150]
- completion of certain additional scientific testing and reports[151]
- compliance with EPA federal regulatory standards for waste repositories[152]
- a program for encouraging and reporting on the hiring of New Mexico residents at the WIPP Project[153]

New Mexico obtained a second modification to the C&C agreement in 1987,[154] after the First Circuit Court of Appeals set aside a segment of EPA's radioactivity standards for nuclear repositories.[155] The initial impetus for the second modification was the need to plug the resulting gap in the environmental regulations governing WIPP. DOE agreed to "continue its performance assessment planning as though the provisions [of EPA's standards] remain applicable."[156] New Mexico also obtained commitments from DOE to take steps to dispose of excavation debris from construction of the repository, especially salt tailings, in an environmentally acceptable manner, and to follow Department of Transportation and NRC regulations in shipping wastes to WIPP.[157] Further, DOE and New Mexico agreed to amend the supplemental stipulated agreement to provide for new sources of funding for waste transportation—specifically for bypasses around cities and for relief routes.

Ambitions to site a repository for commercial nuclear wastes at or near WIPP resurfaced in 1987, when newly elected governor Garrey Carruthers took the idea to New Mexico's congressional delegation and reportedly found that the response "was not just no, but hell no. So I dropped it."[158] Congress's designation in the NWPA amendments of 1987 of Yucca Mountain as a HLW/SNF repository seemed to eliminate the prospect that Congress might provide for SNF disposal at WIPP. However, it was not until 1992 when it enacted WIPPLWA that Congress explicitly limited WIPP to defense TRU alone, finally settling the issue.

In 1988, the C&C agreement was modified a third time to provide for additional testing and geophysical studies to be done on the WIPP site, with a target date set for completion of each test or study.[159] The third modification also provided that New Mexico could take air samplers within the repository.[160]

Controversy over Land Withdrawal for WIPP, 1989–1992

WIPP's surface and underground physical structures were nearly all constructed by 1988, but the facility would not become operational for another eleven years.[161] As of that time, DOE had obtained four land withdrawals under FLPMA, all of them for limited periods. A critical final step in bringing the facility into operation was to obtain authorization for permanent withdrawal of the federal land on which WIPP had been sited. The fourth land withdrawal, granted by Interior in 1983, authorized construction of WIPP and would not expire until 1991 but was conditioned on DOE's not disposing of nuclear waste at the facility without further authorization.[162] Procedurally, in order to make the repository an operational disposal site DOE faced the question of whether to seek another land withdrawal from Interior or to obtain statutory authorization from Congress. Under FLPMA, Interior could grant only temporary withdrawals of federal lands—for up to twenty years.[163] Disposal of waste at WIPP, however, would require permanent withdrawal of federal land for that purpose, which in turn would require congressional legislation. Up until this point, ERDA/DOE had followed a bootstrapping strategy, seeking from Interior a series of administrative land withdrawals for pilot or other short-run activities. This strategy had enabled DOE to keep the project moving forward in low-visibility, small-scale steps without clearly defining the facility's exact mission or development timetable and thereby attracting high-profile scrutiny and controversy.

Now a more difficult decision loomed. A further predisposal stage of the WIPP project might be authorized administratively under FLPMA, but to bring WIPP on line relatively soon as a fully operational repository, DOE seemed to have little choice but to seek authorization for permanent land withdrawal from Congress. Obtaining this authorization, however, would require consideration of legislation by four House and Senate committees and votes by both houses of Congress.[164] This process could involve significant delay, which would undermine DOE's credibility by rendering it unable to deliver on its assurances to Idaho and other western states that it would shortly remove TRU located within their borders.[165] The delay would also give New Mexico's congressional delegation leverage to impose further statutory conditions on WIPP as the price for authorization.

Congress heard testimony on the question of land withdrawal for WIPP from late 1987 through 1988. GAO's associate director, Keith Fultz, expressed doubts regarding the facility that were shared by many observers. Specifically, Fultz stressed the uncertainty surrounding the question of whether brine could seep into WIPP and what the implications of that seepage would be. After considering the scientific opinions of EEG, Sandia, DOE, various NAS committees, and others, Fultz testified that this question "must be addressed before DOE can demonstrate that the facility is suitable as a repository for permanent disposal of TRU waste."[166] Fultz concluded that WIPP should become fully operational only after DOE had developed an operational and experimental plan, confirmed that redesigned shipping containers complied with NRC standards, developed and presented adequate technical justification for disregarding concerns about brine seepage, and determined that WIPP met the standards for nuclear waste repository storage that EPA was then in the process of revising.[167]

News in 1986 of the Chernobyl disaster in Ukraine and of revelations about DOE's dismal record of managing nuclear wastes at its weapons complex facilities in Washington, Ohio, Idaho, and elsewhere reinforced doubts about moving ahead with full-scale operation of WIPP.[168] Congress authorized DOE to store only 25,000 drums of TRU (3 percent of the repository's planned capacity) on-site at WIPP for the duration of a planned five-year test phase, to occur after completion of construction and before full operation; DOE had proposed storing 125,000 drums during the test period.[169]

The WIPP land withdrawal debate drew in a number of affected states in the West, through whose territory waste destined for WIPP would transit, and signaled the increasingly important role of the Western Governors' Association (WGA) in the repository development process. WGA was formally created in 1984[170] and received funding in 1988 from the Department of Transportation to prepare a report for Congress on the "opinions, concerns, and priorities for actions" of the states affected by transport of wastes to WIPP.[171] WGA's 1989 report to Congress emphasized the need for uniformity in the states' transportation safety and emergency preparedness programs, and expressed the willingness of the states to work together on waste transportation matters so long as they received adequate financial support.[172] James Watkins, the newly appointed secretary of energy, responded by drafting the first of a series of cooperative agreements with WGA to fund a transportation safety program specifically for TRU waste.[173]

Western states entered the fray over land withdrawal for WIPP when Idaho governor Cecil Andrus announced in 1988 that Idaho would no longer accept any TRU from Rocky Flats, which was still manufacturing plutonium triggers for nuclear weapons. The TRU waste generated by this production was still being sent to DOE's Idaho facility, well beyond the 1980 deadline agreed upon by DOE and the State of Idaho for removal of all TRU from the state.[174] The Ninth Circuit Court of Appeals eventually decided, in 1991, that Andrus's declaration could not prevent DOE from shipping wastes from Rocky Flats to the Idaho site.[175] Prior to the court's decision, however, Andrus's statement caused substantial political fallout. Governor Roy Romer of Colorado urged that the Rocky Flats wastes go to WIPP. Governor Garrey Carruthers of New Mexico stated that he wanted WIPP to open but sought $250 to $300 million for upgrading the state's highways.[176] DOE eventually patched up the controversy by promising that DOE would continue to push for congressional land withdrawal, would update the 1980 EIS for WIPP, and would find a backup location for TRU outside Idaho if WIPP was not in operation by the new, tentatively scheduled opening date of August 1989.[177] Having emerged as a potent political force, WGA thereafter undertook a nuclear waste transportation initiative, monitored DOE's waste transportation policies and decisions, lobbied DOE, publicly criticized steps with which it disagreed, and took other steps with the stated goal of ensuring that shipments of TRU to WIPP would be "safe and uneventful."[178]

Soon after his February 1989 confirmation, DOE secretary Watkins made it a priority to get waste into WIPP quickly. NEPA required that DOE prepare a supplemental environmental impact statement (SEIS) for WIPP to address the impacts of disposing wastes at the site. Pressed by Watkins, DOE completed the thousand-page draft SEIS in a month, releasing it in April 1989.[179] EEG, however, found serious deficiencies in it: "The document contains mistakes in calculations, reflects an erroneous knowledge of the history of the project, presents tables without units, and displays an indifference to the statistical precision of predictions. . . . It appears that the draft EIS was written with a predetermined conclusion to accept the proposed plan [and that] . . . alternatives were not seriously considered."[180]

Voices in Congress and the national press expressed concern that DOE was trading safety for expediency.[181] Then, on June 6, 1989, EPA and the FBI conducted a widely publicized raid at Rocky Flats, based on allegations of serious environmental regulatory violations.[182] Rockwell International, the DOE contractor that ran Rocky Flats, was later convicted of several criminal violations of environmental laws at the facility. Confidence in DOE sank to new lows, scuttling chances of WIPP being allowed to accept substantial amounts of TRU during its pre-operational test phase.[183] Watkins's energetic efforts at damage control and departmental reform after the FBI/EPA raid failed to satisfy skeptics. Congress refused to grant a permanent land withdrawal for WIPP until DOE had completed a final report on WIPP's safety and developed a plan for retrieving buried waste in the event the site was declared unsuitable.[184] New Mexico's delegation also took the opportunity to demand that DOE make good on its promise in the 1987 second modification to the C&C agreement to help deliver federal funding for transportation infrastructure in New Mexico.[185]

DOE ultimately decided not to seek congressional authorization for a permanent land withdrawal for WIPP but chose instead to apply to Interior for an extension of its current administrative land withdrawal for the site. The extension would provide for construction of "full facilities" for the "WIPP Project"; for conducting a test program using retrievable radioactive material; for removal of the previous restriction adopted by Interior in 1983 that prohibited transportation, storage, or burial of radioactive material; and for expansion of the withdrawn area and extension of withdrawal until 1997.[186] According to the plans set forth in DOE's 1990 FEIS for WIPP, DOE would open WIPP for a five-year test phase after construction of the repository. Once results from the test phase confirmed that WIPP would function safely, as expected, the repository would move into full operation.[187] In a 1990 letter to Interior, DOE stated: "Because of the significance of the WIPP project to the resolution of the radioactive waste disposal problem in this country, [DOE] would prefer to have Congressional authority to use the public land to continue this important project . . . [but seeks the modification and extension] to allow the receipt at WIPP of radioactive waste for the demonstration phase in the event Congress does not act on the bill this year." The letter further asserted that congressional approval was not needed, because this would be a test run of the repository using fully retrievable TRU waste, consistent with the experimental purposes of the current withdrawal, and no final disposal would take place.[188]

Interior put out an official notice of preliminary approval in January 1991.[189] In August 1991, Energy Secretary Watkins confidently predicted that WIPP would open for testing by the end of the year. When asked about the possibility of a court order blocking the opening, he responded: "Sure, we will be sued, but we'll be sued not on the basis of environmental concerns. We already demonstrated we know how to deal with all safety and environmental aspects of WIPP. . . . Only pure emotionalism from [the antinuclear] faction stands between us and accepting wastes at WIPP."[190]

In October 1991, Interior published a notice to proceed, granting DOE authorization to begin the test phase at WIPP.[191] The day that the notice was published, the New Mexico attorney general filed a lawsuit seeking to enjoin DOE from transporting and emplacing waste in WIPP.[192] In November 1991, the District Court for the District of Columbia granted a preliminary injunction, ruling that, under FLPMA, the secretary of interior did not have the authority to extend a withdrawal of federal lands when the purpose of the withdrawal had been changed.[193] In January 1992, the court made the injunction permanent. On appeal, the U.S. Court of Appeals for the District of Columbia Circuit affirmed the permanent injunction, on the ground that the land withdrawal that Interior was seeking to extend specifically precluded DOE from bringing radioactive waste to WIPP, and that Interior could not waive this restriction in the context of a land withdrawal extension proceeding without undertaking further decisional procedures, which it had failed to do.[194]

Restoring WIPP's Credibility, 1992–1996

In 1992, Congress enacted the WIPP Land Withdrawal Act (WIPPLWA), authorizing full-scale operation of WIPP for disposal of defense TRU. The act directed EPA to adopt site-specific radiation standards for the WIPP repository, to certify WIPP's compliance with the EPA standards before it could begin full operation, and to recertify its continuing compliance with those standards every five years thereafter.[195]

WIPPLWA also imposed revisions on DOE's test-phase plans and criteria, reflecting the conclusions of NAS's Committee on the Waste Isolation Pilot Plant. DOE had planned to test TRU's potential for gas generation by bringing TRU on-site and placing it in bins filled with glassware and tools—"dry" materials. The committee found there was "no discernible scientific basis" for using this approach to test the hypothesis that water infiltration of waste containers could promote the generation of explosive hydrogen gas.[196] Based on this, Congress included in WIPPLWA a provision requiring that all tests conducted by DOE during the test phase be "directly relevant to a certification of compliance."[197] The act also included transportation-related provisions codifying many of the undertakings that DOE had previously contractually agreed to with states along the shipping corridor.[198] These provisions included requirements that NRC certify TRU containers shipped to WIPP; that DOE give states and tribes advance notice of shipments that would travel through their territory; and that the federal government provide assistance in accident prevention, transportation safety, and emergency preparedness. WIPPLWA also barred the transportation of TRU from Los Alamos National Lab to WIPP until a Santa Fe bypass highway was constructed or funded by the federal government.[199] Finally, the new law authorized federal grants to New Mexico totaling $300 million over fifteen years.[200] According to a New Mexico official, this was "the cost of doing business in New Mexico."[201] These grants were not specifically earmarked for transportation or any other purpose. While New Mexico had repeatedly invoked waste shipments to WIPP to justify federal funding for highway infrastructure improvement, in the final analysis the state appears to have preferred funding without restrictions.[202]

President Clinton's administration, elected just one week after President Bush signed WIPPLWA into law, took a much more collaborative stance toward WIPP constituencies within New Mexico. Clinton's secretary of energy, Hazel O'Leary, chose as her deputy chief of staff Dan Reicher, an attorney for NRDC who had long advocated for cleanup of DOE's contaminated weapons complex facilities and had been involved in the 1991 lawsuit challenging DOI's extension of the land withdrawal for WIPP. O'Leary also established a DOE field office in Carlsbad in 1993. From 1993 to 1996, the Carlsbad office hosted 128 public meetings.[203] Acceding to calls from EPA, EEG, and its own Task Force on Radioactive Waste Management, the new DOE leadership announced what it called "a major break with the last Administration's approach," namely, that testing to determine how much gas would be generated by TRU waste would be conducted in a laboratory rather than in situ at WIPP.[204] Because the end of the Cold War led to a halt in the production of nuclear weapons—and of the TRU wastes that had, in turn, been accumulating at Rocky Flats—it was perhaps easier at this point for DOE to adopt a more conciliatory approach.[205]

In February 1993, EPA announced that it would promulgate WIPP-specific requirements (40 C.F.R. Part 194) for compliance with EPA's generic radioactive standards for repositories (40 C.F.R. Part 191).[206] In December of that year EPA also published revisions to the Part 191 generic repository standards set aside by the U.S. Court of Appeals for the First Circuit.[207] The revised standards extended to ten thousand years the time period during which the repository must comply with radiation exposure standards for individuals and brought the drinking water contamination portions of the standards into alignment with those of the Safe Drinking Water Act.[208]

In 1996, Congress amended WIPPLWA to ease the way for full operation of WIPP.[209] The amendments reduced from 180 days to 30 days the period between WIPP's certification by EPA and commencement of waste emplacement,[210] exempted WIPP from some of the land disposal restrictions of RCRA,[211] and eliminated some of the facility tests that had been planned but were not yet completed.[212] In addition, the amendments required EPA, every five years after the first receipt of waste, to recertify WIPP's compliance with both sets of EPA repository radiation release standards, those contained in 40 C.F.R. Part 191 (generic) and 40 C.F.R. Part 194 (WIPP-specific).[213]

WIPP Moves into Full-Scale Operation as a TRU Repository

The EPA certification process for WIPP moved forward. In 1995, EPA issued guidelines for DOE's compliance certification application, and DOE submitted a draft of the application.[214] The New Mexico EEG expressed approval of the methodology used in the draft compliance certification application but raised concerns over its lack of detailed findings.[215] EPA made similar criticisms when it reviewed DOE's final application for compliance certification in 1996 and 1997.[216] A November 1996 NAS panel report also noted some deficiencies in DOE's work[217] but provided a generally sunny message about WIPP, declaring that the WIPP repository "appears to be an excellent choice based on geological considerations."[218]

In 1996, New Mexico's attorney general, Tom Udall, challenged EPA's generic and WIPP-specific repository standards in court; parallel suits were brought by the Texas attorney general and environmental groups.[219] The lawsuits gave voice to persistent distrust among many New Mexicans of the federal effort to dispose of nuclear waste at Los Medanos.[220] New Mexico's key arguments were, first, that EPA's generic radioactive repository rules were too vague to constitute "criteria" as required by WIPPLWA, and second, that EPA's collaboration with DOE in crafting the WIPP-specific repository rules was procedurally improper. The U.S. Court of Appeals for the District of Columbia Circuit upheld EPA's standards and its actions against both claims.[221]

New Mexico brought another federal court challenge in May 1998, the same month that EPA was issuing its final certification of WIPP.[222] The state sought to enjoin TRU waste shipments to WIPP on the ground that WIPP did not have a required final RCRA permit from the state. EPA had delegated RCRA authority to New Mexico in 1990, but the RCRA status of WIPP and the corresponding extent of the state's RCRA authority were disputed.[223]

RCRA provides that if a facility qualifies for "interim status," it may temporarily operate in accordance with relatively minimal requirements after applying for, but before receiving, a final RCRA facility permit. A facility that does not qualify for interim status could not begin any operations until after it had received a final RCRA facility permit.[224] New Mexico asserted that WIPP did not qualify for interim status, and that accordingly no shipments of mixed waste to the facility could occur until after the state had issued a final RCRA permit for WIPP. A facility qualifies for interim status if it was "in existence" before 1980 (WIPP was not), or "in existence on the effective date of statutory or regulatory amendments under the Act that render the facility subject to the requirement to have a RCRA permit," subject to timely filing for interim status.[225] In 1989, when WIPP's construction was well advanced, New Mexico adopted legislation that purported to regulate WIPP under RCRA. EPA did not delegate RCRA regulatory authority to New Mexico until 1990. The question was which action qualified as the relevant "statutory or regulatory change" making WIPP subject to RCRA.[226] If it was New Mexico's 1989 action, DOE's filing for interim status in 1990, after EPA's delegation to the state, would have been too late for WIPP to qualify for interim status, and it would require a final RCRA permit before operating. If it was EPA's 1990 delegation to the state, WIPP's filing would be timely and it would enjoy interim status.[227] The court agreed with DOE's position that the "trigger date" for interim status filing was EPA's 1990 delegation, allowing WIPP to operate pending issuance of a final RCRA permit.[228] Accordingly, the court denied the injunction sought by the state.

While awaiting the court's decision on the issue of WIPP's interim status, New Mexico took steps toward issuing a final RCRA facility permit for WIPP.[229] New Mexico stood to gain more by issuing WIPP's RCRA facility permit than by withholding it. Under a RCRA permit issued by the state, WIPP would become subject to the full complement of stringent RCRA operating standards, rather than the less-demanding requirements applicable to interim status facilities. Additionally, New Mexico could then take enforcement action for any permit violations at WIPP, thus giving the state significant leverage in decision making regarding facility operations and the upper hand in assuring the facility's compliance with RCRA.[230]

The end-game interaction between New Mexico and DOE reflected an odd mix of collaboration and contestation. For example, the two sides shared information and staff in order to determine whether New Mexico could refuse outright to allow wastes to be shipped to WIPP.[231] The overall spirit of the enterprise was reflected in remarks by James Channell, a senior member of EEG, who said that EPA's final certification of WIPP in 1998 was "a reasonable decision," and that "you can't wait until everything is perfect, because if you did, you would never start anything."[232]

In May 1998, EPA certified WIPP's compliance with EPA's repository standards pursuant to WIPPLWA.[233] DOE had previously issued a record of decision providing that waste emplacement would proceed after EPA certified the facility.[234] New Mexico governor Gary Johnson was quietly supportive of progress toward an operational WIPP;[235] New Mexico attorney general Tom Udall was not. In July 1998, Udall and citizen groups filed one last volley of lawsuits challenging EPA's certification of WIPP and DOE's plan to accept TRU delivery without awaiting a final RCRA permit from New Mexico.[236] The lawsuits filed against EPA claimed that the 30-day public comment period provided by EPA on its certification decision was inadequate under WIPPLWA and the Administrative Procedure Act. When the court denied an injunction against shipment of wastes to WIPP, on March 26, 1999, Secretary of Energy Richardson started delivery of TRU wastes to the repository.[237] In October 1999, seven months after the first shipment of TRU had been received, New Mexico granted WIPP a final RCRA facility permit.[238]

There are twenty-one DOE facilities at which TRU is now or has been stored, as shown in Map 5.1. As of December 2010, 9,207 shipments of waste had conveyed over 72,000 cubic meters of TRU to WIPP for disposal,[239] completely filling four of the facility's eight planned storage rooms (called "panels"). This represents about a third of WIPP's statutory capacity.[240] WIPP did not begin receiving RH-TRU until seven years after the repository began operation, when five thousand shipments of the less radioactive CH-TRU had already been delivered to the facility.[241] More than 370 deliveries of RH-TRU have arrived at WIPP from six sites since 2007.[242] According to DOE, there have been a relatively small number of truck accidents in transporting wastes to the repository; however, none has resulted in any release of radioactivity.[243]

As provided by WIPPLA, EPA granted WIPP its first five-year recertification of compliance with EPA's radiation standards and other regulatory criteria in 2006;[244] DOE had applied to EPA for certification in 2004.[245] DOE submitted an application to EPA for the second recertification of WIPP on March 26, 2009.[246] EPA granted recertification in 2010.[247] The requirement for recertification by EPA every five years will remain in force until all repository shafts have been filled with TRU waste, backfilled, and closed.[248] EEG activity continued until 2004,[249] when DOE ceased funding the group.[250] Although EEG can well be regarded as the single most effective reviewer, expositor, and facilitator in the project's multiparty dynamic structure, DOE apparently found EEG a thorn in its side and was able, once WIPP was up and running, to get the necessary political support in Congress to terminate its funding.

WIPP has opponents who continue to challenge the safety and environmental soundness of the repository. Among their technical concerns are the potential for migration of radionuclides out of the repository due to extensive resource extraction of oil, natural gas, and potash taking place adjacent to the site; the potential for water intrusion into the repository due to leakage from nearby wells and boreholes; the possible presence of high-pressure brine pockets; and the potential for impurities in the salt to promote the flow of groundwater through the salt formation (contrary to DOE's belief that the salt-bed formation is virtually impermeable).[251]

Accordingly, litigation against WIPP has continued. For example, Citizens for Alternatives to Radioactive Dumping, a local group, challenged DOE's third and final preoperational EIS for WIPP, arguing that it was incomplete and had been doctored to conceal hydrogeological risks.[252] As summarized by the U.S. Court of Appeals for the Tenth Circuit, which affirmed a district court decision in favor of DOE:

> The waste repository for the WIPP is located 2,150 feet underground, in the Salado Formation, a massive salt bed with low permeability that impedes groundwater flow in and out of the WIPP repository. About 1,400 feet above the WIPP is a fractured layer of dolomite rock called the Culebra Dolomite. The Culebra is the first layer above the Salado Formation with a continuous body of groundwater. Above the Culebra sits 86 feet of claystone, mudstone, and siltstone sandwiched between layers of anhydrite called the Tamarisk Member. Above the Tamarisk Member, another layer of dolomite, the Magenta Dolomite, runs from 621 to 596 feet below the surface.[253]

Plaintiffs specifically contended that DOE had acted arbitrarily by concentrating on the risks of radionuclide release posed by the Culebra Dolomite and neglecting risks posed by the Magenta Dolomite, as well as doctoring the record to conceal the latter. The court of appeals upheld the district court's rejection of these contentions and its denial of an injunction.[254]

Controversy also continues over whether certain TRU wastes buried at DOE sites

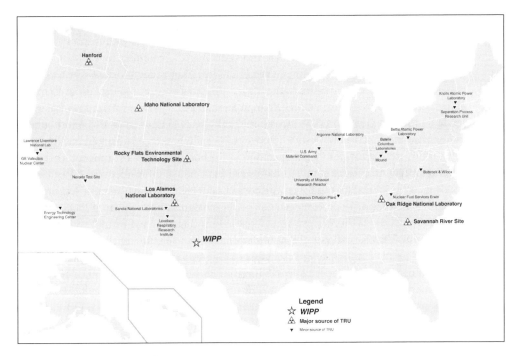

Map 5.1. U.S. TRU waste sites

Source: DOE, *Waste to Be Consolidated at Idaho Site* (2008).

should be unearthed and sent to WIPP. Substantial amounts of defense TRU waste from nuclear weapons production were buried at five major sites in the 1940s and 1950s, and much of it remains buried. Together, the five sites—SRS, Oak Ridge (ORNL), Los Alamos, Hanford, and INL—contain "almost all" the nation's buried TRU.[255] DOE had long considered this waste to be "permanently disposed of," until EPA placed four of the five major sites (all except Los Alamos) on the CERCLA National Priorities List (NPL).[256] DOE began the CERCLA process of assessing the contamination and the risks presented and developing a cleanup plan for the four NPL sites. At ORNL and SRS, which make up about 6 percent and 4 percent by volume, respectively, of all buried TRU, DOE decided to leave the waste in place with a series of engineered barriers to prevent release of radiation.[257] The cleanup decision for the Hanford site, which contains about 53 percent of the buried TRU, is still pending.[258] The second-largest source of the buried TRU waste is INL, which contains 29 percent of the buried waste.[259] Under a 1995 settlement agreement, DOE was required to remove all TRU wastes from INL by 2018.[260] DOE took the position, however, that this did not include removal of TRU wastes buried before 1970, the point at which AEC had decided to stop using burial as the means of disposing of TRU wastes. In 2006, the Idaho federal district court ruled that the settlement agreement covered all TRU wastes, and the Ninth Circuit affirmed in 2008.[261] Despite the court ruling, DOE and Idaho reached an accord whereby DOE committed to remove only about 20 percent of the TRU waste buried at INL.[262] DOE agreed to create engineered barriers at INL to prevent release of the remaining TRU waste into the surrounding environment.[263]

The last site, Los Alamos, which contains about 10 percent of the buried TRU, is not subject to CERCLA, but it is being evaluated under agreements with the State of New Mexico and is regulated under RCRA and state law.[264] According to a 2007 GAO report, the provisions of the agreements regarding Los Alamos are substantively similar to those for sites subject to CERCLA; therefore, the cleanup effort will not differ dramatically when compared to the CERCLA-regulated sites.[265] While no decision has been made on the final disposition of TRU waste buried at Los Alamos, DOE has announced that it believes much of the waste is irretrievable and will therefore remain buried.[266]

Of the two options for final disposition of the buried waste—exhumation and shipment to WIPP, or construction of engineered barriers—DOE prefers the latter, due to cost concerns. DOE estimated that the cost of addressing buried waste at INL, mostly through constructing engineered barriers, would be around $1 billion. If DOE were required to retrieve all the buried TRU, it estimates that the cost would exceed $8 billion.[267] DOE has also estimated that if it were required to exhume all the buried TRU, the total amount requiring geologic disposal would exceed WIPP's statutory cap by sixty thousand cubic meters (i.e., by about 35 percent of WIPP's total capacity).[268] Nevertheless, critics have pressed for exhumation, arguing that it is inconsistent to have different policies for buried TRU and newly generated TRU.[269]

This controversy notwithstanding, WIPP is broadly regarded as functioning successfully. The Obama administration has not changed any of the basic elements of WIPP's operation, but it has participated in the development of a protocol to improve communication between western state governors and DOE on the issue of regional planning for waste shipment.[270] The protocol requires DOE to provide advance notice of TRU shipments to WIPP.[271] In March 2009, $172 million in new funding was allocated for WIPP under the American Recovery and Reinvestment Act.[272] The money was to be used to accelerate preparation and shipment of TRU wastes to WIPP, and to modify the WIPP facility to accommodate different types of waste containers.[273]

Conclusion

The history of WIPP illustrates vividly how a working relationship between the federal government and a host locality and states can evolve in a mixed dynamic of contention and cooperation that ultimately succeeds in satisfying the basic interests of most major stakeholders. The legal and political safeguards of federalism, which New Mexico invoked through litigation, its congressional delegation, the RCRA permit authority that it eventually acquired, and other means, enabled it to gain a voice in decisions about WIPP and to have significant influence over them. The absence of a federal blueprint for the facility set in advance gave the federal government and New Mexico the flexibility to work through issues as they arose and impelled a step-by-step approach to decision making that enhanced the state's ability to influence key features of the facility. The WIPP experience shows that it is both possible and desirable to develop a major nuclear waste facility with the assent of the host locality and state. The background conditions, however, were favorable to a successful resolution. These include Carlsbad's depressed economy, the weakness of the state economy, and New Mexico's generally positive past and then-present experience with federal nuclear activities and facilities within the state.

These favorable conditions may not be replicated in other settings. Other important factors in the eventually successful development of WIPP were the step-by-step evolutionary process by which the facility was developed, and the state's ability to gain leverage in decision making at key stages of the process through successful litigation that challenged DOE actions and through legislation won by its congressional delegation.

The siting and host-state involvement, the negotiated agreement processes, and the adaptive phased decision-making process recommended in Chapter 8 for siting future repositories or consolidated storage facilities reflect the important role played by these mechanisms in facilitating a successful outcome for WIPP.

WIPP is functioning without major incident and is well on its way to safely disposing of the bulk of the nation's TRU waste, relieving thirty-one communities that host DOE facilities with TRU, and the states in which they are located, of the burden of these wastes.[274] WIPP has also helped Carlsbad's economy prosper, earning the town's steadfast support even when New Mexico's state government had, from time to time, been opposed or skeptical.[275] WIPP is viewed locally as the savior of the town of Carlsbad, which even in 2009 boasted an unemployment rate of less than 1 percent and had among New Mexico's highest per capita incomes. It should come as no surprise, then, that Carlsbad is celebrating the cancellation of Yucca—Carlsbad's mayor plans on lobbying Congress to build a HLW repository near his town.[276] The state as a whole has also benefited from WIPP, but less dramatically.

Recently, various New Mexico officials signaled strong interest in expanding WIPP's mission in connection with a visit to WIPP by the members of the Blue Ribbon Commission on America's Nuclear Future in late January 2011. Newly installed Republican governor Susana Martinez stated that the state had had very positive experiences with WIPP and the new URENCO nuclear fuel enrichment facility located nearby in Eunice, New Mexico. She stated that "if the science is sound and the community and state have been consulted and engaged, new nuclear facilities can be built." She added: "If it is done right, New Mexico may be interested in the future." A resolution signed by twenty-eight of the state's forty-two senators that declared support for "the opportunity for other potential missions in southeast New Mexico to adequately address the disposal of defense high-level waste, commercial high-level waste, Greater Than Class C waste and surplus plutonium waste, as well as the interim storage of spent nuclear fuel." New Mexico attorney general Gary King, a Democrat, stated: "WIPP is a great place to permanently dispose of spent fuel, but I'm not sure that permanent disposal is the best idea for spent fuel," noting that salt permanently seals waste. He added that spent fuel should be put into a retrievable facility so that it could possibly be reprocessed in the future. "But I believe southeastern New Mexico would be a good place to put something similar to what we called a monitored retrievable storage facility previously to store waste," he said.[277]

Chapter 6

Yucca Mountain: Blueprint for Failure

The history of the proposed Yucca Mountain, Nevada, repository contrasts sharply with that of WIPP. WIPP was successfully sited and built and has been operating as a repository for TRU waste for more than eleven years. Yucca, after a protracted siting and licensing process, has been abandoned, at least for now, leaving the country to deal with large stockpiles of SNF and HLW that have no other foreseeable or lawful disposal pathway. Since these two projects have emerged as the only significant efforts by the U.S. government to site and operate deep geologic repositories for nuclear wastes, comparisons between them are warranted—and indeed essential. To succeed, future efforts to develop repositories for SNF and HLW must be informed by the lessons learned from these fraternal twins. This chapter examines, in full, the Yucca experience. The lessons learned from the history of WIPP and Yucca are examined in Chapter 8.

After Congress in 1980 restricted the WIPP repository to acceptance of defense wastes only, DOE, Congress, and the nuclear industry began to aggressively promote development of one or more additional federal repositories to take the commercial SNF that WIPP could not. It was concluded that there was an urgent need to develop a repository for defense HLW and SNF, while leaving open the issue of whether these wastes would be co-disposed with civilian wastes in the same facility or facilities. While the possibility remained that WIPP might be opened to defense HLW and SNF in addition to TRU, New Mexico was determined at the time to ward off this eventuality.

The 1982 Nuclear Waste Policy Act strategy—for DOE to select and build in succession two repositories on an accelerated schedule—reflected a broad consensus that the current generation has an imperative ethical obligation to dispose of nuclear wastes permanently so as not to burden future generations with an accumulated waste legacy. The House report on the NWPA bill stated: "As has been emphasized and reiterated over the lifetime of the Federal nuclear program, high level wastes should not be a burden on future generations, and must be disposed of by those who benefited from the energy derived from the nuclear activities which created the wastes."[1] The 1982 NWPA called for DOE to carry out a multisite assessment and selection process leading to construction of two federal repositories, using guidelines for evaluation that would enable DOE to identify and select the best-qualified sites among the candidates. The process was designed to be centralized, technocratic, and meritocratic.

By 1986, DOE had identified as final candidates three potential repository sites in the West, which it originally estimated would cost $60 to $100 million apiece to characterize. The price tag later ballooned to $1 billion per site. In 1987, impatient at the slow progress and high cost of the selection process mandated under the 1982 act, Congress junked it and, through a raw political process, itself designated Yucca as the sole repository.[2] Congress also terminated the search for a second repository in the East, destroying the premise of regional equity underlying enactment of the NWPA.[3]

Congress's selection of Yucca was orchestrated through backroom negotiations that Nevada, then politically weak and vulnerable, has with substantial justification long regarded as deeply unfair. Local or state views played no significant role in the selection of the site. By making the key decisions itself—determining the site location and the wastes that would be stored there—and setting a tight timetable for developing the repository, Congress significantly, and quite deliberately, restricted the scope of Nevada's potential influence over the facility.[4] Nevada's resistance to the Yucca repository was reinforced by its earlier, profoundly negative experience as long-term host to nuclear weapons testing at another federal nuclear facility, the Nuclear Test Site, which forms part of the larger Yucca Mountain site.[5] Moreover, as Nevada itself had no nuclear power plants, most of the waste destined for disposal at Yucca would consist of out-of-state SNF.[6] The state also feared that hosting a nuclear waste repository ninety miles south of Las Vegas would imperil its tourism and gambling industries.[7] These circumstances generated a surge of collective resentment that Nevada had been steamrolled and spawned a campaign of resistance by the state to defeat the Yucca project through all means available. Nevada's unrelenting opposition significantly delayed DOE's efforts to characterize and license the site.

Nevada's political fortunes improved dramatically with Nevada senator Harry Reid's assumption of Senate leadership in January 2007 and the election of President Obama in 2008. The new administration deprived DOE of the funding needed to pursue NRC licensing of the Yucca repository and effectively declared the Yucca project dead. A recent decision by the NRC Atomic Safety and Licensing Board rejecting DOE's motion to withdraw the Yucca license application with prejudice has heartened Yucca proponents. Even if the NRC and the courts were to uphold this ruling, and even if the NRC were later to grant the license and that ruling in turn were upheld by the courts, Congress would still have to fund Yucca's construction. Further turns of the political wheel of fortune—Republican control of Congress, a new president—might resuscitate Yucca from near death. At this juncture, the prospects for Yucca's resurrection are remote, and it is fair to conclude that the elaborate and ambitious NWPA system devised by Congress to force early development of a repository has crashed into bankruptcy, leaving the federal government without any alternative site or statutory authority to develop one.

The 1982 Nuclear Waste Policy Act and Its Implementation by DOE

This section reviews the background of the 1982 NWPA; DOE's implementation of the site selection process that it provides; the problems created by the tight timetable that Congress decreed, compounded by the political resistance in potential host states; and the political calculations and processes that led Congress to amend NWPA in 1987 and ineluctably select Yucca.

Background of NWPA

The Carter administration effectively terminated reprocessing of SNF in the United States, and the ensuing Interagency Review Group (IRG) process and report firmly set the nation on a path whose singular goal was to build, on a sharply accelerated timetable, deep geologic repositories for the most highly radioactive nuclear wastes—namely, SNF and HLW. The 1982 NWPA emerged out of extensive debates and deliberations in Congress over the years 1980–1982. It was strongly shaped by the recommendations expressed in IRG's 1979 draft report, which called for permanent geologic disposal at several repositories located in more than one region of the United States. This plan was designed to reduce the risk and expense of transporting wastes long distances and to provide geographic equity.[8]

IRG conducted technical studies and examined six methods for disposing of HLW and TRU.[9] Of these options, IRG in its final report recommended the use of mined deep geologic repositories for disposal, despite some criticism that the report's earlier draft version did not sufficiently consider alternatives to mined repositories.[10] IRG's recommendation was justified by a scientific consensus, reflected in the report, about the efficacy of deep geologic repositories, and the belief that repositories could be available more quickly than other disposal options.[11] To determine the best place to put the waste, IRG suggested detailed studies "on a variety of geological environments."[12] Moreover, to jump-start the process, it suggested that detailed studies of "specific, potential repository sites," not simply of different media, begin "immediately."[13] IRG envisioned construction of a series of repositories rather than just one.[14] "Ideally," these would be located "in different regions of the country."[15] Finally, in recognition of the extended time it could take to site, build, and open a functioning repository, the report also recommended the construction of interim, away-from-reactor storage facilities.[16] IRG was highly influential, and many of its recommendations became part of the nation's nuclear waste disposal program.[17]

In February 1980, President Carter delivered a message to Congress on radioactive waste management, proposing a comprehensive waste disposal program "consistent with the broad consensus that has evolved from the efforts of the Interagency Review Group on Radioactive Waste Management (IRG) which I established."[18] Carter's plan involved extensive research into different geological media and sites, just as the IRG report had advised.[19] This strategy was later adopted by Congress in NWPA as part of the winnowing process for siting.[20] Carter also recommended interim storage of SNF, preferably at the nuclear power plant sites where it was produced, but called as well for construction of a limited amount of federal, away-from-reactor storage capacity that utilities could use.[21] NWPA incorporated a similar plan, although its federal interim storage program was considerably more limited than that suggested by Carter.[22]

In the 1982 legislation, Congress sought to establish guiding principles and a comprehensive plan and program for evaluating, siting, licensing, constructing, and operating one or more deep geologic repositories for HLW and SNF. In so doing, Congress confronted and addressed a number of key questions: How many repositories would there be? What kinds of nuclear wastes would be disposed of in them? Would the various wastes be co-disposed? How would repositories be financed? How quickly would it be possible to site and build them?[23] How would their safety be assured? Who would define and who would assess that safety? What role would interim storage of nuclear wastes play? How much say would states, tribes, and the public have in the siting and repository development process and decision making? How would conflicts between the federal government and states/tribes be resolved?

Some legislators favored a complex, multiphased siting process with considerable room for state input, like that recommended by IRG and President Carter. Others viewed such a process as an overelaborate waste of time. Congressman Mike McCormack, one of the key opponents of the multiphased approach, lost his seat in 1980; that approach was subsequently adopted.[24] The Ninety-Sixth Congress, however, adjourned in January 1981 before the House and Senate could agree on some key issues, including the extent of state authority in the siting process, and the extent to which consolidated storage should be available as an interim or complementary option to repository disposal.

DOE, states, and interest groups advocated widely divergent answers to the questions involved in designing a comprehensive new approach to dealing with SNF and HLW. The nuclear power industry and DOE wanted maximum expedition and administrative discretion for DOE; they resisted giving states and tribes significant influence over site selection and development.[25] Environmental groups sought to constrain DOE by requiring thoroughgoing scientific selection criteria, articulated through an environmental impact statement.[26] They also sought to exclude any sort of interim storage facility for fear that it would become a de facto permanent repository.[27] This position echoed IRG's strictures regarding the unfairness of burdening future generations. The National Governors Association advocated that states have substantial input to key decisions, although not a veto—a position that reflected its members' diverse perspectives regarding the urgency of disposal, the desire to avoid becoming host to a repository or interim storage facility, and the need for regional balancing.[28] Noting that several potential repository sites were located near Native American lands, tribes urged Congress to assure that they would be able to participate directly in decision making by the federal government about the disposition of nuclear wastes.[29]

In enacting NWPA, Congress sought a compromise among these various interests and considerations, including regional equity, the timetable for repository development, environmental protection, state and tribal rights, and the potential need for interim consolidated storage.[30]

The NWPA Repository-Siting Process

NWPA established a complex, multistage structure and process for siting and constructing repositories, to be implemented on a crash timetable. The act provided for two repositories. The background political understanding was that the first repository would be sited in the West, where DOE had already identified nine candidate sites in six states, and the second in the East, where the bulk of U.S. nuclear waste was generated.[31] At the time of NWPA's passage, approximately 90 percent of SNF and HLW had been produced in states east of the Mississippi River; the regional imbalance in waste generation since then has remained basically unchanged.[32] The processes for selection were the same for both repositories, but they operated on different, staggered timetables. DOE's first responsibility was to develop guidelines on the factors it would consider in recommending a site for a repository,[33] and then use them to identify and winnow candidate sites.

NWPA mandated deep geologic, permanent burial of SNF and HLW on an expedited schedule, as recommended by IRG.[34] IRG had reasoned that this approach was necessary in order not to unfairly burden and constrain subsequent generations: "The responsibility for establishing a waste management program shall not be deferred to future generations. Moreover, the system should not depend on the long-term stability or operation of social or governmental institutions for the security of waste isolation after disposal."[35] NWPA provided that DOE should establish and follow siting guidelines in which "geologic considerations . . . shall be primary criteria for the selection of sites in various geologic media."[36] While site characterization plans and recommendations were to include the "form or packaging" that would be used in connection with waste disposal in the geologic medium, the statute did not mention engineered barriers in conjunction with guidelines for site selection.[37] In addition, NWPA required that SNF placed in a repository be retrievable during the operation of the facility "for any reason pertaining to the public health and safety, or the environment, or for the purpose of permitting the recovery of the economically valuable contents of such spent fuel."[38] The period of retrievability was described as an "appropriate period" and otherwise left undefined.[39] This provision contemplates the possibility of retrieving SNF from a repository in order to deal with problems encountered at the repository or to reprocess emplaced SNF.

NWPA instructed the president, within two years of the passage of the act, to evaluate and determine whether a separate repository was needed for defense HLW. In the event that the president determined that a separate repository was not needed, DOE was instructed to make plans to dispose of defense and civilian wastes in the same repository.[40]

DOE was required to develop, within six months of NWPA's enactment, "general guidelines for the recommendation of sites for repositories," in consultation with the Council on Environmental Quality, EPA, USGS, and state governors, and with the concurrence of NRC.[41] The guidelines were to include such factors as proximity to natural resources, hydrology, seismic activity, proximity to water supplies and populations, geologic diversity among recommended sites, and transportation of SNF and HLW to the site.[42] While NWPA designated geological issues, including factors such as seismic activity, as the most important siting considerations that would qualify or disqualify a site, it also instructed DOE to consider a potential site's transportation requirements and location, including the distance of the site from major population centers, from the existing location of the waste that would be disposed of in the repository, from water supplies, and from natural resources like the National Park System.[43]

DOE was also instructed to produce, within fifteen months of the act's passage, a comprehensive mission plan that included a schedule for compliance with the act; an estimate of costs; a survey of limitations to the department's scientific, engineering, and technical knowledge, and a plan to redress them; its prediction of problems that could occur and strategies for resolving them; its guidelines and methods for evaluating different geological media; and a description of potential sites that it had previously identified.[44]

After issuing the siting guidelines and consulting with governors of affected states, DOE was to nominate at least five sites "suitable for site characterization for selection of the first repository site."[45] Before announcing nomination of the five sites, DOE was required to notify all affected states and tribes of its intent to nominate a site within their lands.[46] For any site nominated, DOE was required to hold public hearings; draft an environmental assessment describing the basis for nomination of the site under its guidelines and the expected impact of site characterization activities; and perform studies of each site based on "available geophysical, geologic, geochemical and hydrologic, and other information." DOE was not to conduct boring, excavation, or other extensive on-site characterization activities unless DOE determined that available information was insufficient.[47] DOE would then evaluate the five nominated sites in order to narrow the pool of five to three sites to undergo extensive and detailed on-site characterization studies.[48] Finally, based on the results of the detailed site characterization studies, DOE was required to select one or more sites from the final three candidates for recommendation to the president.[49]

DOE's selection and recommendation to the president of at least one site for the first repository was to be completed no later than March 31, 1987. Once the president in turn recommended a site to Congress, states or tribes within whose territory the site was located would have the opportunity to submit a notice of disapproval to Congress within sixty days; otherwise, the recommendation would stand as a designation of the site for development of a repository. DOE would thereupon prepare and submit a construction license application to NRC.[50] If a state or tribe issued a notice of disapproval, Congress would have ninety days to pass a joint resolution overriding the state's objection to the site. Otherwise, the notice of disapproval would operate as a veto of the site, and the president would have to recommend another site within a year.[51] Environmental radiation standards would be set by EPA; NRC would determine the repository's compliance with the standards during the licensing process.[52]

Repository Site Selection Timetables

July 7, 1983	DOE to issue proposed site selection guidelines
Jan. 8, 1984	NRC to issue technical guidelines
Apr. 6, 1984	DOE to issue draft mission plan
July 6, 1984	Final mission plan due to Congress (actually issued June 1985)[53]
Jan. 7, 1985	President to decide whether repositories may also be used for defense waste[54]

First Repository

Jan. 1, 1985	Secretary of energy to recommend three sites for the first repository for characterization, Section 112(b)(1)(B)
Mar. 1, 1985	President to approve three sites for characterization for first repository, Section 112(c)(1)
May 1, 1985	Secretary of energy to seek to enter into binding agreement with state authorities, Section 117(c)[55]
(No date)	DOE to recommend one of the three characterized sites to the president as the site for the first repository, Section 114(a)(1)[56]

Mar. 31, 1987	President to recommend to Congress one of the three characterized sites as the site for the first repository, based on DOE's recommendation, Section 114(a)(2)
May 31, 1987	Deadline for host state to submit its notice of disapproval, Section 116(b)(2) (for states), Section 118(b)(2) (for tribes)
(No date)	Deadline for DOE to submit license application to NRC (sixty days after president's recommendation becomes effective), Section 114(b)
Jan. 1, 1989	NRC decision on DOE's application to construct first repository (or within three years of submission, whichever occurs later), Section 114(d)[57]
Jan. 31, 1998	Federal government obliged to begin taking SNF from utilities

Second Repository

July 1, 1989	Secretary of energy to recommend three sites for characterization for the second repository
Sept. 1, 1989	President to approve three sites for characterization for second repository
(No date)	DOE to recommend one of the three characterized sites to the president as the site for the second repository, Section 114(a)(1)
Mar. 31, 1990	President to recommend to Congress the site for the second repository, based on DOE's recommendation, Section 114(a)(2)
May 31, 1990	Deadline for host state to submit a notice of disapproval
(No date)	Deadline for DOE to submit license application to NRC (sixty days after president's recommendation becomes effective), Section 114(b)
Jan. 1, 1992	NRC decision on DOE's application to construct second repository (or within three years of submission, whichever is later), Section 114(d)[58]

Financing for development and construction of repositories would be supplied by a Nuclear Waste Fund to be drawn from fees and taxes on the electricity produced by nuclear power generators.[59] The act provided that "in return for the payment of fees," the secretary of energy would "take title to the high-level radioactive waste or spent nuclear fuel ... upon the request of the generator or owner, [and] dispose of [it] ... not later than January 31, 1998."[60] DOE was to enter into contracts with nuclear power generators that provided for utilities' payment of fees in exchange for government takeover of their SNF; the act instructed NRC to deny or refuse to renew licenses for facilities unless their owners had signed such a contract with DOE or were negotiating a contract in good faith.[61] The House report on the NWPA bill envisaged that the first repository would open by 1995,[62] enabling the federal government to fulfill its undertaking to take charge of SNF by 1998. This expectation proved wildly unrealistic.

NWPA further provided that the Nuclear Waste Fund would be used to provide states and counties proximate to a site with "impact assistance."[63] They would also receive funds in amounts equal to those that they would have been able to raise had they been able to tax the project and property as not owned by the federal government.[64]

The statute capped the total volume of waste that could be disposed of in the first repository at seventy thousand MTHM, with only sixty-three thousand MTHM allocated for commercial SNF.[65] This figure is far less than Yucca Mountain's potential capacity as a repository and also less than needed to meet the nation's actual HLW/SNF disposal needs.[66] The decision to limit Yucca's capacity was deliberate. It was designed to ensure that a second repository would be built in the East, thereby securing regional equity and allaying concerns about the costs and environmental risks of massive east-to-west trans-shipment of SNF.[67] Senator Henry Jackson of Washington stated that "the 70,000 metric ton limit on disposal in the first repository will insure that a second repository is forthcoming. We must not make any one state feel that it is going to be the recipient of all of the Nation's nuclear waste."[68]

NWPA also left open the possibility that WIPP might be selected as a repository for HLW, SNF, or both. The House report (1982) on the bill that became NWPA stated: "The Waste Isolation Pilot project has been authorized for development of a facility for disposal of transuranic wastes, which may eventually be converted to a disposal facility for high level wastes. Such a conversion would presumably subject the facility to requirements for licensing by the Nuclear Regulatory Commission under Section 202 of the Energy Reorganization Act."[69] But the 1980 WIPP Authorization Act had already limited WIPP to disposal of defense waste; without new legislation to remove this limitation, civilian wastes could not be disposed of at WIPP.[70]

MRS Waste Storage Facilities

In addition to the two repositories, NWPA authorized federal development of monitored retrievable storage (MRS) facilities "to accommodate spent nuclear fuel and high-level radioactive waste resulting from civilian nuclear activities."[71] It also provided separately for federal storage for a quite limited amount of SNF (1,900 metric tons), in the event that the owner of a commercial reactor were able to demonstrate hardship because its SNF could not be stored on-site.[72]

The MRS program was envisaged as a complement to, rather than a substitute for, permanent disposal of waste at a repository. A government-run MRS would provide "continuous monitoring, management, and maintenance of such spent fuel and waste for the foreseeable future."[73] While the MRS would be designed "to safely store such spent fuel and waste as long as may be necessary,"[74] it would also "provide for the ready retrieval of such spent fuel and waste for further processing or disposal."[75] NWPA did not require that an MRS be built, only that DOE submit to Congress a study on "the need for and feasibility of" one or more MRS facilities, along with a proposal for the design and construction of such a facility as part of a broader federal program.[76] Congress would then decide whether to authorize construction.[77] While DOE was supposed to provide Congress with alternative MRS sites, the original act did not set out any formal MRS siting process for DOE to follow, nor did it require the preparation of an EIS.[78]

The Role of States and Tribes in Siting

A key feature of the act was a set of provisions defining the role of states and tribes in the process of site selection within their respective territories. Among Congress's NWPA findings was that "state and public participation in the planning and development of repositories is essential in order to promote public confidence in the safety of disposal of such waste and spent fuel."[79] Congress listed among the act's purposes "to define the relationship between the Federal Government and the State governments with respect to the disposal of such waste and spent fuel."[80] The act was also designed to provide for "the participation of an affected Indian tribe in activities proposed to be located on Indian land ... consistent with existing law ... recogniz[ing] Indian tribes as semi-sovereign entities."[81]

NWPA provided for several forms of DOE-state interaction in program formulation, site characterization, and site selection. Section 112(a) of the act required "the Secretary [of Energy], following consultation with ... interested Governors," to develop general guidelines for use in assessing a candidate site's features.[82] The secretary was further directed to consult with governors of affected states when applying the siting guidelines to select candidate sites,[83] as well as to notify a state of the fact that a site within the state had been nominated and the basis for its nomination.[84] In characterizing a selected site, DOE was directed to "consider fully" the affected state's comments and "to the maximum extent practicable and in consultation with the Governor ... conduct site characterization activities."[85] While DOE was not required to consult with affected tribes, it did have to notify a tribe "on whose reservation such a repository is planned to be located."[86] Moreover, once a site was selected, NWPA provided identical means of participation for states and affected tribes. However, unlike states, tribes first had to prove that they were "affected."[87]

The act also provided for public participation in the repository-siting process.[88] Before a site could be recommended to the president, NWPA required that public hearings be held "in the vicinity of such site to inform the residents ... and to receive their comments."[89] It further required DOE to make public environmental assessments of sites recommended for characterization,[90] as well as consultation and cooperation agreements between states or tribes and the federal government.[91]

To help ensure meaningful state involvement in the repository-siting process, NWPA provided for grants from the Nuclear Waste Fund to state governments to sponsor independent studies of DOE's "activities ... for purposes of determining any potential economic, social, public health and safety, and environmental impacts of such repository on the State and its residents."[92] It extended the same grants to affected tribes.[93]

The primary statutory vehicle for DOE-state cooperation and resolution of state and tribal concerns under NWPA was Section 117(c), which provided for negotiation of a formal "consultation and cooperation" (C&C) agreement between a state or tribal governing body and DOE. NWPA provided that "the Secretary [of Energy] shall seek to enter into a binding written agreement" within two months of enactment.[94] The statute did not define "consultation and cooperation"; the agreements negotiated by DOE and the state were expected to specify what this would mean in practice.[95] This provision was similar to the C&C provisions in the 1980 WIPP Authorization Act that resulted in a C&C agreement between New Mexico and DOE in July 1981.[96] Under NWPA, however, no C&C agreements were ever concluded.[97]

The notice-of-disapproval mechanism was another means by which NWPA sought to accommodate the interests of states and tribes that had a candidate repository site within their territory; the 1987 amendments extended this mechanism to MRS sites.[98] The mechanism was not agreed upon until the very end of the legislative process;

amendments to the section were introduced up until the day the NWPA was passed. States with likely sites had advocated a variety of different approaches to securing an effective role in decision making on repository siting.[99] One proposal, by Senator Howard Cannon of Nevada (a potential host state), provided for a process very similar to the disapproval process ultimately included in the legislation. This and other proposals that would have given states some form of veto over repository siting were resisted until the final days of debate,[100] thanks in part to opposition by senators from Washington and Texas, two other potential host states.[101] Because Representative Jim Wright of Texas was the House Speaker, and Representative Tom Foley of Washington was the House majority whip, these two states probably could have blocked any unfavorable siting decision without a state veto provision and were reluctant to let other candidate states have a veto right.[102] The notice-of-disapproval mechanism finally adopted in NWPA was viewed by some members as a sufficiently effective protection for unwilling host states;[103] other members were more skeptical.[104]

Another proposal, by Mississippi, was to add a stringent population criterion that effectively precluded siting in states such as Mississippi with low territory-to-population ratios.[105] While this provision was not adopted, NWPA provided that DOE's siting guidelines were to include "population factors that will disqualify any site from development as a repository if any surface facility of such repository would be located (1) in a highly populated area; or (2) adjacent to an area 1 mile by 1 mile having a population of not less than 1,000 individuals."[106] The final act also required that the guidelines consider "the proximity to sites where high-level radioactive waste and spent nuclear fuel is generated or temporarily stored and the transportation and safety factors involved in moving such waste to a repository," although it did not require the guidelines to limit sites to states where wastes were generated.[107]

DOE's Implementation of the NWPA Siting Process

Notwithstanding NRC's Waste Confidence Decision of 1984, which concluded that permanent disposal of SNF was well enough in hand to justify operating new nuclear power plants, optimism about developing repositories faded in the five years after NWPA's passage.[108] Key events of those years included the winnowing of potential first-repository host states in the West amid increasing controversy, the abrupt termination of DOE's efforts to site a second repository in the face of massive resistance in the East, and the eventual scuttling by Congress of a proposed MRS facility in Tennessee that had been strongly resisted by the state. Also, President Reagan determined in 1985 that defense HLW would not have to be disposed of in a separate repository from that for commercial SNF.[109] Co-disposal of these wastes could not take place at WIPP, however, unless Congress were to amend provisions of the 1980 WIPP Authorization Act that precluded disposal of civilian wastes at the facility.

DOE's efforts to site a first and second repository and an MRS facility elicited strong resistance. By 1986, twenty-five lawsuits had been filed against DOE by states, local communities, environmental groups, and others challenging DOE's implementation of NWPA.[110] While generally unsuccessful, the litigation reflected widespread opposition to the siting program by potential host states.

Pitfalls in Siting the First Repository

Following enactment of NWPA, DOE initiated the first round of siting, which focused on identifying a repository in the West and Gulf South, by designating nine sites in six western and southern states for evaluation.[111] The nine sites were Deaf Smith County and Swisher County in Texas, Davis Canyon and Lavender Canyon in Utah, Cypress Creek Dome and Richton Dome in Mississippi, the Vacherie Dome site in Louisiana, the Hanford site in Washington, and Yucca Mountain in Nevada.[112] These nine sites had been identified through an ERDA program begun in 1975 to identify potential sites for disposal of commercial SNF. This program, known as the National Waste Terminal Storage Program, aimed to survey rock formations in thirty-six states in order to identify suitable sites for six pilot repositories, to be constructed by the year 2000. State opposition in some key states underlain by promising geologic formations, such as Michigan, had led DOE by 1980 to focus solely on Mississippi, Louisiana, Texas, Utah, Nevada, and Washington.[113]

In Utah, Texas, Washington, and Mississippi, there was local community support for DOE to investigate the sites it had previously identified, but support at the state level was tepid at best.[114] Eventually, political pressure against siting mounted from areas of these states that anticipated no direct economic benefit from the construction and operation of a repository.[115] State officials took steps to halt the siting process. The governor of Utah imposed a moratorium on site investigations in the state, enforced by denying requests for rights-of-way across state lands and waters.[116] The governor of Mississippi placed a moratorium on site investigations until the legislature enacted a complex permitting procedure.[117] The Texas legislature passed a bill that would require a permit before an exploratory shaft could be drilled.[118] The Washington legislature, which believed that there were severe geological problems with using the Hanford site for a repository, required independent verification of a federal finding of site suitability, which it believed would make the unsuitability of Hanford's geology clear.[119] Texas sued DOE, challenging the inclusion of two Texas sites among the nine sites under consideration for selection. The Fifth Circuit dismissed the Texas challenge in *Texas v. Department of Energy*, holding that DOE's identification of potential repository sites for was not a "final decision" reviewable by the courts.[120]

In 1984, before the Yucca site had been chosen as one of the three sites for detailed characterization, Nevada requested approximately $2 million for hydrologic and geologic study of the site from DOE under NWPA Section 116(c)(1)(A). DOE denied the request on the ground that the NWPA did not authorize funding until after selection of the final three sites for detailed characterization. It also asserted that the tests Nevada sought to conduct with the funding would be duplicative of data already collected, invoking DOE guidelines which stated that "duplication of data collection efforts and associated activities should be minimized to the maximum extent practicable and avoided if at all possible."[121] Nevada sued.

In December 1985, a Ninth Circuit Court victory for Nevada in *Nevada ex rel. Loux v. Herrington (Nevada v. Herrington* [I]) secured funding for the state to conduct independent studies of the Yucca site and peer reviews of DOE technical findings.[122] The court held that Nevada was entitled to funding for technical study and review activities before selection of the three finalist candidates because the information the state would obtain would be essential to its ability to provide well-supported reasons for noticing disapproval if Yucca were ultimately selected.[123] The court rejected DOE's invocation of its funding guidelines, stating that applying the guidelines to deny funding to Nevada would restrict independent collection of data and thereby "violate the statutory finding that state participation and oversight of DOE is 'essential in order to promote public confidence in the safety of disposal of [nuclear] waste.'"[124]

Nevada's court victory assured dedicated support from the federal Nuclear Waste Fund for the Nevada Commission on Nuclear Projects (NCNP), which had been established by state law earlier that year.[125] NCNP generated ammunition for Nevada to fire at DOE in the form of public opinion polls, technical studies of potential repository impacts (environmental, economic, and health related), and peer reviews critical of DOE's own studies of these impacts.[126] Wisconsin, Mississippi, and Utah also benefited from the 1985 *Nevada v. Herrington* ruling and later joined a suit that sought but failed to compel DOE to defray the expenses of litigation against the federal government under NWPA with monies from the Nuclear Waste Fund.[127]

Other states, tribes, environmental organizations, and private associations initiated dozens of lawsuits challenging DOE's implementation of the 1982 NWPA siting process. According to a 1987 GAO report, the lawsuits "generally involve[d] legal challenges to procedures DOE used to develop its siting guidelines, the contents of the siting guidelines, the decision by DOE to postpone site-specific activities, and the recommendation of first repository sites for detailed testing."[128] Another GAO report that year warned: "Representatives of the states and tribes involved in the first repository program say that if the program's credibility does not improve, they will continue to initiate lawsuits and can be expected to exercise their right to disapprove of the final site selection, forcing the courts, and perhaps ultimately the Congress, to judge whether DOE has adequately ensured the safe disposal of nuclear waste."[129]

Meanwhile, DOE prepared detailed environmental assessments for each of the nine sites it had designated for further evaluation.[130] Based on the siting guidelines it had issued in December 1984,[131] DOE then tentatively nominated for further consideration five sites in a variety of geologic media: Davis Canyon, Utah (salt bed); Richton, Mississippi (salt dome); Hanford, Washington (basalt); Deaf Smith, Texas (salt dome); and Yucca Mountain, Nevada (tuff, that is, volcanic rock). At the same time, it identified three of these, Hanford, Deaf Smith, and Yucca, as the sites it would likely recommend for characterization.[132]

In May 1986, DOE officially nominated five sites for consideration and simultaneously announced, based on ranking criteria it had developed, the three sites it would recommend for site-specific characterization. The three sites were, in order of preference, Yucca Mountain in Nevada, Hanford Reservation in Washington, and Deaf Smith County in Texas.[133] DOE had originally estimated that characterization would cost from $60 to $100 million per site, but by the time DOE was ready to undertake characterization, its revised estimate was $1 billion per site.[134]

Critics sharply questioned the validity of the ranking process by which DOE had selected the final three candidates. The rankings of the five sites chosen in the previous round had been relatively close on most factors.[135] But Hanford ranked last among the five candidate sites and posed widely recognized geological problems, including the technical difficulty of drilling shafts in basalt, a "complex geohydrologic regime," and proximity to the Columbia River;[136] it was nonetheless included in the final three.[137] Local support for a repository at Hanford may well have been a factor.[138] Another factor may have been a desire on DOE's part to have candidate sites in a variety of geologic media. The five sites included three in salt formations. Including Hanford in the final three allowed DOE to select one site in each of three different media (basalt at Hanford, salt at Deaf Smith County, and volcanic rock at Yucca Mountain).[139]

After DOE's final decision was criticized, Ben Rusche of the DOE Office of Civilian Radioactive Waste Management asked the NAS Board on Radioactive Waste Management, formed in 1968 to advise AEC on radioactive waste management, to review and comment on the methodology DOE had employed.[140] The board, which had reviewed and criticized a draft version of the methodology a year earlier, reported that the methodology used to support the final decision was sound; however, it did so without examining either the data DOE had used or the relationship between those data and the final rank order.[141] As a result, its review did not quell critics of DOE's ranking processes.[142]

As the state with the top-ranked site among the three final candidates, Nevada was particularly critical of the site-ranking process.[143] Nevada subsequently (and unsuccessfully) challenged DOE's rankings in litigation.[144] Contemporaneous public opinion surveys of Texans revealed widespread beliefs that a repository at the site in the Texas Panhandle would risk contaminating the aquifer that makes the region agriculturally productive, and that the ranking process demonstrated DOE's indifference to this risk.[145] Nonetheless, on May 28, 1986, the secretary of energy formally recommended to the president that the three sites be further characterized; the president issued his approval the same day.[146]

DOE's subsequent efforts to move forward with selecting a repository site from among the final three candidate sites was clouded by highly publicized events in 1986 that heightened public awareness of and fears about nuclear risks. These included the reactor meltdown at Chernobyl in the Soviet Union, which, according to a DOE official, had a "monumental effect" on DOE's relations with the public on nuclear safety issues.[147] Also in 1986, as a result both of public pressure and a Freedom of Information Act request, DOE released thousands of pages of documents that revealed alarming contamination of the environment around the Hanford Reservation.[148] These revelations created fresh distrust of DOE,[149] as did an audit report by NRC criticizing DOE's quality assurance program for site characterization activities at the three candidate sites.[150] In April 1986, after quality assurance controls had revealed failings in the characterization process, DOE issued stop-work orders to contractors at Yucca.[151] The press reported on a memorandum, authored by scientists at the U.S. Geological Survey, which charged that "the department was using the stop-work orders in attempts to prevent the discovery of problems that would doom the repository."[152] This incident helped fuel Nevada officials' unwillingness to give any appearance of cooperation with DOE; the state declined DOE's invitations, extended in 1986 and 1988, to negotiate a C&C agreement.[153]

In Texas, encouraged by watchdog organizations such as the Southwest Research and Information Center of New Mexico, state officials united to oppose locating a repository site in the Texas Panhandle.[154] Opposition in the Panhandle, where agriculture depended on the Ogallala Aquifer, was based on farmers' concerns that consumers, fearing radioactive contamination, would no longer buy the Panhandle's agricultural products.[155] Political actors in Texas became reasonably confident that Texans' widespread concerns about the Ogallala Aquifer, coupled with the state's large and powerful congressional delegation, could deflect a repository.[156]

The Aborted Second-Repository Program

As the first repository siting process moved forward, DOE also began scouting sites for a second repository. In 1985, it listed 235 potential sites in eastern and midwestern states.[157] The schedule for selecting a site for the second repository was less compressed than that for the first repository; DOE's recommendation to the president of sites for the second repository was not due until July 1989.[158] After further investigations of crystalline rock formations in seventeen states, in January 1986 DOE announced its tentative choice of twelve potential repository sites in seven states for further screening: the states where the sites were located were Minnesota, Wisconsin, Maine, New Hampshire, Virginia, North Carolina, and Georgia.[159]

The announcement of candidate sites for the second repository provoked a political furor in the East. The two New England sites chosen were near highly populated areas and, in the case of the site in Maine, near a major drinking-water source.[160] The selections provoked litigation against DOE. In 1986, Maine and New Hampshire unsuccessfully petitioned the Court of Appeals for the First Circuit for review of DOE's refusal to extend the ninety-day comment period on the department's selection of the twelve potential second-repository sites.[161] The court dismissed the case, saying the decision was not a sufficiently "final" agency action to warrant interlocutory judicial review.[162] When DOE later indefinitely suspended the second-repository program altogether, the case became moot.[163] The same year, the Lakes Environmental Association, a group of local property owners in Maine concerned about the identification of the Sebago Lake area as a potentially acceptable site for the second repository, challenged the screening procedures for the second repository.[164]

The signal sent by the lawsuits was clear: the search for a site for the second repository would be an uphill battle, fought against an already incensed group of states and citizens. Opposition was compounded by DOE's handling of the process. For example, in Wisconsin DOE was blamed for "noncooperation," "thwarting good faith efforts," and "bungling" requests for technical information.[165]

At the same time, election politics came into play. Key Senate races were being held in four of the states with candidate sites, and Republicans were in danger of losing control of the U.S. Senate.[166] The looming presidential election put a premium on New Hampshire's political support. The Republican governor of New Hampshire, John Sununu, argued that the effort to site the second repository was generating huge opposition that was dangerous to the whole nuclear power industry. Sununu worked to mobilize the nuclear power industry to kill the second round of repository siting altogether.[167]

In May 1986, DOE succumbed to the pressure and called a halt to the second repository siting process, but it explicitly premised its decision on the grounds that lowered projections for spent-fuel generation made the second repository unnecessary.[168] Only a month earlier, however, DOE had testified before Congress that a second repository was needed.[169] Almost a year after the decision, the secretary of energy stated publicly that politics had played no role in his decision to defer indefinitely the second repository; a contemporaneous internal DOE memorandum, however, identified "immediate political relief" as an important benefit of the decision.[170]

DOE's cancellation of the second-repository program was deeply unpopular in the West. Key western members of Congress who had supported NWPA denounced DOE's suspension of the program as contrary to the geographic balance built into the act.[171] Western state outrage did not save the second repository program, however. When Congress amended NWPA the next year, it eliminated the timetable for development of a second repository, requiring only that DOE report back to the president and to Congress between January 1, 2007, and January 1, 2010, "on the need for a second repository."[172]

Defeat for Monitored Retrievable Storage

In addition to providing for one or more repositories for SNF and HLW, NWPA called for DOE to evaluate "the need for and feasibility of," and potentially to construct, an MRS facility for SNF.[173] In accordance with the June 1985 deadline in NWPA,[174] DOE reported to Congress that an MRS facility was both warranted and feasible.[175] The department then identified three potential MRS sites for characterization, all of them in Tennessee. The "preferred" site was that of the cancelled Clinch River Breeder Reactor; the two alternative candidate locations were DOE's Oak Ridge Reservation and the site of the cancelled Tennessee Valley Authority Hartsville Nuclear Plant.[176] DOE informed Tennessee governor Lamar Alexander one day before publishing this information in the *Federal Register*.[177] DOE followed up the announcement with offers of grants to assist Tennessee and affected local communities in studying the issues posed by an MRS facility.[178]

Propelled by concern that an MRS for HLW, SNF, or both would become a de facto permanent repository, Tennessee and two of the three site localities firmly opposed DOE's actions.[179] The third locality, Oak Ridge, which was already home to extensive DOE nuclear facilities, indicated openness to hosting an MRS. City and county governments near the Oak Ridge site signaled that DOE's proposal in its current form was unacceptable but that agreement could be reached if certain conditions were met.[180] These included "impact assistance" grants and additional monies equal to the amount Tennessee would receive if it could tax the MRS facility as a private facility; arrangements for local government to share monitoring and enforcement of safety measures with federal authorities; linking construction of the MRS facility to progress on a geologic repository so the MRS facility would not become a de facto substitute for the latter; and accelerating cleanup of existing DOE facilities at Oak Ridge.[181] Further negotiations about the Oak Ridge site were shut down by a state-level decision to oppose any MRS facility within the state.[182] Tennessee also filed a lawsuit against DOE, arguing that NWPA required DOE to initiate consultation and cooperation with a state before, rather than after, designating an MRS site within the state. The Sixth Circuit ruled in favor of DOE.[183] But when Congress amended NWPA in 1987, pressure from the Tennessee delegation, including Senator Al Gore, resulted in enactment of a provision stating that DOE's proposal of MRS sites in Tennessee "is annulled and revoked."[184] It further provided that, in the event that DOE suggested MRS sites in the future, DOE could "make no presumption or preference to such [Tennessee] sites by reason of their previous selection."[185] The 1987 amendments to NWPA also contained other provisions—including one that precluded DOE from characterizing MRS sites until after the secretary of energy had recommended a site for development as a repository[186]—designed to substantially, if not indefinitely, delay the siting of an MRS facility.

Crisis in the NWPA Siting Scheme

By late 1986, it had become clear that the ambitious plan for disposing of SNF and HLW adopted by Congress in 1982 was in very serious trouble.[187] While DOE had succeeded in selecting the final three candidate sites for the first repository, its site selection process was widely criticized, and all three states in which the sites were located strongly opposed a repository within their borders; Texas and Nevada had brought litigation to further that opposition.[188] DOE's estimates of the cost of characterizing the final three candidates had ballooned from $60 to $100 million per site to up to $1 billion per site, as noted earlier.[189] The search for a second repository in the East had been abandoned, which exacerbated the sense of grievance on the part of the western states, as one of them would now be required to handle virtually all the nation's most dangerous nuclear waste, the bulk of it in the form of SNF from nuclear power plants located in the East and Midwest.[190] By the fall of 1986, the NWPA repository-siting effort was in a state of crisis.

As implementation of NWPA faltered, GAO and Congress's Office of Technology Assessment (OTA) identified weaknesses in the 1982 legislative scheme and its implementation by DOE.[191] The most notable were an unrealistically ambitious schedule for investigation and selection of potential repository sites, and inadequate involvement of states and tribes in repository-siting decision making.[192] Other weaknesses in the statutory scheme noted by commentators included poor incentives for states to engage in the repository-siting process, inadequate opportunities for public involvement, and the choice of DOE to lead the program.[193]

Unrealistic Statutory Deadlines

A critical factor impeding NWPA's success, according to GAO, OTA, NAS committees, and others, was the wholly unrealistic timetable dictated for the repository-siting process, which undermined the credibility of the entire scheme. Given that it took almost twenty-five years after the Los Medanos site was selected to take WIPP from preliminary site investigation to operation as a repository, it is evident (at least in retrospect) that the schedule mandated by NWPA was far too ambitious for even a single repository, much less two. The sixteen-year time frame envisioned for completion of the entire process—identifying multiple potential sites for a first and second repository, investigating them, selecting the two sites (after a complex, multistaged winnowing process), further characterizing each site in detail, and licensing, constructing, and beginning operation of at least one repository to fulfill the federal government's obligation to take SNF—was hopelessly unrealistic. This highly compressed timetable, which was a central feature of the NWPA scheme to bring online as quickly as possible repositories for the nation's most dangerous nuclear wastes, worked instead to undermine public confidence in the supposedly meritocratic site selection process.

It was recognized at the time of NWPA's passage that the statutory deadlines were too short to enable DOE to conduct full investigations into brand-new candidate sites.[194] This led DOE at the very outset of its search for a first repository to focus exclusively on the nine sites it had previously studied; a broader search would simply have taken too long.[195] States and tribes asserted, with substantial justification, that this expedient undermined the credibility, fairness, and objectivity of the selection process.[196] They also charged that because the nine sites had been identified and studied prior to enactment of NWPA, those sites—which formed the base set for all further decision making on selection of a site for the first repository—had been chosen without the state/tribe consultation and cooperation process required under the 1982 law, thus undermining the collaborative site selection process envisioned by Congress.[197]

Further, critics charged that the siting guidelines issued by DOE to meet the tight six-month statutory deadline (from January 1983, when NWPA was signed into law, to July 1983, when the guidelines were due) were technically flawed, and that the siting criteria they established were not suitable for assessing nine geologically diverse potential sites in six states. Although considerable research had been done on salt-dome formations, research on the suitability of basalt (Hanford) and tuff (Yucca) for nuclear waste disposal was a relatively new enterprise.[198] When commenters had criticized the technical soundness of its draft guidelines, DOE had responded by taking a few additional months before issuing the final version—not primarily to revise the siting guidelines, however, but to reply more thoroughly to adverse public comments, presumably in anticipation of litigation.[199] Further, the statutory schedule had provided only eighteen months within which DOE was to undertake a detailed characterization, assessment, and comparison of the final three candidate sites. Critics charged that this period was inadequate for conducting a scientifically credible and thorough comparison of the candidates.[200]

These and other factors caused the Office of Technology Assessment (OTA) to conclude in 1985 that the tight NWPA timetable, rather than ensuring speedy development of a functioning repository, had undercut a critical element for success of the siting program—namely, public confidence in it—thus hobbling chances of achieving the goal of developing two repositories within a reasonable time.[201] The NAS Panel on Social and Economic Aspects of Radioactive Waste Management similarly found that public confidence in the ability of DOE and the federal government to deal with nuclear issues was extremely low.[202]

OTA proposed that Congress adopt a less-compressed schedule that would enable DOE to "go beyond the minimum requirements of NWPA" in engaging the public.[203] This schedule, OTA argued, would balance commitments to utilities and nearby communities who wanted SNF to be promptly removed from local reactors with the need for a stable and credible nuclear waste policy supported by adequate scientific evidence demonstrating that public health and safety would be protected. Further, OTA rejected the view that "geologic disposal is a relatively straightforward technical enterprise" and asserted that a more measured decision-making timetable was needed to allow for the degree of public engagement and discussion essential for "securing and sustaining the public's and utilities' confidence."[204] To compensate for extending the timetable, OTA proposed characterization of four sites for each of the two repositories with recommendation of two of the four for licensing, thus ensuring that if one site became unusable, there would be a backup site ready. It also recommended that DOE concentrate its efforts entirely on the repositories and devote attention to MRS and other alternative disposal facilities "only as a last resort, if major problems call into question the feasibility of geologic disposal."[205]

As an additional step, OTA proposed that DOE perform a phased run-up to a fully licensed and operational facility. The first or "demonstration" phase would involve emplacement of small amounts of waste to "show that a licensable disposal technology exists"; the second or "operational" phase would be to "dispose of radioactive waste on a large scale."[206] This phased approach—a process that bears some striking similarities to the evolutionary process that eventually produced WIPP—would, according to OTA, help secure public confidence in the siting program by showing that DOE could meet its scheduled goals.[207]

The Role of States and Tribes in DOE's Siting Processes

Another important source of the siting program's difficulties involved the role of potential host states and tribes and their ability both to influence the site selection process and to opt out of it.

In enacting NWPA, Congress had rejected giving states and tribes an outright veto over siting of a facility within the state. Instead, Congress provided a "checks and balances" mechanism in Section 115 of NWPA to accommodate state and tribal interests by empowering states and tribes to issue a notice disapproving final selection of a repository site within their borders, subject to override by a joint resolution of Congress.[208] Congressional supporters of this approach claimed that it had broad support and presented it as a bona fide opportunity for a host state to challenge DOE's selection.[209] However, its practical political value as a safeguard for a state targeted to host a repository proved limited and was openly questioned at the time by some legislators. By the time a site was recommended by DOE and selected by the president, considerable resources would have been invested. If an unwilling potential host state were to succeed in blocking the decision, the site selection process would have to be restarted, requiring further investments and delays, and necessarily targeting other states as potential (and potentially unwilling) hosts. Any state that might be a repository site candidate, or states storing large amounts of HLW or SNF, would have compelling incentives to vote to override the designated host state's disapproval and uphold the president's selection. As Senator Sonny Montgomery of Mississippi put it bluntly in the floor debate on the NWPA legislation, a state chosen to host a repository faced a "tough time" avoiding override of its notice of disapproval, because "49 other states are going to be ganged up against it."[210] This, indeed, was the experience of Nevada years later, when it issued a notice of disapproval after the president's designation of Yucca for characterization.

Congress also sought to protect the interests of states through a series of federal-state consultation requirements, as well as requirements that DOE fund state studies and independent oversight of DOE's siting activities. NWPA section 117 provided that, within sixty days of either the president's approval of a site for characterization or on request of a potential host state, DOE shall "consult and cooperate" with the state "in an effort to resolve [concerns] regarding the public health and safety, environmental, and economic impacts of any such repository."[211] Further, DOE was obliged to "seek to enter" into a binding written C&C agreement with the state; the C&C agreement was to specify, among other matters, the procedures by which DOE would answer state or tribal requests for information, respond to suggestions and comments, resolve concerns, and implement the repository plan.[212] These would include the procedures by which DOE would keep the affected state informed about "characterization, siting, development, design, licensing, construction, operation, regulation, or decommissioning of such repository."[213] The agreement must also provide how DOE would "resolve the concerns of such State and any affected Indian tribe regarding the public health and safety, environmental, and economic impacts of any such repository."[214] NWPA further required that the C&C agreement be finalized within six months either of the approval of a site for characterization or of a request from a state or tribe for an agreement.[215] If it were not, DOE was required to explain in writing to Congress why the agreement had not been finalized, and to report on the current status of negotiations.[216]

Notwithstanding all these detailed requirements, the statute fails to require DOE to actually conclude C&C agreements with states and fails to make any provision as to what should constitute cooperation and consultation between DOE and an affected state or tribe in the event that a C&C agreement is not concluded.

As noted in *Tennessee v. Herrington*, "the Act's consultation and cooperation requirements present a confused and complicated set of problems."[217] Both sides might in theory gain from a C&C agreement. A state might gain significant concessions, a concrete and enforceable set of its rights, and greater influence than it would otherwise have had by virtue of the NWPA provisions requiring DOE-state consultation and cooperation alone. In the absence of a written agreement with the state, DOE would have the authority unilaterally to interpret what its consultation and cooperation obligations were to the state under relevant provisions of NWPA, subject to only limited judicial review. A C&C agreement could clarify many of the issues that the statute failed to resolve and could also provide a state with a more advantageous dispute settlement procedure than judicial review of DOE's actions might afford. From DOE's perspective, a C&C agreement could be an important tool in securing ongoing dialogue with a state and the prospect of its cooperation, as well as a potential means to avoid an ever-escalating battle waged by a state to kill a potential site.

However, with the exception of Washington, no state with a designated candidate site—Texas, Nevada, Louisiana, and Mississippi—attempted to enter into negotiations with DOE to reach a C&C agreement.[218] These states may have calculated that they would not be able to gain meaningful concessions from DOE, or they may simply have feared that entering into negotiations with DOE would signal to their constituents a willingness to agree to host a repository, inviting challenges by political rivals.[219] Only Washington State and three tribes undertook serious negotiations on a C&C agreement with DOE, but none of them concluded such an agreement.[220] Washington broke off negotiations in 1985 when the state demanded that DOE accept complete liability for any nuclear accident, and DOE claimed it could not, based on the Price-Anderson Act, agree to such a demand.

The explanation proffered by DOE for its failure to conclude any C&C agreements was that negotiations raised complicated issues beyond DOE's authority, such as the amount of federal liability in the case of an accident, as noted. Additionally, states and tribes may not have wanted to risk signaling their approval of siting a repository within their jurisdictions by signing an agreement.[221]

The failure to conclude any C&C agreements left DOE to interpret and implement the statutory arrangements on its own. According to a 1987 GAO report, states felt that DOE had implemented the NWPA consultation requirements in a minimalist and grudging fashion, limited to "keeping the statutorily affected parties informed of the program developments and decisions and allowing them to comment on draft documents."[222] GAO reported a Texas official's view that DOE considered consultation and cooperation to mean "providing selected information, of DOE's choosing, at a time when it felt prepared to present the information."[223] States charged that DOE had excluded them from important aspects of decision making, or had brought them into the process too late for their input to be meaningful. Problems cited by GAO included DOE's failure to share initial working drafts of the repository-siting guidelines, failure to involve affected states or tribes in the development of environmental assessments, and failure to seek state or tribal input prior to preparing its draft (and later, its revised) site-ranking methodologies.[224] Additionally, states claimed that DOE had given them too little time to review complex technical documents. For example, states asserted that the ninety-day comment period for environmental assessments of nine sites, comprising more than ten thousand pages, was too short, and that extension requests should have been granted by DOE.[225] Other problems cited included tardy or inadequate responses to state requests for information, and lack of notice of significant technical or informational meetings.[226]

In *Tennessee v. Herrington,* Tennessee complained that it had not been informed of, much less consulted on, DOE's decision to consider Tennessee as a host for an MRS site. The Department of Energy contended that NWPA did not require DOE to consult and cooperate with a host state to resolve the state's concerns until after Congress had authorized construction of an MRS, which Congress had not done. The court deferred to DOE's interpretation of NWPA and rejected Tennessee's argument.[227]

Officials from Texas, Nevada, and Washington, when interviewed by GAO, asserted that the siting process for the first repository was fatally flawed and that DOE should start over in order to produce a technically sound, objective, and open decision.[228] In response, DOE defended its efforts to involve states and tribes in the repository-siting program, detailing actions that it had undertaken to consult with, and gain early input from, affected parties on many elements of the program. DOE suggested that states' dissatisfaction with the siting process reflected factors other than the adequacy of consultation and cooperation, including the NIMBY syndrome, heated political controversy over nuclear waste disposal generally, and public perceptions that such waste could not be safely managed.[229]

State and Public Participation in Siting

NWPA states that "public participation in the planning and development of repositories is essential in order to promote public confidence in the safety of disposal of such waste and spent fuel."[230] To further this objective, and to help bridge the gulf between the public's views of the risks of nuclear waste and its disposal and the views of government experts,[231] NWPA requires DOE to prepare and make publicly available for comment various documents, including its environmental assessments of potential sites, its nomination of a site for a permanent repository, and any environmental assessments that it prepares on modification or expansion of federal facilities for the storage of spent nuclear fuel;[232] the site characterization plans for the three proposed candidate sites;[233] and its mission plan for implementation of NWPA.[234] The statute further requires DOE to hold public hearings in the vicinity of any potential repository site before it is: (1) nominated from among the initial nine sites for investigation in the round of five sites; (2) selected from among the five sites for further investigation in the round of three sites; and (3) recommended from among three sites to the president for construction.[235] Further, NWPA provides that C&C agreements with states shall make provision for public notification of procedures specified in the C&C agreement.[236]

DOE's implementation of the public participation requirements of NWPA was widely viewed as deficient, undermining the department's credibility and that of the siting program. In its 1985 report on DOE's implementation of NWPA, the Office of Technology Assessment (OTA) concurred with the conclusion reached in July 1980 by key NWPA stakeholders participating in workshops convened by the Keystone Center regarding DOE's approach to public involvement generally: "Current Federal plans for obtaining public participation need substantial improvements" in order to enhance DOE's credibility.[237] OTA suggested that DOE address the issue by including explicit plans for public involvement in its mission plan for implementation of NWPA.[238] DOE issued a revised mission plan in 1988, but by that time opposition to its repository-siting efforts had become widespread.

Weaknesses in DOE's institutional capacities and competence also helped undermine the siting process. DOE was, at the beginning of the 1980s, both relatively new and embattled. It had been strongly criticized for lack of headquarters program management capacity.[239] President Reagan attempted unsuccessfully to abolish DOE and reassign its responsibilities to the Department of Commerce and the Department of the Interior; from 1981 to 1982, however, Reagan did succeed in culling much of its staff.[240] There was widespread skepticism in Washington, D.C., regarding DOE's ability to implement NWPA responsibly; fears were expressed that the department would cater to the interests of the nuclear utilities and seek to develop a repository as quickly as possible, without exercising sufficient care in the siting process.[241] The staff of DOE's new Office of Civilian Radioactive Waste Management (OCRWM) lacked experience. In 1984, the Advisory Panel on Alternative Means of Financing and Managing Radioactive Waste Facilities, appointed by DOE, concluded that a host of failings—chiefly, lack of experience with intergovernmental relations and lack of public credibility—made DOE a second-best choice for the tasks assigned to it by NWPA.[242] The advisory panel concluded that a semi-independent federal government corporation would be preferable.[243] DOE responded to the criticism in a 1985 report; its major contention was that most of the problems identified by the panel were a result of the lack of a clear mandate. DOE claimed that after the enactment of NWPA, its credibility and management improved substantially. DOE cautioned that another change in its mandate would take away from that credibility.[244]

DOE's critics invoked, as evidence of the department's incompetence, the dramatic escalation in DOE estimates of the cost of characterizing the final three candidate sites, from $60 million per site in 1982 to $1 billion per site in 1987.[245] The latter sum is the equivalent of about $2 billion in 2010 dollars.[246] John Bartlett, director of OCRWM, attributed the dramatic rise in cost to the need for "detailed evaluation of the technical realities of meeting the EPA standards and NRC regulations."[247] EPA's 1985 standards were, he asserted, much stricter than anticipated; as a result, DOE had determined that a great deal more work needed to be done than had been accounted for in the original estimate.

The 1987 Nuclear Waste Policy Amendments Act and Resistance to a Yucca Repository

The 1987 NWPA Amendments

In light of the delays in meeting the NWPA siting schedule, escalation in the costs of detailed characterization of the final three candidate sites, mounting state and local opposition to repositories, growing SNF inventories, and the looming 1998 deadline for federal assumption of responsibility for SNF, key members of Congress became convinced that NWPA needed a drastic overhaul. The Nuclear Waste Policy Amendments Act, enacted December 22, 1987, short-circuited the process for technocratic evaluation of multiple candidate sites by DOE and directly designated Yucca Mountain in Nevada as the sole repository site to be evaluated.[248] The 1987 amendments instructed DOE not to conduct site-specific work on a second repository and to terminate its research on granite (i.e., northeastern) sites, unless and until Congress specifically appropriated funds for these activities.[249] No sooner than three decades hence, DOE was to report to Congress on the need for a second repository.[250] Further, DOE was precluded from siting an MRS facility until a permanent repository site had been approved for development by the president.[251]

Congress's decision to abort the repository site selection process that it had established in 1982 and instead to choose Yucca had several plausible justifications. There was sharply mounting resistance, both in the West and in the East, to DOE's implementation of the phased "beauty contest" siting scheme. The multisite evaluation and selection process was proving slow and did not appear likely to produce even a first repository site, much less a licensed repository, any time soon. Characterization of multiple sites would be expensive and time consuming. States and localities with power plants at which SNF was steadily accumulating were agitating for the waste to be removed. Failure to deliver on the federal government's undertaking to assume responsibility for SNF could jeopardize the future of nuclear power.[252]

Senator J. Bennett Johnston of Louisiana, the powerful chair of the Senate Energy and Natural Resources Committee, was influential in engineering Congress's designation of Yucca Mountain as the sole repository site. Johnston was concerned that a repository could be sited in the salt domes of his home state of Louisiana.[253] A Louisiana salt-dome site had been originally identified by DOE as one of nine potential repository sites. Although the Louisiana site was not among the final three recommended for characterization, the nine sites had been relatively close in the rankings, and his concern was thus valid.[254] Johnston also worried about the delays and high costs involved in characterizing multiple sites. Arguing that time was running out on the political window for siting any repository, he pushed for Congress to directly designate a site.[255]

In DOE's rankings, Yucca Mountain was first among the three finalists on technical grounds (the most important factor in its top ranking was that it had the lowest construction cost) and was also estimated to be the least costly to characterize. Although the DOE siting guidelines ranked cost among the least important factors in site selection, all three sites appeared likely to meet EPA's generic radiation standards for repositories; hence, cost estimates became the major discriminating factor.[256] Other perceived advantages of the Yucca Mountain site were its substantial distance from population centers, its immediate adjacency to the Nevada Test Site, its location far above the water table, and the virtual absence of mineral resources and agriculture in the area.[257]

The 1987 amendments were an exercise in raw political power. Johnston handily maneuvered the 1987 NWPA amendments through the Senate. By attaching the provision designating Yucca as a rider to an omnibus budget bill, Johnston averted full floor debate and secured speedy consideration by the conference committee.[258] This enterprise was driven on the House side by Speaker Jim Wright of Texas and Majority Whip Tom Foley of Washington, who each had an obvious interest in designating Yucca in order to foreclose choice of the candidate site in his state.[259] Nevada was in a politically weak position; its congressional delegation was small, and it had no member in the leadership of either house.[260] Once congressional leaders had closed in on Nevada, other states that might otherwise have been targets for a repository had powerful incentives to endorse the Yucca site. As a result of this power play, Nevada acquired a deep and abiding sense of grievance that fueled its unrelenting and finally successful opposition to implementation of the 1987 amendments. Even some representatives of other potential repository states conceded that the selection of Yucca was unfair. Representative Al Swift from Washington State was quoted as saying: "What you are watching is an exercise in pure politics. . . . I am participating in a nonscientific process—sticking it to Nevada."[261] Nevada's opposition to this assault was compounded by a number of other legitimate concerns: fear that nuclear stigma from a repository would severely damage the state's (and especially Las Vegas's) economy; outrage that a state with no nuclear power plants should have to shoulder all the nation's SNF burden; and a deep and abiding mistrust of the federal government's commitment and ability to safely manage SNF at Yucca, a mistrust that had been bred by the state's earlier experience with AEC nuclear bomb testing at NTS and subsequent federal cover-ups of the resulting radiation exposures of its citizens.

The 1987 amendments added several provisions to NWPA purportedly designed to address Nevada's interests and help ensure that the assessment of the Yucca site was technically sound. In addition to the generic C&C agreement, public participation, and study-funding provisions already included in the 1982 NWPA, the amendments authorized a benefits agreement that could provide the state with annual payments of up to $20 million a year.[262] However, the provision conferred broad implementing discretion on DOE, which was required only to "offer to enter into a benefits agreement," the terms of which would not be subject to judicial review.[263] Moreover, in return for the modest payments granted under a benefits agreement, the state would have to give up other financial assistance conferred by NWPA, as well as its right to contest the Yucca repository. Not surprisingly, no such agreement was ever concluded with Nevada. The Nevada Commission on Nuclear Projects asserted that DOE couldn't be trusted to live up to any agreement.[264]

Other provisions of the 1987 amendments were allegedly designed to help assure that DOE's assessment of the Yucca Mountain site was sound. One of these created the Nuclear Waste Technical Review Board (NWTRB), an independent technical organization whose sole purpose was to provide technical oversight of DOE's waste storage and disposal program.[265] NWTRB would be funded by the federal government through appropriations from the Nuclear Waste Fund. The board could demand support, funds, and services from OTA, the Library of Congress, and the comptroller general, and information from any other agency head.[266] In contrast to the WIPP EEB arrangements, there was no provision for any role for Nevada in the NWTB organization. Another provision of the amendments required DOE to prepare an EIS before deciding to recommend Yucca to the president; however, it contained the proviso that DOE "shall not be required in any such environmental impact statement to consider the need for a repository, the alternatives to geological disposal, or alternative sites to the Yucca Mountain site."[267]

By designating Yucca as the sole candidate site for a repository,[268] the amendments relieved DOE of having to follow the siting guideline established by the 1982 NWPA, which required consideration of "proximity to sites where high-level radioactive waste and spent nuclear fuel is generated or temporarily stored and the transportation and safety factors involved in moving such waste to a repository."[269] DOE used this provision to exclude consideration in the Yucca EIS of transportation routes to Yucca through states other than Nevada, enabling it to postpone potential transportation controversies over transit routes with other states and the Western Governors' Association, a move that effectively deprived Nevada of potential support from other Western states.

The 1987 amendments largely tracked the 1982 NWPA host state disclosure and consultation provisions. They required DOE to provide Nevada with "timely and complete information regarding determinations or plans made with respect to the site characterization, siting, development, design, licensing, construction, operation, regulation, or decommissioning of such repository." If DOE failed to make any required information available, Nevada could submit a written request for the information, to which DOE had 30 days to respond in writing with the information requested or the reason for denying the state's request. If DOE failed to respond, the state could notify the president. If neither the president nor DOE responded within thirty days, DOE was required to suspend activities in the state.[270] The act also provided that any C&C agreement would cover procedures "for resolving objections of a State . . . at any stage of the planning, siting, development, construction, operation, or closure of such a facility within such State through negotiation, arbitration, or other appropriate mechanisms."[271] Notably, the requirement in the 1982 NWPA that DOE "seek to enter" into a C&C agreement with a host state was not modified to ensure that DOE would actually conclude such an agreement with Nevada. The 1982 NWPA language had been modeled on a comparable provision in legislation governing WIPP, which, as discussed in Chapter 5, had clearly been premised on Congress's expectation that New Mexico and DOE—which at the time were negotiating such an agreement pursuant to a court order—would soon conclude a C&C agreement. The premise that the parties would have a mutual desire and determination to enter into a C&C agreement was clearly inapplicable to the circumstances surrounding Yucca, and no such agreement was ever concluded.

Nevada's Scorched-Earth Opposition to Yucca

The various provisions added to NWPA in 1987 allegedly to accommodate Nevada's interests did little to persuade the state, its citizens, or the media of the fairness of the Yucca Mountain enterprise, and were widely regarded as window dressing. It was evident that being singled out by Congress as the only candidate made it overwhelmingly likely that Yucca would be selected for construction of a repository unless the site characterization discovered patently disqualifying features. With federal SNF liabilities mounting and no other available site, DOE faced strong pressures to find the Yucca Mountain site suitable. The press soon dubbed the new law "the screw Nevada bill."[272] Richard Bryan won the 1988 race for the U.S. Senate in Nevada by making opposition to Yucca the centerpiece of his campaign.[273]

Nevada had relatively little incentive to enter into a C&C agreement with DOE because Congress in the 1987 NWPA amendments unilaterally determined many of the most fundamental elements of the Yucca repository, including its purpose and location, the type and amount of waste that could be buried there, the terms of compensation for the state and affected localities, and basic arrangements for transportation of wastes.[274] Accordingly, a C&C agreement would not have given Nevada much scope for influencing key decisions. This situation contrasts sharply to that of WIPP, where the C&C and related agreements helped New Mexico to have a say on the key elements of the repository. The 1987 legislation did create a potential opening for Nevada to influence, through a C&C agreement, some repository design and operation details, including the specific location of the repository within Yucca Mountain, the design of the repository chambers and access to them, the extent and type of engineered barriers, and arrangements to ensure retrievability.[275] However, negotiation or collaboration by Nevada officials with DOE would have been seen as tacit acceptance of Yucca, and thus political poison for a Nevada politician. With such a steep political price for negotiating with DOE and only changes at the margins of the project to gain, the state chose not to enter into a C&C agreement or otherwise cooperatively engage with DOE on these issues.

The state instead adopted a strategy of massive resistance to characterization of the site and regulatory approval of the repository. It formed a Nuclear Waste Project Office to oppose the site.[276] Nevada used whatever means it could, including litigation, denial of water access to block DOE efforts to study the site, commissioning of technical studies to question the suitability of the site, challenges to DOE's site characterization methods and procedures, and extensive public relations efforts. Nevada's vehement opposition prompted DOE to complain before Congress of the state's "scorched earth battle plan."[277]

Nevada's Legal Actions to Thwart a Repository at Yucca Mountain

The State of Nevada brought numerous lawsuits to challenge and retard DOE's efforts to site a repository at Yucca Mountain, both under the 1982 NWPA and the 1987 amendments, including suits seeking federal funding for activities that the state could use to defeat the facility and suits raising constitutional challenges and other novel theories to overturn congressional designation of Yucca. The state also used other legal strategies to block siting of the repository and impede DOE's site characterization activities, including state legislation prohibiting the repository and denial of DOE's requests for permits needed to pursue site characterization activities at Yucca. The denial of permits provoked a series of lawsuits by DOE against the state to obtain access to water and rights-of-way relating to the site.

Nevada Suits against DOE to Obtain Funding

Nevada and local government units in the vicinity of Yucca initiated a variety of legal actions to obtain federal funding to support their efforts to oppose development of a repository at the site.

Section 116(c)(1)(B) of the 1982 NWPA mandated funding for states with a site selected by DOE as one the final three candidate sites for a repository.[278] The stated purpose of the funding was to enable recipients

> (i) to review [DOE] activities taken under this subtitle . . . for purposes of determining any potential economic, social, public health and safety, and environmental impacts of such repository on the State and its residents; (ii) to develop a request for impact assistance . . . ; (iii) to engage in any monitoring, testing, or evaluation activities with respect to site characterization programs with regard to such site; (iv) to provide information to its residents regarding any activities . . . with respect to such site; and (v) to request information from, and make comments and recommendations to, [DOE] regarding any activities taken under this subtitle with respect to such site.[279]

Section 116(c)(1)(A) of the act mandated DOE grants to each state notified by DOE that a "potentially acceptable site" for a repository was located within its territory. This included sites where DOE had undertaken only preliminary testing and was not limited to the final three candidate sites. The grants were to cover participation in two types of activities: those "required by" Sections 116 and 117 of the act, and those "authorized by" a C&C agreement under Section 117(c).[280]

As previously noted, in 1985 Nevada won a Ninth Circuit victory in *Nevada v. Herrington* (I), holding that DOE had to provide it funding to conduct its own studies of the Yucca Mountain site even though it had not yet been selected by DOE as one of the three finalist candidates.[281] In doing so, the court determined that in order for Section 116(c)(1)(A) to have an independent effect, it must be interpreted to require some funding for states prior to selection of the three finalist sites. The court then applied the enumerated purposes for which 116(c)(1)(B) funding must be granted to interpret the scope of 116(a)(1)(A) grants for "activities required by sections 116 and 117."[282] It held that these purposes encompassed studies that would inform a state's potential future decision to issue a notice of disapproval under Section 116(b).[283]

In 1987, the Ninth Circuit decided *Nevada v. Herrington* (II), which involved a request to DOE by Nevada, Washington, Utah, Mississippi, and Wisconsin for funding under Section 116(c)(1)(A) for experts and studies to be used in litigation challenging DOE's implementation of NWPA.[284] The states relied upon the Section 116(c)(1)(B) provision authorizing funding "to review activities . . . for purposes of determining any potential economic, social, public health and safety, and environmental impacts of such repository on the State and its residents." They argued that "review" encompassed proceedings for judicial review of DOE's actions and that these were "activities required by sections 116 and 117" under *Nevada v. Herrington* (I).[285] The court rejected this contention and upheld DOE's denial of funding, holding that the provision was intended to cover only review activities by states of the impacts of a repository, not a federal court's review of DOE's actions.[286]

The 1987 amendments to NWPA modified the funding provisions by mandating funding only for "the State of Nevada and any affected unit of local government," in addition to funding for affected tribes.[287] Four counties contiguous to Nye County, Nevada, where Yucca Mountain is located, requested to be designated "affected unit[s] of local government." Such designation would entitle a local government unit to funding and participation rights in the site characterization process.[288] DOE denied the request of two counties—Esmeralda County, Nevada, and Inyo County, California—which sued to overturn the denial.[289] Inyo County contended that it would be affected by socio-economic impacts of a repository and by potential radiation releases. Both counties also argued that they would be affected because there would likely be transportation routes through them. The Ninth Circuit found that DOE's refusal to designate them as affected units of local government was arbitrary and capricious.[290]

As an example of the tactics to which Nevada resorted with regard to federal funding, the state in 1990 created Bullfrog County, population zero, to encompass Yucca in order to maximize the amount of grants that could then be distributed to local communities and counties under state control.[291] DOE countered by granting special status to Clark County, which would be home to much of the workforce for a potential repository, that authorized it to receive funds directly from DOE.[292] However, this ploy backfired; Clark County used DOE funds to sponsor studies showing that Yucca would have deleterious economic effects on the county and the state as a whole.[293] Ultimately, the Nevada state courts ruled that the creation of Bullfrog County violated the state constitution.[294] Further, the Bullfrog County gambit was criticized within the state as implying that Nevada was amenable to the financial benefits of hosting a repository.[295]

In response to Nevada's use of federal funds for its efforts to discredit the Yucca site, Congress in a 1993 appropriations act prohibited the use by states and localities of DOE grants under the NWPA funding provisions to "directly or indirectly ... influence legislative action on any matter pending before Congress or a State legislature," for "lobbying," "for litigation expenses," or "to support multistate efforts or other coalition building activities."[296] And in *Nevada v. Department of Energy*, the Ninth Circuit rejected claims brought by Nevada against DOE that challenged DOE's denial of a funding request for "oversight and related activities."[297] The dispute essentially related to the level of funding that DOE had to provide to Nevada. The court, noting that Congress had generally limited Nevada's funding to $5 million a year and had from time to time imposed restrictions on its use, held that Nevada was not entitled to any specific level of funding, that Nevada had already received sufficient funds (some of which it had yet failed to spend) to enable its participation, and that DOE's denial of additional funding was accordingly not unlawful.

Nevada Lawsuits Raising Constitutional and Statutory Challenges to the Yucca Repository

Nevada also brought repeated litigation, both before and after the 1987 amendments, challenging DOE's implementation of the siting process, the technical and environmental suitability of the Yucca site for a repository, and the adequacy of DOE studies and reports on the site. The resulting court decisions include three Ninth Circuit decisions styled *Nevada v. Watkins*; their titles are numbered based on the date of the court decision rather than the litigation filing date.

In 1985, Nevada challenged the siting guidelines promulgated by DOE in 1984 as inconsistent with NWPA.[298] In *Nevada v. Watkins II*, decided in 1991, the Ninth Circuit dismissed the claim as unripe, holding that the issuance of the guidelines was "preliminary decisionmaking activity" and that judicial review was available only for "final decisions and actions."[299]

In 1986, the state challenged as inadequate DOE's environmental assessment of Yucca, which had formed part of the basis for DOE's recommendation of Yucca to the president as one of the three sites that would receive detailed characterization. The Ninth Circuit did not decide this claim until it issued *Nevada v. Watkins III* in 1991.[300] While the suit was pending, Congress enacted the 1987 NWPA amendments. Concluding that detailed site characterization of the Yucca Mountain site pursuant to the 1987 amendments would lead to a more complete environmental assessment than had the challenged environmental assessment that DOE had conducted in 1986, the court dismissed Nevada's claims as moot.[301]

Nevada in 1988 also brought constitutional challenges to Congress's designation of Yucca as the sole repository site. In 1988, DOE obtained a right-of-way reservation over fifty-one thousand acres of public land from the Bureau of Land Management (BLM) to conduct excavation and site characterization tests. Nevada challenged BLM's action, arguing that the NWPA amendments violated the Tenth Amendment (which reserves rights not granted to the federal government by the Constitution to the states or the people) and the equal footing doctrine (which provides that all states have the same sovereign powers and jurisdiction) by subjecting politically weak Nevada to the will of the other forty-nine states. In 1990, the Ninth Circuit in *Nevada v. Burford* affirmed the district court's dismissal of the case, holding that Nevada could not point to a concrete injury from the BLM right-of-way grant and accordingly lacked standing to sue.[302]

Nevada continued to pursue its constitutional arguments against Yucca and undertook various legal initiatives to block the repository. These included claims that various legislative and executive actions by the state had manifested "disapproval" by Nevada of the Yucca site and that this disapproval had, pursuant to NWPA, become effective in the absence of congressional override, thus vetoing the site. The "disapproval" actions invoked by the state included 1989 Nevada legislation that prohibited storage or disposal of SNF or HLW within the state,[303] which the state had invoked to deny DOE water permits that it needed for site characterization activities.[304] The Nevada attorney general, at the governor's request, had issued an opinion in November 1989, finding that the legislation operated as the state's notice of disapproval of Yucca's selection pursuant to NWPA Section 116(b), and that since Congress had failed to override Nevada's disapproval notice within the sixty days allotted by NWPA, the repository could not lawfully be built at Yucca. The governor invoked this opinion to order various Nevada agencies to cease processing applications by DOE for water and other permits required under state law. Bennett Johnston, chair of the Senate Energy Committee, responded to the state's persistent efforts to obstruct the characterization process in this fashion by securing passage of a provision in the 1990 Energy and Water Development Appropriations Act that halved the state's federal funding until DOE reported that Nevada decision makers were cooperating fully. Johnston's efforts stopped just short of transferring state and local environmental permitting authority from Nevada to the federal government.[305]

In 1990, after DOE refused to terminate site characterization activities, Nevada filed a petition for review in the Ninth Circuit to press its notice-of-disapproval theory, along with arguments that the 1987 amendments were unconstitutional because they exceeded the federal government's authority, invoking the Tenth Amendment, the equal footing doctrine, and a slew of other theories.[306] The state argued that the Constitution did not allow "forty-nine of the republic's members, acting under the pretext of the Supremacy Clause, to destroy the economy or environment of its politically weakest member."[307] Later that year, in *Nevada v. Watkins I*, the Ninth Circuit rejected all Nevada's claims and affirmed Congress's constitutional authority under the Property and Supremacy Clauses to build a repository on federal land at the Yucca Mountain site.[308] In doing so, it rejected Nevada's argument that the courts should act to protect the state because the political safeguards of federalism inherent in the congressional process of legislation, in which states participate through their delegations, had been subverted in this instance by a covert backroom deal among a few powerful members from other states to target Nevada. The court also rejected Nevada's notice-of-disapproval argument, holding that the state could issue such a notice only if the president had designated Yucca as the repository site.[309] After Nevada failed to obtain review by the Supreme Court of the Ninth Circuit's decision, the state grudgingly and haltingly processed DOE's applications for local permits needed to carry out site characterization activities.[310]

Subsequently, however, Nevada state agencies renewed the practice of denying or delaying grants of permits to DOE in connection with site characterization activities, including most notably water permits. DOE believed that water issues would be a "major concern," given growing demand for water in the state and the expectation that repository and railroad construction would impact groundwater levels.[311] In 2000, Nevada denied DOE's application for a water use permit. In 2002, Nevada signed an agreement with DOE allowing DOE to use limited amounts of water solely for showers, restrooms, dust suppression, and fire response. In 2007, after DOE was found to be using water to lubricate drill bits for boreholes and to create mud for samples, Nevada blocked DOE's use of water.[312] DOE parried by bringing suit against Nevada, seeking an injunction against the state's action. In a strongly worded opinion criticizing DOE for creating a "crisis of its own making," the district court denied it relief.[313] The court took issue with DOE's estimates both of the amount of water and the number of boreholes needed for site characterization, stating that "the Court entertains the suspicion that either DOE wants to look busy, or it wants to keep its contractor occupied during its lengthy delays in filing for a license."[314] The court ruled that DOE "has failed to demonstrate the necessity of its voracious water demands. The issue here is not preemption. The issues here are credibility and good faith."[315]

The Battle for Public Opinion

Nevada mounted an extensive public relations campaign against a repository at Yucca Mountain, publicizing unfavorable reports about DOE's management of nuclear wastes and the Yucca site that had been issued by GAO, NRC, the press, and others.[316] These adverse reports included charges that DOE had delayed release of an unfavorable report on seismic activity in the vicinity of Yucca until after passage of the 1987 NWPA amendments,[317] and the statement of DOE geologist Jerry S. Szymanski that DOE's willingness to move ahead with Yucca and ignore evidence of its susceptibility to groundwater upwellings "is prompting the Federal Government to court disaster."[318] The state also used technical reports prepared by its own experts, supported by DOE funds, in an effort to discredit DOE and its technical findings regarding the site.[319]

As part of its effort to discredit DOE, Nevada in 2001 filed an amicus brief in *Leboeuf, Lamb, Green & Macrae, LLP v. Abraham*, an unsuccessful challenge to DOE's award of a contract to another law firm for legal services in connection with Yucca.[320] More important than the litigation was the public relations embarrassment that Nevada sought to fan. Winston & Strawn, the law firm that got the contract, had earlier lobbied for the Yucca Mountain project. Nevada claimed that this lobbying should disqualify Winston & Strawn from working for DOE, which was supposed to evaluate the site impartially.[321] Nevada's congressional delegation used the issue to paint the Yucca process as corrupt.[322] Even after Winston & Strawn withdrew, Senator Harry Reid pushed for congressional hearings, criminal investigations, and bar proceedings against the firm over the alleged conflict of interest.[323]

In an attempt to reverse unfavorable public opinion of Yucca within the state, a nuclear energy industry group, the American Nuclear Energy Council, in 1991 launched a multimillion-dollar pro-Yucca advertising campaign in Nevada.[324] Statewide surveys conducted after the campaign by Decision Research—a nonprofit that received funding from the State of Nevada and the federal government—indicated that the ads had reached almost three-quarters of all Nevadans. The campaign, however, misfired. Just over half of those surveyed reported that their initial position (six to one in opposition) was unchanged. Among the 47 percent whose views did change, twice as many reported being more skeptical of Yucca after seeing the ads than before.[325] Almost half (48.5 percent) of those surveyed said they did not believe the pro-Yucca assertions in the ads, and another 15 percent reported being insulted by the ads or disagreed with them on other grounds.[326] In the weeks after the survey, various documents were leaked that painted the American Nuclear Energy Council in a bad light; these were met with outrage and ridicule from the public.[327] The authors of the follow-up surveys have suggested that part of the reason for the failure of the ads is that the designers misunderstood the issues underlying the controversy. The public opposition to the repository was not driven solely by fear of radiation release. Other important factors, none of which were addressed by the ad campaign, included distrust of DOE and the nuclear industry, concerns over equity, and desire for local input into the program.[328]

Another survey conducted in 1992 by the state found a significant correlation between confidence in DOE and confidence in repository safety. These findings indicate that DOE could not hope to persuade the public of repository safety without first repairing its own credibility.[329] Even the technically focused NWTRB added its voice to the chorus emphasizing the relevance of public confidence to the project's success.[330] Unfortunately for DOE, the opposite happened: from 1993 to 1994, revelations about the legacy of nuclear waste problems at DOE's Hanford and other weapons production facilities made national headlines. Among the news stories was a front-page article in the *New York Times* reporting that "the United States deliberately released large amounts of radiation into the environment in the 1940s and early 1950s as part of a secret program aimed at developing a weapon that would kill enemy soldiers with radioactive fallout."[331] DOE's credibility and reputation for safety plummeted to a new low.

Nevertheless, the ratio of opponents to supporters of the repository in Nevada declined from 1989, when one opinion survey found a ratio of 4.8:1, to 1994, when the ratio was 3.5:1.[332] The results of the various surveys during this period are summarized in

Table 6.1. Surveys of Nevada public opinion regarding a Yucca repository, 1989–1994

	Ratio of opponents to proponents	Neutral or nonanswer (%)
1989	4.8:1	16.2
1991	5.2:1	4.4
Spring 1993	3.8:1	7.8
Autumn 1993	3.0:1	7.7
1994	3.5:1	6.8

Sources: Phone surveys conducted 1989–1994 by Mertz, Flynn, and Slovic, Nevada Agency for Nuclear Projects, Office of the Governor: *1994 Nevada State Telephone Survey*, 4 (1994); *Autumn 1993 Nevada State Telephone Survey*, 3 (1994); *Spring 1993 Nevada State Telephone Survey*, 3 (1993); *1991 Nevada State Telephone Survey*, 2 (1991); *1989 Nevada State Telephone Survey*, 9v (1989).

Table 6.1. More recent surveys have shown inconsistent results: a newspaper-sponsored poll in 2002 indicated that 83 percent of Nevada voters opposed the repository, while a nuclear industry–sponsored poll in 2004 suggested that opposition to the repository had declined to a 2:1 ratio.[333]

One reason some Nevadans have supported the repository is the federal funds they anticipated the state receiving to offset the risks involved in hosting the site.[334] The gradual decrease in opposition may also reflect growing belief that the repository was inevitable. A former governor of Nevada, Robert List, noted in 2004 that "Nevadans feel this project is upon us and ... we as a state and communities need to face the realities that it's going to happen." List also suggested that the state should make it a point to take the benefits that come with the repository; he estimated the project would bring two thousand jobs and $200 million in federal funds.[335]

Also, the prospect of federal funding pursuant to the 1987 NWPA amendments and economic and other benefits from the repository, however, have fostered strong and longstanding support for Yucca among the counties at or near the site, including Nye County, which includes Yucca Mountain, and Esmeralda County. In 1987, a small town near Yucca Mountain proposed annexing the site to capture federal funding. A local battle ensued over the funding, culminating in the creation of Bullfrog County, as described earlier.[336]

Responses to Repository Delay

Notwithstanding Congress's intent to accelerate development of a repository by adopting the 1987 NWPA amendments, it soon became clear that the 1998 target for a federal takeover of SNF would not be met. In its 1989 plan for characterizing Yucca, DOE pushed back the repository's estimated completion date from 2003 to 2010.[337] In 1990, NRC, which had based its 1984 Waste Confidence Decision on the assumption that a federal repository would be built according to the NWPA timeline, issued two new regulations to deal with the delay. One NRC rule authorized (but did not require) the use of dry casks for on-site SNF storage.[338] The other authorized on-site storage of SNF for at least thirty years and potentially for a century, long after reactors had shut down.[339] By using generic rulemaking, NRC was able to authorize plant operators to maintain or expand the volume of their on-site SNF storage without requiring individual plant license amendments, public hearings, environmental impact assessments, or studies.[340] The dry-cask regulations were challenged by Michigan's attorney general and owners of property adjacent to nuclear plants on the grounds that not requiring specific environmental assessments of the effects of the rule at individual plants violated NEPA. The Sixth Circuit rejected these claims, holding that NRC's environmental assessment of the generic rule was sufficient to satisfy NEPA.[341] The *New York Times* reported in 1995 that efforts toward a permanent repository "have failed" and noted that, "for the foreseeable future, more than 70 communities near nuclear generating plants will become repositories for spent nuclear fuel, the most radioactive of all atomic wastes, without any public hearings or environmental studies of the sites."[342] In response to these delays, DOE put together a task force in 1993 to figure out alternative strategies for nuclear waste; the task force recommended shifting the program's goal from rapid, full-scale development of a repository to licensed demonstration of the capability for disposal, since a repository was not urgently needed for safety reasons, and a demonstration project would not foreclose future options.[343]

Congress in 1992 enacted the Energy Policy Act, directing EPA and NRC to adopt Yucca-specific environmental and safety standards in lieu of the generic repository and MRS standards that they had adopted previously.[344] As explained in Chapter 1, Nevada and environmental group representatives criticized this measure as designed to produce weaker standards in order to make it possible to license Yucca. Ironically, the Yucca-specific standards eventually adopted after litigation challenges were substantially more stringent than the previous generic standards, as detailed in later sections.

Twists and Turns in the Technical Debate over Yucca

Sharp technical debates over the Yucca site's characteristics, its suitability for a repository, and the repository design have continued unabated for nearly three decades, from the initial DOE studies of Yucca as part of the NWPA site selection process, to characterization studies and the development of repository designs after the site was selected by Congress in 1987, and since presidential and congressional approval of a repository at Yucca in 2002. Novel questions and lingering uncertainties are to some degree inevitable in a project of Yucca's scope and complexity. The technical debates, moreover, were supercharged by the high stakes in the siting decision: DOE's determination to build a repository at Yucca, the billions of dollars DOE had spent in the effort, Nevada's intense opposition, the state's tactical deployment of its own technical studies, and the alliance between the state and local and national environmental organizations and anti-nuclear groups opposed to the project.

The stepwise ascent along scientific and policy learning curves produced repeated changes in DOE's technical assessments for Yucca and revisions to the project design and objectives. The emergence of additional information and more complete understanding spawned skepticism about Yucca's suitability, and not just from hard-line opponents. Yucca's opponents steadily disseminated all technical discoveries, concerns, criticisms, and project revisions by DOE as they sought to demonstrate the technical unsuitability of the site and to highlight deficiencies in DOE's assessments of and specifications and rationales for the project. Opponents, for example, asserted as evidence of the site's unsuitability that groundwater movement was much higher than expected;[345] that the groundwater under Yucca occasionally rises above its normal level;[346] that 621 small earthquakes had occurred within fifty miles of Yucca between 1977 and 1997;[347] that the Nuclear Waste Technical Review Board had expressed concern that DOE's waste packages would break down too quickly;[348] and that revised estimates showed that volcanic and seismic events in the vicinity were ten times more likely than had been thought earlier.[349]

In 1990, NRC's Board on Radioactive Waste Management criticized DOE's tendency to overpromise in its approach to waste disposal by representing that the Yucca repository would be absolutely safe.[350] According to the board, this approach would jeopardize DOE's credibility when uncertainties and unforeseen issues arose in the course of site characterization, as indeed later occurred.[351]

Undaunted, DOE continued to persevere in its push to build the repository, ultimately assembling a massive license application to NRC documenting its case that Yucca would meet applicable EPA and NRC regulatory standards. If the license application were to go forward in the future under legal compulsion, NRC would have to resolve these technical and regulatory debates, subject to review by the courts and possibly by Congress.

Characteristics of the Yucca Site

Although Yucca Mountain's geology—volcanic rock, also known as tuff—had not been studied as extensively as had salt or other media, DOE had early on included Yucca in its list of candidate sites and subsequently given it a high ranking on the basis of site characteristics thought to be highly favorable. Its location was remote from significant populations. The climate was dry. It was thought to be in an area of low seismic activity. Because Yucca is a tall, expansive mountain, it was believed that the repository could be sited within the mountain at a location far below the upper mountain surface but also far above the water table. The U.S. Geological Survey's initial geological hypotheses, issued in the early 1980s, posited that little water from the relatively scant rainfall at the site (about twelve inches in an average year) would infiltrate Yucca Mountain;[352] further, it predicted that almost no water would infiltrate to the depth of the repository's storage tunnels, one thousand feet beneath the mountain surface.[353] Dry conditions in the mountain interior were expected to slow both the deterioration of the waste package and any eventual escape of radionuclides. Another initial assumption was that, once radionuclides reached the water table beneath the repository, their transport off the mountain would be extremely slow. Subsequently, however, these and a number of other assumptions regarding the site's characteristics were confounded by new studies generated during detailed characterization of the Yucca site following its designation by Congress in 1987 as the sole repository site to be considered. The most notable developments are discussed in this section.

Assumptions that dry conditions would prevail in a repository at Yucca were undermined by studies conducted by DOE and Los Alamos National Laboratory scientists in 1996 and 1997 that discovered traces of chlorine-36, the product of aboveground atomic weapons testing in the 1950s at NTS, in an exploratory tunnel a thousand feet below the surface, where the repository was to be located.[354] This discovery indicated the likelihood of substantially greater and more rapid water infiltration from the mountain surface into the repository, as well as more rapid corrosion of the waste cladding and releases of radioactivity than had previously been predicted.[355]

Also, the assumption that lateral transport of radionuclides through the water table would be very slow was confounded by studies and estimates made in 1999, which predicted water flow and radionuclide mobility to be roughly one hundred times greater than had been estimated in the early 1980s.[356] In an affidavit submitted in 2001 in litigation brought before the D.C. Circuit,[357] John W. Bartlett, former director of OCRWM, concluded, based on his own professional judgment, "that rates of water infiltration into the mountain were on the order of [one hundred] times higher than had been expected; that water flowed very rapidly through fracture pathways in some of the geologic layers (like flow through a pipe rather than dispersed flow through a medium like a bed of sand); and that there appeared to be unexpected 'fast pathways' for movement of radioactivity from the repository to the water table about 1000 feet beneath it."[358] Bartlett concluded that DOE had no sound basis for believing that waste buried at Yucca would remain geologically isolated, stating that such isolation would be contingent on the performance of Yucca's engineered barriers, whose efficacy he characterized as unknowable.[359]

Further, a team of scientists from Cal Tech and the Harvard-Smithsonian Center for Astrophysics reported in 1998 that the likelihood of a seismic or volcanic event at or near Yucca was ten times greater than had been originally estimated.[360]

Regulatory Requirements and Methodologies

The regulatory requirements applicable to the Yucca repository included both generic and Yucca-specific repository requirements adopted by NRC[361] and EPA's Yucca-specific radiation release standards. EPA's initial standards, issued in 2001, required that the Yucca repository limit radiation exposure to individuals in the general environment, individuals exposed to groundwater, and intruders to no more than fifteen millirems annually over a compliance period of ten thousand years.[362]

NRC's 10 C.F.R. Part 60 generic regulations for nuclear repositories, as amended in 1996, had required that each engineered design component of a repository deemed "important to safety"—namely, components that prevent radiation exposure to individuals and the environment—satisfy NRC's standards for protection against radiation, which incorporate EPA's radiations standards.[363] In 2001, as mandated by the 1997 Energy Policy Act, NRC adopted Yucca-specific licensing regulations at 10 C.F.R. Part 63 that incorporated a new approach to evaluating repositories based on a total system performance assessment (TSPA), which integrated performance submodels for separate repository components into a single unified model and system for risk calculation. NRC's stated objective in adopting the TSPA was to "avoid the imposition of unnecessary, ambiguous, or potentially conflicting criteria that could result from the application of some of the Commission's generic requirements."[364]

The regulations provided that NRC would use the TSPA to model individual design components, determine the likelihood of component failures and the effect of those failures in terms of radiation releases, and then integrate the probabilities of different potential failures into a single probable exposure rate for any individual who would come into contact with the Yucca repository. The TSPA allowed DOE to consider how the total design and operation of the repository would affect radionuclide concentrations in the air, soil, and groundwater, rather than examining the contribution of each component separately.[365] In the notice-and-comment rulemaking leading to adoption of the TSPA, comments critical of the proposed rule were filed by Nevada officials, EPA, and the Nuclear Waste Technical Review Board, among others. Commenters criticized a lack of detail in the criteria for determining site suitability and challenged the legal, scientific, and technical basis for the proposal.[366] Yucca opponents charged that the TSPA was calculated to guarantee the site's suitability and to hobble any legal challenge to it.[367] However, the TSPA approach found much support in the scientific community and from expert government bodies.[368]

Prior to NRC's 2001 adoption of the TSPA, Congress in September 1996 enacted the Energy and Water Appropriations Act of 1997, which contained a rider requiring DOE to conduct a viability assessment (VA) for a repository at Yucca Mountain. The VA was required to include: (1) the preliminary design concept for the critical elements for the repository and waste package; (2) a total system performance assessment . . . ; (3) a plan and cost estimate for the remaining work required to complete a license application; and (4) an estimate of the costs to construct and operate the repository in accordance with the design concept."[369] This provision was imported from a proposed but unadopted amendment to NWPA that would not only have required DOE to conduct a VA but also to submit the Yucca VA to the president so that by December 1998, he could determine "in his discretion" whether Yucca "is unsuitable for development as a repository."[370] DOE viewed the appropriations rider as essentially validating its TSPA methodology and used the methodology in the VA it produced for Yucca in 1998.[371] Since NRC's adoption of the 2001 Yucca licensing rule incorporating TSPA, DOE has used TSPA as the primary means for evaluating Yucca's safety and demonstrating compliance with NRC's Yucca-specific requirements.

DOE Modifications of Repository Design and Rationale

DOE had long assumed it would use a "hot" design for repositories that would involve tightly packing the SNF canisters in repository tunnels and limiting cooling ventilation.[372] This arrangement would initially generate significant amounts of heat, with the objective of raising the temperature of the waste container surfaces to around 200°C and the tunnel wall surface to over 100°C in order to evaporate any infiltrated water.[373] Debate among government scientists at Los Alamos National Laboratory (LANL) and SRS over the proposed use of a hot repository design for Yucca surfaced in 1995. Some scientists were concerned that the heat for a hot design would cause degradation of the surrounding rock and the waste packaging, creating possibilities of extreme adverse events, one of which was the occurrence of nuclear explosions if stored plutonium began to interact with the hot surrounding rock.[374] It proved difficult for DOE to entirely negate this possibility. The *New York Times* observed that expert teams at LANL had "uncover[ed] many problems with the [explosion] thesis" but quoted the laboratory's director of energy research on the difficulty of proving that the risk of such an event is zero: "If we knew how to put the stake through [its] heart, we'd do it."[375]

Between 1996 and 1998, as already noted, new information emerged that revealed significant water infiltration at the Yucca repository site. Scientists had previously believed that it would take many thousands of years for rainwater to seep through from the surface to the repository site within the mountain. However, DOE researchers in 1996 and 1997 discovered that water had seeped eight hundred feet from the surface in only forty years.[376] The water infiltration problem contradicted the assumptions in DOE's approach to demonstrating compliance with regulatory standards for the repository, which had been based on the assumption that water traveled slowly from the surface. To address the problem, DOE adopted an even hotter design. Its 1998 repository design relied expressly on temperatures remaining perpetually above the boiling point to keep away almost all moisture for "hundreds to thousands of years."[377] DOE also considered a new waste-packaging design, consisting of a layer each of carbon steel and of high-nickel alloy, to prevent corrosion and releases.[378]

In 1998 DOE prepared a viability assessment based on its TSPA approach in order to comply with the Energy and Water Appropriations Act of 1997 and demonstrate the suitability of Yucca as a repository site.[379] In response to criticism of the TSPA approach, DOE sought a review of the VA by NWTRB and also commissioned an external TSPA peer review panel to provide further oversight.[380] NWTRB endorsed the TSPA methodology and praised DOE's approach in creating both the VA and TSPA.[381] Nevertheless, NWTRB and the external peer reviewers raised concerns about the Yucca VA, in both cases based on fundamental uncertainties about the hot repository design. The board's report stated that "a credible technical basis does not exist for the [high-temperature] repository design described in the VA."[382] The board noted that high temperatures would cause uncertainties about how the site would behave before and after closure.[383] Furthermore, a low-temperature design would require a "less extensive program of studies" because many of the issues would not need to be resolved prior to a suitability decision.[384] The peer review panel declined to make a design recommendation but noted that a hot design was more difficult to model and analyze compared to a "cooler" design.[385] The peer review panel also emphasized the importance of manifest transparency in DOE's development and assessment of the repository design.[386]

In 2002, DOE decided to incorporate additional protective measures into the repository design in the form of arch-shaped titanium drip shields around the canisters to insulate them from infiltrating water.[387] Anticipating that conditions within the repository would be too radioactive and too hot to safely allow humans to enter once the SNF had been emplaced, DOE decided that remote-controlled gantries would be used to install the shields after the SNF had been emplaced and before closure of the repository.[388] In response to repeated warnings by NWTRB that hot storage could make the performance of the natural geological barriers provided by the site itself more uncertain, DOE also adopted a plan that would enable it to adjust thermal conditions in the repository by controlling parameters such as ventilation and by adjusting the packaging of wastes not yet packaged and emplaced, as needed over time.[389] DOE's Yucca license application to NRC called for a hot storage mode to keep the waste packages dry for an initial period, after which the drip shields would be installed by gantries to provide additional protection against water infiltration. The design suggested that the drip shield gantries would operate largely independently of direct human control, with their performance monitored and deployment adjusted if necessary from a control room on the surface outside the repository.[390]

DOE's drip shields plan stirred a great deal of controversy. Even before the gantry-installation element of the plan had been hatched, critics had attacked the cost of the drip shields, estimated to cost at least $5 billion, and the lengthy planned delay in their installation, decades hence, after all SNF had been emplaced. Nevada and other critics also challenged the plan's technical feasibility, arguing that the repository would be too hot and radioactive, clearances would be too tight, and the technology required would have to be invented in order to install the shields once the repository had been filled.[391] Robert Loux, executive director of the Nevada Agency for Nuclear Projects, wryly pointed out that "DOE's proposal presumes the enforceability of a license condition requiring the installation of successfully working drip shields up to three hundred years after waste emplacement, longer than the existence of the United States."[392] After DOE decided installation of the drip shields would be automated, opponents caricatured the notion of using "robots" as evidence of how outlandishly complex and contingent the repository design had become.[393]

Many Yucca skeptics believed that DOE's increasing reliance on engineered barriers, rather than the protective natural features of the site itself, to meet regulatory standards was an implicit acknowledgment of the hydrogeological unsuitability of the site.[394] Victor Gilinsky, a former NRC commissioner who served as a consultant to the State of Nevada, ridiculed DOE's reliance on engineered barriers as implying that highly radioactive waste could, with use of barriers, be appropriately disposed of "in the basement of DOE's headquarters in the Forrestal Building" in Washington, D.C.[395] But others in the technical community viewed reliance on a combination of site conditions and engineered barriers as appropriate or necessary to ensure regulatory compliance, at least at Yucca.[396] Nevada and others challenged NRC's Yucca licensing regulations, claiming that their reliance on engineered barriers violated the NWPA provision that "geologic considerations ... shall be primary criteria for the selection of sites" in DOE's siting guidelines.[397] The D.C. Circuit rejected their claim.[398]

Federal Designation of Yucca for a Repository, and Nevada's Continuing Resistance

In 2002, DOE recommended to President Bush that he designate Yucca as the site for a federal repository, and the president did so.[399] Nevada's governor duly issued a notice of disapproval in April 2002, exercising Nevada's right under Section 116 of NWPA to disapprove of the president's designation of a federal repository site within its borders.[400] In July 2002, both houses of Congress voted to override Nevada's disapproval, and so to authorize DOE to proceed to develop a repository at Yucca; the joint resolution was signed by President Bush.[401] The following Senate floor debate encapsulates the two diametrically opposed positions in Congress:

> Senator Harry Reid (D-NV): Madam President, I will say quickly that this document [a DOE report entitled "The Spent Nuclear Fuel Transportation System"] about which my friend from Alaska refers is not worth the paper on which it is written. . . . This

piece of trash—and that is what it is—is typical of what the Department of Energy has done. It is one big lie after one big lie.[402]

Senator Jeff Bingaman (D-NM): Failure to approve the resolution that we are talking about, S.J. Res. 34, would terminate the Nation's nuclear waste program.[403]

Because of a tireless lobbying effort by Senate Majority Whip Harry Reid, the Democratic leadership in both the Senate and the House backed Nevada in opposition. Nevertheless, ample majorities (60–39 in the Senate, 306–117 in the House) supported the administration's position that Yucca should move forward—and voted, accordingly, to overrule Nevada's notice of disapproval.[404] The delegations from the states that had potential candidate sites, as expected, voted overwhelmingly to override Nevada's notice of disapproval. Key votes supporting construction of the Yucca repository came from other potential host states; the delegations from Washington and Texas, for example, voted overwhelmingly to override Nevada's objections to constructing a repository at Yucca.[405]

After losing the political battle in Washington, Nevada continued its efforts to impede DOE's attempts to move forward with construction of the repository by trying to delay or block DOE's preparation of Yucca's license application to NRC and the commission's licensing of the facility. Governor Kenny Guinn pledged to use all possible tactics to prevent creation of a repository at Yucca Mountain, stating: "We will fight this with every ounce of energy, from the Oval Office to regulatory agencies."[406] Much of Nevada's litigation eventually failed to produce a court victory, but the drag on DOE's progress was considerable.[407] The state's tactics effectively delayed realization of the repository schedule, allowing Nevada eventually to capitalize on a dramatic change in its political fortunes.

In 2004, Nevada won a significant victory before the D.C. Circuit in *Nuclear Energy Institute v. EPA*.[408] In 2002, prior to but anticipating the secretary of energy's recommendation of Yucca to the president and the subsequent actions of Nevada and Congress, Nevada and various other petitioners had brought an array of legal challenges against actions by DOE, NRC, and EPA regarding Yucca; these challenges were consolidated for decision by the D.C. Circuit Court of Appeals into a single, massive proceeding for decision. The court summarized the matter as follows:

> In this case, we consider challenges by the State of Nevada, local communities, several environmental organizations, and the nuclear energy industry to the statutory and regulatory scheme devised to establish and govern a Yucca Mountain nuclear waste repository. Petitioners challenge regulations issued by the three agencies with responsibility for the site: [EPA, NRC, and DOE]. Petitioners also challenge the constitutionality of the joint resolution through which Congress selected Yucca Mountain as the repository site, as well as certain actions of the President and Energy Secretary leading to approval of the Yucca site.[409]

In the case, among its many claims Nevada challenged the validity of DOE/NRC's reliance on engineered barriers at Yucca and resurrected its constitutional argument based on the equal footing doctrine, arguing that the siting scheme violated the state's constitutional rights by allowing the other states, through their representatives in Congress, to impose a repository on Nevada against its consent.[410] The D.C. Circuit's decision, issued after Congress's override of Nevada's notice of disapproval, rejected all the petitioners' claims but one (regarding EPA's Yucca radiation standards) as either without merit or not yet ripe for judicial review.[411] It rejected on the merits Nevada's challenge to DOE/NRC's reliance on engineered barriers. It invoked Congress's override of the state's notice of disapproval in rejecting Nevada's siting challenge:

> Nevada asserted . . . that DOE changed its site-suitability criteria because Yucca could not meet the preexisting criteria. Nevada urged Congress to delay approval of the repository until its legal claims were decided by the courts and stated that direct legislative approval of the Yucca site would mean that "DOE's bogus site suitability determination could never be reviewed on the technical merits." . . .
>
> Everything in the text and legislative history of the Resolution confirms that Congress intended affirmatively to approve the Yucca site. . . . In the absence of any constitutional defect in the Resolution, we have no authority to review the substantive basis for this decision.[412]

The single challenge upheld by the court was petitioners' claim that the Yucca-specific radiation exposure standards adopted by EPA pursuant to the 1992 Energy Policy Act were contrary to that statute.[413] The time at which peak radiation releases from a repository will occur is a function, on the one hand, of the rate of radioactive decay of the wastes and, on the other, of the time required for the waste containment packages and systems to fail (through corrosion, etc.) and for radioactivity to escape from geological containment into the general environment through water flow or the air. In 2001, EPA had issued a standard limiting the peak radioactive exposure to an individual in the vicinity of the site to 15 millirems at any point during a ten-thousand-year compliance period. Section 801(a) of the Energy Policy Act required EPA to set Yucca radiation standards "based upon and consistent with the findings and recommendations of the National Academy of Sciences." In a 1995 report on Yucca radiation standards, an NAS committee had found "no scientific basis for limiting the time period of the individual-risk standard to 10,000 years or any other value" and asserted that it was possible to estimate the risk of releases over a million-year period.[414] The court concluded that the compliance period specified in EPA's standards was inconsistent with the report and accordingly contravened the act. It remanded the matter to EPA to develop new standards consistent with the committee report. The court also vacated NRC's Yucca licensing rule to the extent that it incorporated EPA's ten-thousand-year standard, and required NRC to reconsider the rule after EPA finished its revisions.[415]

Despite having lost with respect to its other claims, Nevada's victory on the EPA radiation standards enabled the state to delay the progress of the repository project. Remand compelled both NRC and EPA to revise their standards; as a result, NRC could not make a final decision on DOE's license application for Yucca until the revisions were complete and the application was revised to show compliance with the new standards. After the decision, Bob Loux, executive director of the Nevada Agency for Nuclear Projects, said the state had been waiting decades for the decision, which he had thought would kill the Yucca Mountain project.[416] Meanwhile, DOE proceeded to develop technical data and analysis in support of the license application to NRC.

In 2006, Nevada again sued DOE, challenging the section of the department's FEIS for Yucca Mountain that dealt with transportation of waste to the repository and DOE's record of decision designating rail as the preferred option. Among the options for transport of waste to Yucca evaluated in the FEIS were transport mostly by truck and transport mostly by rail.[417] DOE's record of decision endorsed a scenario relying primarily on rail transportation along the Caliente Corridor but acknowledged that, if the repository became operational before the rail line was built, trucking would be used while the rail line was being constructed.[418] Nevada challenged the interim trucking scenario as a substantial change from the plan originally proposed and thus requiring a supplemental EIS. In *Nevada v. DOE*, the D.C. Circuit dismissed all Nevada's claims.[419]

In March 2005, Nevada petitioned NRC to revise its 1990 Waste Confidence Rule, which had been based on the assumption that a repository to receive SNF would open by 2025.[420] Nevada argued that if Yucca Mountain were not licensed, the rule would have to be revised because it would not be possible to develop another repository and open it by 2025, and that the commission's interest in avoiding a revision of the Waste Confidence Rule would accordingly bias it in favor of licensing Yucca. It asked the commission to revise the rule to provide that "there is reasonable assurance all licensed reactor spent fuel will be removed from storage sites to some acceptable disposal site well before storage causes any significant safety or environmental impacts."[421] After NRC denied Nevada's petition for rulemaking in August 2005,[422] the state challenged the denial in the D.C. Circuit.[423] In *Nevada v. NRC*, the D.C. Circuit Court dismissed Nevada's challenges for lack of standing, finding that Nevada had suffered no actual injury as a consequence of the Waste Confidence Rule.[424] Further, since the licensing proceeding had not yet begun, the prediction of bias was "neither actual nor imminent."[425] Nevada also initiated an unsuccessful Freedom of Information Act challenge in 2007 seeking the release of draft Yucca license application documents that DOE had sent to NRC.[426] The documents were never made public.

Nevada also resorted to other tactics to obstruct or delay DOE's efforts to characterize the repository site. In 2007, the state sought, for example, a court injunction blocking DOE from using state water to drill boreholes at the Yucca Mountain site.[427] Also in 2007, Nevada claimed that a law firm retained by DOE had a disqualifying conflict of interest because the firm had previously lobbied for the Nuclear Energy Institute and represented nuclear utilities against DOE.[428] The DOE inspector general eventually found that DOE did not properly document the law firm selection process, but he did not void the contract to prevent the firm from representing DOE in the NRC licensing proceedings, as Nevada had requested.[429]

DOE's Yucca License Application to NRC

DOE filed its construction license application for Yucca with NRC on June 3, 2008.[430] On September 8, NRC docketed the application for decision.[431] In October 2008, following a new rulemaking, EPA issued a revised version of the Yucca radiation standards that had been remanded to it by the D.C. Circuit in *Nuclear Energy Institute*.[432] In its August 2005 notice of proposed rulemaking,[433] EPA had proposed a revised standard, which Robert Meyers, a senior EPA official, summarized:

> The proposed rule would limit radiation doses from Yucca Mountain for up to one million years after it closes. No other health and safety rule in the U.S. has ever attempted

to regulate risk for such a long period of time. Within that regulatory time frame, we proposed two dose standards that would apply based on the number of years from the time the facility is closed. For the first 10,000 years, the proposal retained the 2001 final rule's dose limit of 15 millirem per year. This is protection at the level of the most stringent radiation regulations in the U.S. today. From 10,000 to one million years, we proposed a dose limit of 350 millirem per year. The proposed longterm dose standard considered the variation across the country of estimated exposures from natural sources of radiation. Our goal in proposing this level was to ensure that total radiation exposures for people near Yucca Mountain would be no higher than natural levels people live with routinely in other parts of the country today. One million years, which represents 25,000 generations, is consistent with the time period cited by the National Academy of Sciences as providing a reasonable basis for projecting the performance of the disposal system.[434]

EPA received more than two thousand comments on the proposed rule. Its final 2008 rule extended the compliance period from ten thousand to one million years and reduced the dose limit from 350 to 100 millirems. EPA said that the lower dose limit was based on recommendations of NAS and international bodies such as the International Commission on Radiological Protection, the National Council on Radiation Protection and Measurements, the International Atomic Energy Agency, and the United Nations Scientific Committee on the Effects of Atomic Radiation.[435]

DOE submitted a revised license application to NRC demonstrating its compliance with the new EPA standards on February 19, 2009.[436] Nevada in turn filed with NRC a series of more than two hundred legal objections to the revised license application.[437] As required by *Nuclear Energy Institute*, NRC had amended its Yucca licensing regulations at 10 C.F.R. Part 63 to incorporating the new EPA standards as the basis for NRC's licensing decision on Yucca.[438] Nevada promptly filed court challenges to both the revised EPA standards and the revised NRC licensing regulations.[439] Nevada attorney general G. Catherine Masto charged that EPA was again "ignoring science in favor of a project which presents unacceptable risks to the public."[440]

NWPA provides that NRC must decide on DOE's license application for Yucca within three years after the date of submission, with the possibility of a one-year extension.[441] NWPA provides that NRC does not have to prepare a separate EIS for a repository-licensing decision.[442] Such an EIS, had it been required, would have afforded the public an opportunity to review and comment on a draft EIS. However, NRC's Atomic Safety and Licensing Board granted California and Nevada the right to participate in licensing hearings on 299 issues raised by the states.[443]

Obama's Election as President: Political Victory for Nevada

While the regulatory and other legal proceedings on Yucca dragged on, the impending timetable for NRC licensing of Yucca was overtaken by political developments that dramatically changed Nevada's fortunes—namely, the 2008 presidential election and inauguration of the new Obama administration. As a candidate with his eye on the 2008 Nevada Democratic primary (the third such primary in the nation), Obama made numerous public statements calling the repository unsuitable.[444] As early as 2007, he wrote a letter to the *Las Vegas Review-Journal* stating: "I have always opposed using Yucca Mountain as a nuclear waste repository."[445] In a debate leading up to the caucus, Obama said: "Well, I think it's a testimony to my commitment and opposition to Yucca Mountain that, despite the fact that my state has more nuclear power plants than any other state in the country, I've never supported Yucca Mountain, so I just want to make that clear."[446] All the other Democratic presidential contenders also opposed Yucca. Obama's campaign ultimately made a public promise to kill the Yucca project:

> [As president,] Obama will also lead federal efforts to look for a safe, long-term disposal solution based on objective, scientific analysis. In the meantime, Obama will develop requirements to ensure that the waste stored at current reactor sites is contained using the most advanced dry-cask storage technology available. *Barack Obama believes that Yucca Mountain is not an option.* Our government has spent billions of dollars on Yucca Mountain, and yet there are still significant questions about whether nuclear waste can be safely stored there.[447]

Once Obama was in office, the task of delivering the administration's decision on Yucca to the public and Congress fell to DOE secretary Chu. In May 2009, Chu told the Senate Energy and Water Development Subcommittee that the administration would not go forward with Yucca.[448] Thereafter, the administration slashed funding for Yucca, and in March 2010, DOE filed a motion before the Atomic Safety and Licensing Board of NRC to withdraw, on grounds of "policy," its license application for the repository "with prejudice."[449] The decision to cancel Yucca was met with delight across most of Nevada.

Nonetheless, some Nevadans have formed pro-Yucca groups, like the Alliance for Nevada's Economic Prosperity, dedicated to saving Yucca because it will bring jobs to the state.[450] Nevada counties at or near the Yucca site that would stand to gain economic benefits from development of a repository also rallied in its support. According to the *Las Vegas Review-Journal*, these rural counties generally see a Yucca repository "as an economic panacea."[451] Esmeralda County is one of the most sparsely populated counties in the country; its gold rush–era population of twenty thousand has declined over the past one hundred years to about one thousand. The county's most significant economic asset is the only lithium mine in the United States, which recently laid off employees during the economic downturn. The biggest employer is the county government, and one of the most important sources of county revenue is $500,000 in traffic tickets issued along the stretch of U.S. Route 95 that runs through the county. The county has long supported the Yucca repository as a source of jobs and rail infrastructure, which could encourage further mining development in the county.[452]

Nye County, home to Yucca Mountain, has formally supported the Yucca project since August 2002, when its board of county commissioners passed a resolution to "actively and constructively engage with the DOE" on development of the federal repository.[453] Nye County has sought to maximize its involvement in planning, and to maximize the economic benefits from hosting the repository, and describes its approach as "pragmatic."[454] The director of Nye County's Nuclear Waste Project Office strongly voiced the county's support for Yucca in July 2010 in testimony before the Blue Ribbon Commission on America's Nuclear Future, citing the millions of dollars transferred to the county to support oversight activities and in lieu of taxes; he also argued that it is unlikely that the federal government would find a better solution or site for disposal of SNF and HLW than Yucca Mountain.[455]

Despite the Obama administration's steps to terminate the Yucca repository, Nevada officials maintain a public commitment to challenge the project as long as the enabling legislation, NWPAA, is still in place. Nevada deputy attorney general Marta Adams has said: "This beast isn't dead yet . . . [and] in the meantime we can't really stop."[456] Adams's words proved prophetic in light of legal proceedings that could preclude DOE from withdrawing its Yucca license application and force a decision on it by NRC.

In February 2010, various petitions for review were brought before the D.C. Circuit Court challenging DOE's motion to withdraw the license application; petitioners were Aiken County, South Carolina; three business leaders from Hanford, Washington; the State of South Carolina; and the State of Washington.[457] The court originally stayed proceedings, deferring to a parallel NRC licensing proceeding described below. But, following NRC delay, the court ordered the case to proceed and heard oral arguments in March 2011.[458] The petitions claim that NWPA precludes DOE from withdrawing its license application and, further, that it precludes President Obama, Secretary Chu, and DOE from deciding "to unilaterally and irrevocably terminate the Yucca Mountain repository process mandated by NWPA."[459]

Also in February 2010, the Construction Authorization Board of the NRC Atomic Safety and Licensing Board (ASLB) suspended briefing by the parties on DOE's Yucca construction license application and DOE's motion to withdraw the application on "policy" grounds, pending a decision from the D.C. Circuit in *Aiken County*.[460] DOE petitioned NRC for interlocutory review of the board's decision, which NRC then granted. The commission vacated the board's decision to postpone decision on DOE's Yucca application and withdrawal motion, reasoning that a prompt NRC decision on the matter would enable a subsequent reviewing court to take advantage of "the application of our expertise in the interpretation of the AEA, the NWPA, and our own regulations."[461]

The State of Nevada and Clark County (Nevada's most populous county, which includes Las Vegas, one hundred miles from the repository site) intervened in the NRC proceedings in support of DOE's motion to withdraw its application, as did the State of California and the Joint Timbisha Shoshone Tribal Group. Nye County (which would host the repository), White Pine County, and a consortium composed of Churchill, Esmeralda, Lander, and Mineral Counties intervened in opposition to DOE's withdrawal motion, arguing that the NWPA requires completion of the licensing decision process once an application is filed.

In accordance with NRC's order, ASLB proceeded with the matter. On June 29, 2010, it ruled that DOE lacked authority to withdraw the license application. It held that Section 114(d) of NWPA, which provides that NRC "shall consider an application for a construction authorization for all or part of a repository in accordance with the laws applicable to such applications,"[462] passes responsibility for "determining the technical merits of the Application" from DOE to NRC once the application is submitted, and thus deprives DOE of the authority to withdraw that license.[463] The board first considered Congress's intent in passing NWPA and found that, "given the stated purposes of the NWPA and the detailed structure of that legislation, it would be illogical to allow DOE to withdraw the Application without any examination of the merits."[464] The board also pointed out that NWPA Section 113(c)(3) explicitly gives DOE the authority to find

Yucca unsuitable during the site characterization phase and instructs DOE on various steps that must be followed in the event that Yucca is found unsuitable;[465] the absence of similar instructions for DOE in 114(d), which applies once DOE has submitted the license to NRC, "strongly implies that Congress never contemplated that DOE could withdraw the Application before the NRC considered its merits in accordance with section 114(d)."[466] Finally, the board noted that Congress's intent in enacting NWPA was to provide for congressional control of the nuclear waste disposal project and to remove it from the political process; allowing DOE to undermine this careful allocation of authority by unilaterally terminating the project would be contrary to legislative intent.[467]

The board's decision has been appealed to NRC, which instructed the parties to file briefs by June 30. A majority of the five commissioners (four of whom are Obama appointees) seem firmly committed to ending Yucca, but controversy is brewing over which commissioners will be able to participate in deciding the case. In February 2010, during their Senate confirmation hearings, Commissioners George Apostolakis, William Magwood, and William Ostendorff were asked if they would oppose DOE's request to withdraw its Yucca license; all three said they would not.[468] The states of Washington and South Carolina and the counties of Aiken, South Carolina, and White Pine, Nevada, filed a motion for recusal/disqualification for all three commissioners.[469] Commissioner Apostolakis has already recused himself from any adjudicatory proceeding involving Yucca based on his past involvement in Sandia, which manages Yucca, and a Sandia-led review of the adequacy of Yucca's performance assessment.[470] Commissioners Magwood and Ostendorff have not recused themselves. Chair Gregory Jaczko, a former member of Nevada senator Harry Reid's staff, recused himself from matters involving Yucca Mountain during his first year on the commission but has not yet done so for this decision. The only Bush appointee remaining on the commission, Kristine Svinicki, worked for DOE before her appointment.[471]

At the moment, the Yucca repository looks finished. The president, DOE, and a majority of the NRC commissioners seem united in their stance against Yucca Mountain. Reinforced by Reid's Senate leadership role, Congress seems equally firm in its commitment to end Yucca; a proposal to the Senate Appropriations Committee to allocate $200 million for Yucca was voted down sixteen to thirteen on July 22, 2010.[472] Yucca's fate, however, might well change. The NRC Board's view of the law as requiring completion of the Yucca licensing process may eventually be embraced by the courts, whether in review of a contrary Commission decision or in *Aiken County* or both. Yet even if a construction license for Yucca were eventually granted, Congress would still have to provide the money to build it. Despite the Republican majority in the House gained in the November 2010 elections, Senator Reid remains majority leader in the Senate and will continue to oppose funding of Yucca.[473] In addition, there remain many hurdles to be surmounted before Yucca could be licensed and built, and even if all the legal and political stars realign in Yucca's favor, it would not open for decades.

In attacking the constitutionality of the 1987 NWPA amendments, Nevada had argued that the courts should intervene to protect the state's interests because the political safeguards of federalism had failed. Yet, notwithstanding the federal government's plenary constitutional authority to build a repository on federal land in Nevada, and Congress's exercise of that authority in 1987, Nevada succeeded in delaying implementation of NWPA until it was able to take advantage of the shift of political tides in its favor.

The rise and fall of Yucca tracks the swings in Nevada's political influence from its nadir in 1987 to its apogee in the 2008 elections. The long delay in the siting process, even after the 1987 designation of Yucca, ultimately enabled Nevada to take advantage of shifting political contingencies. The delay was the result of the inevitable complexity of the decision process, the division of regulatory authority over Yucca among three federal agencies, and Nevada's determination in using litigation and other available tools to further slow the process. The political contingencies that favored Nevada included the fortuitous early timing of the Nevada Democratic presidential primary, the lack of a commanding Democratic front-runner, and Harry Reid's position in the Senate. But underlying these contingencies lies the powerful if unpredictable role of local issues and interests in U.S. national politics and policies, and the enduring strength of the political safeguards of federalism that ultimately defeated Congress's design, which would impose a repository on an unwilling host state.

Conclusion

Through NWPA, Congress sought to make up for thirty years of abject disregard, including its own, of the nation's nuclear waste stockpiles by mandating a crash program for their permanent deposal. The means Congress chose were, in 1982, a centralized technocratic-meritocratic process administered by DOE and, in 1987, site determination by Congress itself together with the president. Both mechanisms failed. The top-down strategies that Congress selected to address the waste problem relied on the same centralized authoritarian approach that it had followed in creating and sustaining the AEC/ERDA/DOE regime that produced the wastes and then neglected to deal with them. The failure of both the 1982 and 1987 NWPA blueprints, which ultimately relied on sheer federal power to force waste facilities on unwilling states, contrasts with the success of the far more flexible and improvisational approach that led to the successful development of WIPP with the eventual assent of New Mexico.

The data points are too few to draw any definite conclusions from this experience. It is possible that the 1982 NWPA strategy could have succeeded had it provided a more realistic timetable, had it been implemented by DOE in a manner more open to and accommodating of the interests of potential host states, and had Congress stayed the course. It is also possible, although by no means certain, that the 1987 NWPA strategy could have produced a repository at Yucca but for the political contingencies that led to the election of Barack Obama as president and Harry Reid's election to Senate majority leader. Any such success, however, might have been bought at a heavy price. It is doubtful that Congress would have been able to impose a second repository through the same method. Nonetheless, the available evidence indicates that, given the very strong safeguards of federalism in our polity, strategies reliant on unilateral federal imposition run a high risk of failure.

As a result of the failures of both versions of NWPA, the nation again confronts its legacy of highly radioactive defense and civilian waste, this time without a plan or even—under current statutes—legal authority to develop a repository or a consolidated storage facility.

Chapter 7

Options for Orphan Wastes

There are currently more than 62,500 metric tons of commercial SNF stored at reactor sites in thirty-five states around the country, along with large amounts of defense and nondefense HLW and SNF stored at DOE sites; all these wastes currently lack a disposition pathway.[1] There are three basic options for developing a repository to dispose of these wastes: revive Yucca, dispose of them at or near WIPP, or develop a brand-new repository. All these alternatives face serious hurdles, and none could begin to take wastes for many years. Accordingly, the nation must actively pursue one or more of these options and at the same time address what to do, over a long and indefinite time period, with the nation's inventories of SNF and HLW.

The Yucca Mountain repository project has been subjected to sharp technical challenges—by no means confined to those made by Nevada—regarding the suitability of the site and DOE's design for the repository. Yet DOE, in seeking to withdraw its NRC license application for the facility, did not impugn Yucca's ability to safely contain HLW and SNF. Instead, DOE invoked unspecified "policy" considerations. The NRC Atomic Safety and Licensing Board has denied DOE's motion to withdraw the Yucca license application.[2] That denial will be subject to review by the commission, whose decision will almost certainly be challenged in the courts.[3] Even if the Yucca licensing proceeding were to go forward, a final NRC decision would take additional years, followed by review in the courts. And, if a license were eventually granted and sustained in the courts, Congress would still have to appropriate the funds to build the repository, which would then have to be constructed. Based on even the most optimistic assumptions, it would be several decades from now before the facility could open. Even then, according to DOE, it would take forty-six years to dispose of all the SNF that has been and is expected to be generated by existing reactors.[4]

A second option would be to expand the mission of WIPP to accommodate SNF, HLW, or both. Officials of Carlsbad, New Mexico, have indicated that they would be willing to open the WIPP repository to disposal of both SNF and HLW.[5] As a practical political matter, however, this would require the assent of the State of New Mexico, which strongly opposed any SNF disposal at WIPP when the issue was broached more than thirty years ago, and also opposed HLW disposal at the facility. As described in Chapter 5, New Mexico officials have recently signaled interest in expanding WIPP's mission to include disposal of HLW and at least interim or retrievable storage of SNF.[6] Although there are favorable preliminary indications that the WIPP site may be suitable, its ability safely to hold all the nation's SNF and HLW has not been determined and would have to be studied in depth. Also, Congress would have to amend existing legislation for WIPP, which now expressly precludes WIPP from receiving these wastes.

The third option would be to develop and construct a brand-new repository at another site to receive these wastes. This could be a site in areas considered in the past, including the Salado salt formation in which WIPP is located, which covers portions of New Mexico and Texas, or a brand-new site either not identified or not extensively studied in the often closed rounds of siting conducted in the past. The government would have to overcome formidable hurdles in siting any such facility. Even if such a repository could be developed, it would not open for decades.

These three repository options are not necessarily mutually exclusive. But regardless of which option or combination of options is chosen, because the availability and timing of a repository for SNF and HLW is at best highly uncertain, the federal government must adopt and carry out a considered plan for dealing with these waste inventories in the interim—a period of at least the next several decades and possibly much longer. The Blue Ribbon Commission on America's Nuclear Future is charged with helping to develop such a plan.

This chapter deals primarily with interim options for civilian SNF, which presents more urgent and salient issues than defense HLW and SNF currently stored at DOE sites or the small amount of civilian reprocessing HLW currently stored at West Valley. The basic options for SNF now under consideration are continued at-reactor storage for an indefinite period; development of government or private consolidated SNF storage facilities; and reprocessing of SNF, which in turn would require future management of the nuclear wastes generated by reprocessing.[7] Here we examine these alternatives for dealing with SNF, their history to date, and options for the future.

Defense HLW must eventually be disposed of in a repository, and the same is likely true of defense SNF. DOE has signed agreements with the states in which these wastes are currently located, committing to remove HLW from the host states, in some cases under specified timetables. A repository may well not become available in time for DOE to be able to comply with these undertakings by disposing of the wastes in a repository. In that event, the wastes might (and most likely would) continue to be stored at these DOE sites, subject to renegotiated conditions. Alternatively, the wastes could be moved to other DOE sites for storage, subject to agreement by those sites' host states, or to new consolidated storage facilities, if such facilities could be sited and developed. Storage facilities for defense HLW could be independent of, or co-located with, any new consolidated storage facilities for commercial SNF. The issues discussed in this chapter regarding consolidated storage facilities for SNF generally apply to similar storage facilities for defense HLW/SNF.

Continued SNF Storage at Reactors

As of the beginning of 2010, over 62,500 metric tons of civilian SNF were being stored at seventy-seven reactor sites across thirty-five states.[8] Additional SNF is accumulating at the rate of about 2,000 MTHM per year.[9] Decommissioned power plants account for approximately 2,800 metric tons of the current inventory of SNF; the rest of it is located at operating sites.[10] The vast majority of this waste—51,000 metric tons—is currently held in pool storage, the rest in dry-cask storage.[11] NRC authorizes but does not require dry-cask storage, which is more expensive. Use of dry-cask storage has been slowly increasing. Although now responsible for only 15 percent of all SNF,[12] the nuclear power industry estimated in 2007 that by 2017 dry casks will be used at fifty-nine nuclear power sites to store 26,000 metric tons of SNF, 35 percent of the total.[13] Even if dry casks were to become the predominant storage method, pool storage would still be needed, as it generally takes three to five years before SNF cools down sufficiently to be placed in casks.[14] Whether pool or dry-cask storage is used, SNF storage facilities must meet various operational, maintenance, and security requirements, including certain security measures against terrorism.[15] According to the Nuclear Energy Institute, a nuclear industry trade association, there is enough capacity for pool and dry-cask storage at each existing nuclear power plant site to store all SNF that will be produced under each plant's current license.[16] NRC recently directed its staff to propose a rule and develop a draft EIS to "assess the environmental impacts and safety of long-term SNF and HLW storage beyond 120 years."[17]

Pool storage consists of submerging SNF in water-filled pools with concrete floors and walls four to six feet thick.[18] These pools are forty feet deep and range in length from thirty to sixty feet, and from twenty to forty feet in width. SNF is placed at least twenty feet below the water surface to protect workers from radiation and to optimize the cooling effect of the water, which is constantly circulated throughout the pool. Because placing waste in dry storage casks and storing it outside pools is more expensive and labor intensive, plant owners have preferred to expand pool storage capacity in response to increased on-site storage needs as SNF accumulates, rather than transfer all SNF now in pools to dry-cask storage.[19] As SNF accumulates without a repository in sight, NRC has authorized many facilities to add up to five times more fuel rods to the pools than was originally authorized in their operating licenses.[20] The practice of densely storing spent fuel rods in cooling pools poses risks associated with fire and potentially with terrorism.

A complete loss of coolant in a reactor storage pool could push the temperature of the SNF in a fuel assembly above 1650 degrees Fahrenheit, the ignition point of a self-sustaining fire.[21] NRC has found that if such a fire were to occur, it could disperse radioactivity over a large area, causing "thousands of latent cancer fatalities."[22] Storing the fuel assemblies farther apart in a pool reduces the risk of a fire from loss of coolant; if the assemblies are stored far enough apart, air cooling might be sufficient to keep the fuel rods' temperatures below 1650 degrees even if a loss of water coolant were to occur.[23] Currently, however, most storage pools are tightly packed with SNF to minimize cost.[24] The Fukushima crisis has shown the risks of this practice.

According to NRC, the risk that an earthquake or an accidental airplane crash could damage a storage pool, causing loss of water coolant and leading to a fire, is "very low."[25] NRC views the risk as low because of the multiple safety systems that it requires, the design of the pools, and the amount of time required for the fuel to heat up to the point of combustion.[26] But these scenarios involve accidents; a terrorist attack might deliberately seek to cause a complete loss of coolant. The Union of Concerned Scientists (UCS) has expressed serious concern about the risk of a successful terrorist attack on SNF pools and the relative ease with which it might be accomplished under current nuclear power plant security arrangements.[27] The National Academy of Sciences (NAS) Committee on Science and Technology for Countering Terrorism concluded in 2002 that the "potential . . . is high" for "a 9/11 type attack" on spent-fuel storage at nuclear power plants, and that SNF stored in vulnerable storage pools should be moved to dry-cask storage.[28] NRC has responded to various papers evaluating risks from terrorist attacks, and is currently studying the risks associated with some of the more extreme scenarios, such as the deliberate crashing of an airplane into a SNF pool.[29]

The Fukushima crisis has heightened interest in dry-cask storage, although this method cannot be used until after SNF has cooled in pools for three to five years.[30] Dry-cask storage is considered safer than pool storage because it does not pose the risks involved with loss of coolant.[31] To store SNF "dry," spent fuel assemblies are first placed into one-inch-thick steel containers that may house from seven to sixty-eight assemblies, each containing from 91 to 324 fuel rods.[32] These containers are filled with inert gas. The canister is then placed with others either in a large outer container made of steel-reinforced concrete, or in a concrete bunker.[33] The waste inside is passively cooled by heat conduction, natural convection, and thermal radiation.[34] The casks are designed to allow all three cooling processes to take place: the solid materials of the casks themselves allow heat conduction through physical contact, while ducts carved into the concrete, for instance, allow air to reach the edges of the containers and create convection currents.[35] The cooling system prevents the waste from heating to combustion temperature, but the waste inside remains quite hot—around 600 degrees Fahrenheit.[36]

According to NRC, storage of spent fuel in dry casks at reactor sites is safe for a minimum of one hundred years.[37] The Government Accountability Office reported that tests conducted by DOE and the Army Corps of Engineers found that, due to the strength and thickness of the concrete and steel they incorporate, the dry-storage casks currently used would not release significant amounts of radioactivity if struck by crashing airplanes, armor-piercing rounds, or high explosives.[38] The American Physical Society has found that dry-cask storage containers can be securely maintained for at least fifty years with a high degree of confidence, and that replacing aging casks can further extend the secure storage period, making safe on-site storage feasible for as long as adequate resources and attention can be directed to facility maintenance.[39]

These assurances notwithstanding, terrorist attacks on dry-cask storage also have to be considered. NRC evaluates the risk of widespread release of radioactivity from such an attack to be "very low" because of the robust physical properties of transportation and dry storage containers.[40] The NAS Counter Terrorism committee found it unlikely that a terrorist attack would be able to penetrate dry casks or cause a significant release of radiation.[41] UCS, on the other hand, asserts: "Although the dry casks would present less of a hazard than spent fuel pools if attacked, they remain vulnerable to weapons such as rocket-propelled grenades. These weapons could penetrate most dry casks and their vaults, igniting a zirconium fire and resulting in the release of significant amounts of radioactive material."[42] Nonetheless, UCS believes that dry-cask storage can be made an "acceptably safe" and "economically viable option" for the next fifty years by "hardening" such facilities against attack through measures including adding earthen berms and improving security at reactor sites.[43]

According to a GAO report, some experts have suggested that on-site storage is the most equitable solution for nuclear power plant waste. If waste is left at the reactor sites, the risks and responsibilities remain with the communities that benefited most from the electric power generation that created the waste.[44] However, the cancellation of Yucca has begun to stir opposition from local communities and officials at some reactor sites to the option of prolonged at-reactor storage. Some operators of nuclear power plants who have moved to construct dry storage capacity on-site have encountered opposition from the public and from state regulatory agencies.[45] This opposition is likely to grow in coming years as reactors reach the end of their useful lives. No reactor site has yet conducted an environmental review to determine the safety of on-site storage after the end of the license period. GAO found that such reviews will be necessary.[46] However, NRC recently updated its Waste Confidence Rule to allow SNF storage for at least 60 years (up from 30 years) beyond reactors' licensed lives and is considering storage for 120 years or more.[47] SNF's role in the Fukushima crisis will force rethinking this option.

Consolidated SNF Storage

Development of consolidated storage facilities for SNF as an alternative to at-reactor storage has attracted episodic interest over the past thirty-five years. That interest has burgeoned more recently with the delays and uncertainties in developing a repository and with the Fukushima crisis.[48] NWPA contains several provisions for federal interim storage facilities, and leaves open the possibility of private facilities. Indeed, a small private storage facility located at the failed Morris, Illinois, reprocessing plant currently holds around 675 MTHM of SNF from several different reactor sites in pool storage.[49] Yet no federal storage facility has yet been developed, largely because of state opposition. Opposition has also been driven by fears that consolidated storage would drain support for developing a permanent disposal repository and that the storage facilities would become de facto permanent repositories.

Delays in developing Yucca, its cancellation by the Obama administration, and renewed interest in reprocessing have stimulated reexamination of consolidated interim storage options for commercial SNF. The Blue Ribbon Commission on America's Nuclear Future constituted by President Obama is tasked with examining this issue. NRDC supports development of consolidated interim storage capacity for SNF from decommissioned nuclear power plants. But NRDC, along with the Union of Concerned Scientists and others, does not favor this option for SNF at operating reactors, believing it should remain at reactor sites until a repository becomes available.[50] The nuclear industry also favors moving SNF from decommissioned reactors to consolidated storage because safeguarding SNF at various geographically dispersed decommissioned plants would cost more than managing a single consolidated storage facility.[51]

Federal MRS Facilities

NWPA contains several provisions for federal interim storage of SNF. One of these, Section 135, which was never used and is no longer operative, was a hardship provision under which DOE, if requested by utilities, was required to provide interim storage capacity to utilities for up to 1,900 metric tons total of commercial SNF.[52] Utilities never requested this interim storage capacity because they believed that such storage would have cost far more than continued storage at reactor sites, and because the development of dry-cask storage made expanded on-site storage feasible.[53]

NWPA as enacted in 1982 also required DOE to study the need for monitored retrievable storage (MRS) facilities that would temporarily store civilian SNF and HLW and to present to Congress a proposal to construct one or more of these facilities.[54] DOE was authorized to plan, construct, and operate these facilities, which would be designed to "provide for the ready retrieval of such spent fuel and waste for further processing or disposal."[55] DOE eventually developed and submitted to Congress a proposal for an MRS facility at Clinch River, Tennessee, and two alternative sites, also in Tennessee.[56] Although the local communities were receptive, the proposal provoked strong opposition from Tennessee state officials and the state's congressional delegation. The Tennessee delegation won a provision in the 1987 NWPA amendments that prohibited construction of an MRS facility in the state.[57]

The 1987 NWPA amendments greatly constricted the MRS option. The legislation provided that DOE could construct only one MRS facility,[58] which could not be located in Nevada.[59] Congress's concern was that an MRS would be sited near Yucca Mountain and become a de facto repository, so that the Yucca repository would never be built.[60] Further, DOE could not select a potential MRS facility site until the secretary of energy had recommended a site for development of a permanent repository pursuant to the statute—i.e., Yucca Mountain.[61] DOE formally recommended Yucca to the president in February 2002 but has taken no steps to site and develop an MRS facility since that time.[62] The amendments also provided that construction of an MRS facility could not start until NRC had licensed the Yucca Mountain repository.[63] Lastly, Congress in 1987 placed a tight cap on the total amount of SNF that can be stored at an MRS facility: ten thousand metric tons of heavy metal until a permanent repository begins accepting SNF or HLW, and fifteen thousand metric tons of heavy metal thereafter.[64]

The 1987 amendments also created the Office of the Nuclear Waste Negotiator (ONWN) to "attempt to find a State or Indian tribe willing to host a repository or [MRS] facility at a technically qualified site on reasonable terms."[65] This program envisaged a facility built and operated by the federal government.[66] The first negotiator took office in 1990 and established a voluntary siting program that allowed interested host communities to apply for a phased series of study grants to fund research on the effects of hosting an MRS facility.[67] Twenty-four tribes and four counties in four states applied for Phase I grants.[68] NWPA authorized state governors to block counties from participating in the program—which the governors in fact did—but states could not veto participation by tribes.[69]

Environmental justice advocates challenged the program, asserting that it would result in poor communities hosting all of society's dangerous facilities because they would agree to host facilities for less in the way of compensation than would their more affluent counterparts.[70] State officials, on the other hand, protested that sovereign tribes with very small populations and little land should not be able to impose significant negative externalities on their nontribal neighbors in the form of nuclear waste proximity, transportation, and stigma.[71] Opponents invoked a study which found that a nuclear accident at Yucca Mountain could seriously impact convention business in Las Vegas, a hundred miles away.[72]

The ONWN program provided for a series of study grants to jurisdictions that indicated potential interest in hosting an MRS facility. Because of political resistance at the state level, only two applicants reached the third study grant phase, New Mexico's Mescalero Apache Nation and Utah's Skull Valley Band of the Goshute Nation.[73] The Mescalero Apaches came close to agreeing to host an MRS facility but were unable to overcome New Mexico's opposition to the facility.[74] Although the state lacked authority to block the facility, a senator from New Mexico successfully sponsored legislation in Congress that blocked federal funding of MRS study grants in the state.[75] This action by Congress undermined the authority of the negotiator and discouraged tribes from dealing with him further.[76] After Congress failed to reauthorize the Office of Nuclear Waste Negotiator in 1994, the position lapsed in 1995, and the program to site a federal MRS facility died with it.[77]

Private Consolidated Storage Facilities

NWPA provides that the statute shall not "be construed to encourage, authorize, or require the private or Federal use, purchase, lease, or other acquisition of any storage facility located away from the site of any civilian nuclear power plant and not owned by the Federal Government on [January 7, 1983]."[78] The precise legal effect of this provision has not been determined. NRC, however, has used its general authority under the Atomic Energy Act to regulate private uses of nuclear materials to issue regulations for the licensing of private away-from-reactor SNF storage facilities and applied such regulations in licensing the Private Fuel Storage LLC (PFS) SNF storage facility, as discussed below,[79] as well as the Morris facility, discussed above.

After the collapse of discussions with the federal government over construction of a federal MRS facility, both the Mescalero Apaches and several members of the Skull Valley Band of Goshute Indians began negotiating with Private Fuel Storage LLC (PFS), a consortium of nuclear power plant owners.[80] After the Mescalero Apaches withdrew from the negotiations, members of the Skull Valley Band signed an agreement with PFS providing for construction of a private storage facility on the tribe's land.[81] The Skull Valley Band is an offshoot of the Confederated Tribes of the Goshute Nation and consists of 125 members who live on an eighteen-thousand-acre reservation located about fifty miles west of Salt Lake City. The reservation is near a chemical weapons destruction facility that includes an incinerator, a LLW dump, and a coal-burning power plant. Under the plans for the PFS facility, SNF would be stored in dry casks until it could be permanently disposed of in a geologic repository.[82] Some members of the Skull Valley Band, together with the Confederated Tribes of the Goshute Nation, have opposed the facility and contend that the agreement with PFS regarding lease of Goshute lands for an interim storage facility is invalid.[83] The State of Utah has vehemently opposed the PFS facility.[84]

After nearly ten years of effort on the part of PFS and project proponents within the Skull Valley Band, in 2006 NRC issued a twenty-year license to PFS for dry-cask storage of up to forty-four thousand metric tons of SNF at the Skull Valley Goshute Facility.[85] Two bureaus within the Department of Interior subsequently denied PFS's applications for authorizations needed for the project to go forward.[86] The Bureau of Indian Affairs (BIA) disapproved the lease of tribal lands needed to construct the facility on the ground that the storage facility risked becoming a de facto permanent repository.[87] The Bureau of Land Management (BLM) denied rights-of-way across federal land for a truck-to-rail transfer site and for a rail spur needed to transport SNF to the facility—the surface roads in the region were inadequate for SNF transport by truck alone—on the ground that these installations would threaten a wilderness area. The State of Utah opposed the PFS facility for many reasons, including general nuclear stigma and concern that SNF stored at the PFS facility would become orphaned. The state also asserted that PFS lacked the financial and technical capacity to respond effectively to emergencies, including potential terrorist attacks, in connection with temporary storage and transfer of the casks from the intermodal facility where transfer of the SNF from railcars to trucks would take place before the final leg or transportation to the PFS storage facility.[88]

In addition to the reasons BLM and BIA gave for their decisions, their stance may well have reflected pressure exerted by the State of Utah and its congressional delegation, as well as concern on the part of the Bush administration that approval of the PFS facility would undermine the push to develop the Yucca repository. Opposition to the facility by many Goshutes may also have played a role.

In 2007, PFS and the Skull Valley Band of the Goshute Nation filed suit against the Department of Interior (DOI) in federal district court in Utah challenging the BIA and BLM permit denials.[89] The tribe alleged that Interior failed to show good faith in its dealings with the tribe.[90] In July 2010, the court overturned the denial of the permits and remanded the decisions to DOI.[91] At the same time, the State of Utah and a group of Goshutes who oppose the PFS project sought review of NRC's issuance of a license for the facility in the Court of Appeals for the D.C. Circuit.[92] Proceedings in that court were stayed pending the outcome of the case on the BLM and BIA permit denials in the Utah District Court.[93] Now that the district court has overturned and remanded the permit denials, it remains to be seen whether the D.C. Circuit will proceed with the NRC case.

As the PFS story illustrates, private consolidated storage facilities, whether sited on tribal or private lands, are, like federal MRS facilities, vulnerable to opposition by potential host jurisdictions, especially states. Facilities on private lands are, as a purely legal matter, more vulnerable to state and local actions to block them than are facilities on federal and tribal lands. Utah adopted legislation banning the storage of HLW and SNF within the state in an effort to block the PFS facility. The state was also prepared to impose stringent fees and regulations on the PFS facility in the event that the state ban on storage was found by the courts to be preempted by federal law.[94] The ability of a state to impose such measures on a facility located on tribal lands is doubtful, however. And, even in the case of a private facility, the grant by NRC of a license for the facility may have a preemptive effect on state regulation.[95] An unwilling host state could also try to block transportation of SNF to a private facility—as Utah did in opposing the PFS truck transfer station and pressing DOI to deny a right-of-way for it.

Under NWPA, construction of a private storage facility is not dependent on the prior licensing of a permanent repository (as is the case for a federal MRS facility), nor is a private facility subject to the NWPA storage capacity caps applicable to federal MRS facilities.[96] The licensed storage capacity of the PFS facility, forty-four thousand metric tons, is far greater than the cap of ten thousand metric tons imposed on a federal MRS facility under the 1987 NWPA amendments. The absence of restrictions on storage capacity at private facilities makes them more attractive to utilities than a federal MRS,[97] despite the long lead times and uncertainties involved in obtaining government approvals. The absence of such restrictions, however, may increase the risk that a private facility could become a de facto repository, inviting opposition by potential host jurisdictions.[98]

Consolidated Interim SNF Storage at DOE Facilities

DOE has concluded that some of its sites already hosting defense wastes—SRS, Hanford, and possibly INL—could be appropriate for consolidated interim storage of some SNF, but that express congressional authorization would be required. In its 2008 report to Congress on interim storage of SNF from decommissioned power plants, DOE stated that these sites "possess existing infrastructures, including security programs for SNF, operational and regulatory expertise, fully developed environmental baselines and rail access that would facilitate acceptance [of SNF]."[99]

Consolidated interim storage of SNF has attracted growing interest. Attention to this option has mushroomed since the Fukushima Daiichi disaster. A 2006 Senate bill would have authorized DOE to select a site in each state or region to host a federally owned, consolidated interim storage facility for SNF.[100] The bill was not adopted. In 2008, as part of appropriations legislation, Congress directed DOE to "develop a plan to take custody of [SNF] currently stored at decommissioned reactor sites" and to consider consolidated storage of that SNF, managed by DOE, at existing federal sites, operating reactor sites, or at a competitively selected interim storage site.[101] DOE responded with a report finding, as noted earlier, that some DOE sites would be suitable for interim storage of SNF. DOE asserted, however, that the department lacked legal authority to use its sites for this purpose, because NWPA provides that SNF may be stored or disposed of only at a federal repository, a federal MRS facility, or a federal emergency temporary storage facility. Although this interpretation has been questioned and its logic would seem to preclude private SNF storage facilities (which DOE does not regard as prohibited by NPWA), including the PFS facility licensed by NRC, DOE's authority to host civilian SNF at its weapons complex sites is uncertain at best, inviting congressional legislation on the issue.[102] DOE's report recommended that any future legislation to grant it such authority provide additional funding to assure the simultaneous development of a permanent repository.[103]

Evaluating Interim Storage Options

The federal government, with the help of the Blue Ribbon Commission on America's Nuclear Future, will have to confront the question of whether it is preferable to continue to store SNF at reactor sites, to have the federal government develop new consolidated storage facilities (whether at DOE sites or elsewhere), to promote development of private consolidated storage facilities, or some combination of these options. Under existing law, which makes the federal government financially responsible for disposition of SNF after 1998, the federal government would ultimately bear the costs of storing SNF under any of these scenarios. Federal financial responsibility for the costs of managing SNF remains, even though nuclear utilities must continue to pay into the Nuclear Waste Fund based on the extent of electricity they generate from nuclear plants.[104] The alternatives will differ, of course, in terms of cost, transportation, environmental, and security factors, as well as in their broader political and policy implications.

Proponents of nuclear power have tended to favor the option of moving the SNF now being stored at nuclear power plants to one or more consolidated interim storage facilities, based on the belief that doing so will curb opposition to the construction of new nuclear power plants.[105] Other arguments in favor of the consolidated interim storage facility option have included that, in the long run, it may be cheaper than continued at-reactor storage and that it would allow DOE to fulfill its obligation under existing law to take SNF from the utilities, thereby avoiding continued litigation and mitigating the government's potentially enormous financial liability. A 2007 study by the American Physical Society for example, concluded that if Yucca opened by 2017, it would be cheaper for the federal government to pay for at-reactor dry-cask storage than to build an MRS facility.[106] The capital costs of the dry casks are the same no matter where they are put. By contrast, to establish a new facility, DOE would need to provide local and state inducements for siting the facility, acquire land, construct the facility, and probably also improve transportation infrastructure to transfer SNF to the facility. On one hand, these added costs could well offset the savings, due to scale economies, in the operation of a few consolidated interim storage facilities rather than the operation of numerous storage facilities at nuclear power plants dispersed across the country.[107] On the other hand, since so much of the cost of consolidated storage is initial capital investment, if a repository were substantially delayed, substantial savings could result from use of consolidated storage because of its lower operating costs relative to dispersed at-reactor storage.[108] Proponents of consolidated storage contend that SNF will be safer if stored in such a facility, especially if sited in a sparsely populated region. But dry-cask safety studies suggest that the risks from dry-cask storage, especially if hardened against terrorist attack, are very low to negligible, regardless of whether they are located at a consolidated interim storage facility or at many nuclear power plant sites.[109]

There is evidence, moreover, that local communities that have grown used to living with nuclear waste facilities in their midst and have gained significant economic benefits from them may be amenable, if the terms are favorable, to accepting interim storage of some SNF at DOE sites in their vicinity.[110] The interest expressed by Carlsbad, as well as by New Mexico political leaders, in potentially expanding WIPP's mission to include SNF and the fact that a number of local communities volunteered several years ago to take part in the GNEP demonstration program that would have brought SNF into their midst are evidence of such willingness. Such willingness was also reflected in local and tribal participation in the Office of Waste Negotiator's siting program in the 1980s. Local support may not be enough, however, because likely opposition from elsewhere in potential host states could well trigger opposition from state officials. To enhance potential benefits to host-state communities around DOE sites, DOE has floated the possibility that SNF storage at those sites might be coupled with the development of energy parks that could include R&D on advanced fuel cycle reprocessing and reactor technologies, on renewable energy or carbon capture and storage, or on both, along with working links to national laboratories and research universities in the region.[111]

Nonetheless, the difficulties of gaining state-level and broader public acceptance for developing a new federal interim storage facility for SNF are likely to be formidable, as illustrated by the PFS case and the unsuccessful outcome of DOE's attempts to site an MRS in Tennessee in the 1980s.[112] These experiences suggest that attempts to site and license a new consolidated storage facility would take many years, face substantial state and local opposition, and ultimately might very well prove unsuccessful. Strong opposition to consolidated interim storage facilities for SNF is all the more likely to arise when, as is currently the case, there is no viable plan for permanent disposal of SNF, thus making "interim" storage appear inescapably "permanent." The recent expression of interest by New Mexico officials in expanding WIPP's mission to potentially include at least interim or retrievable storage of SNF is, however, a notable new signal of receptivity by a potential host state.

Transportation of SNF is also a likely flash point for public opposition to consolidated interim storage facilities. Having to transport SNF to a consolidated storage site means SNF will have been transported twice before it reaches a permanent repository. As a practical matter, transportation routes through which SNF will pass would have to be negotiated with states in order to gain their cooperation in monitoring and assuring the safety of each shipment. Western governors proved a powerful force in the political battle over transportation and transportation planning for shipment of TRU wastes to WIPP. If transit states were to threaten to block shipments of SNF over their roads—despite federal authority to regulate interstate transportation and the potential for legal challenge to state measures based on the Commerce Clause of the Constitution and the federal transportation regulatory statutes (discussed in Chapter 3)—concerted state opposition could effectively hold up development of a consolidated storage facility.[113]

The public perceives the risk that nuclear waste transportation accidents will cause harmful releases of radiation to be high (see Chapter 3). After study, however, the American Physical Society concluded that the accident risks posed by transporting SNF to consolidated storage facilities are so low as to be *de minimis*.[114] Before spent fuel can be transported, it must be removed from dry or wet storage and placed in NRC-certified and -inspected shipping containers. According to GAO, to threaten residents near a transportation accident site, not only would the containers have to be breached, but also the SNF would have to be "pulverized" by an explosion or high-speed collision into particles small enough to be carried by the air.[115] A 2006 NAS report from a committee that studied the transportation of SNF to Yucca for disposal concluded that existing technologies and regulatory arrangements are generally adequate to ensure safe transport of large volumes of SNF. The American Physical Society and NAS committee findings regarding transport safety are consistent with the extensive record of transportation of TRU by truck to WIPP, with few reported accidents and no releases of radioactivity to the environment. The NAS committee, however, emphasized that large-scale transport of SNF would be a novel and challenging enterprise, and that it would be particularly important to address public concerns about the risks involved and to work closely with state and local authorities.[116]

While the available evidence indicates that transportation accidents do not appear to represent significant risks, deliberate terrorist attacks have not been adequately studied by independent experts because information necessary to evaluate such risks is classified. The NAS committee report cautioned that its generally favorable assessment of SNF transportation risks was subject to a substantial caveat: the committee did not have access to classified information on terrorism risks and was unable to evaluate their significance.[117] In a post-9/11 world, the widely perceived risks of SNF transportation are unlikely to be altered or allayed without a thorough, independent, and credible assessment of the risks posed by deliberate terrorist attack, based on evaluation of all relevant information.

SNF Reprocessing

Another potential response to accumulated SNF is to reprocess it to extract plutonium and uranium to make fresh fuel. Reprocessing, of course, would over time itself produce significant quantities of radioactive wastes, including some quite hazardous wastes, that would require disposal. Past efforts to reprocess civilian SNF in the United States proved a dismal failure. As a result of President Carter's ban on federal assistance, there has been no reprocessing of commercial SNF in the United States for nearly forty years. President George W. Bush advocated revival of reprocessing as part of his administration's GNEP proposal, but the proposal was rejected by Congress and has not been embraced by the Obama administration. GNEP did succeed, however, in putting the reprocessing option back in play, to the point where its inclusion as an element of a U.S. "nuclear renaissance" has been actively debated.

Early Commercial Reprocessing of SNF

In response to Nazi Germany's nuclear weapons program, the United States, in addition to developing uranium enrichment facilities to produce weapons-grade uranium, began constructing reprocessing plants capable of extracting plutonium from irradiated fuel rods for use in nuclear weapons. The intense sense of urgency instilled by World War II and the ensuing Cold War allowed the federal government to pursue reprocessing without regard to cost or, in many cases, environmental consequences. As a result, AEC developed the plutonium uranium extraction (PUREX) reprocessing process and applied it to irradiated fuel rods at reprocessing plants built at Hanford, Savannah River, and INL to obtain plutonium for production of nuclear weapons.[118] The PUREX process requires spent fuel rods to be chopped into small pieces, which are then dissolved in nitric acid. Various solvents are used to remove plutonium, uranium, and other isotopes from the dissolved fuel.[119] This process proved successful and was later used for reprocessing SNF from nuclear power plants with a view to making reactor fuel from the reclaimed plutonium and uranium. AEC justified reprocessing civilian SNF on the ground that uranium supplies were too scarce to support the expected growth of the nuclear power industry. Fuel made from plutonium and residual uranium obtained from reprocessed SNF was expected to replace enriched uranium fuel as the price of virgin uranium mounted.[120] While the price of uranium has fluctuated over time, it has not skyrocketed as AEC had predicted (see Appendix B).

Because fuel used in commercial light-water reactors resides in the reactor substantially longer than fuel for military production reactors, civilian SNF generates significantly greater radioactivity and decay heat than does defense SNF.[121] As a result, reprocessing SNF generated in commercial nuclear reactors poses significantly greater technical challenges than does reprocessing the irradiated fuel rods used in weapons production. Furthermore, the cost of making reprocessed fuel has thus far been significantly higher than that of making fuel from virgin uranium.

The first and only operational commercial reprocessing plant in the United States, located at West Valley, New York, was built to reprocess civilian SNF. It operated from 1966 until 1972. West Valley proved to be a failure on operational, economic, and environmental grounds.[122]

In the early 1960s, efforts to attract nuclear development to the state of New York became a major focus of Governor Nelson Rockefeller's economic development efforts. He formed a financing agency, the Atomic and Space Development Authority, with the charge of promoting new nuclear energy facilities in the state.[123] Negotiations between AEC, the state, and a private company—Nuclear Fuel Services (NFS)—led to an agreement whereby NFS leased land from the state for a reprocessing plant. NFS was responsible for all wastes produced while the facility was operating. At the end of the lease, the state would take over responsibility for the waste. The agreement led to the construction and operation of the Western New York Nuclear Services Center, of which the West Valley reprocessing plant was a part.[124]

AEC had twice considered NFS's initial proposals, and both times determined that there was not "sufficient assurance that the plant could operate efficiently and reliably."[125] Eventually, however, AEC accepted NFS's representations and issued a license for the plant, with the State of New York and NFS as co-licensees.[126] When the plant opened in 1966, there were few commercial nuclear reactors in operation. As a result, AEC agreed to guarantee a minimum amount of defense SNF to be delivered for reprocessing there. About two-thirds of the fuel that was reprocessed at West Valley was defense SNF; the remainder was commercial SNF reprocessed under contract with private utilities.[127] The plant's planned capacity was 300 metric tons of SNF per year. The plant could not meet this level of production, however, because there was not enough SNF available to reprocess.[128] Over its six-year life span, the West Valley plant processed only 640 metric tons of SNF, producing 1,926 kg of plutonium, none of which was used to make fuel for electricity generation in the United States. About 80 percent of the plutonium produced was shipped to Hanford for AEC's use. The remaining 20 percent was sent to domestic and foreign utilities for research purposes and overseas for use in breeder reactors.[129]

The West Valley plant was purportedly designed to prevent any significant release of radiation. However, over the course of its operation, there were numerous instances of failed containment, with at least thirty-eight workers exposed to high concentrations of radioactivity.[130] Although no single incident was judged by AEC to be serious enough to warrant shutting down the plant, the pattern of radiation releases called into question the plant's ability to operate safely in the long term.[131] After AEC had given NFS several ultimatums, threatening closure because of worker safety concerns,[132] NFS closed the facility for what it said was a long-planned temporary shutdown to expand capacity.[133] AEC informed NFS that on reopening, the plant would need to meet newly adopted AEC safety and environmental regulations requiring significant modifications to the plant, including adding a facility for solidifying HLW, a facility for solidifying plutonium, and new protections against natural disasters such as earthquakes and tornadoes.[134] NFS, which had originally estimated the cost of its planned modifications at $15 million, said that it would cost $600 million to meet the new regulatory requirements, $420 million more than its SNF reprocessing contracts with AEC and the private utilities were worth.[135] The company decided not to reopen the plant.

Upon closure of the West Valley plant, NFS exercised its option under its leasing contract with New York to transfer to the state the site's management and the remaining stored waste: a tank filled with more than 600,000 gallons of liquid HLW and 125 SNF assemblies in cooling ponds.[136] Dealing with the HLW posed special problems for cleanup because about half the radioactive material had settled at the bottom of the storage tank in the form of sludge, which could not be pumped out easily.[137] After public outcry from local citizens and environmental groups,[138] Congress enacted the West Valley Demonstration Project Act, which placed responsibility for the cleanup of the West Valley site on DOE. The act specified that the State of New York would be responsible for 10 percent of the cleanup costs, and the federal government would pay for the rest.[139] Cleanup operations at the site—including measures to deal with contaminated buildings, components, equipment, and soils—are far from complete and are expected to cost the State of New York and the federal government around $4.5 billion.[140]

In 1996, as required by the West Valley legislation, DOE began vitrifying the liquid HLW to turn it into glass logs that could be disposed of in a geologic repository. Vitrification of the liquid waste at West Valley was completed in 2002. In total, 275 ten-foot-tall steel canisters of hardened glass are being held in storage on-site.[141] The plan has been to dispose of them at Yucca or another repository when one becomes available.[142] Also, 125 SNF assemblies were transported by rail in 2003 to INL for storage until the SNF could be disposed of at Yucca Mountain or another repository.[143]

Planned commercial reprocessing plants at Morris, Illinois, and Barnwell, South Carolina, also had difficulty meeting regulatory requirements and faced unexpectedly high costs, despite various federal subsidies including donated land (see Chapter 1). The Illinois project was abandoned before construction had begun. The private developers of the Barnwell facility wrote to ERDA in 1975, asking it to construct the reprocessing facilities and have the developers operate the plant as a federally sponsored demonstration project.[144] This request came at a time when SNF reprocessing was viewed in an increasingly negative light because of concerns over proliferation of nuclear materials and weapons. President Carter's termination in 1977 of federal support for civilian SNF reprocessing effectively ended further development of Barnwell.

The 1970s: The End of Commercial SNF Reprocessing

In 1974, India exploded a plutonium-based weapon that owed its completion to a Canadian-built nuclear power reactor and U.S. reprocessing training.[145] Armed with the knowledge that plutonium could be diverted from civilian reprocessing facilities to produce nuclear weapons, Congress grew fearful of the "alarming and almost promiscuous spread of dangerous fuel cycle technologies."[146] At the same time, reports were published arguing that uranium was abundant enough in the earth's crust to ensure that enriched uranium fuel would for the foreseeable future be far cheaper than plutonium-based fuel from reprocessing.[147]

In 1976, President Ford, invoking concerns about nuclear proliferation, directed federal agencies to delay commercialization of SNF reprocessing until these concerns were resolved.[148] On April 7, 1977, President Carter announced his plan to "defer indefinitely the commercial reprocessing and recycling of the plutonium produced in the U.S. nuclear power programs" and halted federal support for civilian SNF reprocessing, including the Barnwell SNF reprocessing plant.[149] He also vetoed congressional authorization for a breeder reactor and associated reprocessing plant. Carter's decision was motivated by a desire to limit the proliferation of nuclear weapons and was based in part on the conclusion that nuclear energy could continue to be produced without plutonium-based fuel created through reprocessing.[150]

Carter's steps did not preclude private development of reprocessing without federal assistance, and for a while industry considered this possibility. The entering Reagan administration took steps to reduce the stringency of regulatory requirements for reprocessing plants and thus reduce some of their costs, but at the same time, it demanded that the private sector pay the full cost of reprocessing.[151] By 1981, none of the concerns about the commercial viability of SNF reprocessing had been resolved, and reprocessing was still politically controversial. In this highly uncertain environment, industry refused to make further investments in the technology. Recognizing that no progress had been made in developing a SNF/HLW repository, South Carolina state leaders feared that Barnwell would suffer the fate of the West Valley plant and be forced to provide indefinite storage for the HLW produced by reprocessing at the site. After unsuccessful efforts to find further financing, the Barnwell plant was terminated in 1983.[152]

Since 1972, when West Valley closed, U.S. nuclear power plants have been based entirely on a once-through nuclear power fuel cycle—the fuel is burned in a reactor only once, rather than reprocessed to be used again—with the expectation that the SNF produced would be disposed of in a repository. This policy was adopted by President Carter in part with the hope of encouraging other countries to abandon reprocessing or to not adopt it at all. The Carter policy's success in this respect was mixed. Various countries, including the United Kingdom, France, and Russia, continued to operate reprocessing plants.[153] At least six other countries, however, terminated plans to construct reprocessing plants, due in part to U.S. diplomacy.[154] The United Kingdom is currently in the process of closing its reprocessing facility on economic grounds, while Japan is struggling to open a troubled reprocessing plant and China is building several reprocessing facilities.[155]

The economics of once-through versus reprocessing fuel cycles is critically dependent on the price of virgin uranium.[156] MOX fuel produced through reprocessing has been, and continues to be, substantially more costly than fuel derived from virgin uranium.[157] Furthermore, only a relatively small number of reactors, mostly in France and Germany, are currently capable of burning MOX fuel.[158] As a result of these factors, plutonium has piled up at the existing reprocessing facilities in other countries, which creates terrorism and proliferation risks and the consequent need to provide security arrangements.[159] In 2000, the United States and Russia signed the Plutonium Management and Disposition Agreement to dispose of at least thirty-four metric tons of surplus weapons-grade plutonium each; the two nations updated that agreement in 2010 to resolve funding issues that had stalled progress. As part of this agreement, the United States pledged to build three major facilities for MOX fuel fabrication at the Savannah River Site (SRS) to turn surplus weapons plutonium into reactor fuel.[160]

To help induce other countries not to develop reprocessing or uranium enrichment facilities, and thereby to reduce the risk of their developing nuclear weapons capability, the United States at one point had a policy of providing other countries with enriched uranium for reactors generating electricity. Until 1973, AEC was required to lease enriched uranium to any country or company that agreed to operate a commercial reactor under international oversight.[161] The program involved close monitoring to ensure that the spent fuel produced was returned to the United States, to prevent its reprocessing by the lessee. This program was designed to discourage the development of reprocessing capabilities in other countries by assuring them of an alternative fuel supply source. But the strategy failed conspicuously in the case of the Indian nuclear bomb, because the Indian reactor used nonenriched uranium as fuel and therefore was not subject to inspection under the program.[162]

The Nixon administration privatized the fuel-leasing system and imposed strict burdens on applicants: fuel demands had to be projected eighteen years into the future, and a cap was placed on the total amount of fuel available.[163] European companies stepped in with supplies of enriched uranium to meet the unfulfilled demand that resulted from limitations of the new U.S. policy.[164] The policy was criticized by leaders in the domestic nuclear industry for "creating doubts about [its] reliability . . . as a nuclear supplier, which not only hurt U.S. commercial interests but weakened U.S. influence on proliferation problems."[165] Today, the United States is a minor player in the uranium fuel market. In fact, the United States depends on foreign sources for more than 80 percent of its uranium requirements, mainly close allies such as Canada and Australia.[166] In January 2007, DOE asserted that "participating fully in [the commercial nuclear] business is essential in order to shape the rules that apply to it."[167] DOE argued that GNEP reprocessing would produce uranium fuel that would promote both U.S. commercial interests and nonproliferation goals because "nuclear material in international commerce . . . of US origin [is] subject to U.S. consent" over its transfer and use.[168]

The Proposed Global Nuclear Energy Partnership and the Potential Revival of Reprocessing

GNEP

In his State of the Union address on January 31, 2006, President George W. Bush outlined his plans for the Advanced Energy Initiative, a program intended to promote various forms of clean energy, including "clean, safe nuclear energy."[169] A week later, DOE secretary Samuel Bodman and deputy secretary Clay Sell announced the Global Nuclear Energy Partnership (GNEP), a part of the Advanced Energy Initiative, which was intended to "bring about a significant, wide-scale use of nuclear energy."[170] At the heart of the program was the Bush administration's decision to resume reprocessing, which DOE claimed would "dramatically reduce the volume and radiotoxicity of the material that ultimately has to be disposed of [in a repository]."[171] In addition to minimizing waste, reprocessing was promoted as more energy-efficient. Because used fuel rods still retain more than 90 percent of their energy value, reprocessing, at least in theory, could retrieve a significant amount of this energy value for reuse.[172] It is not, however, technically possible to recapture and use all the residual energy value of spent fuel. Also, significant cost, environmental, and security issues must be considered in evaluating reprocessing.

As an integral part of GNEP, the United States would have reentered the uranium-leasing business by joining with other fuel cycle countries, such as Russia, the United Kingdom, France, and Japan, to provide enriched uranium obtained through reprocessing SNF to nations that want to develop commercial nuclear power, without their having to construct either uranium enrichment or reprocessing facilities.[173] The lessees would then return the spent fuel to a fuel cycle country, where it could be reprocessed. A number of developed and developing countries agreed to participate in a preliminary prototype for such a regime.

DOE's GNEP proposal did not address whether the waste materials generated by reprocessing would remain in the participating developed countries that would carry out the reprocessing, or turned over to the participating developing countries generating the SNF that was reprocessed.[174] In order to obtain the participation of developing countries seeking to develop nuclear power but unwilling to deal with highly radioactive nuclear wastes, the developed countries might have to agree to take care of all the wastes associated with the program, aggravating the domestic politics of nuclear waste in the developed countries.[175]

GNEP envisaged a new reprocessing separation technology known as UREX+, which would produce a number of products that could be reused as fuel, as well as generate nuclear wastes.[176] Uranium isotopes would be separated out first, then either reused in light-water reactors in the United States or foreign countries or disposed of as LLW. Next, two highly radioactive, thermally hot, moderate-lived fission products, strontium (Sr-90) and cesium (Cs-137), would be separated out as wastes and placed in surface storage for several centuries, most likely in the vicinity of the reprocessing plant, until the waste had decayed enough that it could be disposed of as LLW. Other fission products then would be separated out as wastes (with the exception of the lanthanides, discussed later) and disposed of as HLW in a repository.[177] Finally, the transuranic elements remaining in the reprocessing liquids, including plutonium, would be used to make a fuel mixture that would be burned in new fast reactors to generate electricity.[178] Unlike the PUREX process, then, UREX+ would not separate out pure plutonium but would include other transuranic elements in the fuel. The proposal envisaged several further cycles of electricity generation, production of SNF, and reprocessing, with the result that almost all of these other transuranic elements would be "burned" in the fast reactors.[179] However, a small percentage of TRU would remain in the reprocessing wastes and ultimately require disposal; because of the very large volumes of materials involved in the recycling processes, this small percentage would, as discussed further later, represent a cumulatively large quantity of TRU.[180]

DOE considered two variations of UREX+: UREX+1 and UREX+1a. UREX+1 would leave the lanthanides (fission daughter products with atomic numbers 58–71) in the final reprocessing product along with the plutonium/transuranics fuel mixture, rather than separating them out with the other fission products. The GNEP proposal asserted that the resulting UREX+1 mixture would be more proliferation resistant than MOX, because lanthanides produce significant gamma-radiation levels that would make the mixture more dangerous and difficult to handle and convert to weapons use, and also because a large amount of the mixture would be needed to make bombs.[181] The presence of lanthanides would also prevent fission of the plutonium/transuranics fuel mixture with the aim of making it unusable for weapons, and thereby reduce proliferation risks. However, the lanthanides would have to be removed—and in turn disposed of as HLW—before the reprocessed mixture could be used as fuel in reactors for generating electricity. The UREX+1a process would remove the lanthanides, along with the other fission products, at an earlier stage in the process.[182]

The cesium and strontium in SNF provide the main sources of radioactive decay heat in a repository for the first several hundred years. By removing these fission products from wastes to be disposed of at Yucca, DOE claimed that a twentyfold decrease in the "long-term temperature increase in the rock around [Yucca's] disposal tunnels" would have been achieved.[183] DOE thus touted, as one advantage of reprocessing, the production of a cooler form of waste requiring repository disposal, in contrast to unreprocessed SNF, which according to DOE needed to be disposed in a "hot" repository; DOE's hot repository design for Yucca had stirred up considerable controversy about whether wastes buried at Yucca would be safely contained there over the long-term.[184] Long-lived fission products (other than cesium and strontium) that remained in the waste, including I-129, Cs-135, Tc-99, Sn-126, and Se-79, would still need to be disposed of in a geologic repository, because none of these isotopes has a half-life of less than sixty-five thousand years.[185]

If reprocessing were to be carried out by the private sector, as GNEP contemplated, reprocessing wastes would be regulated solely by NRC; if by the federal government, regulatory authority would have been shared between DOE and NRC—the latter would have had jurisdiction because of the civilian SNF involved. In either case, the waste management and disposal would have been subject to EPA radiation standards.

To implement GNEP, the successful development and construction of three types of facilities would have been necessary:[186] first, a nuclear fuel–recycling center capable of reprocessing spent fuel and fabricating new plutonium-based reactor fuel, which would presumably also store the cesium and strontium removed from reprocessing wastes;[187] second, a new reactor capable of burning the new fuel form containing plutonium and minor transuranics; and third, an advanced fuel cycle research facility that would continue to develop transmutation fuels and improve fuel cell technology.[188]

Criticisms of GNEP: Ill-Founded, Costly, and Counterproductive

GNEP attracted a barrage of criticisms from the NAS Committee on Review of DOE's Nuclear Energy Research and Development Program, GAO, Congress, and outside experts.[189] Congress also expressed great skepticism over the proposal as attempting to accomplish too much too quickly.[190] The principal grounds for criticism included the following.

Technology

A central technological concern raised by the GNEP proposal was that important components of the UREX+ fuel cycle, including the chemical separation processes, had yet to be demonstrated on a commercial scale. In addition, it would have been necessary to develop and qualify a new plutonium-based fuel containing minor actinides, as well as new forms of commercial-scale, sodium-based, fast burner reactors to burn the new fuel.[191] DOE projected that the design, approval, and construction of the GNEP reprocessing facilities would take more than a decade[192]—no doubt a highly optimistic estimate. An expert from Argonne National Laboratory estimated that developing a closed cycle system using UREX+ technology in reprocessing and new fast reactors would take about twenty years to be ready for commercial operations.[193] This estimate also appears optimistic.

DOE circulated a request in August 2006 for expressions of interest from companies willing to build a consolidated fuel treatment center that would include SNF reprocessing and transmutation fuel fabrication capabilities. DOE proposed a two-tiered plan: reprocessing facilities would be built with technology that was currently available or soon to be available on a commercial scale, and R&D would be conducted on advanced recycling processes that had only been tested in the laboratory.[194] This proposal was not adopted.

Wastes

GNEP would have ameliorated some nuclear waste problems while creating others. On the one hand, separating out cesium and strontium for storage and disposal as LLW and then burning actinides in advanced reactors would significantly reduce the volume, heat, and radioactivity of the wastes requiring permanent geologic disposal relative to SNF. On the other hand, facilities for safe and secure storage of cesium and strontium for a period of several centuries would be required. This in turn would require monitoring and other institutional controls to be maintained throughout that period, an unprecedented feat. In addition, the GNEP UREX+ process would generate significant amounts of TRU wastes not found in unreprocessed SNF.[195] DOE projected that the UREX+ process would separate transuranic elements at better than 99 percent efficiency.[196] But even assuming 99 percent efficiency, as much as sixty-three million curies of TRU waste could still be generated; these wastes would be highly radioactive and would require remote handling to process.[197] This estimated amount of the TRU waste that would be generated by UREX+ represents twenty-four times more radioactivity than the TRU waste in current inventories at all DOE sites.[198] The WIPP repository is not currently authorized to receive commercial wastes, including TRU derived from reprocessing commercial SNF. The department, however, did not address where the TRU wastes produced by UREX+ reprocessing would go.

Various groups asserted that, while in theory the GNEP UREX+ process would eliminate the need to dispose of SNF, it would actually increase the total volume of radioactive wastes produced, mostly LLW but also some HLW.[199] DOE itself acknowledged as much in a low-key manner in its draft EIS for GNEP.[200] The current once-through nuclear power fuel cycle in the United States generates 21,950 square meters of nuclear waste per year. Under a reprocessing system with fast reactors, DOE projected that the nation would generate 112,257 square meters of waste annually. Most of this would be LLW, including GTCC waste that may require repository disposal. DOE also projected that a fast reactor system would generate 1,480 square meters of HLW annually that would need to be disposed of in a repository. Also, the international regime envisaged by GNEP might require the United States to take back for disposal SNF from developing nations that the United States had supplied with uranium for nuclear power plants; this would almost certainly stir sharp opposition from states asked to host these wastes.

Cost

Costs merely for fleshing out the GNEP proposal would have been considerable: DOE requested $250 million in FY 2007, and $395 million in FY 2008 for developing mission descriptions, functional requirements, conceptual designs, and life-cycle cost estimates for the proposed GNEP facilities.[201] DOE did not present any life-cycle cost estimate for the project, and no reliable data exist concerning the total price tag for an operational advanced reprocessing and fast burner system. In 1996, the NAS Commission on Geosciences, Environment, and Resources found that the total price tag for a particular advanced reprocessing scenario could reach $500 billion,[202] but changes in technology, the passage of time, and the seemingly inevitable tendency of the costs of ambitious projects like this to increase dramatically over time all sharply limit the usefulness of the commission's estimate.

Significantly, DOE was not willing to assume many of the financial risks involved in developing, constructing, and operating these new technologies and facilities at commercial scale. One of GNEP's central goals was "to develop and implement fuel cycle facilities in a way that will not require a large amount of government construction and operating funding to sustain."[203] A bifurcated strategy was envisioned: industry would take the lead in financing and building the recycling center and new reactors, while DOE would be responsible for establishing a research facility at a government site.[204]

Industry would undoubtedly take a very hard look at the economics of reprocessing, fuel fabrication, and new fast burner reactors before making substantial financial commitments. In the late 1960s and early 1970s, when reprocessing initiatives were underway in the United States, uranium prices were at an all-time high and were expected to keep rising.[205] Also at that time, the United States generated significant amounts of electricity through oil-fired power plants, and the price of oil was peaking. The spike in uranium prices spurred exploration that resulted in the discovery of substantial deposits of uranium ore in Australia and Canada.[206] Increased uranium supplies coupled with declines in fossil fuel prices were a critical factor in U.S. industry's lack of interest in pursuing reprocessing after federal support was halted.[207]

The economics of nuclear power would brighten considerably if climate regulatory requirements effectively imposed a price on emissions of carbon dioxide that would make electricity generation based on coal, or even natural gas, significantly more costly.[208] But a boost for nuclear power resulting from climate regulation would not necessarily translate into more favorable economics for reprocessing, since reprocessing feasibility depends explicitly on the cost of reprocessed fuel versus that of virgin uranium. A report produced by the Belfer Center for Science and International Affairs at Harvard's John F. Kennedy School of Government in 2003 concluded that MOX fuel produced by the existing PUREX process would not become economically competitive until the price of uranium reached \$340 per kilogram, more than \$154 per pound.[209] Because it is a complex technology and has yet to be developed, the break-even point for the GNEP UREX+ process would be much higher.[210] A December 2006 study by DOE's Idaho National Laboratory implied that the price of uranium would have to increase to about \$400 per pound (\$881 per kilogram) in order for advanced reprocessing to become economical.[211] Other cost estimates range even higher.[212] As shown in Appendix B, yearly average prices for uranium have varied for the past several decades but at levels far lower than these figures. A short-term spike in prices, reaching an all-time spot price high of \$140 per pound (\$309 per kilogram), occurred several years ago as a result of market reactions to a 2006 flood at a uranium mine in Saskatchewan, Canada,[213] but by July 2010, prices had declined to \$41.50 per pound (\$91.50 per kilogram).[214] Uranium prices would have to rise for the indefinite future far above their historical levels to make reprocessing economically viable. Although expansion of nuclear power worldwide will increase demand for uranium, abundant supplies of virgin uranium are currently available from many countries, including U.S. allies such as Australia and Canada.[215] Thus, while some energy security benefits may be gained by using current stocks of SNF as a fuel resource, it does not appear that such benefits are likely to be substantial, given significant supplies of virgin uranium from friendly nations.

International Security Issues

The George W. Bush administration contended that GNEP would promote international security and reduce the risk of proliferation by providing developing countries wishing to develop nuclear power with an assured source of uranium, thereby inhibiting the development of uranium enrichment or SNF-reprocessing facilities. A number of experts on nuclear proliferation, however, asserted that the effect on nuclear security would have been the reverse: they strongly attacked the GNEP proposal as significantly increasing proliferation and terrorism risks.[216]

Critics concerned with proliferation and terrorist risks have pointed out that neither UREX+1 nor UREX+1a produces fuel that is "so highly radioactive that it could not be handled directly by human beings."[217] Such critics were also not persuaded that inclusion of lanthanides in the plutonium mixed fuel produced by the UREX+1 process would do much to deter terrorists or others from using the fuel for illicit purposes. Further, they contended that mixing plutonium with other actinides in the new fuel form would make it more difficult to accurately measure the quantity of plutonium present, which might allow undetected diversion to occur for longer periods.[218] To mitigate these risks, the GNEP proposal stipulated that the UREX+ processes would be limited to fuel cycle countries.[219] But since these countries are already able to ensure the safe storage of PUREX-separated plutonium, why, asked the critics, pay for redundant protections by producing, at significantly greater cost, a new form of fuel that contains minor actinides and possibly lanthanides?[220] Moreover, the UREX+1 process would require an extra processing step: before being used as fuel, the plutonium/transuranics mix must be stripped of the lanthanides.[221]

By limiting uranium enrichment and spent-fuel reprocessing to fuel cycle countries, GNEP aimed to dissuade other countries, by providing them with uranium fuel, from developing uranium enrichment or reprocessing capabilities that could be used for nuclear weapons production, as noted.[222] If the fuel cycle nations could cheaply produce reactor-grade uranium through economy-of-scale savings, and thus provide an assured supply of such fuel, then a global consortium might have some success in this aim.

Independent of GNEP, the International Atomic Energy Agency (IAEA) made a proposal in 2006 for a multilateral uranium fuel bank as part of its Nuclear Threat Initiative. As of March 2009, the European Union, Kuwait, Norway, the United Arab Emirates, and the United States had together pledged $100 million to develop such a bank.[223] However, many developing countries have resisted joining the proposal, in part out of protest at the notion that the rich countries should maintain a lock on enrichment or reprocessing technologies. Iran indicated in 2008 that it would join a fuel bank only as a supplier. Countries as diverse as South Africa, Egypt, and Italy have balked at abstaining from uranium enrichment so long as the United States and other nations maintain facilities to do so.[224]

Siting

The GNEP proposal would have required DOE to site reprocessing facilities that would generate substantial amounts of wastes, including HLW, TRU, and GTCC wastes. The federal government could make this prospect more attractive to potential host communities and states by co-locating advanced reactor R&D facilities and other "energy park" elements with reprocessing facilities, and also support related research in universities within the state or region. Whatever the incentives, however, potential hosts, including those that already have DOE weapons facilities, would have to weigh the tradeoffs between the economic and other benefits of such a package against the burden of storing HLW, including storage of cesium and strontium for several centuries if a decision were made to separate out those elements and ultimately dispose of them as LLW.[225] The Hanford site previously mishandled an attempt to hold cesium and strontium in decay storage, resulting in $50 million in cleanup costs.[226] A site that agreed to host a reprocessing plant would also have to face the risk of becoming an indefinite storage facility for the TRU wastes produced by reprocessing, since commercial TRU waste is not currently authorized to be disposed of at WIPP. In addition, any such reprocessing facility would probably include consolidated storage for SNF awaiting reprocessing. If the reprocessing facility failed to open or was shut down—as was the case with West Valley and other abortive attempts to launch commercial reprocessing facilities in the 1960s and 1970s—the SNF might remain at the site indefinitely.[227]

The Cloudy Prospects for Reprocessing

The Obama administration has dropped GNEP.[228] Congress limited FY 2009 funding for GNEP to $3 million.[229] DOE declared in April 2009: "The long-term fuel cycle research and development program will continue but not the near-term deployment of recycling facilities or fast reactors."[230] In June 2009, DOE cancelled GNEP's programmatic environmental impact statement, effectively ending the project.[231]

The U.S. nuclear industry has nonetheless expressed renewed interest in reprocessing, citing expansion of nuclear power production worldwide, resultant increases in demand for nuclear fuels, and SNF disposal concerns.[232] NRC in 2008 asked industry to formally submit comments on a new regulatory framework for licensing reprocessing plants.[233] Within six months of the request, the Nuclear Energy Institute, an industry organization, submitted a white paper to NRC advocating the adoption of a new regulatory framework for SNF-reprocessing facilities.[234] After producing its own regulatory analysis, NRC stated in 2009 that momentum for such a proposal was building and announced that it would next begin working on a technical basis for rulemaking.[235]

GNEP still meets as an international network, but without the full backing of the United States. At a meeting of GNEP in June 2010, the members agreed to change the name of the organization to the International Framework for Nuclear Energy Cooperation. Along with the name change, the organization changed its mission in order to broaden its scope and increase international participation.[236] The framework intends to provide "a forum for cooperation" in the use of nuclear energy for peaceful purposes. The new mission also states that "participating states would not give up any rights and voluntarily engage to share the effort and gain the benefits of economical, peaceful nuclear energy."[237]

The Obama administration has expressed support for a global uranium fuel bank, like that proposed by IAEA, to address the concerns about proliferation. The bank would provide developing countries a supply of low-enriched uranium for nuclear power plants. Although the plan would not prohibit the development of enrichment capabilities, it would create an economic incentive that would discourage their development, because it would be cheaper for countries to purchase uranium fuel from the bank than to develop their own enrichment capacities.[238]

Conclusion

It is highly uncertain when a repository for disposal of SNF and HLW may be available. Prudent planning must proceed on the assumption that it could well be several decades or more. The basic options for dealing with the wastes in the interim are clear: for defense wastes, storage at DOE sites, most probably at the same sites where they are currently stored but also or alternatively, consolidated storage at other sites; for power plant SNF, continued at-reactor storage; consolidated storage at new federal or private facilities, and (although far less promising from today's standpoint) potential reprocessing. It is far from clear, however, which of the available options, or combination of them, should be chosen to deal with power plant SNF. The Blue Ribbon Commission on America's Nuclear Future can contribute greatly to a good decision by gathering information on and analyzing the economic, security, and other policy elements and tradeoffs of the alternatives. In our current state of knowledge, there is much to be said for a diversified strategy, as recommended by IRG, trying out a variety of approaches and mobilizing diverse initiatives. For example, the government could simultaneously ensure that SNF in reactor pools is moved to dry-cask storage as promptly as is feasible while ensuring that both pool and dry-cask storage are hardened against terrorism; and develop at least one federal consolidated interim storage facility for, at a minimum, SNF from decommissioned power plants. The federal government could also facilitate private development of consolidated facilities subject to regulatory safeguards, including public participation requirements, emergency planning and response obligations, and financial assurances, that are equivalent to and as stringent as those applicable to federal facilities. Such a combination of measures would promote more secure storage of SNF, generate valuable experience in siting new waste facilities that could also be of value for future repository siting efforts, and generate information on the performance of different storage methods that would inform future policy. These steps would not foreclose the reprocessing option, but the government should carefully consider and determine the wisdom or folly of SNF reprocessing; the government could, if it thought it prudent, make a limited investment in reprocessing R&D in order to make a more informed future assessment of reprocessing in the future.

Chapter 8

Nuclear Waste in the United States: Lessons Learned and Future Choices

This chapter first briefly recapitulates the highlights of the history of U.S. nuclear waste regulatory law and policy over the past sixty years and summarizes the current status of treatment, storage, and disposal actions and regulatory policies with respect to the categories of nuclear waste that have been the subject of this book. It then identifies and analyzes lessons gleaned from experience to date, especially with regard to siting and developing nuclear waste storage and disposal facilities, with particular attention to the experiences at the Waste Isolation Pilot Plant and Yucca Mountain. These essential "lessons learned" in turn inform the analysis of options for future nuclear waste management in the final section of this book.

Evolution of U.S. Nuclear Waste Policy: Recapitulation

The history of U.S. nuclear waste law and policy begins with the Manhattan Project and the race to make an atomic bomb during World War II. Accumulation of defense nuclear waste, an unaddressed byproduct of that wartime effort, steadily increased, especially after the war, as the Cold War arms race spurred intensive new production of nuclear weapons. Reprocessing of irradiated fuel rods to extract plutonium for weapons production generated large quantities of highly radioactive liquid reprocessing wastes stored in single-hulled tanks. Weapons production and other defense-related activities also generated substantial quantities of liquid HLW from reprocessing, lesser amounts of TRU, and large volumes of LLW. Hasty and often unsafe storage and disposal of these wastes at government facilities left a massive legacy of contamination. These wastes have since been the subject of enormously costly—and far from complete—remediation efforts at DOE's weapons complex sites. Large volumes of these wastes remain in storage at DOE facilities to this day and currently lack any disposal pathway.

Based on studies conducted by various National Academy of Sciences committees and government agencies from the 1950s through the 1970s, a scientific consensus emerged: the best method for safely disposing of highly radioactive nuclear wastes, including SNF and HLW, is burial in geologic repositories sited in dry, stable media far from significant populations. Salt formations have long been considered especially favorable potential sites for such repositories. This consensus in favor of deep geologic repositories has prevailed to the present day. For more than thirty years, it has been the foundation of federal policies for dealing with the most highly radioactive nuclear wastes.

The SNF generated by civilian nuclear power initially presented a distinct set of issues. Beginning in the 1950s, commercial nuclear power was developed with strong federal government promotion and support. AEC's policy was to use uranium fuel in commercial reactors and reprocess the SNF that they produced in order to harvest its constituent uranium and plutonium for reuse as fuel. This policy, and the failure to pursue the alternative of a thorium cycle for electricity production, reflected the federal government's determination to use the same basic fuel cycle for civilian power production that it had developed for weapons production. The expectation that civilian SNF would eventually be reprocessed, coupled with the fact that the civilian nuclear power industry was in the early stages of its development, caused scant attention to be paid to both the HLW that would eventually be produced by reprocessing SNF and the need to dispose of it, and no heed at all to the possibility that large quantities of SNF might one day require disposal if reprocessing did not prove to be economically or otherwise viable in the context of commercial electric generation.

When civilian reprocessing of SNF was eventually undertaken with generous AEC support in the later 1960s, it encountered serious operational, economic, and environmental problems. Two planned reprocessing plants never opened, and the only operating plant—at West Valley, New York—had to be shut down. In the mid-1970s, Presidents Ford and Carter halted federal support for reprocessing because of proliferation concerns associated with the production of plutonium. Since then, there has been no reprocessing of SNF in the United States. As a result, large amounts of commercial SNF have accumulated over the years, stored at the scores of nuclear power plant sites where they were generated.

Also during the late 1960s and the 1970s, public opinion against nuclear power was building, propelled by concerns about reactor safety—later reinforced by the accident at Three Mile Island—and the absence of any plan for disposal of accumulating SNF. Organized opponents of nuclear power, including local and national anti-nuclear, environmental, and scientific advocacy organizations, effectively used litigation to halt or delay construction of new nuclear power plants, while many states passed legislation placing a moratorium on new nuclear plants until a permanent disposal solution could be assured. These developments were among the several factors that led over the past thirty years to a slowdown and then a complete halt in development of new commercial nuclear power plants in the United States. Distrust of the federal government's handling of nuclear wastes was heightened by revelations that the government had withheld information about the deleterious health risks of fallout from its nuclear testing program and also had misled the public about radioactive waste contamination and releases at its weapons production facilities.

Growing concerns by the public and the Carter administration over the environmental risks posed by both civilian and defense wastes, reinforced by the nuclear power industry's belief that the problem of SNF disposal must be solved to avoid imperiling the future of nuclear power, led to Carter's creation in 1978 of the Interagency Review Group on Nuclear Waste Management. IRG's report and recommendations were premised on the imperative obligation of the current generation to achieve speedy burial of nuclear wastes so as not to burden future generations. IRG accordingly proposed a concerted national effort to build, on an expedited basis, several repositories and at least one consolidated storage facility for highly radioactive wastes. It also recommended a stronger federal role to deal with problems in the existing system of LLW disposal, which relied on a handful of commercial disposal facilities that were experiencing environmental and operating problems.

IRG's recommendations, backed by the Carter administration, stimulated congressional legislation in the form of the 1980 Low-Level Radioactive Waste Policy Act (LLRWPA) and the 1982 Nuclear Waste Policy Act (NWPA). Rejecting IRG's proposal for a strong federal role in LLW disposal, Congress in LLRWPA followed the recommendation of the National Governors Association for a system of interstate compacts to spur cooperative state initiatives to develop new disposal sites. This decentralized strategy has proved a failure. By contrast, Congress in NWPA followed IRG's recommendations for a highly centralized federal program for disposal of HLW and SNF. NWPA prescribed a detailed process and tight timetable for DOE to select and develop two repositories through a technocratic process for evaluation of candidate sites. NWPA obligated DOE to take responsibility for power plant SNF beginning in 1998; the highly compressed statutory timetable was designed to enforce development of a repository in time to enable DOE to discharge that commitment.

The issue of disposing of defense TRU, meanwhile, evolved along an entirely different, largely independent pathway as a result of happenstances and resulting controversies and engagements involving AEC/ERDA/DOE and several western states. A 1969 fire at AEC's Rocky Flats, Colorado, weapons plant forced AEC to ship large amounts of TRU wastes from the plant for storage in Idaho; the state subsequently demanded their removal. AEC/ERDA responded with a rushed and clumsy effort to develop a repository at an ill-chosen Kansas site, Project Salt Vault, at which experiments on disposal of HLW were being conducted. When local community leaders in Carlsbad, New Mexico, heard of the incident and expressed interest in hosting a repository, ERDA shifted its efforts to Los Medanos, a salt-bed site on federal lands near Carlsbad that eventually became WIPP.

WIPP was not developed in accordance with an advance plan. It proceeded haltingly, step-by-step, through an iterative process of contention and bargaining involving ERDA/DOE, the State of New Mexico, local community and environmental groups, Congress, the courts, and other western states. Throughout its development, which took more than twenty years, WIPP faced significant opposition within New Mexico. DOE was repeatedly sued in court by the state, environmental groups, and other stakeholders. Major controversies arose over the suitability of the site; what wastes WIPP should hold (SNF and HLW were seriously considered, as well as TRU); the state's right to participate in decision making about, and regulatory authority over, the facility; and transportation issues. The Carlsbad community loyally supported WIPP over the years, eager for the jobs the project would bring to the economically battered town. Over time, the state was able to secure its major objectives concerning WIPP through a series of legally enforceable agreements with DOE and favorable federal legislation that, among other matters, required DOE to enter into an agreement for consultation and cooperation with the state. Throughout the process, the state benefited greatly from the work of an independent expert group, the Environmental Evaluation Group, established by New Mexico with DOE funds to study and evaluate DOE's plans for WIPP. An important part of the eventual deal between the federal government and New Mexico was that WIPP would take only defense TRU wastes.

NWPA's unilateral, detailed blueprint for rapid development of federal SNF and HLW repositories and storage facilities, on the other hand has proved a failure. Despite extensive federal legislation, studies and recommendations by expert bodies, expenditures of billions of dollars (more than $13.5 billion on Yucca alone), and years of work by DOE, neither a repository nor a federal monitored retrievable storage facility has been established. The forced-march scheme devised by Congress in 1982 soon proved to have an overambitious timetable and quickly became bogged down by the complexities of site characterization and by opposition from potential host states.

Within five years, Congress radically changed the siting process by designating Yucca Mountain the sole candidate site for a repository and dropping plans for a second repository in the East that had been called for by the original NWPA legislation. The Yucca plan emerged as the result of a political deal crafted by a small group of powerful members of Congress—including, notably, representatives of the two states with the other top candidate sites. Politically weak Nevada was forced to host all the nation's SNF and HLW. This maneuver created deep and abiding outrage in Nevada; the state has since used every means at hand to block the repository, including attacks on the site's suitability, litigation, public relations efforts, and political mobilization. These efforts failed to stop congressional override in 2002 of Nevada's disapproval of the Yucca site, the limited state veto right provided under NWPA. But litigation by Nevada and environmental groups slowed the process for NRC regulatory approval of the repository. The election as president of Barack Obama, who had stated his opposition to Yucca in the 2008 Nevada Democratic presidential primary, together with Nevada senator Harry Reid's position as Senate majority leader, dramatically lifted Nevada's political fortunes. The new administration quickly terminated Yucca's funding. Further, in March 2010, DOE moved NRC to withdraw its Yucca license application "with prejudice" on "policy" grounds. In June 2010, however, the NRC Atomic Safety and Licensing Board denied DOE's motion on the grounds that, under NWPA, DOE does not have authority to withdraw the license application.[1] That decision has been appealed to NRC, which appears to be deadlocked on the case. Meanwhile, the states of South Carolina and Washington and other petitioners have sued DOE in the Court of Appeals for the D.C. Circuit, challenging DOE's withdrawal of the Yucca license application; the court is proceeding to hear the cases.[2] The license application withdrawal issue will eventually be resolved by the courts, and perhaps ultimately by Congress. If the license application for Yucca were to go forward and be granted by NRC, Congress would still have to decide whether to fund Yucca's construction.

At the same that he has steadfastly opposed Yucca, President Obama has strongly supported expansion of nuclear power in the United States through generous federal subsides. Many in Congress, especially Republicans, heartily endorse this policy. While Obama has justified expanded nuclear power as a means of addressing climate change, many experts disagree that expanding nuclear power through government subsidies is a wise or effective strategy. In any event, many states and citizens will almost certainly oppose allowing SNF to remain at nuclear power plants indefinitely and are unlikely to support dramatic expansion of nuclear power without a credible means for dealing with the most dangerous nuclear wastes. The goal of prompt burial of SNF and HLW embraced by IRG in the late 1970s and by Congress in the 1982 NWPA now appears a mirage: a prospect that, after nearly three decades, is further away than ever. To help solve President Obama's nuclear policy dilemma, the administration has assembled the Blue Ribbon Commission on America's Nuclear Future and charged it with the task, among others, of recommending a plan to deal with the nation's orphaned SNF and HLW.

Current Nuclear Waste Dilemmas and Options

LLW

The Current Situation

Most civilian class A, B, and C LLW is currently disposed of in three privately operated facilities located in Barnwell, South Carolina; Clive, Utah; and Richland, Washington.[3] The Washington and South Carolina facilities, which were established in the 1960s, accept class A, B, and C LLW, but Richland is open only to the eleven states in the Northwest and Rocky Mountain Compact, and Barnwell is open only to the three states in the Atlantic Compact. The Utah facility, which was developed wholly through private initiative, is open to wastes from all states but accepts only class A LLW.[4] It receives large quantities of such wastes, often of very low activity and shipped for long distances. Since passage of the 1980 LLRWPA and the 1985 LLRWPAA, none of the ten approved state compacts, nor any individual state that has not joined a compact, has opened a new LLW waste disposal site.[5] The resulting shortages in disposal capacity have produced higher disposal fees, which in turn have led generators to take measures to reduce the amount of wastes generated and also to reduce the volume of those wastes after they are generated. These steps have reduced the magnitude of the orphan LLW problem, but by no means have they eliminated the problem. Because disposal sites are simply not available to generators of class B and C wastes in many states, and because disposal, when available, is costly, generators are storing many of their class B and C wastes on-site for an indefinite period. The costs and burdens of at-generator storage have created serious problems for many generators.[6] And the projected generation in the next several decades of large amounts of additional LLW—primarily class A LLW but also more hazardous LLW, as a result of decommissioning nuclear power plants—will put additional strain on current arrangements.[7]

A new LLW facility designed to accept class A, B, and C LLW is, however, being constructed at Andrews, Texas, under the auspices of the Texas Compact. Initially it would have accepted waste only from compact members Texas and Vermont, as well the federal government. However, the Texas Low-Level Radioactive Waste Commission recently voted to authorize the Andrews facility to accept class A, B, and C LLW from as many as thirty-six states.[8] The decision may well be subject to legal challenge by local groups, and opening Andrews to noncompact states would require a vote of the Texas Compact Commission. However, the commission, which is controlled by Texas, is expected to approve it and to exploit its market power to extract high fees from noncompact states for use of the facility. If the facility becomes available to out-of-compact class A, B and C wastes, it would relieve existing class B and C disposal problems and provide competition with the Clive, Utah, facility for disposing of class A LLW.

DOE is responsible for disposal of all classes of defense LLW and for civilian GTCC wastes. DOE's own facilities provide a disposal pathway for its class A, B, and C LLW, supplemented by commercial facilities. But DOE still has no disposal plan for defense and civilian GTCC wastes, the most hazardous type of LLW. It remains to be seen whether land disposal of some GTCC wastes, with suitable engineered barriers and institutional controls, at existing DOE or other federal sites, will be environmentally and politically acceptable, or whether GTCC wastes will have to be disposed of in a repository such as WIPP or in a HLW/SNF facility yet to be developed.

Strategies for Expanding LLW Disposal Capacity

The system, adopted by Congress at NGA's urging in 1980, of decentralized state initiatives, structured by a federal system of compacts and incentives, has proven a failure. Congress envisaged a system of regional compacts composed of contiguous states that would cooperatively site new disposal facilities for each region. Most of the compacts that have emerged in shapes of gerrymandered malformation bear no relation to this picture, and none save the Texas Compact has developed a new disposal facility. The Texas Compact, moreover, is really nothing more than an alter ego of the State of Texas, which outvotes Vermont, the only other member. Texas is free to exploit the authority granted by the 1985 LLRWPAA to selectively exclude or admit wastes from other states, and thereby gain significant market power. Texas is now seeking to exploit its market position by opening the Andrews facility to out-of-compact wastes. If this proposal succeeds, the Andrews facility, which is slated to take all classes of LLW, and the Clive facility in Utah, which takes class A wastes from all states, could, together with the existing compact facilities taking compact wastes, provide a near-term remedy to the nation's LLW waste disposal problems. But, as discussed in Chapter 4, this arrangement does not assure a stable, economical, or equitable long-term solution to LLW disposal. The scope for political gamesmanship, NIMBYism, and exploitation of market power under the current regime remains so great that longer-term reforms are needed to develop a more robust LLW disposal system like that envisaged by IRG and the 1980 Congress, consisting of a network of well-regulated regional disposal facilities at environmentally suitable sites.

Congress in 1980 correctly recognized that such a system could not be achieved under the Supreme Court's interpretation of the Commerce Clause, holding that, absent congressional legislation, a state with a disposal facility must open it to wastes from all other states on nondiscriminatory terms.[9] But the solution that Congress subsequently devised has not worked. Part of the blame may be laid on the Supreme Court's *New York* decision, which effectively stripped a key incentive from the scheme by invalidating the LLRWPAA provision requiring a state to take title to LLW generated within its borders unless the state either had a disposal facility or was a member of a compact that did.[10]

Options that might be investigated for developing additional long-term disposal capacity for class A, B, and C civilian LLW include:

- Federal subsidies for existing or new LLW facilities that are open to wastes from all states, especially facilities that accept B and C wastes. Also, if needed, the federal government could provide financial or in-kind benefits to localities and states hosting such facilities.

- A much stronger federal role in civilian LLW disposal, as recommended by IRG, with a federal executive planning body, NRC leadership over siting and regulation, and development of new facilities on federal land. Suitable procedures for host community and state involvement in decision making and incentives for host acceptance of facilities would be needed.
- If serious disposal capacity problems persist or reappear, opening DOE disposal facilities on a commercial basis to civilian wastes. As of 1999, DOE's LLW disposal facilities in Nevada and Washington could accommodate an additional 171 million cubic feet of LLW and were capable of handling class B and C wastes. Only a small fraction of this capacity is needed for DOE's own wastes; the remainder could easily accommodate current and anticipated civilian LLW streams.[11] DOE is already tasked with dealing with civilian GTCC wastes, and there appears to be nothing in federal law prohibiting DOE from accepting commercial LLW.[12]

TRU

The unplanned and often tortuous evolution of WIPP ultimately produced the world's only operating deep geologic repository for highly radioactive wastes. WIPP opened in 1999 and has been steadily receiving defense TRU waste since then. In 1998, EPA initially certified the facility as compliant with environmental requirements. Recertification by EPA is required every five years by the WIPP Land Withdrawal Act. DOE applied to EPA for recertification of WIPP in 2004 and received EPA's recertification in 2006. In 2009, DOE again applied for recertification of WIPP; EPA granted recertification in November 2010.[13] Acting pursuant to its EPA-authorized RCRA permitting authority over mixed wastes, New Mexico in 1999 issued a RCRA facility permit for WIPP that authorized the repository to receive contact-handled TRU waste; the state amended the repository's RCRA permit to authorize WIPP to receive more hazardous, remote-handled TRU waste in 2006.[14] The state granted a ten-year renewal of the permit in December 2010.[15] As of June 2010, WIPP had received more than 8,600 shipments of TRU (about 370 of which were RH-TRU, and the remainder CH-TRU) and had disposed of more than 68,000 cubic meters of TRU waste.[16] The statutory cap on the quantity of TRU wastes that can be disposed at the facility is 175,600 cubic meters;[17] DOE believes this allowance is sufficient to accommodate all current defense TRU, plus all TRU projected to be produced by the federal government's defense activities in the future.[18] DOE currently expects to have disposed of all TRU wastes at WIPP by 2029 and then to close the repository permanently. The small amounts of civilian TRU, statutorily precluded from disposal at WIPP, await another disposal pathway.

HLW and SNF

SNF and HLW present the greatest nuclear waste problems confronting the country.[19] The system of law and policy the United States has followed since 1982 for disposing of these wastes is largely bankrupt. The 1998 target date for opening a repository to receive SNF has long passed. The Obama administration has abandoned Yucca after more than

$13.5 billion has been spent on it. But under NWPA, Yucca is the only repository that can be licensed to receive HLW and SNF. Absent revival of Yucca, siting, licensing, and constructing an entirely new repository will take at least two decades. An alternative is to open WIPP to HLW and SNF, although as a matter of fairness and practical politics this would require New Mexico's assent. As described in Chapter 5, New Mexico officials have recently signaled interest in expanding WIPP's mission to include HLW and possible disposal as well as storage of spent nuclear fuel. Yet, even if a deal could be reached, accomplishing such disposal would take many years. Extended storage of SNF and HLW is thus inevitable—even though this is the very scenario that the NWPA was designed to prevent.

Current HLW/SNF Wastes

Defense HLW and SNF

A midrange estimate of 12,800 MTHM of HLW and SNF are currently stored at DOE facilities around the country; the majority of these wastes consist of HLW.[20] Almost all these wastes are currently stored and managed at one of three DOE sites: Hanford, Savannah River, and INL.[21] DOE has tripartite agreements with the states in which these facilities are located providing for removal of these wastes.[22] Yucca was the intended destination for these wastes. DOE has been in the process of constructing and operating treatment facilities at these sites to solidify the liquid HLW held in storage tanks and to vitrify or otherwise to render them in a form suitable for permanent disposal. The termination of Yucca may make it very difficult for DOE to fulfill its obligations under some of the current agreements.

Civilian SNF

More than 62,500 MTHM of civilian SNF are currently being stored in thirty-five states at seventy-seven different sites, including currently operating reactors, decommissioned reactors, the reprocessing facility at Morris, Illinois, and two DOE sites.[23] Most of the SNF continues to be stored in pools even after it cools down, although the amount of SNF being placed in dry casks and stored on-site has been growing slowly. As of 2009, 85 percent of SNF was still stored in pools.[24] All this SNF was originally destined for disposal at Yucca Mountain. Even if Yucca were to be constructed and begin receiving wastes for disposal in 2020, as was anticipated by DOE in 2009, removal of all SNF from reactor sites would not be completed until 2066; safe and secure storage of some kind would be needed in the interim. In the absence of a repository or consolidated away-from-reactor storage facilities, almost all of the nation's civilian SNF will remain in storage at reactor sites, the great majority of it in pools. NRC has approved on-site storage for SNF in dry casks for extended periods, typically twenty years, subject to renewal.[25] DOE secretary Chu has stated that dry-cask storage of SNF can be relied upon for half a century, perhaps longer.[26] NRC has also begun to examine the possibility of longer-term on-site storage in excess of 120 years.[27] Notwithstanding NRC's approvals of operating-life extensions for many older reactors, a growing number of reactors will be decommissioned in coming years; almost all operating reactors are currently slated for decommissioning by 2030. Local communities and states are likely to strongly oppose indefinite storage of SNF at reactor sites, especially those that are no longer producing electricity or jobs.

Civilian HLW

Between 1966 and 1972, the West Valley facility reprocessed 640 metric tons of spent nuclear fuel. When it closed, the facility left a storage pool full of SNF and underground storage tanks containing 660,460 gallons of liquid HLW. Pursuant to the 1980 West Valley Demonstration Project Act, DOE's contractor completed solidifying and vitrifying the HLW in 275 canisters that are being stored at a new interim storage facility at the site until a repository is available to receive them.[28]

Future Options: Repositories

Developing New Repositories

NWPA currently precludes building a repository at any site other than Yucca and prohibits DOE from characterizing any other site for a repository until Yucca has been licensed. Congress would have to amend NWPA or enact an entirely new scheme in order for another repository or repositories to be developed. Experience has amply demonstrated that siting any kind of new facility for nuclear waste is a hugely challenging enterprise. Virtually every site proposed for disposal or long-term storage of SNF, HLW, TRU, or LLW has had to navigate some or all of the following hurdles: vehement host-state and public opposition, technical problems relating to site suitability, and litigation. And almost all these siting efforts have failed.

Even if these hurdles can be surmounted, the timeline for siting and developing a repository is extremely long. It took more than twenty-five years for WIPP to go from concept to operational repository. Had DOE's license application for Yucca been approved, the repository built, and an operating license granted by NRC, all in accordance with DOE's 2008 timetable, a full thirty-seven years would have elapsed between the time Congress designated Yucca in 1987 and the moment the repository finally became operational; of course, none of this has happened yet.[29] Siting a new repository in the wake of the failed Yucca effort will be all the more difficult. (Principles for addressing these challenges, drawing on siting experience to date with nuclear waste facilities, are discussed later in this chapter.) If a new repository is developed, the experience with the NWPA site selection process indicates that it is well worth the time and resources to consider and characterize in detail a number of candidate sites. Whatever one ultimately concludes about the technical merits or demerits of Yucca, it is painfully obvious in hindsight that Congress would have been well advised to spend $3 billion or more to characterize three sites, as NWPA originally required, rather than fix on a single site that had not been fully investigated and spend $13.5 billion to develop a facility that has since been abandoned.[30]

If new repositories are to be developed, it would make sense to restore some measure of geographic equity by simultaneously considering not only the candidate sites evaluated by DOE to date, most of which are in the West or South, but also midwestern and eastern sites, which were taken off the table at a very early stage of consideration due to political backlash. Other approaches include starting a brand-new nationwide search for sites or conducting a "reverse auction," in which the federal government would allocate a sum of money for the repository and solicit bids from interested communities "detailing how [the communities] would spend it to address the local impacts of and statewide concerns about the proposed facility."[31]

Developers of hazardous waste facilities have successfully used outreach strategies and economic incentives to overcome local opposition to locally undesirable land uses (LULUs). Economic incentives that have facilitated community acceptance of landfills and hazardous waste facilities include in-kind grants for community beautification or development, contingency funds or cleanup bonds, property value guarantees, benefit assurances or employment guarantees, direct monetary payments, and charitable donations.[32] Simply paying communities to accept nuclear waste, however, smacks of bribery unless accompanied by an open participatory process that results in a facility sited in a community that accepts it as a safe and appropriate solution to a waste management problem.[33]

Opening WIPP to SNF and HLW

Given the continuing legal, political, and technical uncertainties over Yucca and the acute difficulties in developing one or more new repositories for disposal of SNF and HLW, opening WIPP to one or both of these wastes as well as GTCC wastes is a serious option for consideration. Earlier DOE studies of the site and facility, as well as the IRG report, indicated that WIPP could be suitable for disposal of SNF, HLW, or both, in addition to TRU.[34] WIPP's technical suitability for disposal of these types of wastes would, however, have to be completely reexamined in depth if there were to be serious thought now of disposing of them at the site. For a variety of political reasons, discussed in Chapter 5, Congress by statute barred commercial wastes from WIPP in 1979, and all but TRU in 1992. Congress could, of course, amend these laws, along with NWPA, to permit HLW/SNF/GTCC disposal at WIPP. The federal government has full legal authority over the site, which is located on federal lands. The Town of Carlsbad's political leaders have indicated that the town is actively interested in expanding the scope of WIPP's mission in order to reap the resulting economic benefits. Carlsbad's mayor has already begun to suggest this possibility to Congress.[35] In light of the experience with Yucca, the notion that the federal government would exercise its legal power to force the repository on an unwilling state seems unlikely, and in any event unwise. New Mexico state officials have recently signaled interest in expanding WIPP's mission to include HLW and either storage or retrievable disposal of SNF, as well as GTCC wastes. Success, then, might well hinge on whether the federal government could provide New Mexico enough reassurances and inducements, including decision-making structures and enforceable commitments as well as economic benefits, to address the state's concerns and win its assent to expansion of WIPP's mission.

Resuscitating Yucca

Despite the taint of the political steamrolling through which the Yucca repository site was chosen, and despite unresolved technical and performance questions regarding the site and repository design, it is still possible that Yucca will be licensed by NRC and eventually built. Even if Yucca were to move forward, the total volume of SNF wastes at reactor sites, as of 2010, is nearing the statutory maximum disposal capacity set by NWPA for a first repository. The overall statutory limit is 70,000 MTHM, of which 63,000 MTHM is allocated to commercial SNF and the remainder to defense HLW and SNF. Accumulated commercial SNF totaled around 62,500 MTHM as of late 2009 and continues to increase at a rate of approximately 2,000 MTHM per year.[36] If this rate of increase holds steady, by 2060 almost 165,000 MTHM would require disposal.[37] Congress would either have to lift the cap to allow substantially more waste to be deposited at Yucca, or a second repository would have to be developed.[38] And more SNF will accumulate—and faster—if nuclear power generation is significantly expanded.

Future Options: Extended Storage

According to DOE's 2008 supplemental EIS for Yucca Mountain, extended surface storage of HLW and SNF can be accomplished safely with adequate institutional controls for up to a hundred years.[39] Beyond that period, the report warns, unforeseen circumstances would be expected to cause security and maintenance to fall below acceptable levels; without institutional controls, releases of radioactivity would be expected to reach the accessible environment within ten thousand years, eventually causing "catastrophic consequences for human health."[40] Other studies have also posited a hundred years as a benchmark for assured safe storage.[41] In late 2010, NRC began the process of a rulemaking and an EIS to update its Waste Confidence Rule; it is considering the possibility of storage in excess of 120 years and up to 200 to 300 years depending upon the agency staff's technical findings. Without endorsing this extended time frame, NRC explained its purpose was "to be fully informed by the current circumstances and scientific knowledge, and also to provide long-term stability" to the Waste Confidence Rule.[42] GNEP proposed that certain reprocessing wastes (strontium and cesium) be stored for several centuries for cooling and radioactive decay, followed by disposal as LLW; but the proposal did not explore the unprecedented institutional challenges in assuring secure storage of highly radioactive wastes for several centuries. As a working premise based on current information and analysis, storage of commercial SNF and HLW at current DOE sites, reactors, or new consolidated storage facilities should not be used for longer than fifty years or so, pending development of one or more repositories for permanent disposal. Designing repositories to ensure retrievability of SNF could extend the availability of the reprocessing option, but robust long-term institutional controls would be necessary to prevent illicit withdrawal or accidental intrusion. Limitations in assuring the viability of such controls for an extended period would imply corresponding limitations on the retrievability period, and the correlative value of maintaining the reprocessing option.

The United States has never designed a repository with long-term retrievability in mind. NWPA provides that a repository should be designed to permit retrieval while the facility is in operation. DOE must identify a retrievability period in the repository design,[43] but NRC can fix the retrievability period in its repository regulations.[44] NRC's general regulations and its Yucca-specific regulations require that repositories allow for retrieval of waste for at least fifty years after emplacement begins.[45] DOE included a plan to comply with this requirement as part of the final supplemental EIS for Yucca.[46] WIPP, by contrast, was designed for retrievability of wastes only during an initial test phase.

Defense HLW and SNF

In the absence of a repository, the most easily implemented alternative strategies are continued storage of defense HLW and SNF at the DOE sites where the wastes are currently located or consolidated storage of some or all of these wastes at fewer sites. Unless Yucca is built and opened soon, relations between the states that host this HLW and SNF will continue to fray. In the case of Idaho and Colorado, DOE has promised to remove all HLW and SNF by 2035.[47] Continued—much less expanded—storage of these wastes at their current sites would require modification of tripartite agreements with these host states to extend deadlines for moving wastes off-site, and might require substantial inducements to states to accept these changes. Potential inducements might include commitments to locate new, job-producing, clean energy facilities and R&D laboratories and facilities within host states. Extended on-site storage could require upgrades of the storage facilities.

Civilian HLW

The FEIS for West Valley, issued in January 2010 by DOE and the New York State Energy Research and Development Authority, provides that the facility's civilian HLW, which has been vitrified and encased in canisters, will be stored on-site, followed by off-site disposal at a federal repository when one becomes available. The expectation had long been that this waste would go to Yucca Mountain. WIPP is not currently authorized to receive HLW or any civilian waste.

Commercial SNF: Indefinite Storage at Reactor Sites

Secretary Chu recently asserted, and NRC had previously determined, that dry-cask storage at reactor sites can be safely maintained for many decades after the expiration of reactor licenses.[48] However, only about 15 percent of the SNF being stored at reactor sites is in dry casks; the remainder is in pool storage.[49] As SNF has accumulated without a repository in sight, NRC has authorized many facilities to add to their storage pools up to five times the number of fuel rods authorized in the original licenses.[50] The Committee on Science and Technology for Countering Terrorism of the National Academy of Sciences concluded in 2002 that the "potential for 9/11-type attacks is high in the near term" for SNF pools, and that SNF should be moved to dry-cask storage.[51] The Union of Concerned Scientists (UCS) contended in a 2007 report that the security of both spent-fuel pools and dry-cask storage was currently "unacceptable."[52] Nonetheless, UCS believes that dry-cask storage can be made an "acceptably safe" and "economically viable option" for the next fifty years, provided that it is "hardened" through relatively simple measures, such as surrounding dry-cask storage areas with earthen berms, and that security at reactor sites is improved.[53]

Communities around a number of existing reactor sites are growing restive at the prospect of indefinite storage of SNF in their vicinities.[54] The six governors of the New England states in December 2009 forwarded a letter to the Department of Energy requesting that it remove spent-fuel and high-level waste from nuclear reactor sites in the region "at the earliest possible date."[55] The governors expressed their concern that the waste had not been removed "as required by law and contract" and was limiting alternative use of three former reactor sites in the region that have been decommissioned.[56] The Prairie Island Indian Community in Minnesota has lobbied Congress and staged protests to have Xcel Energy clear its Red Wing plant of seventeen casks of spent nuclear fuel by 2013.[57] The Fukushima crisis will no doubt fuel more opposition.

Leaving SNF at reactor sites carries still more political costs. The failure to open Yucca is a blow to the federal government's credibility and contradicts its promises to the public and states to begin permanently burying wastes within a matter of years. This continued failure to "solve" the nuclear waste problem as the federal government itself defined it represents a significant: political obstacle to expanded use of nuclear power.

For many nuclear power skeptics, the government's inability to remove and safely dispose of SNF stored at reactor sites represents an inherent and insoluble problem with reliance on nuclear power and a powerful argument against further public investment in the industry.

Commercial SNF: Consolidated Storage

There are no federal or privately owned consolidated storage facilities for SNF at present, and all efforts so far to site such facilities have failed. DOE has taken no steps to site a federal MRS facility. NWPA provides that once a repository site has been designated, DOE can begin siting an MRS facility. Since Yucca has been designated, DOE is free to do so, although the legal implications of withdrawal by DOE of its NRC application to Yucca remain unclear. The Private Fuel Storage (PFS) SNF storage facility on Goshute tribal land in Utah remains in limbo. Although the federal district court in Utah vacated and remanded to DOI for further consideration its denial of two required authorizations for PFS,[58] it is unclear what DOI will decide on remand and whether Private Fuel Storage, LLC, the utility consortium developing the project, will continue to pursue construction of the facility. Although the court's decision could breathe new life into the project, an additional suit by Utah challenging the facility's NRC license would also have to be won. The federal government, notwithstanding Utah's opposition, could decide to encourage rather than block the facility, and could work with opposing members of the Goshute tribes to assure adequate compensation for the tribe and resolve their concerns. However, this by itself would not address Utah's fundamental objections to the site, including its location in a seismically active area adjacent to a military bombing range, and its concerns about the burdens and impacts the facility will impose on communities in the state, including the need to plan for and provide emergency response to a nuclear incident.[59]

The federal government could also encourage consolidated interim storage facility ventures by other private developers; such facilities might be located on existing DOE sites. In order to reduce the federal government's mounting liabilities to utilities for failure to take SNF from reactor sites by 1998, DOE considered the option of consolidated storage of civilian SNF at DOE facilities that now manage defense HLW but concluded that this option is currently precluded by NWPA. Congress could address the issue of consolidated storage at a federal MRS or DOE or other federal installation through changes to NWPA or in a new law. However, host states and nearby local communities might well object to such storage.

A potential advantage of consolidated storage of SNF away from reactor sites is that it would be held at a safer remove from population centers. If, as seems likely, SNF storage will be protracted, proponents of consolidated storage argue that it would be more cost effective and secure to house SNF in state-of-the-art facilities specifically constructed for the purpose, rather than continuing to store it at a much larger number of sites in overcrowded at-reactor storage pools, or requiring utilities to build costly dry-cask facilities. Proponents further assert that once a repository is sited, the logistics of shipping stored waste to the repository from one or a few such consolidated storage locations would be simplified. Obstacles to this option, however, include the high initial capital costs of consolidated storage; likely opposition by potential states to hosting such facilities; and public concerns over the risks posed in transporting large quantities of SNF to consolidated facilities from seventy-seven existing sites around the country, which may also trigger opposition by transit states. The blue-ribbon commission should conduct one or more detailed studies of these and other issues, and carefully and systematically examine the tradeoffs of at-reactor storage versus consolidated SNF storage.

One potential advantage of expanding the mission of the major existing DOE weapons facilities to include consolidated SNF storage, rather than attempting to site a brand-new public or private facility, is suggested by research indicating that host local communities and states, having grown familiar with the presence of nuclear wastes at nearby DOE facilities and the economic benefits those facilities provide, have come to accept them and to welcome those benefits.[60] The experience with siting new nuclear waste facilities, however, shows that even where a local community might be willing to host a facility, potential host states are generally opposed, due the political "doughnut" effect—people in nearby communities tend to perceive the risks posed by the facilities as relatively low, while those farther away (50–150 miles) view them as significantly higher, even though risks in fact decline with distance.[61] This problem might be mitigated by federal inducements for states and localities in the form of co-located job-producing and -sustaining developments at or near the same facility. These developments could include "energy parks" and other energy technology R&D facilities and university-based research programs in renewable energy, energy conservation, and the nuclear fuel cycle, including nuclear waste management. Yet states and local citizens fought long and hard for firm and enforceable federal commitments to remove existing defense wastes from DOE sites that DOE likely cannot meet. In these circumstances, they may be most reluctant to take on the additional burden of long-term storage of large amounts of nuclear power wastes from other states. If consolidated storage is selected as an option for further development, it might be worth considering establishing small demonstration projects of limited duration in three or four parts of the country simultaneously. Consolidated storage projects might proceed in the style of WIPP, with considerable community and state involvement as well as leverage in a decision process that would be carefully staged and include a host-state option to stop or significantly modify a project at each stage. In addition, host states and communities should be provided the resources to engage independent experts to monitor the project throughout its life span, as well as significant and perhaps increasing inducements and benefits to affirm support for the project as it proceeds toward more full-blown status. There might also be a mandatory and enforceable consultation and cooperation agreement. Transit states would also have to be included in planning at an early stage and given an appropriate role in designating transportation routes and planning for emergency preparedness, which would include adequate personnel training and necessary resources.

Commercial SNF and Reprocessing Options

The potential for reprocessing SNF has critical implications for decisions about repository design. The IRG/NWPA goal of developing a repository on an expedited basis was premised on the policy assumption that SNF would not be reprocessed. But if reprocessing were to be regarded as a serious option, then SNF might either be stored or disposed of in a repository designed to ensure retrievability for a substantial period.

Interest in the United States in reprocessing has revived somewhat in recent years as a result of the GNEP proposal (since abandoned); strong support by the Obama administration and in Congress for expansion of nuclear power; heightened concerns with energy security; growing demand for uranium as a result of expansion of nuclear power worldwide, including in China, where a huge buildup of nuclear power is now under way; and decisions by China and Japan to develop reprocessing. Counterfactors, however, include the dismal past record of U.S. reprocessing, the unfavorable economics of fuel produced by reprocessing, and the rise in terrorism worldwide coupled with the success of nations such as Iran and North Korea in developing nuclear weapons despite the international nonproliferation legal regime. Expansion of reprocessing using current technologies would generate more fissionable materials that could be used for making nuclear weapons and terrorism, compounding the challenge of securing such materials.

Advocates for revival of reprocessing in the United States through development of advanced fuel cycle technologies are optimistic that it will unlock the useful energy remaining in our enormous stockpile of SNF, effectively transforming the monumental waste problem into a massive renewable fuel boon, providing a plentiful, secure, long-term source of fuel for low-carbon nuclear power while significantly reducing the volume of, and formidable technical and political challenges posed by, nuclear wastes requiring repository disposal. Advocates further claim that revival of reprocessing will restore U.S. commercial, technological, and strategic leadership in the global nuclear marketplace.

Opponents point to the persistent failure of reprocessing, whether in the United States or abroad, to produce fuel at a price remotely competitive with that of enriched uranium; the huge costs, long lead times, and daunting technological and economic challenges and risks involved in advanced fuel cycle technologies; and the serious risks of proliferation and terrorism associated with the plutonium produced by reprocessing of SNF. Opponents also dispute the claimed environmental and other advantages of reprocessing with respect to waste disposition—arguing that more, not less, radioactive waste requiring disposal will be generated through reprocessing. Further, they sharply question the energy security arguments for reprocessing in light of the fact that the country's primary energy security problems relate to petroleum imports, and that large supplies of uranium are available to the United States from staunch U.S. allies such as Australia and Canada.

Meanwhile, GNEP has been zeroed out of the federal budget and declared dead by the current administration, to be replaced by the proposed global uranium fuel bank.[62] Nonetheless, energy secretary Chu supports research into reprocessing,[63] and the remit of the blue-ribbon commission is broad enough to include consideration of reprocessing.[64]

If a repository were opening soon, it would be important to resolve the reprocessing/retrievability issue promptly. NWPA currently requires that the waste disposed of in a repository be retrievable for fifty years from the time that the repository commences receiving waste for disposal. The capacity to retrieve emplaced waste could be important for at least two reasons: first, if problems occurred early in a repository's operating life, to address them while institutional controls were operational and SNF could be removed if something went wrong; second, in the case of SNF, if a decision were made to reprocess it. But if neither Yucca is revived nor WIPP's mission expanded, developing a new repository will take at least several decades. Such a delay would effectively postpone the need for an immediate, definitive resolution of reprocessing/retrievability issues. Nonetheless, absent a WIPP or Yucca option, the search for a new repository will have to begin soon.

To the extent that ensuring a substantial period of retrievability was to be a goal for a new repository, it could significantly influence the choice of geologic media. In the case of repositories in hard crystalline rock like granite or welded volcanic tuffs as at Yucca, access shafts for retrievability are largely self-supporting.[65] Sedimentary rock formations are less sturdy, followed by clays.[66] Salt is a particularly difficult medium for retrievability purposes. Access tunnels built through salt formations, which gradually close or "heal" over time, require a great deal of engineered support to maintain.[67] Additionally, due to the phenomenon of salt creep, salt can entirely envelop waste containers. To retrieve the emplaced SNF, the repository would have to be mined out, provided this could be done safely. Accordingly, a policy that placed high priority on retrievability would not focus on salt media, even though these media have been heavily favored thus far and are relatively well understood, especially as WIPP has provided valuable experience with actually constructing and operating a repository in a salt formation. Conversely, a policy that disfavored reprocessing or did not put a premium on retrievability could focus on identifying suitable repository sites in salt—for which there are many candidate formations in a number of regions of the country—as well as in a variety of other, less well understood but promising geologic media.

Choices about retrievability could be framed within the decisional theory of adaptive phased management (often called "stepwise decision making" in the European literature).[68] The basic premise of adaptive phased management is that important decisions are made as the project progresses, with an opportunity to evaluate changing circumstances before proceeding from one phase to the next. The difficulty and cost of achieving waste retrieval typically increase as a repository nears completion, when waste packages are buried, backfilling is conducted, and the repository is finally sealed.[69] Thus, if a country decides that it wants to ensure that its SNF is retrievable, prioritizing retrievability in the design and construction phases of the project can reduce the costs of retrieving waste later in the project. More important, given that the United States has made no decision about whether a new repository should enable waste retrievability to maintain a reprocessing option, adaptive phased management would argue for maintaining flexibility by identifying and investigating viable sites in both salt and nonsalt media and generating a set of good repository options, rather than only one.

Various countries have addressed the retrievability question in their repository designs (see Table 8.1). The planned French repository, to be sited within a solid argillite formation, is designed with the goal of ensuring retrievability for one hundred years.[70] Finland, which had previously imposed a clear retrievability requirement, is reportedly now backing away from that position.[71] The planned Swedish repository is not subject to legal requirements for retrievability; however, the Swedish Nuclear Fuel and Waste Management Company (SKB), the body responsible for the repository, has expressed the view that while preclosure retrievability is not a high priority, it should not be made impossible either.[72] Meanwhile, technical retrievability studies are advancing. For example, in 2006, at the Äspö Hard Rock Laboratory in Sweden, SKB successfully retrieved a canister from saturated bentonite clay at a depth of 420 meters.[73]

In the United States, neither reprocessing advocates (who would prefer aboveground storage for SNF) nor opponents of reprocessing strongly favor the retrievability option for an SNF repository. Advocates of reprocessing would argue that once emplaced in a repository, SNF would be politically very difficult to retrieve, even if the technical capacity to do so existed. Opponents of reprocessing generally take the view that it is neither necessary nor wise to ensure SNF retrievability. There is nonetheless a

Table 8.1. Retrievability policies in various countries

Country	Media investigated	Retrievability requirements
United States	Salt, basalt, granite, tuff,* clay, and shale.	NWPA requires repositories to provide for retrievability of SNF and HLW for fifty years from the start of waste emplacement.
Belgium	Clay and shale	No decision made
Canada	Granite and sedimentary rock	The adaptive phased management plan includes "potential for retrievability of the used fuel for an extended period, until such time as a future society makes a determination on the final closure and the appropriate form and duration of post closure monitoring."
China	Granite	No decision made
Finland	Granite, gneiss, grandiorite, and migmatite*	A regulatory requirement for retrievability was eliminated in 2008. However, Posiva is still required under a 2000 decision to present a plan and cost estimate for retrieving the waste when it submits an application for a construction license.
France	Argillite* and granite	The repository must be designed so that it is "reversible" for at least one hundred years. Reversibility is a management concept that requires retrievability.
Germany	Salt	Retrievability of HLW and SNF need not be provided in the disposal concept. However, shielding of the ionizing radiation has to be guaranteed so that the waste will be manageable for possible retrieval for a period of five hundred years after repository closure, taking into account probable developments.
Japan	Granite and sedimentary rock	No decision made
Korea	Granite	No decision made, but in the conceptual-level reference design, waste packages had to be retrievable for an indeterminate period.
Spain	Granite, clay, and salt	No decision made
Sweden	Granite*	None
Switzerland	Clay and granite	The retrievability of HLW has to be considered when designing the repository. The technical feasibility of retrieving the waste has to be demonstrated in experiments on a 1:1 scale before the repository starts operation.
United Kingdom	No decision	No decision made, but planning and guidance require that the option of retrievability not be foreclosed.

*Media at proposed repository site.

Source: Nuclear Waste Technical Review Board, *Survey of National Programs for Managing High-Level Radioactive Waste and Spent Nuclear Fuel* (October 2009).

substantial question of public policy regarding whether or not the nation should incur the costs and risks entailed in ensuring SNF retrievability for a substantial period in order to keep the reprocessing option for future generations. There is a separate question of how far SNF and HLW repositories should be designed for retrievability to address future malfunctions in their performance. There may also be intermediate options that deserve consideration; for example, not all accumulated SNF needs to be deposited in a new repository once it becomes operational—a repository could be opened on a test basis and later receive either the remaining SNF or reprocessing HLW, depending on the decision on reprocessing.

The value placed on maintaining the reprocessing option may also influence the choice between at-reactor versus consolidated storage of SNF for the extended period during which a repository is being developed. Reprocessing proponents might advocate for co-locating reprocessing and other advanced fuel cycle installations near an SNF consolidated storage site, in a package deal that could induce local communities and states to willingly host SNF storage. Yet the history of uneconomic, failed reprocessing facilities—including West Valley, which failed and closed, and the Barnwell facility, which was abandoned when federal subsidies ran out—and the risk that SNF awaiting reprocessing and HLW would remain at reprocessing sites for a long time would give potential host communities serious pause.

Commercial SNF: Resolving Federal Liabilities and Developing New Financial Mechanisms

Clearly, new arrangements are needed to ensure financing for a new repository and to establish stable plans for SNF storage, whether at reactors or at new consolidated facilities. The operative premise thus far has been that utilities (and ratepayers) should bear the costs of SNF storage or disposal. But in charting a new path, it must be recognized that the federal government bears substantial responsibility for the delay in opening a repository and for the seemingly poor investment of many billions of dollars from the Nuclear Waste Fund in a Yucca repository that, as of now, it appears will never open. The federal government has defaulted on its commitments to take SNF by 1998, and Congress has used for other purposes much of the fees paid by the utilities into the NWF. Indeed, the Nuclear Waste Fund is a fund only in name: none of the money paid has been set aside for the fund's legislatively prescribed uses—namely, to take responsibility and ownership of the nation's civilian spent nuclear fuel accumulations. Continued litigation is not a tolerable method for addressing the issues of financing SNF storage and development of a new repository. Congress will sooner or later have to grasp the nettle through legislation that resolves the current liability mess and establishes a new plan for SNF storage coupled with a reconstituted financing mechanism for the future.

The design of new financial mechanisms will have to be adapted to policy choices about what to do with SNF, including at-reactor storage, consolidated storage, or reprocessing, as well as development of a new repository. Financing arrangements will also intersect with potential new institutional arrangements for managing nuclear wastes and siting new storage or disposal facilities.

Meanwhile, utilities are facing serious financial shortfalls in dealing with decommissioning and dismantling nuclear plants at the end of their useful lives, in addition to disposing of the wastes they have produced. Nuclear utilities are required by law to set aside reserves for the costs of decommissioning; nuclear utilities have invested the reserves in stocks and bonds. In theory, these investments should have provided a higher rate of return and increased profits relative to investment in Treasuries, but from 2008 to 2009, because of the recession, decommissioning funds lost an estimated $4.4 billion.[74] It costs, on average, $450 million to decommission a reactor, but most reactors now have only around $300 million saved in their decommissioning funds.[75] And to steepen the losses, these shortfalls have arisen as estimates of decommissioning costs have been skyrocketing due to rising labor and energy prices. In 2009, NRC warned eighteen power plants to address their shortfalls or face legal action.[76] The long-term ramifications of this situation, however, are unclear. NRC is currently exploring the possibility of long-term on-site storage of SNF for over 120 years, long after reactors are shut down. This approach would raise serious questions about assuring operational and financial responsibility for these wastes. Half the nation's nuclear power plants have obtained twenty-year extensions on their licenses, so it remains to be seen if companies will try to delay decommissioning further to build up funds, transfer funds from elsewhere to make up the losses (impeding further nuclear construction), or appeal for government aid.[77] The recession has further pressed the question of what will happen if a nuclear utility goes bankrupt, an issue of particular concern in the case of decommissioned plants where SNF is stored on-site; it is unclear who will be responsible for the waste in the event of a bankruptcy. Nuclear Regulatory Commission deputy general counsel Stephen Burns testified before the Vermont legislature that Vermont Yankee would be required to pay environmental claims "before other claims and other creditors are satisfied." The Vermont Yankee fund has only $347 million, and cleanup would cost between $600 million and $1 billion.[78] Such problems should be addressed in the context of any new financial mechanisms developed to address the SNF liability and financing issues.

Lessons Learned and Future Strategies for U.S. Nuclear Waste Policy

Albeit flawed in some of its premises and assumptions, President Carter's IRG was a thorough and reasoned effort to rethink and restructure a U.S. nuclear waste policy that was in serious disarray. The Obama administration's Blue Ribbon Commission on America's Nuclear Future, and the nation as a whole, faces a similar challenge today (see Table 8.2). U.S. nuclear waste law and policy is currently ruled by statutes, most notably NWPA and LLRWPA and their amendments, that have conspicuously failed to achieve their objectives. Yet political leaders have thus far been unwilling to address this failure and to assume the risks involved in fundamental change. The distinguished members of the commission could help break this political logjam by following the model of IRG, engaging in a total review, stimulating public involvement and public debate, rethinking the country's nuclear waste policy, and charting a new course.[79] The approach taken by such a commission should be ambitious and comprehensive. It should consider options for dealing with current and future wastes of all types in relation to the entire nuclear fuel cycle, potential new fuel cycle technologies, and broader considerations, which include climate change, energy security, and domestic and international security against nuclear proliferation and terrorism. A focus on wastes alone would ignore the several ways in which they are embedded in this larger complex of issues, and would thereby risk promoting adoption of short-sighted approaches that overlook cross-cutting factors, risks, and opportunities.

To frame an agenda for addressing the challenges facing the nation and the blue-ribbon commission, this section summarizes the key features and lessons of U.S. experience to date with siting and developing nuclear waste disposal and management facilities, then proposes five basic steps toward charting a new course, taking into account the experience and lessons learned from the history of nuclear waste law and policy to date. These steps are:

- Rethinking and revising the ethical framework for nuclear waste policy in light of developments over the past half century
- Adopting a new, national nuclear waste facility development strategy that is cooperative, flexible, and adaptive

Table 8.2. Members of the Blue Ribbon Commission on America's Nuclear Future

Lee Hamilton, cochair	President and director, Woodrow Wilson International Center for Scholars; director, Center on Congress, Indiana University; former member, U.S. House of Representatives
Brent Scowcroft, cochair	President, Scowcroft Group; national security advisor to Presidents Gerald Ford and George H. W. Bush
Mark Ayers	President, Building and Construction Trades Department, AFL-CIO
Vicky Bailey	Former commissioner, Federal Energy Regulatory Commission; former Indiana Public Utilities commissioner; former DOE assistant secretary for policy and international affairs
Albert Carnesale	Chancellor emeritus and professor, UCLA
Pete V. Domenici	Senior fellow, Bipartisan Policy Center; former member, U.S. Senate
Susan Eisenhower	President, Eisenhower Group, Inc.
Chuck Hagel	Former member, U.S. Senate
Jonathan Lash	President, World Resources Institute
Allison Macfarlane	Associate professor of environmental science and policy, George Mason University
Richard A. Meserve	President, Carnegie Institution for Science; former chair, U.S. Nuclear Regulatory Commission
Ernie Moniz	Professor of physics, Massachusetts Institute of Technology
Per Peterson	Professor and chair, Department of Nuclear Engineering, University of California–Berkeley
John Rowe	Chair and chief executive officer, Exelon Corporation
Phil Sharp	President, Resources for the Future; former member, U.S. House of Representatives

- Creating and fostering arrangements to gain host and public trust and assent for siting and developing new nuclear waste facilities
- Creating new federal institutions and arrangements for nuclear waste management, disposal, and financing
- Implementing an approach to waste classification that better matches regulation with hazard

Points of Consensus and Disagreement

In thinking about the way forward, it is important to recognize that, even more than half a century after the 1957 NAS Advisory Committee on Nuclear Waste report recommending the development of geologic repository disposal, the scientific and public policy consensus on this issue remains the same: the only viable permanent disposal option for highly radioactive wastes, including HLW and unreprocessed SNF, is a geologic repository. It is widely accepted that at least one deep geologic repository will be needed. This view is shared by the federal government, state governments, the nuclear industry, environmental and public interest organizations, proponents of reprocessing, communities near existing nuclear plants, and most of the informed public. This view is shared not only by those who favor expansion of nuclear power, but also by those who do not. The waste is a fact, and it has to be dealt with. The question is how best to achieve a scientifically and technically sound, secure, and publicly and politically acceptable repository within a realistic time frame.

A second point of consensus—one based on the collective experience of the past thirty years—is that siting an entirely new repository will require decades. Contrary to the assumptions underlying NWPA, it is now clear that attempting to designate a site by fiat tends to lengthen, not shorten, the siting and development process. Indeed, there is growing recognition that successful siting requires state, local, and general public acceptance of a facility, based on trust in the siting and development process and confidence that a proposed repository or interim storage facility will be safe and secure and that those responsible for its operation and financing will meet their responsibilities and commitments. Given acute public concern with the risks posed by nuclear wastes and states' unwillingness to serve as hosts (particularly to wastes generated elsewhere), achieving such acceptance will be difficult and challenging. Achieving acceptance will require a transparent and fair decision-making process that openly promotes a considerable degree of scientific and technical consensus on the merits of sites—a process that not only includes public involvement, but also provides full access to information for the public and host jurisdictions; makes available independent, credible scientific and technical research and expertise; and offers a genuine role for localities, states, tribes, and other important stakeholders in every key aspect of a decision-making process that allows them to raise their concerns and have them addressed and accommodated. Considerable opposition will surely arise and must be addressed and substantially reconciled.

A third important point of consensus—which also represents a shift from past assumptions, due in large part to repository delay—is that safe and secure interim storage of SNF and HLW must be provided for the decades that it will take to develop a repository. IRG thought that implementation of repository disposal could begin within ten to fifteen years.[80] The presumption, held by IRG and many others advocating for a repository and embodied in NWPA, had been that an explicit policy of storing waste for a substantial period, especially consolidated storage in facilities away from power plant sites, would hinder repository development and should therefore be precluded. But such storage has never been seriously regarded as a long-term substitute for a repository, and conversely, the lack of away-from-reactor storage facilities has not been a sufficient incentive to move the repository development process to fruition.

The release of radiation at Fukushima from ignition of pool-stored SNF has greatly strengthened the emerging conclusion that once SNF has cooled sufficiently, it should be moved to more secure dry casks—although questions remain about who should bear the added cost of such storage. The key open question is whether SNF in dry casks should be stored at the scores of reactor sites where the SNF was produced or whether the dry casks should be moved off-site into one or more federal or private consolidated storage facilities until a repository becomes available.[81] Some reprocessing proponents might argue that some if not all SNF should be stored until acceptable reprocessing technologies and facilities are available; other reprocessing proponents contend that retrievable storage in a repository would be the better option. Agreement appears also to have emerged that consolidated, away-from-reactor storage is appropriate for one particular type of SNF—namely, SNF at decommissioned nuclear reactors. Disagreement persists, however, on whether away-from-reactor consolidated storage is appropriate for the rest of the nation's SNF (that is, the vast majority of it)—taking into account factors such as costs, safety and security, transportation risks, and, more controversially, possible synergies with potential reprocessing facilities—or whether the better policy would be to continue with at-reactor storage, while enhancing its security and safety, until a repository opens. There is also disagreement over how the costs of SNF storage should be borne. Another important question is whether defense and civilian HLW should be stored where they are now, at various DOE sites, or put into consolidated storage at one or more such sites or elsewhere.

Many other key points regarding SNF, HLW, and other wastes are still being debated:

- Should Yucca be revived, licensed, and operated?
- Should WIPP be opened to HLW, SNF, or both?
- Should another siting process be initiated to identify a wider range of potential new sites for a permanent repository or repositories? If so, what are the appropriate elements of an effective and fair strategy and process for siting and developing a new repository and (if developed) consolidated interim storage facilities?
- What environmental standards should apply to a new repository or consolidated storage facility, and over what compliance period?
- What, if any, role should reprocessing play, and how should it affect, if at all, the choice between at-reactor and consolidated away-from-reactor interim storage of SNF?
- What strategies should be adopted to respond to current or future civilian LLW disposal problems, especially for GTCC wastes, including potential development of new LLW disposal facilities, expanded access to existing facilities, disposal at DOE facilities, or (in the case of GTCC wastes) repository disposal?

- What existing or new entities (federal, private, or hybrid) should oversee the nuclear waste facility siting process, determine regulatory compliance, manage wastes, and develop and manage waste facilities?
- What financing mechanisms should be used for development and construction of interim storage facilities for SNF, repositories, or both?
- What should be done about federal government liability for the missed NWPA 1998 deadline for DOE assumption of ownership and removal of SNF from power plants?
- What role should states, tribes, localities, and the public at large have in the siting process and in other aspects of developing and operating new facilities?

Rethinking the Normative Foundations of Nuclear Waste Management

This book's review of the nation's nuclear waste law and regulatory policies since World War II demonstrates not only that complete reconsideration of the nation's strategies for dealing with nuclear wastes is long overdue, but that some of the ethical premises underlying those strategies must be rethought and revised. During the first thirty years of the nuclear era, ethics took a backseat to expediency in dealing with wastes. Other goals took priority. Some wastes were buried and others stored, in both cases often without adequate precautions. The future of the most dangerous wastes was largely left to take care of itself. Beginning in the late 1970s, the pendulum swung to the other policy pole. IRG concluded unequivocally: "If present and future generations are to be protected from potential biological damage, a way must be provided either to isolate waste from the biosphere for long periods of time, to remove it entirely from the earth, or to transform it into non-radioactive elements."[82] The highly prescriptive NWPA scenario reflected the ethical premise that the then-current generation, having enjoyed the benefits of nuclear technologies, could not rightfully leave the burden of the waste it had generated to future generations and therefore must secure, as soon as possible, its permanent and perpetual burial in a deep geologic repository. This premise was reinforced by the opposition that emerged during the 1970s to building any more nuclear power plants until the waste problem was solved, and by opposition in some states to continued storage of defense wastes within their borders. Congress, in NWPA, embraced prompt burial of nuclear wastes and promised to achieve it.

In retrospect, it is clear that this position rested, at least in part, on two faulty factual assumptions: that it was feasible to achieve waste burial within twenty years and that arranging for secure interim storage of SNF as well as HLW would undermine, rather than support, achievement of this goal. The normative underpinnings of this position are also flawed: to the extent that nuclear power and nuclear weapons can be said to have benefited past and present generations, these benefits have not accrued solely to them. For instance, as CO_2 emissions reside in the atmosphere for centuries, the carbon emissions avoided by the use of nuclear power to date could well benefit future generations for centuries. At least a part of the national security and economic benefits of past uses of nuclear technology are embedded in the social and economic capital that future generations will inherit, part of which may help support our future energy and environmental security. Moreover, experience has proven unduly absolutist the claim that the benefits conferred on current generations by nuclear technologies, which will themselves evolve in unforeseeable ways in the future, require that we act *immediately* to bury all nuclear wastes "in perpetuity." Though well intentioned, this goal has proven overly simplistic and counterproductive. The record strongly suggests that our responsibilities to future generations can be better fulfilled by using the time we have now to identify a suite of candidate waste sites and measures and to conduct concerted R&D to develop better technologies for nuclear waste management and disposal, building into the process the flexibility to adapt our policies and practices along the way in light of experience and emerging new solutions. A more considered, deliberative (even if lengthier) siting process, combined with accelerated R&D on radioactive waste treatment and permanent disposal techniques, could benefit future generations by assuring better long-term containment of radioactive waste and greater risk reduction.

Yet a core element of the underlying principle driving IRG's recommendations and NWPA remains valid. Having generated a large quantity of long-lived, highly radioactive waste, it is clearly the present generation's responsibility to take prudent actions to ensure safe and effective isolation of this waste from people and the environment now and for as long as it poses a significant risk. This involves designing a fair and viable process for siting and developing one or more permanent repositories; initiating now a search for a set of new repository candidate sites in a variety of geologic media and in different parts of the country; allowing for future flexibility to adjust policies and plans for dealing with wastes in response to new information and improved technology; and assuring safe and secure storage and management of such wastes pending their final disposition.

It is now clear, however, that the work of siting, developing, constructing, and operating a repository for waste burial cannot be accomplished by a single generation and will require partnership across generations. Nor can it simply be assumed that the interests of future generations are best served by burying current waste stockpiles as soon as possible. The experience under NWPA shows that haste can be seriously counterproductive. Further research on using salt and other geologic media as repository sites, as well as initiating a more in-depth, geographically expanded search for suitable sites than was performed in the past, can lead to technologically safer facilities and more equitable distribution of the competing risks and benefits. Indeed, our ability to select appropriate repository sites and the technologies for containing wastes could well improve over time. Further, it may be wise to provide for a significant pilot test phase for a repository, employing a design that enables waste to be retrieved over a substantial period. The precautionary principle for environmental decision making counsels against taking irrevocable actions that pose environmental risks and foreclose options that may be developed and exercised in the future in the light of greater knowledge and better know-how. If we act wisely now, a repository that is built and comes into full operation in the future will likely pose lower total risks to future generations than one built today.[83]

Moreover, despite the formidable economic, environmental, security, and other questions that exist about reprocessing, SNF represents a potential energy resource that future generations might choose to tap using proliferation-resistant advanced fuel cycle technologies that, at least in theory, might be developed in the future, either in the United States or abroad. That SNF will necessarily remain in storage for decades until a repository is opened means in practical terms that the reprocessing option will remain open for a significant period, should future decision makers choose to pursue it. Even after a repository is opened, future generations may choose to maintain the SNF-reprocessing option by designing the repository to ensure SNF retrievability for a substantial period or by choosing not to emplace all or any of the accumulated inventories of SNF in a repository. Alternatively, decision makers in the future may choose to forgo the reprocessing option and to nonretrievably dispose of all SNF in a repository.

Based on these circumstances and considerations, a modified ethic, more nuanced than that embraced by IRG and NWPA, is needed. A useful formulation has been advanced by Tom Isaacs, current director of the Office of Planning and Special Studies at Lawrence Livermore Laboratory and recently appointed lead advisor to the blue-ribbon commission. He has stated that "our obligation is to give succeeding generations a real choice and the opportunity to make their own decisions while at the same time not imposing a burden which future generations may not be able to manage."[84]

Consistent with this formulation, a precautionary approach to environmental decision making would avoid irrevocable decisions and counsel a step-by-step approach that preserves options and the opportunity to make better decisions in the future through an iterative process of research, learning, and public deliberation.[85] This principle points to an incremental and adaptive approach to developing disposal and other waste facilities for nuclear waste (as at WIPP), rather than imposing a single, inflexible solution that has no backup plan (as at Yucca). The Canadian Nuclear Waste Management Organization has recommended a similar stepwise development strategy that effectively explores and maintains various options.[86] Under this approach, several repository options could be explored simultaneously, including WIPP, a new site, and the opening of Yucca on a small-scale test basis (although it would be a daunting challenge to persuade Nevadans to accept this step), along with development of at least one consolidated interim storage facility and a group of regional demonstration projects for consolidated storage.[87]

Developing consensus on the appropriate ethical foundations for nuclear waste policy will require debate and dialogue involving key political decision makers, experts, industry, NGOs, and the general public. The blue-ribbon commission should initiate and promote such discussions.

Lessons from U.S. Experience with Strategies for Siting Nuclear Waste Facilities

The United States has used a wide range of strategies for making nuclear waste facility siting decisions, including:

1. *Storage and, in some cases, disposal of waste at the site where it was generated.* This status quo approach is the default policy. It has been followed more often than any other approach. At-generation storage and disposal follow the path of political and bureaucratic least resistance and avoid the challenges posed by new facility siting. Examples include storage and disposal of defense and HLW at DOE facilities (with some consolidation and reallocation among DOE sites); storage of SNF at reactor sites; and, in many cases, storage or disposal of defense and civilian LLW.
2. *Reliance on market incentives to mobilize private firms to develop new facilities and obtain host-community and host-state acceptance.* This approach succeeded in producing new LLW facilities at Clive, Utah, and Andrews, Texas; its success at

the PFS Skull Goshute Tribe facility in Utah has yet to be determined. The default Commerce Clause rule, which requires states with waste disposal facilities to open them to wastes from other states, has made it difficult to obtain host-state assent to new nuclear waste facilities. Another drawback of this approach is that a site may be chosen not on the basis of its superior technical merits or of systematic identification and evaluation of numerous candidate sites, but on economic and political variables, including the bargaining position of the site host and its need for economic development. Yet, the interest of host jurisdictions in economic benefits is likely to be a key factor under any siting approach, as the WIPP experience exemplifies.

3. *Cooperative siting by groups of states within a federal framework of structures and incentives designed to overcome collective action problems (LLRWPA and LLRWPAA).* In thirty years, this approach has failed to produce a single new LLW disposal facility. Texas has succeeded in developing a new LLW waste facility, but only by virtue of using a two-state "compact" with Vermont, which Texas controls, to manipulate the LLRWPA system to gain market power—hardly the system of regional cooperation envisaged by Congress.

4. *Technocratic top-down selection of facility sites by federal or state administrative bodies.* This strategy was embraced in NWPA in 1982, which gave selection of repositories and storage facilities for HLW and SNF to DOE (subject to certain political review mechanisms in the case of repositories) and by the LLRWPA approach, under which state compact commissions chose sites when no compact state would agree to host a new LLW disposal facility. Yet over the past fifty years, not a single new nuclear waste disposal or storage facility has been developed by this method. Examples of failure include DOE's unsuccessful effort to site an MRS facility in Tennessee, the failures by compact commissions to develop LLW disposal facilities in North Carolina and Nebraska, and Project Salt Vault in Kansas. In many of these cases, the technocratic administrative process was unable to resist subversion by political pressures. Another illustration of this problem is the politically motivated decision of DOE to cancel entirely the second repository due to intense lobbying by eastern states with potential candidate sites, a course that enabled states with potentially highly suitable sites to avoid selection while implicitly discrediting the purportedly technocratic/meritocratic basis for selection of candidate sites for the first repository.

5. *Statutory dictation of a facility site by Congress.* This was the approach of the 1987 NWPA amendments, which also included the ineffectual state-veto-subject-to-congressional-override mechanism for unwilling host states. This approach failed at Yucca. Another failing of purely political top-down decision making is that it may not select the best sites, both technically and in terms of sound overall policy.

6. *Federal or compact initiatives involving dialogue and negotiation with state, tribal, and local authorities, sometimes following litigation, aimed at securing host acceptance of a facility.* This cooperative approach was successfully followed at WIPP, but it made only limited progress (i.e., by successfully inducing interested localities and tribes to study hosting an MRS facility) and ultimately failed when attempted by the Office of Nuclear Waste Negotiator, by DOE when it attempted to persuade states to host a repository under the NWPA, and by California in seeking to develop a LLW disposal facility.

The first approach—storage/disposal at the site of generation—has been the most widely followed, because no alternative was readily available, and cost and disruption were minimized. This approach places the burden of mobilizing to change the status quo on the states and localities hosting the generator site. Host jurisdictions have succeeded in overcoming this burden in a few cases: Idaho's successful demand, for instance, that DOE remove TRU from the state, and host states' use of RCRA regulatory leverage to force DOE to negotiate tripartite agreements providing for waste removal (although most HLW still remains where it was generated). In other cases, the burden of overcoming the status quo and achieving host-jurisdiction acceptance of a new facility lies with the facility developer. Private developers have occasionally succeeded, while state compact mechanisms never have. As for federal development of new facilities, only the last (cooperative) approach has worked.

Experience indicates that local and state host assent is essential for developing new facilities, whether the developer is the federal government, a compact authority, or a private entity. As confirmed by the court's ruling in *Nevada v. Watkins*, the federal government has plenary legal authority to build a nuclear waste facility on federal land, notwithstanding the objections of the state or locality in which the land is located.[88] But as Yucca and the MRS experience show, states have many legal and political weapons available to delay and ultimately defeat efforts by the federal government to impose a facility unilaterally. The legal and political safeguards of federalism are deeply embedded in our fabric of governance. Thus, host-jurisdiction assent must be the essential foundation of future nuclear waste siting strategies.

Securing Informed Public Trust and Host Assent to New Nuclear Waste Facilities

The Necessity of Public Trust and Host Assent

For the reasons just discussed, successful siting of new nuclear waste repositories and storage facilities requires host assent. Host states and communities and their publics must agree on, or be willing to consider, the possibility of a facility on their home turf. DOE recently recognized this imperative when it invited host jurisdictions to express interest in siting GNEP facilities; a number of jurisdictions did so.[89] Achieving assent will require a combination of shared technical expertise and competence; true engagement of local and host state stakeholders in risk assessment and siting and facility design and operation; partnering by facility developers with host states in repository siting, design, and operation planning and decision making; steps to meet host-state and local safety concerns, including host as well as transit states' and localities' concerns relating to waste transportation; and the provision of economic and other benefits to the host locality and state, including investment in physical, economic, or social infrastructure that will support long-term job creation and growth, as well as government services and priority under federal programs. Successful siting will also require a flexible, step-by-step approach by federal or private facility developers that is open to and adept at seizing opportunities, while respectful of and responsive to contestation and doubt. The federal government must abandon the arrogant prescriptions embraced in NWPA and its 1987 amendments and the subsequent dysfunctional approach to their implementation.

The lesson of the U.S. experience to date, confirmed by the record in some other nations, including Finland and Sweden, is that nuclear waste storage and disposal must be based on informed public assent, particularly that of host localities and states. This conclusion has both pragmatic and ethical foundations. As already noted, the federal government's plenary legal power to build a new nuclear waste repository or storage facility on its own lands is counterbalanced by deeply entrenched political and institutional safeguards of federalism that make it very difficult to impose such facilities against the determined opposition of host jurisdictions. And as an ethical matter, such impositions are unfair. Host jurisdictions should not have to bear the burden of other jurisdictions' wastes unless they have shared the benefits (in the form of energy or economic gains) generated along with the waste and have had a true opportunity to fairly contest, influence, and ultimately accept a facility, rather than merely to resist it.

Congress ignored these imperatives in 1987 when it singled out Yucca and imposed its choice on Nevada by brute legal force, and arguably also ignored them in 1982 when it adopted the centralized NWPA administrative-technocratic siting mechanism. Future siting decisions will require federal collaboration with states and localities, open processes, ready public access to information, and public involvement in, or opportunity for, review of data gathering, risk assessment, site evaluation, and facility design. It is only through such open and cooperative processes that informed assent is likely to be secured. The federal government, after considerable prodding, eventually and somewhat grudgingly followed this approach with New Mexico in developing the WIPP facility. WIPP is now the world's only operating deep geologic nuclear waste repository. For more than ten years, it has been viewed by EPA and other oversight bodies as safely, successfully, and permanently disposing of significantly radioactive defense nuclear wastes.[90]

Host assent could take explicit form as an agreement between the federal government or private developer and a state/locality/tribe, such as a memorandum of agreement under which the latter agrees to host the facility on specified terms. A state/locality/tribe may accept a facility on a judgment that the risks and burdens of serving as host are outweighed by economic or other advantages, including the jobs and economic development that the facility will provide and other inducements from the federal government or a private developer, such as funds to upgrade transportation infrastructure, emergency preparedness, or even public education. A similar calculus could operate when the facility developer is a private entity rather than the federal government.

Informed public assent has essential procedural elements. The process for making siting decisions must bring in a potential state/local/tribal host at the early stages of the planning process; the host must be given access to all relevant information about the proposal, be able to voice its concerns and demands, and have the opportunity to resolve them through discussion, deliberation, and negotiation. *Informed* public assent implies a host state or community that has full and accurate information about characteristics of the wastes, the risks posed, the site, and the facility that the government or a private entity proposes to develop, as well as related arrangements such as transportation, and continues to be kept in the loop throughout the project. It also implies that hosts have or are given the resources to hire independent experts to evaluate claims made by the proponents, perform their own investigations, and gather information on issues of importance to them. Correlatively, informed assent also requires public trust in the government or private entity developing the facility, and implicit confidence that decisions will be made on the basis of the information available and for the reasons provided by the developer. Overall, public trust of government is at an all-time low. Trust in DOE is lower still, because of the institutional arrogance, dissembling, and incompetence that have often characterized its—and more generally the federal government's—management of nuclear waste. The government's record has fostered the well-merited view among many segments of the public that waste management decisions are based on expediency and short-term political considerations rather than on the merits of a given site or proposal. This is as true of the decision to cancel Yucca as of the decision to select it. To achieve informed host and public assent to siting major new facilities, the government must rebuild its credibility and nurture public trust through open, informed, and inclusive decision-making processes. In addition, basic changes in the existing federal institutions responsible for nuclear waste management may be necessary. Private facility developers must also follow these procedures if they intend to win host assent and secure regulatory approvals.[91]

Future siting strategies can build on the lessons of WIPP and Yucca. Positive factors identified by a 2001 NAS committee report on WIPP that could be applied elsewhere (or, indeed, to an expansion of WIPP's mission) include the strong scientific and programmatic leadership of the WIPP project; extensive external scientific reviews throughout the project (including reviews by NAS groups); a transparent compliance certification process based on criteria agreed upon in advance; and the flexibility (often stimulated only after contestation) to make significant changes to its original plan. These changes included, for instance, project location and the design of engineered containment systems in response to technical findings by EEG-type expert advisory groups and other government bodies.[92] When combined with a degree of initial local receptivity to a facility, these elements can help build the confidence and trust on the part of the local community, the state or tribe, and other stakeholders that are necessary to support a new facility. However, in the case of WIPP, it is important to note that repeated state litigation against DOE played a major role in the state's leverage and ability to exact the changes and accommodations needed for its eventual assent to the project.

Targeting and Cultivating Receptive Host Jurisdictions

To obtain host-jurisdiction assent, developers of new nuclear waste facilities, whether the federal government, state compacts, or private firms, should first identify and target local jurisdictions that are likely to be receptive to hosting a facility. Experience shows that it is essential to have a strong base of support in the local community where the facility will be sited, and a host state that is not irreversibly opposed.

The local support that was critical at WIPP and the Andrews LLW facility in Texas was conspicuously absent at Yucca.[93] Although early on, some local officials at Yucca "considered with increasing enthusiasm" a plan that would boost a local county's revenue from the repository by creating a special tax zone, that support was swamped by opposition within the state provoked by Congress's heavy-handed approach to the siting process.[94] Yet if properly cultivated, even limited initial local support may grow as a project develops. A facility can provide local economic benefits, including jobs and tax revenue or payments in lieu of such revenue. Support may be especially strong in localities near existing DOE federal nuclear sites or nuclear power plants that have had tangible experience of their economic benefits. At least one survey has found a doughnut effect in public risk perceptions, as mentioned earlier.[95]

Experience reveals a correlative political doughnut effect, with citizens closer to a facility supportive (e.g., Carlsbad and Oak Ridge), and those farther away often opposed (e.g., Santa Fe and Albuquerque, and the rest of Tennessee).[96] This pattern tends to reflect the distribution of the economic benefits from a facility as well as the perceived risks that it poses. In many cases, DOE or ONWN found a supportive local jurisdiction to host a facility only to have state-level officials oppose it. A private developer, PFS, encountered this same pattern of local (tribal) support and state opposition in attempting to develop a private SNF storage facility in Utah. One reason why Sweden and Finland have been able to successfully site new SNF repositories is that they lack politically or legally significant subjurisdictions (such as the states or provinces in federal nations) and are accordingly able to deal directly with local governments and win their assent without strong intermediate state, provincial, or regional authorities that could, as a result of the political doughnut effect, block the facility.[97] The economic situation of a potential host jurisdiction is an important factor in its receptiveness to a nuclear facility. Carlsbad was economically quite depressed, and New Mexico's economy was shaky.[98] Nevada, on the other hand, had a booming economy based on tourism, which it believed would be seriously harmed by the stigma of a nuclear repository at Yucca. The economic factor raises potentially significant environmental justice issues if poor, needy communities end up hosting facilities widely regarded as undesirable. Requiring state-level and federal-level review and assent should help meet this concern; in the case of tribal sites, only federal review would be available.

Local experience with federal nuclear facilities, positive or negative, is also significant. For both Carlsbad and the State of New Mexico, past experiences with federal nuclear facilities and activities had been generally positive, including Project Plowshares and the Los Alamos and Sandia National Laboratories.[99] Nevada's experience was extremely negative—the government's detonation of atomic bombs at the Nevada Test Site, and its later prevarications and cover-up of the serious adverse health impacts caused by the fallout.

A number of local communities near DOE facilities expressed strong interest in hosting GNEP facilities before that program was discontinued.[100] DOE solicited "expressions of interest" from communities willing to host the consolidated fuel treatment center or the advanced burner reactor. Then DOE selected eleven sites from the fourteen applications, which were then eligible to compete for $16 million in grants to provide detailed site-characterization reports to DOE.[101] Six of the eleven sites were owned by DOE. The others included sites in Atomic City, Idaho; Morris, Illinois; Hobbs, New Mexico; Roswell, New Mexico; and Barnwell, South Carolina.[102] Many of these communities were near existing nuclear facilities.

The short-lived Office of Nuclear Waste Negotiator demonstrates that a voluntary approach, in which interested communities are invited to approach DOE, can help federal officials gain and keep the assent of potential local hosts. ONWN was established to negotiate with states and tribes interested in hosting a repository or MRS facility.[103] Congress established ONWN hoping to avoid imposing a facility on a state without its consent.[104] Although ONWN succeeded in locating and working with some local communities interested in serving as hosts, state governments blocked their participation.[105] As a result, ONWN subsequently focused only on interested tribes, which are not subject to direct legal control by the states in which they are located. Ultimately, state opposition to such efforts appears to have played a role in Congress's allowing ONWN to lapse in 1994.[106] But before it was terminated, ONWN offered $100,000 grants to communities interested in studying the feasibility of siting a repository or MRS facility on their land.[107] Twenty communities received initial grants, and several tribes, including the Mescalero Apaches of New Mexico, the Tonkawas in Oklahoma, and the Skull Valley Goshutes in Utah, were still working with ONWN at the time of its termination.[108]

Winning host assent may also require developers to provide economic and development benefits beyond those offered by a waste facility itself. In the case of WIPP, these included federally financed construction of a highway bypass of Santa Fe and $300 million in unrestricted grants to the state. Further inducements could include coupling other facilities or projects with a proposed waste facility, such as energy or other scientific research and development facilities or laboratories, plus support for research, education, and development activities at universities and technical institutes in the region. Facilities and projects that provide benefits that extend beyond construction jobs may be especially valuable. In addition, as again illustrated by WIPP, the federal government can meet state transportation safety concerns by funding training, emergency preparedness, and other initiatives.

Finally, steps to gain host assent must take into account differences in the risk and political legitimacy of the wastes to be stored and disposed of. The public seems to view all forms of nuclear waste as highly hazardous, judged by the opposition to new LLW facilities in California, Nebraska, and North Carolina. Yet it proved possible to site a new LLW facility in Texas. Under the current Texas Compact arrangements, the Andrews facility is licensed to accept class A, B, and C waste from the states in the Texas Compact (Texas and Vermont), as well as federal waste.[109] Very recently, the Texas Low-Level Radioactive Waste Commission voted to authorize the Andrews facility to accept class A, B, and C LLW from as many as thirty-six states.[110] New Mexico initially accepted only the less radioactive CH-TRU at WIPP and accepted the more radioactive RH-TRU only after experience with CH-TRU proved satisfactory.

The difficulties in siting HLW and SNF repositories appear to be far greater than for LLW, as illustrated by New Mexico's past hostility to hosting HLW or SNF. In illustration of the doughnut effect, the New Mexico state government and public have in the past strongly opposed hosting HLW or SNF, while Carlsbad's political leaders have expressed strong interest in the prospect.

The origin of the wastes is also quite relevant in terms of host willingness. As noted, New Mexico and other states have been more willing to accept defense wastes, even from out of state, than SNF from civilian power plants in other states. In the cases of both Nevada and New Mexico, opposition to SNF disposal was heightened by the fact that neither state has a nuclear power plant.

Threshold Prerequisites for Public Trust and Host Acceptance
To win state and local assent to nuclear waste facilities, there must at a minimum be institutional assurances that a project is safe and environmentally sound, funding for independent technical and scientific oversight and review, and effective opportunities for public and host access to information and participation in decisions regarding the project and its implementation.

Institutional Assurances of Project Safety and Environmental Protection
The federal government must set high environmental, health, and safety regulatory standards and stick to them, especially if it means disqualifying a candidate site that

cannot meet them. The Yucca Mountain project, and along with it the entire repository development program, were discredited when Congress enacted amendments to NWPA requiring that specially tailored environmental standards be set for the site, which was widely taken as a signal that the site could not meet the repository standards already set by EPA. To avoid the same fate in future efforts, these high standards must be applied and enforced by federal regulatory authorities (currently NRC and EPA) that are independent of DOE or the project developer, through transparent project-licensing and environmental certification procedures, followed by periodic recertification throughout the life of the facility, on the WIPP model. Also essential are credible institutional and financial assurances of stewardship of a facility after it is closed, as well as during its operational phase. For privately owned facilities, government regulatory requirements to ensure long-term funding, including trust or escrow arrangements, are needed. Clear and adequate monitoring, record keeping, and public disclosure requirements will be necessary to enable host states and localities to track facility performance during and after the period of its operation. There should be mechanisms in place for hosts, as well as regulators, to trigger measures to correct performance problems. New Mexico insisted that the government's assurances and undertakings regarding WIPP be made binding and judicially enforceable; this was critical to the state's ultimate acceptance of the project. Such antecedent and continuous assurances of project safety and environmental protection—and the host jurisdiction's ability to enforce them—are paramount factors in winning host acceptance in the first instance, as well as in ensuring that a project is successfully carried out.

Equipped with the authority to issue and enforce a RCRA permit for WIPP due to the mixed waste character of much of the TRU shipped there and EPA's delegation of RCRA permitting authority to the state, New Mexico has been able to assert a significant measure of control over WIPP's compliance with environmental requirements as well as to exert leverage with federal authorities and contractors running the facility. The RCRA permit includes requirements for monitoring, record keeping, and reporting, as well as for compliance with waste management standards and cleanup requirements; the permit is for ten years, renewable by the state. Under existing law, only a repository or consolidated storage facility that receives mixed waste with the requisite chemically hazardous properties (which might not include SNF or some vitrified HLW) would be subject to RCRA. Accordingly, as an inducement to states to host a HLW or SNF storage facility or repository, Congress might well give host states regulatory authority, including monitoring and environmental compliance authority, over all such facilities.

Independent Technical and Scientific Oversight and Review

Experience shows that gaining public trust and host assent for projects as complex and long-lived and potentially risky as nuclear waste repositories and storage facilities requires not just assurances by regulatory authorities of a project's soundness and safety, but also credible independent oversight and review of the scientific and technical issues relating to the project's siting, design, construction, and operation, as well as its closure and postclosure plans. Assured independent funding should be established for one or more qualified independent, expert bodies, similar to EEG or NWTRB, to operate during the siting, development, operational, and postoperational phases of the project.

Consideration must be given to the relative merits of review bodies created for a specific facility versus those with more general jurisdiction; to whether to create different bodies to address different aspects or phases of a project; and to how best constitute and finance such bodies and their membership so as to assure expertise, independence, and objectivity, and thereby engender trust. There may be a need, in addition, to provide potential host jurisdictions with resources to assemble their own technical and scientific research and review bodies to operate independently of, but concurrently with, similar federally established bodies.

Public Participation, Access to Information, and Transparent Decision Making

A further prerequisite for winning trust and gaining host assent is full disclosure of, and access by the public and host-jurisdiction officials to, information regarding all elements of a project's siting, design, development, and operation, as well as a meaningful opportunity for effective engagement and input into project decision making. The existing procedural apparatus of environmental assessments, environmental impact statements, and *Federal Register* notices inviting public comment, as well as the availability and proven efficacy of the Freedom of Information Act, can further these goals but are not sufficient to establish public trust and gain host assent. The federal government or other project developers must thoroughly embrace an organizational culture and practice of ongoing full disclosure and public engagement. Disclosure, opportunity for written comment, and town hall meetings may not produce the degree and quality of engagement needed to assure host jurisdictions and the public that their concerns have been heard, fully considered, and addressed through a process of genuine dialogue. The successful experience of the Consortium for Risk Evaluation with Stakeholder Participation organization in dealing with local community concerns over waste cleanup at some of DOE's former weapons facilities and over radiological hazards at a former nuclear test site, provides a model that could be usefully followed in the siting context.[111] The CRESP approach is designed to engage, through outreach and open processes, local communities and state authorities at the earliest possible stage in project characterization and risk assessment studies, enabling them to participate in the development of approaches taken and to communicate their knowledge and concerns from the outset. The education process must be two-way, as opposed to outside experts telling the community what the risks are and then attempting to engage the community in discussion about how the risks identified by the project developer should be managed.[112] An open, two-way approach can establish a foundation of mutual understanding that enhances and builds trust in decision making regarding project siting, design, development, and risk management issues. This kind of approach to siting and project development—which may include significant elements of contestation, as WIPP illustrates—is far more effective at building public trust than the top-down NWPA approach and should be made an integral part of any future nuclear waste facility development regime.

Ensuring Potential Host States an Effective Role in Facility Decision Making

The genuine and substantial opportunities that New Mexico won and effectively leveraged to sway important decisions about WIPP proved to be a critical factor in the ultimately successful development of the repository. The extent of these opportunities can be understood as a product of at least three main variables: the kinds of issues that were open (or closed) to meaningful state input and influence, the character of and timetable for decision making on key issues, and the legal and institutional framework for state involvement in decision making.

Flexible, Step-by-Step Project Decision-Making Processes

WIPP was shaped over many years by an evolving process of incremental decision making that sustained its uneven but continuing progress from a pilot project with an undefined mission to successful operation as a TRU repository more than twenty years later. WIPP was in large measure a negotiated package rather than a top-down decision made in Washington, D.C. There was no blueprint drawn up in advance. The key decisions about the location and basic design of the facility and the wastes it would hold were made step-by-step, over a substantial period of time, through an incremental process of contestation and negotiation that directly involved the state, DOE, the federal courts (as a result of litigation brought by the state and environmental and citizens groups), and Congress—and also, although less directly, the Town of Carlsbad, environmental and citizens groups, other western states, and other stakeholders. There was no preset timetable for this process. The step-by-step character of project development, and the considerable flexibility and mutability of the decision-making process—with outcomes often unintended and at times undesired by the various stakeholders—enabled the state to have effective input and influence on key decisions, and to secure many of its most important objectives. The state's ability to exert leverage was strengthened by federal court victories, by a legally enforceable consultation and cooperation agreement negotiated with DOE, by the information and technical expertise provided by EEG, by New Mexico's exercise of RCRA regulatory permit authority over the facility, and through the successful representation of its interests by the state delegation in Congress. The C&C process between the state and DOE was an especially important tool for state influence.

Legislation by Congress for the most part responded to, rather than dictated, developments at WIPP and with the state of New Mexico. Each piece of WIPP legislation corresponded to a particular stage in the project's development.[113] Thus, at each stage New Mexico had an opportunity, through its delegation, to shape congressional decisions and the future direction of the project. As a result, New Mexico succeeded in excluding SNF and HLW from WIPP, and in limiting the amount of RH-TRU to be disposed at the repository. The state also extracted various commitments from federal actors, including greater liability than was provided for by existing statutes for harms arising from nuclear accidents, funding for EEG, training for state emergency services personnel, a facility to monitor and inspect shipments to WIPP, federal construction of a highway bypass around Santa Fe, and $300 million in unrestricted federal grants to the state. These and other substantial changes and concessions, achieved in an incremental fashion by New Mexico throughout the process, helped make the WIPP project ultimately acceptable to the host state and local community.

By contrast, NWPA imposed a blueprint for Yucca that defined at the outset the key elements of the repository project and prescribed a rigid timetable for implementation. Congress determined the facility's location, what wastes it would hold, the amount of wastes that would be disposed of, and the schedule for its completion. Nevada foresaw few benefits from the Yucca Mountain project, only heavy costs, including the potential burden of hosting all the nation's highly radioactive wastes in perpetuity and serious concomitant threats to its tourism and gambling economy. Congress's blueprint provided only limited scope for addressing fundamental state concerns, or for altering course in response to new information. Although arrangements were floated for state input on design, safety, and transportation features, decisions on these issues were made exclusively by DOE and other federal agencies—and on a highly compressed timetable. This rigid top-down scheme ultimately broke against Nevada's adamantine resistance, which drew strength and support within and outside the state from the wide perception of fundamental procedural unfairness in the congressional scheme, multiple substantive concerns regarding the site's potential inability to assure long-term containment of SNF and HLW, and DOE's lack of candor regarding emerging evidence of problems with the site and repository design.

The vital importance of phased, stepwise planning and adaptive development of a repository or other major new nuclear waste facility has been confirmed by both the NAS Committee on Disposition of High-Level Radioactive Waste and the Canadian Nuclear Waste Management Organization (NWMO), as discussed earlier.[114] And experience in other countries supports and confirms that a cooperative, flexible, adaptive phased management approach can help secure public and host-jurisdiction acceptance by encouraging dialogue with the technical community and the public; by expediting correction of weaknesses, uncertainties, and other problems that emerge; and by providing opportunities for stakeholders and host jurisdictions to have input in all stages of project decision making. This approach combines a pragmatic belief that successful engagement and collaborative decision making are key to maintaining momentum and cohesive synergy for a project with an ethical belief that those most exposed to potential harm must be fully informed and involved.[115]

Consultation and Cooperation Agreements

The legally binding C&C agreement and later modifications to it, negotiated by DOE and New Mexico, gave the state significant opportunity to influence key decisions regarding WIPP. The C&C agreement helped define and implement an adaptive, step-by-step decision-making process for development of WIPP and empowered the state to check DOE when it threatened to decide key issues unilaterally. For example, when DOE attempted to move the project from the test phase to repository construction without notifying the state, New Mexico successfully sued DOE to enforce provisions of the agreement guaranteeing notification and prevented DOE from moving ahead with the project. New Mexico was able to use the halt in the project as leverage to obtain important concessions from DOE and Congress on the terms of the construction phase of WIPP. The C&C agreement thus enabled New Mexico to maintain significant control over the pace with which WIPP moved forward and also afforded ongoing opportunities for the state to influence key decisions about the project.

NWPA contained provisions requiring DOE to negotiate a C&C agreement with Nevada. These provisions were apparently modeled on the arrangements that emerged at WIPP.[116] Congress evidently envisioned that the C&C provisions would produce a legally binding agreement between DOE and Nevada that would incorporate dispute settlement and enforcement mechanisms. However, DOE and the state never reached an agreement, and the statute failed to make any provision for this contingency. Even if such an agreement had been reached, its value to Nevada would have been limited, because NWPA took many of the important elements of the Yucca project off the table before any opportunity for negotiation and resolution with the state. The location of the repository, the wastes to be disposed, the timetable for decision making, and the scope of state involvement in decision making had already been determined by Congress. Moreover, NWPA vested the power to make all decisions relating to repository design and transportation exclusively in the federal agencies charged with implementing the statute. Additionally, Congress's imposition on Nevada of a repository at Yucca was made through backroom decisional processes that the state and many others regarded as illegitimate.[117] Under these circumstances, invoking "consultation and cooperation" rang hollow. It would have been political suicide for Nevada officials to be seen as cooperating with DOE in helping move the project forward. Accordingly, even if Congress had included stronger C&C provisions in NWPA, it seems unlikely that these provisions would have produced genuine engagement and productive negotiation between DOE and the state. Legal provisions may facilitate, but cannot mandate or guarantee, genuine consultation and cooperation.

The absence of any genuine engagement and give-and-take between DOE and Nevada may also have flowed from the fact that Nevada's right to issue a notice of disapproval subject to congressional override was a weak instrument that did not give the state significant bargaining leverage with DOE. First, Nevada had made clear at the outset, and DOE was well aware, that the state would file a notice of disapproval if, as Congress clearly intended, the department were to recommend to the president that a repository at Yucca should go forward. The Nevada legislature passed two joint resolutions in January 1989 expressing its opposition to any high-level nuclear waste repository. In July, the governor signed a prohibition of any such facility into law and prompted the Nevada attorney general to issue an opinion that this constituted a valid notice of disapproval under the NWPA.[118] Any concessions that DOE might make to Nevada under NWPA could only be around the margins of the project and thus were highly unlikely to induce Nevada to change its mind. Accordingly, federal officials had no real incentive, other than avoiding adverse court decisions if sued, to accommodate the state's demands, nor did they in any event have great latitude to do so.

Second, because it was evident that members of Congress from other states with potential repository sites and those with stored SNF and HLW had powerful incentives to overturn Nevada's disapproval of Yucca, DOE knew that it was highly likely if not certain that Nevada's veto of Yucca would be overridden by Congress, as in fact occurred. The notice-of-disapproval mechanism, established by Congress as the principal safeguard for the interests of host states, accordingly failed to give Nevada any real leverage with DOE on decisions regarding the facility. Another problem with the disapproval mechanism was that a state could invoke it only once, on an all-or-nothing basis, and only after all the basic decisions regarding the repository had already been made by the federal government. Joseph Canepa, an attorney in the New Mexico attorney general's office from 1978 through 1982 while WIPP was in development, reflected: "A state must impose a phased decision-making process in order to have any type of meaningful role in, and effective control over, such projects. The state must avoid at all cost being put into the position of making a one-time decision which gives the 'green light' to the project forevermore."[119]

Host-State Scientific and Technical Capacity

For a potential host state to have meaningful influence over nuclear waste facility decisions and concurrently to develop confidence in the outcomes, it must have the capacity to understand and assess the scientific and technical questions presented during the project development process, to effectively question and combat erroneous or dubious claims made by the project's proponents, to make knowledgeable judgments about outcomes, and to propose alternatives. Review and oversight by independent expert bodies such as NWTRB or panels or committees put together by NAS, which report to the federal government, can only partially fulfill this need. States will want to either have their own experts (which larger, wealthier states may already to some degree have) or have access to experts they are confident will be responsive to their interests and needs.

At both WIPP and Yucca, the federal government gave grants to the states for this purpose. In the case of WIPP, the state and DOE developed the federally funded but state-appointed and state-run EEG, which evolved into a body that was trusted by the state and was also respected by other stakeholders, enabling it to play an extremely constructive role in the development of the project.[120] In 2004, however, DOE ceased funding EEG, prematurely ending its role in the project.[121] This history indicates the need for assured funding of such bodies throughout the life of a project.

In the case of Yucca, the nominally similar arrangements for federal-state cooperation assumed an opposite, adversarial posture. Nevada used the federal funding to assemble experts to help it attack the suitability of the site and the performance and safety of the repository design. The adversarial use by Nevada of scientific and technical resources is hardly surprising, given the way the project was unilaterally imposed on the state. As with entering into C&C agreements, funding state expertise can work constructively and help promote host acceptance only if the state believes that the overall decision-making process for the facility is fair and truly affords the state a meaningful opportunity to influence decisions and secure considered resolution of its concerns.

Successful Siting Approaches in Other Countries

While a comprehensive examination of the policies and approaches other countries have applied to siting nuclear repositories or consolidated storage facilities is beyond the scope of this book, the successes achieved by Finland and Sweden are notable. Both countries have based their strategies on securing host-jurisdiction assent, using an incremental, adaptive, and collaborative approach that incorporates many of the elements discussed earlier in this chapter.[122]

Finland

The Municipal Council of Eurajoki, Finland, in January 2000 consented to host the world's first deep geological repository for SNF, a decision ratified by the Finnish parliament the next year.[123] Posiva Oy, a company formed by the government-owned Fortum Power and Heat Oy and the privately owned Teollisuuden Voima (TVO) power company to manage the country's SNF, selected the Eurajoki site after a twenty-year site selection process.[124] Initially, TVO identified sixty-five potential sites on the basis of siting criteria that included geological suitability, population density, and land-use planning. Eurajoki was initially excluded because the siting criteria (as was found later) arbitrarily disfavored coastal sites. After this defect in the criteria was corrected and it was determined that they were suitable, sites near the towns of Eurajoki and Loviisa that already hosted nuclear power plants were included in the evaluation process. The candidate sites were then winnowed down to four, based on geology, the local communities' receptiveness to continuing investigations, and further discussions. In 1999, Posiva Oy made its final decision in favor of Eurajoki on the basis of cost, safety, and social acceptability.[125]

Although it allowed early investigations to proceed, the Eurajoki Municipal Council was initially skeptical of the project. An election in 1994 produced a more receptive council, which spent years negotiating with Posiva Oy.[126] Ultimately Eurajoki accepted the project after negotiating a package of assorted economic benefits with Posiva Oy, including a €6.89 million loan for the construction of a new home for the elderly; the renovation of the Vuojoki Manor, an important publicly owned cultural site; and a direct loan of €2.35 million at below-market interest rates.[127] The municipality was especially eager for these capital injections after TVO successfully appealed the 1993–1994 real estate tax rates for its Eurajoki-based nuclear reactors, creating a local fiscal crisis. Eurajoki's acceptance of the project was also conditioned upon Posiva Oy's promise, made binding through enactment in the Finnish Nuclear Energy Act of 1987, that no foreign SNF would be stored in the repository.[128]

After local approval was given, the Finnish Council of State on December 21, 2000, approved the Eurajoki site, a decision ratified by parliament on May 18, 2001.[129] In 2004, Posiva Oy began constructing an underground characterization facility at the site, dubbed ONKALO. As of mid-2010, ONKALO had reached repository depth of 420 meters and excavation was ongoing.[130] The company plans to apply for a construction license in 2012, with an expected operational date of 2020.[131]

Sweden

Efforts to site a HLW repository in Sweden met with similar success after engagement directly with host communities. SKB, a joint venture of the five nuclear-operating utilities in Sweden, is responsible for the planning and construction of SNF facilities.[132] SKB began pursuing a repository in the late 1970s after legislation obliged the nuclear reactor owners to find a long-term solution for SNF disposal.[133] Between 1980 and 1985, SKB drilled a series of test wells throughout the country to inform the site selection process. However, the utilities' failure to engage with the local communities and the widespread perception that a test well was a precursor to a permanent repository engendered a stiff public backlash.[134]

Chastened by this experience, SKB concluded on the basis of the test wells that much of Sweden was geologically suitable for a repository and moved beyond purely technical criteria to emphasize political factors, among other considerations.[135] After efforts between 1992 and 1994 failed to solicit volunteer communities, in 1995 SKB extended invitations to a handful of communities that were already hosting nuclear reactors.[136] By 2002, SKB feasibility studies and local vetoes had narrowed the candidate sites to the municipalities of Östhammar and Oskarshamn, both of which volunteered to continue the siting process via majority vote of their municipal councils.[137] At that point SKB undertook an in-depth site characterization process with a focus on technical criteria.[138] In 2009, SKB selected Östhammar and is currently awaiting approval by the Swedish government from the Swedish Environmental Court and the Radiation Safety Authority.[139]

In contrast to the experience with Eurajoki in Finland, direct financial inducements seem to have played a smaller role in Sweden. Östhammar and Oskarshamn each created its own review process of SKB's investigations, using paid consultants and community boards.[140] These expenses were reimbursed from the Swedish Nuclear Waste Fund, a national fund paid into by the nuclear utilities to cover the costs of nuclear waste disposal.[141] Otherwise, the promise of jobs from the project's construction and operation appear to have been a major inducement for Östhammar, which, as home to the Forsmark Nuclear Power Plant and a LLW repository, is already a nuclear company town.[142]

Although the Finnish and Swedish experiences are instructive, there is no assurance that their strategies would enjoy similar success in the United States. Sweden and Finland are small, relatively homogenous countries that do not have a federal structure that interposes strong jurisdictions, such as states or provinces, between the national government and local governments. Thus national government authorities or private developers with national government support can deal directly and solely with local communities that may (especially if they have existing nuclear facilities) favor new facilities due to anticipated economic or other benefits; this configuration averts the political doughnut effect experienced in the United States whereby states often veto proposals favored by local communities.

Despite its rather unitary governmental structure, France has not progressed nearly as far as Finland and Sweden in developing a repository, which in France's case would be for disposal of HLW from reprocessing. The French National Agency for Nuclear Waste Management (ANDRA) has begun construction of an underground characterization laboratory, the Meuse/Haute-Marne facility, in a clay formation near the village of Bure.[143] France is currently consulting with local communities overlying that 250-square-kilometer formation to determine an acceptable site for a repository.[144]

Canada has a federal structure analogous to that of the United States. Notwithstanding the challenges that such a structure presents, NWMO decided, after careful consideration of siting experiences in a number of other countries, to follow an incremental, adaptive, collaborative approach to siting based on host-jurisdiction assent.[145] NWMO was established in 2002 pursuant to the Canadian Nuclear Fuel Waste Act, which mandated that companies producing nuclear fuel establish an organization to advise the Canadian government on disposal and reprocessing options for Canada's SNF. In 2005, after strong public resistance had scuttled an earlier attempt to site a deep geological repository, NWMO proposed a more collaborative approach.[146] As part of its system of adaptive phased management, NWMO has prioritized public involvement. By publicizing the project's benefits, such as jobs and infrastructural improvements, NWMO hopes that a number of communities will volunteer as candidate sites—most likely in provinces already containing nuclear reactors or uranium mining. The technical review of site suitability would then proceed concurrently with an extended process of community engagement, including education programs, public hearings, and the negotiation of terms and conditions.[147] The strategy, which mirrors that of Finland and Sweden, is just beginning to be implemented in Canada but has already attracted interest from several potential host localities.

New Institutions and Financing Mechanisms for Waste Siting and Management

A further issue that must be addressed by the blue-ribbon commission, Congress, and the Obama administration is future institutional arrangements for nuclear waste management, repository and interim storage facility siting and development, and regulation.

The current arrangements place responsibility for both waste management and waste facility siting and development with DOE and its various offices.

DOE has long been a troubled agency with many disparate missions that are largely unintegrated.[148] There is often lack of communication and cooperation among the internal departmental silos. Top management turns over rapidly. DOE has had difficulty recruiting and retaining qualified personnel to discharge the highly demanding tasks with which it is charged. Although such problems are endemic to the federal government, they are especially serious in the case of DOE, which must manage and execute large-scale, long-term, technology-based construction, management, research, and other projects. The quality and technical capabilities of many DOE personnel, including those overseeing multibillion dollar-cleanup projects at DOE defense facilities, are often weak, and staff are often stretched thin. As a result, great power has devolved to the contractors who execute the cleanup projects, undermining accountability and performance. Moreover, DOE site cleanup projects are rich political prizes: they provide local jobs and huge revenues for contractors. Congressional funding for waste management is intermittent, often driven by constituent and contractor interests. An endemic lack of firm overall management direction has often resulted in overall failure by the agency to set and enforce national priorities amidst the constantly shifting tug and pull over resources and attention among DOE's silos, support contractors, and other local and national constituencies. In addition, DOE is still plagued by an organizational culture of secrecy and isolation inherited from AEC. That culture has been reinforced by government lawyers who have clamped down on any disclosure of records or views that might prejudice the government's liability litigation over waste storage with the utilities, or that could hinder the completion of Yucca.

The most fundamental difficulty with existing arrangements, however, may be that the task of managing nuclear wastes and facilities and the task of siting and developing new waste disposal and storage facilities are fundamentally different and call for distinctive and different organizational skills and attributes. Accordingly, serious consideration should be given to taking both these functions out of DOE's purview and creating two new entities, one responsible for nuclear waste facility siting and development, the other for nuclear waste facility construction and operation and nuclear waste management. The choice of an institutional structure must also take into account its relation with arrangements to assure long-term financing for development and operation of new and existing nuclear waste facilities, discussed below.

Waste Management

A new entity for waste management could be responsible for waste storage, treatment, and transportation; development and use of waste containers; construction and operation of interim consolidated storage facilities; and construction, operation, closure, and postclosure monitoring of a repository—but not for siting and development of such facilities. Its responsibilities not only could include management of HLW and defense SNF, but also could encompass management of the civilian SNF currently managed by utilities, pursuant to new federal legislation that would also have to resolve issues relating to disposition of and liability for power plant SNF. The institutional requisites for such a waste management entity should include a clearly defined mission; a business model of management; high-quality, technically adept personnel; and assured, long-term, stable funding. Such an entity might take one of several institutional forms:

- A federal agency with a single head who reports to the president
- A federal agency with a single head who reports to the secretary of DOE but is located outside DOE (on the model of the Bonneville Power Authority)[149]
- A federal corporation owned by the federal government with a presidentially appointed board that selects a CEO to manage its operations (on the model of the Tennessee Valley Authority)[150]
- A hybrid federal corporation owned in part by the federal government and in part by the nuclear utilities, with a board selected by both (a model adopted in a number of western European countries)

Another option is a multimember independent federal agency similar to NRC or FCC. Such an organization would, because of its plural governance structure, tend to lack the unity of leadership and clear lines of control and accountability needed for the nuclear waste management enterprise, which involves large-scale, capital-intensive, long-term, technically complex projects of a type that requires a more hierarchical business or military model of organization. These requisites point to a further institutional option: a private corporation owned by the nuclear electric utilities and regulated by the government. Although this model has, as discussed later, been adopted by several Scandinavian countries, it might be too different from existing U.S. arrangements to be politically palatable.[151]

The advantages of a corporate form, whether public or private, are that it would promote and help realize a business model, implying integrated planning and execution, efficiency, accountability, and bottom-line functionality. It would also free the entity from federal personnel and procurement requirements, thus promoting flexibility and efficiency, as well as enabling it to hire and retain highly qualified personnel.[152] A variety of possible arrangements could provide more assured continuity of financing for facility construction and management than under the NWF scheme and existing practices for funding DOE that have been available to date. Some of these mechanisms would be easier to establish in connection with the corporate form of organization than with the traditional model of a government department or agency funded through annual appropriations by Congress.

A further advantage of the corporate form, if coupled with appropriate financing arrangements, is that it could build on the commonality of interests between government and the utilities in successful waste management; the current NWPA waste management and financing scheme makes these parties into adversaries. The model adopted by Sweden and Finland gives operational responsibility for nuclear waste management to a utility-owned corporate entity, while Belgium, Germany, France, and the United Kingdom have opted for various forms of hybrid public-private governance of waste management.[153] Canada has established the Nuclear Waste Management Organization to manage and dispose of SNF. NWMO's shareholders are three Canadian nuclear utilities; they have appointed several outside directors to represent broader perspectives.[154] In the U.S. context, a hybrid federal corporation, co-owned by the government and the nuclear utilities, could represent a sensible compromise arrangement. The waste management entity could assume ownership of SNF or reprocessing wastes once they left the site of a reactor or a privately owned reprocessing facility and, as in a number of European countries, assume responsibility for managing, treating, and disposing of them, subject to independent federal and, as appropriate, state regulation.

A federal or hybrid corporate form, however, risks certain disadvantages. It would be subject to both political and private capital market incentives and influences, potentially blurring responsibility and undermining accountability in its policies and finances.[155] Just such an eventuality was illustrated by the recent saga of Fanny Mae and Freddie Mac, in which incentives created by management compensation based on relatively short-term financial performance, together with political influences from Congress and elsewhere, promoted excessive risk taking and other distortions and resulted in dire mission failure.[156] Also, a waste management entity in the form of a federal or hybrid public-private corporation might intrinsically resist cooperation with, or find it difficult to coordinate waste management decisions with, the functionally related policies and decisions of existing federal agencies with a role in the field. A separate federal agency with a single head reporting to the president would ameliorate these problems but would sacrifice many of the attributes and advantages of the business model, including freedom from federal personnel requirements and independence from distorting short-term political pressures. A separate agency that reports to the secretary of energy could promote coordination but might sacrifice even more of the advantages of independence. Whichever were chosen, assured funding for any new waste management entity would have to be developed through a revolving fund, fees paid directly by the utilities for SNF services, or otherwise. The use of a corporate form could allow the entity to tap private capital markets for funding capital-intensive, long-term projects.

The IRG report of the late 1970s concluded that the operational tasks of SNF waste management required a managerial structure with well-defined program authority, efficient (i.e., businesslike) operations, and a predictable, transparent funding mechanism.[157] A 1982 congressional Office of Technology Assessment report examined a number of options, including a new executive branch agency similar to NASA; an "independent" agency with loose ties to DOE, such as the Bonneville Power Administration; a government corporation, such as TVA; and a federally chartered nonfederal entity such as Comsat, the global telecommunications Communications Satellite Corporation. The report indicated that a corporate structure might be most desirable because it would allow for direct funding through user fees and greater flexibility in personnel policies, and would therefore "increase accountability for the achievement of the goals of the program."[158] A 1984 report to the secretary of energy by his Advisory Panel on Alternative Means of Financing and Managing Radioactive Waste Facilities followed a similar analysis, stressing the need for a financing mechanism outside the normal congressional budget process.[159] The advisory panel recommended creating a federal corporation to take over nuclear waste management functions.[160] An internal DOE group established to review the panel's recommendations concluded, however, that they should not be followed, and that internal DOE reforms would suffice.[161] Neither the president nor Congress has taken steps to implement these options or recommendations. Experience since these outside reports were published strongly suggests that they were on the right track and that fundamentally new institutional arrangements indeed are needed. DOE has recently floated the idea of a government-chartered corporation like TVA to assume responsibility for nuclear waste management, storage, and disposal.[162]

Siting and Development of New Waste Facilities

Siting and development of new HLW/SNF waste repositories and consolidated storage facilities and LLW disposal facilities call for different institutional requisites than those fitted for waste management. While technical competence is essential, experience indicates that a purely technocratic model is too narrow and rigid to be successful. As discussed earlier, successful development of new storage facilities or repositories will require close engagement with states and localities, with the utility and nuclear industry, and with environmental and local NGOs, accompanied by a flexible capacity for constructive negotiation with those many different types of stakeholders. This will require an institution that is relatively open, can represent or accommodate different viewpoints and stakeholder interests, is politically attuned, and can develop good connections with Congress, other federal agencies, the states, and localities. The multimember independent commission form may best suit these specifications. The multimember structure facilitates representation of different viewpoints in its governance. Such agencies have typically had closer ties with Congress (and, through Congress, to state and local governmental and other interests) than have agencies with single heads who report to the president. An office of waste negotiator could be included as a component within the commission to take the entrepreneurial and operational lead in exploring and negotiating siting opportunities, building on experience gained under the now-expired ONWN, as well on the successful experience at WIPP and with private LLW and hazardous waste facilities.

The creation of several new entities to assume functions now housed in DOE will undoubtedly create problems of coordination between the new entities and DOE, as well as with NRC, other federal agencies, and states and localities. Such coordination problems might make it more difficult than does even the current dysfunctional arrangement to achieve an integrated approach to waste management that takes account of the entire nuclear fuel cycle and concerns over environmental protection, energy security, and security against terrorism and nuclear proliferation But the limitations of the current arrangement indicate that a new approach, relying on more independent and functionally specialized bodies and the flexibility that they would provide, is likely warranted.

Financing Nuclear Waste Management, Storage, and Disposal

With leadership from the blue-ribbon commission and the administration, Congress will have to grasp the nettle of DOE's liability for failure to meet its contractual obligations, pursuant to NWPA, to take ownership of utility SNF beginning in 1998. Continuing to leave the issue to be resolved through scores of lawsuits is not an effective or tolerable basis for financing nuclear waste management. Congress should resolve the government's past and future liabilities through legislation that will at the same time provide a more secure system of financing for SNF management, storage, and disposal than was achieved under the NWPA Nuclear Waste Fund. As previously explained, the term Nuclear Waste Fund is misleading, because it implies that there is a reserve of money set aside in the U.S. Treasury earmarked for dealing with SNF. In fact, no such reserve exists. Only a fraction of the revenues collected on nuclear electricity generation has been spent on dealing with SNF; the balance has been effectively used by Congress for other purposes. Of the billions collected—ultimately from electricity users—a great deal is long gone.

To effectively address the SNF disposition problem and help instill industry and public confidence in an orderly and responsible system for financing the new enterprise, a variety of mechanisms could provide more secure, dedicated funding for SNF management, storage, and disposal:

- Industry-financed storage and disposal through a corporation owned and operated by the utilities, with some government/public representation in its governance. This is the NWMO model being followed in Canada.
- A federal corporation with utility representation in its governance that would have authority to finance its operations by fees on nuclear electricity generation. Either of these first two forms of corporation could borrow or raise funds in private financial markets to help finance capital projects and other activities.
- Funding for a special-purpose government agency, or an SNF management and disposal program within a government agency, such as OCRWM/DOE, through dedicated revenues from nuclear electricity fees and congressional appropriations placed in an escrow account in the treasury.

States and some experts have suggested using a system of escrow accounts to provide more secure and assured financing for SNF management, storage, and disposal. Under this approach, one or a number of escrow accounts would be established within the Treasury to receive and hold funds, which could be withdrawn only by specified entities for these purposes.[163] The financing arrangement would be established through congressional legislation that would settle the government's SNF liabilities and establish a system for future financing of SNF repositories and storage facilities. The financing could include monies appropriated by Congress in liquidation of its liabilities under existing law, and fees on future nuclear electricity generation, including generation by new nuclear power plants. A general escrow account could be established, or a separate account for each plant. To the extent that utilities continue to hold title to and manage the waste, they could draw funds from the escrow account to cover specified expenses. If a government agency assumed responsibility, it also could draw funds for designated purposes. Transitional measures would have to be taken to phase in these new arrangements.[164]

Another option is reclassification of revenues from the nuclear generation fee as offsetting collections and receipts. Under this system, expenditures for SNF management would not be subject to the overall federal-spending budget cap, and the SNF program would not have to compete with other federal programs for limited resources.[165]

As a further option, some Republican members of Congress have introduced legislation that would liquidate the Nuclear Waste Fund and return unspent collected revenues to taxpayers if the government does not proceed to develop Yucca Mountain for receiving SNF.[166] But as the revenues in question do not really exist, this proposal appears to be a transparent ploy to open Yucca, rather than a genuine attempt at resolution of the liability/financing problem.

Any congressional resolution of DOE's SNF contractual liabilities should seek to give nuclear electricity ratepayers some form of credit for NWF revenues collected that were not spent for SNF disposition. In addition, the government should bear at least some of the increased costs of dealing with SNF that result from the abandonment of Yucca.

The discussion here has thus far focused on commercial SNF. Attention must also focus on defense wastes, providing more predictable and assured funding for managing and cleaning up defense wastes than is provided under the current somewhat erratic annual appropriations process. If the government corporation model were used for defense wastes as well as for SNF, a system of five-year or ten-year government contracts and grants might be used, with arrangements to give the entity's management considerable discretion to set cleanup priorities, subject to arrangements to secure accountability for performance. Alternatively, in the more likely event that a government agency model were used, financing could be provided through long-term (e.g., five- or ten-year) congressional appropriations for specific capital and operating projects, possibly including a revolving fund separate from the unified federal budget. In this case, too, discretionary arrangements for management to set cleanup priorities and ensure performance accountability should be developed.

Environmental Health and Safety Regulation of Highly Radioactive Nuclear Wastes

The institutional arrangements for environmental health and safety (EHS) regulation and management of nuclear fuel cycle activities and the SNF, HLW, and TRU they produce have evolved in patchwork fashion. Nonetheless, it appears that the existing institutional arrangements work tolerably well, notwithstanding regulatory duplication at the federal level with respect to nuclear waste repositories. Serious consideration should be given, however, to empowering states to regulate all forms of the most highly radioactive wastes at storage and disposal facilities that they host, not just (as currently) those high-activity radioactive wastes that happen to contain chemically hazardous constituents and can accordingly be regulated as mixed wastes under RCRA by states that have been delegated that regulatory authority by EPA.

Civilian nuclear fuel cycle activities have always been regulated by a federal government regulator—initially by AEC and later by NRC. NRC was created pursuant to the 1974 Energy Reorganization Act (ERA) based on the conclusion that AEC's role as regulator could be undermined by its role as developer and promoter of nuclear technologies. But the decision to have an independent agency charged only with regulatory responsibilities applied only to civilian nuclear operations and wastes. AEC had regulated its own weapons production and other nuclear programs, and this system of self-regulation was continued after 1974 when those programs were transferred to ERDA/DOE, notwithstanding the concerns that had promoted the creation of NRC to regulate civilian activities. No doubt the decision to continue the practice of government self-regulation reflected the power in Congress of the Joint Committee on Atomic Energy and of the armed services committees, and the concern that independent regulation could undermine national security.[167] Finally, the 1954 AEA gave AEC power to create radiation standards and regulations for all nuclear facilities; the authority to set generic radiation standards was transferred to EPA when the new agency was created in 1970.

But the Office of Management and Budget in 1973 determined that EPA, which proposed to adopt regulatory requirements for specific types of civilian nuclear facilities, lacked authority to implement the generic standards through such measures and also lacked enforcement authority.[168] After 1974, NRC had the responsibility for ensuring compliance by civilian nuclear power plants with EPA's generic radiation standards, while ERDA/DOE was responsible for assuring compliance by federal facilities.

Subsequent legislation added even more regulatory complexity to this mix. First, the 1982 NWPA provided for NRC regulation and licensing of SNF/HLW repositories and MRS facilities, and adoption by EPA of generic radiation release standards for repositories. Following President Reagan's 1985 decision not to dispose of defense HLW and SNF separately from civilian wastes, disposition of the government's nuclear wastes became subject to NRC regulation.[169] Second, the 1987 NWPA amendments provided specifically for NRC licensing of a repository at Yucca Mountain, and for EPA adoption of Yucca-specific environmental radiation standards. Third, the 1992 Federal Facility Compliance Act (FFCA) made DOE facilities subject to state and local regulation of mixed wastes, including RCRA regulation by those states exercising RCRA authority delegated to them by EPA. Fourth and finally, the 1992 WIPP Land Withdrawal Act directed EPA to set WIPP-specific environmental radiation standards, and to certify WIPP's compliance with them.

NRC also regulates commercial landfills where DOE has disposed of some of its LLW, and would regulate any civilian wastes stored, managed, or disposed of at DOE sites (if, for example, a decision were made to store commercial SNF there).[170] DOE, however, continues to self-regulate its on-site management and disposition of its own wastes pursuant to AEA, except for mixed wastes, which are subject to federal, state, and local regulation of their chemically hazardous constituents. In addition, there is the question of the extent to which NWPA limits DOE's authority to classify and dispose of certain reprocessing wastes as other than HLW, and the potential role of NRC in determining any such limitations (see Chapter 2).

This maze of intersecting institutional arrangements reflects successive congressional judgments that one or more independent authorities should regulate DOE's off-site disposition of TRU, HLW, and SNF (see Table 8.3). The existing regulatory system in a number of instances provides for two regulators responsible for regulating the same activities. Both EPA and New Mexico regulate WIPP. Both EPA and NRC regulate HLW/SNF repositories. Having more than one regulator can guard against the sloth, incompetence, or regulated industry capture of an agency that can arise in the case of regulation by a single authority. But having two regulators of the same activity can muddle accountability while creating costly and dysfunctional duplication, conflict, and delay. Is this pattern, then, simply a crazy-quilt and haphazard result of repeated, uncoordinated microdecisions—or does it work as a functional regulatory regime?

Dual state-federal regulation—a form of vertical regulatory duplication—is a commonplace in U.S. environmental regulation, with the federal government setting minimum standards and the states generally at liberty to adopt more demanding requirements. But AEA made federal regulation of nuclear EHS risks exclusive, preempting state regulation. FFCA does not formally disturb this arrangement, but it does so functionally because the chemical and radiological elements of mixed wastes cannot feasibly be separated; in regulating the chemical elements of mixed wastes, states are effectively

Table 8.3. Disposition of regulatory authority, 1970–1992

	Executive action/Legislation	Effect on regulatory authorities
1970	Reorganization Plan No. 3	EPA created and given AEC's authority to set general radiation exposure standards
1974	Energy Reorganization Act	AEC divided into NRC and ERDA/DOE; NRC given responsibility for setting facility-specific requirements to secure compliance with EPA's general radiation exposure standards and for enforcing such standards
1982	Nuclear Waste Policy Act	NRC given regulatory, including licensing, authority over SNF/HLW repositories and MRS facilities. EPA given authority to set radiation release standards for repositories; NRC to ensure compliance
1986	Superfund Amendments and Reauthorization Act	EPA given oversight role in implementing CERCLA at federal facilities
1987	Nuclear Waste Policy Act Amendments	NRC given regulatory, including licensing, authority over Yucca Mountain repository subject to EPA/Yucca-specific radiation release standards
1992	Federal Facility Compliance Act	DOE facilities subject to RCRA regulation and cleanup requirements by EPA and delegated states; state enforcement of RCRA; and state and local regulation of chemically hazardous wastes
1992	WIPP Land Withdrawal Act	EPA authorized to set WIPP-specific radiation release standards and to certify facility compliance; recertification required every five years

regulating nuclear wastes as well. Since the application of RCRA to mixed wastes was established well before 1992, Congress in enacting FFCA was presumably aware that it was effectively amending AEA to give states regulatory authority over nuclear wastes that contained chemically hazardous elements. This mobilization of state regulation was critical in prodding DOE to clean up its weapons sites and begin steps for disposition of its HLW wastes. Further, RCRA regulation by New Mexico gave the state a significant role in regulating and enforcing environmental requirements at the WIPP facility. This state authority, along with EPA's role as an independent federal-level regulator of DOE's actions at WIPP, responsible for initially certifying and periodically recertifying WIPP's compliance with EPA's generic environmental radiation standards, was important in promoting the state's confidence in, and acceptance of, the facility and its agreement to dispose of RH-TRU as well as CH-TRU at the facility. The WIPP experience points to the desirability of giving potential host states EHS regulatory authority over nuclear wastes at future federal repositories or consolidated storage facilities, even in the case of wastes (such as SNF or vitrified HLW) that are not chemically hazardous.

Much more unusual than the vertical regulatory duplication involved in dual state-federal regulation is the system of horizontal regulatory duplication created by NRC and EPA regulation of civilian nuclear power plants and of HLW/SNF repositories.[171] Indeed, in the case of NWPA repositories there are effectively three federal agencies regulating repositories: NRC, EPA, and DOE, which regulates its own activities in developing and constructing repositories. Moreover, NWPA, as construed by the court in *Nuclear Energy Institute*, effectively establishes NAS as a *fourth* federal-level regulator by requiring EPA's radioactive exposure standards, which must be applied by NRC in licensing requirements and decisions, to be "consistent" with NAS committee findings and recommendations.[172]

Congress has for good reasons declined to rely solely on DOE self-regulation in developing repositories. At the same time, it has not reexamined its 1974 decision in the Energy Reorganization Act to authorize ERDA/DOE to self-regulate management and storage of its nuclear wastes at its facilities and disposal of its LLW at those facilities, an arrangement arguably inconsistent with its embrace in NWPA of an independent regulator for repositories.[173] But why have two independent regulators—both NRC and EPA—to oversee DOE's development and construction of HLW/SNF repositories?

There is a potential case for making NRC, as the specialized and correspondingly expert agency regulating nuclear technologies and wastes, the sole independent federal-level regulator of highly radioactive nuclear wastes and their disposition. EPA's nuclear regulatory authority is a vestige of the Nixon administration's consolidation of environmental regulatory authority within EPA. The transfer of AEC authority to EPA occurred, however, some years before the creation of NRC as an independent regulator; this allocation of authority was apparently not reexamined when NRC was created in 1974. Moreover, EPA assumed only a limited part of AEC's regulatory authority: the power to set general radiation standards, without the authority to implement them through facility-specific standards or licensing. Congress's action in giving EPA certification authority over WIPP in 1992 appears to have been based on a judgment that some independent federal regulator was required, as well as on the fact that EPA was a competent and experienced environmental and radiation regulator trusted by New Mexico; New Mexico, lacking nuclear power plants, had had little prior relationship with NRC. Moreover, as described in Chapter 5, the House Armed Services Committee adamantly and successfully resisted giving NRC jurisdiction over WIPP for reasons having nothing to do with the respective competences of NRC and EPA.

Going forward, an argument for maintaining EPA's role as an independent federal watchdog of repositories is that specialized nuclear expertise is largely irrelevant to EPA's role, which is to set generic fencepost radiation exposure standards to protect public health and the environment, a subject falling within its general remit. Moreover, EPA has experience with regulating risks similar to those presented by nuclear wastes disposition, although for shorter periods. EPA regulates hazardous waste disposal facilities and underground injection of hazardous waste in order to ensure secure containment of the wastes' chemically hazardous components and thereby protect public health and the environment. Such regulation involves use of models to project containment performance, in the case of underground injection, to more than ten thousand years. Nonetheless, in the case of nuclear repositories, the risks (which involve scenarios projecting hundreds of thousands of years into the future) are a function of the characteristics of nuclear wastes and of packaging and containment technologies. NRC has substantially more experience and technical capacity with respect to these matters than does EPA.

NRC's record of regulatory performance, however, has been strongly criticized by many constituencies with long experience with the agency, including environmental organizations and communities and tribes near nuclear power plants. GAO has also reported significant deficiencies in NRC regulation, citing failure to secure and track nuclear materials. For example, GAO employees posed as a buyer and were able to obtain a radioactive materials license from NRC.[174] Other experts, including those at environmental groups, have criticized NRC's safety record at aging nuclear power plants.[175] Outside experts have also charged that NRC has paid insufficient attention to terrorist risks at nuclear plants. In one particularly troubling incident at the Davis-Besse reactor in Ohio, leaking coolant nearly triggered a meltdown after NRC management deferred an inspection of the aging reactor vessel despite severe warnings from agency staff.[176] Citizen groups have also criticized the effectiveness and sincerity of NRC's efforts to engage the public.[177] The commission, however, has recently taken serious steps to improve its capacities and performance.[178]

One theme in much of the criticism is that NRC has been "captured" by the nuclear industry. A less pejorative diagnosis, long ago developed by Louis Jaffe, is that agencies charged with regulating a specific industry tend, for structural reasons, to adopt in greater or lesser degree the orientation of that industry. They necessarily interact on a regular basis with the industry that they regulate, depend on it for much of the information needed for regulatory decisions, and may face heavy criticism if their regulatory policies significantly damage the industry.[179] EPA, by contrast to NRC, has a general environmental protection mission; its purview extends over many different industries. EPA is accordingly less likely than a single-industry regulator to develop an industry orientation and more likely to take a viewpoint that is responsive to environmental concerns. But by the same token, EPA often lacks close knowledge of many of the industries (including the nuclear industry) that it regulates, including sector-specific issues of technology, economics, and practicality that such industries face; this lack may weaken or impair the functionality of EPA's regulation. As a general environmental regulatory agency with an environmentalist mission, EPA may slight the costs and other burdens imposed by the requirements that it adopts for specific sectors.

Regardless of which agency or agencies exercise regulatory authority, forecasting compliance with radiation exposure standards eons hence in order to determine nuclear waste repository locations and make design decisions today is a deeply problematic, indeed visionary, enterprise. EPA's repository radiation standards have implicitly acknowledged the difficulties presented by providing that DOE/NRC need establish only a "reasonable expectation" of compliance with the release and exposure limits, which were originally set for ten thousand years.[180] Even using a fuzzy and diluted "reasonable expectation" test, determining compliance with a standard as low as fifteen millirems ten thousand years into the future is an exquisite exercise in imagination. To put this standard in context, the annual radiation dose received by the average TV viewer is thirty millirems, while a single cross-country plane flight gives a passenger an eight-millirems dose.[181] The NAS Board on Radioactive Waste Management, when reviewing EPA's initial ten-thousand-year standard, found that the compliance period should be for a vastly longer period, and the *Nuclear Energy Institute* court agreed.[182] On remand, EPA promulgated a revised individual exposure limit of fifteen millirems per year for ten thousand years and one hundred millirems per year for a million years. Determining compliance a million years hence is an exercise in pure speculation. Yet the challenge of determining the performance of a repository over thousands of years in the future is ineluctable.

However these conundrums might be resolved if the regulatory system were being designed from scratch, at this juncture EPA regulation through generic radiation standards is long-standing and has become accepted as an integral part of the regulatory process for disposal of the most highly radioactive wastes. President Clinton in 1997 vetoed legislation eliminating EPA regulation of a proposed interim SNF storage facility near Yucca on the grounds that EPA oversight was necessary for safety and credibility.[183] Notwithstanding the problems of duplication, EPA and NRC appear to have worked together successfully over the years to manage the problems created by divided regulatory authority. EPA's role as a guarantor of environmental integrity in this context seems too firmly settled to be overcome by arguments from first principles regarding institutional design.[184] Fundamental changes in existing institutional and regulatory structures might emerge, however, if the blue-ribbon commission produces a package of measures that could win support from a broad range of political constituencies, including the White House, DOE, NRC, Congress, industry, environmental groups, and a large number of states.

A Waste Classification Approach

A further step in rethinking nuclear waste law and policy that should be examined by the blue-ribbon commission is to follow a more consistently risk-informed approach to regulatory classification of wastes in order to better align regulatory requirements with the hazards presented by different wastes. The existing U.S. waste classification system represents an amalgam of provisions in statutes and regulations that has been built up in patchwork fashion over many years. It was not developed in accordance with, nor does it reflect, any consistent set of classificatory principles. Among the factors that have determined the present classifications are the nature of the processes that produced the wastes, the ownership of the facilities that produced them, their disposal destination, their radioactive characteristics, and the circumstance (exemplified by the LLW category) that the wastes do not fall within any other classification. In a number of significant instances, the categorization of a waste does not match the relative hazards that it presents, and the regulatory requirements that apply (or fail to apply) as a result are not appropriate for the waste. In these instances, classification departs from sound waste management policies and skews regulatory priorities. NAS committees and other independent experts have recognized this problem and proposed the adoption of more consistently risk-informed classifications.[185] The International Atomic Energy Agency has developed a generic framework for waste classification that reflects practices in a wide range of countries and provides a useful point of reference for such efforts.

Despite instances of arbitrariness in the current U.S. system for classifying nuclear wastes, many existing classifications, including those for SNF, HLW, TRU, and class B and C LLW, achieve a reasonably good fit between the hazards that they present and regulatory requirements. But there are also cases of serious mismatch. Examples of existing waste classifications that should be reexamined include the following:

- With some ad hoc exceptions, all reprocessing wastes are classified as HLW that must be disposed of in a geologic repository after being pretreated for such disposition. Some components of reprocessing wastes that have LLW radioactivity characteristics should appropriately be classified and disposed of as LLW.
- Disposition of GTCC wastes remains in limbo at DOE. A firm program for their disposal, including, where appropriate, disposal in a repository, must be developed and adopted on a priority basis. Additionally, the classification of GTCC wastes

in a single regulatory category, solely because their radiation levels exceed those of class C LLW and because they fall in no other waste classification, will not enable a regulatory regime for these wastes to reflect the varying levels of hazard of different GTCC wastes. Clearly, higher-activity GTCC, which can reach TRU radioactivity levels, should be regulated as stringently as TRU and disposed of in a repository. By contrast, different requirements for management and disposal might appropriately apply to the lowest-activity GTCC wastes.

- The NAS Committee on Improving Practices for Regulating and Managing Low-Activity Radioactive Waste found that there are very large volumes of very low-activity waste that must, under current law, be classified and disposed of as LLW, at a cost of many billions of dollars. The committee concluded that there is no justification for treating these wastes as LLW, and that other means for disposing of them would be adequately protective and appropriate. Other developed countries classify very low-activity radioactive wastes as below the level of regulatory concern (BRC) and allow their disposal in ordinary landfills.[186] NRC and EPA have recognized that it would be desirable and appropriate to create one or more regulatory classifications that would allow such wastes to be disposed of in RCRA or ordinary landfills, but their efforts have been blocked by Congress or have otherwise been stalled.
- The NAS Committee on Improving Practices for Regulating and Managing Low-Activity Radioactive Waste found the existing U.S. programs for regulating devices used for medical, industrial, and resource applications containing highly concentrated radioactive elements to be deficient and inadequately protective. These devices are classified as LLW when retired from use, but they often have no assured disposition pathway and therefore present potentially serious risks. The committee recommended adopting a strengthened regulatory program tailored to the specific radioactive characteristics of this category of wastes in order to ensure their safe management and disposal, rather than continuing to treat them as generic LLW.[187]

In addition, the differences in regulatory treatment of wastes depending on whether they were created by civilian or defense authorities are a historical artifact that should be reexamined in connection with the development of any new framework for waste classification and regulation.

As some of these examples indicate, steps toward a more risk-informed approach to waste classification and regulation would, quite appropriately, produce more rigorous regulation for some wastes, and less burdensome and costly regulation for others. Experience suggests that legislation will probably be necessary to make basic changes in the existing classification system. Congress might adopt, on a case-by-case basis, new classifications and regulatory requirements for specific types of wastes in order to address specific problems, including those discussed here and others identified by the blue-ribbon commission. In lieu of such a waste-by-waste approach, which could invite political influences leading to ad hoc results, Congress could consider a systematic reworking of the existing classification system as a whole on the basis of a hazard-based and risk-informed framework like that of IAEA. This wholesale approach, in addition to being more ambitious and politically difficult than a "retail" waste-by-waste approach focused on particular problem areas in isolation, does not seem necessary, since most of the key categories are working fairly well; however, it is possible that it might yield higher payoffs in terms of consistency and sound regulatory policy. Whatever changes are made to existing classifications and regulatory requirements, however, should be decided through procedures that allow full opportunity for public participation and judicial review. This approach should take into account societal views of risk as well as scientific ones.[188] The Blue Ribbon Commission on America's Nuclear Future should consider these and other approaches to addressing problems with the existing waste classifications, as well as reconsider the validity of the longstanding divide in waste regulation and management for civilian versus defense wastes.

Moving toward a more rational system of waste classification and regulation will by no means solve the most fundamental problems of nuclear waste in the United States, but such an effort would make significant contributions toward a more balanced and effective system of nuclear waste regulation and policy.

Appendix A

Operating U.S. Nuclear Power Units by Year

Year	Operating units
1955	0
1956	0
1957	1
1958	1
1959	2
1960	3
1961	3
1962	9
1963	11
1964	13
1965	13
1966	14
1967	15
1968	13
1969	17
1970	20
1971	22
1972	27
1973	42

Year	Operating units
1974	55
1975	57
1976	63
1977	67
1978	70
1979	69
1980	71
1981	75
1982	78
1983	81
1984	87
1985	96
1986	101
1987	107
1988	109
1989	111
1990	112
1991	111
1992	109

Year	Operating units
1993	110
1994	109
1995	109
1996	109
1997	107
1998	104
1999	104
2000	104
2001	104
2002	104
2003	104
2004	104
2005	104
2006	104
2007	104
2008	104
2009	104
2010	104

Source: Energy Information Agency, "Annual Energy Review: Nuclear Energy," *www.eia.doe.gov/emeu/aer/nuclear.html*.

Appendix B

Uranium Oxide Spot Prices (per pound)

Year	Price (US$)	Inflation-adjusted price (US$)*	Year	Price (US$)	Inflation-adjusted price (US$)*
1968	6.50	40.48	1990	9.70	16.09
1969	6.20	36.62	1991	8.75	13.92
1970	6.15	34.35	1992	7.85	12.13
1971	5.95	31.84	1993	6.90	10.35
1972	5.95	30.85	1994	7.20	10.53
1973	7.00	34.17	1995	10.00	14.22
1974	15.00	65.95	1996	13.75	18.99
1975	35.00	141.00	1997	9.65	13.03
1976	41.00	156.18	1998	8.45	11.24
1977	43.20	154.51	1999	7.60	9.89
1978	43.25	143.77	2000	6.40	8.06
1979	40.75	121.66	2001	9.50	11.63
1980	27.00	71.02	2002	10.20	12.29
1981	23.50	56.03	2003	14.40	16.96
1982	20.25	45.48	2004	20.50	23.52
1983	22.00	47.87	2005	36.25	40.23
1984	15.25	31.81	2006	72.00	77.41
1985	17.00	34.24	2007	90.00	94.08
1986	16.75	33.12	2008	52.50	52.85
1987	16.55	31.58	2009	44.50	44.96
1988	11.75	21.53	2010	47.67	47.67
1989	9.00	15.73			

Source: Global Financial Database, 2010.

*Adjusted based on Consumer Price Index, with March 2010 as base.

Appendix C

The Hanford Waste Cleanup Agreement and Program

This overview of the Hanford cleanup agreement and program for the Hanford facility, which contains the largest volume of HLW of any DOE facility, provides a concrete picture of the legal and programmatic elements of DOE cleanup efforts.

The Hanford Tri-Party Agreement

DOE, EPA, and the Washington State Department of Ecology entered into the Hanford Federal Facility Agreement and Consent Order, commonly known as the Tri-Party Agreement (TPA), for cleanup at the Hanford site in 1989.[1] The first such agreement for a DOE facility, it was promoted as the flagship of DOE's waste cleanup efforts.[2] Since the original accord, the agreement has been modified more than four hundred times by mutual agreement of the parties.[3] Further cleanup requirements are specified in the Hanford Federal Facility Site-Wide Permit, the facility's RCRA permit, jointly issued by EPA and the Washington Department of Ecology.[4] The Hanford TPA defines and ranks CERCLA and RCRA cleanup commitments, defines the responsibilities of the parties, sets forth a budgeting process, and provides arrangements for public information and involvement.[5]

The TPA also provides an elaborate dispute settlement procedure with set timetables. For CERCLA disputes, the final decision is made by the regional EPA administrator with the possibility of review by the EPA administrator in Washington, D.C., if DOE determines that there are "significant national policy implications."[6] Disputes relating to RCRA are referred to the director of the Washington State Department of Ecology for a final decision. DOE can appeal such a decision to the courts.[7] The TPA also stipulates money damages for breaches by DOE of the requirements that it establishes. For example, if DOE misses a deadline, the agreement stipulates a $5,000 penalty for the first week and $10,000 for each additional week. In the event of a CERCLA-related breach, the fine is paid to EPA. Fines related to RCRA are paid to the State of Washington.[8] Citizens may also seek to enforce the agreement through judicial actions under RCRA Section 7002(a)(1)(A).[9]

The TPA enforcement provisions have been invoked many times. DOE has been assessed stipulated penalties on several occasions.[10] For example, it was assessed penalties of $270,000 in 2004 and $1.1 million in 2007.[11] The dispute settlement process has been invoked in a few instances.[12] For example, in 1998, the procedures for settlement of a dispute over one of the TPA cleanup schedules resulted in schedule modifications that were then noticed for public comment.[13] Even with these procedures, several disputes have ended with lawsuits. In one prominent example, the State of Washington sued DOE in 2008 for failing to meet some forty deadlines it had agreed to under the TPA.[14] The parties eventually settled and agreed to extend many of the deadlines in a consent decree entered with the court.[15]

The Hanford Cleanup Program

DOE has split the Hanford cleanup project into two operationally separate areas: the Richlands Operations Office area, which comprises the original Hanford facilities; and the River Corridor Site, an eighty-square-mile area contaminated by discharged reactor-cooling water. The most recent estimates from the DOE for the total estimated life-cycle cleanup costs for Hanford are between $113 billion and $134 billion, or approximately one-third of the Office of Environmental Management's (EM) cleanup cost estimates for all the sites for which it is responsible.[16] Different components of the project deal with reprocessing wastes, most of which are stored in liquid form in tanks; SNF; collection, management, and disposal of solid wastes; and remediation of soils and groundwater. The overall goal for Hanford is to shrink its contamination area from 581 square miles to 75 square miles by 2015. Other goals include completing the pretreatment and vitrification of 10 percent of tank waste by mass by 2018 and vitrification of all HLW by 2028.[17] DOE estimates that cleanup at Hanford will finally be completed between 2050 and 2062.[18] For the moment, cleanup is on schedule,[19] but DOE has admitted that 44 of 146 major enforceable milestones are at risk of not being completed on time.[20] Table C.1, from the December 2009 version of the Action Plan in the TPA, provides an overview of major milestones for the cleanup.

One of the major challenges faced by DOE at the Hanford site is the stabilization, treatment, and storage of tank waste, that is, 53 million gallons of reprocessing and other liquid wastes now stored in 177 tanks;[21] these wastes account for 176 million curies of radioactivity.[22] Some of the tanks date back to the 1940s,[23] and many have exceeded their design lives.[24] EM devotes substantial efforts to achieving more secure storage of these wastes.[25] As part of these efforts, approximately 3 million gallons of liquids have been removed from single-shell tanks and transferred into double-shell tanks.[26] The Waste Treatment and Immobilization Plant (WTP), currently under construction and scheduled to begin operation in 2019, will separate and vitrify the low-activity wastes and high-activity fractions of the reprocessing wastes in tanks.[27] Vitrification is designed to immobilize the wastes to promote secure disposal and eliminate chemically hazardous constituents. As mentioned, delays in the development and construction of the WTP resulted in failures to meet schedules in the Tri-Party Agreement, litigation by the States of Washington and Oregon, and negotiated modifications of the agreement.[28] The WTP is currently projected to cost $12.3 billion and begin operation in 2019.[29]

The WTP is designed to treat only about half the total volume of the low-activity fractions of this waste; for now, EM has postponed firm plans for dealing with the remainder, based on reports from an independent review that it will not be necessary to have plans until 2015–2017.[30] For tank wastes that were not produced in connection with reprocessing, EM plans to determine other treatment and disposal methods on a case-by-case basis, including potential disposal at WIPP.[31]

Table C.1. Major milestones from the Hanford Site Compliance Agreement Action Plan

Milestone*	Date
Complete remedial actions for all nontank farm operable units	9/30/2024
Complete closure of all single-shell tank farms (follows retrieval of as much tank waste as technically possible with set maximum residue limits)	9/30/2024
Complete pretreatment processing of Hanford tank waste	12/31/2028
Complete vitrification of Hanford high-level tank waste and high-level waste	12/31/2028
Complete pretreatment and immobilization of Hanford low-activity waste	12/31/2028
Complete the acquisition of new facilities, modification of existing facilities, and modification of planned facilities necessary for retrieval, storage, and disposal/processing of all Hanford RCRA mixed and suspect mixed low-level waste and RCRA mixed and suspect mixed transuranic waste	To be determined
Complete retrieval of contact-handled RSW** (to be completed by the targeted removal of 10,700 cumulative cubic meters by 9/30/2010, and the targeted removal of 1,000 cubic meters per year thereafter)***	To be determined

*As of December 29, 2009.

**RSW is retrievable stored waste (waste that is or was believed to be contaminated with significant concentrations of transuranic isotopes when it was placed in burial trenches after May 6, 1970).

***For retrieved contact-handled transuranic waste, DOE may treat such wastes to meet the RCRA Land Disposal Requirements (LDR) or, alternatively, complete certification of the waste for disposal at WIPP.

EM is also responsible for disposing of the SNF located at Hanford, approximately 2,100 MTHM.[32] Much of the SNF had originally been stored in cooling basins approximately a quarter mile from the Columbia River.[33] Since 2002, EM has moved all SNF to dry storage on a plateau away from the Columbia River. DOE has begun removing debris, sludge, and concrete from the basins. These wastes, which include radioactive components, must be characterized before disposal. DOE is considering treating the wastes for disposal at WIPP.[34]

Other nuclear wastes present at Hanford include TRU, which is being shipped to WIPP for disposal; very large quantities of many different types of LLW and mixed LLW; and surplus plutonium. Much of the LLW and mixed LLW is being disposed of at Hanford in landfills, including (in the case of mixed LLW) landfills meeting RCRA requirements.[35] DOE had used Hanford as a regional disposal facility for LLW and mixed low-level waste from other sites but was forced to stop this practice, pending issuance of a new EIS, as a result of a suit brought by the State of Washington.[36]

EM has reported that it initiated the stabilization, packaging, and shipping of all surplus plutonium stored at Hanford to the Savannah River Site in 2008, and this work was to be completed by September 2009.[37] However, there is disagreement over whether DOE had located all the plutonium for which it was responsible. A 2010 report from the

Institute for Policy Studies found that as of 2009, approximately four tons of plutonium, or one-third of DOE's plutonium wastes, had been reclassified, underestimated, or simply lost at Hanford.[38]

Further, there is widespread and significant plutonium contamination of the soil and groundwater resulting from the practice, continued through the 1980s, of discharging wastes containing plutonium directly into the soil or burying them in surface trenches.[39] While DOE plans to clean up approximately 4 percent of the total acreage at Hanford containing buried plutonium by 2025, at an estimated cost of $320 million, critics claim this would still leave about 0.7 metric tons of plutonium in the ground.[40] If, as critics believe, DOE has seriously underestimated the amount of plutonium wastes or simply does not know the amount of such wastes and where they were disposed of, significant plutonium contamination may remain at Hanford indefinitely. One estimate concluded that plutonium in the ground "could reach the near shore of the Columbia River in less than one thousand years at concentrations 283 times greater than the federal drinking water standard."[41]

In addition, EM has concluded that approximately 450 billion gallons of contaminated water was discharged into the ground at the site, resulting in "approximately 80 square miles of groundwater being contaminated with radionuclides, metals, and organic chemicals above Government standards for drinking water." Some of the groundwater plumes are discharging to the Columbia River.[42] To date, EM has treated around four billion gallons of contaminated groundwater.[43]

The scale of the work involved in dealing with this massive cleanup program is illustrated by just one component of the Hanford cleanup, the River Corridor project. It has deactivated, decommissioned, and demolished three nuclear facilities, twenty-nine radiological facilities, and seventy-nine industrial facilities. In addition, five of the nine reactors were placed into interim safe storage. The reactor support facilities have been demolished, and the reactor blocks are being cocooned for interim storage. The cocooning process involves removing all of the reactor building except the five-foot-thick shield walls surrounding the reactor core. Openings and penetrations are sealed with corrosion resistant materials, and a seventy-five-year roof is placed over the remaining structure.

The lack of a repository for disposal of the nuclear waste at Hanford has raised concern in the local communities. When the blue-ribbon commission recently visited Hanford, they attended a meeting with approximately 150 people. Susan Leckband, chair of the Hanford Advisory Board, expressed concern that Hanford "will become a de facto high-level waste repository for vitrified waste." Tribal nations also voiced concern. Stuart Harris, a member of the Confederated Tribes of the Umatilla Indian Reservation stated: "My people won't remember your names in 1,000 years, but they will live your decisions." Similarly, Brooklyn Baptiste of the Nez Perce tribe worried that "science always seems to be topped by politics in the end."[44]

Notes

Acknowledgments
1. *See www.CRESP.org*. CRESP is funded through a Cooperative Agreement with the U.S. Department of Energy.
2. This report is partially based on work supported by the U.S. Department of Energy, under Cooperative Agreement Number DE-FC01-06EW07053 entitled "The Consortium for Risk Evaluation with Stakeholder Participation III" awarded to Vanderbilt University. The opinions, findings, conclusions, or recommendations expressed herein are those of the authors and do not necessarily represent the views of the Department of Energy, CRESP, or Vanderbilt University. Any errors contained in the book are attributable solely to the authors, not to them.

    Disclaimer: This report was prepared as an account of work sponsored by an Agency of the United States Government. Neither the United States Government nor any agency thereof, nor any of their employees, makes any warranty, express or implied, or assumes any legal liability or responsibility for the accuracy, completeness, or usefulness of any information, apparatus, product, or process disclosed, or represents that its use would not infringe privately owned rights. Reference herein to any specific commercial product, process, or service by trade name, trademark, manufacturer, or otherwise does not necessarily constitute or imply its endorsement, recommendation, or favoring by the United States Government or any agency thereof.

Introduction
1. Blue Ribbon Commission on America's Nuclear Future: Memorandum for the Secretary of Energy, 75 Fed. Reg. 5,485, at 5,485 (Feb. 3, 2010).
2. The Consortium on Risk Based Evaluation with Stakeholder Participation (CRESP) has successfully developed and applies strategies to engage local communities and groups and other stakeholders in the assessment of nuclear waste risks and in risk management decision making for those wastes. *See www.cresp.org*.
3. The hazardous waste regulatory and cleanup laws passed by Congress in the 1970s and 1980s were later made applicable to federal government facilities, providing the legal impetus for a massive effort, begun by DOE in the late 1980s and still ongoing, to remediate radioactive and chemical contamination at its weapons sites. The total cleanup bill is projected to run to several hundred billion dollars. The Chernobyl accident had helped trigger the DOE initiative by increasing awareness and concerns in the United States about radioactive hazards.
4. We are indebted to Tom Isaacs, Director of Policy and Planning, Lawrence Livermore National Laboratories and lead adviser to the staff of the Blue Ribbon Commission, for this formulation.
5. IRG held an extensive number of public meetings in many states over the course of the short period in which it developed its recommendations. A significant number of public meetings held by IRG addressed the issues raised by WIPP.

Chapter 1
1. Atomic Energy Act of 1946, Pub. L. No. 79-585, 60 Stat. 755 (1946).
2. The Idaho National Laboratory (INL) was known as the National Reactor Testing Station from 1949 to 1977, the Idaho National Engineering Laboratory from 1977 to 1997, and the Idaho National Engineering and Environmental Laboratory from 1997 to 2005. For consistency, it will be referred to as INL throughout this book.
3. Walker, *Road to Yucca*, 3 (2009); *see also* Office of Environmental Management, DOE, *Linking Legacies*, 71–93 (1997).
4. Halgraves and Mohr, "Liquid Fluoride Thorium Reactors" (2010). For discussion of the thorium fuel cycle, *see* International Atomic Energy Agency, *Thorium Fuel Cycle: Potential Benefits and Challenges* (2005).
5. Atomic Energy Act of 1946 §§ 2(a), 5, as enacted, Pub. L. No. 79-585, 60 Stat. 755, at 756, 760–763 (1946).
6. *Id.* §§ 5(a)(4), 5(c)(2), 12(a)(2), as enacted, Pub. L. No. 79-585, 60 Stat. 755, at 760–761, 763, 770.
7. Atomic Energy Act of 1954, as enacted, Pub. L.

No. 83-703, 68 Stat. 919 (1954). Congress further enabled expansion of nuclear power generation by the private sector in 1964 by enacting legislation to allow private ownership of uranium fuel rods by the nuclear power industry. Private Ownership of Special Nuclear Materials Act, Pub. L. No. 88-489, 78 Stat. 602 (1964), codified as amended in scattered sections of 42 U.S.C. (2010).

8. Atomic Energy Act of 1954 § 3(d), as enacted, Pub. L. No. 83-703, 68 Stat. 919, at 922 (1954).
9. *Id.* §§ 53(f), 103–104, as enacted, Pub. L. No. 83-703, 68 Stat. 919, at 931, 936–938; Yates, "Preemption under the Atomic Energy Act of 1954," 399–400 (1976). In distributing fissile materials, AEC was to give priority to parties proposing "research and development activities . . . which are most likely . . . to contribute to basic research, to the development of peacetime uses of atomic energy, or to the economic and military strength of the Nation." Atomic Energy Act of 1954 § 53(f), as enacted, Pub. L. No. 83-703, 68 Stat. 919, at 931 (1954).
10. Atomic Energy Act of 1954 §§ 2(d), 161, as enacted, Pub. L. No. 83-703, 68 Stat. 919, at 921, 923, 924 (1954).
11. *Id.* §§ (11)(e), 11(s), 161, as enacted, Pub. L. No. 83-703, 68 Stat. 919, at 921, 948–951 (1954).
12. These included Westinghouse, which worked on the Shippingport, Pennsylvania reactor; General Electric, which worked on the Dresden reactor project; and Babcock and Wilcox, which worked on the Lynchburg Research Center, the first privately financed U.S. nuclear facility.
13. Greenberg et al., *Reporter's Handbook*, 228 (2009). The differences between a breeder reactor, which is designed to produce plutonium for further use as fuel, and the light-water reactors used for commercial nuclear energy generation, are discussed later.
14. *Id.*
15. *Id.* at 229.
16. Energy Policy Act of 2005 § 170H, as enacted, Pub. L. No. 109-58, 119 Stat. 594, at 806, amending the Atomic Energy Act of 1954 §11(e)(3)–(4), 42 U.S.C. § 2014(e)(3)–(4) (2006), and implementing NRC regulations.
17. Act of Sep. 23, 1959 § 274(a)(1), as enacted, Pub. L. No. 86-373, 73 Stat. 688, at 688 (1959). "Until the passage of the 1959 amendment, reference to the states' authority respecting [radiation hazards] is conspicuously absent." Yates, "Preemption under the Atomic Energy Act of 1954," 401 (1976).
18. Joint Committee on Atomic Energy, *Amendments to the Atomic Energy Act of 1954*, at 8 (1959).
19. *Id.* at 8–9.
20. Yates, "Preemption under the Atomic Energy Act of 1954," 401 (1976).
21. Act of Sep. 23, 1959 § 274(e)(2)(k), as enacted, Pub. L. No. 86-373, 73 Stat. 688, at 691 (1959).
22. See *N. States Power Co. v. Minnesota*, 447 F.2d 1143, 1154 (8th Cir. 1971), which held that the act preempted state efforts to regulate operation and construction of nuclear power plants; *see also Pac. Gas and Elec. v. State Energy Res. Conservation and Dev. Comm'n*, 461 U.S. 190, 212 (1983), which found that the AEA preempts all nuclear safety concerns not expressly ceded to states, and *United States v. Kentucky*, 252 F.3d 816, 823 (6th Cir. 2001), which determined that the AEA preempts state attempts to regulate, for safety purposes, materials covered by the act.
23. *N. States Power Co. v. Minnesota*, 447 F.2d 1143, 1149–1150 (8th Cir. 1971), *summarily aff'd*, 405 U.S. 1035 (1972).
24. *P. Gas & Elec. Co. v. State Energy Res. Conservation & Develop. Comm'n*, 461 U.S. 190, 207 (1983). The Court invoked the savings clause in Section 271 of AEA to uphold California's moratorium, finding that the measure was based not on the radiation hazards of the waste but on the potential economic burdens on ratepayers, a traditional subject of state regulation that was not preempted by AEA. The Court found that federal law did not preempt the states' authority to regulate power generation, including the power to determine "the need for additional generating capacity, the type of generating facilities to be licensed, land use, ratemaking, and the like." *Id.* at 212.
25. Atomic Energy Act of 1946 § 271, 42 U.S.C. §2018.
26. Atomic Energy Act of 1946 § 5(b)(1), as enacted, Pub. L. No. 79-585, 60 Stat. 755, 761 (1946); Atomic Energy Act of 1954 § 11(s), as enacted, Pub. L. No. 83-703, 68 Stat. 919, 924 (1954).
27. Atomic Energy Act of 1946 § 5(a)(1), as enacted, Pub. L. No. 79-585, 60 Stat. 755, 759 (1946); Atomic Energy Act of 1954 § 11(t), as enacted, Pub. L. No. 83-703, 68 Stat. 919, 924 (1954).
28. Atomic Energy Act of 1946 § 5(c)(1), as enacted, Pub. L. No. 79-585, 60 Stat. 755, 763 (1946); Atomic Energy Act of 1954 § 11(e), Pub. L. No. 83-703, 68 Stat. 919, 923 (1954).
29. Atomic Energy Act of 1954 § 161(b), as enacted, Pub. L. No. 83-703, 68 Stat. 919, 948 (1954).
30. Standards for Protection Against Radiation, 22 Fed. Reg. 548 (Jan. 29, 1957), codified as amended at 40 C.F.R. pt. 20 (2010); Standards for Protection against Radiation: Miscellaneous Changes, 22 Fed. Reg. 3,005 (Apr. 27, 1957), codified as amended at 40 C.F.R. pt. 20 (2010).
31. Advisory Committee on Nuclear Waste, NRC, *History and Framework of Low-Level Radioactive Waste Management*, 9 (2007); *see also* Robertson, "Geologic Problems at Low-Level Disposal Sites," 105–107 (1984).
32. Office of Environmental Management, DOE, *Linking Legacies*, 41 (1997).
33. Walker, *Road to Yucca*, 28–31 (2009); Office of Environmental Management, DOE, *Linking Legacies*, 89 (1997).
34. Walker, *Road to Yucca*, 18–19 (2009).
35. Committee on Waste Disposal, National Research Council of the National Academies, *Disposal of Radioactive Waste*, 1 (1957).
36. Walker, *Road to Yucca*, 20 (2009).

37. *Id.* at 21.
38. Glenn T. Seaborg, "Peacetime Uses of Atomic Energy," *Seaborg Papers: Box 862 (1955–1960 Speeches)* (July 30, 1955), as quoted in Walker, *Road to Yucca*, 19 (2009).
39. Walker, *Road to Yucca*, 18 (2009).
40. Committee on Waste Disposal, National Research Council of the National Academies, *Disposal of Radioactive Waste*, 2 (1957). The report states that the NAS was asked to report on "high-level radioactive waste." *Id.* at 1. Although this term is never defined, appendices to the report indicate that already generated and future defense HLW generated at the AEC weapons complex and HLW from anticipated reprocessing of spent fuel from the new nuclear power industry were the major focus of this effort. *Id.* app. B at 16–17; *id.* app. F at 132–133.
41. *Id.* at 3 (emphasis in original).
42. *Id.* app. F at 132.
43. *Id.* app. F at 133. Under the metric system, this works out to approximately 197 million liters total annual volume of liquid and 198,000 cubic meters of space underground.
44. Most federal disposal programs generally use 55-gallon (208-liter) drums. DOE, "WIPP Fact Sheet: The Remote-Handled Transuranic Waste Program," *www.wipp.energy.gov/fctshts/RH_TRU.pdf*.
45. Committee on Waste Disposal, National Research Council of the National Academies, *Disposal of Radioactive Waste*, app. B at 25–31 (1957).
46. *Id.* at 4.
47. *Id.* app. D at 101.
48. *Id.*
49. *Id.* app. F at 108–142.
50. *Id.* at 6.
51. *Id.* at 5.
52. See *id.* app. C, D, and F for discussions of these alternatives.
53. *Id.* at 5.
54. *Id.*
55. Committee on Geologic Aspects of Radioactive Waste Disposal, National Research Council of the National Academies, *Report to the Division of Reactor Development and Technology*, 23 (1966).
56. Commission on Geosciences, Environment, and Resources, National Research Council of the National Academies, *Disposition of High-Level Radioactive Waste*, 10 (1999); Hollister and Nadis, "Burial of Radioactive Waste under the Seabed," 60 (Jan. 1998).
57. Advisory Committee on Nuclear Waste, National Research Council of the National Academies, *Low-Level Radioactive Waste Management*, 7 (2007).
58. Walker, *Road to Yucca*, 24 (2009).
59. Berlin and Stanton, *Radioactive Waste Management*, 293 (1989).
60. Calmet, "Ocean Disposal of Radioactive Waste: Status Report," 48 (1989).
61. Rice and Priest, "Overview of Nuclear Waste Disposal in Space," 370 (1981).
62. Berlin and Stanton, *Radioactive Waste Management*, 296 (1989).
63. Zeller, Saunders, and Angino, "Putting Radioactive Waste on Ice," 4 (1973).
64. Berlin and Stanton, *Radioactive Waste Management*, 301 (1989).
65. Antarctic Treaty art. 5, Dec. 5, 1959, 12 U.S.T. 794, 402 U.N.T.S.5778.
66. Commission on Geosciences, Environment, and Resources, Nuclear Research Council of the National Academies, *Nuclear Wastes*, 2 (1996).
67. *Id.*
68. Organisation for Economic Cooperation and Development, Nuclear Energy Agency, *Geological Disposal of Radioactive Waste: Review of Developments in the Last Decade*, 38 (1999).
69. Commission on Geosciences, Environment, and Resources, National Research Council of the National Academies, Board on Radioactive Waste Management, *Rethinking High-Level Radioactive Waste Disposal*, 2 (1990).
70. Commission on Geosciences, Environment, and Resources, National Research Council of the National Academies, *Disposition of High-Level Radioactive Waste* (1999). The 1999 report noted that the Commission had moved from a prescriptive approach to waste disposal to an approach that recognized and allowed for uncertainty. The 1957 committee report concluded that based on the half-lives of the key isotopes, containment of high-level waste was necessary for only 600 years. The 1999 report dramatically increased the recommended containment period to thousands of years. *Id.* at 7.
71. Organisation for Economic Co-operation and Development, Nuclear Energy Agency, *Can Long-Term Safety Be Evaluated?* (1991).
72. Committee on Waste Disposal, National Research Council of the National Academies, *Disposal of Radioactive Waste*, 4 (1957).
73. Pierce and Rich, U.S. Geological Survey, *Summary of Rock Salt Deposits*, 64–82 (1958).
74. Christopher J. Wentz, State of New Mexico Radioactive Waste Consultation Task Force, "Chronology of WIPP," *www.emnrd.state.nm.us/WIPP/chronolo.htm*.
75. Letter from Harry Hess, Chairman, National Research Council Committee on Waste Disposal, to John A. McCone, Chairman, Atomic Energy Commission (June 21, 1960), as reprinted in Hubbard, National Research Council of the National Academies, *Energy Resources*, 117 (1962).
76. Committee on Geologic Aspects of Radioactive Waste Disposal, National Research Council of the National Academies, *Report to the Division of Reactor Development and Technology*, 76–77 (1966).
77. Cotton, "Nuclear Waste Story," 30–33 (2006).
78. *Id.* at 30; Siting of Fuel Reprocessing Plants and Related Waste Management Facilities, 35 Fed. Reg. 17,530, at 17,531 (Nov. 14, 1970).
79. Siting of Fuel Reprocessing Plants and Related Waste Management Facilities, 35 Fed. Reg. 17,530,

at 17,531 (Nov. 14, 1970). The consensus on use of geologic repositories (but not necessarily of salt beds) was reaffirmed yet again later in the decade, when the issue was reexamined by federal regulators, experts, and the public as part of a nationwide debate orchestrated by President Carter. Interagency Review Group on Nuclear Waste Management, *Report to the President*, 40–43 (1979)
80. Siting of Fuel Reprocessing Plants and Related Waste Management Facilities, 35 Fed. Reg. 17,530, at 17,533 (Nov. 14, 1970), codified at 10 C.F.R. pt. 50 app. F (2010).
81. *Id.* at 17,531.
82. Office of Technology Assessment, *Managing the Nation's Commercial High-Level Radioactive Waste*, app. A-2 at 239 (1985). The Radioactive Waste Management Committee is "an international committee made up of senior representatives from regulatory authorities, radioactive waste management agencies, policy making bodies and research and development institutions" from the Nuclear Energy Agency's twenty-nine member states. Organisation for Economic Co-operation and Development, Nuclear Energy Agency, "Radioactive Waste Management Committee," www.nea.fr/rwm/rwmc.html. The member states are: Australia, Austria, Belgium, Canada, Czech Republic, Denmark, Finland, France, Germany, Greece, Hungary, Iceland, Ireland, Italy, Japan, Korea, Luxembourg, Mexico, Netherlands, Norway, Poland, Portugal, Slovak Republic, Spain, Sweden, Switzerland, Turkey, the United Kingdom, and the United States.
83. According to OTA, AEC cited political opposition and technical uncertainties as the reasons for scuttling the bedrock option.
84. Office of Environmental Management, DOE, *Linking Legacies*, 41, 43 (1997).
85. *Id.* at 42.
86. Lawrence E. Davies, "Fire Cleanup Keeps Plutonium Plant Busy," *New York Times*, 10 (June 27, 1969).
87. Ackland, *Making a Real Killing*, 157 (1999).
88. *Id.* at 158.
89. Organisation for Economic Co-operation and Development, Nuclear Energy Agency, *Stakeholder Participation in Radiological Decision Making*, 22 (2003).
90. Ackland, *Making a Real Killing*, 156–176 (1999).
91. Lawrence E. Davies, "Fire Cleanup Keeps Plutonium Plant Busy," *New York Times*, 10 (June 17, 1969).
92. *Id.*
93. Gary Richardson, "Perspective of a Former Idaho Trout Farmer" (2001), 10 *Science for Democratic Action* (Nov. 2001), www.ieer.org/sdafiles/vol_10/10-1/index.html.
94. Stacy, *Proving the Principle*, 197–199 (2000).
95. *Id.* at 202.
96. Carter, *Nuclear Imperatives*, 67 (1987).
97. Office of Technology Assessment, *Managing the Nation's Commercial High-Level Radioactive Waste*, app. A-1 at 222 (1985); Carter, *Nuclear Imperatives*, 67 (1987).
98. Carter, *Nuclear Imperatives*, 66 (1987).
99. *Id.* at 61–62; Siting of Fuel Reprocessing Plants and Related Waste Management Facilities, 35 Fed. Reg. 17,530, at 17,531 (Nov. 14, 1970). Previously, AEC had considered disposing of defense reprocessing wastes at Hanford and Savannah River in underground caverns at those sites, but the Commission later abandoned this idea in favor of disposal in a repository. Carter, *Nuclear Imperatives*, 60–61 (1987).
100. Siting of Fuel Reprocessing Plants and Related Waste Management Facilities, 35 Fed. Reg. 17,530, at 17,531 (Nov. 14, 1970). Early tests at Lyons found no structural changes from the tests. Office of Technology Assessment, *Managing the Nation's Commercial High-Level Radioactive Waste*, app. A-1 at 210 (1985).
101. Frank Parker, Distinguished Professor of Environmental and Water Resources, Professor of Engineering Management, Vanderbilt University (former head of Radioactive Waste Disposal Research at Oak Ridge National Laboratory), telephone interview by the author (Sep. 27, 2007).
102. Office of Technology Assessment, *Managing the Nation's Commercial High-Level Radioactive Waste*, app. A-1 at 222–223 (1985).
103. Anthony Ripley, "Kansas Geologists Oppose a Nuclear Waste Dump," *New York Times*, 27 (Feb. 17, 1971); "The Kansas Geologists and the AEC," *Science News* 161 (1971).
104. Boffey, "Waste Site Search Gets into Deep Water," 361 (1975).
105. "The Kansas Geologists and the AEC," 161 (1971).
106. Kasperson et al., "Public Opposition to Nuclear Energy," 12–15 (1980).
107. Light, "The Hidden Dimension of the National Energy Plan," 177 (1979) (describing reaction in Michigan); Mazur and Conant, "Controversy over a Local Nuclear Waste Repository," 236 (1978) (describing reaction in New York).
108. Office of Technology Assessment, *Managing the Nation's Commercial High-Level Radioactive Waste*, app. A-2 at 239 (1985).
109. McCutcheon, *Nuclear Reactions*, 12 (2002); Carlsbad Area Office, DOE, *Pioneering Nuclear Waste Disposal*, 6–7 (2000).
110. Downey, "Politics and Technology in Repository Siting," 53 (1985).
111. Rechard, WIPP Performance Assessment Department, DOE, *Milestones for Disposal at WIPP*, 2 (1998).
112. Downey, "Politics and Technology in Repository Siting," 56 (1985).
113. New Mexico: Proposed Withdrawal and Reservation of Lands, 41 Fed. Reg. 54,994, at 54,994–54,995 (Dec. 16, 1976).
114. On use of WIPP for civilian and defense wastes, statement of John M. Deutch, Acting Assistant Secretary for Energy Technology, DOE *Report of the Interagency Review Group on Nuclear Waste Management: Hearing before the Senate Subcommittee on Energy, Nuclear Proliferation, and*

Federal Services of the Committee on Governmental Affairs, 96th Cong. 154 (1979). On lack of a decision regarding type and amount of wastes, McCutcheon, *Nuclear Reactions*, 61, 70 (2002). On lack of a decision on safety standards and criteria, Office of Technology Assessment, *Managing the Nation's Commercial High-Level Radioactive Waste*, 86 (1985). On lack of decisions on transportation, roles of federal and state actors, and decision-making procedures, Office of Technology Assessment, *Comparative Analysis of ERDA Plan*, 20 (1976). The Office of Technology Assessment highlighted this lack of procedural and organizational definition in federal-state roles in its 1976 report criticizing ERDA's program plan.
115. For a general discussion of the rise of environmentalism, *see* Lazarus, *The Making of Environmental Law* (2004).
116. *Id.* at 67–68, 81–84.
117. Pollack, "Time for a Test-Ban Bargain" (2009).
118. Nevada Site Office, DOE, *Nuclear Timeline*, 1–3 (2008).
119. Lawrence Livermore National Laboratory, "Marshall Islands Program: Bikini Atoll," *marshallislands.llnl.gov/bikini.php*.
120. National Cancer Institute, National Institutes of Health, *Estimated Exposures and Thyroid Doses following Nevada Atmospheric Nuclear Bomb Tests*, 8.46–8.47 (1999).
121. Utah State Historical Society, "Utah State Historical Society," *historytogo.utah.gov/utah_chapters/utah_today/nucleartestingandthedownwinders.html*.
122. *Bulloch v. United States*, 145 F. Supp. 824, 828 (C.D. Utah 1956), *vacated*, 95 F.R.D. 123 (D. Utah 1982), *rev'd*, 721 F.2d 713 (10th Cir. 1983), *cert. denied*, 474 U.S. 1086 (1986).
123. Titus, "Governmental Responsibility for Victims of Atomic Testing," 282–284 (1983).
124. *Bulloch v. United States*, 95 F.R.D. 123, 124 (D. Utah 1982), *rev'd*, 721 F.2d 713 (10th Cir. 1983), *cert. denied*, 474 U.S. 1086 (1986)
125. AEC conducted a public relations campaign to combat what it perceived as public misunderstanding about nuclear power that resulted in favorable news articles. A 1965 poll of three cities showed that participants had positive views of nuclear power. Walker, *Containing the Atom*, 391–92 (1992).
126. The Three Mile Island accident, which occurred on March 28, 1979, was the worst commercial nuclear power accident in the United States. There was a severe core meltdown of the Three Mile Island 2 reactor; but the radiation did not breach the containment walls, which would have released large amounts of radiation into the environment. For a definitive history of the causes and consequences of the accident, *see* Walker, *Three Mile Island* (2004). Chernobyl occurred on April 26, 1986, and is the worst nuclear power accident that the world has seen. A test of the badly designed power system led to a meltdown of the entire Chernobyl 4 reactor. About 5 percent of the radioactive material was released in the accident along with radioactive gasses. Yablokov, Nesterenko, and Nesterenko, *Chernobyl* (2009), provides a complete account of the short- and long-term consequences of the accident.
127. Mayer et al., *Challenges of Nuclear Weapons Waste*, 36 (2010).
128. National Environmental Policy Act § 102(2)(c), 42 U.S.C. § 4332(2)(c) (2006).
129. Murchison, "Does NEPA Matter?" 574 (1983).
130. Buccino, "NEPA under Assault," 51–52 (2003).
131. *Kleppe v. Sierra Club*, 427 U.S. 390, 407 (1976).
132. See discussion on Calvert Cliffs in Chapter 1, in the fifth paragraph after the heading "Litigation Challenging Reactor Safety."
133. Freedom of Information Act of 1966, 5 U.S.C. § 552 (2006).
134. *Id.*, §§ 552(b), 552(a)(4)(B).
135. Downey, "Politics and Technology in Repository Siting," 56 (1985).
136. Hanford Health Information Network, Washington State Department of Health, *The Release of Radioactive Material from Hanford: 1944–1972* (2000).
137. *In re Hanford Nuclear Reservation Litigation*, 780 F. Supp. 1551 (E.D. Wash. 1991).
138. Titus, "Governmental Responsibility for Victims of Atomic Testing," 282–283 (1983).
139. Special Message to the Congress about Reorganization Plans to Establish the Environmental Protection Agency and the National Oceanic and Atmospheric Administration, 1970 PUB. PAPERS 578 (July 9, 1970).
140. Reorganization Plan No. 3 of 1970: Environmental Protection Agency, 84 Stat. 2086.
141. *Id.* § 2(a)(6), 84 Stat. 2088.
142. *Id.*
143. Walker, *Permissible Dose*, 69–74 (2000).
144. Office of Radiation and Indoor Air, EPA, *Radiation Protection at EPA*, 15 (2000).
145. Roy L. Ash, Director, Office of Management and Budget, "Memorandum to Administrator Train and Chairman Ray: Responsibility for Setting Radiation Protection Standards" (Dec. 7, 1973), as reprinted in Office of the General Counsel, NRC, *Nuclear Regulatory Legislation*, 266 (2009).
146. Environmental Protection Agency, Radiation Protection Programs: Environmental Radiation Protection Standards for Nuclear Power Operations, 42 FED. REG. 2857 (Jan. 13, 1977); 40 C.F.R. § 190.10(a) (1977).
147. Walker, *Permissible Dose*, 77–78 (2000); 40 C.F.R. § 190.02(c), (k) (1978).
148. 40 C.F.R. § 190.10(a) (2009).
149. It was recognized that EPA might be expressly authorized by statute to enforce its generic standards. Office of Radiation and Indoor Air, EPA, *Radiation Protection at EPA*, 5 (2000). But no such authority has been conferred on EPA.
150. Energy Reorganization Act, as enacted, Pub. L. No. 93-438, 88 Stat. 1233 (1974).

151. *Id.* § 202, as enacted, Pub. L. No. 93-438, 88 Stat. 1233, at 1244.
152. Department of Energy Organization Act § 201, 42 U.S.C. § 7131 (2006).
153. "Energy Research Agency Replaces AEC," *Science News*, 248 (1974).
154. On the AEC's struggles, Buck, DOE, *A History of the Atomic Energy Commission*, 6–7 (1983). On public concern about radioactive releases, Walker, *Permissible Dose*, 29 (2000).
155. "Passing of the AEC," *Science News*, 55 (1975).
156. Energy Reorganization Act § 2(c), 42 U.S.C. § 5801 (2006).
157. Resource Conservation and Recovery Act, 42 U.S.C. §§ 6901–6992 (2006); Comprehensive Environmental Response, Compensation, and Liability Act of 1980, 42 U.S.C. §§ 9601–9675 (2006).
158. Resource Conservation and Recovery Act § 1003, 42 U.S.C. § 6902(a) (2006); Applegate and Laitos, *Management of Hazardous Waste*, 16–17 (2006).
159. Hess, "Hanford," 205 (1996).
160. Hazardous and Solid Waste Amendments of 1984 §§ 206–207, 42 U.S.C. § 6924(u)–(v).
161. Resource Conservation and Recovery Act § 3006(b), 42 U.S.C. § 6926(b) (2006). RCRA provides, however, that "delegated" state RCRA programs may be more stringent than EPA's RCRA program, so long as they are substantially equivalent to EPA's own program. States are not preempted from establishing their own hazardous waste programs, even if those differ substantially from EPA's; in those cases, both EPA's and the state's hazardous waste programs apply within those states. Most, but not all, states have in fact chosen to adopt part or all of the federal RCRA program and have been authorized by EPA to do so.
162. Superfund Amendments and Reauthorization Act of 1986 § 120, 42 U.S.C. § 9620 (2006).
163. The unitary executive theory holds that conflict between federal agencies must be resolved by the president rather than the courts, and that such a conflict does not present a "case or controversy" within the jurisdiction of the federal courts under Article III of the Constitution. For a discussion of this theory, *see* Calabresi and Rhodes, "The Structural Constitution" (1992).
164. EPA, "Federal Facilities Cleanup Enforcement," *www.epa.gov/oecaerth/federalfacilities/enforcement/cleanup/index.html*.
165. These sites are Weldon Spring Quarry, MO; Wayne Interim Storage Site, NJ; Savannah River Site, SC; Rocky Flats Plant, CO; Pantex Plant, TX; Paducah Gaseous Diffusion Plant, KY; Oak Ridge Reservation, TN; Mound Plant, OH; Monticello Mill Tailings, UT; Middlesex Sampling Plant, NJ; Lawrence Livermore National Laboratory Main Site, CA; Lawrence Livermore National Laboratory Site 300, CA; Laboratory for Energy-Related Health Research, CA; Idaho National Laboratory, ID; Hanford 300, WA; Hanford 200, WA; Hanford 100, WA; Feed Materials Production Center, OH; and Brookhaven National Laboratory, NY. EPA, "National Priorities List (NPL)/Base Realignment and Closure (BRAC) Sites," *www.epa.gov/fedfac/ff/nplbracsites.htm*.
166. Federal Facility Compliance Act of 1992, as enacted, Pub. L. No. 102-386, 106 Stat. 1505 (1992).
167. Statement of Representative Bill Richardson, 137 Cong. Rec. H16034 (June 24, 1991).
168. Statement of Senator David Durenberger,137 Cong. Rec. S28428 (Oct. 24, 1991).
169. Statement of Representative Charles Luken, 137 Cong. Rec. H16031 (June 24, 1991).
170. Representative Bob Schaefer noted that the federal government was allowed to "violate environmental laws relatively free from retribution." 137 Cong. Rec. H16028 (June 24, 1991). Senator Howard Metzenbaum said he believed "Congress ought to step in and tell DOE once and for all that it is not above Federal law. It cannot be allowed to ignore the harmful pollution problems it helped create." 137 Cong. Rec. S28429 (Oct. 24, 1991).
171. Statement of Senator Howard Metzenbaum, 137 Cong. Rec. S28429 (Oct. 24, 1991).
172. The Ninth Circuit, for example, refused to conclude that RCRA unambiguously waived sovereign immunity from civil suits by the states. Such a waiver would allow states to enforce their own hazardous waste statues against the federal government. *United States v. Washington*, 872 F.2d 874 (9th Cir. 1989).
173. Statement of Representative Al Swift, 137 Cong. Rec. H16028 (June 24, 1991); statement of Representative Donald Ritter, 137 Cong. Rec. H16026 (June 24, 1991); statement of Representative Brian Bilbray, 137 Cong. Rec. H16032 (June 24, 1991); statement of Senator Sam Nunn, 137 Cong. Rec. S28421 (Oct. 24, 1991).
174. Federal Facility Compliance Act of 1992 § 103, as enacted, Pub. L. No. 102-386, 106 Stat. 1505, at 1507 (1992).
175. *Id.* § 102(a), as enacted, Pub. L. No. 102-386, 106 Stat. 1505, at 1505.
176. Resource Conservation and Recovery Act § 7002, 42 U.S.C. § 6972 (2006).
177. *Id.* § 3021, 42 U.S.C. § 6939(c) (2006).
178. For example, the agreement governing INL, to which Idaho, EPA, and DOE are parties, states that "all parties shall have the right to enforce the terms of this agreement." EPA, DOE, Idaho, *Federal Facility Agreement and Consent Order*, 27 (1991). Authorized states can also enter federal facilities to inspect for compliance with state hazardous waste programs. Federal Facility Compliance Act of 1992 § 103, as enacted, Pub. L. No. 102-386, 106 Stat. 1505, at 1507 (1992).
179. Citizens and groups can bring civil suits pursuant to Comprehensive Environmental Response, Compensation, and Liability Act of 1980 § 310, 42 U.S.C. § 9659 (2006). The enforceability is reiterated in the agreements. For example, the agreement governing INL states that "any standard, regulation, condition, requirement, or order which has become

effective under CERCLA and is incorporated into this Agreement is enforceable by any person." EPA, DOE, Idaho, *Federal Facility Agreement and Consent Order*, 26 (1991).
180. See Appendix C for a discussion of dispute settlement and stipulated damages provisions in the Hanford TPA.
181. Lambright, "Changing Course: Admiral James Watkins and the DOE Nuclear Weapons Complex," (2002).
182. Mayer et al., *Challenges of Nuclear Weapons Wastes*, 10–12 (2010).
183. *Id.* at 35.
184. *Id.* at 28.
185. For example, DOE is obligated to excavate and clean up the TRU waste it buried at Hanford, INL, Oak Ridge, and SRS under CERCLA. Additionally, DOE is required to do the same at Los Alamos National Laboratory under other environmental laws. Government Accountability Office, *Plans for Addressing Most Buried TRU Wastes*, 3–4 (2007).
186. Office of Environmental Management, DOE, *Report to Congress: Status of Environmental Management Initiatives*, 69–71 tbl.2.3 (2009).
187. *Id.* at 67.
188. *Id.* at 21–22.
189. *Id.* at 68–69.
190. *Id.* at 71–73.
191. General Accounting Office, *Waste Cleanup*, 9–11 (2002).
192. *Id.* at 12 (2002).
193. Holt, Congressional Research Service, *Nuclear Weapons Production Complex* (1997); General Accounting Office, *Waste Cleanup*, 2–3 (2002).
194. Office of Environmental Management, DOE, *Linking Legacies*, 31 (1997).
195. Office of Environmental Management, DOE, *Report to Congress: Status of Environmental Management Initiatives*, at i–iii (2009).
196. *Id.* at 3–7.
197. *Id.* at 32.
198. Office of Environmental Management, DOE, *Linking Legacies*, 7, 34 (1997).
199. Office of Environmental Management, DOE, *Report to Congress: Status of Environmental Management Initiatives*, 24 (2009).
200. *Id.* at 24–25.
201. *Id.* at 23.
202. *Id.* at 24.
203. Picket and Norford, Westinghouse Savannah River Co., *First Commercial U.S. Mixed Waste Vitrification Facility*, 7 (1998).
204. Office of Environmental Management, DOE, *Report to Congress: Status of Environmental Management Initiatives*, 23–25 (2009).
205. *Id.* at 25–26.
206. *Id.*
207. *Id.* at 26–27.
208. *Id.* at 29.
209. *Id.* at 30–31.
210. *Id.* at 31–32.
211. *Id.* at 33–34. TRU waste is being stored at Rocky Flats, Fernald, Missouri University Research Reactor, Energy Technology Engineering Center, Lovelace Respiratory Research Inc., Mound, Battelle Columbus, Brookhaven, Teledyne–Brown, Knolls Atomic Power Laboratory, USAMC, Arco Medical Products, Lawrence Berkeley Laboratory, and Framatome.
212. *Id.* at 36–37.
213. *Id.* at 38.
214. *Id.* at 41.
215. *Id.* at 48.
216. *Id.* at 52.
217. *Id.* at 49–52. Rocky Flats was closed in 2005, while Fernald and Mound were completed in 2006. All were completed ahead of schedule below budget.
218. *Id.* at 54–55.
219. *Id.*
220. Mayer et al., *Progress and Challenges: The Cost of Cleaning Up the Nation's Nuclear Weapons Legacy Wastes*, 80 (2010).
221. *Id.* at 79.
222. Office of Environmental Management, DOE, *Report to Congress: Status of Environmental Management Initiatives*, 2010, Appendices, 1 (2010).
223. Mayer et al., *Progress and Challenges*.
224. Carter, *Nuclear Imperatives*, 62 (1987).
225. *Id.* at 92.
226. On technology for private firms, Campbell, "The State and the Nuclear Waste Crisis," 21 (1987). On AEC success with reprocessing, Walker, *Road to Yucca*, 83 (2009).
227. Campbell, "The State and the Nuclear Waste Crisis," 21–22 (1987).
228. Metz, "Reprocessing: How Necessary Is It for the Near Term?" 44 (1977).
229. DOE, *Plutonium Recovery at West Valley* (1996).
230. Campbell, "The State and the Nuclear Waste Crisis," 24 (1987). When AEC passed the rule, there was not a dominant solidification technology. At the time, AEC was sponsoring research into three solidification technologies: phosphate glass, spray solidification, and pot calcination. Battelle Memorial Institute, *Waste Solidification Demonstration Program*, 2 (1970).
231. DOE, *Plutonium Recovery at West Valley* (1996); Natural Resources Defense Council, *DOE's Nuclear Energy Research Programs Threaten National Security*, www.nrdc.org/nuclear/bush/freprocessing.asp.
232. DOE, *Plutonium Recovery at West Valley* (1996).
233. Douglas Dales, "Pioneer Atom Site Pushed by State," *New York Times*, 46 (June 14, 1963); *New York State Energy Research and Dev. Auth. v. Nuclear Fuel Services, Inc.*, 666 F.2d 787, 788 (2d Cir. 1981).
234. Douglas Dales, "Pioneer Atom Site Pushed by State," *New York Times*, 46 (June 14, 1963).
235. Wohlstetter, "Spreading the Bomb without Quite Breaking the Rules," 173 (1976).
236. West Valley Demonstration Project Act § 2(b)(4)(C), as enacted, Pub. L. No. 96-368, 94 Stat. 1347, at 1348 (1980).

237. DOE, "West Valley Demonstration Project Nuclear Timeline," www.wv.doe.gov/Site_History.html.
238. DOE, "Waste Management," www.wv.doe.gov (follow Waste Management link under General Links).
239. "The parties agree that the NRC 'induced' private industry to undertake the awesome reprocessing task, the motive of the NRC being, of course, its commitment to the power producers to dispose of their spent fuel. It believed that this operation, too, could be best performed by private industry with private capital. Besides jawboning, the 'inducement' took the form of free land as well as technical assistance." *Allied-General Nuclear Services v. United States*, 839 F.2d 1572, 1573 (Fed. Cir. 1988).
240. Associated Press, "Commercial Reprocessing of Nuclear Fuel Dead," *Tuscaloosa News*, 8 (Apr. 26, 1977).
241. DOE, *Plutonium Recovery at West Valley* (1996).
242. Gerald Ford, Statement on Nuclear Policy, 3 Pub. Papers 2763 (Oct. 28, 1976).
243. Bergeron, *Tritium on Ice*, 30 (2002).
244. Gerald Ford, Statement on Nuclear Policy, 3 Pub. Papers 2763, 2776 (Oct. 28, 1976); Don Oberdorfer, "Nuclear Policy Study Set, Tougher Stance Hinted," *Washington Post*, A3 (Jan. 28, 1977).
245. Gerald Ford, Statement on Nuclear Policy, 3 Pub. Papers 2763, 2776 (Oct. 28, 1976); Office of Technology Assessment, *Managing the Nation's Commercial High-Level Radioactive Waste*, 93 (1985).
246. Jimmy Carter, Nuclear Power Policy: Statement on Decisions Reached Following Review, 1 Pub. Papers 587, 588 (Apr. 7, 1978).
247. Andrews, Congressional Research Service, *Nuclear Fuel Reprocessing*, 4 (2008); Brenner, *Nuclear Power and Non-Proliferation*, 142–153 (1981).
248. Andrews, Congressional Research Service, *Nuclear Fuel Reprocessing*, 5 (2008).
249. Stucker, "Nuclear White Elephant," 15–18 (1982).
250. Cotton, "Nuclear Waste Story," 31 (2006).
251. Union of Concerned Scientists, "Fact Sheet: Reprocessing Would Increase Total Volume of Radioactive Waste," 1 (2009).
252. Institute for Policy Studies, *Radioactive Wastes and GNEP*, 14–15 (2007)
253. Downey, "Politics and Technology in Repository Siting," 57 (1985). As discussed earlier in the chapter, ERDA reached out to thirty-six states to avoid the communication problems it had faced in the past. Many of those states responded by enacting legislation to block waste repositories. Light, "The Hidden Dimension of the National Energy Plan," 177 (1979).
254. General Accounting Office, *Liquid Metal Fast Breeder Reactor*, 3 (1982).
255. *Id.*; General Accounting Office, *U.S. Fast Breeder Reactor Program Needs Direction*, 2 (1980).
256. Act of June 2, 1970, as enacted, Pub. L. No 91-273, 84 Stat. 299 (1970).
257. General Accounting Office, *Liquid Metal Fast Breeder Reactor*, 11–12 (1982).
258. General Accounting Office, *U.S. Fast Breeder Reactor Program Needs Direction*, 16 (1980).
259. General Accounting Office, *Liquid Metal Fast Breeder Reactor*, 3 (1982).
260. Katz, "Clinch River Breeder Reactor," 54 (1984).
261. General Accounting Office, *Liquid Metal Fast Breeder Reactor*, 22–23 (1982); *see* Appendix B for more information on the price of uranium.
262. General Accounting Office, *U.S. Fast Breeder Reactor Program Needs Direction*, 32 (1980).
263. Jimmy Carter, Veto of Department of Energy Authorization Bill: Message to the Senate Returning S. 1811 without Approval, 2 Pub. Papers 1972, 1972 (Nov. 5, 1977).
264. General Accounting Office, *Liquid Metal Fast Breeder Reactor*, 3 (1982).
265. *Natural Resources Defense Council, Inc. v. Zeller*, 688 F.2d 706, 711 (11th Cir. 1982).
266. Marshall, "Clinch River Dies," 590 (1983).
267. Jack Anderson, "Nuclear Plant Goof Blamed on Fraud, Mismanagement," *Spokane Daily Chronicle*, 4 (July 21, 1981).
268. Marshall, "Clinch River Dies," 590 (1983).
269. *Id.*; statement of Rudolph G. Penner, Director, Congressional Budget Office, *Clinch River–An Alternate Financing Plan: Hearing before the House Committee on Energy and Commerce*, 98th Cong. 56 (1983).
270. Statement of Shelby T. Brewer, Assistant Secretary for Nuclear Energy, Department of Energy, *id.* at 226.
271. For an account of the successful Environmental Defense Fund effort to stop PG&E from building new nuclear power plants in California, *see* Roe, *Dynamos and Virgins*, 157–196 (1984). EDF lobbied the California Public Utilities Commission and conducted a public relations campaign against new nuclear and coal-fired power plants on economic grounds.
272. Atomic Energy Act of 1954 § 185, as enacted, Pub. L. 83-703, 68 Stat. 919, at 955 (1954).
273. Boyle, "The Case of Nuclear Power in Four Countries," 151 (1998).
274. Advisory Committee on Reactor Safeguards, *Report on Power Reactor Development Corporation Applications for Construction Permit* (1956), as appears in *Enrico Fermi Reactor–Use for Irradiation Testing and Background Information: Hearing before the Joint Committee on Atomic Energy*, 89th Cong. app. 4 at 77–78 (1966).
275. *Power Reactor Dev. Co. v. Int'l Union of Elec. Workers*, 367 U.S. 396, 411–416 (1961); Boyle, "The Case of Nuclear Power in Four Countries," 151 (1998).
276. *Nader v. Ray*, 363 F. Supp. 946, 949 (D.D.C. 1973).
277. *Id.* at 953.
278. *Union of Concerned Scientists v. Atomic Energy Comm'n*, 499 F.2d 1069, 1094 (D.C. Cir. 1974); *Nader v. Nuclear Reg. Comm'n*, 513 F.2d 1045, 1056 (D.C. Cir. 1975).
279. *York Comm. for a Safe Environment v. Nuclear Reg. Comm'n*, 527 F.2d 812, 814 (D.C. Cir. 1975).

280. *Id.* at 816.
281. *Train v. Colorado Public Interest Research Group, Inc.*, 426 U.S. 1, 24–25 (1975).
282. *Calvert Cliffs' Coordinating Comm., Inc. v. Atomic Energy Comm'n*, 449 F.2d 1109, 1119 (D.C. Cir. 1971).
283. Cohen, McCubbins, and Rosenbluth, "The Politics of Nuclear Power," 184 (1995).
284. Boyle, "The Case of Nuclear Power in Four Countries," 151 (1998).
285. Zinberg, "The Public and Nuclear Waste Management," 36 (1979).
286. *Vermont Yankee Nuclear Power Corp. v. Natural Res. Def. Council*, 435 U.S. 519, 543 (1978).
287. Cohen, McCubbins, and Rosenbluth, "The Politics of Nuclear Power," 184 (1995).
288. Cook, *Nuclear Power and Legal Advocacy*, 46–48 (1980).
289. *Id.* at 48.
290. Cohen, McCubbins, and Rosenbluth, "The Politics of Nuclear Power," 184 (1995).
291. Keating, "The Role of Technology Assessment," 61 (1975).
292. Cohen, McCubbins, and Rosenbluth, "The Politics of Nuclear Power," 185 (1995).
293. Environmental Effects of the Uranium Fuel Cycle, 37 Fed. Reg. 24,191, at 24,192 (Nov. 15, 1972); Environmental Effects of the Uranium Fuel Cycle, 39 Fed. Reg. 14,188, at 14,189–14,190 (Apr. 22, 1974); Uranium Fuel Cycle Impacts from Spent Fuel Reprocessing and Radioactive Waste Management, 41 Fed. Reg. 45,849, at 45,850 (Oct. 18, 1976); Licensing and Regulatory Policy and Procedures for Environmental Protection; Uranium Fuel Cycle Impacts from Spent Fuel Reprocessing and Radioactive Waste Management, 44 Fed. Reg. 45,362, at 45,372 (Aug. 2, 1979).
294. Licensing and Regulatory Policy and Procedures for Environmental Protection; Uranium Fuel Cycle Impacts from Spent Fuel Reprocessing and Radioactive Waste Management, 44 Fed. Reg. 45,362, at 45,369 (Aug. 2, 1979); *Baltimore Gas & Elec. v. NRDC*, 462 U.S. 87, 96 (1983).
295. Environmental Effects of the Uranium Fuel Cycle, 37 Fed. Reg. 24,191, at 24,191–24,193 (Nov. 15, 1972).
296. Environmental Effects of the Uranium Fuel Cycle, 39 Fed. Reg. 14,188, at 14,190 (Apr. 22, 1974).
297. *Natural Res. Def. Council v. Nuclear Reg. Comm'n*, 547 F.2d 633 (D.C. Cir. 1976) at 653–654.
298. In 1976, NRC published a revised survey of environmental impacts from August 16 to October 18 of that year. NRC, *Environmental Survey of Reprocessing and Waste Management* (1976). For the interim rule, Uranium Fuel Cycle Impacts from Spent Fuel Reprocessing and Radioactive Waste Management, 41 Fed. Reg. 45,849, at 45,852 (Oct. 18, 1976).
299. *Vermont Yankee Nuclear Power Corp. v. Natural Res. Def. Council*, 435 U.S. 519, 525 (1978). In the same opinion, the court reversed a related decision by the D.C. Circuit Court of Appeals. *Aeschliman v. Nuclear Reg. Comm'n*, 547 F.2d 622 (D.C. Cir. 1976). *Id.*
300. Licensing and Regulatory Policy and Procedures for Environmental Protection; Uranium Fuel Cycle Impacts from Spent Fuel Reprocessing and Radioactive Waste Management, 44 Fed. Reg. 45,362 (Aug. 2, 1979).
301. *Id.*
302. *Id.* at 45,367–45,368.
303. *Natural Res. Def. Council v. Nuclear Reg. Comm'n*, 685 F.2d 459, 467 (D.C. Cir. 1982). The court's decision resolved four related cases that had been consolidated. Case No. 74-1586 was the remand from the Supreme Court's *Vermont Yankee* decision to determine if the Table S-3 rule was supported by the administrative record. In case No. 77-1448, NRDC challenged the interim Table S-3 rule. In case No. 79-2110, New York challenged the final S-3 rule on the grounds that NRC's findings regarding economic feasibility were arbitrary and capricious. In case No. 79-2131, NRDC challenged the final rule on the grounds that it failed to disclose and allow consideration of the uncertainties underlying the environmental impacts listed in the final Table S-3. The D.C. Circuit also found that the original and interim table S-3 rules were arbitrary and capricious before amendment because they failed to allow consideration of the health, socioeconomic, and cumulative effects of fuel-cycle activities. It upheld the Commission's findings regarding economic feasibility. *Id.* at 477–478.
304. *Id.* at 478, 481–485.
305. *Baltimore Gas & Elec. Co. v. Natural Res. Def. Council*, 462 U.S. 87, 106–108 (1983).
306. *Id.* at 91–95.
307. *Northern States Power Co.*, 7 N.R.C. 41, 49–51 (1978).
308. *Minnesota v. Nuclear Reg. Comm'n*, 602 F.2d 412, 418–419 (D.C. Cir. 1979).
309. Storage and Disposal of Nuclear Waste, 44 Fed. Reg. 61,372, at 61,372–61,373 (Oct. 25, 1979).
310. *Minnesota v. Nuclear Reg. Comm'n*, 602 F.2d 412, 414–415 (D.C. Cir. 1979); *Vermont Yankee Nuclear Power Corp.*, 6 N.R.C. 436 (1977); *Northern States Power Co.*, 6 N.R.C. 265 (1977).
311. Waste Confidence Decision, 49 Fed. Reg. 34,658, at 34,658 (Aug. 31, 1984). A singular, basic solution emerged as its mandate: a federal repository.
312. NRC, "Dry Cask Storage," www.nrc.gov/waste/spent-fuel-storage/dry-cask-storage.html.
313. Government Accountability Office, *Nuclear Waste Management*, 10 (2009).
314. Of a total of 60,000 metric tons of commercial SNF, 9,000 metric tons were in dry-cask storage; the rest were held in wet storage. Institute for 21st Century Energy, U.S. Chamber of Commerce, *Revisiting America's Nuclear Waste Policy*, 4, 8 (2009).
315. Murphy and La Pierre, "Nuclear Moratorium Legislation in the States," 392 (1976).
316. Cal. Pub. Res. Code § 25524.2 (1977).
317. Cal. Pub. Res. Code § 25524.1(b) (1977) requires demonstrated adequate storage capacity for the

wastes that would be generated by new plants before they can be built.
318. On the demonstrations, Giugni, *Social Protest and Policy Change*, 43 fig.3.1b (2004). On falling public support, Walker, *Three Mile Island*, 197–198 (2004).
319. Wayne King, "Concern Rises in South Carolina, Home of Many Nuclear Reactors," *New York Times*, 30 (Apr. 1, 1979). Connecticut's governor spoke out against construction of any additional nuclear reactors. Richard Severo, "Debate on Safety of Nuclear Plants Intensifies in the Tristate Area," *New York Times*, 31 (Apr. 1, 1979).
320. *Pac. Gas and Elec. v. State Energy Res. Conservation and Dev. Comm'n*, 461 U.S. 190, 194–195 (1983).
321. *Id*. at 216. For general discussion of the issues presented, *see* Murphy and La Pierre, "Nuclear Moratorium Legislation in the States," 421–424 (1976).
322. Or. Rev. Stat. Ann. § 469.595 (West 2010); Me. Rev. Stat. Ann. tit. 35-A, § 4374 (2010) (West); Mont. Code Ann. § 75-20-1203 (1993); Conn. Gen. Stat. Ann. § 22a-136 (West 2010); Mass. Ann. Laws ch. S63, § 3 (LexisNexis 2010); W. Va. Code Ann. § 16-27A-2 (LexisNexis 2010); Wis. Stat. Ann. § 196.493 (West 2010); Ky. Rev. Stat. Ann. § 278.610 (LexisNexis 2010); Kan. Stat. Ann. § 66-128h (2009); 220 Ill. Comp. Stat. Ann. 5/8-406 (LexisNexis 2010); N.J. Stat. Ann. § 13:19-11 (West 2010); Minn. Stat. Ann. § 216B.243 (West 2010); 66 Pa. Cons. Stat. Ann. § 518 (West 2010). *See also* Parker and Holt, Congressional Research Service, *Outlook for New U.S. Reactors*, 23 (2007); Blake, "Where Nuclear Reactors Can (and Can't) Be Built," 23 (Nov. 2006); Farsetta, "Nuclear Industry Targets State Laws" (2009).
323. Farsetta, "Nuclear Industry Targets State Laws" (2009).
324. *Id*.
325. Donovan, "Kentucky Bills Would End Moratorium on Nuclear Power Plant Construction" (2008); Steven Walters, "Doyle Favors Lifting Nuclear Moratorium," *Milwaukee Journal Sentinel*, D3 (Aug. 6, 2008); Blake, "Where Nuclear Reactors Can (and Can't) Be Built," 23–25 (Nov. 2006); Mark Brunswick, "Nuclear Plant Moratorium Upheld," *St. Paul Star Tribune*, B5 (May 1, 2009).
326. Interagency Review Group on Nuclear Waste Management, *Report to the President* (1979). The members of the IRG commission were largely assistant secretaries or their equivalents.
327. *Id*. at 3–5. The IRG defined TRU as "low-level waste with concentrations of transuranic elements in excess of 10 nanocuries/gram" but noted that the definition was under review by NRC and could be revised upward. *Id*. at D-11.
328. *Id*. at 2.
329. *Id*. at C-1.
330. *Id*. at preface, C-3.
331. *Id*. at 16. The ethical premises of the statement are discussed in Chapter 8.
332. *Id*. at H-9.
333. *Id*.
334. *Id*. at 100–101.
335. *Id*. at 35.
336. *Id*. at 70.
337. *Id*. at 73.
338. *Id*. The IRG report defined an ISF as a facility where a smaller amount of waste could be emplaced with the possibility, but not the expectation, of retrievability that would provide data helpful for a future repository and knowledge helpful for future siting processes. *Id*. at 54.
339. *Id*.
340. *Id*. at 92–94.
341. *Id*. at 106.
342. Nuclear Waste Policy Act, H.R. 8378, 96th Cong. (1980); Nuclear Waste Policy Act, S. 2189, 96th Cong. (1980).
343. Joseph M. Hendrie, NRC chair, suggested that Congress authorize NRC to license DOE facilities for TRU storage. *Nuclear Waste and Facility Siting Policy, Part 1: Hearing before the Senate Committee on Energy and Natural Resources*, 96th Cong. 87 (1979). Congress did not follow this suggestion.
344. Low-Level Radioactive Waste Policy Act of 1980, 42 U.S.C. § 2021(b)-(d) (2006).
345. DOE, *FEIS: Management of Commercial Radioactive Waste* (1980). In the FEIS, DOE included SNF in its definition of HLW. *Id*. at 1.1.
346. *Id*. at 1.17.
347. *Id*. at 1.16.
348. *Id*.
349. *Id*. at 1.18–1.19.
350. *Id*. at 1.17.
351. *Id*. at 1.18.
352. *Id*. at 1.20.
353. *Id*.
354. Pelham, "Government Groping with Waste Disposal," 2555 (1977).
355. *Id*.
356. Light, "The Hidden Dimension of the National Energy Plan," 176–178 (1979); General Accounting Office, *Institutional Relations under NWPA*, 10 (1987). States approached by ERDA were Louisiana, Mississippi, Texas, Utah, Nevada, and Washington.
357. Downey, "Politics and Technology in Repository Siting," 58 (1985). The proffered explanation for this move was that an NRC license was required under the Energy Reorganization Act of 1974. *Id*.
358. *Id*.
359. *Id*. at 58.
360. Staats, *Nuclear Energy's Dilemma: Disposing of Hazardous Radioactive Waste Safely*, xi (1977). GAO's name was changed in 2004 from the General Accounting Office to the Government Accountability Office as part of the agency's reorganization under the GAO Human Capital Reform Act of 2004, Pub. L. 108-271, 118 Stat. 811 (2004). For the purposes of this book, the organization will be referred to as GAO, regardless of date.
361. DOE's *Final Environmental Impact Statement for Management of Commercially Generated Radioactive Waste*, issued in October 1980, identified no bar

to co-locating HLW or SNF at WIPP. DOE, *FEIS: Management of Commercial Radioactive Waste* (1980). In addition, DOE's FEIS for WIPP, also issued in October 1980, stated that in addition to achieving safe disposal, co-location of these wastes could reduce the administrative work required, the amount of land required, and the overall environmental impacts of a federal repository. DOE, *Final EIS for WIPP*, at 1-7-1-9, 3-16 (1980).
362. WIPP Authorization Act, Department of Energy National Security and Military Applications of Nuclear Energy Authorization Act of 1980 § 213(a), as enacted, Pub. L. No. 96-164, 93 Stat. 1259, 1265 (1979).
363. Waste Isolation Pilot Plant Land Withdrawal Act § 12, as enacted, Pub. L. No. 102-579, 106 Stat. 4777, 4791 (1992).
364. Carter, "Congressional Committees Ponder Veto," 1137 (1978).
365. See Committee on Armed Services, *National Defense Authorization Act for FY89*, 160 (1988) for a discussion of the history of the EEG.
366. WIPP Authorization Act, Department of Energy National Security and Military Applications of Nuclear Energy Authorization Act of 1980 § 213(a)–(b)(1), as enacted, Pub. L. No. 96-164, 93 Stat. 1259, 1265–1266 (1979).
367. Complaint, *New Mexico ex rel. Bingaman v. Department of Energy*, No. 81-0363 JB (D.N.M. 1981).
368. "Agreement for Consultation and Cooperation," as appears in Appendix A, Stipulated Agreement, *New Mexico ex rel. Bingaman v. Department of Energy*, No. 81-0363 JB (D.N.M. 1981).
369. The agreements were approved and entered in a decree by the court in litigation brought by New Mexico against DOE. Stipulated Agreement, *New Mexico ex rel. Bingaman v. Department of Energy*, No. 81-0363 JB (D.N.M. 1981).
370. McCutcheon, *Nuclear Reactions*, 100–04 (2002). The waste was a result of ongoing production of nuclear weapons triggers at Rocky Flats. *Id.*
371. Ackland, *Making a Real Killing*, 213 (1999).
372. Matthew L. Wald, "3 States Ask Waste Cleanup as Price of Atomic Operation," *New York Times*, 11 (Dec. 17, 1988).
373. McCutcheon, *Nuclear Reactions*, 103 (2002). In 1995, Idaho and DOE reached a settlement agreement that required DOE to remove all TRU waste from the state by 2018. The agreement settled two cases filed by Idaho that sought to prevent DOE from shipping SNF to INL. Idaho Department of Environmental Quality, "Contamination at INL: Text of the 1995 Settlement Agreement," *www.deq.state.id.us/inl_oversight/contamination/settlement_agreement_entire.cfm*. Shipment of TRU to WIPP began in March 1999. Carlsbad Field Office, DOE, "DOE Announces First Shipment of Transuranic Waste from INEEL to WIPP on April 27, 1999," press release (Apr. 13, 1999).
374. Ackland, *Making a Real Killing*, 221 (1999).
375. *Id.* at 222–227.
376. McCutcheon, *Nuclear Reactions*, 123–124 (2002).
377. Leftover TRU began shipping to WIPP from Rocky Flats when WIPP opened in 1999. Buffer, DOE, *Rocky Flats History* (2003).
378. *New Mexico ex rel. Udall v. Watkins*, 783 F. Supp. 633, 638 (D.D.C. 1992), aff'd 969 F.2d 1122, 1126 (D.C. Cir. 1991).
379. Waste Isolation Pilot Plant Land Withdrawal Act, Pub. L. No. 102-579, 106 Stat. 4777 (1992).
380. *Id.* § 8(b)(3), 42 U.S.C. § 10107(b)(3) (2006). The act also directed EPA to issue criteria for final disposal as a further basis for certification. *Id.* §8(c). EPA did so. Criteria for the Certification and Recertification of the Waste Isolation Pilot Plant's Compliance with the 40 CFR Part 191 Disposal Regulations, 61 Fed. Reg. 5,224 (Feb. 9, 1996) (codified at 40 C.F.R. pt. 194).
381. Committee on Energy and Resources, *Waste Isolation Pilot Plant Land Withdrawal Act*, 12 (1991).
382. Committee on Armed Services, *Waste Isolation Pilot Plant Land Withdrawal Act of 1991*, at 16 (1991).
383. Committee on Energy and Resources, *Waste Isolation Pilot Plant Land Withdrawal Act*, 12 (1991).
384. Environmental Radiation Protection Standards for the Management and Disposal of Spent Nuclear Fuel, High-Level and Transuranic Radioactive Wastes, 58 Fed. Reg. 66,398 (Dec. 20, 1993); 40 C.F.R. pt. 191 (1994).
385. The amendments were designed to fix only the three pieces of sections 191.15 and 191.16 that WIPPLWA did not reinstate.
386. 40 C.F.R. pt. 191.15 (1994).
387. *Id.* pt. 191.124. The applicable groundwater standards can be found at 40 C.F.R. Part 141 (1994).
388. *Id.* pt. 191.12.
389. Maggiore, New Mexico Environment Department, *Final Permit Issued to DOE for WIPP*, 4 (1999).
390. Department of Energy, "Waste Isolation Pilot Plant," *www.wipp.energy.gov/index.htm*.
391. Criteria for the Certification and Recertification of the Waste Isolation Pilot Plant's Compliance with the Disposal Regulations: Certification Decision, 63 Fed. Reg. 27,354 (May 18, 1998); Criteria for the Certification and Recertification of the Waste Isolation Pilot Plant's Compliance with the Disposal Regulations: Recertification Decision, 71 Fed. Reg. 18,010 (Apr. 10, 2006); EPA, "Radiation Protection," *www.epa.gov/radiation/wipp/*.
392. Maggiore, New Mexico Environment Department, *Final Permit Issued to DOE for WIPP*, 4 (1999); Carlsbad Field Office, DOE, "State Renews WIPP Facility Permit," press release (Dec. 1, 2010).
393. *City of Philadelphia v. New Jersey*, 437 U.S. 617 (1978).
394. Low-Level Radioactive Waste Policy Act of 1980, Pub. L. No. 96-573, 94 Stat. 3347, codified as amended at 42 U.S.C. § 2021b et seq. (2010); Government Accountability Office, *LLRW: Disposal Availability Adequate in the Short Term,* 1 (2004).

395. Low-Level Radioactive Waste Policy Act of 1980 § 4(a)(1)(A), 42 U.S.C. 2021d(a)(1)(A) (2006).
396. *Id.* at § 4(a)(2)(A), 42 U.S.C. § 2021d(a)(2)(A).
397. Low-Level Radioactive Waste Policy Amendments Act of 1985, Pub. L. No. 99-240, codified at 42 U.S.C. § 2021b et seq. (2006) at § 4(a)(2)(B). *See also* Chapter 4.
398. Chuang, "Who Should Win the Garbage Wars?" 2456 (2004).
399. Government Accountability Office, *LLRW: Disposal Availability Adequate in the Short Term,* 20 (2004).
400. Government Accountability Office, *LLRW: Status of Disposal Availability,* 5 (2008).
401. Advisory Committee on Nuclear Waste, NRC, *History and Framework of Low-Level Radioactive Waste Management,* 17 (2007).
402. Government Accountability Office, *LLRW: Disposal Availability Adequate in the Short Term,* 42 (2004).
403. *Id.* at 21.
404. C. Ramit Plushnick-Masti, "Commission Lets 36 States Dump Nuke Waste in Texas," Associated Press (Jan. 4, 2011).
405. Government Accountability Office, *LLRW: Status of Disposal Availability,* 1–2 (2008).
406. Nuclear Waste Policy Act of 1982, 42 U.S.C. §§ 10101–10270 (2006).
407. Statement of Sherwood H. Smith, Chairman and CEO, Carolina Power and Light Co., *Nuclear Waste Disposal: Joint Hearing before the House Committee on Energy and Resources and the Senate Committee on Environment and Public Works,* 97th Cong. 330 (1981).
408. "[NWPA] made clear the government's commitment to deal with a complex and controversial issue" Walker, *Road to Yucca,* 181 (2009) The bill was specifically designed to promote fairness among the states. Easterling, "Fair Rules for Siting a Repository," 461 n.19 (1992). The bill was not an unqualified success; Walker, *Road to Yucca,* 181 (2009), discusses doubts some groups had about the bill. But it did seem to strike a mostly satisfactory balance. NRC, for instance, approved of NWPA. "Although the NWPA is intrinsically incapable of resolving technical issues, it will establish necessary programs, milestones, and funding mechanisms to enable their resolution in the years ahead," NRC stated in its issuance of a preliminary finding of "confidence" that nuclear waste disposal needs could be met, grounded in the passage of the act in 1982. Waste Confidence Decision, 49 Fed. Reg. 34,658, at 34,659 (Aug. 31, 1984). *But see* Carter, *Nuclear Imperatives,* 228 (1987), which discusses reservations held by NRC commissioner Ahearne.
409. Nuclear Waste Policy Act of 1982 § 111(b), 42 U.S.C. § 10131(b) (2006).
410. *Id.* § 8(b), 42 U.S.C. § 10107(b). It is unclear whether defense SNF was also eligible for disposal in the commercial repository under the act; Section 8(b) does not mention defense SNF. *Id.*
411. New Mexico Radioactive Waste Consultation Task Force, New Mexico Energy, Minerals, and Natural Resources Department, "Chronology of WIPP," *www.emnrd.state.nm.us/WIPP/chronolo.htm*; Flynn and Slovic, "Yucca Mountain," 90 (1995).
412. Nuclear Waste Policy Act of 1982 § 114(a)(2)(A), as enacted, Pub. L. No. 97-425, 96 Stat. 2201, 2214 (1983).
413. Nuclear Waste Policy Act of 1982 § 302(a)(5)(B), 42 U.S.C. § 10222(a)(5)(B) (2006).
414. *Id.* § 114(d), 42 U.S.C. § 10134(d). The limit was determined by the tonnage of SNF disposed or, in the case of HLW, the tonnage of SNF from which the HLW derived. Thus if one ton of SNF were reprocessed, leaving 0.5 tons of solidified HLW reprocessing waste, the 0.5 tons of solidified HLW deposited in Yucca would be counted as 1 ton for the purposes of the limit. *Id.*
415. Senator Henry Jackson stated that "the 70,000 metric ton limit on disposal in the first repository will insure that a second repository is forthcoming. We must not make any one state feel that it is going to be the recipient of all the Nation's nuclear waste." 128 Cong. Rec. 32,568 (Dec. 20, 1982).
416. "After submission of the second such recommendation, the President may submit to the Congress recommendations for other sites, in accordance with the provisions of this subtitle." Nuclear Waste Policy Act of 1982 § 114(a)(2)(A), as enacted, Pub. L. No. 97-425, 96 Stat. 2201, 2214 (1982).
417. *Id.* § 116(a), 42 U.S.C. § 10136(a) (2006).
418. *Id.* § 112(b)(1)(A)–(B), 42 U.S.C. § 10132(b)(1)(A)–(B).
419. *Id.* § 112(c)(1)–(2), 42 U.S.C. § 10132(c)(1)–(2).
420. *Id.* § 114(a)(2)(A), as enacted, Pub. L. No. 97-425, 96 Stat. 2201, at 2214.
421. *Id.* § 116(b)(2), 42 U.S.C. § 10136(b)(2); *id.* § 135(d)(6)(D), 42 U.S.C. § 10155(d)(6)(D).
422. *Id.* § 302(a), 42 U.S.C. § 10222(a) (2006). Section 302 provides that "in return for the payment of fees established by this section, the Secretary, beginning not later than January 31, 1998, will dispose of the high-level radioactive waste or spent nuclear fuel involved as provided in this subtitle." *Id.* § 302(a)(5)(B), 42 U.S.C. § 10222(a)(5)(B).
423. 10 C.F.R. § 961.11(2010); Nuclear Waste Policy Act of 1982 § 302(a)(5)(B), 42 U.S.C. § 10222(a)(5)(B) (2006). Section 123 further provides that "delivery and acceptance by the Secretary, of [HLW or SNF] for a repository constructed under [the act] shall constitute a transfer to the Secretary of title to such [HLW or SNF]." *Id.*, 42 U.S.C. § 10143.
424. *See* Office of Technology Assessment, *Managing the Nation's Commercial High-Level Waste,* 102–103 (1985). The Office of Technology Assessment's 1985 report indicates that there was some debate over whether the act required the federal government to take title to waste in 1998 only if a functional repository existed. *Id.* at 125–126. It was also thought that to satisfy the statute and contracts, DOE need accept for initial disposal only a small amount of waste by the deadline. *Id.* at 4.

425. Nuclear Waste Policy Act of 1982 § 304, 42 U.S.C. § 10224 (2006).
426. Nuclear Waste Policy Act of 1982 § 121(b), 42 U.S.C. § 10141(b) (2006).
427. Nuclear Waste Policy Act of 1982 § 114(b), as enacted, Pub. L. No. 97-425, 96 Stat. 2201, at 2213 (1983). After the act as amended in 1987 statutorily designated Yucca Mountain as the site for the first repository, DOE was still required to submit the same three applications to NRC. Nuclear Waste Policy Act of 1982 § 114(b), 42 U.S.C. § 10134(b) (2006).
428. Nuclear Waste Policy Act of 1982 § 121(a), 42 U.S.C. § 10141(a) (2006). EPA promulgated radiation protection standards generally for all geologic repositories and standards specific to Yucca Mountain. 40 C.F.R. pts. 191, 197 (2009). EPA also promulgated certification and recertification criteria for WIPP. 40 C.F.R. pt. 194 (2009). Section 121 of the act requires EPA and NRC to promulgate standards only for repositories, not for MRS facilities. Nuclear Waste Policy Act of 1982 § 121(a)–(b), 42 U.S.C. § 10141(a) (2006). However, before sending a proposal to Congress for an MRS facility, DOE was required to consult with EPA and NRC and publish the results of the consultation. Id. § 141(b)(3), 42 U.S.C. § 10161(b)(3).
429. 10 C.F.R. pt. 960 (2010).
430. Id. § 960.3-1-1 to 3-1-2, 960.3-2-2-4.
431. The sites were located in Mississippi (two sites), Louisiana, Texas (two sites), Utah (two sites), Nevada, and Washington. They had previously been identified as promising candidates through a 1975 ERDA program. General Accounting Office, Institutional Relations under NWPA, 10 (1987). Given the short time frame (sixty days) within which DOE was required to identify candidate sites pursuant to NWPA Section 116(a), it has been argued that these nine ERDA sites were effectively grandfathered into the site selection process by NWPA. Carter, "The Path to Yucca Mountain and Beyond," 383 (2006); Cotton, "Nuclear Waste Story," 35 (2006).
432. According to a GAO report issued several years later, the lawsuits "generally involve[d] legal challenges to procedures DOE used to develop its siting guidelines, the contents of the siting guidelines, the decision by DOE to postpone site-specific activities, and the recommendation of first repository sites for detailed testing." General Accounting Office, NWPA Implementation as of 1987, 80 (1987). The report stated: "Representatives of the states and tribes involved in the first repository program say that if the program's credibility does not improve, they will continue to initiate lawsuits and can be expected to exercise their right to disapprove of the final site selection, forcing the courts, and perhaps ultimately the Congress, to judge whether DOE has adequately ensured the safe disposal of nuclear waste." Id. at 81.
433. Keeney, "Analysis of Portfolio of Sites for Repository," 196 (1987).
434. On the ranking of the five sites, Office of Civilian Radioactive Waste Management, DOE, *Recommendation by the Secretary of Energy of Candidate Sites for Site Characterization* (1986). On the relative scores, Frank Parker, distinguished professor of environmental and water resources, professor of engineering management, Vanderbilt University (former head of Radioactive Waste Disposal Research at Oak Ridge National Laboratory), telephone interview by Jane Stewart (Sep. 27, 2007). (Professor Parker was directly involved in the DOE siting process.)
435. Office of Civilian Radioactive Waste Management, DOE, *Recommendation by the Secretary of Energy of Candidate Sites for Site Characterization*, 9 (1986).
436. Robert D. Hershey Jr., "U.S. Suspends Plan for Nuclear Dump in East or Midwest," *New York Times*, A1 (May 29, 1986). For more information on the reasons for the decision, *see* Chapter 6.
437. Nuclear Waste Policy Act of 1982 § 141(b)(1)(B)–(C), 42 U.S.C. § 10161(b)(1)(B)–(C) (2006). NWPA required the secretary to submit a study on "the need for and feasibility of . . . the construction of one or more monitored retrievable storage facilities for high-level radioactive waste and spent nuclear fuel" and to submit a proposal to Congress, which Congress would then consider. Id. § 141(b)(1), 42 U.S.C. § 10161(b)(1). It did not, however, require the construction of an MRS.
438. In the event that an MRS was planned to be located within fifty miles of the first repository, waste stored therein would count toward the cap of 70,000 metric tons established for the first repository until a second repository was in operation. Nuclear Waste Policy Act of 1982 § 114(d), 42 U.S.C. § 10134(d) (2006).
439. Office of Civilian Waste Management, DOE, *Integrated Monitored Retrievable Storage/Repository Comparative Study* (1985).
440. See Chapter 6 for a full discussion of the failure of the federal MRS program.
441. Nuclear Waste Policy Act of 1982 § 135(a), 42 U.S.C. § 10155(a) (2006).
442. Id. § 136(a)(1), 42 U.S.C. § 10156(a)(1).
443. General Accounting Office, *Quarterly Report on DOE's Nuclear Waste Program as of December 31, 1986* at 7 (1987); statement of John W. Bartlett, Director, Office of Civilian Radioactive Waste Management, DOE, *Nuclear Waste Disposal Issues: Hearing before the Subcommittee on Nuclear Regulation of the Senate Committee on Environment and Public Works*, 102nd Cong. 48 (1991).
444. Nuclear Waste Policy Amendments Act of 1987 § 5011(a), Pub. L. No. 100-203, 101 Stat. 1330-227, 1330-228 (1987).
445. Id. § 5012. The 1987 Amendments required, however, that between 2007 and 2011, DOE report to Congress on the need for a second repository. Id. at 1330-231. DOE did so in 2008, saying that a second repository would indeed be necessary. Office of Civilian Radioactive Waste Management, DOE,

Report to the President on the Need for a Second Repository (2008).

446. Rissmiller, "Equality of Status," 104–105 (1993), discusses ways in which the NWPA was designed to promote geographic equality.
447. Susan Rasky, "The Nation: Nevada May End Up Holding the Bag," *New York Times*, 4 (Dec. 20, 1987); Rissmiller, "Equality of Status," 105–108 (1993).
448. Carter, *Nuclear Imperatives*, 201 (1987). Although the Louisiana site originally identified by DOE as one of nine potential sites was not among the final three recommended for characterization, all the sites were relatively equally ranked, and there was no assurance that DOE would not decide to revisit the other sites at a later time.
449. Easterling and Kunreuther, *Dilemma of Siting a Repository*, 42–43, 63n.45 (1995).
450. Carter, "The Path to Yucca Mountain and Beyond," 384 (2006), notes the relative political strength of Texas with respect to the Deaf Smith site and acknowledges that Nevada's weak political position was a factor in the decision to choose Yucca Mountain. Nevada's inability to wield any influence over the 1987 Amendments was partly because its representatives were not well placed on the relevant committees. Nevada senator Harry Reid's attempts to kill the provision on the floor were overcome by a cloture vote, and none of Nevada's representatives were invited to the conference committee that eventually singled out Yucca Mountain as the sole candidate site. Rissmiller, "Equality of Status," 107 (1993).
451. *See* Solomon, "HLW Management in the US" (2009).
452. Nuclear Waste Policy Amendments Act of 1987 § 5011(h)(1), Pub. L. No. 100-203, 101 Stat. 1330-227, 1330-229 (1987).
453. *Id.* § 5011(h)(2), Pub. L. No. 100-203, 101 Stat. 1330-227, 1330-230.
454. Nuclear Waste Policy Act of 1982 § 116(b)(2), 42 U.S.C. § 10136(b)(2) (2006); *id.* § 135(d)(6)(D), 42 U.S.C. § 10155(d)(6)(D).
455. Nuclear Waste Policy Amendments Act of 1987 § 5021, Pub. L. No. 100-203, 101 Stat. 1330-227, 1330-236 (1987).
456. *Id.* § 4021. Once a repository opened and had begun accepting waste, the capacity limit for the MRS increased to 15,000 MTHM. *Id.*
457. *Id.* § 4021, Pub. L. No. 100-203, 101 Stat. 1330-227, 1330-232.
458. Environmental Standards for the Management and Disposal of Spent Nuclear Fuel, High-Level and Transuranic Radioactive Wastes, 50 Fed. Reg. 38,065 (Sep. 19, 1985); 40 C.F.R. pt. 191 (1986). NWPA set a deadline of January 7, 1983, for EPA to promulgate generally applicable radiation standards as required by AEA. Nuclear Waste Policy Act of 1982 § 121(a), 42 U.S.C. § 10141(a) (2006). These standards applied to facilities for both civilian and defense wastes. Environmental Standards for the Management and Disposal of Spent Nuclear Fuel, High-Level and Transuranic Radioactive Wastes, 50 Fed. Reg. 38,065 (Sep. 19, 1985); 40 C.F.R. pt. 191 (1986).
459. The regulations limited the total annual exposure to any individual member of the public to twenty-five millirems for the whole body and seventy-five millirems to any critical organ. 40 C.F.R. § 191.03 (1986).
460. *Id.* §§ 191.12; 191.14; 191.15; 191.16. The regulations applied only to groundwater sources designated "special source[s] of groundwater." These were defined as "those Class 1 groundwaters identified in accordance with the agency's Ground-Water Protection Strategy published in August 1984 that: (1) Are within the controlled area . . . or are less than 5 kilometers beyond the controlled area; (2) are supplying drinking water for thousands of persons . . . (3) are irreplaceable in that no reasonable alternative source of drinking water is available to that population." *Id.* § 191.12(n).
461. *NRDC v. Envtl. Prot. Agency*, 824 F.2d 1258 (1st Cir. 1987). The court found that NWPA allowed contamination of waters within the controlled area, but that the Safe Drinking Water Act still applied outside the controlled area, and EPA's NWPA standards for directly adjacent groundwaters are less stringent than those found in EPA's SDWA regulations. *Id.* at 1276–1277.
462. *Id.* at 1293–1294.
463. Waste Isolation Pilot Plant Withdrawal Act § 8(b)(1)–(2), Pub. L. No. 102-579, 106 Stat. 4777, 4786 (1992).
464. Committee on Interior and Insular Affairs, *Comprehensive National Energy Policy*, 110 (1991).
465. *See* Cochran, NRDC, *How Safe Is Yucca Mountain?* (2008); *see also* Cochran and Fettus, "NRDC's Perspective on the Nuclear Waste Dilemma" (2010).
466. Energy Policy Act of 1992 § 801, Pub. L. No. 102-486, 106 Stat. 2776, 2921 (1992); Conference Committee, *Energy Policy Act of 1991: Conference Report (to accompany H.R. 776)*, 390 (1991).
467. Holt, Congressional Research Service, *Civilian Nuclear Waste Disposal*, 11 (2009).
468. 138 Cong. Rec. H11420 (Oct. 5, 1992) (Statement of Representative James Bilbray).
469. *Id.* (Statement of Representative Barbara Vucanovich). "The conferees propose to make this call, because the nuclear power lobby is concerned that Yucca Mountain, NV, might not qualify as a repository unless the rules are relaxed quite a bit." *Id.*
470. Cochran, NRDC, *How Safe Is Yucca Mountain?* 8 (2008).
471. Committee on Technical Bases for Yucca Mountain Standards, National Research Council of the National Academies, *Technical Bases for Yucca Mountain Standards* (1995).
472. Public Health and Environmental Radiation Protection Standards for Yucca Mountain, NV, 66 Fed. Reg. 32,074 (Jun. 13, 2001); 40 C.F.R. pt. 197 (2001).
473. 40 C.F.R. § 197.4 (2001).

474. *Id.* § 197.20.
475. *Id.* § 197.21.
476. The human intrusion standard required that if DOE determines that within 10,000 years, human intrusion by drilling will occur without recognition by the drillers, DOE must assure that the "reasonably maximally exposed individual" will not receive an annual dose higher than fifteen millirems as a result. If the intrusion is not projected to occur within ten thousand years, DOE is required only to include an analysis of this risk in the EIS for Yucca. *Id.* § 197.25.
477. *Id.* § 197.30.
478. *Id.* § 197.12.
479. *Id.* § 197.36.
480. *Nuclear Energy Inst. v. Envtl. Prot. Agency*, 373 F.3d 1251 (D.C. Cir. 2004). *See* Chapter 6 for a discussion of other challenges in the case, including a challenge to EPA's controlled area.
481. DOE, *Recommendation by the Secretary of Energy Regarding the Suitability of Yucca Mountain* (2002); President George W. Bush, Letter to Congressional Leaders Recommending the Yucca Mountain Site for the Disposal of Spent Nuclear Fuel and Nuclear Waste, 1 Pub. Papers 234 (Feb. 15, 2002).
482. Kenny C. Guinn, "Statement of Reasons Supporting the Governor of Nevada's Notice of Disapproval of the Proposed Yucca Mountain Project" (Apr. 8, 2002), www.state.nv.us/nucwaste/news2002/nn11650.pdf.
483. Yucca Mountain Repository Site Approval Act, Pub. L. No. 107-200, 116 Stat. 735 (2002), "Approving the Site at Yucca Mountain, Nevada, for the Development of a Repository for the Disposal of High-Level Radioactive Waste and Spent Nuclear Fuel, Pursuant to the Nuclear Waste Policy Act of 1982."
484. DOE, *Yucca Mountain Repository License Application for Construction Authorization* (2008). *See* Chapter 6 for a discussion of the lawsuits brought by Nevada and its denial of a water permit to DOE.
485. NRC, "Fact Sheet on Licensing Yucca Mountain," www.nrc.gov/reading-rm/doc-collections/fact-sheets/fs-yucca-license-review.html.
486. *Nuclear Energy Inst. v. Envtl. Prot. Agency*, 373 F.3d 1251, 1261-62 (D.C. Cir. 2004).
487. Committee on Technical Bases for Yucca Mountain Standards, National Research Council of the National Academies, *Technical Bases for Yucca Mountain Standards*, 6–7, 55–56 (1995).
488. *Nuclear Energy Inst. v. Envtl. Prot. Agency*, 373 F.3d 1251, 1273 (D.C. Cir. 2004).
489. *Id.* at 1315. The court held that, with the exception of the incorporated EPA ten-thousand-year compliance period, NRC's licensing requirements were lawful.
490. Public Health and Environmental Radiation Protection Standards for Yucca Mountain, NV, 73 Fed. Reg. 61,255 (Oct. 15, 2008); 40 C.F.R. pt. 197 (2010).
491. 40 C.F.R. § 197.12 (2009).
492. *Id.* § 197.20.
493. *Id.* § 197.25.
494. *Id.* § 197.36.
495. Nuclear Waste Policy Amendments Act of 1987 § 5011(j), Pub. L. No. 100-203, 101 Stat. 1330-227, 1330-230 (1987).
496. Lisa Mascaro, "Once Flatlining, Now on Life Support," *Las Vegas Sun*, 1 (Feb. 10, 2009).
497. Steve Tetrault, "Yucca Funding at Lowest Level Ever," *Las Vegas Review-Journal*, B3 (May 7, 2009).
498. Steve Tetrault, "Chu Promises to Develop Yucca Alternative Quickly," *Las Vegas Review-Journal*, B1 (Mar. 12, 2009).
499. Office of Management and Budget, *Budget of the U.S. Govt., FY 2011 Budget DOE Overview*, 437 (2010).
500. Motion to Stay the Proceedings, *U.S. Dep't of Energy (High-Level Waste Repository)*, No. 63-001-HLW (ASLB, Nuclear Regulatory Comm'n Feb. 1, 2010).
501. Motion to Withdraw, *U.S. Dep't of Energy (High-Level Waste Repository)*, No. 63-001-HLW (ASLB, Nuclear Regulatory Comm'n, Mar. 3, 2010).
502. U.S. Dep't of Energy's Reply to the Motion to Withdraw at 31n.102, *U.S. Dep't of Energy (High Level Waste Repository)*, No. 63-001-HLW (ASLB, Nuclear Regulatory Comm'n, May 27, 2010).
503. Nuclear Waste Policy Act of 1982 § 114, 42 U.S.C. § 10134 (2006).
504. *U.S. Dep't of Energy (High-Level Waste Repository)*, slip op. at 3, No. 63-001 (ASLB Nuclear Regulatory Comm'n, June 29, 2010).
505. *U.S. Dep't of Energy (High-Level Waste Repository)*, slip op. at 3, No. 63-001-HLW (ASLB, Nuclear Regulatory Comm'n, June 29, 2010); U.S. Dep't of Energy's Reply to the Motion to Withdraw at 31n.102, *U.S. Dep't of Energy (High-Level Waste Repository)*, No. 63-001-HLW (ASLB, Nuclear Regulatory Comm'n, May 27, 2010).
506. *In re Aiken County*, No. 10-1050 (D.C. Cir. filed Feb. 19, 2010); *Ferguson v. U.S. Dep't of Energy*, No. 10-1052 (D.C. Cir., filed Feb. 25, 2010); *South Carolina v. U.S. Dep't of Energy*, No. 10-1069 (D.C. Cir., transferred Mar. 25, 2010); *Washington v. Dep't of Energy*, No. 10-1082 (D.C. Cir., filed Apr. 13, 2010). These cases were consolidated as *In re Aiken County*, No. 10-1050.
507. Order, *In re Aiken County*, No. 10-1050 (D.C. Cir., filed Jan. 10, 2011).
508. State of the Union Address by President Obama, 156 Cong. Rec. H414 (daily ed. Jan. 27, 2010).
509. Julie Pace, "Obama Touts Nuclear Loan Guarantee as Just a Start," Associated Press (Feb. 16, 2010).
510. GreenHouse Blog, "Obama's Nuclear Loan Guarantees Draw Opposition," *USA Today* (Feb. 1, 2010), content.usatoday.com/communities/greenhouse/post/2010/02/obamas-nuclear-loan-guarantees-draw-broad-opposition/1.
511. Mark Holt, Congressional Research Service, "Nuclear Waste Policy: How We Got Here," presentation, 33 (Mar. 25, 2010). brc.gov/pdfFiles/

CRS_BlueRibbonCommissionWastePolicyHistory .pdf (62,500 MTHM); John Kessler, Electric Power Research Institute, "Used Nuclear Fuel: Inventory Projections," presentation to the Blue Ribbon Commission, 7 (Aug. 19, 2010) (65,000 MTHM). *See also* Van Ness Feldman, P.C., *Federal Commitments Regarding Used Fuel and High-Level Wastes*, 1 (2010), *brc.gov/commissioned_papers.html*.

512. Mark Holt, Congressional Research Service, "Nuclear Waste Policy: How We Got Here," presentation, 40 (Mar. 25, 2010), brc.gov/pdfFiles/ CRS_BlueRibbonCommissionWastePolicyHistory .pdf. There are 104 operating commercial nuclear reactors at sixty-five sites across the country. In addition, there are nine decommissioned reactor sites that store SNF. Also, commercial SNF is stored at the uncompleted commercial reprocessing site in Morris, Illinois, and at two DOE sites. NRC, "Operating Nuclear Power Reactors by Location or Name," www.nrc.gov/info-finder/reactor; Office of Civilian Radioactive Waste Management, DOE, *Report to Congress on Demonstration of Interim Storage of Fuel from Decommissioned Reactors*, 1 (2008); Nuclear Energy Study Group, American Physical Society, *Consolidated Interim Storage of Commercial Spent Nuclear Fuel*, 6 tbl.1 (2007).

513. Mark Holt, Congressional Research Service, "Nuclear Waste Policy: How We Got Here," presentation, 33 (Mar. 25, 2010), brc.gov/pdfFiles/ CRS_BlueRibbonCommissionWastePolicyHistory.pdf.

514. A steady generation rate assumes all existing reactors request extensions from NRC to operate for sixty years. This also does not take into account any new reactors that may be licensed by NRC, though the nuclear utilities had told NRC they intended to file license applications for thirty-four new nuclear reactors by the end of 2010. *See* Office of Civilian Radioactive Waste Management, DOE, *Report to the President on the Need for a Second Repository*, 2 (2008).

515. Idaho National Laboratory, "2008 DOE Spent Nuclear Fuel and High Level Waste Inventory," *inlportal.inl.gov/portal/server.pt?tbb=siteindex*. This includes 2,129 MTHM at Hanford, 280 MTHM at INL, 28 MTHM at SRS, 15 MTHM at Fort St. Vrain, and 3 MTHM at other sites.

516. *Id.*; Office of Environmental Management, "DOE Spent Nuclear Fuel," www.em.doe.gov/pages/ spentfuel.aspx; Andrews, Congressional Research Service, *Spent Nuclear Fuel Storage Locations and Inventory*, 2 (2004).

517. Idaho National Laboratory, "2008 DOE Spent Nuclear Fuel and High Level Waste Inventory," *inlportal.inl.gov/portal/server.pt?tbb=siteindex*; Office of Environmental Management, DOE, *Status of Environmental Management Initiatives to Accelerate the Reduction of Environmental Risks and Challenges Posed by the Legacy of the Cold War*, 24 (2009).

518. Office of Environmental Management, DOE, Status of Environmental Management Initiatives to Accelerate the Reduction of Environmental Risks and Challenges Posed by the Legacy of the Cold War, 29 (2009); Office of Civilian Radioactive Waste Management, DOE, *Report to the President on the Need for a Second Repository*, 2 (2008). For difficulties in conversion to MTHM, *see* Knecht et al., INL, Options for Determining Equivalent MTHM for DOE High-Level Waste (1999) (discussing the four methods of conversion).

519. Blue Ribbon Commission on America's Nuclear Future: Memorandum for the Secretary of Energy, 75 Fed. Reg. 5,485, at 5,485 (Feb. 3, 2010). The commission is chaired by former congressman Lee Hamilton and former national security advisor Brent Scowcroft. Its other members are Mark Ayers, Vicky Bailey, Albert Carnesale, Pete V. Domenici, Susan Eisenhower, Chuck Hagel, Jonathan Lash, Allison Macfarlane, Richard A. Meserve, Ernie Moniz, Per Peterson, John Rowe, and Phil Sharp.

520. Nuclear Waste Policy Act of 1982 § 161(a)–(b), 42 U.S.C. § 10172a (2006).

521. Office of Civilian Radioactive Waste Management, DOE, *Report to the President on the Need for a Second Repository*, 2 (2008). This statutory cap is still in place.

Following President Reagan's 1985 decision, pursuant to Section 8 of NWPA, to use the first repository for disposal of defense HLW/SNF, DOE established a policy to allocate 90 percent of the repository's capacity to civilian SNF and 10 percent for military HLW. This means that sixty-three thousand MTHM of Yucca's statutorily limited capacity is allocated to civilian waste and seven thousand MTHM to defense wastes.

522. *Id.*

523. A 2009 survey found that two-thirds of people around the world believe that the use of nuclear power should increase in their countries. Accenture, "Consumers Warm to Nuclear Power in Fight against Fossil Fuel Dependency, Accenture Survey Finds," press release (Mar. 17, 2009). The 2011 Japan nuclear crisis may change this.

524. Five applications for six new power plants have been suspended for reasons unrelated to the NRC regulatory process. Kristine L. Svinicki, Commissioner, U.S. Nuclear Regulatory Commission, "State of the Nuclear Renaissance— A Regulatory Perspective," speech (Jun. 8, 2010).

525. Joskow and Parsons, "The Economic Future of Nuclear Power," 51 (2009).

526. *Id.*

527. Hertzberg, "Some Nukes," 19 (2010).

528. Socolow and Glaser, "Balancing Risks," 35 (2009). The indirect emissions associated with nuclear power are "nuclear plant construction, operation, uranium mining and milling, and plant decommissioning." Sovacool, "Greenhouse Gas Emissions from Nuclear Power," 2951 (2008).

529. Feiveson, "A Skeptic's View of Nuclear Energy," 67 (2009); Socolow and Glaser, "Balancing Risks," 32 (2009); Gronlund, Lochbaum and Lyman, Union of Concerned Scientists, *Nuclear Power in a Warming World*, 11–12 (2007).

530. Socolow and Pacala, "Stabilization Wedges: Solving the Climate Change Problem for the Next 50 Years with Current Technologies," 969 (2004); Socolow and Glaser, "Balancing Risks," 31 (Fall 2009).
531. Tom Meersman, "Supporters of Nuclear Power Push to Lift Ban," *Minneapolis Star Tribune* (Mar. 3, 2010); Wisconsin Legislative Reference Bureau, *Rethinking the Moratorium on Nuclear Energy* (2006).
532. On improved operating efficiency, Joskow and Parsons, "The Economic Future of Nuclear Power," 46 (2009). NRC allows extensions of up to twenty years past the expiration of the original license, with the total length of the renewal not to exceed forty years. Those standards can be found at Issuance of a Renewed License, 10 C.F.R. § 54.31 (2010).
533. DOE, *Generation IV Nuclear Energy Systems Initiatives*, 2 (2006).
534. Gronlund, Lochbaum, and Lyman, Union of Concerned Scientists, *Nuclear Power in a Warming World*, 1 (2007); Natural Resources Defense Council, *Position Paper: Commercial Nuclear Power*, 10–15 (2005).
535. Matthew L. Wald, "Vermont Senate Votes to Close Nuclear Power Plant," *New York Times*, A14 (Feb. 24, 2010).
536. Joskow and Parsons, "Economic Future of Nuclear Power," 52 (2009).
537. Without adding a charge for carbon dioxide emissions, Joskow and Parsons calculate that the levelized cost of nuclear energy from new plants is 8.4 cents per kilowatt-hour, which is more expensive than coal in a low- (5.2), moderate- (6.2), and high- (7.2) priced scenario and more expensive than gas in a low- (4.2) and moderate- (6.5) priced scenario; only gas in a high-priced scenario (8.7) would cost more than nuclear power. *Id.*
538. For the most recent Energy Information Agency study of subsidies, *see* Energy Information Agency, DOE, *Federal Financial Interventions and Subsidies in Energy Markets*, xii (2008).
539. Feiveson, "A Skeptic's View of Nuclear Energy," 65 (2009); Accenture, "Consumers Warm to Nuclear Power in Fight against Fossil Fuel Dependency, Accenture Survey Finds," press release (Mar. 17, 2009).
540. Feiveson, "A Skeptic's View of Nuclear Energy," 53 (2009).
541. Nuclear Waste Policy Act of 1982 § 302, as enacted, 97 Pub. L. 425, 96 Stat. 2201, 2257–2261 (1982). Section 302 effectively makes entry into such contracts by utilities mandatory by prohibiting NRC from issuing licenses to an operator who had not entered into such a contract or was not actively negotiating such a contract with the secretary of energy. *Maine Yankee Atomic Power Co. v. United States*, 225 F.3d 1336, 1337 (Fed. Cir. 2000). Pursuant to section 302, DOE issued a Standard Contract, which can be found at 10 C.F.R. § 961.11 (2010).
542. Statement of Kim Cawley, Unit Chief, Natural and Physical Resources Cost Estimates Unit, Congressional Budget Office, *Budgeting for Nuclear Waste Management: Hearing before the House Committee on the Budget*, 111th Cong. (2009).
543. Nuclear Waste Policy Act of 1982 § 302(e)(2), as enacted, 97 Pub. L. 425, 96 Stat. 2201, 2260 (1982).
544. Statement of Edward F. Sproat III, Director, Office of Civilian Radioactive Waste Management, DOE, *Issues in Federal Government Financial Liabilities–Commercial Nuclear Waste: Hearing before the House Committee on the Budget*, 110th Cong. 7–8 (2007).
545. Institute for 21st Century Energy, U.S. Chamber of Commerce, *Revisiting America's Nuclear Waste Policy*, 10 (2009).
546. Standard Contract, 10 C.F.R. § 961.11 (2010). The contract also provides that in the event of delays in the delivery, acceptance, or transport of waste, the charges will be equitably adjusted.
547. Notice of Inquiry: Waste Acceptance Issues, 59 Fed. Reg. 27,007, at 27,008 (May 25, 1994).
548. *Id*. at 27,008–27,009. One congressional observer had summed up the practical nature of DOE's commitment in 1985: "The utility and nuclear industries assume that the NWPA of 1982 commits the Federal government to accept SNF for disposal beginning in 1998. *DOE reiterates this commitment at every opportunity*." (Emphasis added.) Benjamin Cooper, Senate Committee on Energy and Natural Resources, "Monitored Retrievable Storage," 378 (1985).
549. *See N. States Power Co. v. U.S. Dep't of Energy*, 128 F.3d 754, 760 (D.C. Cir. 1997); *Neb. Pub. Power Dist. v. United States*, 590 F.3d 1357, 1376 (Fed. Cir. 2010).
550. *Maine Yankee Atomic Power Co. v. United States*, 225 F.3d 1336, 1342 (Fed. Cir. 2000); *Commonwealth Edison Co. v. United States*, No. 98-483 (Fed. Cl. 2001); *Fla. Power & Light Co. v. United States*, No. 98-486 (Fed. Cl. 2002); *Wis. Elec. Power Co. v. United States*, No. 00-697 (Fed. Cl. 2004); *Tenn. Valley Auth. v. United States*, 60 Fed. Cl. 665, 679 (Fed. Cl. 2004); *S. Nuclear Operating Co. v. United States*, No. 98-614 (Fed. Cl. 2004); *Sacramento Mun. Util. Dist. v. United States*, 65 Fed. Cl. 180, 181 (Fed. Cl. 2005). *Sys. Fuel Inc. v. United States*, 79 Fed. Cl. 37, 125–126 (2007) required DOE to compensate a utility for costs of expanding on-site SNF storage. The U.S. Court of Federal Claims recently awarded more than $50 million to Wisconsin Electric Power Company for the partial breach by DOE of its contract to accept and dispose of wastes from the company's Point Beach Nuclear Power Plant. *Wis. Elec. Power Co. v. United States*, 90 Fed. Cl. 714, 803 (2009). Damages awarded through settlements typically include an initial payment to compensate for past storage costs as well as annual reimbursement for future storage costs. Damages awarded by courts through litigation cover only past costs, so utilities must continue filing claims as they accrue more costs. Holt, Congressional Research Service, *Civilian Nuclear Waste Disposal*, 4 (2008). *See also* Thies, "The Decline of the Court

of Federal Claims" (2010), for discussion of recent court decisions finding that NWPA unconditionally required DOE to accept SNF from utilities by Jan. 31, 1998, and rejecting DOE's argument that failure to accept waste from the utilities resulted from "unavoidable delays" that would absolve the government of liability.

551. *Sacramento Mun. Util. Dist. v. United States*, 70 Fed. Cl. 332, 378 (Fed. Cl. 2006).
552. For example, *see Sys. Fuels, Inc. v. United States,* 79 Fed. Cl. 37, 74 (2007).
553. Statement of Michael Hertz, Deputy Assistant Attorney General, Department of Justice, *Budgeting for Nuclear Waste Management: Hearing before the House Committee on the Budget*, 111th Cong. (2009).
554. See note 557.
555. This is the cost only to the government, paid out of the general Treasury Judgment Fund.
556. In *Ala. Power Co. v. Dep't of Energy*, 307 F.3d 1300, 1315 (11th Cir. 2002), the court held that the claim could not be settled through fee adjustments.
557. Martha Coyle, "Breach Cases Could Cost U.S. Government Billions," *National Law Journal* (Sep. 8, 2008), www.law.com/jsp/article.jsp?id=1202424340326.
558. Holt, Congressional Research Service, *Civilian Nuclear Waste Disposal*, 4 (2009); Government Accountability Office, *Challenges and Alternatives to Yucca*, 16 (2009).
559. *Sacramento Mun. Util. Dist. v. United States*, 70 Fed. Cl. 332, 378 (Fed. Cl. 2006).
560. Rebating America's Deposits Act, S. 861, 111th Cong. (2009).
561. Steve Tetrault, "States Resist Adding to Nuclear Waste Construction Fund," *Las Vegas Review-Journal*, A7 (Apr 9, 2009).
562. Jones, "FY 2010 Energy and Water Development Appropriations Bill: Nuclear Waste" (2009).
563. Letter from the National Association of Regulatory Utility Commissioners to Steven Chu, Secretary, Department of Energy (Jul. 8, 2009), as reprinted at *www.naruc.org/Testimony/NARUC percent20ltr percent20on percent20NWF percent20fees percent20070809.pdf*.
564. Ben German, "Utilities Sue Energy Department to Stop Nuclear Waste Management Fees," *The Hill* (Mar. 4, 2010).
565. While DOE has promised to remove HLW and SNF from various sites around the country, the only agreements with deadlines are a settlement with Idaho to remove some HLW and all SNF by January 1, 2035, and a commitment to Colorado to remove SNF from the state by the same date. Frank Marcinowski, Office of Environmental Management, DOE, "Overview of DOE's Spent Nuclear Fuel & High-Level Waste," presentation, 5, 8 (Mar. 25, 2010), brc.gov/pdfFiles/Environmental_Management_BRC_03252010.pdf.
566. Mark Holt, Congressional Research Service, "Nuclear Waste Policy: How We Got Here," presentation, 40 (Mar. 25, 2010), brc.gov/pdfFiles/ *CRS_BlueRibbonCommissionWastePolicyHistory .pdf*. These sites are: sixty-five sites comprising 104 operating commercial nuclear reactors, nine decommissioned reactor sites, the uncompleted commercial reprocessing site in Morris, Illinois, and two DOE sites. NRC, "Operating Nuclear Power Reactors by Location or Name," *www.nrc.gov/info-finder/reactor*; Office of Civilian Radioactive Waste Management, DOE, Report to Congress on Demonstration of Interim Storage of Fuel from Decommissioned Reactors, 1 (2008); Nuclear Energy Study Group, American Physical Society, Consolidated Interim Storage of Commercial Spent Nuclear Fuel, 6 tbl.1 (2007).
567. Institute for 21st Century Energy, U.S. Chamber of Commerce, *Revisiting America's Nuclear Waste Policy*, 4, 8 (2009).
568. General Accounting Office, *Spent Nuclear Fuel*, 13–16 (2003). NRC stated that several factors combine to make a pool fire unlikely, including pool design and construction and the long time required to heat up the fuel. The Japan nuclear crisis has led NRC to reevaluate the issue. *Id.* at 17.
569. Gronlund, Lochbaum, and Lyman, Union of Concerned Scientists, *Nuclear Power in a Warming World*, 4 (2007).
570. *Id.* at 47.
571. Board on Radioactive Waste Management, National Research Council of the National Academies, *Safety and Security of Commercial SNF Storage*, 71 (2006).
572. Institute for 21st Century Energy, U.S. Chamber of Commerce, *Revisiting America's Nuclear Waste Policy*, 5 (2009).
573. Nuclear Regulatory Commission, *Fact Sheet on Dry Cask Storage of Spent Nuclear Fuel*, www.nrc.gov/reading-rm/doc-collections/fact-sheets/dry-cask-storage.html; Nuclear Regulatory Commission, *Pilot Probabilistic Risk Assessment of a Dry Cask Storage System*, 2–3 (2006). Experiments have shown that dry storage casks are capable of resisting crashing airplanes, armor-piercing rounds, and high explosives. General Accounting Office, *Spent Nuclear Fuel*, 16–17 (2003).
574. 10 C.F.R. § 72.42 (2010). NRC is currently contemplating extending licenses for an additional twenty to one hundred years, although it has yet to make a determination on the issue. Keanneally and Kessler, "Behavior of Spent Fuel in Dry Cask Storage Systems," 1 (2009).
575. Peter Behr, "The Administration Puts Its Own Stamp on a Possible Nuclear Revival," *New York Times* (Feb. 2, 2010).
576. Nuclear Regulatory Commission, *Fact Sheet on Dry Cask Storage of Spent Nuclear Fuel*, www.nrc.gov/reading-rm/doc-collections/fact-sheets/dry-cask-storage.html.
577. Office of Civilian Radioactive Waste Management, DOE, *Report to Congress on Demonstration of Interim Storage of Fuel from Decommissioned Reactors*, 15–16 (2008).
578. Cotton, "Nuclear Waste Story," 33–37 (2006).

579. NWPA appears to allow private parties to establish consolidated SNF storage facilities (Nuclear Waste Policy Act of 1982 § 131, 42 U.S.C. § 10151); NRC, "NRC Issues License to Private Fuel Storage for Spent Nuclear Fuel Storage Facility In Utah," press release (Feb. 22, 2006).
580. Office of Senator Bob Bennett, "Bennett Hails News That PFS Loses Final Push to Bring Nuclear Waste to Utah," press release (Sep. 7, 2006); Paul Foy, "Feds Reject Plan to Create Utah Nuclear Waste Stockpile," Associated Press (Sep. 8, 2006); *Devia v. Nuclear Regulatory Comm'n*, 492 F.3d 421, 423 (D.C. Cir. 2007).
581. *Skull Valley Band of Goshute Indians v. Cason*, No. 2:07cv00526 (D. Utah 2007).
582. *Skull Valley Band of Goshute Indians v. Davis*, No. 07-cv-0526-DME-DON, slip op. at 1 (D. Utah July 26, 2010).
583. Clay Sell, Deputy Secretary, DOE, "Announcing the Global Nuclear Energy Partnership," press briefing (Feb. 6, 2006), *www.energy.gov/print/3171.htm*.
584. France and Russia currently operate reprocessing facilities. China and Japan (which in the past had sent SNF to be reprocessed in France or the United Kingdom) are building facilities. The United Kingdom is in the process of closing its reprocessing plant, primarily on economic grounds. Belgium, Finland, Sweden, Germany, and Switzerland at various points in the past sent SNF for reprocessing to one or more of the countries with facilities (France, Russia, United Kingdom) but no longer do so.
585. Institute for Policy Studies, *Radioactive Wastes and GNEP*, 8 (2007).
586. For criticism from experts, Committee on Review of DOE's Nuclear Energy Research and Development Program, National Research Council of the National Academies, *Review of DOE's Nuclear R&D Program*, 47–57 (2008).
587. Rob Pavey, "Nuclear Reuse Initiative Scrapped," *Augusta [Georgia] Chronicle*, B1 (Apr. 16, 2009).
588. Notice of Cancellation of Environmental Impact Statement Process, 74 Fed. Reg. 31,018 (Jun. 29, 2009).
589. Statement of Dr. Steven Chu, Secretary of Energy, Department of Energy, *New Directions for Energy Research and Development at the U.S. Department of Energy: Hearing before the House Committee on Science and Technology*, 111th Cong. 30–32 (2009).
590. Institute for Policy Studies, *Radioactive Wastes and GNEP*, 3 (2007); von Hippel, *Managing Spent Fuel*, 12 (2007); Government Accountability Office, *Global Nuclear Energy Partnership*, 15 (2008).
591. Global Nuclear Energy Partnership, *GNEP Action Plan*, 2 (2007).
592. Institute for Policy Studies, *Radioactive Wastes and GNEP*, 8 (2007).
593. Global Nuclear Energy Partnership, *GNEP Action Plan*, 1 (2007). These countries include Russia, the United Kingdom, France, and Japan.
594. Office of Fuel Cycle Management, DOE, *Global Nuclear Energy Partnership Strategic Plan*, 5 (2007).
595. Clay Sell, Deputy Secretary, DOE, "Announcing the Global Nuclear Energy Partnership," press briefing (Feb. 6, 2006), *www.energy.gov/print/3171.htm*.
596. Gronlund, Lochbaum, and Lyman, Union of Concerned Scientists, *Nuclear Power in a Warming World*, 68–74 (2007); Thomas B. Cochran, Director, Nuclear Program, Natural Resources Defense Council, statement on "Environmental, Safety, and Economic Implications of Nuclear Power" (2007); Committee on Review of DOE's Nuclear Energy Research and Development Program, National Research Council of the National Academies, *Review of DOE's Nuclear Energy Research and Development Program*, 4–6 (2008). In 1996, an NAS committee estimated the cost of technologies like those proposed for GNEP could range from $50 to $100 billion. Committee on Separations Technology and Transmutation Systems, National Research Council of the National Academies, *Nuclear Wastes: Technologies for Separation and Transmutation*, 7 (1996).
597. Committee on Review of DOE's Nuclear Energy Research and Development Program, National Research Council of the National Academies, *Review of DOE's Nuclear Energy Research and Development Program*, 5 (2008).
598. Government Accountability Office, *Global Nuclear Energy Partnership*, 34–35 (2008).
599. For example, the Duncan Hunter National Defense Authorization Act for Fiscal Year 2009 § 3117, 110 P.L. 417, 122 Stat. 4356, 4758 (2008), mandates that of the amounts authorized to be appropriated for FY 2009 by section 3101(a)(2) for defense nuclear nonproliferation activities, not more than $3 million "may be used for projects that are specifically designed for the Global Nuclear Energy Partnership."
600. On proliferation risks, Gronlund, Lochbaum, and Lyman, Union of Concerned Scientists, *Nuclear Power in a Warming World*, 1 (2007); Natural Resources Defense Council, *Position Paper: Commercial Nuclear Power*, 10–15 (2005). Although GNEP's proposed reprocessing method would not produce pure plutonium, critics warned that "the separation technologies proposed under GNEP are far from being 'proliferation-resistant' and, in fact, would significantly reduce the time needed to acquire pure plutonium," asserting that they would significantly reduce the number of reprocessing steps those seeking to extract plutonium would have to perform on their own. Lindemyer, "GNEP Proliferation Concerns," 89.
601. Steve Christ, "Investing in Uranium: America's 'Other' Energy Crisis," *Wealth Daily* (Aug. 14, 2008). "Some 44 percent of the reactors in the United States are fueled in part with cheap uranium coming from the Megatons to Megawatts programme [an agreement with Russia to import downblended Russian weapons uranium for use as fuel]." Charles Digges, "US OKs Some Uranium Imports from Russia—But Figures Set to Jump within Decade,"

(Aug. 12, 2007), *www.bellona.org/articles/articles_2007/Uranium_USimports*.
602. Organisation for Economic Co-operation and Development Nuclear Energy Agency, International Atomic Energy Agency, *Uranium 2007*, 82–84 (2008).
603. Gronlund, Lochbaum, and Lyman, Union of Concerned Scientists, *Nuclear Power in a Warming World*, 49, 51 (2007). *See* discussion of GNEP proposal in Chapter 7 here.
604. Institute for Policy Studies, *Radioactive Wastes and GNEP*, 14–15 (2007).
605. *See* Union of Concerned Scientists, "Reprocessing and Nuclear Waste" (2009), *www.ucsusa.org/nuclear_weapons_and_global_security/nuclear_terrorism/technical_issues/reprocessing-and-nuclear.html*.
606. Office of Nuclear Energy, DOE, *Draft GNEP Programmatic EIS*, at 4-46, 4-54 (2008).
607. The *Draft GNEP Programmatic EIS* estimates total volume at 158,000 MTHM. *Id.* Most sources assume a generation rate of 2,000 MTHM per year, resulting in 100,000 MTHM over fifty years. *See, e.g.*, Mark Holt, Congressional Research Service, "Nuclear Waste Policy: How We Got Here," presentation, 40 (Mar. 25, 2010), *brc.gov/pdfFiles/CRS_BlueRibbonCommissionWastePolicyHistory.pdf* (estimating rate of generation at 2,000 MTHM).
608. NRC regulations "require disposal of [GTCC] wastes in a deep geologic repository unless disposal elsewhere has been approved by the Commission." 54 Fed. Reg. 22,578 (May 25, 1989). DOE is considering several alternatives for disposal of GTCC waste, including reclassifying it as HLW so it could be disposed of at a federal repository under NWPA. Notice of Intent to Prepare an Environmental Impact Statement for the Disposal of Greater-Than-Class-C Low-Level Radioactive Waste, 72 Fed. Reg. 40,135, at 40,138 (Jul. 23, 2007).
609. Office of Nuclear Energy, DOE, *Draft GNEP Programmatic EIS*, at 4-139 (2008). DOE estimates that reprocessing plans will increase the latent cancer fatalities attributable to transportation by 66 percent to 584 percent. *Id.*
610. *See, e.g.*, Schneider and Marignac, International Panel on Fissile Materials, *Spent Nuclear Fuel Reprocessing in France*, 8–11 (2008); Bunn et al., Harvard Kennedy School, *The Economics of Reprocessing* (2003).
611. Bunn et al., Harvard Kennedy School, *The Economics of Reprocessing*, ix, 87 (2003).
612. Wang Ying, John Duce, and Anna Stablum, "Uranium Bottoming as China Boosts Stockpiles," *Bloomberg News* (Jul. 12, 2010), quotes uranium prices at $41.75 per pound, or $91.85 per kilo on July 5, 2010.
613. Schneider and Marignac, International Panel on Fissile Materials, *Reprocessing in France*, 8–11 (2008).
614. *Id.* at 10. In 2003, the French government's General Directorate of Energy and Primary Materials estimated that, after accounting for anticipated changes in the uranium market, the gap between once-through and reprocessed fuel would fall from $1.60 to $0.13 per megawatt hour. However, this report is considered optimistic by some members of France's nuclear power industry. *Id.* at 11.
615. Bryan Bender, "Obama Seeks Uranium Fuel Bank," *Boston Globe*, A1 (Jun. 8, 2009).
616. International Atomic Energy Agency, "Multinational Fuel Bank Proposal Reaches Key Milestone" (2009).
617. Marc Champion, "Iran's Uranium Stance Keeps a Deal at Bay," *Wall Street Journal*, A3 (Jan. 28, 2008); Charles Hanley, "Big Names, Bucks Bet Behind Nuclear Fuel Bank," *Pittsburgh Tribune* (May 24, 2009).
618. Institute for Policy Studies, *Radioactive Wastes and GNEP*, 8 (2007).

Chapter 2

1. These particles have a mass about 10,000 times smaller than that of an alpha particle and are capable of approximately a hundredfold deeper penetration into matter.
2. The becquerel is a very small amount of radioactivity; for instance, the International Atomic Energy Agency suggests that waste material with radioactivity on the order of 1 kBq/g be considered below regulatory concern. The curie is equivalent to about 37 GBq.
3. The relative damage of various forms of radiation is a complex calculation: alpha radiation is weighted high but is more easily mitigated through shielding; beta and gamma radiation have lower weighting factors but are harder to block.
4. One joule of energy per kilogram of matter dosed, multiplied by the radiation type's weighting factor. Hence a piece of waste emitting alpha particles would contribute more to equivalent dose rate than a piece of waste emitting the same number of beta particles, even though the alpha particles are much easier to block.
5. One hundred rem is equivalent to one Sv.
6. The percentage of the original quantity of a radioactive substance remaining after n half-lives have elapsed is given by $P(remaining) = 100/(2^n)$. To illustrate: a substance will decay to 1 percent of its original concentration after 6.64 of its half-lives have passed.
7. Vitrified waste is suspended in borosilicate glass. Suspension of waste in Synroc, an advanced ceramic, is an alternative immobilization technology.
8. A typical fuel assembly will be placed in the core for three 1.5-year cycles before it becomes spent fuel.
9. For example, in the centrifuge process, uranium is first chemically bonded with fluorine to create uranium hexafluoride. The gas is then spun at high speeds, causing the heavier U-238 to move toward the outside of the cylinder, and the U-235 to stay toward the center, where it can be collected. This

9. process is repeated as necessary until the desired enrichment is achieved. Typical enrichment values are 3 percent–5 percent for PWRs, 1–2 percent for CANDU reactors, and 85 percent for typical uranium nuclear devices, though crude nuclear weapons are possible at 20 percent enrichment.
10. The length of the burn depends on the aim of the burner: in general, fuel used for defense purposes has a higher proportion of actinides because the fuel is meant to produce a "younger" form of plutonium (Pu-239), i.e., a plutonium isotope which is generated earlier in the fission cycle and which gets diluted with different isotopes (mainly Pu-240) the longer the fuel remains in the reactor. In contrast, civilian reactor fuel is burned longer to extract the maximum amount of energy possible.
11. World Nuclear Association, *Radioactive Waste Management* (2009).
12. Note that 0.8 percent is slightly above the naturally occurring ratio of U-235, though well below the 3–5 percent ratio of enriched uranium.
13. Pu-240 can be used as fuel for reactors and for weapons, but fast neutron reactors are the only currently available reactor design that burns it, and they are not widely used. Five fast neutron reactors (FNRs) have operated in the United States: all were built for experimental or demonstration purposes and have ceased operation. The technology has seen wider use internationally, with four FNRs currently in operation and five more planned or under construction in Europe and Asia. Still, because the economics of reprocessing depend on fuel reprocessing, the technology is not likely to see widespread use until the price of natural uranium increases significantly. World Nuclear Association, *Fast Neutron Reactors* (2010), www.world-nuclear.org/info/inf98.html. Transuranic elements produce decay chains. As they decay, they often in turn produce other long-lived, alpha-emitting wastes, posing further challenges for disposal.
14. Nuclear Energy Institute, *Nuclear Waste Disposal for the Future*, 1 (2006).
15. Both Cs-137 and Sr-90 have medical and industrial uses and are disposed of as LLW once they are no longer useful in that capacity, pursuant to NRC's LLW regulations. 10 C.F.R. § 61.55 (2010).
16. Nuclear Energy Institute, *Nuclear Waste Disposal for the Future*, 2–3 (2006).
17. Regarding the one-year storage period, Nuclear Regulatory Commission, "Dry Cask Storage," www.nrc.gov/waste/spent-fuel-storage/dry-cask-storage.html.
18. Dry casks use steel containers containing inert gas to prevent chemical reactions with the fuel. They also contain a number of passive cooling devices. Pool and dry-cask storage are discussed more extensively in Chapter 7.
19. The chemical extraction techniques used in reprocessing cannot separate out specific isotopes. The uranium and plutonium output from reprocessing may require a separate enrichment cycle or have to be fabricated into advanced fuel types before it can be reused in reactors. World Nuclear Association, *Processing of Used Nuclear Fuel* (2010).
20. *Id.*
21. World Nuclear Association, *Mixed Oxide Fuel* (2009). Fuel reprocessing has outpaced MOX conversion, however, and European countries have started stockpiling civilian plutonium.
22. Areva, "Transport of MOX fuel from Europe to Japan," *areva.com/EN/operations-1391/transport-of-mox-fuel-from-europe-to-japan-the-stakes.html*.
23. Areva, "Recycling Benefits," *areva.com/EN/operations-3028/recycling-used-fuel-closed-cycle-management.html*.
24. *E.g.*, Union of Concerned Scientists, "Reprocessing and Nuclear Waste" (2009). For a detailed assessment of waste generation under the once-through fuel cycle and various reprocessing proposals, *see* Office of Nuclear Energy, DOE, *Draft GNEP Programmatic EIS*, at 4-46, 4-54 (2008).
25. Office of the Spokesman, U.S. Department of State, "2000 Plutonium Management and Disposition Agreement," press release (Apr. 13, 2010).
26. United States Enrichment Corporation, "Megatons to Megawatts" (2010), www.usec.com/megatonstomegawatts.htm.
27. Public Affairs Department, DOE, "U.S. Removes 9 Metric Tons of Plutonium from Nuclear Weapons Stockpile," press release (Sep. 17, 2007).
28. Scott Miller, "Duke Energy Interested in MOX Fuel, Despite Dropping Contract," *Charleston Regional Business Journal* (Mar. 17, 2009).
29. Robert Alvarez, Institute for Policy Studies, *Radioactive Wastes and GNEP*, 13–15 (2007); DOE, *Spent Nuclear Fuel Recycling Program Plan: Report to Congress*, 12–13 (2006).
30. DOE, *Spent Nuclear Fuel Recycling Program Plan: Report to Congress*, 12 (2006).
31. Fast neutron reactors rely on higher-energy neutrons to sustain their nuclear reaction than do those used in the more common thermal reactor. Fast reactors require fuel with higher concentrations of fissile material but are able to fission more nuclides than thermal reactors can. Thus, they are able to split a higher proportion of the heavier actinides into smaller fission products than a thermal reactor can. See World Nuclear Association, *Fast Neutron Reactors* (2010), www.world-nuclear.org/info/inf98.html.
32. *Id.* at 13.
33. As further discussed in Chapter 8, DOE considered two variations of UREX+—UREX+1 and UREX+1a—that would extract the lanthanides (fission daughter products with atomic number 58-71) from SNF at different stages in the process. *Id.* at 13–14.
34. Office of Nuclear Energy, DOE, *Draft GNEP Programmatic EIS*, at 4-46, 4-54 (2008).
35. NRC, Criteria Relating to Uranium Mills, 10 C.F.R. pt. 40 app. A (2010).
36. NRC, "Oversight of Materials and Reactor

Decommissioning," *www.nrc.gov/about-nrc/regulatory/decommissioning/oversight.html*.
37. Office of Radiation Protection, Washington State Department of Health, "Naturally Occurring and Accelerator Produced Radioactive Materials," *www.doh.wa.gov/ehp/rp/waste/llwnarm.htm*.
38. The 2005 Energy Policy Act expanded the definition of "byproduct material" to include "(4) any discrete source of naturally occurring radioactive material, other than source material, that—(A) the Commission . . . determines would pose a threat similar to the threat posed by a discrete source of radium-226 to the public health and safety or the common defense and security; and (B) before, on, or after the date of enactment . . . is extracted or converted after extraction for use in a commercial, medical, or research activity." Energy Policy Act of 2005 § 651, 42 U.S.C. § 2014(e) (2006). Pursuant to the act, NRC modified its regulations to bring radium 226 and some accelerator-produced wastes under NRC regulations. 72 Fed. Reg. 55,864, at 55,867–55,869 (Oct. 1, 2007). However, since the expanded definition applies only to NARM that has been packaged for medical and commercial use, most NARM remains under state authority.
39. *See* NRC Criteria Relating to the Operation of Uranium Mills and the Disposition of Tailings or Wastes Produced by the Extraction or Concentration of Source Material from Ores Processed Primarily for Their Source Material, 10 C.F.R. pt. 40 app. A (2010).
40. Some residual U-235 remains. NRC defines DU as any uranium with a concentration of U-235 below the naturally occurring rate of 0.711 percent. 10 C.F.R. § 40.4 (2010). The Department of Defense uses only DU with less than 0.2 percent U-235. Army Environmental Policy Institute, U.S. Army, *Health and Environmental Consequences of Depleted Uranium Use*, 2.2.3 (1995).
41. Craft et al., "Depleted and Natural Uranium," 306–313 (2004).
42. Louisiana Energy Serv., L.P., CLI-05-05 (Jan. 18, 2005) (NRC Memorandum and Order).
43. U-233 is produced by the thorium fuel cycle, which has been used only for research in the United States, so only very small quantities have been generated.
44. Office of Environmental Management, DOE, *Report to Congress: Status of Environmental Management Initiatives*, at iii (2009).
45. *Id*. at 41–45.
46. "The term 'low-level radioactive waste' means radioactive material that (i) is not high-level radioactive waste, spent nuclear fuel, or byproduct material [as defined in section 11(e)(2) of the Atomic Energy Act of 1954 (42 U.S.C. § 2014(e)(2))]." Low-Level Radioactive Waste Policy Act § 2, 42 U.S.C. § 2021b(9)(A) (2006). The excluded materials are specified in AEA sections 11e(2)–e(4) as materials that are nonreactor wastes, including accelerator-produced and naturally occurring materials (ARM/NORM) and uranium mill tailings (UMT). Atomic Energy Act of 1954 § 11(e), 42 U.S.C. § 2014(e)(2)–(4) (2006). Additionally, LLW includes waste that "(ii) the Nuclear Regulatory Commission, consistent with existing law and in accordance with paragraph (A), classifies as low-level radioactive waste." Low-Level Radioactive Waste Policy Act § 2, 42 U.S.C. § 2021b(9)(A) (2006). However, this has had little effect, because the NRC's regulatory definition is almost the same as that of LLRWPAA. The only difference is that NRC excludes transuranic waste in addition to the other excluded wastes. 10 C.F.R. § 61.2 (2010). However, there is no practical difference in terms of regulatory treatment whether TRU is included or excluded from the definition of LLW.
47. 10 C.F.R. § 61.55 (2010).
48. Low-Level Radioactive Waste Policy § 3, 42 U.S.C. § 2021c (2006).
49. *See generally* Chapter 4 of DOE, *Implementation Guide for Use with DOE M 435.1-1* (1999).
50. 10 C.F.R. § 61.55 tbls.1 and 2 (2010). The data in the table are taken from 10 C.F.R. § 61.55 (2010). The entries with "no limit" indicate that any concentration of these wastes is considered class B, unless "practical considerations such as the effects of external radiation and internal heat generation on transportation, handling, and disposal . . . limit the concentrations for these wastes." The entries with a n/a in the class B limit column indicate that no concentration of these isotopes are considered class B; anything above the class A limit is considered class C or GTCC waste.
51. *See* Committee on Improving Practices for Regulating and Managing Low-Activity Radioactive Waste, National Research Council, *Improving the Regulation of Low-Activity Wastes*, 38 (2006).
52. *Id*.; 10 C.F.R. § 61.55 (2010) establishes the nuclide limits for each LLW class and requires that class B and class C waste meet packaging requirements delineated in 10 C.F.R. § 61.56.
53. Low-Level Radioactive Waste Policy Amendments Act of 1985 § 10, 42 U.S.C. § 2021j (2006). The waste may still have disposal restrictions; for instance, if the waste contains toxic material, then it would still be subject to RCRA disposal requirements.
54. 10 C.F.R. § 30.11 (2010). NRC has exempted smoke and gas detectors, some radiologic drugs, glow-in-the-dark watch and compass hands, and vacuum tubes, provided that they contain less than specified concentrations of nuclides. 10 C.F.R. §§ 30.15, at 19–20 (2010). The power to regulate any waste that NRC exempts was given to the states by Section 276 of the Energy Policy Act of 1992, 28 U.S.C. § 2023 (2006).
55. 10 C.F.R. § 61.3 (2010); Low-Level Radioactive Waste Policy Amendments Act of 1985 § 102, as enacted, Pub. L. No. 99-240, 99 Stat. 1842, at 1843–1844 (1985), as codified, 42 U.S.C. § 2021c (2006). No GTCC waste disposal facility has been submitted to NRC for approval.
56. The regulations cap maximum exposure to the "reasonable maximally exposed individual"

57. 47 Fed. Reg. 57,446 (Dec. 27, 1982).
58. 10 C.F.R. §§ 61.7, at 40–44 (2010).
59. 54 Fed. Reg. 22,578 (May 25, 1989); DOE, *Implementation Guide for Use with DOE M 435.1-1*, at IV-16 (1999).
60. Swan and Lakes, *Challenges with Retrieving TRU*, 1–2 (2007).
61. 40 C.F.R. § 191.02 (2010); DOE *Implementation Guide for Use with DOE M 435.1-1*, at III-1 (1999). The DOE regulations except waste for which DOE has received EPA approval to exempt from 40 C.F.R. Part 191, or NRC approval to exempt from 10 C.F.R. Part 61. *See also* Waste Isolation Pilot Plant Land Withdrawal Act § 2(20), as enacted, Pub. L. No. 102-579, 106 Stat. 4777, at 4779 (1992).
62. The border between the two categories is waste that would expose an unprotected worker handling the waste package to a dose rate of two hundred millirems per hour. *Id.* § 2(3), (12), as enacted, Pub. L. No. 102-579, 106 Stat. 4777, at 4777–4778.
63. 10 C.F.R. Part 61 specifies transuranics with greater than a five year half-life to be subject to 10 C.F.R. Section 61.55 table 1. In that table, a concentration in excess of ten nanocuries per gram would be class C LLW, while a concentration in excess of one hundred nanocuries per gram would be deemed GTCC LLW. AEA specifies as transuranic wastes those with a concentration of ten nanocuries per gram, without any half-life threshold, but does not provide any distinctive regulatory requirements regarding their management or disposal. 42 U.S.C. § 2014(ee) (2006). The AEA definition is apparently left over from initial proposals for TRU standards and was never updated to reflect the final implemented DOE and EPA standards for TRU.
64. Committee on the Waste Isolation Plant, National Research Council of the National Academies, *The Waste Isolation Pilot Plant: A Potential Solution for the Disposal of TRU Waste*, 7 (1996).
65. *E.g.*, 10 C.F.R. § 61.55 (2010); DOE, *Implementation Guide for Use with DOE M 435.1-1*, at III-7 (1999), barring TRU from near-surface disposal via reference to 40 C.F.R. Part 191.
66. Alvarez, *Transuranic Wastes at Hanford*, 5 (2009)
67. Transuranic Waste Disposal, Proposed Standards for Protection against Radiation, 39 Fed. Reg. 32,921, at 32,921 (Sep. 12, 1974). *See also* EPA Environmental Standards for the Management of Spent Nuclear Fuel, 47 Fed. Reg. 58,196, at 58,197 (Dec. 29, 1982).
68. See, for example, 47 Fed. Reg. 58,196, at 58,197 (Dec. 29, 1982), proposing EPA's 40 C.F.R. Part 191 definition of transuranic waste, and Licensing Requirements for Land Disposal of Radioactive Waste, 47 Fed. Reg. 57,446, at 57,473 (Dec. 27, 1982), implementing NRC's 10 C.F.R. regulations, including 10 C.F.R. Part 61.
69. Waste Isolation Pilot Plant Land Withdrawal Act § 2, as enacted, Pub. L. No. 102-579, 106 Stat. 4777, 4778 (1992).
70. Nuclear Waste Policy Act § 2, 42 U.S.C. § 10101(23) (2006); *see also* Atomic Energy Act § 11, 42 U.S.C. § 2014(dd), which defines SNF by reference to NWPA. The NRC definition of HLW, at 10 C.F.R. Section 60.2, classifies spent nuclear fuel as HLW. In this book, however, we use the term "SNF" to include both civilian and defense used nuclear fuel, and "HLW" to include both defense and civilian reprocessing wastes that have not been reclassified into another regulatory category.
71. Commercial fuel tends to be irradiated up to the point at which the fission products have accumulated enough to reduce the fuel's efficiency below that necessary to maintain the self-sustaining chain reaction.
72. *See* Marcinowski, EM, DOE, *Overview of DOE SNF and HLW*, 3 (2010); Nuclear Energy Institute, *U.S. State-by State Commercial Nuclear Used Fuel* (2010).
73. *See* Nuclear Waste Policy Act § 111, 42 U.S.C. § 10131 (2006).
74. Marine Protection, Research, and Sanctuaries Act of 1972 §§ 101–02, 33 U.S.C. §§ 1411–12 (2006), as amended, Pub. L. No. 93-254 § 1(4)(A) (1974) (forbidding waivers of the anti-dumping provision).
75. The Energy Reorganization Act of 1974 §203(3), 42 U.S.C. § 5842(3) (2006), grants NRC authority to regulate Energy Research and Development Administration (later DOE) facilities related to storage of HLW.
76. 46 Fed. Reg. 13.971 (1981). See also 52 Fed. Reg. 5992, 5993 (Feb. 27, 1987) (summarizing the evolution of the term "HLW").
77. West Valley Demonstration Project Act § 6(4), as enacted, Pub. L. No. 96-368, 94 Stat. 1347, at 1350 (1980).
78. Nuclear Waste Policy Act of 1982 § 2(12), 42 U.S.C. § 10101(12) (2006).
79. Committee on Armed Services, *Nuclear Waste Policy Act of 1982*, 4 (1982).
80. *Chevron U.S.A., Inc. v. Natural Res. Def. Council, Inc.*, 467 U.S. 837 (1984). *See* Breyer et al., *Administrative Law and Regulatory Policy*, 342–344 (2006). *See generally id.* at 242–346.
81. The interagency allocation of authority is complicated further by the fact that NWPA applies only to wastes destined for Yucca, not to defense wastes otherwise possessed and managed by DOE.
82. 10 C.F.R. § 60.2 (2010).
83. 10 C.F.R. § 63.2 (2010). Note that 10 C.F.R. pt. 50 app. F (2010) is still in force as well.
84. DOE, *Implementation Guide for Use with DOE M 435.1-1*, at II-1 (1999). As a technical matter, DOE is subject to the AEA, which incorporates NWPA's definition, which in turn incorporates whatever NRC regulations are "consistent with existing law." Atomic Energy Act of 1954 § 11, 42 U.S.C. § 2014(dd) (2006); Nuclear Waste Policy Act of 1982, 42 U.S.C. § 10101(12)(B) (2006).
85. Nuclear Waste Policy Act of 1982 § 137, 42 U.S.C. § 10157 (2006).

86. Atomic Energy Act of 1954 §11, 42 U.S.C. § 2014(dd) (2006), as amended; Price-Anderson Amendments Act of 1988 § 4(b), as enacted, Pub. L. No. 100-408, 102 Stat. 1066, at 1069 (1988).
87. Committee on Risk-Based Approaches for Disposition of Transuranic and High-Level Radioactive Waste, National Research Council of the National Academies, *Risk and Decisions about Disposition of Transuranic and High-Level Radioactive Waste*, 45–55 (2005).
88. Department of Environmental Management, DOE, "Recovery Act Funds $24 Million in Technology Projects," *Environmental Management Recovery Act News Flash*, 1 (June 30, 2010), describes the Mobile Arm Retrieval System, a robotic arm designed to maximize recovery of Hanford waste tanks.
89. Allyn Boldt, "Hanford Waste Treatment Plant—History and Technical Issues," 2, *www.hanfordchallenge.org/cmsAdmin/uploads/WTP_History_and_Technical_Issues.pdf*.
90. Committee on the Management of Certain Radioactive Waste Streams Stored in Tanks at Three Department of Energy Sites, National Research Council of the National Academies, *Tank Waste Retrieval*, 29–30, 64 (2006).
91. See 52 Fed. Reg. 5992, 5993 (Feb. 27, 1987); NRC Disposal of High-Level Wastes in Geologic Repositories, 46 Fed. Reg. 13,971, at 13,980 (Feb. 25, 1981). NRC proposed changing the 10 C.F.R. § 60.2 definition of HLW, which was "liquid wastes resulting from the operation of the first cycle solvent extraction system, or equivalent, and the concentrated wastes from subsequent extraction cycles, or equivalent, in a facility for reprocessing irradiated reactor fuel, and . . . solids into which such liquid wastes have been converted."
92. 52 Fed. Reg. 5992, at 5992 (Feb. 27, 1987).
93. *Id.* at 5994.
94. *Id.*
95. NRC Disposal of Radioactive Wastes, 53 Fed. Reg. 17,709, at 17,709 (May 18, 1988).
96. *Id.*
97. "These facilities might make use of intermediate depth burial or various engineering measures, such as intruder barriers, to accommodate wastes with radionuclide concentrations unsuitable for disposal by shallow land burial." 52 Fed. Reg. 5992, at 5995 (Feb. 27, 1987).
98. Kocher and Croff, Oak Ridge National Laboratory, *A Proposed Classification System for High-Level and Other Radioactive Wastes* (1987).
99. *Id.*
100. 53 Fed. Reg. 17,709, at 17,709 (May 18, 1988).
101. *E.g.*, Callan, Executive Director of Operations, NRC, *Classification of Hanford Waste as Incidental* (1997).
102. NRC, *NRC Staff Guidance for DOE Waste Determinations*, xiii (2007).
103. Office of Environmental Management, DOE, *Report to Congress: Status of Environmental Management Initiatives*, 24–25 (2009). Some of the INL tank closures were conducted pursuant to § 3116 of the 2005 Defense Authorization Act, discussed later. *See* Ronald W. Reagan National Defense Authorization Act for Fiscal Year 2005 § 3116, as enacted, Pub. L. No. 108-375, 118 Stat. 1811, at 2162–2163 (2005).
104. NRC, *NRC Staff Guidance for DOE Waste Determinations*, xiii (2007).
105. *See* DOE, *Implementation Guide for Use with DOE M 435.1-1*, at II-16 (1999).
106. 55 Fed. Reg. 51,732, at 51,732 (Dec. 17, 1990).
107. *Id.*
108. *Id.*
109. States of Washington and Oregon: Denial of Petition for Rulemaking, 58 Fed. Reg. 12,342, at 12,342 (Mar. 4, 1993).
110. *Id.* at 12,345.
111. *Id.* at 12,343.
112. *Id.* at 12,344, which cites Letter from Robert M. Bernero, Director, Office of Nuclear Material Safety and Safeguards, NRC, to A. J. Rizzo, Assistant Manager for Operations, Richland Operations Office, DOE, September 25, 1989. However, the NRC did not take a completely hands-off approach: "The letter also called upon DOE to advise NRC periodically of the analytical results of samples of key radionuclides entering the grout facility, so that the classification of the waste might be reconsidered if the inventories were significantly higher than DOE had estimated." *Id.*
113. There is no indication that the petitioners sought judicial review of NRC's denial of their petition.
114. *See* DOE, *Radioactive Waste Management Manual for 435.1-1* (1999); DOE, *Implementation Guide for Use with DOE M 435.1-1* (1999). For a complete list of DOE 435.1 *Implementation Guide* sections, *www.directives.doe.gov*.
115. DOE, *Radioactive Waste Management Manual for 435.1-1*, at II-1 (1999). DOE explained the purpose of these provisions in the *Manual* as follows: "It is not the Department's intent to use the waste incidental to reprocessing process to circumvent high-level waste disposal standards by not disposing of high-level waste in the NRC-licensed geologic repository. The goal of the waste incidental to reprocessing determination process is to safely manage and dispose of a limited number of reprocessing waste streams that do not warrant geologic repository disposal because of their lack of long-term threats to the environment and man." DOE, *Implementation Guide for Use with DOE M 435.1-1*, at II-18 (1999).
116. DOE, *Radioactive Waste Management Manual for 435.1-1*, at II-14 to II-16 (1999).
117. *Id.* at II-16 to II-17 (1999). DOE asserted that "the NRC staff also confirmed that it supports the position that DOE has authority to make incidental waste determinations that involve waste streams that are incidental by use of the citation process." DOE, *Implementation Guide for Use with DOE M 435.1-1*, at II-17 (1999). Under the citation process, DOE field managers are authorized to make WIR determinations without consulting DOE headquarters. AEC had proposed in 1969 that materials such as tools and gloves be listed

in paragraphs (6) and (7) of Appendix D of 10 C.F.R. pt. 50 and not treated as highly radioactive reprocessing wastes. Atomic Energy Commission Siting of Commercial Fuel Reprocessing Plants and Related Waste Management Facilities: Statement of Proposed Policy, 34 Fed. Reg. 8712, at 8712 (June 3, 1969). Although Appendix D was never finalized into regulation, NRC has traditionally regarded such materials as "incidental to reprocessing." *See* 52 Fed. Reg. 5992, at 5993 (Feb. 27, 1987).
118. DOE, *Radioactive Waste Management Manual for 435.1-1*, at II-1 to II-2 (1999).
119. "The basis for the Commission's conclusion is that the reprocessing wastes disposed of in the grout facility would be 'incidental' wastes because of DOE's assurance that they. . . (2) will be incorporated in a solid physical form at a concentration that does not exceed the applicable concentration limits for Class C LLW as set out in 10 CFR Part 61." 58 Fed. Reg. 12,342, at 12,345 (Mar. 4, 1993).
120. DOE *Implementation Guide for Use with DOE M 435.1-1* states at II-16: "The question of whether the NRC or DOE has the authority to make incidental waste determinations (using the evaluation process) was raised by NRC Commissioner Curtiss in December 1992 (SECY-92-391), as a precursor to the Commission's action on the 1993 Denial of Petition for Rulemaking [filed by Washington and Oregon]. In response, the NRC staff (memo for Commissioner Curtiss from J. M. Taylor, 1/14/93) stated that DOE has the responsibility to make an initial determination, and if DOE concludes that the action is not subject to NRC jurisdiction, then DOE can undertake the activity without involving the NRC in any manner. However, if DOE concludes that NRC jurisdiction is unclear (i.e., the waste may be high-level waste and therefore potentially subject to NRC licensing), then DOE has two options: (1) consult with the NRC and then make a decision based on the results of the consultation; or (2) proceed without communication with the NRC." DOE, *Implementation Guide for Use with DOE M 435.1-1*, at II-16 (1999).
121. *See* NRC, *NRC Staff Guidance for DOE Waste Determinations* (2007).
122. *Natural Res. Def. Council, Inc. v. Abraham*, 271 F. Supp. 2d 1260 (D. Idaho 2003).
123. Plaintiffs' Response to Defendants' Motion to Dismiss, 8, *National Res. Def. Council, Inc. v. Abraham*, 271 F. Supp. 2d 1260 (D. Idaho 2003) (No. 01-CV-413), 2002 WL 32972157.
124. Memorandum of Points and Authorities in Support of Federal Defendants' Motion to Dismiss, 20–22, *National Res. Def. Council, Inc. v. Abraham*, 271 F. Supp. 2d 1260 (D. Idaho 2003) (No. 01-CV-413), 2002 WL 34351695.
125. *Natural Res. Def. Council, Inc. v. Abraham*, 271 F. Supp. 2d 1260 at 1264 (D. Idaho 2003). NRDC had initially sought direct review of DOE's issuance of the *Manual* in the court of appeals, which held that jurisdiction lay in the district court. *Natural Res.*
Def. Council, Inc. v. Abraham, 244 F.3d 742 (9th Cir. 2001).
126. *Natural Res. Def. Council, Inc. v. Abraham*, 271 F. Supp. 2d 1260 at 1265 (D. Idaho 2003).
127. *Id.*
128. The court noted that NWPA allowed for some DOE reclassification, because it "allows DOE to treat the solids to remove fission products, thereby permitting reclassification of the waste" provided that the solid waste no longer had "sufficient concentrations" of fission products. *Id.*
129. *Natural Res. Def. Council, Inc. v. Abraham*, 388 F.3d 701, 705 (9th Cir. 2004).
130. *Id.* at 703, 706.
131. The lower court's ruling that the *Manual*'s provisions conflict with NWPA arguably disregards the logic of the evaluation criteria in Order 435.1. The second and third evaluation criteria—namely, that wastes must meet 10 CFR Part 61, Subpart C, Performance Objectives for disposal and must be class C or lower LLW or meet alternative requirements established by DOE—are designed to ensure that wastes classified as WIR have radioactivity levels below the levels of "sufficient concentrations" requiring repository disposal. Arguably, that is all that the NWPA HLW definition requires for wastes to be classified as other than HLW. If so, the first DOE evaluation criterion—maximum "technically and economically practical" radionuclide removal—can be interpreted as serving to reduce radioactivity levels even further below the level of "sufficient concentrations." The order does not purport to authorize classification of wastes as non-HLW simply because they have been treated to the extent technically and economic feasible, even though the other evaluation criteria are not met; on their face, the three requirements in the evaluation provisions are cumulative, not alternative. But the lower court apparently believed that the inclusion of any factors outside the NWPA HLW definition in the WIR process would make that process invalid.

The court also asserted that DOE's third criterion, providing for DOE to set "alternative requirements" for radioactivity levels, was arbitrary and capricious because DOE could, under this provision, set requirements based on "whim." The Ninth Circuit concluded that the courts should consider this issue in the context of a concrete DOE decision, rather than as an abstract matter.
132. Ronald W. Reagan National Defense Authorization Act for Fiscal Year 2005 § 3116, as enacted, Pub. L. No. 108-375, 118 Stat. 1811, at 2162–2163 (2005).
133. *Id.*
134. *Id.* at 2163.
135. Senator Graham praised DOE's involvement with South Carolina and stated that he looked forward to creating a framework for federal-state cooperation for the remaining tanks at the Savannah River Site, 150 Cong. Rec. S5902, at S5903–S5904 (May 20, 2004).
136. *Id.* at S5904.

137. *Id.* at S5903.
138. *E.g.*, Statement of Senator Patti Murray, 150 CONG. REC. S6411 (June 3, 2004).
139. Statement of Senator Hillary Clinton, 150 CONG. REC. S6410 (June 3, 2004).
140. *See id.* at S6410–S6412.
141. Ronald W. Reagan National Defense Authorization Act for Fiscal Year 2005 §3116(b), as enacted, Pub. L. No. 108-375, 118 Stat. 1811, at 2163 (2005).
142. *Id.* § 3146, as enacted, Pub. L. No. 108-375, 118 Stat. 1811, at 2173–2175.
143. Committee on the Management of Certain Radioactive Waste Streams Stored in Tanks at Three Department of Energy Sites, National Research Council of the National Academies, *Tank Waste Retrieval*, 30 (2006). It also noted that section 3116(b) requires NRC to monitor DOE's implementation of this authority, in coordination with Idaho and South Carolina.
144. *Id.* at ix.
145. The conference report for the act stated: "Section 3116 does not establish any precedent for and is not binding on the States of Washington, Oregon or any other state that is not a covered state for the management, storage, treatment, and disposition of radioactive and hazardous material." Conference Committee, *Ronald W. Reagan National Defense Authorization Act for FY05*, 884–885, 2004. The legislative background indicates a purpose to validate HLW reclassification agreements between DOE and South Carolina and Idaho, respectively.
146. Committee on the Management of Certain Radioactive Waste Streams Stored in Tanks at Three Department of Energy Sites, National Research Council of the National Academies, *Tank Waste Retrieval*, 68–71, 110–111 (2006).
147. *Id.* at 68–71.
148. *Id.* at 60–61.
149. *Id.* at 51–52.
150. Committee on the Management of Certain Radioactive Waste Streams Stored in Tanks at Three Department of Energy Sites, National Research Council of the National Academies, *Tank Waste Retrieval*, 68–71, 110–111 (2006).
151. *Id.* at 60–61.
152. *Id.* at 56–60. For specifics regarding aluminum removal at Hanford, *see id.* at 58; DOE, *Tank Waste System Integrated Project Team Feasibility Report* (2009).
153. *Id.* Saltstone is a method of immobilizing the salt washes from HLW treatment by mixing the salts with cement and other solids. The resulting slurry is then pumped into storage vaults, where it sets to an immobile form. Savannah River Remediation, LLC., "Saltstone Facilities," *www.srs.gov/general/news/factsheets/salt.pdf*. Steam reformation is another solidification process that destroys any organic compounds in the waste stream, then renders the radioactive component for solidification. The process is less costly than other methods and was chosen for use at INL because the homogeneous nature of tank waste there did not require the more extensive and costly treatments that were necessary at SRS and Hanford. Mason, et al., THOR Treatment Technologies, LLC., *Steam Reforming at INL*, 1–4 (2006).
154. Office of Environmental Management, DOE, *Report to Congress: Status of Environmental Management Initiatives*, 25–26 (2009).
155. Federal Facility Compliance Act of 1992 § 105(a), 42 U.S.C. § 6939c (2006).
156. EPA, "About Mixed-Waste," *www.epa.gov/radiation/mixed-waste/about.html*.
157. *Id.*
158. Resource Conservation and Recovery Act of 1976 § 1004(27), 42 U.S.C. § 6903(27) (2006).
159. Bonstead, "EPA's Mixed Approach to Mixed Waste," 541 (2002).
160. *See, e.g., Legal Envtl. Assistance Found. v. Hodel*, 586 F. Supp. 1163 (E.D. Tenn. 1984).
161. *Id.* at 1167.
162. *Sierra Club v. U.S. Dep't of Energy*, 734 F. Supp. 946, 949 (D. Colo. 1990).
163. Bonstead, "EPA's Mixed Approach to Mixed Waste," 539–542 (2002).
164. EPA, "About Mixed-Waste," *www.epa.gov/radiation/mixed-waste/about.html*; Bonstead, "EPA's Mixed Approach to Mixed Waste," 541 (2002).
165. Hazardous and Solid Waste Amendments of 1984 § 201, 42 U.S.C. § 6924(d)-(k) (2006).
166. Leonard, "All Mixed Up about Mixed Waste," 236 (2002).
167. Urban, "EPA's Hazardous Waste Rule Gone Haywire," 111–112 (1998); Hazardous and Solid Waste Amendments of 1984 § 201, 42 U.S.C. § 6924(d) (2006).
168. Hazardous and Solid Waste Amendments of 1984 § 201, 42 U.S.C. § 6924(j) (2006). After one year of storage, the waste's owner bears the burden of proving that storage is necessary to achieve an approved purpose. *Id. See also* Bonstead, "EPA's Mixed Approach to Mixed Waste," 548 (2002).
169. Leonard, "All Mixed Up about Mixed Waste," 226–227 (2002).
170. Resource Conservation and Recovery Act of 1976 § 6001, 42 U.S.C. § 6961 (2006); 55 FED. REG. 22,520 (June 1, 1990); Bonstead, "EPA's Mixed Approach to Mixed Waste," 552 (2002).
171. Federal Facility Compliance Act of 1992 § 102(c)3, as enacted, Pub. L. No. 102-386, 106 Stat. 1505, at 1506–1507 (1992).
172. 66 FED. REG. 27,218 (May 16, 2001).
173. *Id.* at 27,221–27,222. *See also* Bonstead, "EPA's Mixed Approach to Mixed Waste," 561 (2002). For low-level mixed waste to be eligible for the exception, first, it must be "generated and managed . . . under a single NRC or NRC Agreement State license." 40 C.F.R. § 266.225 (2010). Second, the waste must be stored "in tanks or containers in compliance with the

requirements of [the NRC] license that apply to the proper storage of low-level radioactive waste." 40 C.F.R. § 266.230(b) (2010). Third, these tanks or containers must meet "chemical compatibility requirements." *Id.* Fourth, "facility personnel who manage stored conditionally exempt LLMW [must be] trained in a manner that ensures that the conditionally exempt waste is safely managed and includes training in chemical waste management and hazardous materials incidents response." *Id.* Fifth, the conditionally exempt low-level waste must be inspected quarterly and inventoried annually. *Id.* Last, the operator of the storage facility must "maintain an accurate emergency plan and provide it to all local authorities who may have to respond to a fire, explosion, or release of hazardous waste or hazardous constituents." *Id.* RCRA restrictions resume if owners fail to comply with these conditions, or if waste decays to a point that its radioactivity is "below regulatory concern." 40 C.F.R. § 261.3 (2010). *See also* Leonard, "All Mixed Up about Mixed Waste," 228 (2002).

174. "To qualify for the transportation and disposal exemption, a generator must satisfy four conditions. First, the eligible waste must be treated to meet . . . LDR (land disposal restriction) standards. . . . Treatment of waste to the LDR standards removes any flammable, corrosive, and reactive waste characteristics. The only possible remaining hazardous characteristic is the toxicity of the waste. . . . The second condition requires that NRC or NRC agreement state manifest and transportation regulations be followed for all mixed waste shipments, even by generators who are not already subject to such regulation. . . . The third condition requires that all exempted waste be in containers for disposal. This condition responds to the differences in design requirements for RCRA disposal facilities and NRC disposal facilities. All RCRA disposal facilities are required to have a synthetic liner, whereas NRC disposal facilities are not. EPA determined that the protection provided by the containers specified in the condition was comparable to the protection provided by synthetic liners at hazardous waste disposal facilities with respect to retaining the integrity of the waste in the disposal cell. The final condition requires that waste be disposed at a LLRWDF that is licensed by the NRC or a NRC agreement state." Bonstead, "EPA's Mixed Approach to Mixed Waste," 565–566 (2002), analyzes the requirements of 40 C.F.R. § 266.315(a)–(d).
175. *Id.* at 561.
176. 66 Fed. Reg. 27,218, at 27,221 (May 16, 2001).
177. Bonstead, "EPA's Mixed Approach to Mixed Waste," 577 (2002).
178. *Id.* at 579.
179. National Defense Authorization Act for Fiscal Year 1997 § 3188(a), as enacted, Pub. L. No. 104-201, 110 Stat. 2422, at 2853 (1997).
180. *Waste Isolation Pilot Plant Land Withdrawal Amendment Act*, 11 (1996).
181. *Id.* at 18–19.
182. *Id.* at 18.
183. Nuclear Fuel Management and Disposal Act, S. 2589, 109th Cong. (2006); A Bill to Enhance the Management and Disposal of Spent Nuclear Fuel and High-Level Radioactive Waste, and for Other Purposes, S. 2610, 109th Cong. (2006); Nuclear Fuel Management and Disposal Act, H.R. 5360, 109th Cong. (2006). The same exemption would be extended to material "located at the Yucca Mountain site for disposal" if that material was subject to a NRC license. *See, e.g.,* Nuclear Fuel Management and Disposal Act, S. 2589, 109th Cong. § 2 (2006).
184. Resource Conservation and Recovery Act of 1976 § 6001, 42 U.S.C. § 6961(a) (2006).
185. Public Citizen, "Yucca Mountain Bills (S. 2589 and S. 2610): A Wholesale Rollback of Public Health and Safety Laws; States' Rights," www.citizen.org/documents/YuccaBillSummary.pdf.
186. Samuel W. Bodman, Secretary of Energy, "Legislative Proposal for the Nuclear Fuel Management and Disposal Act," 12 (Mar. 6, 2007), www.energy.gov/media/BodmanLetterToPelosi.pdf.
187. *Id.*
188. *See* Lowenthal, *Radioactive-Waste Classification in the United States*, 1 (1997).
189. *See id.*, at 14; Croff, "Risk-Informed Waste Classification," 450–451 (2006); Garrick, "Contemporary Issues in Risk-Informed Decision Making," 431–432 (2006).
190. Croff, "Risk-Informed Waste Classification," 457 (2006).
191. Esh et al., *Risks and Uncertainties with HLW Tank Closure* (2002).
192. Walker, *Road to Yucca*, 7 (2009).
193. International Atomic Energy Agency, *Classification of Radioactive Waste* (1994).
194. Croff, "Risk-Informed Waste Classification," 452 (2006).
195. International Atomic Energy Agency, *Radioactive Waste Management Profiles*, 75, 250, 486, 876, 917 (2008).
196. International Atomic Energy Agency, *Classification of Radioactive Waste*, 12 (1994).
197. Of eighteen foreign countries surveyed by the GAO, fifteen either exempt very low-level radioactive waste or allow for disposal with minimal regulations. Government Accountability Office, *Low-Level Radioactive Waste Management: Approaches Used by Foreign Countries*, 26 (2007).
198. International Atomic Energy Agency, *Classification of Radioactive Waste*, 13, 15–17 (1994). There is no official radiation cap for LILW waste. Rather, the waste must generate less than two kilowatts per cubic meter of thermal energy. *Id.*
199. *Id.* at 14–15.
200. International Atomic Energy Agency, *Predisposal Management of HLW*, 1–2 (2003).

201. DOE's attempts to dispose of the low-activity fractions of Hanford and SRS tank wastes highlight this problem: although the low-activity fractions had activity levels which met LLW requirements, the source-based definition created legal barriers to disposing the waste as LLW.
202. International Atomic Energy Agency, *Classification of Radioactive Waste*, 16–17 (1994).
203. *See* 10 C.F.R. § 61.55 (2010). *See also* National Council on Radiation Protection and Measurements, *Risk-Based Classification of Wastes*, 7.2.2 (2002).
204. Atomic Energy Act of 1954 § 81, 42 U.S.C. § 2111 (2006). *See also* Low-Level Radioactive Waste Policy Act, 42 U.S.C. § 2021j (2006), authorizing NRC to exempt particular LLW categories from disposal requirements.
205. 10 C.F.R pt. 30.20 (2010) exempts smoke detectors; 10 C.F.R. § 30.21 (2010) exempts carbon-14 containing drugs when used for diagnostic purposes.
206. 55 Fed. Reg. 27,522 (July 3, 1990), 51 Fed. Reg. 30,839 (Aug. 29, 1986).
207. *Public Citizen, Inc. vs. NRC*, 940 F.2d 679 (D.C. Cir. 1991). Public Citizen, a nonprofit group, challenged the proposal, charging that it had been issued without sufficient notice or opportunity for comment, that it was beyond NRC's statutory authority, and that NEPA required an EIS, which was not done. The court rejected the claims as unripe.
208. Energy Policy Act of 1992 § 276, 42 U.S.C. § 2023 (2006).
209. 68 Fed. Reg. 65,120 (Nov. 18, 2003). There has been no action on such a rule since the publication of its advance notice of proposed rulemaking.
210. *See* Croff, "Risk-Informed Waste Classification," 455 (2006)
211. Committee on Improving Practices for Regulating and Managing Low-Activity Radioactive Waste, National Research Council of the National Academies, *Improving the Regulation of Low-Activity Wastes*, app. A, at 3 (2006).
212. *Id.*, at 5, 44.
213. Committee on Improving Practices for Regulating and Managing Low-Activity Radioactive Waste, National Research Council of the National Academies, *Improving the Regulation of Low-Activity Wastes*, 9 (2006).
214. *See id.* at 59, 93–94.
215. *Id.* at 14.
216. *Id.* at 20–21.
217. *See* Committee on Alternatives for Controlling the Release of Solid Materials from Nuclear Regulatory Commission-Licensed Facilities, National Research Council of the National Academies, *The Disposition Dilemma*, 2–7 (2002).
218. Committee on Improving Practices for Regulating and Managing Low-Activity Wastes, National Research Council of the National Academies, *Improving the Regulation of Low-Activity Wastes*, 25–26. The committee described the "Off-Site Source Recovery Program," a DOE program to recover high-risk sources.
219. National Council on Radiation Protection and Measurements, *Risk-Based Classification of Wastes*, 6.2.2 (2002).
220. *Id.* at 7.2.1.
221. *Id.* at 7.2.2.
222. *Id.* at 7.2.1–7.2.2.

Chapter 3

1. Committee on Transportation of Radioactive Waste, National Research Council of the National Academies, *Going the Distance?* (2006). *See* Electric Power Research Institute, *Spent Nuclear Fuel Transportation* (2004), for another comprehensive overview of nuclear waste transportation, which also concluded that existing transportation regulations were adequate and that risks associated with transporting SNF were small.
2. Committee on Transportation of Radioactive Waste, National Research Council of the National Academies, *Going the Distance?* 117–122 (2006); DOE, "WIPP Chronology," www.wipp.energy.gov/fctshts/chronology.pdf.
3. Haggerty, "'TRU' Cooperative Regulatory Federalism," 55–56 (2002), reporting, based on a "Fact Sheet" by the "Radioactive Waste Campaign," that from 1971 to 1985, 1,034 LLW transport accidents occurred, resulting in some release of radioactivity from ninety containers. Haggerty also finds that coordination among states, and between DOE and transit states, is currently poor. *Id.* at 56–57. He argues that delegating regulatory authority to "quasi-governmental" organizations such as the Western Governors' Association and the Commercial Vehicle Safety Alliance is a "dangerous experiment" that avoids public accountability for nuclear waste transportation management. *Id.* at 59–71.
4. Committee on Transportation of Radioactive Waste, National Research Council of the National Academies, *Going the Distance?* 21 (2006).
5. For example, the American Physical Society described the NAS Committee report as "comprehensive" and stated that an "overwhelming consensus" of independent research supports the NAS Committee's conclusion that there are no technical barriers to the safe transportation of SNF and HLW. Nuclear Energy Study Group, American Physical Society, Consolidated Interim Storage of Commercial Spent Nuclear Fuel: A Technical and Programmatic Assessment, 4 (Feb. 2007), www.aps.org/policy/reports/popa-reports/upload/Energy-2007-Report-InterimStorage.pdf.
6. Surveys show significant public concern over the possibility of releases and contamination from transportation-related accidents. Greenberg et. al., "Nuclear Waste and Public Worries," 5 (2007). In 1992 the New Mexico Supreme Court ordered compensation for the decreased value of land located along the proposed transport route to WIPP,

regardless of the actual contamination risks. *City of Santa Fe v. Komis*, 845 P.2d 753, 760 (N.M. 1992).
7. Atomic Energy Act of 1954 §§ 61–66, 81, 91, 42 U.S.C. §§ 2091–2096, 2111, 2121 (2006).
8. Energy Reorganization Act of 1974 §§ 2, 101, 201, 42 U.S.C. §§ 5801, 5811, 5841 (2006). ERDA/DOE was given the authority to arrange for, and manage transportation of, wastes produced from nuclear materials used for defense-related activities. *Id.* § 104, 42 U.S.C. § 5814 (2006), assigns responsibilities of the AEC Military Liaison Committee to ERDA/DOE.
9. Act of Sept. 6, 1960 §§ 832(b), 834(a), (c), Pub. L. No. 86-710, 74 Stat. 808, at 809–810 (1960).
10. *Id.* § 834(b), Pub. L. No. 86-710, 74 Stat. 808, at 810.
11. *Id.* § 832(c), Pub. L. No. 86-710, 74 Stat. 808, at 809.
12. Department of Transportation Act § 6(e)–(f), Pub. L. No. 89-670, 80 Stat. 931, at 939–940 (1966) (identifying powers transferred from ICC to DOT).
13. Federal Railroad Safety Act of 1970 § 202, Pub. L. No. 91-458, 84 Stat. 971, at 971 (1970); Hazardous Materials Transportation Control Act of 1970, Pub. L. No. 91-458, 84 Stat. 977 (1970) (requiring DOT to study risks of hazardous materials transportation, to review existing federal regulation, and to advise president and Congress on need for additional legislation). Other agencies with regulatory roles over hazardous materials transportation included the Coast Guard, the Federal Aviation Administration, and the Civil Aeronautics Board. Millan and Harrison, "Primer on Hazardous Materials Transportation Law of the 1990s," at 10,584 (1992). Congress transferred FAA and CAB authority over hazardous materials transportation by air to DOT in 1975. Hazardous Materials Transportation Act § 113, Pub. L. No. 93-633, 88 Stat. 2156, at 2162 (1975). The Coast Guard's authority currently includes enforcing DOT regulations on water-based shipments of radioactive materials and advising shippers and carriers on selecting routes. DOE, *Transporting Radioactive Materials*, 10, 20 (1999).
14. AEC's authority under the AEA was broad enough to include the power to regulate transportation of radioactive materials, but it declined to exercise this authority, pursuant to interagency agreements with DOT. Yates, "Preemption under the Atomic Energy Act of 1954," 413n.113 (1976).
15. Millan and Harrison, "Primer on Hazardous Materials Transportation Law of the 1990s," at 10,585 (1992); Thompson, "HMTA: Chemicals at Uncertain Crossroads," 415 (1987).
16. Federal Railroad Safety Act of 1970 §§ 101, 205, Pub. L. No. 91-458, 84 Stat. 971, at 971–972 (1970).
17. *Id.* § 202, Pub. L. No. 91-458, 84 Stat. 971, at 971. FRSA did not directly address whether DOT railroad safety regulations applied to rail shipments by AEC or other government agencies.
18. *Id.* § 205, Pub. L. No. 91-458, 84 Stat. 971, at 971.
19. Hazardous Materials Transportation Act §§ 103(2), 104, as enacted, Pub. L. No. 93-633, 88 Stat. 2156, at 2156 (1975), codified as amended at 49 U.S.C. § 5103(a) (2006).
20. *Id.* §§ 104–105, as enacted, Pub. L. No. 93-633, 88 Stat. 2156, at 2156–2157.
21. *Id.* § 105(b), as enacted, Pub. L. No. 93-633, 88 Stat. 2156, at 2156–2157.
22. *Id.* § 112(a), as enacted, Pub. L. No. 93-633, 88 Stat. 2156, at 2161.
23. *Id.* § 112(b), as enacted, Pub. L. No. 93-633, 88 Stat. 2156, at 2161.
24. Millan and Harrison, "Primer on Hazardous Materials Transportation Law of the 1990s," at 10,585 (1992).
25. Committee on Energy and Commerce, *HMTUSA*, 20–21 (1990).
26. *Id.*
27. *City of New York v. Dept. of Transportation*, 700 F. Supp. 1294, 1296 (S.D.N.Y. 1988).
28. *Id.*; Highway Routing of Radioactive Materials: Inquiry, 43 Fed. Reg. 36,492 (Aug. 17, 1978).
29. Highway Routing of Radioactive Materials, 45 Fed. Reg. 7140 (Jan. 31, 1980); Radioactive Materials: Routing and Driver Training Requirements, 46 Fed. Reg. 5298 (Jan. 19, 1981).
30. *See, e.g., City of New York v. Dept. of Transportation*, 700 F. Supp. 1294 (S.D.N.Y. 1988).
31. Low-Level Radioactive Waste Policy Act § 4(b)(3), as amended, 42 U.S.C. § 2021d(b)(3) (2006).
32. Hazardous Materials Transportation Uniform Safety Act of 1990 § 2(3), Pub. L. No. 101-615, 104 Stat. 3244, at 3245 (1990).
33. *Id.* § 15, Pub. L. No. 101-615, 104 Stat. 3244, at 3261 (1990), codified as amended at 49 U.S.C. § 5105 (2006); *see also* Nuclear Waste Policy Act of 1982 § 180, 42 U.S.C. § 10175 (2006). DOT originally created the Research and Special Programs Administration to carry out the oversight of hazardous materials transportation, but in 2004 Congress transferred this authority to the Pipeline and Hazardous Materials Safety Administration. Norman Y. Mineta Research and Special Improvements Act § 2(b), Pub. L. No. 108-426, 118 Stat. 2423, at 2424 (2004); 49 U.S.C. § 108 note (2006).
34. Hazardous Materials Transportation Uniform Safety Act of 1990 § 15, Pub. L. No. 101-615, 104 Stat. 3244, at 3261 (1990).
35. *Id.* HMTUSA did not authorize DOT to *require* carriers to base routing decisions on the factors identified in the study. *Id.*
36. *Id.* § 4, Pub. L. No. 101-615, 104 Stat. 3244, at 3248 (amending Hazardous Materials Transportation Act § 105(a), Pub. L. No. 93-633, 88 Stat. 2156 (1975)).
37. *Id.* § 13, Pub. L. No. 101-615, 104 Stat. 3244, at 3259–3260 (amending Hazardous Materials Transportation Act § 112(a), Pub. L. No. 93-633, 88 Stat. 2156 (1975)).
38. *Id.*, Pub. L. No. 101-615, 104 Stat. 3244, at 3260 (amending Hazardous Materials Transportation Act § 112(c), Pub. L. No. 93-633, 88 Stat. 2156 (1975)).
39. Hazardous Materials Transportation Uniform Safety Act of 1990 § 4, Pub. L. No. 101-615, 104 Stat. 3244,

at 3248–3249 (amending Hazardous Materials Transportation Act § 105(b)(1)–(2), Pub. L. No. 93-633, 88 Stat. 2156 (1975)).
40. *Id.*, Pub. L. No. 101-615, 104 Stat. 3244, at 3249–3250 (amending Hazardous Materials Transportation Act § 105(b)(3), Pub. L. No. 93-633, 88 Stat. 2156 (1975)).
41. *Id.*, Pub. L. No. 101-615, 104 Stat. 3244, at 3249–3250 (amending Hazardous Materials Transportation Act § 105(b)(2), Pub. L. No. 93-633, 88 Stat. 2156 (1975)).
42. *Id.*; *see* Southern States Energy Board, "Radioactive Materials: Emergency Response and Transportation Planning," *www.sseb.org/radioactive-materials.php*; Western Governors' Association, "WGA WIPP Fact Sheet," 1, *www.westgov.org/wga_governors.htm*.
43. Hazardous Materials Transportation Uniform Safety Act of 1990 § 15, Pub. L. No. 101-615, 104 Stat. 3244, at 3261 (1990).
44. Implementing Recommendations of the 9/11 Commission Act of 2007, Pub. L. No. 110-53, 121 Stat. 266 (to be codified in scattered sections of 5, 6, 8, 14, 22, 42, 46–47, 49, 50 U.S.C.).
45. *Id.* § 1501(13), 6 U.S.C.A. § 1151 (West 2010).
46. Rail Transportation Security, 73 Fed. Reg. 72,132 (Nov. 26, 2008), 49 C.F.R. pts. 1520 and 1580 (2009).
47. Implementing Recommendations of the 9/11 Commission Act of 2007 § 1551(b), 6 U.S.C.A. § 1201 (West 2010).
48. Hazardous Materials: Enhancing Rail Transportation Safety and Security for Hazardous Materials Shipments, 73 Fed. Reg. 72,182 (Nov. 26, 2008), 49 C.F.R. pts. 172, 174 (2009). The Transportation Security Administration (a component of DHS), the Pipeline and Hazardous Materials Safety Administration (part of DOT), and the Federal Railroad Administration (also part of DOT) coordinated closely to develop the regulations required by the 9/11 Commission Act and issuing final rules.
49. Implementing Recommendations of the 9/11 Commission Act of 2007 § 1551(c)–(g), 6 U.S.C.A. § 1201(c)–(g) (West 2010).
50. *Id.* § 1551(h), 6 U.S.C.A. § 1201(h) (West 2010).
51. Compare the Hazardous Materials Transportation Uniform Safety Act of 1990 § 4, Pub. L. No. 101-615, 104 Stat. 3244, at 3247, which amends the Hazardous Materials Transportation Act § 105(b)(3), Pub. L. No. 93-633, 88 Stat. 2156 (1975), providing that DOT "shall issue regulations for the safe transportation of hazardous material (including radioactive material)," with the Atomic Energy Act of 1954 §§ 57, 62, 81, Pub. L. No. 83-703, 68 Stat. 919, at 932, 935 (1954), prohibiting transfer of special, source, or byproduct radioactive material except as authorized by AEC; *see also* Committee on Transportation of Radioactive Waste, National Research Council of the National Academies, *Going the Distance?* 50 (2006).
52. Transportation of Radioactive Materials: Memorandum of Understanding, 44 Fed. Reg. 38,690, at 38,690 (July 2, 1979).
53. *Id.* at 38,690–38,691.
54. Act of Sept. 6, 1960 § 832(c), Pub. L. No. 86-710, 74 Stat. 808, at 809 (1960).
55. Hazardous Materials Transportation Act §105(a), Pub. L. No. 93-633, 88 Stat. 2156, at 2157 (1975).
56. Government Operations and Materials, 49 C.F.R. 173.7(b) (2009).
57. Committee on Energy and Commerce, *HMTUSA of 1990*, 12 (1990).
58. Nuclear Waste Policy Act of 1982 §§ 117(c)(7), 137, 180, 42 U.S.C. §§ 10137(c)(7), 10157, 10175 (2006).
59. Waste Isolation Pilot Plant Land Withdrawal Act § 16, Pub. L. No. 102-579, 106 Stat. 4777, at 4792–4794 (1992). DOE determined in its 1990 record of decision on its FEIS for WIPP that its TRU shipments would comply with NRC packaging requirements. DOE, *WIPP Disposal Phase: Final Supplemental EIS*, S-18 (1997).
60. "Second Modification to the July 1, 1981 'Agreement for Consultation and Cooperation' on WIPP by the State of New Mexico and the U.S. Department of Energy," ¶¶ 2, 4 (1987).
61. Committee on Transportation of Radioactive Waste, National Research Council of the National Academies, *Going the Distance?* 46, 52 n. a (2006).
62. *E.g.*, Southern States Energy Board, "Radioactive Materials: Emergency Response and Transportation Planning," *www.sseb.org/radioactive-materials.php*; *2009 Memorandum Between WGA and DOE* (2009).
63. Office of Environmental Management, DOE, *Radioactive Transportation Practices Manual*, § 1.3 (2008).
64. *Id.*
65. Standards for Protection Against Radiation, 10 C.F.R. pt. 20 (2010); Packaging and Transportation of Radioactive Material, 10 C.F.R. pt. 71 (2010).
66. Rem is a measure of equivalent dose rate, used to relate the radioactivity of a substance to its biological effects by measuring the amount of radioactive energy per unit of organic mass exposed, multiplied by a factor that depends on the type of radiation in question and the relative damage that it causes. *See* 10 C.F.R. Section 20.1004 (2009) (Table 1004(b),1). Dose limits for various categories of exposures are set in Occupational Dose Limits, 10 C.F.R. pt. 20, subpt. C (2010); Radiation Dose Limits for Individual Members of the Public, 10 C.F.R. pt. 20, subpt. D (2010); Application for Package Approval, 10 C.F.R. pt. 71, subpt. D (2010); Package Approval Standards, 10 C.F.R. pt. 71, subpt. E (2010); Physical Protection of Plants and Materials, 10 C.F.R. §§ 73.25–73.37 (2010).
67. Physical Protection of Plants and Materials, 10 C.F.R. §§ 73.25–73.37 (2010). A_2 is a measure of the maximum activity of radioactive materials that is either listed in 10 C.F.R. Part 71, appendix A, Table A-1 (2010), or may be derived in accordance with the procedures prescribed in 10 C.F.R. Part 71 Appendix A (2010).
68. Package, Special Form, and LSA-III Tests, 10 C.F.R. pt. 71, subpt. F (2010). Subpart F establishes additional requirements for special-

form (encapsulated) radioactive materials and for LSA-III materials (solid objects contaminated with insoluble low-level radioactive materials, such as contaminated concrete). *Id.* §§ 71.75–71.77; *see id.* § 71.4 for definitions of special-form radioactive materials and LSA-III materials.
69. Operating Controls and Procedures, 10 C.F.R. pt. 71, subpt. G (2010).
70. Physical Protection of Special Nuclear Materials in Transit, 10 C.F.R. §§ 73.25–73.37 (2010).
71. Committee on Transportation of Radioactive Waste, National Research Council of the National Academies, *Going the Distance?* 12–15 (2006). The committee noted that concerns about the risk of engulfing fires had increased after several train accidents that resulted in long-burning fires, including derailment of a train carrying non-radioactive hazardous materials in a tunnel in Baltimore in July 2001; the resulting fire burned for up to 12 hours with temperatures in the tunnel reaching an estimated 1000°C. *Id.* at 13, 78–80.
72. Hazardous Materials Program Procedures, 49 C.F.R. pt. 107 (2009); Hazardous Materials Public Sector Training and Planning Grants, 49 C.F.R. pt. 110 (2009); Carriage by Rail, 49 C.F.R. pt. 174 (2009); Transportation of Hazardous Materials: Driving and Parking Rules, 49 C.F.R. pt. 397 (2009).
73. Hazardous Materials Program Procedures, 49 C.F.R. pt. 107 (2009).
74. Hazardous Materials Public Sector Training and Planning Grants, 49 C.F.R. pt. 110 (2009).
75. Safety and Security Plans, 49 C.F.R. pt. 172, subpt. I (2009); Rail Risk Analysis Factors, 49 C.F.R. pt. 172, app. D (2009).
76. Carriage by Rail, 49 C.F.R. pt. 174 (2009).
77. Transportation of Hazardous Materials: Driving and Parking Rules, 49 C.F.R. pt. 397 (2009).
78. Routing of Class 7 (Radioactive) Materials, 49 C.F.R pt. 397, subpt. D (2009).
79. *Id.*
80. DOT, Guidelines for Selecting Preferred Routes for Shipments of Radioactive Materials, 16–31 (1992); *see also* Hazardous Materials Transportation Uniform Safety Act of 1990 § 4, Pub. L. No. 101-615, 104 Stat. 3244, at 3248–3250 (1990).
81. 49 C.F.R. § 397.103(c) (2009).
82. Hazardous Materials Transportation Act § 117a(a), 49 U.S.C. § 5116(a) (2006).
83. DOE, "Questions and Answers about Transportation of Radioactive Materials by DOE," 5, *www.em.doe.gov/PDFs/transPDFs/Questions_Answers_About_Transportation_Radioactive.pdf*.
84. Southern States Energy Board, "Radioactive Materials: Emergency Response and Transportation Planning," *www.sseb.org/radioactive-materials.php*.
85. Western Governors' Association, "WGA WIPP Fact Sheet," *www.westgov.org/wga_governors.htm*.
86. Council of State Governments, Midwestern Office, "Midwestern Radioactive Materials Transportation Committee," *www.csgmidwest.org*.
87. Council of State Governments—Eastern Regional Conference, "Northeast High-Level Radioactive Waste Transportation Project," *www.csgeast.org*.
88. *See, e.g.*, Office of Environmental Management, DOE, *Radioactive Material Transportation Practices Manual*, 2–3 (2008).
89. Larry Stern, Commercial Vehicle Safety Alliance, Presentation at the 2009 Waste Management Conference: "North American Standard Level VI Inspection Program—Ensuring Safe Transportation of Radioactive Materials," 1–5 (Mar. 3, 2009).
90. Implementing Recommendations of the 9/11 Commission Act of 2007 § 1551, 6 U.S.C.A. § 1201 (West 2010).
91. Hazardous Materials: Enhancing Rail Transportation Safety and Security for Hazardous Materials Shipments, 73 Fed. Reg. 72,182 (Nov. 26, 2008), 49 C.F.R. pts. 172 (2009).
92. Additional Planning Requirements for Transportation by Rail, 49 C.F.R. § 172.820 (2009).
93. Components of a Security Plan, 49 C.F.R. § 172.802 (2009).
94. Rail Risk Analysis Factors, 49 C.F.R. pt. 172, app. D (2009).
95. Implementing Recommendations of the 9/11 Commission Act of 2007 § 1551, 6 U.S.C.A. § 1201 (West 2010).
96. Office of Environmental Management, DOE, *Radioactive Material Transportation Practices Manual*, § 1.3 (2008).
97. DOE, *Final Supplemental EIS for Yucca*, 2-48 to 2-50 (2008); Record of Decision on Mode of Transportation and Nevada Rail Corridor for the Disposal of Spent Nuclear Fuel and High-Level Radioactive Waste at Yucca Mountain, Nye County, NV, 69 Fed. Reg. 18,557, at 18,562 (Apr. 8, 2004).
98. *Nevada v. Dept. of Energy*, 457 F.3d 78, 86–87 (D.C. Cir. 2006). The court rejected Nevada's other challenges to DOE's plans with regard to the Yucca repository as either unripe or failing to show that DOE made "arbitrary and capricious" decisions. *Id.* at 81, 92–93.
99. *Id.* at 93; Record of Decision on Mode of Transportation and Nevada Rail Corridor for the Disposal of Spent Nuclear Fuel and High-Level Radioactive Waste at Yucca Mountain, Nye County, NV, 68 Fed. Reg. 18,557, at 18,858 (Apr. 8, 2004).
100. Nuclear Waste Policy Act of 1982 §§ 114(a)(1)(D), 141(c)(2), 407, 42 U.S.C. §§ 10134(a)(1)(D), 10161(c)(2), 10247 (2006); Office of Environmental Management, DOE, *Radioactive Material Transportation Practices Manual*, § 1.3 (2008).
101. See the Notice of Intent to Expand the Scope of the Environmental Impact Statement for the Alignment, Construction, and Operation of a Rail Line to a Geologic Repository at Yucca Mountain, Nye County, NV, 69 Fed. Reg. 18,565, at 18,565 (Apr. 8, 2004), for DOE's determination that an additional rail corridor EIS was necessary.
102. Hazardous Materials Transportation Act § 105(b), 49 U.S.C. § 5112 (2006);
103. Committee on Transportation of Radioactive

Waste, National Research Council of the National Academies, *Going the Distance?* 192 (2006).
104. Office of Environmental Management, DOE, *Radioactive Material Transportation Practices Manual*, §§ 5.2–5.3 (2008). When transporting LLW, DOE allows carriers to determine routing. *Id.*
105. Waste Isolation Pilot Plant Land Withdrawal Act § 16, Pub. L. No. 102-579, 106 Stat. 4777, at 4792–4793 (1992); Nuclear Waste Policy Act of 1982 §§ 175, 180, 42 U.S.C. §§ 10174a, 10175 (2006).
106. Noyes, *Nuclear Waste Cleanup Technology and Opportunities*, 85 (1995).
107. DOE accounted for 6.6 percent (by volume) of the class A waste shipped to the Clive, Utah facility in 1994, and 77.8 percent in 2003. Government Accountability Office, *LLRW: Disposal Availability Adequate in the Short Term*, 13–14 (2004). GAO noted that the DOE's central LLRW database was unreliable, and that it had to estimate LLRW volumes based on data from three commercial disposal facilities. *Id.* at 14–15.
108. Current arrangements are summarized in Committee on Transportation of Radioactive Waste, National Research Council of the National Academies, *Going the Distance?* 52 (2006).
109. *Id.* at 117; NRC, "Transportation of Spent Nuclear Fuel" (Apr. 8, 2010), www.nrc.gov/waste/spent-fuel-transp.html.
110. Idaho Department of Environmental Quality, "Waste at INL: Spent Nuclear Fuel," www.deq.idaho.gov/inl_oversight/waste/spent_nuclear_fuel.cfm.
111. *Going the Distance?* at 40–41.
112. DOE, "National Transportation Program: Spent Nuclear Fuel and High-Level Radioactive Waste Transportation," www.nti.org/e_research/official_docs/doe/spent_nuc_feul.pdf; Committee on Transportation of Radioactive Waste, National Research Council of the National Academies, *Going the Distance?* 186 (2006).
113. Office of Environmental Management, DOE, *Radioactive Transportation Practices Manual*, § 5.2.1.a(4) (2008).
114. Committee on Transportation of Radioactive Waste, National Research Council of the National Academies, *Going the Distance?* 187, 203 (2006).
115. *Id.* at 186–187; Office of Environmental Management, DOE, *Radioactive Transportation Practices Manual*, § 5.2.1.a(4) (2008).
116. *Going the Distance?* at 46.
117. Committee on Transportation of Radioactive Waste, National Research Council of the National Academies, *Going the Distance?* 118 (2006).
118. *Id.*
119. Idaho National Laboratory, "2008 DOE Spent Nuclear Fuel and High Level Waste Inventory," inlportal.inl.gov/portal/server.pt?tbb=siteindex; Gary DeLeon, DOE, "Nuclear Waste Technical Review Board Meeting: Status of DOE Spent Nuclear Fuel and High Level Waste," 3 (June 11, 2009), www.nwtrb.gov/meetings/2009/june/deleon.pdf.
120. Committee on Transportation of Radioactive Waste, National Research Council of the National Academies, *Going the Distance?* 32 (2006)
121. Gary DeLeon, DOE, "Nuclear Waste Technical Review Board Meeting: Status of DOE Spent Nuclear Fuel and High Level Waste," 3 (June 11, 2009), www.nwtrb.gov/meetings/2009/june/deleon.pdf.
122. Committee on Transportation of Radioactive Waste, National Research Council of the National Academies, *Going the Distance?* 31–32 (2006).
123. DOE, "WIPP Shipment and Disposal Information," www.wipp.energy.gov/shipments.htm.
124. Record of Decision for the Department of Energy's Waste Management Program: Treatment and Storage of Transuranic Waste, 63 Fed. Reg. 3629, 3629 (Jan. 23, 1998). DOE's preferred practice, however, is to prepare TRU for transportation to WIPP at the site where the TRU is produced. *Id.*
125. DOE, "Waste to be Consolidated at Idaho Site Before Shipment to WIPP," www.wipp.energy.gov/fctshts/Consolidation.pdf.
126. CH2MHILL, "Miamisburg Closure," 2, www.ch2m.com/corporate/services/decontamination_and_decommissioning/assets/ProjectPortfolio/Miamisburg.pdf. DOE noted that building repackaging facilities at Mound Plant was not practical because it was closing the facility. DOE, *Supplement Analysis for Transportation of Transuranic Waste from the Mound Plant to Savannah River Site for Storage, Characterization, and Repackaging*, DOE/EIS-0200-SA02, at 1.0 (2001).
127. Committee on Transportation of Radioactive Waste, National Research Council of the National Academies, *Going the Distance?* 15–16, 50, 200–202 (2006).
128. *Id.* at 247–257.
129. Committee on Transportation of Radioactive Waste, National Research Council of the National Academies, *Going the Distance?* 122 (2006); NRC, *Safety of Spent Nuclear Fuel* (2003).
130. Committee on Transportation of Radioactive Waste, National Research Council of the National Academies, *Going the Distance?* 122 (2006).
131. The NAS Committee noted that few details are available about these incidents, but concluded that some probably were small leaks of water, which was then used for cooling and protection of the SNF during transport. Under current practice, SNF is cooled for 5 years before shipping, and then is transported in a dry state. *Id.* at 121. Most of the reported incidents were not leaks of radioactive material from the packages, but surface contamination on the package resulting from incomplete decontamination after loading of SNF. *Id.*
132. Haggerty, "'TRU' Cooperative Regulatory Federalism," 55–56 (2002), reports, based on a "Fact Sheet" by the "Radioactive Waste Campaign," that from 1971 to 1985, 1,034 LLW transport accidents occurred, resulting in some release of radioactivity from ninety containers.

133. Committee on Transportation of Radioactive Waste, National Research Council of the National Academies, *Going the Distance?* 32, 118–119 (2006).
134. *Id.* at 118.
135. *See Skull Valley Band of Goshute Indians v. Davis*, No. 07-cv-0526-DME-DON, at 4 (D. Utah July 26, 2010), for the order vacating and remanding DOI's decision denying PFS's right-of-way application.
136. *Id.* at 19, 28–29, 35. The history of the PFS facility is discussed further in Chapter 7.
137. Committee on Transportation of Radioactive Waste, National Research Council of the National Academies, *Going the Distance?* 253 (2006).
138. *Id.* at 232, 253.
139. "WIPP Shipment and Disposal Information," *www.wipp.energy.gov/shipments.htm*.
140. Committee on Transportation of Radioactive Waste, National Research Council of the National Academies, *Going the Distance?* 142 (2006).
141. Office of Nuclear Energy, DOE, *Draft GNEP Programmatic EIS*, app. E, at E-18 to E-21, E-39 to E-41 (2008).
142. The agreements consisted of a stipulated agreement (and later supplements to it) settling litigation brought by New Mexico against DOE and a consultation and cooperation agreement (and later modifications to it) incorporated as an appendix to the stipulated agreement. *See* Chapter 5.
143. Stipulated Agreement, at ¶ 7, *New Mexico ex rel. Bingaman v. Dept. of Energy*, No. 81-0363 JB (D.N.M. July 1, 1981).
144. Supplemental Stipulated Agreement Resolving Certain State Off-Site Concerns over WIPP, at Art. III §§ A(2)(c), (C)(2)(a)–(e), (h)–(i), *New Mexico ex rel. Bingaman v. Dep't of Energy*, No. 81-0363 JB (D.N.M. Dec. 27, 1982).
145. *Id.* at Art. III § C(2)(f).
146. *Id.* at Art. III § E(2)(a)–(b).
147. "Second Modification to the July 1, 1981 'Agreement for Consultation and Cooperation' on WIPP by the State of New Mexico and U.S. Department of Energy," ¶¶ 2, 4 (1987).
148. Waste Isolation Pilot Plant Land Withdrawal Act §§ 15–16, Pub. L. No. 102-579, 106 Stat. 4777, at 4791–4793 (1992).
149. *Id.* § 16(e), Pub. L. No. 102-579, 106 Stat. 4777, at 4793 (1992).
150. *Id.* § 16(c)–(d), Pub. L. No. 102-579, 106 Stat. 4777, at 4792–4793 (1992).
151. John Gervers, *The Record of U.S. Department of Energy Commitments to the State of New Mexico Regarding the Waste Isolation Pilot Plant* (1990), as quoted in Urban Environmental Research, LLC, *Lessons Learned from WIPP*, 13 (2001).
152. Waste Isolation Pilot Plant: Availability of Draft Supplement to the Final Environmental Impact Statement, 54 Fed. Reg. 16,350, at 16,350 (Apr. 21, 1989).
153. *Id.*
154. DOE, *WIPP Disposal Phase: Final Supplemental EIS*, S-42 (1997).
155. *Id.* at S-18, S-44, S-46.
156. *Id.* at S-19.
157. *Id.* at 3-6 to 3-7.
158. *Id.* at 19-73 to 19-83. The New Mexico Environment Department, while accepting the decision to use truck transportation initially, urged DOE to continue studying the rail option; DOE responded that it would continue studying the possibility of switching to rail transport. *Id.* at S-18, Comment A-002.
159. Record of Decision for the Department of Energy's Waste Isolation Pilot Plant Disposal Phase, 63 Fed. Reg. 3624, 3624 (Jan. 23, 1998).
160. Designation of Highway Routes for Transport of Radioactive Materials, 10 N.M. Reg. 231 (Apr. 30, 1999).
161. DOE, "WIPP TRU Waste Transportation Routes," *www.wipp.energy.gov/routes.htm*.
162. Western Governors' Association, *WIPP Transportation Safety Program Implementation Guide*, XII-1 (2003).
163. *Id.* at XII-2. So long as state-selected routes meet DOT criteria, they become part of the national waste transportation routes established by DOE.
164. Waste Isolation Pilot Plant: Availability of Draft Supplement to the Final Environmental Impact Statement, 54 Fed. Reg. 16,350, at 16,350 (Apr. 21, 1989).
165. McCutcheon, *Nuclear Reactions*, 91 (2002).
166. *Id.*
167. *Id.*; DOE, "WIPP Chronology," *www.wipp.energy.gov/fctshts/chronology.pdf*.
168. McCutcheon, *Nuclear Reactions*, 91 (2002). A smaller version of TRUPACT-II, the HalfPACT, was certified by the NRC in 2000. DOE, "Transuranic Waste Transportation Containers," *www.wipp.energy.gov/fctshts/truwastecontainers.pdf*. The RH-72B container for RH-TRU that needed to be remotely handled due to higher radioactivity was certified by NRC in 2000 as well. Office of Nuclear Material Safety and Safeguards, NRC, "Certificate of Compliance for RH-TRU 72B Transport Package Issued," press release (Mar. 10, 2000).
169. DOE, "WIPP Chronology," *www.wipp.energy.gov/fctshts/chronology.pdf*.
170. *Id.*; Office of Nuclear Material Safety and Safeguards, NRC, "Certificate of Compliance for RH-TRU 72B Transport Package Issued," press release (Mar. 10, 2000).
171. Sue Vorenberg, "Ten Years in Operation, WIPP Boasts Sterling Safety Record, Continued Support," Santa Fe New Mexican, A1 (Mar. 29, 2009). Tests after an accident on August 25, 2002, indicated that radioactivity had released inside the TRUPACT-II container. The shipment was sent back to Idaho to be examined. DOE eventually determined that the release was not caused by the traffic accident, but rather by improper loading of the container. Don Hancock, Director, Nuclear Waste Safety Program, Southwest Research and Information Center, e-mail correspondence on file with the authors (Mar. 2, 2010).

172. For WIPP's capacity cap, Waste Isolation Pilot Plant Land Withdrawal Act § 7(a)(3), Pub. L. No. 102-579, 106 Stat. 4777, at 4785 (1992).
173. DOE, "DOE Looks at New Shipping Package for WIPP Use," press release (Mar. 15, 2004), www.wipp.energy.gov/pr/2004/TRUPACT-III.pdf.
174. NRC, "Certificate of Compliance for Radioactive Material Packages: No. 9305" (June 1, 2010).
175. Task Force on Nuclear Waste, Western Governors' Association, *Transport of TRU to WIPP*, 13 (1989); *see also* 2009 *Memorandum between WGA and DOE* (2009); WIPP Transportation Technical Advisory Group, Western Governors' Association, *WIPP Rail Transportation Safety Program*, 6 (2004).
176. WIPP Transportation Technical Advisory Group, Western Governors' Association, *WIPP Rail Transportation Safety Program*, 6 (2004).
177. *Id.* at 6–7.
178. Waste Isolation Pilot Plant Land Withdrawal Act § 16(d), Pub. L. No. 102-579, 106 Stat. 4777, at 4793 (1992).
179. The *Memorandum between WGA and DOE*, 2–3 (2009), references the 1995 and 2003 WGA-DOE Memoranda of Agreement. For the current version of the guide, *see* Western Governors' Association, *WIPP Program Implementation Guide*, www.westgov.org/index.php?option=com_content&view=article&catid=102:initiatives&id=226:wga-wipp-program-implementation-guide&Itemid=79.
180. *See* DOE, *WIPP Transportation Plan*, Revision 1, at 1 (2002). The WIPP Transportation Plan also incorporated protocols developed by the Southern States Energy Board. *Id.*
181. *Memorandum between WGA and DOE*, 2–3 (2009).
182. Western Governors' Association, "Western Governors Applaud 10 Years of Safe, Uneventful Shipments to WIPP," press release (Mar. 26, 2009).
183. *Memorandum between WGA and DOE*, 2 (2003).
184. WIPP Transportation Technical Advisory Group, Western Governors' Association, *WIPP Rail Transportation Safety Program Implementation Guide*, ii (2004).
185. Western Governors' Association, "Letter to Steven Chu, Secretary of Energy" (May 21, 2009), www.westgov.org/wieb/radioact/doi/05-21-09WGA-Chu.pdf.
186. *Memorandum between WGA and DOE* (2009).
187. Haggerty, "'TRU' Cooperative Regulatory Federalism," 70–71 (2002).
188. Committee on Transportation of Radioactive Waste, National Research Council of the National Academies, *Going the Distance?* 228 (2006).
189. *Id.*
190. DOE, *Final EIS for Yucca*, at 2-2 (2002). DOE estimated that rail transport would have many advantages over truck transport, including lower occupational and nonoccupational exposure levels, fewer accidents, fewer fatalities from accidents, and less risk of, and lower levels of exposure from, accidents. *Id.* at 6-38 to 6-39, 6-43, 6-47 to 6-50.
191. Record of Decision on Mode of Transportation and Nevada Rail Corridor for the Disposal of Spent Nuclear Fuel and High-Level Radioactive Waste at Yucca Mountain, Nye County, NV, 69 Fed. Reg. 18,557, at 18,561 (Apr. 8, 2004).
192. *Id.* at 18,559.
193. *Id.* at 18,561.
194. Committee on Transportation of Radioactive Waste, National Research Council of the National Academies, *Going the Distance?* 223 (2006).
195. DOE, *Final EIS for Yucca*, 2-47 (2002); DOE, "Statement for Use of Dedicated Trains for Waste Shipments to Yucca Mountain" (2005).
196. Committee on Transportation of Radioactive Waste, National Research Council of the National Academies, *Going the Distance?* 217 (2006).
197. *Id.* at 235, citing General Accounting Office, *Spent Nuclear Fuel* (2003).
198. Record of Decision on Mode of Transportation and Nevada Rail Corridor for the Disposal of Spent Nuclear Fuel and High-Level Radioactive Waste at Yucca Mountain, Nye County, NV, 69 Fed. Reg. 18,557, at 18,564 (Apr. 8, 2004); *see* Nuclear Waste Repository Project Office, Nye County, Nevada, Rail Transportation Economic Impact Study (2007).
199. *Nevada v. Dep't of Energy*, 457 F.3d 78, 81 (D.C. Cir. 2006).
200. *Id.* at 86, 88–89.
201. *Id.* at 81, 89, citing 40 C.F.R. § 1503.1 (2009).
202. DOE, *Final EIS for Railroad to Yucca* (2008).
203. Record of Decision and Floodplain Statement of Findings—Nevada Rail Alignment for the Disposal of Spent Nuclear Fuel and High-Level Radioactive Waste at Yucca Mountain, Nye County, NV, 73 Fed. Reg. 60,247, at 60,248 (Oct. 10, 2008).
204. Nevada, Motion to Suspend Further Proceedings, Or in the Alternative, to Reopen the Procedural Schedule and Record Previously Established for Public Comment on Public Convenience and Necessity (PCN) Issues Related to the Application Filed by the United States Department of Energy under 49 U.S.C. 10901, STB Finance Docket No. 35106, at 5 (Apr. 7, 2009).
205. Committee on Transportation of Radioactive Waste, National Research Council of the National Academies, *Going the Distance?* 13–14, 21–22, 38 (2006).
206. *See* the statements of Senators Barbara Boxer and Harry Reid. *Status of the Yucca Mountain Project: Hearing before the Senate Committee on Environment and Public Works*, 109th Cong., 2d Sess., 10, 48–49 (Mar. 1, 2006).
207. *Going the Distance?*, 7–8, 247–248.
208. *Id.* at 11–12.
209. Western Governors' Association, "Assessing the Risks of Terrorism and Sabotage against Spent Nuclear Fuel High-Level Waste Shipments," Policy Resolution 10-2 (2010), www.westgov.org/index.php?option=com_wga&view=resolutions&Itemid=53.
210. Office of Civilian Radioactive Waste Management, DOE, National Transportation Plan, DOE/RW-0603, at 25 (Jan. 2009).
211. *Going the Distance?*, 21–22.

Chapter 4

1. Atomic Energy Act of 1954 Amendments, Pub. L. No. 86-373, 73 Stat. 688 (1959).
2. NRC, "Integrated Materials Performance Evaluation Program (IMPEP)," *nrc-stp.ornl.gov/impeptoolbox/impep.html*; Criteria for Guidance of States and NRC in Discontinuance of NRC Regulatory Authority and Assumption thereof by States through Agreement, 46 FED. REG. 7540 (Jan. 23, 1981); Discontinuance of NRC Authority and Assumption Thereof by States through Agreement; Criteria for Guidance of States and NRC, 48 FED. REG. 33,376 (July 21, 1983); NRC, "Becoming an Agreement State," *www.nrc.gov/about-nrc/state-tribal/become-agreement.html*.
3. Organization of Agreement States, "Agreement State Links," *www.agreementstates.org/page/agreement-state-links*.
4. These standards can be found at 10 C.F.R. pt. 61 (2010).
5. This book uses the term "low-level waste" or "LLW" to include both what NRC terms LLRW and what DOE terms LLW-like waste.
6. Advisory Committee on Nuclear Waste, NRC, *History and Framework of Low-Level Radioactive Waste Management*, 1 (2007); Low-Level Radioactive Waste Policy Amendments Act of 1985 § 3(b)(1), 42 U.S.C. § 2021c(b)(1) (2006).
7. Advisory Committee on Nuclear Waste, NRC, *History and Framework of Low-Level Radioactive Waste Management*, 2 (2007).
8. Government Accountability Office, *Low-Level Radioactive Waste Management: Approaches Used by Foreign Countries*, 4 (2007).
9. Advisory Committee on Nuclear Waste, NRC, *History and Framework of Low-Level Radioactive Waste Management*, 2 (2007). This does not include GTCC; NRC estimates include only waste that is under its control.
10. Government Accountability Office, *LLRW: Disposal Availability Adequate in the Short Term*, 13 (2004)
11. *Id.*, 14.
12. *Id.*
13. English, *Siting Low-Level Radioactive Waste Disposal Facilities*, 6 (1992).
14. Because no regulations existed dealing specifically with LLW disposal, AEC licensed these facilities according to the generic radiation protection standards then available under 10 C.F.R. Part 20 (1971). Advisory Committee on Nuclear Waste, NRC, *History and Framework of Low-Level Radioactive Waste Management*, 9 (2007).
15. Along with shallow land burial, other types of near-surface land disposal include belowground vaults, earth-mounded concrete bunkers, and augured holes. These alternate disposal methods are regarded as technically sound but are significantly more expensive. Bell, *Engineering Geology and Construction*, 671 (2004).
16. Kiefer, "Low-Level Radioactive Waste Issues in Michigan," 351 (2002). It is likely not a coincidence that the Barnwell and Beatty facilities were sited on state land adjacent to major DOE facilities. Government Accountability Office, *LLRW: Disposal Availability Adequate in the Short Term*, 9 (2004).
17. Advisory Committee on Nuclear Waste, NRC, *History and Framework of Low-Level Radioactive Waste Management*, 9 (2007).
18. EPA, *Record of Decision: Maxey Flats Disposal Site*, 11 (1991); New York State Department of Environmental Conservation, *West Valley: History and Future*, 1 (2008).
19. EPA, *Record of Decision: Maxey Flats Disposal Site*, 21 (1991).
20. New York State Department of Environmental Conservation, *West Valley: History and Future*, 2 (2008).
21. 10 C.F.R. pt. 61.1 (2010); Licensing Requirements for the Disposal of Radioactive Waste, 47 FED. REG. 57,446, at 57,463 (Dec. 27, 1982).
22. Office of Technology Assessment, *Partnerships under Pressure*, 136–137 (1989); *see also* South Carolina Department of Health and Environmental Control, *Commercial Low-Level Radioactive Waste Disposal in South Carolina*, 7 (2007).
23. Advisory Committee on Nuclear Waste, NRC, *History and Framework of Low-Level Radioactive Waste Management*, 11 (2007).
24. English, *Siting Low-Level Radioactive Waste Disposal Facilities*, 6 (1992); Chuang, "Who Should Win the Garbage Wars?" 2434 (2004); Kiefer, "Low-Level Radioactive Waste Issues in Michigan," 351 (2002).
25. South Carolina Department of Health and Environmental Control, *Commercial Low-Level Radioactive Waste Disposal in South Carolina*, 2 (2007).
26. *City of Philadelphia v. New Jersey*, 437 U.S. 617 (1978).
27. *Id.* at 629.
28. Interagency Review Group on Nuclear Waste Management, *Report to the President*, 106–107 (1979).
29. *Id.* at 90.
30. *Id.* at 94.
31. National Governors Association, *Low-Level Waste*, 5 (1980).
32. General Accounting Office, *Radioactive Waste: Status of Commercial Low-Level Waste Facilities*, 15 (1995).
33. English, *Siting Low-Level Radioactive Waste Disposal Facilities*, 7 (1992).
34. *Id.*
35. Nuclear Waste Policy Act, H.R. 8378, 96th Cong. (1980).
36. Rabe et al., "NIMBY and Maybe," 68 (1994).
37. *Id.* at 78.
38. H.R. 8378, 96th Cong. § 201(a)(1) (1980).
39. *Id.* § 2021(a)(2).
40. *See, e.g.*, National Governors Association, *Low-Level Waste*, 6; *cf.* Low-Level Radioactive Waste Policy Act of 1980 § 4(a)(1) (1981).

41. H.R. 8378, 96th Cong. § 201(a)(2) (1986).
42. Rabe et al., "NIMBY and Maybe," 78–79 (1994).
43. H.R. 8378, 96th Cong. § 201(a)(2)(B); *see also* Mostaghel, "The Low-Level Radioactive Waste Policy Amendments Act," 389 (1994).
44. Committee on Energy and Commerce, *Low-Level Radioactive Waste Policy Amendments Act of 1985*, 18 (1985).
45. Low-Level Radioactive Waste Policy Act of 1980, Pub. L. No. 96-573, 94 Stat. 3347 (1980); Energy and Natural Resources Committee, *Low-Level Radioactive Waste Policy Act*, at 13–14 (1980).
46. Chuang, "Who Should Win the Garbage Wars?" 2435 (2004).
47. H.R. Rep. No. 99-314, 13–14 (1985) ("In July and August of 1984 the Governor of South Carolina and the Washington State Legislature substantially improved the prospects for ratification of compacts by announcing that they could consider accepting low-level waste from outside their compact regions for some period after January 1, 1986, if Congress would ratify their compacts.").
48. *Id.*; *see also* Rabe et al., "NIMBY and Maybe," 85 (1994).
49. Vari, Reagan-Cirincione, and Mumpower, *LLRW Disposal Facility Siting*, 21 (1994).
50. Advisory Committee on Nuclear Waste, NRC, *History and Framework of Low-Level Radioactive Waste Management*, 17 (2007).
51. Low-Level Radioactive Waste Policy Amendments Act of 1985, Pub. L. No. 99-240, 99 Stat. 1842 (1986).
52. Low-Level Radioactive Waste Policy Amendments Act of 1985 § 5, 42 U.S.C. § 2021e (2009); *see also* Committee on Energy and Commerce, *Low-Level Radioactive Waste Policy Amendments Act of 1985*, at 18–19 (1985).
53. Low-Level Radioactive Waste Policy Amendments Act of 1985 § 6, 42 U.S.C. § 2021f (2010).
54. *Id.* § 5(d)(2), 42 U.S.C. § 2021e(d)(2) (2010); *see also* Advisory Committee on Nuclear Waste, NRC, *History and Framework of Low-Level Radioactive Waste Management*, 17 (2007).
55. Low-Level Radioactive Waste Policy Amendments Act of 1985 § 5(d)(2), 42 U.S.C. § 2021e(d)(2) (2010). English, *Siting Low-Level Radioactive Waste Disposal Facilities*, 6 (1992); Chuang, "Who Should Win the Garbage Wars?" 2434n.28 (2004).
56. Low-Level Radioactive Waste Policy Amendments Act of 1985 § 5(d)(2)(C)(i), 42 U.S.C. § 2021e (d)(2)(C)(i) (2010).
57. *Id.*
58. Committee on Energy and Commerce, *Low-Level Radioactive Waste Policy Amendments Act of 1985*, 18–19 (1985).
59. *New York v. United States*, 505 U.S. 144, 148 (1992). Seventeen states submitted amicus briefs on behalf of New York; the three states with LLW facilities, Washington, South Carolina and Nevada, were named as respondents. *Id.* at 148. *See* Rabe et al., "NIMBY and Maybe," 88–99 (1994).
60. *Id.* at 166–167.
61. For a post hoc explanation of California's motivations for building a disposal facility in the first place, *see* Office of Assembly Speaker Pro Tem Fred Keeley, "AB 2214 Factsheet 2," press release (May 28, 2002).
62. *See* Committee on Energy and Commerce, *Low-Level Radioactive Waste Policy Amendments Act of 1985*, 23 (1985).
63. Chuang, "Who Should Win the Garbage Wars?" 2435 (2004).
64. Texas would employ this strategy in the late 1990s and early 2000s after deciding to build its Andrews County disposal site. AB 2214, 2002 Cal. Legis. Serv. Ch. 513 § 4, codified at Cal. Health & Safety Code § 115261 (2010); *see also* Office of Assembly Speaker Pro Tem Fred Keeley, "AB 2214 Factsheet 2," press release (May 28, 2002).
65. Atlantic Interstate Low-Level Radioactive Waste Compact Implementation Act, 2000 S.C. Acts 357 § 1, codified at S.C. Laws § 48-46-40(A)(6).
66. Committee on Energy and Commerce, *Texas Low-Level Radioactive Waste Disposal Compact Consent Act*, 8; Cragin, *Nuclear Nebraska*, 253 (2007).
67. Texas Low-Level Radioactive Waste Disposal Compact Consent Act § 5, codified at 25 U.S.C. § 2021d note.
68. Vari, Reagan-Cirincione, and Mumpower, LLRW Disposal Facility Siting, 68, 136 (1994).
69. In the 1990s, the Central Compact unsuccessfully offered Boyd County, Nebraska, the Central Compact's proposed host site, $120 million over 40 years in community development grants. As of the 2000 Census, Boyd County had a population of 2,438.
70. *See, e.g.*, Lone Star Chapter, Sierra Club, "Disaster in the Making: Nuke Waste Bill Passes Texas House," press release (Apr. 23, 2003).
71. Newberry, "Rise and Fall and Rise and Fall of American Public Policy on Disposal of Low-Level Radioactive Waste," *Southeastern Environmental Law Journal* 69 (1993).
72. *Id.*
73. *See* Kiefer, "Low-Level Radioactive Waste Issues in Michigan," 358 (2002); Mich. Comp. Laws §§ 333.26201 to 333.26226 (2010).
74. South Carolina Department of Health and Environmental Control, *Commercial Low-Level Radioactive Waste Disposal in South Carolina*, 3 (2007).
75. *Alabama v. North Carolina*, No. 22O132-ORG (U.S. Jan. 11, 2010).
76. *See Alabama v. North Carolina*, 560 U.S. ___, slip op. at 26 (2010), rejecting motions for summary judgment.
77. Cragin, *Nuclear Nebraska*, 131 (2007).
78. *Id.* at 159–160.
79. *Entergy Arkansas, Inc. v. Nebraska*, 226 F.Supp.2d 1047, 1102–1104 and 1140–1142 (D. Neb. 2002), *aff'd*, 358 F.3d 528 (8th Cir. 2004).

80. *Entergy Arkansas, Inc. v. Nebraska*, 358 F.3d 528, 558 (8th Cir. 2004).
81. Chuang, "Who Should Win the Garbage Wars?" 2438–2439 (2004).
82. The Andrews site is near the New Mexico–Texas border and is located in the same geological formation as WIPP. Texas Commission on Environmental Quality, *Radioactive Material License No. R04100* (2009).
83. Chuang, "Who Should Win the Garbage Wars?" 2452 (2004).
84. Cragin, *Nuclear Nebraska*, 210–211 (2007); AB 2214, 2002 Cal. Legis. Serv. Ch. 513 § 4, codified at Cal. Health & Safety Code § 115261 (West 2010); *see also* Office of Assembly Speaker Pro Tem Fred Keeley, "AB 2214 Factsheet 2," press release (May 28, 2002).
85. Vari, Reagan-Cirincione, and Mumpower, *LLRW Disposal Facility Siting*, 98, 114 (1994).
86. Rabe et al., "NIMBY and Maybe," 82 (1994).
87. Cragin, *Nuclear Nebraska*, xxiii (2007).
88. *Id.* at 50–51, 60.
89. *Id.* at 18, 38.
90. *Id.* at 120–123, 138–140.
91. Chuang, "Who Should Win the Garbage Wars?" 2435 (2004).
92. The Beatty site is currently accepting RCRA and polychlorinated biphenyl waste. Advisory Committee on Nuclear Waste, NRC, *History and Framework of Low-Level Radioactive Waste Management*, 30 (2007).
93. Chuang, "Who Should Win the Garbage Wars?" 2435 (2004); Government Accountability Office, *Low-Level Radioactive Waste Management: Approaches Used by Foreign Countries*, 70 (2007).
94. Holt, Congressional Research Service, Civilian Nuclear Waste Disposal, 17 (2007).
95. Chuang, "Who Should Win the Garbage Wars?" 2435 (2004). South Carolina unilaterally excluded North Carolina waste from Barnwell in response to North Carolina's failure to build a new Southeast Compact facility.
96. *Id.* at 2455.
97. Government Accountability Office, *Low-Level Radioactive Waste Management: Approaches Used by Foreign Countries*, 70 (2007).
98. In 2007, the Atlantic Compact produced only a little over a quarter of all waste accepted by Barnwell. *Id.* at 81.
99. Texas Commission on Environmental Quality, *Radioactive Material License No. R04100* (2009).
100. "Commission Lets 36 States Dump Nuke Waste in Texas," Associated Press (Jan. 4, 2011).
101. Nick Lawton, "Andrews Waste Rule Approved," NewsWest9 (January 5, 2011), www.newswest9.com/Global/story.asp?S=13783892.
102. Ramit Plushnick-Masti, "Commission Lets 36 States Dump Nuke Waste in Texas," *Rutland [Vermont] Herald* (January 5, 2011), www.vermonttoday.com/apps/pbcs.dll/article?AID=/RH/20110105/BUSINESS/711239927/-1/CBJ.
103. Government Accountability Office, *LLRW: Disposal Availability Adequate in the Short Term*, 8–9 (2004).
104. Government Accountability Office, *Low-Level Radioactive Waste Management: Approaches Used by Foreign Countries*, 70 (2007).
105. EnergySolutions Inc., *Form 10-K*, F-27 (2008).
106. *EnergySolutions, LLC v. Northwest Interstate Compact*, 2009 U.S. Dist. LEXIS 41209 (May 15, 2009); Advisory Committee on Nuclear Waste, NRC, *History and Framework of Low-Level Radioactive Waste Management*, 30 (2007); Government Accountability Office, *LLRW: Disposal Availability Adequate in the Short Term*, 8–9 (2004).
107. SB 24, 2005 Utah Laws ch. 10 § 2, codified at Utah Code § 19-3-103.7.
108. Lisa Riley Roche, "EnergySolutions Abandons Plan to Import Italian Nuclear Waste to Utah," *Deseret News* (Salt Lake City) (July 14, 2010).
109. EnergySolutions, LLC v. Northwest Interstate Compact, 2009 U.S. Dist. LEXIS 41209, 47 (May 15, 2009).
110. *Id.* at 13.
111. *Id.* Utah's license for the Clive facility stipulated that the Northwest Compact had to approve storage at Clive of waste generated within the compact's boundaries.
112. Amy O'Donoghue, "Utah Appeals Ruling on Italian Nuclear Waste," *Deseret News* (Salt Lake City), B1 (June 24, 2009).
113. Radioactive Import Deterrence Act, H.R. 515, 111th Cong. (2009).
114. Judy Fahys and Thomas Burr, "EnergySolutions Dumps Italian Waste Bid for Utah," *Salt Lake Tribune* (Sept. 21, 2010).
115. Advisory Committee on Nuclear Waste, NRC, *History and Framework of Low-Level Radioactive Waste Management*, 42–43 (2007).
116. *Id.* For a relatively recent discussion of the LLW disposal situation and policy options, *see* Committee on Improving Practices for Regulating and Managing Low-Activity Radioactive Waste, National Research Council of the National Academies, Improving the Regulation and Management of Low-Activity Wastes (2006).
117. *Id.*
118. Sandia National Laboratories, Greater-Than-Class C Low-Level Waste: Inventory Estimates, iii (2007). Large amounts of activated metal waste must be disposed of when reactors are decommissioned. Nuclear Waste Project Office, State of Nevada, "Greater-than-Class-C Radioactive Waste" www.state.nv.us/nucwaste/gtcc/gtcc.htm.
119. DOE, *Recommendations for the Management of Greater-Than-Class-C Low-Level Radioactive Waste*, 3-2 (1987).
120. Corrections: Notice of Intent to Prepare an Environmental Impact Statement for the Disposal of Greater-Than-Class-C Low-Level Radioactive Waste, 72 Fed. Reg. 41,819 (Jul. 31, 2007).

121. Waste Isolation Pilot Plant Land Withdrawal Act § 12, as enacted, Pub. L. No. 102-579, 106 Stat. 4777, 4791 (1992).
122. Nuclear Waste Policy Act of 1982 § 2(12)(b), 42 U.S.C. 10101(12)(b) (2010). Yucca's purpose is to provide for a "repositor[y] that will provide a reasonable assurance that the public and the environment will be adequately protected from the hazards posed by high-level radioactive waste and such spent nuclear fuel as may be disposed of in a repository." Nuclear Waste Policy Act of 1982 § 111, 42 U.S.C. 10131 (2010). Thus, if NRC chose to exercise its authority to name GTCC waste as HLW, there is little doubt that Yucca would be open to such waste.
123. 10 C.F.R. § 61.55(a)(2)(iv) (2010).
124. DOE, Environmental Impact Statement for the Disposal of Greater-Than-Class-C (GTCC) Low-Level Waste and GTCC-Like Waste (2011).
125. *Id.* at 40,138.
126. The seven proposed federal sites are Hanford Reservation (Washington), Idaho National Laboratory (Idaho), Los Alamos National Laboratory (New Mexico), the Nevada Test Site (Nevada), Oak Ridge Reservation (Tennessee), the Savannah River Site (South Carolina), and on federal lands adjacent to WIPP (New Mexico). *Id.*
127. Government Accountability Office, *LLRW: Disposal Availability Adequate in the Short Term*, 8–9 (2004).
128. English, *Siting Low-Level Radioactive Waste Disposal Facilities*, 21 (1992).
129. If a particular long-lived radionuclide has a concentration below 0.1 times the amount listed in table 1 in 10 C.F.R. Section 61.55, then it is in class A; if the concentration exceeds 0.1 times the amount listed in table 1 but is less than the value in that table, then it is in class B. The class of a particular short-lived radionuclide is determined by whether its concentration is lower than those values listed in table 2, 10 C.F.R. Section 61.55.
130. Government Accountability Office, *LLRW: Disposal Availability Adequate in the Short Term*, 20 (2004).
131. *Id.* at 10. Storage of these wastes at generator sites is regulated under 10 CFR Part 20.
132. Committee on Improving Practices for Regulating and Managing Low-Activity Radioactive Waste, National Research Council of the National Academies, *Improving the Regulation of Low-Activity Wastes*, 25 (2006).
133. Advisory Committee on Nuclear Waste, NRC, *History and Framework of Low-Level Radioactive Waste Management*, 32 (2007).
134. *Id.*
135. *Id.* at 6.
136. Government Accountability Office, *LLRW: Disposal Availability Adequate in the Short Term*, 42 (2004).
137. Government Accountability Office, *Low-Level Radioactive Waste Management: Approaches Used by Foreign Countries*, 4 (2007).
138. Low-Level Radioactive Waste Policy Amendments Act § 3(b), 42 U.S.C. § 2021c(b) (2006). Between 1959 and 1963, AEC accepted commercial LLW for disposal at Oak Ridge and Idaho National Laboratories. Kiefer, "Low-Level Radioactive Waste Issues in Michigan," 346 (2002).
139. AIFs act as retrievable storage for LLW, rather than as the permanent burial that shallow land disposal would be. Advisory Committee on Nuclear Waste, NRC, *History and Framework of Low-Level Radioactive Waste Management*, 32 (2007).
140. Cragin, *Nuclear Nebraska*, 261–262 (2007).
141. *Id.* at 33.
142. NRC, *Blending of Low-Level Radioactive Waste*, 36 (2010).
143. Thomas Burr, "EnergySolutions Pitches 'Blending' Hotter Radioactive Waste," *Salt Lake Tribune* (Dec. 15, 2009).
144. Katherine Ling, "NRC Weighs 'Blending' Regs for Low-Level Waste," *Greenwire* (Jan. 13, 2010), www.eenews.net/Greenwire/2010/01/13/13/.
145. Judy Fahys, "Deal Would Bring Blended Radioactive Waste to Utah," *Salt Lake Tribune* (Feb. 8, 2011).
146. The LAMW that the 2004 proposals sought to bury in RCRA landfills would be classified as very low level waste by European standards; such RCRA disposal was proposed because LLW disposal is expensive compared to RCRA disposal. Adequate capacity exists for LAMW at RCRA Subtitle C (i.e., hazardous) waste landfills. Approaches to an Integrated Framework for Management and Disposal of Low-Activity Radioactive Waste, 68 Fed. Reg. 65,120, at 65,123 (Nov. 18, 2003).
147. See, for example, the statement of Raymond A. Guilmette, President, Health Physics Society, *Hearing on Low Level Waste Oversight before the Senate Committee on Energy and Natural Resources*, 108th Cong. 5–6 (2004); 10 C.F.R. § 20.2002 (2010); NRC, "Low-Level Waste Disposal under 10 C.F.R. 20.2002" (Mar. 21, 2007), www.nrc.gov/waste/llw-disposal/10cfr20-2002-info.html.
148. Approaches to an Integrated Framework for Management and Disposal of Low-Activity Radioactive Waste, 68 Fed. Reg. 65,120, at 65,123 (Nov. 18, 2003).
149. *See* "EPA Reviving RCRA Proposal to Landfill Low-Level Nuclear Waste," Environmental NewsStand (Mar. 4, 2008), *environmentalnewsstand.com/ Defense-Environment-Alert/Defense-Environment-Alert-03/04/2008/menu-id-307.html*.
150. Committee on Improving Practices for Regulating and Managing Low-Activity Radioactive Wastes, National Research Council of the National Academies, *Improving the Regulation and Management of Low-Activity Radioactive Wastes*, 20 (2006).
151. *Id.*
152. *Id.*
153. *Id.* at 14, 15, 24.
154. Low-Level Radioactive Waste Policy Amendments Act of 1985 § 6, Pub. L. No. 99-240, 99 Stat. 1842, 1843 (1986), codified as amended at 42 U.S.C. § 2021f.

Chapter 5

1. See Chapter 1 for a discussion of the technical issues that led the AEC to abandon Project Salt Vault. The biggest problems were unplugged boreholes in the salt formation, the mysterious disappearance of 175,000 gallons of water from a nearby mine, and the resulting disputation of AEC's original belief that it could successfully plug twenty nearby oil and gas boreholes that might compromise containment of radioactive wastes in the salt formation.
2. Downey, "Politics and Technology in Repository Siting," 53–56 (1985).
3. Office of Technology Assessment, *Managing the Nation's Commercial High-Level Radioactive Waste*, 86 (1985). The Office of Technology Assessment highlighted this lack of procedural and organizational definition in federal-state roles in a 1976 report critiquing ERDA's program plan. Office of Technology Assessment, *Comparative Analysis of ERDA Plan*, 20 (1976).
4. Office of Technology Assessment, *Managing the Nation's Commercial High-Level Radioactive Waste*, 86 (1985); Statement of Representative Harold Runnels, Chairman, Subcommittee on Oversight and Investigations, *Nuclear Waste Isolation Pilot Plant* (WIPP): *Oversight Hearings before the Subcommittee on Oversight and Investigations of the House Committee on Interior and Insular Affairs*, 96th Cong. 1 (1979).
5. WIPP Authorization Act, Department of Energy National Security and Military Applications of Nuclear Energy Authorization Act of 1980 § 213(a), Pub. L. No. 96-164, 93 Stat. 1259, 1265 (1979).
6. Waste Isolation Pilot Plant Land Withdrawal Act § 12, Pub. L. No. 102-579, 106 Stat. 4777, 4791 (1992).
7. Downey, "Politics and Technology in Repository Siting," 56 (1985).
8. *Id.* at 56–57.
9. The first WIPP borehole hit unexpected brine reservoirs, releasing noxious gases that endangered the drilling crew. The prospective site for the plant was then relocated six miles to the southwest. Rechard, *Milestones for Disposal at WIPP*, 3 (1998); Carter, *Nuclear Imperatives*, 181 (1987). Problems associated with brine pockets are discussed in greater detail later in this chapter.
10. New Mexico: Proposed Withdrawal and Reservation of Lands, 41 Fed. Reg. 54,994, at 54,994–54,995 (Dec. 16, 1976).
11. Federal Land Policy and Management Act of 1976 § 204(b)(1), 43 U.S.C. 1714(b)(1) (2006).
12. New Mexico: Notice of Proposed Withdrawal and Reservation of Lands, 43 Fed. Reg. 53,063, at 53,063 (Nov. 15, 1978).
13. Downey, "Politics and Technology in Repository Siting," 57 (1985).
14. This is the account of Owen Gormley, an AEC/ERDA project engineer for the nuclear waste disposal effort. Carter, *Nuclear Imperatives*, 179 (1987).
15. *Id.* at 58.
16. *Id.*
17. McCutcheon, *Nuclear Reactions*, 69 (2002).
18. Southwest Research and Information Center, Nuclear Waste Safety Program, "Chronology of WIPP Events: 1972 to 2000," www.sric.org/nuclear/docs/WIPPCHRON.html.
19. Nuclear Power Policy: Statements on Decisions Reached Following a Review, 1 Pub. Papers 587, 588 (Apr. 7, 1977).
20. *Id.*
21. Andrews, Congressional Research Service, *Nuclear Fuel Reprocessing*, 4 (2008).
22. Light, "The Hidden Dimension of the National Energy Plan," 176–178 (1979); Office of Technology Assessment, *Managing the Nation's Commercial High-Level Radioactive Waste*, 87 (1985); General Accounting Office, *Institutional Relations under NWPA*, 10 (1987), notes that, among the six states that considered ERDA's overture, only Washington—home of the Hanford Reservation site—had been explored with nuclear waste storage in mind.
23. The *Congressional Quarterly* took note of the situation in December: "The swimming pools in the backyards of the nation's nuclear reactors are filling up with used fuel, and there's no firm plan to dispose of the deadly radioactive waste." Pelham, "Government Groping with Waste Disposal," *Congressional Quarterly*, 2555 (1977).
24. Downey, "Politics and Technology in Repository Siting," 58 (1985).
25. Staats, Comptroller General, General Accounting Office, *Nuclear Energy's Dilemma: Disposing of Hazardous Radioactive Waste Safely*, xi (1977).
26. Downey, "Politics and Technology in Repository Siting," 59 (1985).
27. *Id.* Had New Mexico amended the state constitution in this manner, its action might have later been invalidated by a court citing *City of Philadelphia v. New Jersey*, 437 U.S. 617 (1978). Nevertheless, politically, the threat of a constitutional amendment was an effective strategy for rebuffing the initiative to expand WIPP's mission to include storage of SNF.
28. Senator Pete V. Domenici, Senator Harrison H. Schmidt, Congressman Manuel Lujan Jr., and Congressman Harold Runnels, "Letter to James R. Schlesinger, Secretary, Department of Energy" (Nov. 1, 1977), as quoted in Downey, "Politics and Technology in Repository Siting," 58 (1985).
29. *Id.*
30. *Id.*
31. Downey, "Politics and Technology in Repository Siting," 59 (1985).
32. Light, "The Hidden Dimension of the National Energy Plan," 179 (1979).
33. Statement of James R. Schlesinger, Secretary of Energy, *Authorizing Appropriations to the Department of Energy: Hearing before the Subcommittee on Energy and the Environment of the House Committee on Interior and Insular Affairs*, 95th Cong. 40 (1978), as quoted in Carter, "Congressional Committees Ponder Veto," 1136 (1978).

34. Thomas C. Newkirk, Assistant General Counsel for Legal Counsel, Department of Energy, to Worth Bateman, Deputy Under Secretary, "Memorandum on State Participation in Nuclear Waste Siting," as reprinted in *Nuclear Waste Disposal: Hearing Before the Subcommittee on Nuclear Regulation of the Senate Committee on Environment and Public Works*, 96th Cong. 73–92 (1979); Keller, Comptroller General, General Accounting Office, *DOE Policies on Construction and Operation of WIPP*, 6 (1979).

35. Urban Environmental Research, LLC, *Lessons Learned from WIPP*, 6–7 (2001); Downey, "Politics and Technology in Repository Siting," 62–63 (1979).

36. The task force's report was intended to be a first step in the creation of a national nuclear waste management policy. After the report was released, President Carter, upon the recommendation of Secretary Schlesinger, formed the Interagency Review Group on Nuclear Waste Management in March 1978. Carter, *Nuclear Imperatives*, 134–135 (1987); Greenwood, "Nuclear Waste Management," 18–20 (1982).

37. DOE, *Report on Nuclear Waste Management, Draft*, 16–17 (1978).

38. See Greenwood, "Nuclear Waste Management," 20 (1982), for a discussion of DOE's lack of credibility on nuclear waste management.

39. Downey, "Politics and Technology in Repository Siting," 54 (1985).

40. Carter, "Nuclear Wastes: Science of Geologic Disposal Seen as Weak," 1136–1137 (1978).

41. *Id.* at 1135.

42. Anderson, *Report to Sandia Laboratories on Deep Dissolution of Salt* (1978); Mercer and Orr, U.S. Geological Survey, *Review of Hydrogeologic Conditions near Repository Site* (1977); Bredehoeft et al., U.S. Geological Survey, *Geologic Disposal of High-Level Radioactive Wastes* (1978).

43. Carter, *Nuclear Imperatives*, 183 (1987).

44. Carter, "Nuclear Wastes: Science of Geologic Disposal Seen as Weak," 1135–37 (1978).

45. The Southwest Research Information Center, an Albuquerque-based group that continues to monitor and criticize WIPP's operation, led the charge against the repository. *See* Southwest Research Information Center, www.sric.org.

46. Coverage of newly elected New Mexico governor Toney Anaya in December 1978 in the *Albuquerque Journal* included the headline, "Anaya Says DOE Lied about Veto Power." Downey, "Politics and Technology in Repository Siting," 72n.38 (1985).

47. Sun, "Radwaste Dump Controversy," 1484 (1982).

48. Carter, "Congressional Committees Ponder Veto," 1137 (1978).

49. DOE, *Draft EIS for WIPP*, iv–v (1979). The DEIS stated that WIPP would be used "for the disposal of defense transuranic nuclear wastes (TRU), experimental research and development with high-level waste forms and for the potential disposal of up to a thousand spent fuel assemblies in an intermediate scale facility (ISF)." *Id.*

50. McCutcheon, *Nuclear Reactions*, 70 (2002).

51. Downey, "Politics and Technology in Repository Siting," 61 (1985).

52. *Id.*

53. McCutcheon, *Nuclear Reactions*, 70 (2002).

54. Downey, "Politics and Technology in Repository Siting," 62 (1985)

55. Interagency Review Group on Nuclear Waste Management, *Report to the President*, 69–73 (1979). The IRG report contained separate discussions of TRU and of a facility for SNF, but it noted that a TRU repository (which "could conceivably be . . . the Waste Isolation Pilot Plant") could be "co-located" with an intermediate-scale facility—a demonstration facility that would provide disposal for 1,000 SNF assemblies. *Id.* at 54–55, 70, 73. The Three Mile Island core meltdown in the same month, while not directly relevant to the ongoing debates over WIPP, certainly raised the profile of all issues touching U.S. nuclear energy, WIPP among them. Carter, *Nuclear Imperatives*, 88–89 (1987).

56. *Id.* at 54–58.

57. DOE, *Draft EIS for WIPP*, 2-1 (1979).

58. Senator Pete Domenici's intervention averted the referendum: the senator assured state legislators that DOE would not ignore New Mexico's objections to WIPP and encouraged the state to consider more cooperative approaches to dealing with DOE. Urban Environmental Research, LLC, *Lessons Learned from WIPP*, 8 (2001).

59. Radioactive Waste Consultation Act § 4, 1979 N.M. Laws 1636, 1637. According to the DOJ Office of Legal Counsel legal memorandum on the question, written in 1978 but not made public until 1979, it is unlikely that the provision would have survived a court challenge. But it was, nonetheless, similar to provisions adopted by at least nine other states at the time. These included Colorado, Louisiana, Michigan, Minnesota, Montana, New Jersey (ban on transport), Oregon, South Dakota (legislative concurrence required), and Vermont (legislative concurrence required). *See* Office of State Programs, NRC, *Means for Improving State Participation*, D-1 to D-5 (discussing DOJ memorandum). New Mexico subsequently amended the legislation in 1981 to read "until the state has concurred . . . except as specifically preempted by federal law." Radioactive Materials Act § 6, 1981 N.M. Laws 2121, 2125, N.M. STAT. § 74-4A-11.1 (2009).

60. Committee on Armed Services, *Department of Energy National Security and Military Applications of Nuclear Authorization Act*, 24 (1979): "In consideration of the above factors [including the assistant secretary of energy technology's testimony that DOE wished to submit WIPP for licensing as an SNF repository] the Committee recommends the deletion of the $20.8 million requested for operating expenses for the Waste Isolation Pilot Plant for fiscal year 1980."

61. Conference Committee, *DOE National Security and Military Applications of Nuclear Energy Authorization Act*, 18 (1979).

62. Downey, "Politics and Technology in Repository Siting," 62–63 (1985).
63. Carter, *Nuclear Imperatives*, 187 (1987).
64. Conference Committee, *DOE National Security and Military Applications of Nuclear Energy Authorization Act*, 18 (1979).
65. WIPP Authorization Act, Department of Energy National Security and Military Applications of Nuclear Energy Authorization Act of 1980 § 213(a), Pub. L. No. 96-164, 93 Stat. 1259, 1265 (1979). (emphasis added).
66. DOE, *Final EIS for WIPP*, 1-5 to 1-6 (1980). While DOE described this defense-waste-only plan as an "authorized alternative," under its "preferred alternative" "the Los Medanos site could become a potential site for a commercial-high-level-waste (HLW) repository that would include the disposal of defense TRU waste." *Id.* at 1–8. The FEIS did not address how this alternative might be lawfully accomplished; congressional legislation would be required. Despite DOE's preference for co-locating a commercial HLW and defense TRU repository, this alternative was never developed. As it turned out, the cessation of commercial reprocessing meant that there would be very little commercial HLW to dispose of. See Chapter 1 for a discussion of the end of commercial reprocessing.
67. Message to Congress: Radioactive Waste Management Program, 1 Pub. Papers 296, 298 (Feb. 12, 1980).
68. *Id.*
69. Carter, *Nuclear Imperatives*, 186–187 (1987).
70. Nuclear Waste Policy Act of 1982 § 114(a)(2)(A), as enacted, Pub. L. No. 97-425, 96 Stat. 2201, 2214 (1983). DOE continued to conduct experiments at WIPP to determine the suitability of bedded salt as an appropriate geologic setting for an SNF repository. After enactment of the 1987 Amendments designating Yucca as the sole HLW/SNF repository candidate site, DOE canceled the experiments. Rechard, *Milestones for Disposal at WIPP*, 3 (1998).
71. Nuclear Waste Policy Amendments Act of 1987 § 5011(a), Pub. L. No. 100-203, 101 Stat. 1330-227, 1330-228 (1987), as codified at 42 U.S.C. § 10172(a) (2006).
72. James Watkins, Secretary, Department of Energy, "Letter to Bruce King, Governor of New Mexico," (Oct. 4, 1991), as reprinted in *WIPP Land Withdrawal Act: Hearing before the Subcommittee on Energy and Power of the House Committee on Energy and Commerce*, 102nd Cong. 71–72 (1992).
73. Waste Isolation Pilot Plant Land Withdrawal Act § 12, as enacted, Pub. L. No. 102-579, 106 Stat. 4777, 4791 (1992), states: "The Secretary shall not transport high-level radioactive waste or spent nuclear fuel to WIPP or emplace or dispose of such waste or fuel at WIPP."
74. Radioactive Waste Consultation Act § 2, 5, 1979 N.M. Laws 1636, 1636–1637, authorizes creation of the New Mexico Radioactive Waste Consultation Task Force in answer to the "need to centralize and coordinate information on the plant concerns and to develop recommendations for action by the state." The New Mexico Radioactive Waste Consultation Task Force, New Mexico Energy, Minerals, and Natural Resources Department, "The Radioactive Waste Consultation Task Force," discusses the members and duties of the task force, at www.emnrd.state.nm.us/Wipp/TaskForce.htm.
75. Environmental Evaluation Group, *Review of WIPP Draft Application*, iii (1996).
76. *Contract between Department of Energy and New Mexico Health and Environment Department*, Contract No. ET-78-C-04-5344 (July 10, 1978), subsequently redesignated Contract No. DE-AC04-79AL107 (Oct. 1, 1978); Rechard, *Milestones for Disposal at WIPP*, 8 (1998).
77. National Defense Authorization Act Fiscal Year 1989 §1433(a), as enacted, Pub. L. No. 100-456, 102 Stat. 1918, 2073 (1988).
78. *Id.* § 1433(b)(1), as enacted, Pub. L. No. 100-456, 102 Stat. 1918, 2073.
79. Rechard, *Milestones for Disposal at WIPP*, 8 (1998).
80. National Defense Authorization Act Fiscal Year 1989 § 1433(a), as enacted, Pub. L. No. 100-456, 102 Stat. 1918, 2073 (1988); *Contract between Department of Energy and New Mexico Health and Environment Department*, Modification A030 to Contract No. DE-AC04-79AL10752 (May 7, 1988).
81. McCutcheon, *Nuclear Reactions*, 64–66 (2002).
82. Kerr, "For Radioactive Waste from Weapons," 1628 (1999); McCutcheon, *Nuclear Reactions*, 66–69 (2002).
83. Robert H. Neill, foreword to Channell and Walker, EEG, *Evaluation of Risks and Waste Characterization Requirements*, 1 (2000).
84. Patricia A. Madrid, Office of the Attorney General, State of New Mexico, "Attorney General Urges Continued Funding of WIPP Watchdog EEG," press release (May 5, 2004). In 2004, DOE cut EEG's budget, which had reached $1.5 million in 2004; despite New Mexico attorney general Madrid's attempts to convince then-governor Bill Richardson to continue funding the group, EEG was never revived.
85. Office of Technology Assessment, *Long-Lived Legacy: Managing High-Level and Transuranic Waste*, 57 (1991).
86. McCutcheon, *Nuclear Reactions*, 91–92 (2002); *see also* EEG, *Radiological Health Review of the WIPP FEIS*, iv (1981).
87. Channell, Rodgers, and Neill, EEG, *Adequacy of TRUPACT-I for Transporting CH-TRU* (1986); Colglazier and Langum, "Policy Conflicts in the Process for Siting Repositories," 342 (1988).
88. *See, e.g.*, Silva, Rucker, and Chaturvedi, "Resolution of Long-Term Performance Issues at WIPP," 1003 (1999).
89. McCutcheon, *Nuclear Reactions*, 68, 93 (2002).
90. In a 1982 memo, Larry Barrows, a geophysicist working with Sandia National Laboratories, argued that karst-type conditions might exist near WIPP. Larry Barrows "WIPP Geohydrology—The

Implications of Karst" (unpublished manuscript dated May 20, 1982), as reprinted in Chaturvedi and Channell, EEG, *Rustler Formation as a Medium for Contaminated Groundwater*, app. A (1985); Chaturvedi, *Karst and Related Issues at WIPP*, 1 (2009).

91. EEG concluded that there was not enough information to determine if karst would be a problem at the site. It noted two possible options: conduct more studies on the hydrology of the site, or assume that the natural geology would not act as a natural barrier and add engineered barriers instead. The report recommended pursuing both pathways simultaneously. Chaturvedi, and Channell, EEG, *Rustler Formation as a Medium for Contaminated Groundwater*, v, 72–74 (1985).

92. EPA, *Evaluation of Karst at WIPP Site* (2006); Committee on the Waste Isolation Pilot Plant, National Research Council of the National Academies, *Improving Operations and Safety of WIPP* (2001).

93. Chaturvedi, *Karst and Related Issues at WIPP* (2009).

94. McCutcheon, *Nuclear Reactions*, 64–65, 71, 73–74 (2002).

95. For example, Senator Domenici stated: "I still say this concurrence process, to be distinguished from veto, this concurrence process is . . . the keystone to a successful national State policy on nuclear disposal." 125 Cong. Rec. 18,691 (July 16, 1979).

96. The Office of Technology Assessment's 1985 report notes that IRG made particular effort in 1978 and 1979 to clarify that concurrence differed from veto in that the latter created a one-shot game, whereas the former fostered ongoing dialogue. According to OTA, however, most in Congress believed IRG was laboring over a distinction without a difference. Office of Technology Assessment, *Managing the Nation's Commercial High-Level Waste*, 224 (1985). Congress addressed this point after Senator George McGovern entered into the record the comptroller general's 1978 letter reporting that DOE's own Office of General Counsel shared the comptroller's legal opinion that a state lacked the legal authority to veto or concur regarding construction of a repository within its borders. R. F. Keller, Acting Comptroller General, "Letter to Congressman John D. Dingell, Chairman, Subcommittee on Energy and Power of the House Interstate and Foreign Commerce Committee" (June 19, 1978), as reprinted in 125 Cong. Rec. 18,685 (July 16, 1979).

97. Downey, "Politics and Technology in Repository Siting," 63 (1985).

98. WIPP Authorization Act, Department of Energy National Security and Military Applications of Nuclear Energy Authorization Act of 1980 § 213(b)(1)–(2), as enacted, Pub. L. No. 96-164, 93 Stat. 1259, 1265–1266 (1979).

99. Conference Committee, *DOE National Security and Military Applications of Nuclear Energy Authorization Act*, 19, 21–22 (1979).

100. The agreement was entered by the federal district court as part of the settlement in the litigation brought by New Mexico against DOE, complaining, among other matters, that DOE had violated the WIPP Authorization Act by commencing construction at WIPP without having first entered into a consultation and cooperation agreement with the state. Stipulated Agreement, *New Mexico ex rel. Bingaman v. Dep't of Energy*, No. 81-0363 JB (D.N.M. 1981).

101. The Agreement for Consultation and Cooperation was first entered on July 1, 1981. It was amended three times: "First Modification to the July 1, 1981 'Agreement for Consultation and Cooperation' on WIPP by the State of New Mexico and U.S. Department of Energy" (Nov. 30, 1984); "Second Modification to the July 1, 1981 'Agreement for Consultation and Cooperation' on WIPP by the State of New Mexico and U.S. Department of Energy" (Aug. 4, 1987); "1988 Modification to the Working Agreement of the Consultation and Cooperation Agreement between the Department of Energy and the State of New Mexico on the Waste Isolation Pilot Plant" (Mar. 18, 1988).

102. "Agreement for Consultation and Cooperation," as appears in Appendix A, Stipulated Agreement, *New Mexico ex rel. Bingaman v. Dep't of Energy*, No. 81-0363 JB (D.N.M. 1981).

103. WIPP Authorization Act, Department of Energy National Security and Military Applications of Nuclear Energy Authorization Act of 1980 § 213(b)(1)–(2), as enacted, Pub. L. No. 96-164, 93 Stat. 1259, 1265–1266 (1979).

104. The EEG's first meeting, in January 1980, convened thirty-five scientists and other interested parties to discuss technical issues. New Mexico Radioactive Waste Consultation Task Force, New Mexico Energy, Minerals, and Natural Resources Department, "Chronology of WIPP," www.emnrd.state.nm.us/WIPP/chronolo.htm. Between 1980 and 2004, EEG issued more than eighty technical reports on aspects of WIPP.

105. McCutcheon, *Nuclear Reactions*, 77–79 (2002).

106. Downey, "Politics and Technology in Repository Siting," 64 (1985).

107. McCutcheon, *Nuclear Reactions*, 77 (2002).

108. Complaint at ¶¶ 5, 27–30, 42–43, 52, 59, 65, 107, *New Mexico ex rel. Bingaman v. U.S. Dep't of Energy*, No. 81-0363 JB (D.N.M. 1981).

109. Don Hancock, Director, Nuclear Waste Safety Program, Southwest Research and Information Center, telephone interview by Justin Gundlack (Sep. 2, 2008).

110. *Id.*

111. *Id.*

112. WIPP's precedent-setting character was explicitly discussed in the deliberations of the House-Senate Conference on the 1980 WIPP Authorization Act. Conference Committee, *DOE National Security and Military Applications of Nuclear Energy Authorization Act*, 20 (1979): "The conferees believed that the 'precedent setting' nature of WIPP has been overstated."

113. Don Hancock, Director, Nuclear Waste Safety Program, Southwest Research and Information Center, telephone interview by Justin Gundlack (Sep. 2, 2008).
114. "Agreement for Consultation and Cooperation," as appears in Appendix A, Stipulated Agreement, *New Mexico ex rel. Bingaman v. Dep't of Energy*, No. 81-0363 JB (D.N.M. 1981).
115. *Id.* at Art. VII § C, which identifies "Key Events" whose commencement was to follow written notice by DOE to New Mexico.
116. The C&C agreement allowed only temporary storage of high-level waste for experiments. *Id.* at Art. VI § B. Although DOE committed to exclude HLW from permanent disposal at WIPP in a legally binding contract, Congress could of course override this commitment legislatively. SNF had already been banned from disposal at WIPP by this time.
117. Stipulated Agreement at ¶¶ 1, 7, *New Mexico ex rel. Bingaman v. Dep't of Energy*, No. 81-0363 JB (D.N.M. 1981).
118. Order, *New Mexico ex rel. Bingaman v. U.S. Dep't of Energy*, No. 81-0363 JB (D.N.M. 1981).
119. Order, *New Mexico ex rel. Bardacke v. U.S. Dep't of Energy*, No. 81-0363 JB (D.N.M. 1985). The supplemental stipulated agreement is discussed in greater detail in the next subsection of this chapter.
120. Stipulated Agreement at ¶¶ 1–2, 5, *New Mexico ex rel. Bingaman v. Dep't of Energy*, No. 81-0363 JB (D.N.M. 1981).
121. "Agreement for Consultation and Cooperation," at Art. IX, as appears in Appendix A, Stipulated Agreement, *New Mexico ex rel. Bingaman v. Dep't of Energy*, No. 81-0363 JB (D.N.M. 1981).
122. *Id.* at Art. IX §§ B, E, G.
123. *Id.* at § I.
124. Urban Environmental Research, LLC, *Lessons Learned from WIPP*, 11–12 (2001).
125. Gervers, "The NIMBY Syndrome," 20 (1987).
126. Supplemental Stipulated Agreement Resolving Certain State Off-Site Concerns over WIPP, *New Mexico ex rel. Bingaman v. U.S. Dep't of Energy*, No. 81-0363 JB (D.N.M. 1982).
127. *Id.* at 1–2.
128. Gervers, "The NIMBY Syndrome," 20 (1987).
129. Molecke, Sandia National Laboratories, *A Comparison of Brines*, 14 (1983); Carter, *Nuclear Imperatives*, 183 (1987). Measurements taken in the weeks after the brine pocket was discovered generated estimates that 350 gallons of brine per minute were flowing from the brine pocket into the WIPP-12 shaft. New Mexico Radioactive Waste Consultation Task Force, New Mexico Energy, Minerals, and Natural Resources Department, "Chronology of WIPP," *www.emnrd.state.nm.us/WIPP/chronolo.htm*.
130. Carter, *Nuclear Imperatives*, 183–184 (1987).
131. Supplemental Stipulated Agreement Resolving Certain State Off-Site Concerns over WIPP at 10–11, *New Mexico ex rel. Bingaman v. Dep't of Energy*, No. 81-0363 JB (D.N.M. 1982). The Price-Anderson Act of 1957 was up for amendment in August 1983, and the supplemental stipulated agreement called upon DOE to advocate for various changes sought by the state, including, for instance, to "increase substantially the amount of protection from the present 500 million dollar level." *Id.* at 10.
132. *Id.* at 30–31 (highway upgrades); *id.* at 19–20 (state monitoring of transportation); *id.* at 11–18 (emergency response preparedness).
133. *Id.* at 28–29.
134. "This Supplemental Stipulated Agreement and the previous Stipulated Agreement of July 1, 1981, are binding contractual agreements the compliance with which is subject to the appropriate oversight jurisdiction of this court." *Id.* at 32. Like DOE, the state was bound to comply with the terms of the supplemental agreement, and noncompliance on its part was also subject to judicial review.
135. *See* Urban Environmental Research, LLC, *Lessons Learned from WIPP*, 13–14 (2001).
136. New Mexico: Proposed Withdrawal and Reservation of Lands, 45 Fed. Reg. 75,768 (Nov. 17, 1980); Downey, "Politics and Technology in Repository Siting," 64–65 (1985).
137. New Mexico: Withdrawal of Lands, 47 Fed. Reg. 13,340 (Mar. 30, 1982) (as codified at 43 C.F.R. Public Land Order 6432).
138. New Mexico: Proposed Withdrawal and Reservation of Lands, 48 Fed. Reg. 3878 (Jan. 27, 1983).
139. New Mexico: Withdrawal of Lands, 48 Fed. Reg. 31,038 (July 6, 1983) (as codified at 43 C.F.R. Public Land Order 6402).
140. Stipulated Agreement at 1, *New Mexico ex rel. Bingaman v. Dep't of Energy*, No. 81-0363 JB (D.N.M. 1981).
141. Carter, "WIPP Goes Ahead," 1104 (1983). One cause for concern was the possibility, left open under NWPA, that the president could decide not to co-locate defense HLW in the NWPA repository (along with commercial HLW and SNF), thus introducing the prospect of either construction of a new defense waste repository or expansion of WIPP's mission to include defense HLW. *Id.*
142. McCutcheon, *Nuclear Reactions*, 90 (2002).
143. Announcement of Decision to Proceed with Construction of the Waste Isolation Pilot Plant, 48 Fed. Reg. 30,427 (July 1, 1983).
144. The list of key events contained several milestones, which all required notice to the state. "Agreement for Consultation and Cooperation," at Art. VII § C, app. B at 9–12, as appears in Appendix A, Stipulated Agreement, *New Mexico ex rel. Bingaman v. Dep't of Energy*, No. 81-0363 JB (D.N.M. 1981). But Urban Environmental Research, LLC, *Lessons Learned from WIPP*, 12 (2001), notes that construction was not listed as a key event.
145. Urban Environmental Research, LLC, *Lessons Learned from WIPP*, 11–12 (2001).
146. "First Modification to the July 1, 1981 'Agreement for Consultation and Cooperation' on WIPP by the State of New Mexico and U.S. Department of Energy" (Nov. 30, 1984).

147. *Id.* at 3–4. The modification made legally binding a limit of 250,000 cubic feet of RH-TRU (approximately 5 percent of WIPP's capacity); this limit had been set earlier by DOE in a record of decision dated January 22, 1981. *Id.* at 3.
148. *Id.* at 2, 5.
149. *Id.* at 9–11. The WIPP Land Withdrawal Act of 1992 subsequently excluded HLW from WIPP for any purpose, even experimentation. Waste Isolation Pilot Plant Land Withdrawal Act § 12, as enacted, Pub. L. No. 102-579, 106 Stat. 4777, 4791 (1992).
150. "First Modification to the July 1, 1981 'Agreement for Consultation and Cooperation' on WIPP by the State of New Mexico and U.S. Department of Energy," 5, 13 (Nov. 30, 1984).
151. *Id.* at 11–12.
152. *Id.* at 6–7. This provision was chiefly intended to make WIPP subject to requirements equivalent to those provided in Section 121 of NWPA, which directed repositories to comply with environmental standards set by EPA. McCutcheon, *Nuclear Reactions*, 93 (2002); Nuclear Waste Policy Act § 121(a), 42 U.S.C. § 10141(a) (2006).
153. "First Modification to the July 1, 1981 'Agreement for Consultation and Cooperation' on WIPP by the State of New Mexico and U.S. Department of Energy," 8–9 (Nov. 30, 1984); Montange, "Federal Nuclear Waste Disposal Policy," 393 (1987).
154. "Second Modification to the July 1, 1981 'Agreement for Consultation and Cooperation' on WIPP by the State of New Mexico and U.S. Department of Energy" (Aug. 4, 1987).
155. *NRDC v. Envtl. Prot. Agency*, 824 F.2d 1258 (1st Cir. 1987), upholds 40 CFR Part 191.14, Subpart A, but remands to EPA the long-term "Environmental Standards for Disposal" in Subpart B of Part 191.14 for noncompliance with the Safe Drinking Water Act; "Second Modification to the July 1, 1981 'Agreement for Consultation and Cooperation' on WIPP by the State of New Mexico and U.S. Department of Energy," 5 (Aug. 4, 1987), states: "The parties are aware of the opinion issued by the United States Court of Appeals for the First Circuit in *Natural Resources Defense Council, et al. v. United States Environmental Protection Agency, et al.*"
156. "Second Modification to the July 1, 1981 'Agreement for Consultation and Cooperation' on WIPP by the State of New Mexico and U.S. Department of Energy," 5 (Aug. 4, 1987).
157. *Id.* at 3.
158. McCutcheon, *Nuclear Reactions*, 88 (2002).
159. "1988 Modification to the Working Agreement of the Consultation and Cooperation Agreement between the Department of Energy and the State of New Mexico on the Waste Isolation Pilot Plant" (March 18, 1988).
160. *Id.* at 12.
161. Nelson, Carlsbad Field Office, DOE, "WIPP Site Selection and Early Site Studies," 12 (2008).
162. New Mexico: Withdrawal of Lands, 48 Fed. Reg. 31,038 (July 6, 1983) (as codified at 43 C.F.R. Public Land Order 6402).
163. Federal Land Policy and Management Act of 1976 § 204(c)(1), 43 U.S.C. 1714(c)(1) (2006).
164. These included the House Committee on Energy and Commerce, the House Committee on Interior and Insular Affairs, the House Committee on Armed Services, and the Senate Committee on Energy and Natural Resources.
165. McCutcheon, *Nuclear Reactions*, 122, 126 (2002), discusses the importance of DOE's ability to open WIPP to DOE's overall credibility, and at 82, 123–124, discusses the problems DOE experienced in finding a site for interim storage of TRU.
166. Statement of Keith O. Fultz, Senior Associate Director, General Accounting Office, *Status of the Waste Isolation Pilot Plant Project: Hearing before the Environment, Energy, and Natural Resources Subcommittee of the House Committee on Government Operations*, 100th Cong. 6 (1988).
167. *Id.* at 9, 18–19.
168. McCutcheon, *Nuclear Reactions*, 102–104, 117–118 (2002).
169. *Id.* at 98.
170. "Western States Form Group to Gain Greater U.S. Impact," *New York Times*, 50 (Mar. 11, 1984).
171. WIPP Transportation Technical Advisory Group, Western Governors' Association, *WIPP Rail Transportation Safety Program*, 6 (2004).
172. *Id.*
173. *Id.* at 6–7; *Memorandum between WGA and DOE*, 2 (2009). Participating in WGA's WIPP Transportation Technical Advisory Group were Arizona, California, Colorado, Idaho (cochair), Nebraska, Nevada (cochair), New Mexico, Oregon, Texas, Utah, Washington, and Wyoming. Western Governors' Association, "WGA WIPP Fact Sheet," 1, www.westgov.org/index.php?option=com_joomdoc&task=doc_download&gid=259. WGA now includes the governors of nineteen of the fifty states, as well as U.S. territories and possessions: Alaska, American Samoa, Arizona, California, Colorado, Guam, Hawaii, Idaho, Kansas, Montana, Nebraska, Nevada, New Mexico, North Dakota, Northern Mariana Islands, Oklahoma, Oregon, South Dakota, Texas, Utah, Washington, and Wyoming. Western Governors Association, *www.westgov.org*.
174. McCutcheon, *Nuclear Reactions*, 100–104 (2002); *see also* Office of Technology Assessment, *Managing the Nation's Commercial High-Level Radioactive Waste*, app. A-1, at 222 (1985).
175. In February 1991, DOE notified Governor Andrus that DOE intended to accept commercial waste from Colorado at its Idaho facility, despite Andrus's declaration. Idaho filed suit the next day to overturn DOE's decision to accept the waste. *Idaho v. Dep't of Energy*, 945 F.2d 295 (9th Cir. 1991).
176. McCutcheon, *Nuclear Reactions*, 103 (2002); Matthew L. Wald, "3 States Ask Waste Cleanup as Price of Atomic Operation," *New York Times*, 11 (Dec. 17, 1988).
177. McCutcheon, *Nuclear Reactions*, 103 (2002).

Shipments to DOE's Idaho site from Rocky Flats continued until August 1989. Thereafter, a search began for an interim storage location for the TRU that was still being generated. The search was eventually given up when Secretary James Watkins announced in 1989 that all production at Rocky Flats would halt indefinitely, obviating the need for a new storage location. *Id.* at 123–124. Due to the end of the Cold War, weapons production at the plant was never restarted. Ackland, *Making a Real Killing*, 222–227 (1999). TRU waste generated by the decommissioning phase of Rocky Flats has been held on-site for eventual shipment to WIPP. DOE, *Rocky Flats Fact Sheet* (2010).

178. *See, e.g.*, Western Governors' Association, "WIPP Transportation Safety Program Initiative," www.westgov.org/index.php?option=com_content&view=article&id=136&Itemid=79.

179. DOE, *Final Supplement EIS for WIPP* (1990); Rechard, *Milestones for Disposal at* WIPP, 11 (1998); Waste Isolation Pilot Plant; Availability of Draft Supplement to the Final Environmental Impact Statement, 54 Fed. Reg. 16,350 (Apr. 29, 1989).

180. Environmental Evaluation Group, *Review of the Draft SEIS for WIPP*, 1, 10 (1989).

181. McCutcheon, *Nuclear Reactions*, 119–120 (2002).

182. *Id.* at 12; David Johnston, "Criminal Investigation Is Begun at Arms Plant," *New York Times*, A1 (June 7, 1989).

183. McCutcheon, *Nuclear Reactions*, 121–122 (2002)

184. *Id.* at 128.

185. *Id.* at 128–129.

186. Proposed Modification and Partial Termination of Public Land Order No. 6403 and Public Hearings; New Mexico, 54 Fed. Reg. 15,814 (Apr. 19, 1989).

187. DOE, *WIPP Disposal Phase: Final Supplemental EIS*, vol. 1, at iv to v, 1-5 to 1-7 (1990).

188. DOE, "Letter to Department of Interior" (1989), as quoted in *New Mexico v. Watkins*, 969 F.2d 1122, 1126 and n.7 (D.C. Cir. 1992).

189. Interior issued a record of decision accepting DOE's recently published FSEIS and entered a modification to Public Land Order 6403 as requested by DOE in April 1989. Record of Decision, Waste Isolation Pilot Plant Project; New Mexico, 56 Fed. Reg. 3114 (Jan. 28, 1991); Modification of Public Land Order No. 6503; New Mexico, 56 Fed. Reg. 3038 (Jan. 28, 1991); note that the title of the notice was corrected in a later notice, Modification of Public Land Order No. 6403; New Mexico, 56 Fed. Reg. 5731 (Feb. 12, 1991).

190. Gregg Easterbrook, "James Watkins for the Energy Department, a Man Willing to Call Them as He Sees Them," *Los Angeles Times*, M3 (Aug. 11, 1991).

191. Notice to Proceed, Waste Isolation Pilot Plant Project, New Mexico, 56 Fed. Reg. 50,923 (Oct. 9, 1991).

192. New Mexico Radioactive Waste Consultation Task Force, New Mexico Energy, Minerals, and Natural Resources Department, "Chronology of WIPP," www.emnrd.state.nm.us/WIPP/chronolo.htm.

193. *New Mexico ex rel. Udall v. Watkins*, 783 F. Supp. 633, 638 (D.D.C. 1992), aff'd 969 F.2d 1122, 1126 (D.C. Cir. 1992).

194. *New Mexico v. Watkins*, 969 F.2d 1122 (D.C. Cir. 1992); WIPP Authorization Act, Department of Energy National Security and Military Applications of Nuclear Energy Authorization Act of 1980 § 213(a), as enacted, Pub. L. No. 96-164, 93 Stat. 1259, 1265 (1979).

195. Waste Isolation Pilot Plant Land Withdrawal Act, Pub. L. No. 102-579, 106 Stat. 4777 (1992).

196. Committee on the Waste Isolation Pilot Plant, National Research Council of the National Academies, "Letter to Leo Duffy, Assistant Secretary for Environmental Restoration and Waste, Department of Energy" (June 17, 1992), as quoted in Task Force on Radioactive Waste Management, DOE, *Earning Public Trust and Confidence*, 34 (1993).

197. Waste Isolation Pilot Plant Land Withdrawal Act § 5, as enacted, Pub. L. No. 102-579, 106 Stat. 4777, 4782 (1992).

198. Committee on Energy and Resources, *Waste Isolation Pilot Plant Land Withdrawal Act*, at 48, 50 (1991).

199. *Id.* § 16(a)–(e), as enacted, Pub. L. No. 102-579, 106 Stat. 4777, 4792–4793.

200. "There are authorized to be appropriated to the Secretary for payments to the State $20,000,000 for each of the 15 fiscal years beginning with the fiscal year in which the transport of transuranic waste to WIPP is initiated." Waste Isolation Pilot Plant Land Withdrawal Act § 15(a), as enacted, Pub. L. No. 102-579, 106 Stat. 4777, 4791 (1992).

201. *Regarding the Waste Isolation Pilot Plant* (1990), as quoted in Urban Environmental Research, LLC, *Lessons Learned from WIPP*, 13 (2001).

202. Urban Environmental Research, LLC, *Lessons Learned from WIPP*, 24 (2001).

203. Carlsbad Area Office, DOE, "DOE's Carlsbad Area Office Stresses Importance of Public Involvement," press release (Aug. 14, 1996).

204. John H. Cushman Jr., "U.S. Drops Test Plan at Bomb Waste Site," *New York Times*, A16 (Oct. 22, 1993), quoting DOE secretary Hazel O'Leary. For demands that DOE make this change, *see* McCutcheon, *Nuclear Reactions*, 154–155 (2002); Robert H. Neill, Director, EEG, "Letter to Tom Grumbly, Undersecretary, DOE" (Sep. 21, 1993), cited in McCutcheon, *Nuclear Reactions*, 155n.17 (2002); Task Force on Radioactive Waste Management, DOE, *Earning Public Trust and Confidence*, 34 (1993).

205. McCutcheon, *Nuclear Reactions*, 156 (2002).

206. Criteria for the Certification of Compliance with Environmental Radiation Protection Standards for the Management and Disposal of Spent Nuclear Fuel, High-Level and Transuranic Radioactive Wastes, 58 Fed. Reg. 8,029 (Feb. 11, 1993). EPA's generic radioactivity standards for repositories require compliance with specified exposure limits over a specified compliance period. 40 C.F.R. § 191.13 (2009).

207. *Natural Res. Def. Council v EPA*, 824 F.2d 1258 (1st Cir. 1987).

208. Environmental Radiation Protection Standards for

the Management and Disposal of Spent Nuclear Fuel, High-Level and Transuranic Radioactive Wastes, 58 FED. REG. 66,398, 66,405 (Dec. 20, 1993) (codified at 40 C.F.R. pt. 191).
209. Waste Isolation Pilot Plant Land Withdrawal Amendment Act, National Defense Authorization Act for Fiscal Year 1997, Pub. L. No. 104-201, 110 Stat. 2422, 2851 (Sep. 23, 1996).
210. *Compare* Waste Isolation Pilot Plant Land Withdrawal Act § 7(b)(3), as enacted, Pub. L. No. 102-579, 106 Stat. 4777, 4785 (1992), *with* Waste Isolation Pilot Plant Land Withdrawal Amendment Act, National Defense Authorization Act for Fiscal Year 1997 § 3186(b)(3), as enacted, Pub. L. No. 104-201, 110 Stat. 2422, 2852 (1996).
211. Resource Conservation and Recovery Act § 3004(d)–(g), (m), 42 U.S.C. § 6924(d)–(g), (m) (2006); Waste Isolation Pilot Plant Land Withdrawal Amendment Act, National Defense Authorization Act for Fiscal Year 1997 § 3188(a), as enacted, Pub. L. No. 104-201, 110 Stat. 2422, 2853 (1996). Notably, however, the amendments did not exempt WIPP from RCRA permitting and other requirements. *Compare* Waste Isolation Pilot Plant Land Withdrawal Act § 9(a)(1)(H), as enacted, Pub. L. No. 102-579, 106 Stat. 4777, 4788 (1992), *with* Waste Isolation Pilot Plant Land Withdrawal Amendment Act, National Defense Authorization Act for Fiscal Year 1997 § 3188(a), as enacted, Pub. L. No. 104-201, 110 Stat. 2422, 2853 (1996).
212. Waste Isolation Pilot Plant Land Withdrawal Amendment Act, National Defense Authorization Act for Fiscal Year 1997 § 3184, as enacted, Pub. L. No. 104-201, 110 Stat. 2422, 2851–2852 (1996). The cancelled tests had been devised to measure how much pressurized gas would be generated by TRU emplaced at WIPP. DOE planned to conduct tests at three scales—"laboratory tests" (smallest), "bin tests," and "alcove tests." DOE's 1990 plan called for emplacing drums full of TRU at WIPP for five years, and monitoring both the rate of gas release from waste stored in those drums and the effects of that release on the surrounding storage spaces. Neill and Chaturvedi, EEG, *Status of WIPP*, 7–9 (1991).
213. Waste Isolation Pilot Plant Land Withdrawal Act § 8(f), as enacted, Pub. L. No. 102-579, 106 Stat. 4777, 4787–4788 (1992), as amended by Pub. L. No. 104-201, 110 Stat. 2422 (1996).
214. Draft Compliance Application Guidance Document: Notice of Availability, 60 FED. REG. 53,921 (Oct. 18, 1995); DOE, *Draft Compliance Certification Application for WIPP* (1995).
215. "The EEG finds that even the most basic information is lacking in this draft." Neill et al., EEG, *Review of the WIPP Draft Application* (1995).
216. Rechard, *Milestones for Disposal at WIPP*, 5, 13 (1998). After granting EPA access to its input data, DOE waited for all of 1997 while EPA thoroughly reviewed its compliance certification application. *Id.*
217. "The degree and comprehensiveness of the understanding of some aspects of the local and regional ground-water flow systems are lower than desirable relative to the importance of the repository and to the time and money spent studying the WIPP site and constructing the repository." Committee on the Waste Isolation Pilot Plant, National Research Council of the National Academies, *The Waste Isolation Pilot Plant: A Potential Solution for the Disposal of TRU Waste*, 69–70 (1996).
218. *Id.* at 80.
219. Udall's petition, Civil Action No. 96-1107, was consolidated by the U.S. Court of Appeals for the District of Columbia Circuit with two others brought against EPA by environmental organizations, Civil Action No. 96-1108, and the Texas attorney general, Civil Action No. 96-1109. New Mexico Radioactive Waste Consultation Task Force, New Mexico Energy, Minerals, and Natural Resources Department, "Chronology of WIPP," *www.emnrd.state.nm.us/WIPP/chronolo.htm*.
220. McCutcheon, *Nuclear Reactions*, 169 (2002).
221. *New Mexico v. Envtl. Prot. Agency*, 114 F.3d 290, 293, 295 (D.C. Cir. 1997), determined that EPA was owed deference in its interpretation of "criteria," pursuant to § 8(c)(2) of the WIPP Authorization Act; and that off-the-record communications with nonagency personnel are permissible, so long as EPA could justify decisions made in its rulemaking based on the existing record.
222. *New Mexico ex rel. Madrid v. Richardson*, 39 F. Supp. 2d 48 (D.D.C. 1999).
223. McCutcheon, *Nuclear Reactions*, 179–183 (2002).
224. 40 C.F.R. § 270.70 (2009) provides for interim status qualification process and criteria; 40 C.F.R. pt. 265 (2009) enumerates operating standards for interim status facilities. The standards imposed by a final RCRA facility permit at 40 C.F.R. Part 264 are more stringent than those in effect at interim status facilities. *See, e.g.*, Sadler, "Standards for Containers, Tanks, Incinerators, Boilers, and Furnaces," 217, 219 (2004).
225. *New Mexico ex rel. Madrid v. Richardson*, 39 F. Supp. 2d 48, 52 (D.D.C. 1999).
226. Resource Conservation and Recovery Act § 3005(e)(A)(i)–(ii), 42 U.S.C. § 6925(e)(A)(i)–(ii) (2006); *see also* 40 C.F.R.§ 270.2 (2009), which defines "existing hazardous waste management (HWM) facility" to mean one "in operation or for which construction commenced on or before November 19, 1980"; 40 C.F.R. § 270.70 (2009), which qualifies a facility for interim status if it is either an "'existing HWM facility' or . . . in existence on the effective date of statutory or regulatory amendments under the Act that render the facility subject to the requirement to have a RCRA permit."
227. The issue of WIPP's interim status under RCRA had been noted but not addressed or resolved in the litigation over administrative land withdrawal for WIPP. *New Mexico v. Watkins*, 969 F.2d 1122, 1132 (D.C. Cir. 1992).
228. *Id.* at 52–53. The court held that WIPP had interim status under RCRA, and so the waste in question could be shipped to WIPP. As there was

no definition in RCRA itself of the "regulatory or statutory change" that would trigger such interim status, the court in so ruling had deferred to EPA's interpretation. EPA had concluded that its January 1990 authorization of New Mexico to implement the RCRA permit program sufficed as a "change" that qualified WIPP for interim status. *Id.*
229. Southwest Research and Information Center, Nuclear Waste Safety Program, "Chronology of WIPP Events: 1972 to 2000," *www.sric.org/nuclear/docs/WIPPCHRON.html.*
230. Once a state has been authorized by EPA to administer the RCRA permit program (or some portion of it), the state becomes the primary enforcement authority; while EPA thereafter retains its enforcement authority, the agency does not generally take enforcement action unless the state is not adequately enforcing the program (or relevant portion of it). EPA, *RCRA Orientation Manual*, III-143 (2002).
231. McCutcheon, *Nuclear Reactions*, 183 (2002).
232. Mike Taugher, "WIPP Receives Federal Go-Ahead," *Albuquerque Journal*, A1 (May 14, 1998).
233. Criteria for the Certification and Recertification of the Waste Isolation Pilot Plant's Compliance With the Disposal Regulations: Certification Decision, 63 Fed. Reg. 27,354 (May 18, 1998). EPA examined WIPP for compliance with both the general radioactive waste repository standards, 40 C.F.R. §§ 191.B–191.C (2009), and WIPP-specific standards, 40 C.F.R. pt. 194 (2009). *Id.* at 27,354. EPA found that WIPP had complied with both sets of regulations; however, EPA required that DOE meet certain additional requirements to maintain certification and in advance of shipping waste to WIPP for disposal. *Id.* at 27,355.
234. Record of Decision for the Department of Energy's Waste Isolation Pilot Plant Disposal Phase, 63 Fed. Reg. 3,623, 3,624 (Jan. 23, 1998). The record of decision was issued after EPA proposed to certify that WIPP would comply with the applicable regulations. Criteria for the Certification and Recertification of the Waste Isolation Pilot Plant's Compliance With the 40 C.F.R. § 191 Disposal Regulations: Certification Decision, 62 Fed. Reg. 58,792 (Oct. 30, 1997).
235. McCutcheon, *Nuclear Reactions*, 186 (2002).
236. For lawsuits against EPA, *see Southwest Research and Information Center v. Envtl. Prot. Agency*, 194 F.3d 175 (D.C. Cir. 1999). For lawsuits against DOE, *see New Mexico ex rel. Madrid v. Richardson*, 39 F. Supp. 2d 48 (D.C. Cir. 1999). *See also* Associated Press, "Opponents: WIPP Must Wait," *Albuquerque Journal*, D1 (Jan. 26, 1999).
237. McCutcheon, *Nuclear Reactions*, 185–186 (2002); Carlsbad Field Office, DOE, "World's First Underground Waste Repository Begins Operations," press release (Mar. 26, 1999).
238. Maggiore, Secretary, New Mexico Environment Department, *Final Order Granting a Hazardous Waste Permit to DOE for WIPP*, 4 (1999).
239. DOE, "WIPP Shipment and Disposal Information," *www.wipp.energy.gov/shipments.htm.*
240. Statement of James A. Rispoli, Assistant Secretary for Environmental Management, Department of Energy, *Energy and Water Development Appropriations for Fiscal Year 2009: Hearing before the Energy and Water Development Subcommittee of the Senate Committee on Appropriations*, 110th Cong. 135 (2008); Carlsbad Field Office, DOE, "WIPP Marks a Decade of Safe Disposal," press release (Mar. 25, 2009); Oregon Department of Energy, *Radioactive Material Transport in Oregon*, 5 (2010); Nelson, Carlsbad Field Office, DOE, "WIPP-DOE's TRU Strategy," 6 (2008).
241. DOE, *The Remote-Handled Transuranic Waste Program*, 1 (2007).
242. DOE, "WIPP Quick Facts," *www.wipp.energy.gov/TeamWorks/TRUTeamWorksArchives/TTW%20 6-16-10.pdf*; Carlsbad Field Office, DOE, "WIPP Completes First RH-TRU Shipment from VNC," press release (Sep. 18, 2009).
243. In its first ten years of operation, there were fourteen traffic incidents involving WIPP trucks, according to officials at WIPP. Sue Vorenberg, "Ten Years in Operation, WIPP Boasts Sterling Safety Record, Continued Support," *Santa Fe New Mexican*, A1 (Mar. 29, 2009). Tests after an accident on August 25, 2002 indicated that radioactivity had released inside the TRUPACT-II container. The shipment was sent back to Idaho to be examined. DOE eventually determined that the release was not caused by the traffic accident, but rather by improper loading of the container. Don Hancock, Director, Nuclear Waste Safety Program, Southwest Research and Information Center, e-mail correspondence on file with the authors (Mar. 2, 2010).
244. Criteria for the Certification and Recertification of the Waste Isolation Pilot Plant's Compliance with the Disposal Regulations: Recertification Decision, 71 Fed. Reg. 18,010, 18,010 (Apr. 10, 2006).
245. Intent to Evaluate Whether the Waste Isolation Pilot Plant Continues to Comply with the Disposal Regulations and Compliance Criteria, 69 Fed. Reg. 29,646, 29,646 (May 24, 2004).
246. DOE, Compliance Recertification Application for WIPP (2009).
247. EPA, "Radiation Protection," *www.epa.gov/radiation/wipp/.*
248. Waste Isolation Pilot Plant Land Withdrawal Act §§ 2(4), 8(f), as enacted, Pub. L. No. 102-579, 106 Stat. 4777, 4787–4788 (1992), as amended by Pub. L. No. 104-201, 110 Stat. 2422 (1996), which provides that recertification will continue until the end of the "decommissioning phase," defined as the backfilling of all shafts.
249. *See, e.g.,* Silva, Rucker, and Chaturvedi, "Resolution of Long-Term Performance Issues at WIPP," 1003 (1999); Rucker, Silva, and Chaturvedi, EEG, *Performance Assessment Issues to be Resolved at WIPP* (2000); Allen, Silva, and Channell,

EEG, *Identification of Issues Relevant to First Recertification of WIPP* (2002).
250. Associated Press, "DOE Eliminates Oversight Group's Funding," *Associated Press Online* (Apr. 15, 2004); John Fleck, "DOE Cuts Funding to WIPP Watchdog," *Albuquerque Journal*, A1 (Apr. 15, 2004).
251. Citizens for Alternatives to Radioactive Dumping, "WIPP Background: The Site," www.cardnm.org/backfrm_a.html.
252. DOE, *WIPP Disposal Phase: Final Supplemental EIS* (1997).
253. *Citizens for Alternatives to Radioactive Dumping v. Dept. of Energy*, 485 F.3d 1091, 1094 (10th Cir. 2007).
254. *Id.* at 1095, 1100.
255. Government Accountability Office, *Plans for Addressing Most Buried TRU Wastes*, 2–3 (2007). DOE also buried TRU at a handful of smaller sites. For comparison, one such smaller site, the Nevada Test Site (NTS), contains less than a tenth of a percent (by total radioactivity) of the total amount of TRU wastes. NTS is, like Los Alamos, subject only to RCRA cleanup requirements, and the cleanup decision for TRU buried at NTS is still pending. Neill and Neill, *Shallow Buried TRU*, 158, 161 (2009).
256. Government Accountability Office, *Plans for Addressing Most Buried TRU Wastes*, 3 (2007).
257. *Id.* at 4, 20.
258. Neill and Neill, *Shallow Buried TRU*, 162 (2009); Government Accountability Office, *Plans for Addressing Most Buried TRU Wastes*, 16 (2007).
259. Government Accountability Office, *Plans for Addressing Most Buried TRU Wastes*, 16 (2007).
260. Idaho Department of Environmental Quality, "Contamination at INL: Text of the 1995 Settlement Agreement," www.deq.state.id.us/inl_oversight/contamination/settlement_agreement_entire.cfm. See Chapter 1 for discussion of Rocky Flats waste buried at INL.
261. *Public Service Co. v. Kempthorne*, No. CV 91-035-S-EJL, 2006 U.S. Dist. LEXIS 34584 (D. Idaho May 26, 2006); *United States v. Otter*, 270 Fed. Appx. 568, 569–570 (9th Cir. 2008).
262. Neill and Neill, *Shallow Buried TRU*, 158 (2009).
263. *Id.*
264. Government Accountability Office, *Plans for Addressing Most Buried TRU Wastes*, 17–18 (2007).
265. *Id.* at 9.
266. Neill and Neill, *Shallow Buried TRU*, 165 (2009).
267. Government Accountability Office, *Plans for Addressing Most Buried TRU Wastes*, 21 (2007).
268. *Id.* at 21–22.
269. *See, e.g.*, Neill and Neill, *Shallow Buried TRU*, 187–188 (2009).
270. *Memorandum between WGA and DOE*, 2 (2009).
271. *Id.*
272. DOE, "Energy Secretary Chu Announces $6 Billion in Recovery Act Funding for Environmental Cleanup," press release (Mar. 31, 2009); American Recovery and Reinvestment Act of 2009, Pub. L. No. 111-5, 123 Stat. 115 (2009).
273. *Id.*; Office of Congressman Harry Teague, "Congressman Harry Teague Announces $172 Million in Recovery Funding for WIPP," press release (Mar. 2009); Matthew Reichbach, "On the Nuclear Waste Beat, Should We WIPP It Good?," *New Mexico Independent* (Mar. 31, 2009), newmexicoindependent.com/23653/on-the-nuclear-waste-beat-should-we-wipp-it-good.
274. TRU remains at 17 of the 31 TRU sites throughout the United States. DOE, *Waste to Be Consolidated at Idaho Site* (2008).
275. McCutcheon, *Nuclear Reactions*, 158 (2002).
276. Sue Vorenberg, "Ten Years in Operation, WIPP Boasts Sterling Safety Record, Continued Support," *Santa Fe New Mexican*, A1 (Mar. 29, 2009); Sue Major Holmes, Associated Press, "Some Suggest NM Area Could Replace Yucca Mountain," KLAS-TV Channel 8 News Las Vegas (Mar. 26, 2009) www.lasvegasnow.com/Global/story.asp?S=10079606.
277. "In Comments to Blue Ribbon Panel, N.M. Officials Open to More Waste," 22 *Weapons Complex Monitor*, no. 4–5, 11 (Jan. 28, 2011).

Chapter 6
1. Committee on Interior and Insular Affairs, *NWPA of 1982*, at 29 (1982).
2. Nuclear Waste Policy Amendments Act of 1987 § 5011, Pub. L. 100-203, 101 Stat. 1330-227, at 1330-228 (1987), 42 U.S.C. § 10172(a)(1) (2006). Even before the 1987 Amendments to NWPA were enacted, DOE secretary John S. Herrington had indefinitely postponed siting of the second repository. Thomas Isaacs, Director for the Office of Planning and Special Studies, Lawrence Livermore National Laboratory, telephone interview by Jane Stewart (Aug. 2008).
3. "The Secretary may not conduct site-specific activities with respect to a second repository unless Congress has specifically authorized and appropriated funds for such activities." Nuclear Waste Policy Amendments Act of 1987 § 5012, Pub. L. No. 100-203, 101 Stat. 1330-227, at 1330-231 (1987), 42 U.S.C. § 10172a(a) (2006). The second repository was relegated to a requirement that DOE evaluate the need for additional disposal space two decades later. *Id.*, Pub. L. No. 100-203, 101 Stat. 1330-227, at 1330-231, 42 U.S.C. § 10172a(b).
4. *Id.* § 10172.
5. Kemp, *Exploring Environmental Issues*, 213 (2004).
6. *Id.*
7. The State of Nevada commissioned numerous studies on the socioeconomic impacts that Yucca would have on the local economy, starting in 1986. *See* Nevada Agency for Nuclear Projects, Office of the Governor, *A Mountain of Trouble*, 51 (2002). The state studies consistently found that "impacts of the proposed Yucca Mountain project and

related high-level nuclear waste transportation from around the country to a repository would be ubiquitous, major in scale, and long-lasting." *Id.*
8. Interagency Review Group on Nuclear Waste Management, *Report to the President*, 51–52 (1979).
9. These included, among others, emplacement in mined repositories, deep ocean sediments, deep drill holes, and outer space. *Id.* at 35.
10. *Id.* at 35–37; Carter, *Nuclear Imperatives*, 136–139 (1987), discusses the behind-the-scenes discussions at IRG about the validity of different types of geologic settings, especially salt deposits. In contrast, some who commented on the draft argued that alternatives to mined repositories were "too speculative and exotic to warrant funding." Interagency Review Group on Nuclear Waste Management, *Report to the President*, 36 (1979).
11. Interagency Review Group on Nuclear Waste Management, *Report to the President*, 35 (1979).
12. *Id.* at H-9. The IRG report recommended looking at "a number of geologic environments possessing a wide variety of emplacement media such as salt, shale, or granite." *Id.* at H-8.
13. *Id.* at 38.
14. *Id.*
15. *Id.* at H-9.
16. *Id.* at 37.
17. For example, the two-repository siting scheme in NWPA can be traced back to the IRG report.
18. Message to Congress: Radioactive Waste Management Program, 1 Pub. Papers 296, 297 (Feb. 12, 1980).
19. *Id.*
20. Carter, *Nuclear Imperatives*, 200 (1987); Nuclear Waste Policy Act of 1982 § 112(a), 42 U.S.C. § 10132(a) (2006).
21. Message to Congress: Radioactive Waste Management Program, 1 Pub. Papers 296, 299 (Feb. 12, 1980).
22. *See* Nuclear Waste Policy Act of 1982 §§ 131–137, 42 U.S.C. §§ 10151–10157 (2006). In fact, Congress did debate, but eventually discarded, the idea of an away-from-reactor storage program. Carter, *Nuclear Imperatives*, 201 (1987).
23. In 1982, NRC announced the estimate that one-third of U.S. SNF storage would be filled to capacity by 1990. Ben A. Franklin, "Atom Waste Disposal Issue Still Unwelcome in Congress," *New York Times*, A12 (Aug. 25, 1982).
24. Carter, *Nuclear Imperatives*, 199–200 and n.6 (1987).
25. Jacob, *Site Unseen*, 84–86 (1990); Carter, *Nuclear Imperatives*, 195–196, 206, 212 (1987), describes the nuclear power industry's "delight" with the accelerated siting schedule included in the Senate bill.
26. Testimony of Brooks Yeager, Washington Representative of the Sierra Club, *Nuclear Waste Disposal Policy: Hearings on H.R. 1993, H.R. 2881, H.R. 3809, and H.R. 5016 before the Subcommittee on Energy Conservation and Power of the House Committee on Energy and Commerce*, 97th Cong., 2d Sess. 550–551 (1982).
27. Testimony of Renee Parsons, Legislative Representative of Friends of the Earth, *id.* at 559; Carter, *Nuclear Imperatives*, 204 (1987).
28. Testimony of Robert List, Governor, Nevada, and member, Subcommittee on Nuclear Power of the National Governors Association, *Nuclear Waste Disposal Policy: Hearings on H.R. 1993, H.R. 2881, H.R. 3809, and H.R. 5016 before the Subcommittee on Energy Conservation and Power of the House Committee on Energy and Commerce*, 97th Cong., 2d Sess. 392 (1982); Jacob, *Site Unseen*, 63, 86–87 (1990).
29. Statement of Russell Jim, member, Yakima Indian Nation, and representative, Council of Energy Resource Tribes, *Nuclear Waste Disposal: Joint Hearings Before the Senate Committee on Natural Resources and the Subcommittee on Nuclear Regulation of the Senate Committee on Environment and Public Works*, 97th Cong., 1st Sess. 117–119 (1981).
30. Carter, *Nuclear Imperatives*, 199–200 (1987).
31. "Following the issuance of guidelines under subsection (a) and consultation with the Governors of affected States, the Secretary shall nominate at least 5 sites that he determines suitable for site characterization for selection of the first repository site." Nuclear Waste Policy Act of 1982 § 112(b)(1)(A), 42 U.S.C. § 10132(b)(1)(A) (2006). "Not later than July 1, 1989, the Secretary shall nominate 5 sites, which shall include at least 3 additional sites not nominated under subparagraph (A), and recommend by such date to the President from such 5 nominated sites 3 candidate sites the Secretary determines suitable for site characterization for selection of the second repository." Nuclear Waste Policy Act of 1982 § 112(b)(1)(C), as enacted, Pub. L. No. 97-425, 96 Stat. 2201, at 2208 (1983).
32. Barlett and Steele, *Forevermore: Nuclear Waste in America*, 153–154 (1985); Andrews, Congressional Research Service, *Spent Nuclear Fuel Storage Locations and Inventory*, 4 (2004).
33. Nuclear Waste Policy Act of 1982 § 112(a), 42 U.S.C. § 10132(a) (2006).
34. Interagency Review Group on Nuclear Waste Management, *Report to the President*, 16 (1979).
35. *Id.*
36. Nuclear Waste Policy Act of 1982 § 112(a), 42 U.S.C. § 10132(a) (2006).
37. Nuclear Waste Policy Act of 1982 § 113(b)(1)(B), as enacted, Pub. L. No. 97-425, 96 Stat. 2201, at 2211 (1983); Nuclear Waste Policy Act of 1982 § 114(a)(1)(B), 42 U.S.C. § 10134(a)(1)(B) (2006).
38. Nuclear Waste Policy Act of 1982 § 122, 42 U.S.C. § 10142 (2006).
39. *Id.*
40. *Id.* § 8(b)(2), 42 U.S.C. § 10107(b)(1)–(2).
41. *Id.* § 112(a), 42 U.S.C. § 10132(a).
42. *Id.* § 112(a)–(b), 42 U.S.C. § 10132(a)–(b). The guidelines were published at Nuclear Waste Policy Act of 1992: General Guidelines for the Recommendation of Sites for the Nuclear Waste

Repositories, 49 FED. REG. 47,714, at 47,714 (Dec. 6, 1984).
43. Nuclear Waste Policy Act of 1982 § 213(a), 42 U.S.C. § 10193(a) (2006).
44. *Id.* § 301, 42 U.S.C. § 10221.
45. *Id.* § 112(b)(1)(A), 42 U.S.C. § 10132(b)(1)(A).
46. *Id.* § 112(b)(1)(H), 42 U.S.C. § 10132 (b)(1)(G).
47. *Id.* § 112(b)(1)(E), (b)(2), (b)(3), 42 U.S.C. § 10132(b)(1)(D), (b)(2), (b)(3).
48. NWPA did not prescribe how DOE was to determine which three of the five nominated sites it would recommend undergo site characterization; however, DOE announced that "the decision to recommend a site will be based on (1) the available geophysical, geologic, geochemical, and hydrologic data, . . . (2) other information, and (3) the associated evaluations and findings reported in the environmental assessments." Nuclear Waste Policy Act of 1992: General Guidelines for the Recommendation of Sites for the Nuclear Waste Repositories, 49 FED. REG. 47,714, at 47,717 (Dec. 6, 1984).
49. Nuclear Waste Policy Act of 1982 § 112(b)(1)(A)–(B), 42 U.S.C. § 10132(b)(1)(A)–(B) (2006).
50. Nuclear Waste Policy Act of 1982 § 114(a)(1), 42 U.S.C. § 10135(b) (2006).
51. *Id.* § 115(c), 42 U.S.C. § 10135(c); Nuclear Waste Policy Act of 1982 § 114(a)(3), as enacted, Pub. L. No. 97-425, 96 Stat. 2201, at 2214 (1983).
52. Nuclear Waste Policy Act of 1982 § 114(b), as enacted, Pub. L. No. 97-425, 96 Stat. 2201, at 2214–2215 (1983); Nuclear Waste Policy Act of 1982 § 121(a)–(b), 42 U.S.C. § 10141(a)–(b) (2006). The act called for DOE to undertake environmental assessments of sites nominated for characterization; DOE was not required to prepare environmental impact statements, which are far more in-depth studies, until it had completed site characterization and was ready to recommend selected sites for construction of a repository. *Id.* §§ 112(b)(1)(E)–(G), (e), 113(d), 42 U.S.C. §§ 10132(b)(1)(D)–(F), (e), 10133(d); Nuclear Waste Policy Act of 1982 § 114(a)(1)(D), as enacted, Pub. L. No. 97-425, 96 Stat. 2201, at 2213 (1983).
53. Office of Civilian Radioactive Waste Management, DOE, *Mission Plan for the Civilian Radioactive Waste Management Program*, Vol. 1, at ii (1985).
54. President Reagan decided that there was no need to dispose of defense wastes separately, in another repository, in April 1985. WIPP Transportation Safety Program, New Mexico Energy, Minerals, and Natural Resources Department, "Chronology of WIPP," *www.emnrd.state.nm.us/wipp/chronolo.htm*.
55. Nuclear Waste Policy Act of 1982, 42 U.S.C. § 10137(c) (2006).
56. NWPA does not include specific deadlines for DOE's recommendation to the president on the site for a repository; NWPA's deadline for DOE to submit a license application to NRC for a repository site depends on the date the president's recommendation of a repository site becomes effective, rather than falling on a specific date.
57. DOE was required to submit its application for construction of a repository to the NRC within 90 days of the site recommendation for designation becoming official. Nuclear Waste Policy Act of 1982 § 114(b), as enacted, Pub. L. No. 97-425, 96 Stat. 2201, at 2214–2215 (1983). A recommendation for designation became official within 60 days of the president's recommendation of a repository site if there was no notice of disapproval, or upon Congress' passing a joint resolution overriding the notice. If a state or tribe did issue a notice of disapproval, the president's recommendation became effective only if Congress overrode the disapproval during the first ninety days of its first session after receiving the notice. Nuclear Waste Policy Act of 1982 § 115(b)–(c), 42 U.S.C. § 10135(b)–(c) (2006).
58. Nuclear Waste Policy Act of 1982 §§ 112–118, as enacted, Pub. L. No. 97-425, 96 Stat. 2201, at 2208–2227 (1983); Jacob, *Site Unseen*, 96 (1990).
59. Nuclear Waste Policy Act of 1982 §§ 111(a)(4), 302(a), (c)–(d), 42 U.S.C. §§ 10131(a)(4), 10222(a), (c)–(d) (2006).
60. *Id.* § 302(a)(5)(B), 42 U.S.C. § 10222(a)(5)(B). As discussed in Chapter 3, the nature and extent of the government's liability for its failure to meet this obligation has been much disputed and was the subject of lawsuits in the mid-1990s. *See* Wall, *Going Nowhere: Breach of Yucca Contract* (2007). Court decisions in those suits confirmed that DOE was liable for monetary damages to utilities for their detrimental reliance upon the federal government's obligation to take title to SNF stored at reactor sites by 1998. *Id.* at 165.
61. Nuclear Waste Policy Act of 1982 §§ 111(a)(4), 302(b), 42 U.S.C. §§ 1013(a)(4), 10222(b) (2006). Disposal of defense or other federal HLW in a NWPA repository was allowed by the act provided that the agency responsible for the waste transferred funds into the Nuclear Waste Fund at rates equivalent to the utilities. *Id.* The act also instructed DOE to charge nuclear electric utilities a fee for delivery of SNF, or HLW derived from SNF, to a federal repository if that SNF had been produced in a power reactor that was not yet subject to a contract with DOE. *Id.* § 302(a)(3), 42 U.S.C. § 10222(a)(3).
62. Committee on Interior and Insular Affairs, *NWPA of 1982*, at 31 (1982).
63. Nuclear Waste Policy Act of 1982 § 116(c), as enacted, Pub. L. No. 97-425, 96 Stat. 2201, at 2221–2222 (1983). Impact assistance was intended to be "designed to mitigate the impact on such State of the development of such repository" by enabling the state to protect its interests, including "any economic, social, public health and safety, and environmental impacts that are likely as a result of the development of a repository." *Id.* § 116(c)(2)(A)–(B), as enacted, Pub. L. No. 97-425, 96 Stat.

2201, at 2221. The state was to submit its request for funding to DOE before the site was recommended to the president; once the construction license was granted by NRC, the act provided that "the Secretary shall seek to enter into a binding agreement [regarding such funding] with the State involved." *Id.* § 116(c)(2)(B), as enacted, Pub. L. No. 97-425, 96 Stat. 2201, at 2221.

64. *Id.* § 116(c)(3), as enacted, Pub. L. No. 97-425, 96 Stat. 2201, at 2222.

65. Nuclear Waste Policy Act of 1982 § 114(d), as enacted, Pub. L. No. 97-425, 96 Stat. 2201, at 2215.

66. Mark Holt, Congressional Research Service, "Nuclear Waste Policy: How We Got Here," presentation, 40 (Mar. 25, 2010), *brc.gov/pdfFiles/CRS_BlueRibbonCommissionWastePolicyHistory.pdf*.

67. Statement of Senator Slade Gorton: "This [emplacement] limitation . . . discourages long or frequent transportation of wastes throughout the country." 128 CONG. REC. 8213 (1982); *see also* Office of Technology Assessment, *Managing the Nation's Commercial High-Level Radioactive Waste*, 102 (1985).

68. Statement of Senator Henry Jackson, 128 CONG. REC. 32,568 (Dec. 20, 1982). Washington's two senators, Slade Gorton and Henry Jackson, cosponsored the amendment on emplacement limitations; it was passed without objection and enacted as part of Nuclear Waste Policy Act of 1982 § 114(d), as enacted Pub. L. No. 97-425, 96 Stat. 2201, 2215 (1983).

69. Committee on Interior and Insular Affairs, *NWPA of 1982*, at 44–45 (1982).

70. WIPP Authorization Act, Department of Energy National Security and Military Applications of Nuclear Energy Authorization Act of 1980 § 213(a), Pub. L. No. 96-164, 93 Stat. 1259, at 1265 (1979).

71. Nuclear Waste Policy Act of 1982 § 141(b)(1)(A), 42 U.S.C. § 10161(b)(1)(A) (2006).

72. *Id.* § 131(b)(2), 42 U.S.C. § 10151(b)(2).

73. *Id.* § 141(b)(1)(B), 42 U.S.C. § 10161(b)(1)(B).

74. *Id.* § 141(b)(1)(D), 42 U.S.C. § 10161(b)(1)(D).

75. *Id.* § 141(b)(1)(C), 42 U.S.C. § 10161(b)(1)(C).

76. *Id.* § 141(b)(1)–(2), 42 U.S.C. § 10161(b)(1)–(2).

77. *Id.* § 141(b)(2)(C)(ii), 42 U.S.C. § 10161(b)(2)(C)(ii).

78. *Id.* § 141(c), 42 U.S.C. § 10161(c). NWPA did, however, give states and tribes the same participation rights as under the repository siting system. *Id.* § 141(h), 42 U.S.C. § 10161(h). The 1987 Amendments added a formal MRS-siting procedure. They also established an MRS commission, which would report on need for an MRS. Nuclear Waste Policy Amendments Act of 1987 § 5021, Pub. L. No. 100-203, 101 Stat. 1330-227, at 1330-232 (1987), 42 U.S.C. § 10163 (2006). The MRS commission would survey and evaluate possible sites. *Id.*, Pub. L. No. 100-203, 101 Stat. 1330-227, at 1330-234, 42 U.S.C. § 10164. The energy secretary could then recommend a site to the president, but only after DOE had recommended a site for development of a repository. *Id.*, Pub. L. No. 100-203, 101 Stat. 1330-227, at 1330-234, 42 U.S.C. § 10165.

79. Nuclear Waste Policy Act of 1982 § 111(a)(6), 42 U.S.C. § 10131(a)(6) (2006).

80. *Id.* § 111(b)(3), 42 U.S.C. § 10131(b)(3).

81. Committee on Interior and Insular Affairs, *NWPA of 1982*, at 46 (1982).

82. Nuclear Waste Policy Act of 1982 § 112(a), 42 U.S.C. § 10132(a) (2006).

83. *Id.* § 112(b)(1)(A), 42 U.S.C. § 10132(b)(1)(A).

84. *Id.* § 112(b)(1)(H), 42 U.S.C. § 10132(b)(1)(G).

85. Nuclear Waste Policy Act of 1982 § 113(a), as enacted, Pub. L. No. 97-425, 96 Stat. 2201, at 2211 (1983). Similarly, section 113 provided that "before proceeding to sink shafts . . . the Secretary shall submit . . . [to] the Governor and legislature of the State . . . for their review and comment (A) a general plan for site characterization activities . . . includ[ing] (i) a description of such candidate site; (ii) a description of . . . site characterization activities . . . ; [and] (iv) criteria to be used to determine the suitability of such candidate site for the location of a repository." *Id.* § 113(b)(1), as enacted, Pub. L. No. 97-425, 96 Stat. 2201, at 2211.

86. Nuclear Waste Policy Act of 1982 § 101(a), 42 U.S.C. § 10121(a) (2006).

87. *Id.* § 101(b), 42 U.S.C. § 10121(b). A tribe holding title to the land on which a proposed repository would be sited could only be considered "affected" for the purposes of the act if it proved by petition to the secretary of the interior that the construction of a facility on its land would have effects "both substantial and adverse to the tribe." *Id.* § 2(2)(B), 42 U.S.C. § 10101(2)(B).

88. *Id.* § 111(a)(6), 42 U.S.C. § 10131(a)(6). The original 1982 NWPA did not provide for public participation in the MRS-siting process. The 1987 amendments to NWPA instead required public hearings before MRS site selection. Nuclear Waste Policy Amendments Act of 1987 § 5021, Pub. L. No. 100-203, 101 Stat. 1330-227, at 1330-235 (1987), 42 U.S.C. § 10165(e)(2) (2006).

89. Nuclear Waste Policy Act of 1982 § 112(b)(2), 42 U.S.C. § 10132(b)(2) (2006). The 1982 act also provided for public input into negotiations between a state or tribe and DOE for consultation and cooperation in the construction of an interim storage facility. *Id.* § 135(d)(2), 42 U.S.C. § 10155(d)(2).

90. *Id.* § 112(b)(1)(G), 42 U.S.C. § 10132(b)(1)(F). It provided the same for environmental impact statements prepared for possible interim storage facility sites. *Id.* § 135(c)(2)(A), 42 U.S.C. § 10155(c)(2)(A).

91. *Id.* § 117(c)(10), 42 U.S.C. § 10137(c)(10).

92. Nuclear Waste Policy Act of 1982 § 116(c), as enacted, Pub. L. No. 97-425, 96 Stat. 2201, at 2207 (1983).

93. *Id.* § 118(b), as enacted, Pub. L. No. 97-425, 96 Stat. 2201, at 2225–2227.

94. Nuclear Waste Policy Act of 1982 § 117(c), 42 U.S.C. § 10137(c) (2006). A Senate version proposed

requiring the parties to an MRS siting to "seek to conclude" an agreement, 128 Cong. Rec. 32,545 (1982), but at no point did the legislature discuss or propose requiring the parties "to conclude" an agreement.
95. *See* Committee on Interior and Insular Affairs, *NWPA of 1982*, at 12 (1982). GAO summarized it this way in a 1987 report detailing problems with implementation of NWPA of 1982: "Rather than specifying the level of participation expected, the act stated that the Secretary of Energy shall consult and cooperate with the governor and legislature of affected states . . . to try to resolve their concerns regarding public health and safety, environmental, and economic impacts of a repository." General Accounting Office, *Institutional Relations under NWPA*, 14 (1987).
96. Nuclear Waste Policy Act of 1982, 42 U.S.C. § 10137(c) (2006). Section 117(c) is similar to the consultation provision of the WIPP Authorization Act § 213(b), Pub. L. No. 96-164, 93 Stat. 1259, at 1265–1266 (1979). *Compare* "the Secretary shall seek to enter into a binding written agreement, and shall begin negotiations, with such State and, where appropriate, to enter into a separate binding agreement with the governing body of any affected Indian tribe, setting forth (but not limited to) the procedures under which the requirements of subsections (a) and (b)," which describe requirements for provision of information and consultation and cooperation, "and the provisions of such written agreement, shall be carried out," Nuclear Waste Policy Act of 1982 § 117(c), 42 U.S.C. § 10137(c) (2006), *with* "(1) The Secretary shall consult and cooperate with the appropriate officials of the State of New Mexico, with respect to the public health and safety concerns of such State in regard to such project and shall . . . give consideration to such concerns and cooperate with such officials in resolving such concerns; (2) The Secretary shall seek to enter into a written agreement with the appropriate officials of the State of New Mexico, as provided by the laws of the State of New Mexico, not later than September 30, 1980, setting forth the procedures under which the consultation and cooperation required by paragraph (1) shall be carried out," WIPP Authorization Act § 213(b), Pub. L. No. 96-164, 93 Stat. 1259, at 1265–1266 (1979).
97. General Accounting Office, *Nuclear Waste: Status of DOE's Implementation of NWPA*, 36 (1987). The State of Washington attempted to negotiate a C&C agreement with DOE but suspended negotiations after the parties disputed whether the federal government would assume unlimited liability for potential accidents at a repository. *Id.* The DOE Office of Civilian Radioactive Waste Management (OCRWM) contends that it offered to begin negotiations with Nevada in November 1986. Office of Civilian Radioactive Waste Management, DOE, *Annual Report to Congress*, 22 (1989). Also, according to OCRWM, in April 1988, DOE once more offered to negotiate a formal C&C agreement with Nevada, which declined, saying it preferred to rely on informal procedures. *Id.* Nevada, however, asserts that "no such 'C&C' agreement has ever been seriously pursued by federal officials." Nevada Agency for Nuclear Projects, Office of the Governor, *Executive Summary of Nevada Socioeconomic Studies Biannual Report* (1995).
98. The Nuclear Waste Policy Act of 1982 § 116(b), 42 U.S.C. § 10136(b) (2006), allows a state to issue a notice of disapproval with respect to a repository site in the state; *id.* § 118(a), 42 U.S.C. § 10138(a), allows a tribe to issue a notice of disapproval with respect to a repository site on tribal lands; and the Nuclear Waste Policy Amendments Act of 1987 § 5021, Pub. L. No. 100-203, 101 Stat. 1330-227, at 1330-235 (1987), 42 U.S.C. § 10166 (2006), allows a state or tribe to issue a notice of disapproval on an MRS site within its territory.
99. Carter, *Nuclear Imperatives*, 209–210 (1987); Jacob, *Site Unseen*, 87–89 (1990).
100. 128 Cong. Reg. 8209 (Apr. 15, 1982).
101. Carter, *Nuclear Imperatives*, 210 (1987).
102. Easterling and Kunreuther, *The Dilemma of Siting a Repository*, 42 and n.45 (1995).
103. *See* the statement of Congressman Toby Roth: "I want to urge my colleagues to give their support to the provision that requires that both Houses of Congress vote in the affirmative before the decision of a State Governor can be overridden. Every major organization representing the States—National Governors Association, National Conference of State Legislatures, and State Planning Council—have [sic] endorsed the two-House procedure." 128 Cong. Rec. 27,766 (Nov. 29, 1982).
104. *See* the statement of Senator William Proxmire, 128 Cong. Rec. 8207 (Apr. 29, 1982).
105. Senator Trent Lott of Mississippi failed in his gambit to exclude his state from consideration by inserting a stiffer population-proximity criterion than the one ultimately codified in the act. 128 Cong. Rec. 27,787 (Nov. 29, 1982). *Compare* Lott's amendment that "such guidelines shall specify population factors that will disqualify any site from development as a repository *if any such site* would be located . . . adjacent to an area 1 mile by 1 mile square having a population of not less than 1,000 individuals" (emphasis added) *with* NWPA's enacted provision that "such guidelines shall specify population factors that will disqualify any site from development as a repository *if any surface facility of such repository* would be located . . . adjacent to an area 1 mile by 1 mile having a population of not less than 1,000 individuals," Nuclear Waste Policy Act of 1982 § 112(a), 42 U.S.C. § 10132(a) (2006) (emphasis added). Since the actual sites could extend for miles underground, the difference was an appreciable one.
106. Nuclear Waste Policy Act of 1982 § 112(a), 42 U.S.C. § 10132(a) (2006).
107. *Id.*
108. Final Waste Confidence Decision, 49 Fed. Reg. 34,658 (Aug. 31, 1984). The decision itself addressed

the narrow question of whether nuclear waste storage and disposal could be accomplished safely, but its key effect was to counter efforts by states, suspicious that the federal government would not proceed with waste disposal plans in a reasonable time frame, to block construction of additional nuclear power plants. *See Pac. Gas & Elec. Co. v. State Energy Res. Conservation and Dev. Comm'n*, 461 U.S. 190 (1983).

109. WIPP Transportation Safety Program, New Mexico Energy, Minerals, and Natural Resources Department, "Chronology of WIPP," *www.emnrd.state.nm.us/wipp/chronolo.htm*; Flynn and Slovic, "Yucca Mountain," 90 (1995). This determination was made pursuant to the Nuclear Waste Policy Act of 1982 § 8, 42 U.S.C. § 10107 (2006), which required the president to evaluate the possibility of using a NWPA repository for co-disposal of defense HLW.

110. General Accounting Office, *Institutional Relations under NWPA*, 31 (1987).

111. Nuclear Waste Policy Act of 1982: Availability of Draft Environmental Assessments for Proposed Site Nominations and Announcement of Public Information Meetings and Hearings, 49 Fed. Reg. 49,540, 49,540 (Dec. 20, 1984).

112. *Id.*

113. General Accounting Office, *Institutional Relations under NWPA*, 10 (1987).

114. *See* Carter, *Nuclear Imperatives*, 151–161, 165–170 (1987). Building a repository in Paradox Basin, Utah, for instance, was popular in San Juan County, where the basin is located, but strongly opposed by most Utahans and national environmental groups. *Id.* at 151–153.

115. *See id.* at 156–157, 161, 166.

116. *Id.* at 153.

117. *Id.* at 161.

118. *Id.* at 157.

119. *Id.* at 166.

120. *Texas v. Dept. of Energy*, 764 F.2d 278, 285 (5th Cir. 1985). Texas challenged as defective DOE's notification to the state that DOE was considering the two sites as locations for a federal repository. *Id.* at 281. Nuclear Waste Policy Act § 116(a), 42 U.S.C. § 10136(a) (2006), governs DOE's identification of potentially acceptable sites.

121. *Nevada ex rel. Loux v. Herrington*, 777 F.2d 529, 535 (9th Cir. 1985).

122. *Nevada ex rel. Loux v. Herrington*, 777 F.2d 529 (9th Cir. 1985). The court held that Nevada was entitled to receive funds from DOE to study issues relevant to site characterization for a repository at Yucca. *Id.* at 531.

123. *Id.* at 529, 534, 536.

124. *Id.* at 536.

125. Nev. Rev. Stat. § 459.0091 (2009). The NCNP succeeded the state's Nuclear Waste Projects Office in its watchdog and advocacy functions.

126. *See* Agency for Nuclear Projects, State of Nevada, *www.state.nv.us/nucwaste/*.

127. *Nevada v. Herrington* [II], 827 F.2d 1394 (9th Cir. 1987).

128. General Accounting Office, *Nuclear Waste: Status of DOE's Implementation of NWPA*, 80 (1987).

129. General Accounting Office, *Institutional Relations under NWPA*, 2–3 (1987).

130. Nuclear Waste Policy Act of 1982: Availability of Draft Environmental Assessments for Proposed Site Nominations and Announcement of Public Information Meetings and Hearings, 49 Fed. Reg. 49,540, 49,540 (Dec. 20, 1984).

131. These guidelines considered separately the site's features for preclosure (construction, operation, sealing) and postclosure (100,000 years after sealing). Nuclear Waste Policy Act of 1982: General Guidelines for the Recommendation of Sites for the Nuclear Waste Repositories, 49 Fed. Reg. 47,714, 47,723 (Dec. 6, 1984), 10 C.F.R. pt. 960 (2010).

132. Carter, *Nuclear Imperatives*, 403–404 (1987).

133. The secretary of energy announced on May 28, 1986, that DOE had recommended, and the president had approved, Yucca, Hanford, and Deaf County as the three sites that would undergo site characterization. Carter, *Nuclear Imperatives*, 401 (1987); *see also* Office of Civilian Radioactive Waste Management, DOE, *Recommendation of Candidate Sites for First Repository*, 9 (1986). DOE apparently had not yet officially nominated the five sites it had considered for characterization; it did so several days later. Nuclear Waste Policy Act of 1982: Nomination of Five Sites for the First High-Level Nuclear Waste Repository, and Availability of Accompanying Environmental Assessments, 51 Fed. Reg. 19,783, 19,783 (June 2, 1986).

134. Statement of Senator Daniel J. Evans, *Civilian Radioactive Waste Disposal: Hearings before the Senate Committee on Energy and Natural Resources*, 100th Cong., 1st Sess. 236 (1987); statement of John W. Bartlett, Director, Office of Civilian Radioactive Waste Management, DOE, *Nuclear Waste Disposal Issues: Hearing before the Subcommittee on Nuclear Regulation of the Senate Committee on Environment and Public Works*, 102nd Cong. 48 (1991); General Accounting Office, *Quarterly Report on DOE's Nuclear Waste Program as of December 31, 1986*, at 7 (1987).

135. DOE noted that the ranking reflected the large differences in costs; the differences among the sites based on other factors were very small. If repository and transportation costs had been excluded, Hanford would have ranked highest among the five sites. Office of Civilian Radioactive Waste Management, DOE, *Recommendation of Candidate Sites for First Repository*, 5 (1986); Office of Civilian Radioactive Waste Management, DOE, *A Multiattribute Utility Analysis*, 5–16 (1986).

136. *See* Carter, *Nuclear Imperatives*, 166–170, 403, 405–406 (1987).

137. *Id.* at 405.

138. *Id.* at 165.

139. Office of Civilian Radioactive Waste Management,

DOE, *Recommendation of Candidate Sites for First Repository*, 7–8 (1986).
140. Flynn and Slovic, "Yucca Mountain," 89 (1995); Frank L. Parker, Chairman, Board on Radioactive Waste Management, National Research Council, "Letter to Ben C. Rusche, Director, Office of Civilian Radioactive Waste Management, Department of Energy," 1 (Apr. 10, 1986) (on file with authors). The board was originally organized as the Committee on Radioactive Waste Management and has since been reorganized as the Nuclear and Radiation Studies Board. National Research Council of the National Academies, *Annual Report, Fiscal Years 1973 and 1974*, 83 (1975); Committee on Transportation of Radioactive Waste, National Research Council of the National Academies, *Going the Distance?*, iv n.1 (2006).
141. Frank L. Parker, Chairman, Board on Radioactive Waste Management, National Research Council of the National Academies, wrote Ben C. Rusche, Director, Office of Civilian Radioactive Waste Management, Department of Energy: "It should be noted, however, that the Board's focus was on methodology and its implementation and that the Board has not reviewed in detail the data and judgments on which the conclusions from the multi-attribute procedure are based." "Letter to Ben C. Rusche, Director, Office of Civilian Radioactive Waste Management, Department of Energy," 1–2, 4 (Apr. 10, 1986) (on file with authors); testimony of Raphael G. Kasper, Executive Director, Commission on Physical Sciences, Mathematics, and Resources, National Research Council, *Fiscal Year 1988 Department of Energy Authorization (Nuclear Waste; Uranium Enrichment): Hearings before the Subcommittee on Energy Research and Development of the House Committee on Science, Space, and Technology*, 100th Cong., 1st Sess., 127–129 (Mar. 19, 1987).
142. Flynn and Slovic, "Yucca Mountain," 89 and nn.11–14 (1995).
143. The spokeswoman for a Nevada representative explained her boss's opposition to Yucca as follows: "If all those [siting] criteria had led to an orderly process that (Yucca) tuff was the safest in the United States, she probably wouldn't object. . . . However, that is not what has happened." Mary Manning, "GOP, Demos Agree: DOE Boss Wrong," *Las Vegas Sun* (May 31, 1986).
144. *See, e.g., Nevada v. Watkins*, 943 F.2d 1080 (9th Cir. 1991), in which the court rejected as moot Nevada's claims that DOE's environmental assessment for Yucca was deficient.
145. Brody and Fleishman, "Sources of Public Concern," 120, 132–133 (1993).
146. Office of Civilian Radioactive Waste Management, DOE, *Recommendation of Candidate Sites for First Repository*, 9 (1986); Lomenick, Oak Ridge National Laboratory, *The Siting Record*, 52 (1996).
147. McCutcheon, *Nuclear Reactions*, 117 (2002).
148. Edelstein, "Hanford: the Closed City," 267 (2007).
149. *Id.*
150. *See* General Accounting Office, *Nuclear Waste: Repository Work Should Not Proceed*, 24–27 (1988).
151. *Id.* at 54–57 documents stop-work orders issued to DOE contractors upon findings of poor quality assurance.
152. Matthew L. Wald, "Work Is Faltering on U.S. Repository for Atomic Waste," *New York Times*, A1 (Jan. 17, 1989). The USGS memorandum specifically alleged that scientists had not been allowed to sample potentially radioactive test hole gas emissions at Yucca because of purported "paperwork" problems. *Id.*
153. Office of Civilian Radioactive Waste Management, DOE, *Annual Report to Congress*, 22 (1989), reports that Nevada declined DOE's invitations in November 1986 and April 1988 to negotiate a C&C agreement.
154. Don Hancock, Nuclear Waste Safety Program, Southwest Research and Information Center, telephone interview by Alice Byowitz (Sept. 2, 2008).
155. Quotes from those participating in 1984 and 1986 public opinion surveys of the counties under consideration are illustrative. One respondent said: "It is beyond me why the government would even consider placing the repository in one of the most agriculturally productive areas of the United States." Brody and Fleishman, "Sources of Public Concern," 132 (1993). Another reported that "some expectations [of stigma] have already been translated into actions, including concrete changes in agricultural investments." *Id.* at 133.
156. Don Hancock, Nuclear Waste Safety Program, Southwest Research and Information Center, telephone interview by Alice Byowitz (Sept. 2, 2008); Carter, "The Path to Yucca Mountain and Beyond," 384 (2006).
157. Kraft and Clary, "Citizen Participation and NIMBY," 304 (1991), cites DOE's 1985 mission plan.
158. Nuclear Waste Policy Act of 1982 § 112(b)(1)(C), as enacted, Pub. L. No. 97-425, 96 Stat. 2201, at 2208 (1983).
159. Nuclear Waste Policy Act of 1982: Availability of Crystalline Repository Project Draft Area Recommendation Report and Announcement of Public Information Meetings and Hearings, 51 Fed. Reg. 2420, 2420 (Jan. 16, 1986), reports on sites identified in seventeen states; *see also* Carter, *Nuclear Imperatives*, 410 (1987), describing a narrowed list of twelve preferred sites in seven states.
160. Carter, *Nuclear Imperatives*, 410 (1987).
161. *Maine v. Herrington*, 790 F.2d 8, 9 (1st Cir. 1986).
162. *Id.* at 10.
163. General Accounting Office, *Nuclear Waste: Quarterly Report, June 30, 1986*, 14 (1986).
164. *Id.* at 17.
165. Schaefer, *State Opposition to a Federal Nuclear Waste Repository Siting* (1988).
166. Carter, *Nuclear Imperatives*, 411 (1987).
167. *Id.*
168. Robert D. Hershey Jr., "U.S. Suspends Plan

for Nuclear Dump in East or Midwest," *New York Times*, A1 (May 29, 1986); Carter, *Nuclear Imperatives*, 412 (1987), quotes from Secretary Herrington's May 28, 1986, press conference: "To go ahead and spend hundreds of millions of dollars on site identification now would be both premature and unsound fiscal management."
169. Carter, *Nuclear Imperatives*, 412n.34 (1987).
170. *Id.* at 412 and n.32 cites Crystalline Rock Program Chicago Office, DOE, "Crystalline Options" (May 13, 1986).
171. Carter, *Nuclear Imperatives*, 413–414 (1987).
172. Nuclear Waste Policy Amendments Act of 1987 § 5012, Pub. L. No. 100-203, 101 Stat. 1330-227, at 1330-231 (1987), 42 U.S.C. § 10172a(b) (2006).
173. Nuclear Waste Policy Act of 1982 § 141(b)(1), 42 U.S.C. § 10141(b)(1) (2006).
174. "On or before June 1, 1985, the Secretary shall complete a detailed study of the need for and feasibility of, and shall submit to the Congress a proposal for, the construction of one or more monitored retrievable storage facilities for high-level radioactive waste and spent nuclear fuel." *Id.* § 161(b)(1), 42 U.S.C. § 10161(b)(1).
175. DOE, *A Monitored Retrievable Storage Facility: Technical Background Information*, 4 (1991).
176. Announcement of Identification of Candidate Sites for a Proposal to Congress for a Monitored Retrievable Storage Facility and Availability of a Preliminary Analysis of the Need for and Feasibility of the Facility, 50 Fed. Reg. 16,536, at 16,537 (Apr. 26, 1985).
177. *Tennessee v. Herrington*, 806 F.2d 642, 645–646 (6th Cir. 1986).
178. Sigmon, "Achieving a Negotiated Compensation Agreement," 172–174 (1987).
179. Colglazier and Langum, "Policy Conflicts in the Process for Siting Repositories," 335 (1988).
180. Sigmon, "Achieving a Negotiated Compensation Agreement," 172–173 (1987).
181. *Id.*
182. Governor Lamar Alexander of Tennessee said that he opposed the project because he felt the MRS was unnecessary for the country's nuclear waste program, would raise electricity rates in the state, and would negatively impact public attitudes (especially those of tourists and business leaders) toward the region. *Id.* at 174.
183. *Tennessee v. Herrington*, 806 F.2d 642, 643 (6th Cir. 1986).
184. Nuclear Waste Policy Amendments Act of 1987 § 5021, Pub. L. No. 100-203, 101 Stat. 1330-227, at 1330-232 (1987), 42 U.S.C. § 10162(a) (2006).
185. *Id.* The purpose of this provision was to prevent DOE from catapulting the Tennessee sites that it had already studied and found potentially suitable to the top of the list of any restarted search for MRS sites in the future. Both DOE and Congress had used this approach in the repository-siting process. Congress chose Yucca in the 1987 NWPA amendments in part based on its top ranking among sites already studied by DOE, and DOE had begun its search for first-repository sites under the 1982 NWPA with sites the federal government had studied in an earlier repository-siting attempt.
186. Nuclear Waste Policy Amendments Act of 1987 § 145(b), as enacted, Pub. L. No. 100-203, 101 Stat. 1330-227, 1330-234 (1987).
187. *See* Carter, *Nuclear Imperatives*, 414 (1987).
188. *See Texas v. Department of Energy*, 764 F.2d 278 (5th Cir. 1985), which found Texas's challenges to site characterization unripe; *Nevada ex rel Loux v. Herrington*, 777 F.2d 529 (9th Cir. 1985), in which Nevada won DOE funds to conduct the state's repository-related research agenda; Carter, *Nuclear Imperatives*, 166, 170 (1987), which notes political antipathy in Washington State to a potential repository site at Hanford and quotes multiple observations about the site's technical unsuitability; and Colglazier and Langnum, "Policy Conflicts in the Process for Siting Repositories," 331 (1988), noting the Washington governor's firm opposition on technical grounds to siting a repository at Hanford.
189. Statement of Senator Daniel J. Evans, *Civilian Radioactive Waste Disposal: Hearings before the Senate Committee on Energy and Natural Resources*, 100th Cong., 1st Sess. 236 (1987); statement of John W. Bartlett, Director, Office of Civilian Radioactive Waste Management, DOE *Nuclear Waste Disposal Issues: Hearing before the Subcommittee on Nuclear Regulation of the Senate Committee on Environment and Public Works*, 102nd Cong. 48 (1991); General Accounting Office, *Quarterly Report on DOE's Nuclear Waste Program as of December 31, 1986*, 7 (1987).
190. *See* Chapter 1, Map 1.2, for locations of operating reactor sites, the great majority of which are in the East and Midwest.
191. General Accounting Office, *Institutional Relations under NWPA*, 2–3 (1987); testimony of Keith O. Fultz, Associate Director, Resources, Community, and Economic Development Division, General Accounting Office, *Nuclear Waste Policy Act of 1982: Progress and Problems: Hearing before the Subcommittee on Energy Research and Production of the House Committee on Science and Technology*, 99th Cong., 1st Sess. 172–174 (1985); General Accounting Office, *Nuclear Waste: Repository Work Should Not Proceed*, 54–57 (1988); Office of Technology Assessment, *Managing the Nation's Commercial High-Level Radioactive Waste*, 3–6 (1985).
192. General Accounting Office, *Institutional Relations under NWPA*, 48–49 (1987), notes that DOE had done a poor job of interacting with states and tribes; Office of Technology Assessment, *Managing the Nation's Commercial High-Level Radioactive Waste*, 4 (1985), notes DOE's overly optimistic schedule; *id.* at 160 and n.20 reports that DOE's approach to intergovernmental relations had garnered criticism from officials of various states, including those whose states were not being considered for a permanent repository.

193. *See, e.g.*, Flynn and Slovic, "Yucca Mountain," 91–92 (1995), for criticism of Congress's decision to designate DOE as the agency in charge of implementing NWPA.
194. According to one commentator, this meant that potential sites "had been essentially decided before the act was passed." *See* Carter, *Nuclear Imperatives*, 403 (1987). The tight statutory timetable also meant that DOE and others missed some of the early NWPA deadlines. For example, DOE was supposed to submit to Congress its final mission plan for evaluating and selecting repository sites on July 6, 1984; the mission plan was actually submitted in February 1985. *See* General Accounting Office, *Status of DOE's Implementation of NWPA as of 1984*, at ii (1984). The president's decision on whether the repository could also be used for military waste, due January 7, 1985, was actually made in April 1985. *See* WIPP Transportation Safety Program, New Mexico Energy, Minerals, and Natural Resources Department, "Chronology of WIPP," www.emnrd.state.nm.us/wipp/chronolo.htm.
195. The nine potential sites lay in Texas (two: Deaf Smith County, Swisher County), Utah (two: Davis Canyon, Lavender Canyon), Mississippi (two: Cypress Creek Dome, Richton Dome), Louisiana (one: Vacherie Dome), Washington (one: Hanford Reservation), Nevada (one: Yucca Mountain). Nuclear Waste Policy Act of 1982: Availability of Draft Environmental Assessments for Proposed Site Nominations and Announcement of Public Information Meetings and Hearings, 49 Fed. Reg. 49,540, 49,540 (Dec. 20, 1984).
196. General Accounting Office, *Institutional Relations under NWPA*, 21, 49 (1987).
197. *Id.* at 21; *see also* Gervers, "The NIMBY Syndrome" (1987), which describes how DOE's poor coordination with states led state governors to oppose the department's nuclear repository-siting decisions.
198. Macfarlane, "Underlying Yucca Mountain," 785–786 (2003).
199. Nuclear Waste Policy Act of 1982: General Guidelines for the Recommendation of Sites for the Nuclear Waste Repositories, 49 Fed. Reg. 47,714 (Dec. 6, 1984).
200. Governor Booth Gardner of Washington, for example, testified before Congress asking for a "pause" in the siting process to assure the public that "there has been adequate consideration for safety in this process." His prepared statement criticized the 1998 target date for opening a repository as "unrealistic" given the time needed to adequately perform site characterization. *DOE Radioactive Waste Repository Program: Hearings before the Subcommittee on Energy Conservation and Power of the House Committee on Energy and Commerce*, 99th Cong. 12, 16 (1985).
201. *See* Office of Technology Assessment, *Managing the Nation's Commercial High-Level Radioactive Waste*, 3–4 (1985); *see also* Crawford, "DOE, States Reheat Nuclear Waste Debate," 151 (1985).
202. Panel on Social and Economic Aspects of Radioactive Waste Management, Board on Radioactive Waste Management, National Research Council of the National Academies, *Social and Economic Aspects*, 38 (1984).
203. Office of Technology Assessment, *Managing the Nation's Commercial High-Level Radioactive Waste*, 116 (1985).
204. *Id.* at 116, 103.
205. *Id.* at 4. This recommendation was seconded by the Monitored Retrievable Storage Review Commission established under NWPA, which found in 1989 that construction of an MRS could not be justified under the existing legal constraints. Monitored Retrievable Storage Review Commission, *Nuclear Waste: Is There a Need for Federal Interim Storage?*, xvi (1989).
206. Office of Technology Assessment, *Managing the Nation's Commercial High-Level Radioactive Waste*, 5, 14–15 (1985).
207. *Id.* at 14.
208. Nuclear Waste Policy Act of 1982 § 115, 42 U.S.C. § 10135 (2006).
209. For example, Congressman Toby Roth declared: "Every major organization representing the States—National Governor's Association, National Conference of State Legislatures, and State Planning Council—have [sic] endorsed the procedure [requiring House and Senate majorities to override a State veto]." 128 Cong. Rec. 27,766 (Nov. 29, 1982).
210. Remarks of Senator Sonny Montgomery, 128 Cong. Rec. 27,797 (Nov. 29, 1982).
211. Nuclear Waste Policy Act of 1982 § 117(b), 42 U.S.C. § 10137(b) (2006).
212. *Id.* § 117(c), 42 U.S.C. § 10137(c).
213. *Id.* § 117(a)(1), 42 U.S.C. § 10137(a)(1).
214. *Id.* § 117(b), 42 U.S.C. § 10137(b).
215. *Id.* § 117(c), 42 U.S.C. § 10137(c).
216. *Id. See id.*, § 117(c)(1)–(11), 42 U.S.C. § 10137(c)(1)–(11), for all matters that the C&C agreement must cover.
217. *Tennessee v. Herrington*, 806 F.2d 642, 654 (6th Cir. 1986) (Wellford, J., dissenting).
218. Colglazier, "Evidential, Ethical, and Policy Disputes," 144 (1991); General Accounting Office, *Institutional Relations under NWPA*, 42–43 (1987).
219. In the midst of a 1983 election campaign in which Yucca became an issue, Nevada's governor initially rejected Nuclear Waste Fund dollars available to support a state review of DOE's DEIS at Yucca, "out of concern that acceptance would be interpreted as support for the program." Gervers, "The NIMBY Syndrome," 40 (1986).
220. General Accounting Office, *Nuclear Waste: Status of DOE's Implementation of NWPA*, 42–43 (1987). GAO recommended that DOE make up for the lack of C&C agreements by elaborating the meaning of C&C in its mission plan, the overall program blueprint that DOE was required, per Nuclear Waste Policy Act § 301, 42 U.S.C. § 10221 (2006), to submit for congressional review. General

Accounting Office, *Institutional Relations under NWPA*, 50 (1987).
221. General Accounting Office, *Institutional Relations under NWPA*, 51 (1987).
222. *Id.* at 20.
223. *Id.* at 21.
224. *Id.* at 20–22.
225. *Id.* at 25.
226. *Id.* at 26.
227. *Tennessee v. Herrington*, 806 F.2d 642, 646, 651–653 (6th Cir. 1986).
228. General Accounting Office, *Institutional Relations under NWPA*, 54–55 (1987).
229. *Id.* at 34–41. It should be noted that those advocating greater state participation in siting decisions disputed the NIMBY characterization. John Gervers, a longtime consultant to state and tribal governments in New Mexico and Nevada on repository issues, argued in 1989: "Almost every governor of every state under consideration . . . was initially open to a demonstration of the merits of such facilities by the [DOE]. In each case, what could have evolved into a reasonably cooperative relationship between the state and DOE gradually eroded through disappointments in the manner of DOE's implementation of the program. Once this erosion had occurred and the public became aware of the failings of the program, any accommodation with the repository would spell political suicide for an elected official." Gervers, "The NIMBY Syndrome," 18 (1987).
230. Nuclear Waste Policy Act of 1982 § 111(a)(6), 42 U.S.C. § 10131(a)(6) (2006).
231. *See, generally*, Dunlap, Kraft, and Rosa, eds., *Public Reactions to Nuclear Waste* (1993); Boholm and Löfstedt, eds., *Facility Siting* (2004).
232. Nuclear Waste Policy Act of 1982 §§ 112(b)(1)(G), 135(c)(2)(A), 42 U.S.C. §§ 10132(b)(1)(F), 10155(c)(2)(A) (2006).
233. Nuclear Waste Policy Act of 1982 § 113(b)(2), as enacted, Pub. L. No. 97-425, 96 Stat. 2201, at 2212 (1983).
234. Nuclear Waste Policy Act of 1982 § 301(b)(2), 42 U.S.C. § 10221(b)(2) (2006).
235. *See* Nuclear Waste Policy Act of 1982 §§ 112(b), 112(f), 113(b)(2), 114(a), as enacted, Pub. L. No. 97-425, 96 Stat. 2201, at 2208–2213 (1983).
236. Nuclear Waste Policy Act of 1982 § 117(c)(10), 42 U.S.C. § 10137(c)(10) (2006).
237. Office of Technology Assessment, *Managing the Nation's High-Level Radioactive Waste*, 194 (1985), quotes Keystone Center, "Public Participation in Developing National Plans for Radioactive Waste Management: Summary Report of the Second Keystone Conference on Public Participation in Radioactive Waste Management Decision Making," iv (Oct. 1980).
238. *Id.* at 194.
239. Statement of Elmer B. Staats, Comptroller General, *Oversight of the Structure and Management of the Department of Energy: Hearings before the Senate Committee on Governmental Affairs and Its Subcommittee on Energy, Nuclear Proliferation, and Federal Services*, 96th Cong., 1st Sess. 151–152 (1979).
240. *See* Tugwell, *The Energy Crisis and American Political Economy*, 131 (1988); Robert D. Hershey Jr., "Department of Energy Stays Alive," *New York Times*, D1 (Feb. 15, 1982).
241. "The conventional wisdom which prevailed in many corridors at the time of the adoption of the NWPA was that DOE, perhaps at the urging of nuclear utilities, would cut corners and engage in slip-shod practices in a head-long dash to bring a repository on line." Montange, "The Initial Environmental Assessments for the Nuclear Waste Repository," 195 (1985).
242. Advisory Panel on Alternative Means of Financing and Managing Radioactive Waste Facilities, DOE, *Managing Nuclear Waste*, XII-2 (1984).
243. *Id.* at XI-1 to XI-9.
244. DOE, *Report to the Secretary on the Conclusions of the Advisory Panel*, 1 (1985).
245. Statement of Senator Daniel J. Evans, *Civilian Radioactive Waste Disposal: Hearings before the Senate Committee on Energy and Natural Resources*, 100th Cong., 1st Sess. 236 (1987).
246. Bureau of Labor Statistics, "CPI Inflation Calculator," www.bls.gov/data/inflation_calculator.htm.
247. Statement of John W. Bartlett, Director, Office of Civilian Radioactive Waste Management, DOE, *Nuclear Waste Disposal Issues: Hearing before the Subcommittee on Nuclear Regulation of the Senate Committee on Environment and Public Works*, 102nd Cong. 48 (1991).
248. Nuclear Waste Policy Amendments Act of 1987 § 5011, Pub. L. No. 100-203, 101 Stat. 1330-227, at 1330-228 (1987), 42 U.S.C. § 10172 (2006).
249. The Nuclear Waste Policy Amendments Act of 1987 enacted the following subsections:
§ 10172a(a): "Congressional action required. The Secretary may not conduct site-specific activities with respect to a second repository unless Congress has specifically authorized and appropriated funds for such activities."
§ 10172a(c): "Termination of granite research. Not later than 6 months after the date of the enactment of the Nuclear Waste Policy Amendments Act of 1987, the Secretary shall phase out in an orderly manner funding for all research programs in existence on such date of enactment designed to evaluate the suitability of crystalline rock as a potential repository host medium." *Id.* § 5012, Pub. L. No. 100-203, 101 Stat. 1330-227, at 1330-231, 42 U.S.C. §§ 10172a(a), 10172a(c).
250. "The Secretary shall report to the President and to Congress on or after January 1, 2007, but not later than January 1, 2010, on the need for a second repository." *Id.*, § 5012, Pub. L. No. 100-203, 101 Stat. 1330-227, 1330-231, 42 U.S.C. § 10172a(b).
251. "Limitation. The Secretary may not select a [MRS] site under subsection (a) until the Secretary recommends to the President the approval of a

site for development as a repository under section [10134(a)]." *Id.* § 5021, Pub. L. No. 100-203, 101 Stat. 1330-227, 1330-234, 42 U.S.C. § 10165(b). In addition, Section 5021 of the amendments required that an NRC construction license for an MRS facility must contain the following conditions: construction of the facility cannot begin until NRC has issued a construction license for a permanent repository; MRS facility construction must cease if the repository license is revoked or if repository construction ceases; the MRS facility cannot accept more than ten thousand MTHM until a repository first accepts SNF or HLW; and the facility cannot contain more than fifteen thousand MTHM at any given time. *Id.*, Pub. L. No. 100-203, 101 Stat. 1330-227, 1330-266, 42 U.S.C. § 10168(d). These strictures were meant to reassure a state hosting an MRS site that waste being stored there would not remain there permanently. *See* Colglazier and Langum, "Policy Conflicts in the Process for Siting Repositories," 333–334 (1988).

Benefits Schedule, in Millions of Dollars		
	MRS	Repository
Annual payments prior to first spent-fuel receipt	5	10
Upon first spent-fuel receipt	10	20
Annual payments after first spent-fuel receipt until closure of the facility	10	20

252. Cotton, "Nuclear Waste Story," 36 (2006).
253. Carter, *Nuclear Imperatives*, 201 (1987).
254. Frank Parker, Distinguished Professor of Environmental and Water Resources, Professor of Engineering Management, Vanderbilt University, telephone interview by Jane Stewart (Sept. 27, 2007). Parker was directly involved in the DOE siting process.
255. Easterling and Kunreuther, *The Dilemma of Siting a Repository*, 41–42 (1995).
256. Thomas Isaacs, Director for the Office of Planning and Special Studies, Lawrence Livermore National Laboratory, telephone interview by Jane Stewart (Aug. 2008). Isaacs, a longtime employee of DOE, served as the director of repository coordination in the late 1980s.
257. Carter, *Nuclear Imperatives*, 175 (1987).
258. Easterling and Kunreuther, *The Dilemma of Siting a Repository*, 42 (1995).
259. *Id.* at 42–43 and n.64.
260. *Id.* at 42.
261. Swainston, "The Characterization of Yucca Mountain," 152 (1991).
262. "The Secretary shall offer to enter into a benefits agreement with the Governor of Nevada. Any benefits agreement with a State under this subsection shall be negotiated in consultation with any affected units of local government in such State." Nuclear Waste Policy Amendments Act of 1987 § 5031, Pub. L. No. 100-203, 101 Stat. 1330-227, at 1330-237 (1987), 42 U.S.C. § 10173(c) (2006). "In addition to the benefits to which a State, an affected unit of local government or Indian tribe is entitled under title I, the Secretary shall make payments to a State or Indian tribe that is a party to a benefits agreement under section 170 in accordance with the schedule shown in the table. *Id.*, Pub. L. No. 100-203, 101 Stat. 1330-227, at 1330-237, 42 U.S.C. § 10173a(a)(1). In exchange for annual payments under a benefits agreement, Nevada would have had to agree to waive its right to disapprove of the president's recommendation of the Yucca site for a repository. Pub. L. No. 100-203, 101 Stat. 1330-227, at 1330-238, 42 U.S.C. § 10173a(b)(2). DOE and Nevada would also have had to share technical information and cooperate in the design for Yucca and preparation of necessary documents. *Id.*, 42 U.S.C. § 10173a(b)(3)–(4). 101 Stat. 1330-238. Payments made under a benefits agreement would also be in lieu of, and not in addition to, impact assistance provided under NWPA. *Id.*, § 10173a(b)(5). 101 Stat. 1330-238.
263. *Id.* § 5031, 101 Stat. 1330-228, at 1330-237, 42 U.S.C. §§ 10173(c), 10173(f).
264. Nevada Commission on Nuclear Projects, *Report and Recommendations*, 15 (2009).
265. Nuclear Waste Policy Amendments Act of 1987 § 5051, Pub. L. No. 100-203, 101 Stat. 1330-227, at 1330-248 to 1330-251 (1987), 42 U.S.C. §§ 10261–10270 (2006).
266. *Id.*
267. *Id.* § 5011(a), Pub. L. No. 100-203, 101 Stat. 1330-227, at 1330-229, 42 U.S.C. § 10134(a)(1)(D).
268. Nuclear Waste Policy Amendments Act of 1987 § 5011(a), Pub. L. No. 100-203, 101 Stat. 1330-227, 1330-228, 42 U.S.C. § 10172(a)(2).
269. Nuclear Waste Policy Act of 1982 § 112(a), 42 U.S.C. § 10132(a) (2006).
270. *Id.* § 117(a)(1)–(2), 42 U.S.C. § 10137(a)(1)–(2) (2006).
271. *Id.* § 117(c)(11), 42 U.S.C. § 10137(c)(11).
272. *See, e.g.*, Thomas Lippman, "Nevada's Objections Stall Plan for Nuclear Waste Repository; Alternative Sites Lacking as Setbacks Mount," *Washington Post*, A1 (Oct. 3, 1989).
273. Stuart D. Waymire, "Senator Bryan," *www.yuccamountainexpose.com/Y38.htm*.
274. *See* Garrick and Kaplan, "A Decision Theory Perspective" (1999).
275. *Id.*
276. *See* Nuclear Waste Project Office, State of Nevada, "Why Does the State Oppose Yucca Mountain?" *www.state.nv.us/nucwaste/yucca/state01.htm*.
277. The phrase was used by the head of OCRWM and by energy secretary Watkins in March 1991. Flynn and Slovic, "Yucca Mountain," 97 (1995). Swainston catalogues Nevada's pre- and post-1987

lawsuits to illustrate the "gloves off" approach that followed enactment of NWPAA. Swainston, "The Characterization of Yucca Mountain," 154–155 (1991).
278. Nuclear Waste Policy Act of 1982. § 116(c)(1)(B), as enacted, Pub. L. No. 97-425, 96 Stat. 2201, 2221 (1983).
279. *Id.* § 116(c)(1)(B)(i)–(v), as enacted, Pub. L. No. 97-425, 96 Stat. 2201, 2221 (1983). There was a nearly identical provision to provide funding for tribes. Nuclear Waste Policy Act of 1982 § 118(b)(2)(A), 42 U.S.C. § 10138(b)(2)(A) (2006)
280. *Id.* § 116(c)(1)(A), as enacted, Pub. L. No. 97-425, 96 Stat. 2201, 2221 (1983).
281. *Nevada v. Herrington* [I], 777 F.2d 529, 531, 535 (9th Cir. 1985).
282. *Nevada v. Herrington* [I], 777 F.2d 529, 531, 534 (9th Cir. 1985).
283. Nuclear Waste Policy Act of 1982. § 116(b), as enacted, Pub. L. No. 97-425, 96 Stat. 2201, 2220 (1983).
284. *Nevada v. Herrington* [II], 827 F.2d 1394 (9th Cir. 1987).
285. Nuclear Waste Policy Act of 1982 § 116(c)(1)(B)(i), 42 U.S.C. § 10136(c)(1)(B)(i) (2006); *Nevada v. Herrington* [II], 827 F.2d 1394, 1399 (9th Cir. 1987)
286. *Nevada v. Herrington* [II], 827 F.2d 1394, 1399–1400 (9th Cir. 1987). The court stated: "NWPA's state participation provisions authorize independent oversight and peer review by states. Neither the language of section 10136 nor its legislative history makes any reference to judicial review." *Id.* at 1400 (internal quotation marks and citation omitted).
287. Nuclear Waste Policy Amendments Act of 1987 § 5032(a), Pub. L. No. 100-203, 101 Stat. 1330-227, 1330-241 (1987), 42 U.S.C. § 10116(c)(1)(B) (2006).
288. Nuclear Waste Policy Act of 1982 § 116(c), 42 U.S.C. § 10136(c) (2006).
289. *County of Esmeralda v. Dep't of Energy*, 925 F.2d 1216, 1217–1218 (9th Cir. 1991).
290. *Id.* at 1220–1221.
291. Titus, "Bullfrog County: A Nevada Response," 128 (1990). The grants, under Section 116(c)(3) of NWPA, were intended to compensate the local community for tax revenue that would have been collected if the facility were a private facility. Bullfrog County was created with a much higher property tax and thus would receive larger grants. *Id.* at 125, 128.
292. *Id.* at 127; Jacob, *Site Unseen*, 171 (1990).
293. *See* Rosa and Short, "Importance of Context in Siting Controversies," 5, 10 (2004).
294. *Nye County, Nevada v. Nevada*, No. 4606 (5th Jud. Dist. Ct. Nev. 1988); *see also* Associated Press, "Nevada County Is Held Illegal," *New York Times*, § 1 at 7 (Feb. 13, 1988).
295. *See* Titus, "Bullfrog County: A Nevada Response," 133–134 (1990).
296. Energy and Water Development Appropriations Act of 1993, Pub. L. No. 102-377, 106 Stat. 1315, at 1334 (1992); *see also* Hancock, "Symposium: Nuclear West: Which Road?" 33 (2004).
297. *Nevada v. Dep't of Energy*, 133 F.3d 1201, 1203, 1205 (9th Cir. 1998).
298. "Nuclear Waste Policy Act of 1982: General Guidelines for the Recommendation of Sites for the Nuclear Waste Repositories," 49 Fed. Reg. 47,714 (Dec. 9, 1984); 10 C.F.R. pt. 960 (1985).
299. *Nevada v. Watkins II*, 939 F.2d 710, 713, 715 (9th Cir. 1991).
300. *Nevada v. Watkins III*, 943 F.2d 1080, 1083–1084 (9th Cir. 1991).
301. *Id.* at 1087.
302. *Nevada v. Burford*, 918 F.2d 854, 857 (9th Cir. 1990).
303. Nev. Rev. Stat. § 459.910 (2009).
304. *Nevada v. Watkins I*, 914 F.2d 1545, 1551 (9th Cir. 1990), *cert. denied*, 499 U.S. 906 (1991).
305. Jacob, *Site Unseen*, 171–172 (1990). The Energy and Water Development Appropriations Act of 1990 provided that "not more than $6,000,000 may be provided to the State of Nevada, at the discretion of the Secretary of Energy, to conduct appropriate activities pursuant to the Act." Energy and Water Development Appropriations Act of 1990, Pub. L. No. 101-101, 103 Stat. 641, at 658 (1989). The Senate report on the bill made it clear that the secretary was only to exercise this discretion "based upon his certification to Congress of good faith efforts and cooperation on the part of the State in allowing technical field work, including at-depth testing in an exploratory shaft facility, to proceed at the Yucca Mountain site." Committee on Appropriations, Energy and Water Development Appropriation Bill, 1990, at 132 (1989); *see also* the statement of Senator Richard Bryan: "In addition to this provision of the act, report language, contained at page 132, makes it abundantly clear that until such time as the State ceases its opposition to the high-level [waste] site—until such time as the State capitulates, it concludes, it agrees, it acknowledges that Yucca Mountain shall be the site—that $6 million provided in the act will be withheld. The Secretary of Energy is directed to make a certification that indeed that Nevada's cooperation is forthcoming before that money can be released." 135 Cong. Rec. 16,608 (July 27, 1989). This provision was also a response to Nevada's state legislature, which passed legislation the same year making it "unlawful for any person or governmental entity to store high-level radioactive waste in Nevada." Nev. Rev. Stat. § 459.910 (2009); Jacob, *Site Unseen*, 171 (1990).
306. The case was consolidated with another case that had been filed on the same day as the petition for review in *Nevada v. Watkins III*.
307. *Nevada v. Watkins I*, 914 F.2d 1545, 1555 (9th Cir. 1990), *cert. denied*, 499 U.S. 906 (1991).
308. *Id.* at 1552–1558. The Property Clause of the Constitution provides: "The Congress shall have Power to dispose of and make all needful Rules and Regulations respecting the Territory or other Property belonging to the United States; and nothing in this Constitution shall be so construed as to Prejudice any Claims of the United States, or of any particular State." U.S. Const. art. IV, § 3, cl. 2.

The Supremacy Clause of the Constitution provides: "This Constitution, and the Laws of the United States which shall be made in Pursuance thereof; and all Treaties made, or which shall be made, under the Authority of the United States, shall be the supreme Law of the Land; and the Judges in every State shall be bound thereby, any Thing in the Constitution or Laws of any State to the Contrary notwithstanding." U.S. CONST. art. VI., cl. 2.

309. *Nevada v. Watkins I*, 914 F.2d 1545, 1558–1559 (9th Cir. 1990), *cert. denied*, 499 U.S. 906 (1991).
310. Flynn and Slovic, "Yucca Mountain," 98 (1995).
311. DOE, *Final Supplemental EIS for Yucca Mountain*, S-23, S-59 to S-60, S-66 (2008).
312. Keith Rogers, "Government Challenges Order Halting Water Use at Yucca Site," *Las Vegas Review-Journal*, 1B (July 26, 2007).
313. *United States v. Nevada*, 2007 U.S. Dist. LEXIS 69177, at 26, 31–32 (D. Nev. Aug. 31, 2007). The district court had initially dismissed the case on abstention grounds, but the Ninth Circuit held that it should hear and decide it on the merits. *United States v. Morros*, 268 F.3d 695, 697 (9th Cir. 2001). Nevada officials hoped that the district court's favorable ruling in 2007 would prevent DOE from collecting data it needed to obtain an NRC license for the repository. Ralph Vartabedian, "Judge Denies Water to Nuclear Dump," *Los Angeles Times*, 12 (Sept. 5, 2007). In the months after the order was issued, a battle ensued over whether the order covered all water use or whether it was more limited. DOE agreed to stop using the water for its phase 2 borehole drilling but refused to stop using it for its phase 1 drilling, which was nearly complete. Keith Rogers, "Judge Refuses to Stop Yucca Water Use," *Las Vegas Review-Journal* (Sept. 21, 2007).
314. *United States v. Nevada*, 2007 U.S. Dist. LEXIS 69177, at 23 (D. Nev. Aug. 31, 2007).
315. *Id.* at 32.
316. *See, e.g.*, General Accounting Office, *Nuclear Waste: Repository Work Should Not Proceed*, 3 (1988), noting NRC's lack of confidence in DOE's quality assurance program, as reported formally in March 1988.
317. Jacob, *Site Unseen*, 173–174 (1990).
318. Broad, "A Mountain of Trouble," 37 (1990).
319. *See, e.g.*, Chalmers et al., *State of Nevada Socioeconomic Studies 1986-1992* (1993).
320. *Leboeuf, Lamb, Green & Macrae, LLP v. Abraham*, 180 F. Supp. 2d 65 (D.D.C. 2001).
321. Benjamin Grove and Mary Manning, "Law Firm for DOE Lobbied for Yucca," *Las Vegas Sun* (July 27, 2001).
322. Benjamin Grove, "Berkley Seeks Probe of Yucca Law Firm," *Las Vegas Sun* (Oct. 12, 2001).
323. Benjamin Grove, "Nevadans Plan to File Complaint in Yucca Case," *Las Vegas Sun* (Dec. 7, 2001).
324. Keith Schneider, "Nuclear Industry Plans Ads to Counter Critics," *New York Times*, A18 (Nov. 13, 1991).
325. Rosa, Dunlap, and Kraft, "Prospects for Public Acceptance," 312–315 (1993).
326. Flynn, Slovic, and Mertz, "The Nevada Initiative," 498 (1993).
327. *Id.* at 499–500
328. *Id.* at 501.
329. Mushkatel et al., Nevada Agency for Nuclear Projects, Office of the Governor, *Governmental Trust and Risk Perceptions* (1992). Notably, a grant from DOE funded the study.
330. Nuclear Waste Technical Review Board, *Sixth Report to Congress and Secretary of Energy*, 43 (1992).
331. Keith Schneider, "Trying to Build Secret Weapons, U.S. Spread Radiation in 1950s," *New York Times*, A1 (Dec. 16, 1993).
332. Nevada Agency for Nuclear Projects, Office of the Governor, *1989 Nevada State Telephone Survey* v (1989), reported 14.4 percent in favor of Yucca Mountain repository, 69.4 percent opposed; Mertz, Flynn, and Slovic, Nevada Agency for Nuclear Projects, Office of the Governor, *1994 Nevada State Telephone Survey*, 4 (1994), reported 20.9 percent in favor, 72.3 percent opposed.
333. Keith Rogers, "Poll Finds Slight Rise in Acceptance of Yucca Plans," *Las Vegas Review-Journal*, 3B (June 18, 2004).
334. Flynn, Slovic, and Mertz, Nevada Agency for Nuclear Projects, Office of the Governor, *Autumn 1993 Nevada State Telephone Survey*, 23–26 (1989).
335. *Id.*
336. Titus, "Bullfrog County: A Nevada Response," 128 (1990)
337. DOE, *Reassessment of Civilian Radioactive Waste Program*, vii, 10 (1989).
338. Storage of Spent Fuel in NRC-Approved Storage Casks at Power Reactor Sites, 55 FED. REG. 29,181, 29,182 (July 18, 1990), 10 C.F.R. pt. 72, subpts. K–L (2010).
339. Consideration of Environmental Impacts of Temporary Storage of Spent Fuel after Cessation of Reactor Operation, 55 FED. REG. 38,472, at 38,472–38,473 (Sept. 18, 1990), 10 C.F.R. pt. 51.23(a) (2010).
340. As described in Chapter 1, NRC used this rulemaking technique in the 1970s to avoid having to address on a site-by-site basis the environmental impacts of the additional SNF that would be generated by licensing new nuclear power plants, and successfully defended legal challenges to it.
341. *Kelley v. Selin*, 42 F.3d 1501, 1521 (6th Cir. 1995).
342. Keith Schneider, "Nuclear Plants to Become de Facto Radioactive Dumps," *New York Times*, A19 (Feb. 15, 1995).
343. Task Force on an Alternative Program Strategy, *A Proposed Alternative Strategy for DOE's Radioactive Waste Program*, 2–3 (1993).
344. Energy Policy Act of 1992 § 801(a)–(b), Pub. L. No. 102-486, 106 Stat. 2921 (1992); for generic standards, *see* 40 C.F.R. pt. 191 (2009) (EPA) and 10 C.F.R. pt. 60 (2010) (NRC).
345. M. Manning, "Tainted Rocks Found," *Las Vegas Sun* (Apr. 26, 1996).
346. "AG Turns Up the Heat on Yucca," *Las Vegas Sun* (Mar. 26, 1997).

347. Nuclear Waste Project Office, State of Nevada, "Earthquakes in the Vicinity of Yucca Mountain," *www.state.nv.us/nucwaste/yucca/seismo01.htm*; "Yucca Shaking with Quakes Last Two Decades," *Las Vegas Sun* (Apr. 28, 1997).
348. Suzanne Struglinski, "DOE Predicts Nuke Reactions in Casks—Nevadans Worry about Danger at Yucca," *Las Vegas Sun* (Nov. 26, 2003); Scott Sonner, "Panel: Data Back Yucca Concerns," *Las Vegas Sun* (Feb. 20, 2004).
349. Wernicke et al., "Anomalous Strain Accumulation," 2096, 2099 (1998).
350. Commission on Geosciences, Environment, and Resources, National Research Council of the National Academies, *Rethinking High-Level Radioactive Waste Disposal*, 3–6 (1990).
351. *Id.* at 6. As has been observed, DOE received much criticism for policies that Congress had directed it to implement.
352. DOE, *Viability Assessment of a Repository at Yucca*, Overview, 10–11 (1998).
353. Craig, "High-Level Nuclear Waste," 471 (1999). Craig indicates that two U.S. Geological Survey scientists in particular were instrumental in developing this theory: Winograd, "Radioactive Waste Disposal in Thick Unsaturated Zones" (1981), and Roseboom, U.S. Geological Survey, *Disposal of High-Level Nuclear Waste above the Water Table in Arid Regions* (1983).
354. Bodvarsson et al., "Overview of Scientific Investigations at Yucca," 11 (1999),
355. Craig, "High-Level Nuclear Waste," 474 (1999).
356. *Id.*
357. *See Nuclear Energy Inst. v. Envtl. Prot. Agency*, 373 F.3d 1251 (D.C. Cir., 2004), discussed later.
358. Affidavit of Dr. John W. Bartlett, at ¶ 22, *Nevada v. Department of Energy*, No. 01-1516 (D.C. Cir. Feb. 6, 2002), as republished in Nevada Agency for Nuclear Projects, Office of the Governor, "Affidavit of Dr. John W. Bartlett" (2002), *www.state.nv.us/nucwaste/news2002/nn11580.pdf*.
359. *Id.* at ¶ 35–37.
360. Wernicke et al., "Anomalous Strain Accumulation," 2096, 2099 (1998).
361. *See* 10 C.F.R. pt. 60 (2010), which governs geologic repositories generally, and 10 C.F.R. pt. 63 (2010), which governs the repository at Yucca specifically.
362. 66 Fed. Reg. 32,074 (June 13, 2001). As described later, the standards were revised in 2008 in order to require maximum dosages for periods over ten thousand years, as required by court order. 73 Fed. Reg. 61,256 (Oct. 15, 2008). For criticisms of the new standards and EPA's response to these criticisms, *see* Commission on Geosciences, Environment, and Resources, National Research Council of the National Academies, *Radioactive Waste Repository Licensing* (1992).
363. *See* Disposal of High-Level Radioactive Wastes in Geologic Repositories: Design Basis Events, 61 Fed. Reg. 64,257, at 64,267–64,270 (Dec. 4, 1996), 10 C.F.R. §§ 60.2, .21, .111, .136 (2010).
364. Disposal of High-Level Radioactive Wastes in a Proposed Geologic Repository at Yucca Mountain, NV, 66 Fed. Reg. 55,732, at 55,807 (Nov. 2, 2001), 10 C.F.R. §§ 63.114 (2010). DOE promptly brought its own regulations into line with NRC's by adopting TSPA into its guidelines. Office of Civilian Radioactive Waste Management: General Guidelines for the Recommendation of Sites for Nuclear Waste Repositories—Yucca Mountain Site Suitability Guidelines, 66 Fed. Reg. 57,298 (Nov. 14, 2001).
365. Office of Civilian Radioactive Waste Management: General Guidelines for the Recommendation of Sites for Nuclear Waste Repositories—Yucca Mountain Site Suitability Guidelines, 66 Fed. Reg. 57,298, 57,305 (Nov. 14, 2001).
366. *Id.* at 57,309 acknowledges criticism of the rule as proposed by DOE in December of 1996.
367. *See, e.g.*, Nuclear Information and Resource Service, "DOE Attempting to Change (Yet Again!) Rules for Yucca Mountain," press release (Jan. 16 1997); Kevin Kamps, "Yucca Heats Up," Nuclear Monitor Online (Dec. 20, 2001), *www.nirs.org/mononline/yuccaheatsup122001.htm*.
368. DOE reported: "On a national level, the NRC, the NAS and the Nuclear Waste Technical Review Board ("NWTRB") (a Congressionally mandated committee of experts chartered to evaluate . . . Yucca Mountain['s] . . . suitability as a location for a repository) have acknowledged the value of [the TSPA] for evaluating post-closure performance for a repository at Yucca Mountain." 66 Fed. Reg. 57,298, at 57,305 (Nov. 14, 2001). In endorsing use of the TSPA for evaluating the performance of Yucca, NRC stated that "experience and improvements in the technology of performance assessment, acquired over more than 15 years, now provide significantly greater confidence in the technical ability to assess comprehensively overall repository performance, and to address and quantify the corresponding uncertainty." 64 Fed. Reg. 8640 (Feb. 22, 1999). In a 1995 report, NAS noted that "because it is the performance of the total system in light of the risk-based standard that is crucial, imposing subsystem performance requirements might result in suboptimal repository design." Committee on Technical Bases for Yucca Mountain Standards, National Research Council, *Technical Bases for Yucca Mountain Standards*, 13 (1995); *see also* Nuclear Waste Technical Review Board, *1997 Findings and Recommendations*, 8 (1998); Kessler and McGuire, "TSPA for Waste Disposal" (1999).
369. Energy and Water Development Appropriations Act of 1997, Pub. L. No. 104-206, 110 Stat. 2984, at 2995–2996 (1996).
370. Nuclear Waste Policy Act of 1996, S. 1936, 104th Cong., 2d Sess. § 204(b)(1)(B) (1996).
371. DOE, *Viability Assessment of a Repository at Yucca*, Vol. 1, O1 (1998). DOE set forth the TSPA approach for the first time in 1991 (TSPA-1991). DOE developed the approach over time, issuing TSPAs with improved models in 1993 and 1995 (TSPA-1993 and TSPA-1995 respectively). *See* Robert Andrews, Bechtel SAIC Co., "Overview of U.S. Department of

Energy Total System Performance Assessment for the Yucca Mountain Repository," presentation, 9 (Sept. 20, 2004), *www.nwtrb.gov/meetings/2004/sept/andrews.pdf*. DOE used the TSPA for the 2001 FEIS and for the June 3, 2008 NRC license application (TSPA-LA) for Yucca.
372. "Important documents relating to the evolution of the U.S. program clearly suggest that the DOE has always assumed high thermal loads for repositories in this country, regardless of the disposal environment." Nuclear Waste Technical Review Board, *Fifth Report to Congress and Secretary of Energy*, xii–xiii (1992). NWTRB went on to point out concerns over a hot repository design, as discussed later.
373. *See* DOE, *Viability Assessment of a Repository at Yucca*, Vol. 2, at 1-3, 3-4 (1998); Nuclear Waste Technical Review Board, *Moving beyond Yucca Mountain Viability Assessment*, 6 (1999).
374. William J. Broad, "Theory on Threat of Blast at Nuclear Waste Site Gains Support," *New York Times*, A18 (Mar. 23, 1995). The article reported that the main concern voiced by the scientists was with respect to burial of vitrified nuclear weapons waste; it stated: "[d]anger could arise many thousands of years from now after repository tunnels had collapsed, after geologic erosion and flows of underground water had dissolved the steel canisters holding wastes and after plutonium had slowly dispersed into surrounding rock. In turn, the physical properties of the rock could help set off a nuclear chain reaction and explosion, according to the [scientists'] thesis." *Id.*
375. William J. Broad, "Scientists Fear Atomic Explosion of Buried Waste," *New York Times*, 11 (Mar. 5, 1995).
376. Scientists dated the rainwater by the presence of the isotope chlorine-36, a byproduct of nuclear weapons tests in the 1950s and 1960s. Matthew L. Wald, "Doubt Cast on Prime Site as Nuclear Waste Dump," *New York Times* (Mar. 5, 1995); *see also* Campbell et al., "Chlorine-36 Data at Yucca Mountain," 44 (2003).
377. DOE, *Yucca Mountain Viability Assessment*, Overview, 20 (1998). *See also* Nuclear Waste Technical Review Board, *Moving Beyond Yucca Mountain Viability Assessment*, 6–7 (1999).
378. DOE, *Yucca Mountain Viability Assessment*, Overview, 20–21 (1998). DOE later decided to switch the corrosion-resistant nickel alloy from the inner layer to the outer layer. Office of Civilian Radioactive Waste Management, DOE, *Yucca Mountain Science and Engineering Report*, 3-2, 3-3 (2002).
379. DOE, *Yucca Mountain Viability Assessment*, Overview 1.
380. Nuclear Waste Technical Review Board, *1997 Findings and Recommendations*, 8 (1998), notes the formation by DOE of a six-member "external TSPA peer review panel" in anticipation of the VA's completion.
381. *Id.* "The DOE has devoted significant and laudable effort to achieving the goal of developing a credible TSPA." *Id.*
382. Nuclear Waste Technical Review Board, *Report to Congress and Secretary of Energy*, 3 (2000).
383. *Id.*
384. Nuclear Waste Technical Review Board, *Moving beyond Yucca Viability Assessment*, 2, 8 (1999).
385. Total System Performance Assessment Peer Review Panel, DOE, *Final Report TSPA Peer Review Panel*, § III.D (1999).
386. The peer review panel also emphasized the importance of manifest transparency: "In the Panel's view, the confidence that the public can have in the TSPA results will, to a large degree, depend on how the analyses of the major attributes of the repository system are conducted and presented." Total System Performance Assessment Peer Review Panel, DOE, *Final Report TSPA Peer Review Panel*, Introduction, § I (1999).
387. Office of Civilian Radioactive Waste Management, DOE, *Yucca Mountain Science and Engineering Report*, 2-153 (2002).
388. "The drip shield emplacement gantry is designed to operate in the harsh environment inside the emplacement drifts, taking into account the moderately high temperature of 50°C (122°F), a relative humidity ranging from 10 to 100 percent, and high radiation levels." Office of Civilian Radioactive Waste Management, DOE, *Yucca Mountain Science and Engineering Report*, xxxi, xxxv, 2-153 to 2-158 (2002). The gantry was envisioned as a modification of the emplacement railcars, which would clasp the shields and maneuver them into place over the waste packages. *See id.* for detailed description and proposal drawings for the gantry system.
389. *Id.* at xxxi–xxxii.
390. Office of Civilian Radioactive Waste Management, DOE, *Yucca Mountain Repository License Application*, §§ 1.3.4.7–1.3.4.7.2 (2008).
391. *See, e.g.*, Robert Loux, Executive Director, Nevada Agency for Nuclear Projects, Office of the Governor, "Letter to Dale Klein, Chairman, Nuclear Regulatory Commission," 2 (Apr. 19, 2008), *www.state.nv.us/nucwaste/news2008/pdf/nv070419klein.pdf*.
392. *Id.* (emphasis in original).
393. For example, Nevada congresswoman Shelley Berkley was quoted as saying that the plan to use "robots" was "straight out of a bad science fiction plot." Mary Manning, "Feds Withheld Negative Yucca Data, Say Nevada Officials," *Las Vegas Sun* (Aug. 11, 2009).
394. *See, e.g.*, Nuclear Information and Resource Service, "U.S. Department of Energy (DOE) Changes Its Own Rules in the Middle of the Game," press release (Nov. 30, 2001).
395. Johnson, "Yucca Mountain," 20 (2002).
396. David Stahl, a nuclear materials expert and adjunct professor at the University of Nevada, stated: "A repository at Yucca Mountain will depend on both the geologic and engineered barriers for safety."

Stahl, "Drip Shield and Backfill," 301 (2006). David Shoesmith, a professor of chemistry and consultant to Bechtel on the Yucca project, summarized the "integral role" of engineered waste containers to "overall vault performance" as follows: "Retarding the release of radionuclides from the proposed nuclear waste repository at Yucca Mountain will require a system of multiple barriers. . . . As such, the design and fabrication [of the metal package that encases the waste] have been adapted in an attempt to compensate for the uncertainties inherent in natural geologic barriers and for the uncontrollable features of barriers. . . . This approach has led to the perception that the accumulation of uncertainties in performance in all pre- and post-waste package barriers is best dealt with by improving the design and durability of the waste package." Shoesmith, "Waste Package Corrosion," 287 (2006).

397. "The Secretary . . . shall issue general guidelines for the recommendation of sites for repositories [which] shall specify detailed geologic considerations that shall be primary criteria for the selection of sites in various geologic media. Such guidelines shall specify factors that qualify or disqualify any site from development as a repository, including factors pertaining to the location of valuable natural resources, hydrology, geophysics, seismic activity, and atomic energy defense activities." Nuclear Waste Policy Act of 1982 § 112(a), 42 U.S.C. § 10132(a) (2006). While the act does mention engineered barriers, defined as "manmade components of a disposal system designed to prevent the release of radionuclides into the geologic medium involved," *id.* § 2(11), 42 U.S.C. § 10101(11), it does so in relation to EPA's standards, *id.* § 121, 42 U.S.C. § 10141, and DOE's R&D facilities, *id.* § 217(b)(3), (c)(1)(B), (c)(1)(D), (e), 42 U.S.C. § 10197(b)(3), (c)(1)(B), (c)(1)(D), (e), but not in relation to the DOE siting guidelines.

398. *Nuclear Energy Inst. v. Envtl. Prot. Agency*, 373 F.3d 1251, 1289–1297 (D.C. Cir. 2004).

399. DOE, *Recommendation by the Secretary of Energy Regarding the Suitability of Yucca Mountain*, 6 (2002).

400. Eric Pianin, "Mountain of Controversy Brews over Nuclear Waste Site; Nevada Veto Shifts Debate to Congress," *Washington Post*, A03 (Apr. 9, 2002); *see also* Kenny C. Guinn, Governor of Nevada, "Statement of Reasons Supporting the Governor of Nevada's Notice of Disapproval of the Proposed Yucca Mountain Project (Apr. 8, 2002), www.state.nv.us/nucwaste/news2002/nn11650.pdf.

401. Yucca Mountain Repository Site Approval Act, Pub. L. No. 107-200, 116 Stat. 735 (2002), 42 U.S.C. § 10135 note (2006).

402. Statement of Senator Harry Reid, 148 Cong. Rec. S6452 (July 9, 2002).

403. Statement of Senator Jeff Bingaman, 148 Cong. Rec. S6454 (July 9, 2002).

404. 148 Cong. Rec. H2204 (May 8, 2002); 148 Cong. Rec. S6478 (July 9, 2002).

405. The results in the House for the states that had potential candidate sites are as follows (yea indicates a vote to override the veto): Louisiana, 7 yea, 0 nay; Mississippi, 5 yea, 0 nay; Texas, 22 yea, 8 nay; Utah, 2 yea, 1 nay; Washington, 7 yea, 2 nay. 148 Cong. Rec. H2204 (May 8, 2002). The Senate voted by voice vote, so individual votes were not recorded. However, a roll-call vote held on a related procedural measure passed immediately before the resolution. The results of that procedural vote are as follows: Louisiana; 1 yea, 1 nay; Mississippi, 2 yea, 0 nay; Texas, 2 yea, 0 nay; Utah, 2 yea, 0 nay; Washington, 2 yea, 0 nay. 148 Cong. Rec. S6490 (July 9, 2002); Steve Tetreault, "Yucca Mountain: Senate OKs Dump," *Las Vegas Review-Journal*, 1A (July 10, 2002).

406. Mary Manning, "State Is Geared for Years of Opposition," *Las Vegas Sun* (Jan. 15, 2002).

407. See, for example, *Nevada v. Nuclear Regulatory Comm'n*, 199 Fed. App'x 1 (D.C. Cir. 2006), dismissed for petitioner-Nevada's lack of standing; *Nevada v. Dep't of Energy*, 457 F.3d 78 (D.C. Cir. 2006), dismissed for lack of ripeness; and *Nevada v. Dep't of Energy*, 517 F. Supp. 2d 1245 (D.C. Cir. 2007), dismissed after finding that the DOE documents sought by Nevada were protected from disclosure. DOE did suffer a notable courtroom loss in 2007 over the issue of the extent of its access to water resources in Nevada. Before rejecting DOE's request for an injunction against the Nevada Department of Conservation and Natural Resources that would have provided carte blanche access to Nevada water supplies, the Ninth Circuit panel opined that DOE had been the agent of its own misfortune: "None of this presents a conflict between state and federal law or regulation. Years ago the DOE sought the Court's help in maintaining the status quo at Yucca Mountain. It has, since that time, engaged in activities which dramatically alter the status quo." *United States v. Nevada*, 2:00-CV-0268-RLH-LRL, 2007 U.S. Dist. LEXIS 69177, 31–32 (9th Cir. Aug. 31, 2007).

408. *Nuclear Energy Inst. v. Envtl. Prot. Agency*, 373 F.3d 1251 (D.C. Cir. 2004).

409. *Id.* at 1257.

410. *Id.* at 1305.

411. *Id.* at 1315.

412. *Id.* at 1310–1311. The "DOE's bogus site . . ." phrase in the court's opinion is a quotation from Kenny C. Guinn, Governor of Nevada, "Statement of Reasons Supporting the Governor of Nevada's Notice of Disapproval of the Proposed Yucca Mountain Project," 5–6 (Apr. 8, 2002).

413. *Id.* at 1315.

414. *Id.* at 1257, quoting the Energy Policy Act of 1992 § 801(a)(1), Pub. L. No. 102-486, 106 Stat. 2776, at 2921 (1992); *id.* at 1267, quoting Committee on Technical Bases for Yucca Mountain Standards, National Research Council, *Technical Bases for Yucca Mountain Standards*, 55 (1995).

415. *Id.* at 1299.

416. Suzanne Struglinski, "Yucca in for Long Delay;

Radiation Standard Too Low," *Las Vegas Sun* (July 9, 2004).
417. DOE, *Final EIS for Yucca*, 6-32 to 6-37, 6-54 to 6-62 (2002).
418. Record of Decision on Mode of Transportation and Nevada Rail Corridor for the Disposal of Spent Nuclear Fuel and High-Level Radioactive Waste at Yucca Mountain, Nye County, NV, 69 Fed. Reg. 18,557, at 18,558, 18,561 (Apr. 8, 2004).
419. *Nevada v. Dep't of Energy*, 457 F.3d 78, 81–83 (D.C. Cir. 2006).
420. Martin G. Malsh, Counsel for the State of Nevada, "Letter to Annette L. Vietti-Cook, Secretary, Nuclear Regulatory Commission," (Mar. 1, 2005), *www.state.nv.us/nucwaste/news2005/pdf/nv050228diaz.pdf*.
421. Robert R. Loux, Executive Director, Nevada Agency for Nuclear Projects, "Letter to Nils Diaz, Chairman, Nuclear Regulatory Commission," (Feb. 28, 2005), *www.state.nv.us/nucwaste/news2005/pdf/nv050228diaz.pdf*.
422. State of Nevada; Denial of a Petition for Rulemaking, 70 Fed. Reg. 48,329 (Aug. 17, 2005).
423. Ken Ritter, "Nevada Files Suit against NRC over Yucca License Process," *Associated Press Newswires* (Sept. 1, 2005).
424. *Nevada v. Nuclear Regulatory Comm'n*, 199 Fed. App'x 1 (D.C. Cir. 2006).
425. The court noted that "[b]ias in a proceeding that *might* take place *years* from now is not 'actual or imminent.'" *Id.*
426. *Nevada v. Dep't of Energy*, 517 F. Supp. 2d 1245, 1265 (D. Nev. 2007). The court held that the draft license application documents were covered by the deliberative process privilege, and thus exempted from FOIA. Under this exemption, predecisional materials are exempt from FOIA if disclosure would "discourage candid discussion within the agency and thereby undermine the agency's ability to perform its functions." *Carter v. Dep't of Commerce*, 307 F.3d 1084, 1090 (9th Cir. 2002), as quoted in *Nevada v. Dep't of Energy*, 517 F. Supp. 2d 1245, 1262 (D. Nev. 2007). The purpose of Nevada's FOIA request was to find discrepancies between the draft and the final application, the latter of which would, of course, be publicly available. *Nevada v. Dep't of Energy*, 517 F. Supp. 2d 1245, 1264 (D. Nev. 2007).
427. Keith Rogers, "Judge Refuses to Stop Yucca Water Use: Energy Department Can Finish Phase 1 Drilling," *Las Vegas Review-Journal*, B7 (Sept. 21, 2007); *see also* DOE, *Final Supplemental EIS for Yucca Mountain*, S-23, S-59 to S-60, S-66 (2008), which discusses DOE water requirements.
428. Steve Tetreault, "Nevada Fights Law Firm's Deal at Yucca Site," *Las Vegas Review-Journal*, 3B (Dec. 6, 2007).
429. Steve Tetreault, "Law Firm's Yucca Pact with DOE Criticized," *Las Vegas Review-Journal*, 3B (Apr. 4, 2008).
430. Edward F. Sproat, III, Director, Office of Civilian Radioactive Waste Management, Department of Energy, "Letter to Michael F. Weber, Director, Office of Nuclear Material Safety and Safeguards, Nuclear Regulatory Commission" (June 3, 2008), available at Nuclear Regulatory Commission, "DOE's License Application for a High-Level Waste Geologic Repository at Yucca Mountain" (2010), *www.nrc.gov/waste/hlw-disposal/yucca-lic-app.html*: follow link for "Transmittal Letter (June 3, 2008)." The license application is available at *www.nrc.gov/waste/hlw-disposal/yucca-lic-app/yucca-lic-app-general-info.html*.
431. Department of Energy, Notice of Acceptance for Docketing of a License Application for Authority to Construct a Geologic Repository at a Geologic Repository Operations Area at Yucca Mountain, NV, 73 Fed. Reg. 53,284 (Sept. 15, 2008).
432. Public Health and Environmental Radiation Protection Standards for Yucca Mountain, 73 Fed. Reg. 61,256 (Oct. 15, 2008).
433. Public Health and Environmental Radiation Protection Standards for Yucca Mountain, NV, 70 Fed. Reg. 49,014 (Aug. 22, 2005); *see also* NRC's proposed rule implementing the EPA standards, Implementation of a Dose Standard after 10,000 Years, 70 Fed. Reg. 53,313 (Sept. 8, 2005).
434. Testimony of Robert Meyers, Principal Deputy Assistant Administrator for the Office of Air and Radiation, EPA, *Next Steps toward Permanent Nuclear Waste Disposal: Hearing before the Subcommittee on Energy and Air Quality of the House Committee on Energy and Commerce*, 110th Cong. (2008).
435. Public Health and Environmental Radiation Protection Standards for Yucca Mountain, 73 Fed. Reg. 61,256, 61,264 (Oct. 15, 2008), 40 C.F.R. pt. 197 (2009).
436. William J. Boyle, Director, Regulatory Affairs Division, Department of Energy, "Letter to Michael F. Weber, Director, Office of Nuclear Material Safety and Safeguards, Nuclear Regulatory Commission" (Feb. 19, 2009), available at Nuclear Regulatory Commission, "DOE's License Application for a High-Level Waste Geologic Repository at Yucca Mountain" (2010), *www.nrc.gov/waste/hlw-disposal/yucca-lic-app.html*: follow link for "Letter to NRC re: Update to the Yucca Mountain Repository License Application (LA) for Construction Authorization. (February 19, 2009)." Licensing application update is available at *www.nrc.gov/waste/hlw-disposal/yucca-lic-app/yucca-lic-app-general-info.html*.
437. Phoebe Sweet, "Bryan: Dump Plan Demise Is Not a Lock," *Las Vegas Sun* 1 (Jan. 23, 2009). Nevada initially filed more than two hundred contentions to the original license application. *See* "State of Nevada's Petition to Intervene as a Full Party," U.S. Dep't of Energy (High Level Waste Repository), No. 63-001-HLW (ASLB, Nuclear Regulatory Comm'n December 19, 2008), *www.state.nv.us/nucwaste/licensing/Contentions_NV.pdf*. After DOE submitted its license application update, Nevada filed two additional contentions. *See* "State of Nevada's New Contentions Based on DOE's February 19, 2009 License Application Update," U.S. Dep't of Energy (High Level Waste Repository), No. 63-

001-HLW (ASLB, Nuclear Regulatory Comm'n June 8, 2009), *www.state.nv.us/nucwaste/licensing/nv090608contentions.pdf*.
438. Implementation of a Dose Standard after 10,000 Years, 74 Fed. Reg. 10,811 (Mar. 13, 2009), 10 C.F.R. pt. 63 (2010); *see also* Nuclear Regulatory Commission, "DOE's License Application for a High-Level Waste Geologic Repository at Yucca Mountain" (2010), *www.nrc.gov/waste/hlw-disposal/yucca-lic-app.html*.
439. As of June 2010, both cases were being held in abeyance pending review of DOE's motion to withdraw the license application for Yucca, discussed later. Joint Status Report, *Nevada v. Envtl. Prot. Agency*, No. 08-1327 (D.C. Cir. 2008), *www.state.nv.us/nucwaste/licensing/appeal100610epa.pdf*; Joint Status Report, *Nevada v. Nuclear Regulatory Comm'n*, No, 09-1133 (D.C. Cir. 2009), *www.state.nv.us/nucwaste/licensing/appeal100610nrc.pdf*.
440. Mary Manning, "A.G. Files Challenge to Yucca Radiation Rule," *Las Vegas Sun* (Oct. 10, 2008).
441. Nuclear Waste Policy Act of 1982 § 114(d), 42 U.S.C. § 10134(d) (2006). Based on the submission date of June 3, 2008, NRC must decide on the license application by June 3, 2011, with a possible extension until June 3, 2012.
442. *Id.* § 114(f)(4), 42 U.S.C. § 10134(f)(4). The act provides that, in connection with its licensing decision, the NRC shall adopt the EIS prepared by DOE for the repository and that no consideration beyond DOE's EIS shall be required. *Id.*
443. Keith Rogers and Steve Tetreault, "Yucca Opponents to Raise 299 Concerns to Federal Regulators," *Las Vegas Review-Journal*, B1 (May 12, 2009).
444. *See, e.g.*, Mark Holt, Congressional Research Service, "Nuclear Waste Policy: Alternatives to Yucca Mountain," presentation, 4 (June 11, 2009), available at Nuclear Waste Technical Review Board, *www.nwtrb.gov/meetings/2009/june/holt.pdf* ("Obama-Biden Campaign Called Yucca Mountain Not a '*Suitable Site*'"); Alec MacGillis, "Candidates Running Closest in the West; Demographics Shifting in Mountain States," *Washington Post*, A26 (Aug. 25, 2008), describes the Obama campaign's political advertisements in Nevada criticizing opponent McCain's support for Yucca.
445. Senator Barack Obama, "Barack Obama Explains Yucca Mountain Stance," Letter to the Editor, *Las Vegas Review-Journal*, 4D (May 20, 2007).
446. "The Democratic Debate in Las Vegas," *New York Times* (Jan. 15, 2008), *www.nytimes.com/2008/01/15/politics/15demdebate-transcript.html*.
447. Obama Campaign, "Barack Obama's Plan to Make America a Global Energy Leader," 4–5, press release (2008) (emphasis added).
448. Steven Chu, the secretary of energy, testified: "The FY 2010 budget request of $197 million for OCRWM implements the Administration's decision to terminate the Yucca Mountain program while developing nuclear waste disposal alternatives." *Funding and Oversight of the Department of Energy: Hearing before the Subcommittee on Energy and Water Development of the Senate Committee on Appropriations*, 111th Cong., 1st Sess. (May 19, 2009).
449. Motion to Withdraw, *U.S. Dept. of Energy (High-Level Waste Repository)*, No. 63-001 (ASLB, Nuclear Regulatory Comm'n March 3, 2010).
450. Alliance for Nevada's Economic Prosperity, "The Other Side of Yucca Mt.: The Encyclopedia on Yucca Mt.," *www.yuccapedia.com*.
451. Steve Tetreault, "Nevada and Counties Split over Ending Yucca," *Las Vegas Review-Journal* (May 18, 2010), *www.lvrj.com/news/nevada-and-counties-split-over-ending-yucca-94211039.html*. In June 2009, six rural Nevada counties argued before the NRC Atomic Safety and Licensing Board that DOE does not have the authority to withdraw the Yucca project. The motions of the four-county consortium are available at *esmeraldanvnuke.com/motions.html*.
452. Henry Brean, "Esmeralda County Has Few People, Fewer Jobs, but Don't Talk about Consolidation," *Las Vegas Review-Journal* (June 6, 2010), *www.lvrj.com/news/esmeralda-county-has-few-people-fewer-jobs-but-don-t-talk-about-consolidation-95716319.html*; Esmeralda County, "Resolution of the Esmeralda County Board of County Commissioners to Support Realignment of the Caliente Rail Line to a Geologic Repository at Yucca Mountain in Nye County, Nevada" (June 1, 2004), *esmeraldanvnuke.com/documents/Esm_Com.pdf*.
453. Nye County Board of County Commissioners, "Nye County, Nevada, Community Protection Plan," 3 (2006).
454. *Id.* at v.
455. Mark Waite, "Lacy Strong for Yucca Mountain," *Pahrump Valley Times* (July 14, 2010), *pvtimes.com/news/lacy-strong-for-yucca-mountain-2/*.
456. Phoebe Sweet, "Bryan: Dump Plan Demise Is Not a Lock," *Las Vegas Sun*, 1 (Jan. 23, 2009).
457. *In re Aiken County*, No. 10-1050 (D.C. Cir. filed Feb. 19, 2010); *Ferguson v. Dep't of Energy*, No. 10-1052 (D.C. Cir. filed Feb. 25, 2010); *South Carolina v. Dep't of Energy*, No. 10-1069 (D.C. Cir. transferred Mar. 25, 2010); *Washington v. Dep't of Energy*, No. 10-1082 (D.C. Cir. filed Apr. 13, 2010).
458. Naureen S. Malik, "Federal Appeals Court to Begin Yucca Mountain Hearings in 2011," *Wall Street Journal* (Dec. 10, 2010).
459. Brief of Petitioners at ii, *In re Aiken County*, No. 10-1050 (D.C. Cir. filed June 18, 2010).
460. Memorandum and Order (Suspending Briefing and Consideration of Withdrawal Motion), *U.S. Dept. of Energy (High-Level Waste Repository)*, No. 63-001 (ASLB, Nuclear Regulatory Comm'n April 6, 2010).
461. Memorandum and Order (Vacating the Board's Suspension Order and Remanding the Matter to the Board for Prompt Resolution of DOE's Motion to Withdraw) at 4, No. 63-001 (Nuclear Regulatory Comm'n April 23, 2010).
462. Nuclear Waste Policy Act of 1982 § 114(d), 42 U.S.C. § 10134(d) (2006).

463. *U.S. Dep't of Energy (High Level Waste Repository)*, slip op. at 5, No. 63-001-HLW (ASLB, Nuclear Regulatory Comm'n, June 29, 2010).
464. *Id.* at 7.
465. Nuclear Waste Policy Act of 1982 § 113(c)(3), 42 U.S.C. § 10133(c)(3) (2006).
466. *U.S. Dep't of Energy (High Level Waste Repository)*, slip op. at 8–9, No. 63-001-HLW (ASLB, Nuclear Regulatory Comm'n, June 29, 2010).
467. *See id.* at 9–10.
468. Steve Tetreault, "NRC Nominees Won't Stand in Way of Yucca Mountain Shutdown," *Las Vegas Review-Journal* (Feb. 11, 2010).
469. Motion of Recusal/Disqualification, *U.S. Dep't of Energy (High-Level Waste Repository)*, No. 63-001-HLW (NRC, July 9, 2010).
470. Notice of Recusal, *U.S. Dep't of Energy (High-Level Waste Repository)*, No. 63-001-HLW (NRC, July 9, 2010).
471. NRC, "Commissioner Kristine L. Svinicki," www.nrc.gov/about-nrc/organization/commission/svinickibio.pdf.
472. Steve Tetreault, "Bid to Revive Yucca Mountain Project Defeated in Senate," *Las Vegas Review-Journal* (July 22, 2010).
473. Tony Batt, "Yucca Mountain: Survey: Project Has Big Support," *Las Vegas Review-Journal* (Apr. 28, 2002).

Chapter 7

1. Holt, Congressional Research Service, "Nuclear Waste Policy: How We Got Here," presentation, 40 (Mar. 25, 2010), brc.gov/pdfFiles/CRS_BlueRibbonCommissionWastePolicyHistory.pdf. One recent presentation to the Blue Ribbon Commission estimated commercial SNF at 65,000 MTHM. Kessler, Electric Power Research Institute, "Used Nuclear Fuel: Inventory Projections," presentation to the Blue Ribbon Commission, 7 (Aug. 19, 2010).
2. Steve Tetreault, "Ruling Keeps Yucca Mountain Alive," *Las Vegas Review-Journal* (June 30, 2010); Memorandum and Order at 3, U.S. Dep't of Energy (High Level Waste Repository), No. 63-001-HLW (ASLB NRC 2010).
3. Steve Tetreault, "NRC Sets Schedule for Yucca Mountain Appeals," *Las Vegas Review-Journal* (June 30, 2010).
4. Holt, Congressional Research Service, *Nuclear Waste Disposal: Alternatives to Yucca*, 3 (2009). As explained by Holt, DOE's scenario assumes that the amount of additional SNF generated at existing reactors continues to be 2,000 metric tons of heavy metal (MTHM) per year. It does not account for any net increase in SNF resulting from an increase in the number of U.S. nuclear power plants, a goal the Obama administration and nuclear industry are actively promoting. Further, DOE's scenario assumes that Congress will lift the current statutory cap on disposal of SNF at Yucca. There is already almost enough SNF at existing reactors to fill Yucca to its statutory capacity. In addition to 63,000 MTHM of civilian SNF, the repository is slated to accept 7,000 MTHM of the total 12,800 MTHM of defense SNF and HLW from DOE sites. If Congress were to leave the cap in place, by the time Yucca opened, there would be a substantial amount of SNF that could not lawfully be accommodated at the repository. DOE believes, however, that Yucca's potential physical capacity as a repository far exceeds the NWPA cap, which could be raised by Congress. *Id.*
5. *Id.* at 23; *see, e.g.*, Rick Michal, "James Conca: On WIPP and Other Things Nuclear," *Nuclear News*, 44 (Feb. 2008).
6. "In Comments to Blue Ribbon Panel, N.M. Officials Open to More Waste," 22 *Weapons Complex Monitor*, no. 4–5, 11 (Jan. 28, 2011).
7. Holt, Congressional Research Service, *Nuclear Waste Disposal: Alternatives to Yucca*, 12–17 (2009).
8. Holt, Congressional Research Service, "Nuclear Waste Policy: How We Got Here," presentation, 40 (Mar. 25, 2010), brc.gov/pdfFiles/CRS_BlueRibbonCommissionWastePolicyHistory.pdf. There are 104 operating commercial nuclear reactors at sixty-five sites across the country that store SNF, in addition to nine decommissioned reactor sites. Commercial SNF is also stored at the uncompleted commercial reprocessing site in Morris, Illinois, and at two DOE sites. NRC, "Operating Nuclear Power Reactors by Location or Name," www.nrc.gov/info-finder/reactor; Office of Civilian Radioactive Waste Management, DOE, *Report to Congress on Demonstration of Interim Storage of Fuel from Decommissioned Reactors*, 1 (2008); Nuclear Energy Study Group, American Physical Society, *Consolidated Interim Storage of Commercial Spent Nuclear Fuel*, 6tbl.1 (2007).
9. Holt, Congressional Research Service, "Nuclear Waste Policy: How We Got Here," presentation, 43 (Mar. 25, 2010), brc.gov/pdfFiles/CRS_BlueRibbonCommissionWastePolicyHistory.pdf.
10. Office of Civilian Radioactive Waste Management, DOE, *Report to Congress on Demonstration of Interim Storage of Fuel from Decommissioned Reactors*, 1–2 (2008).
11. Institute for Twenty-First Century Energy, U.S. Chamber of Commerce, *Revisiting America's Nuclear Waste Policy*, 5, 7 (2009).
12. *Id.* at 4, 8.
13. Nuclear Energy Study Group, American Physical Society, *Consolidated Interim Storage of Commercial Spent Nuclear Fuel*, 7 (2007).
14. From a technical standpoint, heat and radiation have decreased enough after three years for the fuel to be passively cooled in dry storage. However, most dry-storage systems in the United States have been licensed only for fuel that has been cooled for five years. *Id.* at 1.
15. General Accounting Office, *Spent Nuclear Fuel*, 7 (2003), notes that NRC requires security systems and armed guards at nuclear power plants; the document also describes the security measures required at facilities with spent fuel pools; *id.*,

32–33; Nuclear Energy Study Group, American Physical Society, *Consolidated Interim Storage of Commercial Spent Nuclear Fuel*, 3–5 (2007), assesses terrorist risks at interim storage facilities. *See also* Gronlund, Lochbaum, and Lyman, Union of Concerned Scientists, *Nuclear Power in a Warming World*, 47 (2007).
16. Nuclear Energy Study Group, American Physical Society, *Consolidated Interim Storage of Spent Fuel* 7 (2007).
17. Waste Confidence Decision Update, 75 Fed. Reg. 81,037, at 81,040 (Dec. 23, 2010).
18. General Accounting Office, *Spent Nuclear Fuel*, 31 (2003).
19. *Id.* at 4–5, 32.
20. Gronlund, Lochbaum, and Lyman, Union of Concerned Scientists, *Nuclear Power in a Warming World*, 47 (2007).
21. General Accounting Office, *Spent Nuclear Fuel*, 15 (2003).
22. *Id.* at 13.
23. *Id.* at 15.
24. *Id.* at 15, 22.
25. *Id.* at 13–16; Collins and Hubbard, NRC, *Technical Study of Spent Fuel Pool Accident Risk* (2001).
26. General Accounting Office, *Spent Nuclear Fuel*, 14 (2003).
27. Gronlund, Lochbaum, and Lyman, Union of Concerned Scientists, *Nuclear Power in a Warming World*, 47 (2007).
28. Committee on Science and Technology for Countering Terrorism, National Research Council of the National Academies, *Making the Nation Safer*, 46–47 (2002).
29. *See* Nuclear Regulatory Commission, "Fact Sheet on NRC Review of Paper on Reducing Hazards from Stored Spent Nuclear Fuel" (Aug. 2003), *www.nrc.gov/reading-rm/doc-collections/fact-sheets/reducing-hazards-spent-fuel.html*; *see also* Waste Confidence Decision Update, 75 Fed. Reg. 81,037, at 81,051 (Dec. 23, 2010).
30. Nuclear Energy Study Group, American Physical Society, *Consolidated Interim Storage of Commercial Spent Nuclear Fuel*, 1 (2007).
31. Holt, Congressional Research Service, *Civilian Nuclear Spent Fuel: Temporary Storage Options for Spent Fuel*, 16 (1998).
32. General Accounting Office, *Spent Nuclear Fuel*, 34 (2003); Keheley, *AREVA Use of CFD in Fuel Assembly Design and Licensing*, 3–4 (2006).
33. General Accounting Office, *Spent Nuclear Fuel*, 34 (2003).
34. Board on Radioactive Waste Management, National Research Council of the National Academies, *Safety and Security of Commercial SNF Storage*, 61 (2006).
35. *Id.*; Committee on Transportation of Radioactive Waste, National Research Council of the National Academies, *Going the Distance?* 42 (2006).
36. Kravets, "With No Long-Term Solution, Nuclear Pallbearers Bury Waste," *Wired* (Mar. 16, 2009).
37. Waste Confidence Decision Review, 55 Fed. Reg. 38,474, 38,482 (Sep. 18, 1990).
38. General Accounting Office, *Spent Nuclear Fuel*, 16–17 (2003).
39. Nuclear Energy Study Group, American Physical Society, *Consolidated Interim Storage of Commercial Spent Nuclear Fuel*, 3 (2007).
40. *Id.* at 8.
41. Committee on Science and Technology for Countering Terrorism, National Research Council of the National Academies, *Making the Nation Safer*, 46–47 (2002).
42. Gronlund, Lochbaum, and Lyman, Union of Concerned Scientists, *Nuclear Power in a Warming World*, 47 (2007).
43. *Id.*, 45, 47.
44. Government Accountability Office, *Nuclear Waste Management*, 38 (2009).
45. Committee on Transportation of Radioactive Waste, National Research Council of the National Academies, *Going the Distance?* 44 (2006).
46. Government Accountability Office, *Nuclear Waste Management*, 38 (2009).
47. Waste Confidence Decision Update, 75 Fed. Reg. 81,037, at 81,038 (Dec. 23, 2010).
48. Holt, Congressional Research Service, Nuclear Waste Disposal: Alternatives to Yucca, 13–15 (2009); Nuclear Energy Study Group, American Physical Society, Consolidated Interim Storage of Commercial Spent Nuclear Fuel, 1–2 (2007); *see also* Holt, Congressional Research Service, "Nuclear Waste Policy: How We Got Here," presentation, 40 (Mar. 25, 2010), brc.gov/pdfFiles/CRS_BlueRibbonCommissionWastePolicyHistory.pdf.
49. Andrews, Congressional Research Service, Spent Nuclear Fuel Storage Locations and Inventory, 5 (2004). In 2004, the NRC renewed the Morris facility's storage license for a further twenty years. "NRC Renews License for Interim Spent Fuel Storage Installation at G.E. Morris Facility in Illinois," press release (Dec. 30, 2004), www.nrc.gov/reading-rm/doc-collections/news/2004/04-166.html.
50. Statement of Thomas B. Cochran, Ph.D., Senior Scientist, Natural Resources Defense Council, Inc., Opportunities and Challenges for Nuclear Power: Hearing before the House Committee on Science and Technology, 110th Cong. 41 (2008).
51. Holt, Congressional Research Service, *Nuclear Waste Disposal: Alternatives to Yucca*, 14 (2009).
52. Nuclear Waste Policy Act of 1982 § 135(a), 42 U.S.C. § 10155(a) (2006). Utilities were required to pay for this service according to a price set by the secretary of energy after notice and comment in the *Federal Register*. *Id.* § 136(a)(2)–(3), 42 U.S.C. § 10156(a)(2)–(3). Under the hardship provision, DOE was required to provide storage capacity to a utility after a successful demonstration by the utility and an NRC determination that the utility was pursuing all other alternatives but could not reasonably provide the storage capacity itself; the hardship provision required environmental review and public, state, and tribal involvement. *Id.* § 135(b)–(d), 42 U.S.C. § 10155(b)–(d). DOE

remained responsible for providing interim storage capacity for a utility that had made the requisite demonstration until three years after the opening of a repository or an MRS facility. *Id.* § 135(e), 42 U.S.C. § 10155(e). DOE's authority to enter into contracts to provide interim storage under the hardship provision of NWPA expired in 1990 and was never renewed by Congress. *Id.* § 136(a)(1), 42 U.S.C. § 10156(a)(1).

53. Holt, Congressional Research Service, *Civilian Nuclear Spent Fuel: Temporary Storage Options for Spent Fuel*, 37–38 (1998).
54. Nuclear Waste Policy Act of 1982 § 141(b)(1), 42 U.S.C. § 10161(b)(1) (2006).
55. Gross, "Nuclear Native America," 154 (2001); Nuclear Waste Policy Act of 1982 § 141(b)(1)(C), 42 U.S.C. § 10161(b)(1)(C) (2006). Section 141(b)(1) requires that any MRS facility be designed (1) to accommodate spent nuclear fuel and high-level radioactive waste resulting from civilian nuclear activities; (2) to permit continuous monitoring, management, and maintenance of such spent fuel and high-level waste for the foreseeable future; (3) to provide for the ready retrieval of such spent fuel and high-level waste for further processing or disposal; and (4) to safely store such spent fuel and high-level waste as long as may be necessary by maintaining such facility through appropriate means, including any required replacement of such facility. *Id.* § 141(b)(1), 42 U.S.C. § 10161(b)(1).
56. Ross, "Yucca Mountain: The Need for Monitored Retrievable Storage," 844 (2001). The three candidate sites were selected by the DOE after examining eleven sites in detail. Office of Civilian Radioactive Waste Management, DOE, *Screening and Identification of Sites for a Proposed Monitored Retrievable Storage Facility*, 1 (1985).
57. Nuclear Waste Policy Amendments Act of 1987 § 5021, Pub. L. No. 100-203, 101 Stat. 1330-227, 1330-232 (1987), as codified at 42 U.S.C. § 10162(a) (2006). The amendments annulled the original DOE proposal in its entirety, including the contingency plans for siting an MRS facility at one of the two alternative sites in Tennessee, but stopped short of precluding the siting of a future MRS facility at any of the locations considered in the original proposal: "[In evaluating the suitability of sites for an MRS facility], the Secretary shall make no presumption or preference to such sites by reason of their previous selection." *Id.*
58. *Id.* § 5021, Pub. L. No. 100-203, 101 Stat. 1330-227, 1330-232 (1987), as codified at 42 U.S.C. § 10162(b) (2006). The amended act authorizes federal interim storage only at an MRS facility or as provided under the hardship provision. Office of Civilian Radioactive Waste Management, DOE, *Report to Congress on Demonstration of Interim Storage of Fuel from Decommissioned Reactors*, 7–8 (2008).
59. Nuclear Waste Policy Amendments Act of 1987 § 5021, Pub. L. No. 100-203, 101 Stat. 1330-227, 1330-235 (1987), as codified at 42 U.S.C. § 10165(g) (2006).
60. Hardin, "Tipping the Scales: Create a Federal Interim Storage Facility," 297–298 (1999).
61. Nuclear Waste Policy Amendments Act of 1987 § 5021, Pub. L. No. 100-203, 101 Stat. 1330-227, 1330-234 (1987), as codified at 42 U.S.C. § 10165(b) (2006).
62. Nuclear Waste Repository Program: Yucca Mountain Site Recommendation to the President and Availability of Supporting Documents, 67 FED. REG. 9048 (Feb. 27, 2002); Holt, Congressional Research Service, *Civilian Nuclear Waste Disposal*, 7, 10 (2009); Cynkar, "Constitutional Conflicts on Public Land," 1277 (2004).
63. Nuclear Waste Policy Amendments Act of 1987 § 5021, Pub. L. No. 100-203, 101 Stat. 1330-227, 1330-236 (1987), as codified at 42 U.S.C. § 10168(d)(1) (2006).
64. *Id.* § 5021, Pub. L. No. 100-203, 101 Stat. 1330-227, 1330-236 (1987), as codified at 42 U.S.C. § 10168(d)(3)–(4) (2006).
65. *Id.* § 5041, Pub. L. No. 100-203, 101 Stat. 1330-227, 1330-244 (1987), as codified at 42 U.S.C. § 10242(b)(2) (2006); Gowda and Easterling, "Nuclear Waste and Native America," 231 (1998).
66. Nuclear Waste Policy Amendments Act of 1987 § 5041, Pub. L. No. 100-203, 101 Stat. 1330-227, 1330-246 (1987), as codified at 42 U.S.C. § 10246(a) (2006).
67. Grants were divided into three phases: I, II-A, and II-B. Phase I granted applicants $100,000 to begin independently studying MRS facilities; Phase II-A granted applicants $200,000 for continued education and feasibility studies; Phase II-B granted applicants $2.8 million to continue steps taken in previous phases, enter formal negotiations, identify potential sites, and commence an environmental assessment. Gowda and Easterling, "Nuclear Waste and Native America," 232–233 (1998); Erickson, Chapman, and Johnny, "MRS in Indian Country," 79–80 (1994); Sachs, "The Mescalero Apache Indians and MRS," 884 (1996).
68. Gowda and Easterling, "Nuclear Waste and Native America," 233, 235 (1998).
69. Nuclear Waste Policy Act of 1982 § 141(h), 42 U.S.C. § 10161(h) (2006); Gowda and Easterling, "Nuclear Waste and Native America," 231–232, 234 (1998). Although state governors were required to provide a statement of reasons for disapproval, no specific criteria for disapproval were established by the act. Nuclear Waste Policy Act of 1982 § 116(b)(2), 42 U.S.C. § 10136(b)(2) (2006). The scope of state veto authority explicitly excluded sites on Indian reservations. *Id.* § 116(b)(3), 42 U.S.C. § 10136(b)(3).
70. Sachs, "The Mescalero Apache Indians and MRS," 897–898 (1996).
71. *Id.* at 907–910.
72. *Id.* at 907.
73. Gowda and Easterling, "Nuclear Waste and Native America," 235 (1998); Erickson, Chapman, and Johnny, "MRS in Indian Country," 80 (1994).

74. Gowda and Easterling, "Nuclear Waste and Native America," 236 (1998).
75. *Id.*; Sachs, "The Mescalero Apache Indians and MRS," 884–885 (1996).
76. Gowda and Easterling, "Nuclear Waste and Native America," 236 (1998).
77. Skibine, "High Level Nuclear Waste on Indian Reservations," 291 (2001).
78. Nuclear Waste Policy Act of 1982 § 135(h), 42 U.S.C. § 10155(h) (2006).
79. Section 202 of the Energy Reorganization Act of 1974 grants NRC "licensing and related regulatory authority pursuant to chapters 6, 7, 8, and 10 of the Atomic Energy Act of 1954" with respect to "facilities used primarily for the receipt and storage of high-level radioactive wastes resulting from activities licensed under such Act." Energy Reorganization Act of 1974 § 202, 42 U.S.C. § 5842 (2006). Skibine, "High Level Nuclear Waste on Indian Reservations," 292–294 (2001); Hardin, "Tipping the Scales: Create a Federal Interim Storage Facility," 306–307 (1999). Private interim storage facilities are governed by 10 C.F.R. Part 72, the same regulations that would apply to a DOE-owned MRS facility. Notice of Issuance of Materials License Snm-2513 for the Private Fuel Storage Facility, 71 Fed. Reg. 10,068 (Feb. 28, 2006); Holt, Congressional Research Service, *Civilian Nuclear Spent Fuel: Temporary Storage Options for Spent Fuel*, 6 (1998).
80. Gross, "Nuclear Native America," 151 (2001); Gowda and Easterling, "Nuclear Waste and Native America," 236 (1998).
81. Gross, "Nuclear Native America," 150–153 (2001); Skibine, "High Level Nuclear Waste on Indian Reservations," 287 (2001); Holt, Congressional Research Service, *Civilian Nuclear Spent Fuel: Temporary Storage Options for Spent Fuel*, 6–7 (1998).
82. Gross, "Nuclear Native America," 150 (2001); No! The Coalition Opposed to High-Level Nuclear Waste, *White Paper Opposing PFS Facility*, 4 (2000).
83. No! The Coalition Opposed to High-Level Nuclear Waste, *White Paper Opposing PFS Facility*, 6–9 (2000).
84. *Id.* at 7–8; Gross, "Nuclear Native America," 150–153 (2001).
85. Notice of Issuance of Materials License Snm-2513 for the Private Fuel Storage Facility, 71 Fed. Reg. 10,068 (Feb. 28, 2006); Utah Department of Environmental Quality, *High Level Nuclear Waste Storage in Utah*, 1 (2006).
86. Notice of Availability of the Record of Decision for the Right-of-Way Applications Filed by Private Fuel Storage, L.L.C., for an Independent Spent Fuel Storage Installation on the Reservation of the Skull Valley Band of Goshute Indians and the Related Transportation Facility in Tooele County, UT, 71 Fed. Reg. 57,005 (Sep. 28, 2006); Notice of Availability of the Record of Decision for a Proposed Lease of Tribal Trust Lands between Private Fuel Storage, L.L.C., and Skull Valley Band of Goshute Indians in Tooele County, UT, 71 Fed. Reg. 58,629 (Oct. 4, 2006); Bureau of Indian Affairs, *Record of Decision for the Construction and Operation of an Independent Spent Fuel Storage Installation (ISFSI) on the Reservation of the Skull Valley Band of Goshute Indians (Band) in Tooele, Utah* (2006); Bureau of Land Management, *Record of Decision Addressing Right-of-Way Applications U76985 and U76896 to Transport Spent Nuclear Fuel to the Reservation of the Skull Valley Band of Goshute Indians* (2006); Holt, Congressional Research Service, *Nuclear Waste Disposal: Alternatives to Yucca*, 15 (2009).
87. Bureau of Indian Affairs, *Record of Decision for the Construction and Operation of an Independent Spent Fuel Storage Installation (ISFSI) on the Reservation of the Skull Valley Band of Goshute Indians (Band) in Tooele, Utah*, 26–29 (2006); Holt, Congressional Research Service, *Nuclear Waste Disposal: Alternatives to Yucca*, 15 (2009).
88. A state fact sheet on the right-of-way issue elaborated the latter concern:
    "PFS, incapable of responding to an accident or other emergency, will rely on emergency responders from State and local government. PFS will not train, equip, or fund Utah emergency responders. PFS's reliance on Utahans to respond to a high level nuclear waste incident creates an unacceptable health and safety risk and an uncompensated economic burden.
    "NRC will not regulate the intermodal as a storage site, has made no site-specific evaluation of risks at the site, and requires no funding plan to cleanup accidents or to terminate use of the site.
    "PFS, a shell limited liability company, has no capital backing. Thus, its financial and technical capability to safely operate and maintain the intermodal site is questionable. In addition, PFS has no funds escrowed to cleanup or terminate its use of the site.
    Utah Department of Environmental Quality, *Utah Fact Sheet on PFS's Right-of-Way Application*, 2 (2006).
89. *Skull Valley Band of Goshute Indians v. Cason*, No. 2:07cv00526 (D. Utah 2007).
90. Judy Fahys, "Goshutes Fight N-Waste Rulings," *Salt Lake Tribune* (Mar. 23, 2010). Additional controversy surrounds the case. While DOI was deliberating, a former deputy secretary of the interior approached the tribe about dropping the project in exchange for additional land and hunting and fishing rights. "Judge Holds Hearing over Rejected Goshute Nuclear Dump" (Mar. 23, 2010), www.Indianz.com/News/2010/018901.asp.
91. *Skull Valley Band of Goshute Indians v. Davis*, No. 07-cv-0526-DME-DON, slip op. at 1 (D. Utah July 26, 2010).
92. *Devia v. Nuclear Regulatory Comm'n*, 492 F.3d 421 (D.C. Cir. 2007).
93. Jay E. Silberg, Partner, Pillsbury Winthrop Shaw Pittman, LLP, representing PFS in the Court of

Appeals for the D.C. Circuit, e-mail correspondence with the authors (Mar. 21, 2010). *See also Devia v. Nuclear Regulatory Comm'n*, 492 F.3d 421, 422 (D.C. Cir. 2007) ("Because it is speculative whether the project will ever be able to proceed, we find the petitioners' challenge unripe and direct that the case be held in abeyance.").

94. Skibine, "High Level Nuclear Waste on Indian Reservations," 287–288 (2001).
95. According to one commentator, NRC's issuance of a license for a private storage facility "come[s] attached with a comprehensive federal regulatory scheme which will preempt most state regulations"—but the full reach of such preemption is not clear. *Id.* at 315. For an analysis of the constitutional ramifications of state attempts to regulate nuclear waste storage on Indian reservations, *see id.*
96. Nuclear Waste Policy Act of 1982 § 148(d), 42 U.S.C. § 10168(d) (2006).
97. Holt, Congressional Research Service, *Nuclear Waste Disposal: Alternatives to Yucca*, 15 (2009).
98. Gross, "Nuclear Native America," 158 (2001).
99. Office of Civilian Radioactive Waste Management, DOE, *Report to Congress on Demonstration of Interim Storage of Fuel from Decommissioned Reactors*, 11 (2008).
100. Committee on Appropriations, *2007 Energy and Water Appropriations Bill*, 126–127 (2006).
101. Committee on Appropriations, *Consolidated Appropriations Act, 2008: Committee Print*, 581 (2008). The law as enacted appears as Consolidated Appropriations Act, 2008, Pub. L. No. 110-161, 121 Stat. 1844 (2007).
102. DOE's acceptance of SNF from Three Mile Island at INL in 1979 suggests that DOE might have general authority under the Atomic Energy Act to accept larger quantities of SNF at one or more of its weapons complex sites. Holt, Congressional Research Service, *Nuclear Waste Disposal: Alternatives to Yucca*, 14 (2009).
103. Office of Civilian Radioactive Waste Management, DOE, *Report to Congress on Demonstration of Interim Storage of Fuel from Decommissioned Reactors*, 15–16 (2008).
104. Nuclear Energy Study Group, American Physical Society, *Consolidated Interim Storage of Commercial Spent Nuclear Fuel*, 1–2 (2007); Hardin, "Tipping the Scales: Create a Federal Interim Storage Facility," 308–311 (1999).
105. *See* Private Fuel Storage, LLC, "The PFS Project Benefits Nuclear Power Generators and Customers," www.privatefuelstorage.com/benefit/industry.html.
106. Nuclear Energy Study Group, American Physical Society, *Consolidated Interim Storage of Commercial Spent Nuclear Fuel*, 11 (2007).
107. Von Hippel, International Panel on Fissile Materials, *Managing Spent Fuel*, 26 (2007).
108. The Nuclear Energy Study Group of the American Physical Society noted that it was reasonable to expect that consolidation would lead to some cost savings due to economies of scale but emphasized that no comprehensive studies on the matter were available. *See* Nuclear Energy Study Group, American Physical Society, *Consolidated Interim Storage of Commercial Spent Nuclear Fuel*, 10 (2007).
109. *Id.*, 3, 5–6 (2007); Holt, Congressional Research Service, *Civilian Nuclear Spent Fuel: Temporary Storage Options for Spent Fuel*, 15, 17 (1998). Dry-cask storage facilities can be maintained for at least fifty years with a high degree of confidence; replacing casks can further extend their life, making safe on-site storage feasible for as long as adequate resources and attention can be directed to facility maintenance. Nuclear Energy Study Group, American Physical Society, *Consolidated Interim Storage of Commercial Spent Nuclear Fuel*, 3 (2007). Security requirements are likely to be the same regardless of whether SNF is stored in a consolidated facility or in on-site storage facilities. *Id.*, 5.
110. *See* Greenberg, "NIMBY, CLAMP, and the Location of New Nuclear Facilities," 1242 (2009); *see also* Greenberg et al., "Nuclear Waste and Public Worries," 346 (2007).
111. The DOE Office of Environmental Management terms the concept the Energy Parks Initiative. Gilbertson, Deputy Assistant Secretary for Engineering and Technology, DOE, "Energy Parks Initiative" (2009); Office of Environmental Management, DOE, *Reduction of EM Footprint and Establishment of Energy Parks* (2008); *see also* Campbell and Haried, "Financing and Licensing Nuclear Energy Parks" (2009). The Office of Environmental Management has conducted several public workshops on the issue, including one discussing the possibility of energy park development at the Savannah River Site in Georgia. *See* Savannah River Operations Office, DOE, "Energy Park Concept at SRS," *sro.srs.gov/energypark.htm*.
112. Holt, Congressional Research Service, *Civilian Nuclear Spent Fuel: Temporary Storage Options for Spent Fuel*, 36 (1998).
113. The Western Governors' Association insists that DOE must coordinate and cooperate with the states in identifying routes for shipment of SNF and HLW. Western Governors' Association, *Transportation of SNF and HLW*, 2–4 (2008).
114. Nuclear Energy Study Group, American Physical Society, *Consolidated Interim Storage of Commercial Spent Nuclear Fuel*, 5 (2007).
115. General Accounting Office, *Spent Nuclear Fuel*, 7–8 (2003).
116. Committee on Transportation of Radioactive Waste, National Research Council of the National Academies, *Going the Distance?* 7–9 (2006).
117. *Id.*
118. Carter, *Nuclear Imperatives*, 93–95, 107–108 (1987).
119. Alvarez, Institute for Policy Studies, *Radioactive Wastes and GNEP*, 6 (2007).
120. Brenner, *Nuclear Power and Non-Proliferation*, 77 (1974).
121. *Id.* at 78; Carter, *Nuclear Imperatives*, 18 (1987).

122. Carter, *Nuclear Imperatives*, 98–105 (1987); Alvarez, Institute for Policy Studies, *Radioactive Wastes and GNEP*, 6 (2007).
123. Mauro and Yago, "Government Targeting in Economic Development," 72–73 (1989).
124. Committee on Science and Technology, *West Valley Demonstration Project Act*, 6 (1980).
125. General Accounting Office, *Issues Related to the Closing of the West Valley Plant*, Enclosure at 2 (1977).
126. *Id.*
127. DOE, *Plutonium Recovery at West Valley* (1996).
128. Carter, *Nuclear Imperatives*, 98 (1987).
129. DOE, *Plutonium Recovery at West Valley* (1996).
130. Carter, *Nuclear Imperatives*, 99–101 (1987); Gillette, "Plutonium: Questions of Health in a New Industry," 1031 (1974).
131. Carter, *Nuclear Imperatives*, 101 (1987).
132. Gillette, "Plutonium: Questions of Health in a New Industry," 1032 (1974). AEC officials accused the company of a "failure to make reasonable efforts to maintain the lowest levels of contamination and radiation" and of a "failure to adequately instruct or effectively train employees . . . in the radiation hazards involved in their job assignments." *Id.*
133. Carter, *Nuclear Imperatives*, 102 (1987).
134. General Accounting Office, *Issues Related to the Closing of the West Valley Plant*, Encl. at 2 (1977).
135. Carter, *Nuclear Imperatives*, 102 (1987); Wohlstetter, "Spreading the Bomb without Quite Breaking the Rules," 173 (1976).
136. DOE, *Plutonium Recovery at West Valley* (1996).
137. Carter, *Nuclear Imperatives*, 102–103 (1987).
138. *Id.* at 103.
139. West Valley Demonstration Project Act § 2(b)(4)(C), as enacted, Pub. L. No. 96-368, 94 Stat. 1347, at 1348 (1980).
140. Alvarez, Institute for Policy Studies, *Radioactive Wastes and GNEP*, 6 (2007).
141. DOE, "West Valley Demonstration Project Nuclear Timeline," www.wv.doe.gov/Site_History.html.
142. Alvarez, Institute for Policy Studies, *Radioactive Wastes and GNEP*, 10 (2007).
143. Williams, DOE, "West Valley Spent Nuclear Fuel Shipment," 7 (2004). A 1995 West Valley settlement agreement also required DOE to remove all TRU waste from the state. *Id.* at 3. *See* Chapter 1 for a discussion of the agreement.
144. Carter, *Nuclear Imperatives*, 112 (1987).
145. Bergeron, *Tritium on Ice*, 30–34 (2002).
146. Carter, *Nuclear Imperatives*, 117 (1987).
147. *Id.*; Bergeron, *Tritium on Ice*, 32 (2002).
148. Statement on Nuclear Policy, 3 Pub. Papers 2763, 2769 (Oct. 28, 1976).
149. Nuclear Power Policy: Statements on Decisions Reached Following a Review, 1 Pub. Papers 587, 588 (Apr. 7, 1977).
150. Nuclear Power Policy: The President's Remarks and a Question-and-Answer Session with Reporters on Decisions Following a Review of U.S. Policy and a Question-and-Answer Session with Reporters, 1 Pub. Papers 581 (Apr. 7, 1977).
151. Ronald Reagan, Statement Announcing a Series of Policy Initiatives on Nuclear Energy, 1981 Pub. Papers 903, 904 (Oct. 8, 1981).
152. Carter, *Nuclear Imperatives*, 120–125 (1987).
153. Von Hippel, International Panel on Fissile Materials, *Managing Spent Fuel*, 3, 13 (2007).
154. Lyman, Union of Concerned Scientists, *The GNEP: Will It Advance or Undermine Nonproliferation?* 2 (2006); Von Hippel, International Panel on Fissile Materials, *Managing Spent Fuel*, 4 (2007). The six countries are Argentina, Belgium, Brazil, Italy, Sweden, and Germany.
155. Caroline Merrell, "Sellafield Reprocessing Plant to Close by 2010," *Times*, 19 (Aug. 26, 2003); Von Hippel, International Panel on Fissile Materials, *Managing Spent Fuel*, 11 (2007); World Nuclear Association, "China's Nuclear Fuel Cycle," www.world-nuclear.org/info/inf63b_china_nuclearfuelcycle.html.
156. Bunn et al., Belfer Center, *The Economics of Reprocessing*, 23–26, 105–117 (2003), discusses uranium prices in both the near term and the long term, and the extent of worldwide uranium resources.
157. Von Hippel, International Panel on Fissile Materials, *Managing Spent Fuel*, 9–10 (2007).
158. *Id.* As of 2009, there were only thirty-five operating reactors that used MOX fuel (twenty in France, ten in Germany, two in Belgium, and three in Switzerland). Jarret Adams, "Areva Has Many Satisfied MOX Fuel Customers," Areva North America: Next Energy Blog (May 11, 2009), us.arevablog.com/2009/05/11/areva-has-many-satisfied-mox-fuel-customers. Areva, a French nuclear power conglomerate that is 90 percent state owned, signed contracts with Japanese utilities to provide MOX fuel for sixteen to eighteen reactors in Japan starting in 2010. *Id.*; *see, e.g.*, Areva, "Recycling: Areva Signs an Agreement to Supply MOX Fuel to Japanese Utility Electric Power Development," press release (Apr 3, 2009).
159. Lyman, Union of Concerned Scientists, *The GNEP: Will It Advance or Undermine Nonproliferation?*, 2 (2006).
160. National Nuclear Security Administration, "Plutonium Disposition," nnsa.energy.gov/aboutus/ourprograms/nonproliferation/programoffices/fissilematerialsdisposition/plutoniumdisposition; Office of the Spokesman, U.S. Department of State, "2000 Plutonium Management and Disposition Agreement," press release (Apr. 13, 2010).
161. See Bergeron, *Tritium on Ice*, 17–20 (2002), which describes Eisenhower's Atoms for Peace initiative.
162. *Id.*, 29–30.
163. *Id.*, 26–27.
164. *Id.*, 28–29.
165. Carter, *Nuclear Imperatives*, 119 (1987).
166. Office of Fuel Cycle Management, DOE, *GNEP Strategic Plan*, 5 (2007).
167. *Id.*
168. *Id.*
169. Address before a Joint Session of the Congress on

the State of the Union, 42 WEEKLY COMP. PRES. DOC. 145, 150 (Jan. 31, 2006).

170. Clay Sell, Deputy Secretary, DOE, "Announcing the Global Nuclear Energy Partnership," press briefing (Feb. 6, 2006), *www.energy.gov/print/3171.htm*; Office of Fuel Cycle Management, DOE, *GNEP Strategic Plan*, 1 (2007).
171. Clay Sell, Deputy Secretary, DOE, "Announcing the Global Nuclear Energy Partnership," press briefing (Feb. 6, 2006), *www.energy.gov/print/3171.htm*.
172. *Id.*
173. Office of Fuel Cycle Management, DOE, *GNEP Strategic Plan*, 4–5 (2007).
174. Clay Sell, Deputy Secretary, DOE, "Announcing the Global Nuclear Energy Partnership," press briefing (Feb. 6, 2006), *www.energy.gov/print/3171.htm*.
175. *Id.* China, France, Japan, and Russia signed on in May 2007, and eleven other countries became partners in September 2007. Statement of Dennis R. Spurgeon, Assistant Secretary for Nuclear Energy, Department of Energy, *Global Nuclear Energy Partnership: Hearing before the Senate Committee on Energy and Natural Resources*, 110th Cong. 12 (2007).
176. Alvarez, Institute for Policy Studies, *Radioactive Wastes and GNEP*, 13–14 (2007). As part of GNEP, the United States would have encouraged other fuel cycle countries using reprocessing to replace PUREX with UREX+. Clay Sell, Deputy Secretary, DOE, "Announcing the Global Nuclear Energy Partnership," press briefing (Feb. 6, 2006), *www.energy.gov/print/3171.htm*.
177. DOE, *Spent Fuel Recycling Plan*, 11–13 (2006); Alvarez, Institute for Policy Studies, *Radioactive Wastes and GNEP*, 3, 14–17 (2007).
178. A fast reactor differs from a breeder reactor. "Fast" signifies a reactor that does not have a moderator and thus uses neutrons of higher energy than the usual thermal reactors to produce fission, while "breeder" refers to a reactor that produces more fissile material than it consumes. As discussed in this chapter, GNEP proposed to use a fast reactor that could burn reprocessed fuel produced through the UREX+ process.
179. DOE, *Spent Fuel Recycling Plan*, 12 (2006).
180. Alvarez, Institute for Policy Studies, *Radioactive Wastes and GNEP*, 3, 14–17 (2007).
181. DOE, *Spent Fuel Recycling Plan*, 12 (2006); Clay Sell, Deputy Secretary, DOE, "Announcing the Global Nuclear Energy Partnership," press briefing (Feb. 6, 2006), *www.energy.gov/print/3171.htm*. For a skeptical view of the extent to which this arrangement would reduce security risks, *see* Von Hippel, International Panel on Fissile Materials, *Managing Spent Fuel*, 22 (2007).
182. DOE, *Spent Fuel Recycling Plan*, 13 (2006); Von Hippel, International Panel on Fissile Materials, *Managing Spent Fuel*, 22 (2007).
183. Wigeland et al., "Separations and Transmutation Criteria to Improve Utilization of a Geologic Repository" (2006), as cited in Von Hippel, International Panel on Fissile Materials, *Managing Spent Fuel*, 12 (2007). Under this scenario, the cesium and strontium would be separately stored for several centuries until the point that their radioactivity was low enough to permit disposal in near-surface facilities.
184. In the Yucca context, NWTRB found that a "hot" repository design would increase uncertainty about site performance before and after closure and would make modeling more difficult. Nuclear Waste Technical Review Board, *1999 Report to Congress and Secretary of Energy*, 3 (2000). A cooler waste would presumably mitigate some or all of the concerns expressed by the NWTRB about the potential adverse effects of a hot design on the performance of the Yucca repository (see Chapter 6).
185. Alvarez, Institute for Policy Studies, *Radioactive Wastes and GNEP*, 16–17 (2007).
186. Office of Fuel Cycle Management, DOE, *GNEP Strategic Plan*, 6 (2007).
187. Alvarez, Institute for Policy Studies, *Radioactive Wastes and GNEP*, 12, 15–16 (2007).
188. Office of Fuel Cycle Management, DOE, *GNEP Strategic Plan*, 6 (2007).
189. Committee on Review of DOE's Nuclear Energy Research and Development Program, National Research Council of the National Academies, *Review of DOE's Nuclear Energy R&D Program*, 4–6 (2008); Government Accountability Office, *GNEP: DOE Should Reassess Its Approach*, 9–11 (2008); Von Hippel, International Panel on Fissile Materials, *Managing Spent Fuel*, 4 (2007).
190. "The Committee does not support the Department's rushed, poorly defined, expansive, and expensive Global Nuclear Energy Partnership (GNEP) proposal." Committee on Appropriations, *Energy and Water Development Appropriations Bill, 2009*, at 93 (2008).
191. Bunn, *Assessing the Benefits, Costs, and Risks of Reprocessing*, 6–7 (2006).
192. Office of Fuel Cycle Management, DOE, *GNEP Strategic Plan*, 10 (2007).
193. Phillip J. Finck, Deputy Associate Laboratory Director, Argonne National Laboratory, stated at a House subcommittee hearing: "Assuming that demonstrations of these processes are started by 2007, commercial operations are possible starting in 2025." *Nuclear Fuel Reprocessing: Hearing before the Subcommittee on Energy of the House Committee on Science*, 109th Cong. 70–71 (2005).
194. Notice of Request for Expressions of Interest in a Consolidated Fuel Treatment Center to Support the Global Nuclear Energy Partnership, 71 FED. REG. 44,676, at 44,677 (Aug. 7, 2006).
195. Alvarez, Institute for Policy Studies, *Radioactive Wastes and GNEP*, 16 (2007).
196. DOE, *Spent Fuel Recycling Plan*, 13 (2006).
197. Alvarez, Institute for Policy Studies, *Radioactive Wastes and GNEP*, 16 (2007).
198. *Id.*
199. Statement of Thomas B. Cochran, Ph.D., Senior Scientist, Natural Resources Defense Council, Inc.,

Opportunities and Challenges for Nuclear Power: Hearing before the House Committee on Science and Technology, 110th Cong. 38 (2008); Union of Concerned Scientists, *Reprocessing Would Increase Volume of Radioactive Waste* (2009).
200. DOE, *Draft EIS for GNEP*, 4-138 (2008).
201. Office of the Chief Financial Officer, DOE, *FY 2007 Congressional Budget Request: Highlights*, 1–2, 6 (2006); Office of the Chief Financial Officer, DOE, *FY 2008 Congressional Budget Request: Highlights*, 8, 40–41 (2007).
202. Commission on Geosciences, Environment, and Resources, National Research Council of the National Academies, *Nuclear Wastes*, 82 (1996).
203. Office of Fuel Cycle Management, DOE, *GNEP Strategic Plan*, 6 (2007).
204. *Id.* at 6–7, 9.
205. Carter, *Nuclear Imperatives*, 113 (1987).
206. Von Hippel, International Panel on Fissile Materials, *Managing Spent Fuel*, 3, 7–8 (2007).
207. Carter, *Nuclear Imperatives*, 388–389 (1987).
208. Joskow and Parsons, "The Economic Future of Nuclear Power," 47 (2009).
209. Bunn et al., Belfer Center, *The Economics of Reprocessing*, 72 (2003).
210. Alvarez, Institute for Policy Studies, *Radioactive Wastes and GNEP*, 18 (2007).
211. Shropshire et al., *Advanced Fuel Cycle Economic Sensitivity Analysis*, 14 (2006); Alvarez, Institute for Policy Studies, *Radioactive Wastes and GNEP*, 18 (2007).
212. One study estimated that the cost of reprocessing could range from about $1350/kgHM to over $3100/kgHM, depending upon the cost of capital. Bunn et al., Belfer Center, *The Economics of Reprocessing*, 28 (2003).
213. See Andrew Mickey, Chief Investment Strategist, Q1 Publishing, "Uranium has Bottomed: Two Uranium Bulls to Jump on Now" (Aug. 22, 2008), *www.uraniumseek.com/news/UraniumSeek/1219431716.php*.
214. Ux Consulting Company, LLC, "UxC Nuclear Fuel Price Indicators," *www.uxc.com/review/uxc_Prices.aspx*. Note that figures in Appendix B are based on yearly average prices; the average price in 2007, the year in which the spike in uranium prices occurred, was $90 per pound ($198 per kilogram) in 2007 dollars.
215. Von Hippel, International Panel on Fissile Materials, *Managing Spent Fuel*, 2–3 (2007).
216. *See, e.g.*, Von Hippel, International Panel on Fissile Materials, *Managing Spent Fuel* (2007); Lyman, Union of Concerned Scientists, *The GNEP: Will It Advance or Undermine Nonproliferation?* (2006); Bunn, *Assessing the Benefits, Costs, and Risks of Reprocessing* (2006).
217. Lyman, Union of Concerned Scientists, *The GNEP: Will It Advance or Undermine Nonproliferation?*, 5 (2006).
218. Von Hippel, International Panel on Fissile Materials, *Managing Spent Fuel*, 23 (2007).
219. Clay Sell, Deputy Secretary, DOE, "Announcing the Global Nuclear Energy Partnership," press briefing (Feb. 6, 2006), *www.energy.gov/print/3171.htm*; DOE, *Spent Fuel Recycling Plan*, 10 (2006).
220. See Bunn, *Assessing the Benefits, Costs, and Risks of Reprocessing*, 2 (2006).
221. Von Hippel, International Panel on Fissile Materials, *Managing Spent Fuel*, 22 (2007).
222. Fred Weir and Howard LaFranchi, "Russia and US as Global Nuclear Waste Collectors?" *Christian Science Monitor*, 1 (Feb. 7, 2006).
223. International Atomic Energy Agency, "Agency Multinational Fuel Bank Reaches Key Milestone: Kuwait Pledge of $10 Million Secures International Funding for Next Steps" (Mar. 6, 2009), *www.iaea.org/NewsCenter/News/2009/fbankmilestone.html*; Nuclear Threat Initiative, "NTI/IAEA Fuel Bank Hits $100 Million Milestone: Kuwaiti Contribution Fulfills Buffett Monetary Condition," press release (Mar. 5, 2009).
224. Marc Champion, "Iran's Uranium Stance Keeps a Deal at Bay," *Wall Street Journal*, A3 (Jan. 28, 2008); Charles Hanley, "Big Names, Bucks Bet behind Nuclear Fuel Bank," *Pittsburgh Tribune-Review* (May 24, 2009).
225. Alvarez, Institute for Policy Studies, *Radioactive Wastes and GNEP*, 15 (2007).
226. *Id.*
227. Von Hippel, International Panel on Fissile Materials, *Managing Spent Fuel*, 3 (2007).
228. Energy secretary Chu, however, might have been hinting otherwise when he stated at his Senate confirmation hearing that he was "confident the Department of Energy, perhaps in collaboration with other countries, can get a solution to the nuclear waste problem." *Chu Nomination: Hearing before the Committee on Energy and Natural Resources*, 111th Cong. 24 (2009). Equally, he might have been thinking of the global uranium fuel-bank proposal.
229. "Of the amounts authorized to be appropriated for fiscal year 2009 by section 3101(a)(2) for defense nuclear nonproliferation activities, not more than $3,000,000 may be used for projects that are specifically designed for the Global Nuclear Energy Partnership." Duncan Hunter National Defense Authorization Act for Fiscal Year 2009 § 3117, Pub. L. No. 110-417, 122 Stat. 4356, 4758 (2008).
230. Dalrymple, "GNEP Turns to the World," 9 (2009).
231. Notice of Cancellation of the Global Nuclear Energy Partnership (GNEP) Programmatic Environmental Impact Statement (PEIS), 74 Fed. Reg. 31,017 (June 29, 2009).
232. Nuclear Energy Institute, *The Potential of Reprocessing and Recycling*, 3 (2006); Dale E. Klein, Chair, NRC, "Some Views on Closing the Fuel Cycle," speech at Fuel Cycle Information Exchange, S-08-025 (June 17, 2008).
233. Dale E. Klein, Chair, NRC, "Some Views on Closing the Fuel Cycle," speech at Fuel Cycle Information Exchange, S-08-025 (June 17, 2008).
234. Nuclear Energy Institute, *Regulatory Framework for Recycling Nuclear Fuel*, 2–4 (2008). The proposal

called for creation of a set of regulations to be codified in a new 10 C.F.R. pt. 7X. Absent the creation of Part 7X, recycling facilities would be governed by the same regulations that apply to the licensing of light-water reactors, 10 C.F.R. pt. 50. *Id.* The white paper argues, and NRC concedes, that recycling facilities and wastes have characteristics very different from those of light water reactors, and suggests that revision of Part 50 to reflect these differences would be desirable. *Id.*; NRC, *Regulatory Structure for Reprocessing* (2008); NRC, *Update on Reprocessing Regulatory Framework* (2009).
235. NRC, *Update on Reprocessing Regulatory Framework* (2009); Dale E. Klein, Chair, NRC, "Some Views on Closing the Fuel Cycle," speech at Fuel Cycle Information Exchange, S-08-025 (June 17, 2008).
236. Department of Energy, "Global Nuclear Energy Partnership Steering Group Members Approve Transformation to the International Framework for Nuclear Energy Cooperation," press release (June 21, 2010).
237. International Framework for Nuclear Energy Cooperation, *Statement of Mission* (2010), *www.gneppartnership.org/docs/IFNEC_ StatementofMission.pdf*.
238. Bryan Bender, "Obama Seeks Global Uranium Fuel Bank," *Boston Globe*, A1 (June 8, 2009).

Chapter 8

1. U.S. Dep't of Energy (High Level Waste Repository), slip op. at 3, No. 63-001-HLW (ASLB, Nuclear Regulatory Comm'n, June 29, 2010); U.S. Dep't of Energy's Reply to the Motion to Withdraw at 31n.102, U.S. Dep't of Energy (High Level Waste Repository), No. 63-001-HLW (ASLB, Nuclear Regulatory Comm'n, May 27, 2010).
2. *In re Aiken County*, No. 10-1050 (D.C. Cir., filed Feb. 19, 2010); *Ferguson v. U.S. Dep't of Energy*, No. 10-1052 (D.C. Cir., filed Feb. 25, 2010); *South Carolina v. U.S. Dep't of Energy*, No. 10-1069 (D.C. Cir., transferred Mar. 25, 2010); *Washington v. Dep't of Energy*, No. 10-1082 (D.C. Cir., filed Apr. 13, 2010). These cases have been consolidated as *In re Aiken County*, No. 10-1050. The court heard oral arguments March 22, 2011. Order, *In re Aiken County*, No. 10-1050 (D.C. Cir., filed Jan. 10, 2011).
3. Government Accountability Office, *LLRW: Status of Disposal Availability*, 3–5 (2008).
4. *Id.*, 5.
5. Nuclear Regulatory Commission, *Radioactive Waste: Production, Storage, Disposal*, 27–29 (2002); Government Accountability Office, *LLRW: Disposal Availability Adequate in the Short Term*, 9 (2004).
6. *See, e.g.*, Robert Knox, "Low-level Waste to Be Stored at Pilgrim Nuclear Plant," *Boston Globe*, 3 (June 8, 2009).
7. For information on the startup and operation of each reactor in the United States, see Energy Information Agency, "Nuclear Power Plants Operating in the United States as of December 31, 2008," *www.eia.doe.gov/cneaf/nuclear/page/at_a_ glance/reactors/nuke1.html*. Assuming that a nuclear reactor has forty years of useful operational life, nearly 100 of the 104 running reactors in the United States will be ripe for decommissioning by 2030. With recent NRC operating license extensions, it has become increasingly common to assume a sixty-year operational life, which would still necessitate retiring one-third of existing U.S. reactors by 2035. Energy Information Agency, "U.S. Nuclear Power Plants: Continued Life or Replacement after 60?" *www.eia.doe.gov/oiaf/aeo/otheranalysis/ aeo_2010analysispapers/nuclear_power.html*.
8. "Commission Lets 36 States Dump Nuke Waste in Texas," *Associated Press* (Jan. 4, 2011).
9. *City of Philadelphia v. New Jersey*, 437 U.S. 617 (1978).
10. *New York v. United States*, 505 U.S. 144 (1992).
11. Government Accountability Office, *LLRW: Disposal Availability Adequate in the Short Term*, app. 2, at 42 (2004).
12. The only provision in LLRWPA that addresses the federal government's disposition of LLW states: "The Federal Government shall be responsible for the disposal of low-level radioactive waste owned or generated by the Department of Energy; low-level radioactive waste owned or generated by the United States Navy as a result of the decommissioning of vessels of the United States Navy; low-level radioactive waste owned or generated by the Federal Government as a result of any research, development, testing, or production of any atomic weapon; and any other low-level radioactive waste with concentrations of radionuclides that exceed the limits established by the Commission for class C radioactive waste, as defined by section 61.55 of title 10, Code of Federal Regulations, as in effect on January 26, 1983." Low-Level Radioactive Waste Policy Amendments Act of 1985 § 3(b)(1), 42 U.S.C. § 2021c(b)(1) (2006). Neither this provision nor any other in the statute explicitly permits or prohibits DOE from voluntarily entering into arrangements to dispose of civilian LLW other than GTCC.
13. EPA, "Radiation Protection," *www.epa.gov/ radiation/wipp/*.
14. Until that time, WIPP had been receiving only CH-TRU waste, which can be directly handled by workers because of its relatively low level of hazard. RH-TRU waste must be handled remotely using special equipment, because direct exposure of workers to the waste is not safe.
15. Carlsbad Field Office, DOE, "State Renews WIPP Facility Permit," press release (Dec. 1, 2010).
16. Department of Energy, "WIPP Quick Facts," *www .wipp.energy.gov/TeamWorks/index.htm*.
17. WIPP Land Withdrawal Act of 1992 § 7(a)(3), as enacted, Pub. L. No. 102-579, 106 Stat. 4777, 4785 (1992).
18. DOE estimates the total volume of existing and anticipated defense TRU as 171,970 cubic meters. DOE, *WIPP Disposal Phase: Final Supplemental*

EIS, vol. 1, at 1–4 (1997). This estimate does not assume reprocessing of civilian SNF, which, if allowed in the future using advanced reprocessing technologies like the UREX system proposed by GNEP, would be expected to generate significant quantities of TRU waste.

19. Another important nuclear waste problem, though somewhat less publicized, is how to handle stocks of "waste" plutonium that are no longer needed for weapons production. This problem is not addressed here; *but see* Committee on End Points for Spent Nuclear Fuel and High-Level Radioactive Waste in Russia and the United States, National Research Council, *End Points for SNF and HLW in Russia and the US* (2003).

20. This number is broken down into nearly 2,500 MTHM of defense SNF and an estimated 10,300 MTHM of HLW, when rendered in a form suitable for repository disposal. Office of Civilian Radioactive Waste Management, DOE, *Report to the President on the Need for a Second Repository*, 2, 5 (2008). The HLW is currently in the form of 90 megagallons primarily located at Hanford and SRS and 4,400 cubic meters at INL. Idaho National Laboratory, "2008 DOE Spent Nuclear Fuel and High Level Waste Inventory," *inlportal.inl.gov/portal/server.pt?tbb=siteindex*. Conversion to MTHM is an inexact science. *See* Knecht et al., INL, *Options for Determining Equivalent MTHM for DOE High-Level Waste* (1999). While OCRWM gives the number at 10,300 MTHM, EM estimates that it could be anywhere between 8,000 and 17,000 MTHM. Office of Environmental Management, DOE, *Report to Congress: Status of Environmental Management Initiatives*, 27 (2009).

21. Office of Civilian Radioactive Waste Management, DOE, *Report to the President on the Need for a Second Repository*, 2 (2008).

22. *See, e.g.,* DOE, "Tri-Party Agreement," www.hanford.gov/page.cfm/TriParty. Although DOE has agreed to remove HLW and SNF from all the states, it has only agreed to specified timetables in the cases of Idaho (INL) and Colorado (Fort St. Vrain). Frank Marcinowski, Office of Environmental Management, DOE, "Overview of DOE's Spent Nuclear Fuel and High-Level Waste," presentation, 5, 18 (Mar. 25, 2010), *brc.gov/pdfFiles/Environmental_Management_BRC_03252010.pdf*.

23. NRC, "Operating Nuclear Power Reactors by Location or Name," www.nrc.gov/info-finder/reactor; Office of Civilian Radioactive Waste Management, DOE, *Report to Congress on Demonstration of Interim Storage of Fuel from Decommissioned Reactors*, 1 (2008); Nuclear Energy Study Group, American Physical Society, *Consolidated Interim Storage of Commercial Spent Nuclear Fuel*, 6 tbl.1 (2007).

24. Institute for Twenty-First Century Energy, U.S. Chamber of Commerce, *Revisiting America's Nuclear Waste Policy*, 4, 8 (2009).

25. Nuclear Regulatory Commission, *Pilot Probabilistic Risk Assessment of a Dry Cask Storage System*, 2–3 (2006).

26. Peter Behr, "The Administration Puts Its Own Stamp on a Possible Nuclear Revival," *New York Times* (Feb. 2, 2010).

27. Barbara Miller, "NRC Extends Time That Radioactive Waste Can Be Stored at Nuclear Plants," *Patriot-News* (Mechanicsburg, PA) (Jan. 10, 2011).

28. DOE, *Revised DEIS for Decommissioning West Valley*, 1 (2008).

29. Office of Civilian Radioactive Waste Management, DOE, *Report to the President on the Need for a Second Repository*, 10 (2008).

30. DOE, "U.S. Department of Energy Releases Revised Total System Life Cycle Cost Estimate and Fee Adequacy Report for Yucca Mountain Project," press release (Aug. 5, 2008). The Obama FY 2010 budget virtually eliminated Yucca's funding.

31. David S. Kosson and Charles W. Powers, "The US Nuclear Waste Issue—Solved," *Christian Science Monitor* (Nov. 12, 2008).

32. Kunreuther, "Voluntary Procedures for Siting Noxious Facilities," 344–345 (2006).

33. *Id.*, 354–364.

34. DOE recommended using WIPP as an SNF demonstration site. DOE, *Report of Task Force for Review of Nuclear Waste Management*, 17–18 (1978); Interagency Review Group on Nuclear Waste Management, *Report to the President*, 58 (1979).

35. Sue Vorenberg, "WIPP Wonders," *Santa Fe New Mexican*, A1 (Mar. 29, 2009).

36. Holt, Congressional Research Service, "Nuclear Waste Policy: How We Got Here," presentation, 33 (Mar. 25, 2010), brc.gov/pdfFiles/CRS_BlueRibbonCommissionWastePolicyHistory.pdf (62,500 MTHM); John Kessler, Electric Power Research Institute, "Used Nuclear Fuel: Inventory Projections," presentation to the Blue Ribbon Commission, 7 (Aug. 19, 2010) (65,000 MTHM).

37. There are no guarantees that the rate of generation will hold steady. It assumes, first, that all existing nuclear power plants request and are granted a sixty-year extension of their operating licenses and, second, that no new plants begin operation.

38. Office of Civilian Radioactive Waste Management, DOE, *Report to the President on the Need for a Second Repository*, 12–13 (2008). DOE examined these two options as well as a third, deferring decision on a second repository. DOE notes that Yucca's physical capacity has been determined to be much greater than that allowed to be used for disposal of nuclear wastes under NWPA. *Id.* at 8–9.

39. Office of Civilian Radioactive Waste Management, DOE, *Final Supplemental EIS for Yucca*, S-50 (2008).

40. *Id.* at S-51.

41. Committee on Disposition of High-Level Radioactive Waste through Geological Isolation, National Research Council, *Disposition of HLW and SNF*, 115 (2001); Hoffman and Stacey, "Nuclear and Fuel Cycle Analysis for a Fusion Transmutation of Waste Reactor," 90 (2002).

42. NRC, *Final Update of the Commission's Waste Confidence Decision*, 2 (2010).
43. Nuclear Waste Policy Act of 1982 § 122, 42 U.S.C. § 10142 (2006).
44. *Id.* § 121(b)(1)(B), 42 U.S.C. § 10141(b)(1)(B).
45. 10 C.F.R. § 60.111 (2010); 10 C.F.R. § 63.111 (2010).
46. "Under the Proposed Action, DOE could retrieve emplaced waste for at least 50 years after the start of emplacement." DOE, *Final Supplemental EIS for Yucca Mountain*, 2–5 (2008).
47. Frank Marcinowski, Office of Environmental Management, DOE, "Overview of DOE's Spent Nuclear Fuel and High-Level Waste," presentation, 5, 18 (Mar. 25, 2010), *brc.gov/pdfFiles/Environmental_Management_BRC_03252010.pdf*.
48. NRC, in its updated Waste Confidence Decision in 2008, found that at-reactor storage of SNF in dry casks can be done safely for up to sixty years following expiration of the reactor license, though it was unwilling to make a similar finding for pool storage beyond thirty years. Waste Confidence Decision Update, 73 Fed. Reg. 59,551, at 59,568 (2008); statement of Robert W. Fri, Visiting Scholar, Resources for the Future, *Opportunities and Challenges for Nuclear Power: Hearing before the Committee on Science and Technology*, 100th Cong. 70 (2008). NRC has recently begun to examine the possibility of long-term on-site storage in excess of 120 years. NRC, *Final Update of the Commission's Waste Confidence Decision*, 2 (2010).
49. Institute for Twenty-First Century Energy, U.S. Chamber of Commerce, *Revisiting America's Nuclear Waste Policy*, 4, 8 (2009).
50. Gronlund, Lochbaum, and Lyman, Union of Concerned Scientists, *Nuclear Power in a Warming World*, 47 (2007).
51. Committee on Science and Technology for Countering Terrorism, National Research Council of the National Academies, *Making the Nation Safer*, 47 and tbl.2.1C (2002).
52. Gronlund, Lochbaum, and Lyman, Union of Concerned Scientists, *Nuclear Power in a Warming World*, 47 (2007). A court required NRC to prepare an EIS on terrorist risks for dry-cask storage. *San Luis Obispo Mothers for Peace v. NRC*, 449 F.3d 1016 (2005).
53. *Id.*
54. Tux Turkel, "Spent Fuel Could Stay in Maine for Decade," *Portland Press Herald*, A1 (Mar. 10, 2009); Thomas Burr and Judy Fahys, "After Yucca: America's Homeless Nuclear Waste," *Salt Lake Tribune* (Mar. 21, 2009).
55. Letter from New England Governors' Conference, Inc., to Steven Chu, Secretary, DOE (Dec. 16, 2009). *www.sustainablefuelcycle.com/resources/20091216GovernorsNEGCNuclearWasteLetter.pdf*.
56. *Id.*
57. John Croman, "Minn. Tribe Rallies against Nuke Plant's Expansion," KARE-11 News (Minneapolis/St. Paul) (Nov. 27, 2009); Nina Petersen-Perlman, "Prairie Island Indians Seek Removal of Nuclear Waste," *Minneapolis Star Tribune* (Nov. 1, 2007).
58. *Skull Valley Band of Goshute Indians v. Davis*, No. 07-cv-0526-DME-DON, slip op. at 1 (D. Utah, July 26, 2010).
59. *See, e.g.*, Utah Dept. of Environmental Quality, "High Level Nuclear Waste Storage in Utah," *www.deq.utah.gov/Issues/no_high_level_waste/documents/pdocs/06_PFS_Factsheet.pdf*. Utah also believes that if PFS were to go forward, this would increase the likelihood that DOE will site a second repository at one of the former repository candidate sites located in the state. *Id.*
60. A survey of the public in various communities with former or current nuclear facilities on attitudes toward hosting new nuclear waste facilities found that, "by facility type, respondents at the four older nuclear weapons sites (Hanford, Idaho, Oak Ridge, Savannah River) were the strongest supporters of new waste management facilities." Greenberg, "NIMBY, CLAMP, and the Location of New Nuclear Facilities," 1246–1247 (2009).
61. For a discussion of the doughnut effect, see Greenberg et al., "Preferences for Alternative Risk Management Policies at the United States Major Nuclear Weapons Legacy Sites" (2007).
62. "US GNEP Program Dead, DOE Confirms," *Nuclear Engineering International Magazine* (Apr. 15, 2009), *www.neimagazine.com/story.asp?storyCode=2052719*.
63. *Proposals on Energy Research and Development: Hearing before the Committee on Energy and National Resources*, 111th Cong. 14 (2009) (Testimony of Secretary Stephen Chu).
64. The commission was charged with, among other tasks, considering options for management of SNF that take the "current and potential full fuel cycles into account." Department of Energy, *Blue Ribbon Commission on America's Nuclear Future Charter* (2010), *www.energy.gov/news/8698.htm*. The blue-ribbon commission is not expected to release a draft report until mid-2011; however, it has heard testimony from politicians, industry representatives, and the public on the subject of reprocessing. *See, e.g.*, Rob Pavey, "Don't Make SRS Nuke Dump, Panel Told," *Augusta Chronicle* (Jan. 7, 2011). Materials submitted to the commission can be found at *brc.gov*.
65. International Atomic Energy Agency, *Geological Disposal of Radioactive Waste*, 19 (2009).
66. *Id.* at 19–20.
67. *Id.* at 20.
68. Swedish National Council for Nuclear Waste, *Nuclear Waste Report*, 29 (2010).
69. Radioactive Waste Management Committee, Nuclear Energy Agency, *Reversibility and Retrievability for the Deep Disposal of High-Level Radioactive Waste and Spent Fuel*, 21 (2010).
70. ANDRA, *2009 Activity Report*, 32 (2010).
71. Lehtonen, "Opening up or Closing Down Radioactive Waste Management Policy?" 153–55 (2010).
72. Swedish National Council for Nuclear Waste, *Nuclear Waste Report*, 38 (2010).

73. SKB, *Äspö Hard Rock Laboratory Annual Report 2006*, 83 (2007).
74. Dave Gram, "Dismantling Funds Short," *Brattleboro Reformer* (June 17, 2009). Gram looked at the figures for three reactors, which lost between 7 percent and 12 percent of their decommissioning funds.
75. *Id.*
76. *Id.*; Nuclear Regulatory Commission, "NRC Requests Plans from 18 Nuclear Power Plants to Address Apparent Decommissioning Funding Assurance Shortfalls," press release (June 19, 2009).
77. Rebecca Smith, "NRC Cites Utility Shortfalls," *Wall Street Journal*, B5 (June 20, 2009).
78. Bob Audette, "Bankrupt Nuke Plant Owners Can't 'Walk Away' from Site Clean Up," *Brattleboro Reformer* (Mar. 13, 2009).
79. The blue-ribbon commission's charge is to "conduct a comprehensive review of policies for managing the back end of the nuclear fuel cycle, including all alternatives for the storage, processing, and disposal of civilian and defense used nuclear fuel and nuclear waste." Blue Ribbon Commission on America's Nuclear Future: Memorandum for the Secretary of Energy, 75 Fed. Reg. 5,485, at 5,485 (Feb. 3, 2010).
80. Interagency Review Group on Nuclear Waste Management, *Report to the President*, 35 (1979).
81. If and to the extent that reprocessing of SNF is instituted, a repository would be needed for disposal of the reprocessing wastes, as well as any SNF not reprocessed. Reprocessing proponents would argue that some if not all SNF should be stored until acceptable reprocessing technologies and facilities have been developed.
82. Interagency Review Group on Nuclear Waste Management, *Report to the President*, 9 (1979).
83. Stewart, "Environmental Regulation under Uncertainty" (2002).
84. The authors are indebted to Tom Isaacs, Director of Policy and Planning, Lawrence Livermore National Laboratories, for this formulation and other invaluable counsel.
85. Stewart, "Environmental Regulation under Uncertainty" (2002).
86. NWMO recommended an "adaptive phased management" approach in 2005. Nuclear Waste Management Organization, *Choosing a Way Forward: Management of Canada's Used Nuclear Fuel*, 23–29 (2005). Canada is now implementing adaptive phased management in nuclear waste facility siting. Nuclear Waste Management Organization, *Moving Forward Together: Designing the Process for Selecting a Site*, 10 (2009).
87. Tests at Yucca could involve emplacing, retrievably and on an experimental basis, only a small portion of the statutorily authorized quantity of SNF and HLW. Such testing could be helpful in demonstrating the suitability of tufa geology, and the Yucca site specifically, for permanent disposal of these wastes; if the tests are successful, the experimental phase might be followed by operation at Yucca of a pilot-scale repository. These steps would make it possible to better determine whether a repository can be built and successfully operated for waste disposal at the site, reserving for the future the decision to fill and close the repository. Successful demonstration of the Yucca repository on this model might also ultimately help persuade the public that a permanent HLW/SNF waste disposal option is available, set a precedent for consolidated interim storage, and help reduce opposition to a second repository. These steps might alleviate somewhat the intensity of Nevadans' concerns with and opposition to the site, although, given the politically tainted history of the facility, this outcome would be very hard to vouchsafe. NWPA would not have to be modified to accommodate this result; it provides only that the federal government must begin sending SNF to Yucca in 1998 and does not appear to preclude pilot-scale emplacement to test the repository performance and the retrievability feature required by law. NWPA also requires DOE to seek NRC approval for test emplacements of radioactive material, and conditions NRC approval on the necessity of such emplacement for generating useful data. Nuclear Waste Policy Act of 1982 § 113(c)(2), 42 U.S.C. § 10133(c)(2).
88. *Nevada v. Watkins* 914 F.2d 1545, 1552–1553 (9th Cir. 1990).
89. Under the program, DOE offered $16 million in grants to communities that volunteered to host a GNEP site. Tomero, "The Future of GNEP: Domestic Stakeholders" (Aug. 8, 2008).
90. After DOE submissions, independent technical review, and public comment, EPA recertified WIPP's compliance on November 18, 2010. Criteria for the Certification and Recertification of the Waste Isolation Pilot Plant's Compliance with the Disposal Regulations: Recertification Decision, 75 Fed. Reg. 70,584 (Nov. 18, 2010), www.epa.gov/rpdweb00/wipp/2010recertification.html#fact. As further evidence of WIPP's success, see Sue Vorenberg, "Ten Years in Operation, WIPP Boasts Sterling Safety Record, Continued Support," *Santa Fe New Mexican* (Mar. 29, 2009). For the view that the jury is out on the long-term "success" of WIPP as a permanent disposal site for nuclear wastes, see Dernbach, *Stumbling toward Sustainability*, 499 (2002).
91. For an overview of the experience and practices followed by private developers of hazardous waste facilities and similar projects, see Munton, *Hazardous Waste Siting and Democratic Choice* (1996).
92. Committee on Disposition of High-Level Radioactive Waste through Geological Isolation, National Research Council, *Disposition of HLW and SNF*, 65–66 (2001).
93. WIPP had a strongly supportive local leadership. Carlsbad community leaders initiated and promoted their area as the home for WIPP, and they continued to stand by the project over time. The

local newspaper was also supportive of the project. Carter, *Nuclear Imperatives*, 176–179 (1987).
94. Titus, "Bullfrog County: A Nevada Response," 126–127 (1990).
95. For a discussion of the doughnut effect, see Greenberg et al., "Preferences for Alternative Risk Management Policies at the United States Major Nuclear Weapons Legacy Sites" (2007).
96. SRIC and the New Mexico groups affiliated with it were based in Albuquerque and Santa Fe rather than at Carlsbad. Mora, Sandia National Laboratories, *Sandia and WIPP*, 6–8 (2000).
97. John Tagliabue, "A Town Says 'Yes, in Our Backyard' to Nuclear Site," *New York Times* (Apr. 5, 2010); Kojo, Kari, and Litmanen, "Socio-Economic and Communication Challenges of SNF Management in Finland," 171-172 (2010).
98. Downey, "Politics and Technology in Repository Siting," 54–55 (1985). WIPP employs about eight hundred people in New Mexico. McCutcheon, *Nuclear Reactions*, 12 (2002).
99. Project Plowshares was an Eisenhower administration effort to find peaceful uses for nuclear weapons. Black, *Nature and the Environment in 20th Century American Life*, 117 (2006).
100. Tomero, "The Future of GNEP: Domestic Stakeholders" (Aug. 8, 2008). DOE discusses the receipt of comments in response to a request for indications of interest in hosting GNEP-related research and reprocessing facilities in Notice of Intent to Prepare a Programmatic Environmental Impact Statement for the Global Nuclear Energy Partnership, 72 Fed. Reg. 331, at 333 (Jan. 4, 2007).
101. DOE, "Department of Energy Selects Recipients of GNEP Siting Grants," press release (Nov. 29, 2006).
102. "U.S. Moves to Become Global Nuclear Fuel Supplier," Environment News Service (Jan. 8, 2008), www.ens-newswire.com/ens/jan2007/2007-01-08-05.asp; Idaho National Laboratory, "GNEP: Regional Development Alliance to Receive Grant for Siting Study," www.inl.gov/featurestories/2006-11-30.shtml.
103. Nuclear Waste Policy Act Amendments of 1987 § 402, 42 U.S.C. § 10242 (2006).
104. Gross, "Nuclear Native America," 148 (2001).
105. Holt, Congressional Research Service, "Nuclear Waste Policy: How We Got Here," presentation, 33 (Mar. 25, 2010), brc.gov/pdfFiles/CRS_BlueRibbonCommissionWastePolicyHistory.pdf.
106. The Nuclear Waste Policy Act Amendments of 1987 § 410, 42 U.S.C. § 10250 (2006), stated that, unless reauthorized by Congress, ONWN would expire seven years after the passage of the NWPA amendments. No reauthorization ever occurred.
107. Easterling and Kunreuther, *Dilemma of Siting a Repository*, 72 (1995). As communities moved further along in the investigation and siting process, the potential amount of the grants increased to $200,000 and eventually to $2,800,000. Id. at 72–73.
108. *Id.* at 75; Gross, "Nuclear Native America," 150–151 (2001).
109. Texas Commission on Environmental Quality, Radioactive Material License No. R04100 (2009).
110. C. Ramit Plushnick-Masti, "Commission Lets 36 States Dump Nuke Waste in Texas," *Associated Press* (Jan. 4, 2011).
111. A description of, and materials on, the CRESP approach to engaging local communities in risk assessment are available at CRESP, www.cresp.org.
112. Especially relevant is CRESP's work at Amchitka, Alaska. Powers et al., CRESP, *Amchitka Independent Science Assessment* (2005).
113. For instance, the Department of Energy National Security and Military Applications of Nuclear Energy Authorization Act of 1980 § 213, as enacted, Pub. L. 96-164, 93 Stat. 1259, 1265–1266 (1980), addressed WIPP in its test phase. Construction and full-scale operation of the WIPP repository were addressed in WIPPLWA and the 1996 Amendments to WIPPLWA, respectively. WIPP Land Withdrawal Act of 1992, Pub. L. No. 102-579, 106 Stat. 4777 (1992), amended by Pub. L. No. 104-201, 110 Stat. 2422 (1996).
114. Committee on Disposition of High-Level Radioactive Waste through Geological Isolation, National Research Council of the National Academies, *Disposition of High-Level Radioactive Waste*, 9–10 (1999); Nuclear Waste Management Organization, *Choosing a Way Forward: Management of Canada's Used Nuclear Fuel*, 23–29 (2005); Nuclear Waste Management Organization, *Implementing Adaptive Phased Management: 2010–2014* (2010).
115. Nuclear Waste Management Organization, *Choosing a Way Forward: Management of Canada's Used Nuclear Fuel*, 42–43 (2005).
116. Herman Roser, Assistant Secretary for Defense Programs, DOE, addressed consultation and cooperation requirements proposed for inclusion in NWPA during congressional hearings, testifying based on his experience negotiating with New Mexico on WIPP: "It is imperative that in the development of any repository that we work very closely with the State and be in a position to assure the State that the concerns of the citizens of the area have been adequately addressed and that when the facility, whatever it may be, is completed that their safety or well-being will not be affected." *Hearing before the Subcommittee on Strategic and Theater Nuclear Forces of the Committee on Armed Services*, 97th Cong. 22 (1982).
117. Flynn and Slovic, "Yucca Mountain," 92 (1995).
118. *See Nevada v. Watkins*, 914 F.2d 1545, 1551 (9th Cir. 1990).
119. Canepa, "Implementation of State-Federal Agreements," 27 (1985).
120. International Atomic Energy Agency, "EGG: The Environmental Evaluation Group," www.iaea.org/inisnkm/nkm/ws/d2/r1588.html.
121. John Fleck, "WIPP Watchdog Gets No Reprieve," *Albuquerque Journal* (Apr. 24, 2004).
122. John Tagliabue, "A Town Says 'Yes, in Our Backyard' to Nuclear Site," *New York Times* (Apr. 5, 2010);

Posiva Oy, "Posiva's ONKALO at the Final Disposal Depth," press release (June 17, 2010).
123. Kojo, "The Strategy of Site Selection for the SNF Repository in Finland," 161 (2009).
124. Organisation for Economic Cooperation and Development, *Nuclear Legislation in OECD Countries*, 3.
125. Kojo, "The Strategy of Site Selection for the SNF Repository in Finland," 169–177 (2009).
126. *Id.* at 175–176.
127. *Id.* at 183–184.
128. Kojo, Kari, and Litmanen, *The Socio-Economic and Communication Challenges of SNF Management in Finland*, 172 (2010).
129. Kojo, "The Strategy of Site Selection for the SNF Repository in Finland," 161 (2009).
130. Posiva, "Posiva's ONKALO at the Final Repository Depth," www.posiva.fi/en/news/press_releases/posiva_s_onkalo_at_the_final_disposal_depth.html.
131. Posiva, "General Time Schedule for Final Disposal," www.posiva.fi/en/final_disposal/general_time_schedule_for_final_disposal.
132. SKB, "SKB's Owners and Board of Directors," www.skb.se/Templates/Standard_17436.aspx.
133. Elam and Sundqvist, *Stakeholder Involvement in Swedish Nuclear Waste Management*, 12 (2007).
134. *Id.* at 19.
135. *Id.* at 20.
136. *Id.* at 31.
137. *Id.* at 48.
138. Elam and Sundqvist, "The Swedish KBS Project," 979 (2009).
139. Swedish National Council for Nuclear Waste, *Nuclear Waste State-of-the-Art Report 2010—Challenges for the Final Repository Programme*, Swedish Government Official Reports, SOU 2010:6, 7 (2010).
140. OECD, *Partnering for Long-Term Management of Radioactive Waste—Evolution and Current Practice in Thirteen Countries*, 85–89 (2010), www.nea.fr/pub/ret.cgi?div=RWM#6823.
141. *Id.* at 91.
142. John Tagliabue, "A Town Says 'Yes, in Our Backyard' to Nuclear Site," *New York Times* (Apr. 5, 2010).
143. World Nuclear Association, "Nuclear Power in France," www.world-nuclear.org/info/inf40.html.
144. Nuclear Waste Technical Review Board, *Survey of National Programs for Managing HLW and SNF*, 30 (2009).
145. Nuclear Waste Management Organization, *Choosing a Way Forward: Management of Canada's Used Nuclear Fuel*, 40–41 (2005).
146. *Id.* at 16–19.
147. Nuclear Waste Management Organization, *Moving Forward Together: Designing the Process for Selecting a Site*, 19 (2009).
148. The problems identified in this paragraph are well documented in the following and other sources: Government Accountability Office, *Department of Energy: Further Actions Are Needed to Strengthen Contract Management for Major Projects*, 4–6 (2005); General Accounting Office, *Department of Energy: Fundamental Reassessment Needed to Address Major Mission, Structure, and Accountability Problems*, 2–4 (2001); and General Accounting Office, *Nuclear Waste: Comprehensive Review of the Disposal Program Is Needed*, 4–7 (1994).
149. The Bonneville Power Administration is a federal entity established in 1937 to deliver and sell the power generated at the Bonneville Dam. It has since been designated by Congress to be the wholesale marketing agent for power produced by federally owned dams in the Northwest. It receives no tax revenues or appropriations but pays its employees and expenses out of revenues from the sale of electricity; its rates of sale are at cost. Bonneville Power Administration, "About BPA," www.bpa.gov/corporate/About_BPA.
150. TVA, a federal corporation, was famously established during the New Deal by the Tennessee Valley Authority Act ch. 32, Pub. L. No. 73-17, 48 Stat. 58, 58 (1933). It has since undertaken a variety of energy generation and supply projects, and continues to operate now that wholesale and retail energy markets in the region are divided and regulated differently.
151. The Posiva Oy was founded by Finnish utilities in 1995 and has, with the approval of the Finnish government, conducted research, development, and implementation of a strategy for nuclear waste disposal. Rogers, "Fire in the Hole: Review of National SNF Disposal Policy," 283–284 (2008); Posiva Oy, "Posiva's Operating Policy," www.posiva.fi/en/posiva/posiva_s_operating_policy.
152. Froomkin, "Reinventing the Government Corporation," 557–558, 582 (1995).
153. On Sweden, Rogers, "Fire in the Hole: Review of National SNF Disposal Policy," 285 (2008); Swedish Nuclear Fuel and Waste Management Company, www.skb.se. On Finland, British Nuclear Fuels Limited, "BNFL," www.bnfl.com. On Belgium, Rogers, "Fire in the Hole: Review of National SNF Disposal Policy," 282 (2008); Belgian Agency for Radioactive Waste and Enriched Fissile Materials, "Collaboration," www.nirond.be/engels/9_samenwerking_eng.html. On Germany, Rogers, "Fire in the Hole: Review of National SNF Disposal Policy," 284 (2008). The German Company for the Construction and Operation of Waste Repositories (DBE) was created by the Federal Office for Radiation Protection. DBE, "Activities," www.dbe.de/en/about-dbe/activities-of-dbe/index.php. German utilities hold the majority of shares in DBE. On France, Nuclear Waste Technical Review Board, *Survey of National Programs for Managing HLW and SNF*, 26 (2009).
154. Rogers, "Fire in the Hole: Review of National SNF Disposal Policy," 282–283 (2008); Nuclear Waste Management Organization, "NWMO Board of Directors: Update on Board Appointments" (2007), www.nwmo.ca.
155. Froomkin, "Reinventing the Government Corporation," 560, 594–595, 607–608 (1995).

156. Emmons and Sierra, "Incentives Askew?" 22 (Winter 2004).
157. Interagency Review Group on Nuclear Waste Management, *Report to the President*, 118, 122 (1979).
158. Office of Technology Assessment, *Managing Commercial High-Level Radioactive Waste*, 57 (1982).
159. Advisory Panel on Alternative Means of Financing and Managing Radioactive Waste Facilities, DOE, *Managing Nuclear Waste*, XI-4 (1984).
160. *Id.* at XI-1.
161. DOE Review Group, *Report to the Secretary of Energy on the Conclusions of and Recommendations of the Advisory Panel on Alternative Means of Financing and Managing Radioactive Waste Management Facilities* (1984).
162. Clinton E. Crackel, Nuclear Fuels Reprocessing Coalition, "The United States Nuclear Waste Management Agency," presentation, 7 (May 18, 2010 ed.), *aaenvironment.com/NWMAPowerpoint.ppt*; "US DOE Considers TVA-like Entity for Nuclear Waste Program," *Platts Commodity News* (Feb 28, 2008).
163. Rosenstein and Roy, Program in Arms Control, Disarmament, and International Security, *"Plan D" for Spent Nuclear Fuel*, 3 (2009).
164. *See id.*
165. The Bush administration introduced legislation in 2004 that would have adopted this approach for nuclear generation revenues. H.R. 3981, 108th Cong. (2004).
166. Rebating America's Deposits Act, H.R. 2372, 111th Cong. (2009); Rebating America's Deposits Act, S. 861, 111th Cong. (2009).
167. The White House issued a press release explaining the national security justification for keeping weapons development within the purview of the NRC: "It will continue to serve our national security needs by carrying on AEC's responsibility for the design, development and fabrication of weapons systems for the Department of Defense." Office of the Press Secretary, "Statement by the President on Signing the Energy Reorganization Act of 1974," press release (Oct. 11, 1974).
168. Office of Radiation and Indoor Air, EPA, *Radiation Protection at EPA*, 15 (2000).
169. Flynn and Slovic, "Yucca Mountain," 90 (1995).
170. NRC has jurisdiction over DOE waste disposed at commercial landfills under the Low-Level Radioactive Waste Policy Amendments Act of 1985 § 3(b), 42 U.S.C. § 2021c(b) (2006). For a more complete discussion of LLW disposal, see Chapter 4.
171. Another instance of dual regulation at the federal level is EPA and the Army Corps of Engineers regulation of wetlands development pursuant to Section 404 of the Clean Water Act. *See* Clean Water Act § 404, 33 U.S.C. 1344 (2006).
172. *Nuclear Energy Inst., Inc. v. Envtl. Prot. Agency*, 373 F.3d 1251, 1257 (D.C. Cir. 2004).
173. A potential justification for requiring independent NRC regulation of repositories is that civilian SNF may be stored in them. But the decision for NRC regulation in the 1982 NWPA was made before it was decided that SNF and defense HLW would be co-disposed.
174. Government Accountability Office, *Nuclear Security: Actions Taken by NRC to Strengthen Its Licensing Process for Sealed Radioactive Sources Are Not Effective*, 2–3 (2007).
175. Thomas B. Cochran, Natural Resources Defense Council, "Statement on the Environmental, Safety and Economic Implications of Nuclear Power," No. 06-IEP-1N (California Energy Commission June 28, 2007); Gunter, Beyond Nuclear, *Leak First, Fix Later: Radioactive Releases from Nuclear Power Plants*, 7 (2010); Gronlund, Lochbaum, and Lyman, Union of Concerned Scientists, *Nuclear Power in a Warming World*, 3–4 (2007).
176. Gronlund, Lochbaum, and Lyman, Union of Concerned Scientists, *Nuclear Power in a Warming World*, 17 (2007).
177. North Carolina Waste Awareness and Reduction Network, "No Confidence in NRC: 100s of Citizen Groups Led by UCS' Lochbaum" (Sep. 15, 2003), *www.ncwarn.org/?p=1261*.
178. Kristine L. Svinicki, Commissioner, NRC, "Prepared Remarks at the U.S. WIN and WIN Global 2009 Conference, Washington D.C." (July 22, 2009).
179. Jaffe, "The Effective Limits of the Administrative Process: A Reevaluation," 1113–1119 (1954); Jaffe, *Judicial Control of Administrative Actions*, 11–14 (1965).
180. In its Yucca-specific standards, EPA also established an irregularly shaped compliance boundary, which has been challenged as a gerrymander designed to ensure Yucca's compliance. Thomas B. Cochran, Senior Scientist, Natural Resources Defense Council, "How Safe Is Yucca Mountain?" speech (Jan. 7, 2008).
181. University of Iowa Health Care, "Health Topics: Radiation Exposure—The Facts vs. Fiction," *www.uihealthcare.com/topics/medicaldepartments/cancercenter/prevention/preventionradiation.html*.
182. Committee on Technical Bases for Yucca Mountain Standards, National Research Council of the National Academies, *Technical Bases for Yucca Mountain Standards*, 6, 55 (1995); *Nuclear Energy Inst., Inc. v. Envtl. Prot. Agency*, 373 F.3d at 1267 (D.C. Cir. 2004).
183. Holt, Congressional Research Service, *Nuclear Waste Disposal: Alternatives to Yucca*, 8 (2009); President William Clinton, Message to the Senate Returning without Approval Legislation on Nuclear Waste Policy, 1 Pub. Papers 776 (Apr. 25, 2000).
184. However, the experiment in the 1987 NWPA Amendments, construed in *Nuclear Energy Institute* to give effective regulatory authority to a panel of NAS scientists raises significant concerns. This arrangement undermines NAS's role as an independent, technically expert advisory body, inappropriately makes it a regulator, and creates counterproductive duplication of authority and diffusion of responsibility.

185. Committee on Improving Practices of Regulating and Managing Low Activity Radioactive Waste, National Research Council, *Improving the Regulation and Management of Low Activity Waste*, 44 (2006); Croff, "Risk-Informed Waste Classification" (2006); Garrick, "Contemporary Issues in Risk-Informed Decision Making" (2006); Linsley, "International Standards Related to the Classification and Deregulation of Radioactive Waste" (2006); Rechard, "Performance and Risk Assessment for Radioactive Waste Disposal," 799–801 (1999).
186. These countries include Japan, Sweden, Spain, and France. Y. A. Sokolov, "Opening Address at the Disposal of Low Level Radioactive Waste International Symposium at Cordoba, Spain," speech (Dec. 13, 2004).
187. *Id.* at 39–40.
188. *Id.* at 67.

Appendix C

1. General Accounting Office, *Waste Cleanup*, 42 (2002).
2. Hess, "Hanford," 212, 219 (1996).
3. Shannon Dininny, "Washington, DOE Agree on Hanford Cleanup Deadlines," Associated Press (Aug. 11, 2009).
4. EPA, "Hanford Federal Facility RCRA and TSCA Cleanup Activities," yosemite.epa.gov/R10/OWCM.NSF/webpage/Hanford+Federal+Facility+RCRA+and+TSCA+Cleanup+Activities?OpenDocument#overview.
5. DOE, "Tri-Party Agreement," www.hanford.gov/page.cfm/TriParty.
6. *Id.*, 38–39.
7. *Id.*, 21.
8. *Id.*, 25, 44–45.
9. Resource Conservation and Recovery Act, 42 U.S.C. §6972(a)(1)(A) (2006).
10. Hawkins, "Regulation at Hanford: A Case Study," 123 (2008).
11. *Id.*, 122–123.
12. *See, e.g.*, EPA, "Extension to Dispute Resolution for Hanford Federal Facility Agreement and Consent Decree Milestone M-40-07," www5.hanford.gov/pdw/fsd/AR/FSD0001/FSD0028/D197225601/D197225601_16230_1.pdf; *see also* EPA, "Sixty Day Extension to Dispute Resolution for Hanford Federal Facility Agreement and Consent Decree Milestone M-44-02C" www5.hanford.gov/pdw/fsd/AR/FSD0001/FSD0028/D197225407/D197225407_16212_1.pdf.
13. EPA, "Hanford—Resolution of Dispute" yosemite.epa.gov/r10/cleanup.nsf/9f3c21896330b4898825687b007a0f33/4e51a59b6a088bf68825667a0066db89; EPA, "Sixty Day Extension to Dispute Resolution for Hanford Federal Facility Agreement and Consent Decree Milestone M-44-02C" www5.hanford.gov/pdw/fsd/AR/FSD0001/FSD0028/D197225407/D197225407_16212_1.pdf.
14. Complaint, *Washington v. Dep't of Energy*, No. CV-08-5085-FVS (E.D. Wa. 2008); Chris Gregoire and Rob McKenna, "Washington's Lawsuit over Hanford Cleanup Is Timely and Measured Response to Delays," *Seattle Times* (Dec. 15, 2008).
15. Consent Decree, *Washington v. Dep't of Energy*, No. CV-08-5085-FVS (E.D. Wa. 2009).
16. Mayer et al., *Challenges of Nuclear Weapons Waste* (CRESP Report).
17. Office of Environmental Management, DOE, *Updated Appendices to Status of Environmental Management Initiatives*, 19–20 (2010).
18. Office of Environmental Management, DOE, *Report to Congress: Status of Environmental Management Initiatives*, 79 (2009).
19. DOE estimates that it will be able to finish the River Corridor cleanup twenty years early and $1 billion under budget. *Id.*, 55.
20. DOE, *Updated Appendices to the Status of Environmental Management Initiatives*, 3 app. A (2010); *Id.*, 18–21, 23–30, 71–73.
21. DOE is storing 88 million gallons of liquid HLW in 230 underground tanks located at Hanford, INL, and SRS. Office of Environmental Management, DOE, *Report to Congress: Status of Environmental Management Initiatives*, 23 (2009).
22. Annette Cary, "Blue Ribbon Commission Sees Hanford Waste," *Tri-City Herald* [Kennewick, Washington] (Jul. 15, 2010).
23. *Id.*
24. Office of Environmental Management, DOE, *Report to Congress: Status of Environmental Management Initiatives*, 24 (2009).
25. *Id.*
26. *Id.*
27. *Id.*, 26 (figure of the construction project, 26).
28. Department of Energy, "Energy Secretary Chu, EPA Administrator Jackson, Washington State Governor Gregoire and Oregon Governor Kulongoski Join Elected Officials in Announcing Agreement on New Commitments for Hanford Cleanup," press release (Aug. 11, 2009). When completed, the Hanford Tank Waste Treatment and Immobilization Plant will be the largest vitrification facility in the world. It will treat the fifty-three million gallons of waste currently stored underground at Hanford. The process will take several decades. Office of River Protection, DOE, *High Aluminum HLW Glasses for Hanford's WTP*, 4 (2009).
29. Office of Environmental Management, DOE, *Report to Congress: Status of Environmental Management Initiatives*, 25–26 (2009).
30. *Id.*, 29. DOE wants to wait until around 2015 in order to defray the cost of construction, and additionally for the contingency that treatment technology develops in the meantime. Until then, the low-activity fraction will begin treatment at WTP or continue to be stored on-site in the newer double-hulled tanks.
31. *Id.*, 28.
32. *Id.*, 30.

33. Office of Environmental Management, DOE, *Report to Congress: Status of Environmental Management Initiatives*, 30 (2009).
34. *Id.*, 30–31.
35. These landfills employ disposal trenches covered by a barrier consisting of layers of "three feet of soil/gravel mix, three feet of soil, a six-inch sand filter, a one-foot gravel filter, five feet of fractured basalt, another one-foot gravel layer, six inches of asphalt, and a compacted soil foundation." There are also warning markers above and below ground. Hess, "Hanford," 219–222 (1996).
36. *Id.*, 40. The State of Washington complained that the decision to accept outside wastes was in violation of NEPA, because NEPA requires that "all federal agencies prepare a detailed EIS on every proposal for a major federal action significantly affecting the quality of the human environment." The state further argued that DOE's EIS regarding solid waste disposal at Hanford covered disposal only of wastes generated on-site at Hanford. *See* State of Washington's Amended Complaint, *Washington v. Bodman*, Civil No. 2:03-cv-05018-AAM, 2003 U.S. Dist. Ct. Pleadings 398697, ¶¶ 103–107 (Aug. 19, 2004). The litigation was settled, and the court dismissed the NEPA claims without prejudice. DOE agreed that shipments would be halted pending a new EIS, and also that certain milestones would be legally binding. A copy of the settlement agreement is available at *www.hanford.gov/orp/uploadfiles/settlement-agreement.pdf*.
37. Office of Environmental Management, DOE, *Report to Congress: Status of Environmental Management Initiatives*, 42 (2009).
38. Robert Alvarez, Institute for Policy Studies, *Plutonium Wastes from the U.S. Nuclear Weapons Complex*, 4–5 (2010).
39. *Id.*, 8–9.
40. *Id.*, 109.
41. *Id.*, 110.
42. Office of Environmental Management, DOE, *Report to Congress: Status of Environmental Management Initiatives*, 53 (2009).
43. *Id.*, 55. (Figure 1.11 shows two photos of a reactor, before and after cocooning.)
44. Annette Cary, "Blue Ribbon Commission Sees Hanford Waste," *Tri-City Herald* [Kennewick, Washington] (Jul. 15, 2010).

Bibliography

Statutes: Federal

Act of July 5, 1994, Pub. L. No. 103-272, 108 Stat. 745 (1994).
Act of June 2, 1970, Pub. L. No. 91-273, 84 Stat. 299 (1970).
Act of September 6, 1960, Pub. L. No. 89-670, 80 Stat. 931 (1996).
Act of September 23, 1959, Pub. L. No. 86-373, 73 Stat. 688 (1959).
American Recovery and Reinvestment Act of 2009, Pub. L. No. 111-5, 123 Stat. 115 (2009).
Antarctic Treaty, Dec. 5, 1959, 12 U.S.T. 794, 402 U.N.T.S. 5778.
Atomic Energy Act of 1946, Pub. L. No. 79-585, 60 Stat. 755 (1946).
Atomic Energy Act of 1954, Pub. L. No. 83-703, 68 Stat. 919 (1954), codified at 42 U.S.C. §§ 2011–2297 (2006).
Atomic Energy Act of 1954 Amendments, Pub. L. No. 86-373, 73 Stat. 688 (1959).
Atomic Energy Commission Appropriation Authorization Act, Pub. L. No. 91-273, 84 Stat. 299 (1970).
Clean Air Act, 42 U.S.C. §§ 7401–7515 (2006).
Comprehensive Environmental Response, Compensation, and Liability Act of 1980, 42 U.S.C. §§ 9601–9675 (2006).
Department of Energy Organization Act, Pub. L. No. 95-91, 91 Stat. 565 (1977), codified at 42 U.S.C. §§ 7101–7352 (2006).
Department of Transportation Act, Pub. L. No. 89-670, 80 Stat. 931 (1966).
Duncan Hunter National Defense Authorization Act for Fiscal Year 2009, Pub. L. No. 110-417, 122 Stat. 4356 (2008).
Energy and Water Development Appropriations Act of 1990, Pub. L. No. 101-101, 103 Stat. 641 (1989).
Energy and Water Development Appropriations Act of 1993, Pub. L. No. 102-377, 106 Stat. 1315 (1992).
Energy and Water Development Appropriations Act of 1997, Pub. L. No. 104-206, 110 Stat. 2984 (1996).
Energy Policy Act of 1992, Pub. L. No. 102-486, 106 Stat. 2776 (1992).
Energy Policy Act of 2005, Pub. L. No. 109-58, 119 Stat. 594 (2005).
Energy Reorganization Act of 1974, 42 U.S.C. §§ 5801–5891 (2006).
Federal Facility Compliance Act of 1992, Pub. L. No. 102-386, 106 Stat. 1505 (1992).
Federal Land Policy and Management Act of 1976, 43 U.S.C. §§ 1701–1787 (2006).
Federal Railroad Safety Act of 1970, Pub. L. No. 91-458, 84 Stat. 971 (1970).
Freedom of Information Act of 1966, 5 U.S.C. § 552 (2006).
Hazardous and Solid Waste Amendments of 1984, Pub. L. No. 98-616, 98 Stat. 3221 (1984).
Hazardous Materials Transportation Act, Pub. L. No. 93-633, 88 Stat. 2156 (1975).
Hazardous Materials Transportation Control Act of 1970, Pub. L. No. 91-458, 84 Stat. 977 (1970).
Hazardous Materials Transportation Uniform Safety Act of 1990, Pub. L. No. 101-615, 104 Stat. 3244 (1990).
Implementing Recommendations of the 9/11 Commission Act of 2007, Pub. L. No. 110-53, 121 Stat. 508 (2007) (to be codified in scattered sections of the U.S.C.).
Low-Level Radioactive Waste Policy Act of 1980, 42 U.S.C. §§ 2021b–2021i (2006).
Low-Level Radioactive Waste Policy Amendments Act of 1985, Pub. L. No. 99-240, 99 Stat. 1842 (1986).
Marine Protection, Research, and Sanctuaries Act of 1972, 33 U.S.C. §§ 1401–1445 (2006).
National Defense Authorization Act, Fiscal Year 89, Pub. L. No. 100-456, 102 Stat. 1918 (1988).
National Defense Authorization Act for Fiscal Year 1997, Pub. L. No. 104-201, 110 Stat. 2422 (1997).
National Environmental Policy Act, 42 U.S.C. §§ 4321–4335 (2006).
Norman Y. Mineta Research and Special Improvements Act, Pub. L. No. 108-426, 118 Stat. 2423 (2004), codified at 49 U.S.C. § 108 note (2006).
Nuclear Waste Policy Amendments Act of 1987, Pub. L. No. 100–203, 101 Stat. 1330 (1986).
Nuclear Waste Policy Act of 1982, 42 U.S.C. §§ 10101–10270 (2006).
Nuclear Waste Policy Act of 1996, S. 1936, 104th Cong., 2d Sess. (1996) (unenacted).
Price-Anderson Amendments Act of 1988, Pub. L. No. 100-408, 102 Stat. 1066 (1988).

Private Ownership of Special Nuclear Materials Act, Pub. L. No. 88-489, 78 Stat. 602 (1964).

Reorganization Plan No. 3 of 1970: Environmental Protection Agency, 84 Stat. 2086 (1970).

Resource Conservation and Recovery Act, 42 U.S.C. §§ 6901–6992 (2006).

Ronald W. Reagan National Defense Authorization Act for Fiscal Year 2005, Pub. L. No. 108-375, 118 Stat. 1811 (2005).

Safe Drinking Water Act, 42 U.S.C. §§ 300f–300j-25 (2006).

Superfund Amendments and Reauthorization Act of 1986, Pub. L. No. 99–499, 100 Stat. 1613 (1986).

Texas Low-Level Radioactive Waste Disposal Compact Consent Act § 5, codified at 25 U.S.C. § 2021d note.

Waste Isolation Pilot Plant Land Withdrawal Act, Pub. L. No. 102-579, 106 Stat. 4777 (1992), amended by Pub. L. No. 104-201, 110 Stat. 2422 (1996).

Waste Isolation Pilot Plant Land Withdrawal Amendment Act, enacted as part of National Defense Authorization Act for Fiscal Year 1997, Pub. L. No. 104-201, 110 Stat. 2422, 2851 (1996).

West Valley Demonstration Project Act, Pub. L. No. 96-368, 94 Stat. 1347 (1980).

WIPP Authorization Act, Department of Energy National Security and Military Applications of Nuclear Energy Authorization Act of 1980, Pub. L. No. 96-164, 93 Stat. 1259, 1265 (1979).

Yucca Mountain Repository Site Approval Act, Pub. L. No. 107-200, 116 Stat. 735 (2002).

Statutes: State

AB 2214, 2002 Cal. Legis. Serv. Ch. 513 § 4, codified at Cal. Health and Safety Code § 115261 (2010).

Atlantic Interstate Low-Level Radioactive Waste Compact Implementation Act, 2000 S.C. Acts 357 § 1, codified at S.C. Laws § 48-46-40(A)(6).

H.B. 1567, 2003 Tex. Sess. Law Serv. Ch. 1067, § 1.

Nev. Rev. Stat. § 459.0091 (2009).

Radioactive Materials Act, 1981 N.M. Laws 2121, N.M. STAT. § 74-4A-11.1 (2006).

Radioactive Waste Consultation Act § 4, 1979 N.M. Laws 1636, 1637.

Books, Articles, and Reports

Ackland, Len. *Making a Real Killing: Rocky Flats and the Nuclear West*. Albuquerque: University of New Mexico Press, 1999.

Adams, Marta. "Yucca Mountain: Nevada's Perspective." 46 *Idaho Law Review* 1 (2010).

Alvarez, Robert, Environmental Defense Institute. *Transuranic Wastes at Hanford*. 2009.

Alvarez, Robert, Institute for Policy Studies. *Plutonium Wastes from the U.S. Nuclear Weapons Complex*. 2010.

Alvarez, Robert, Institute for Policy Studies. *Radioactive Wastes and the Global Nuclear Energy Partnership*. 2007.

ANDRA. *2009 Activity Report*. 2010.

Applegate, John S., and Jan G. Laitos. *Environmental Law: RCRA, CERCLA, and the Management of Hazardous Waste*. New York: Foundation Press, 2006.

Barlett, Donald L., and James B. Steele. *Forevermore: Nuclear Waste in America*. New York: W. W. Norton, 1985.

Battelle Memorial Institute. *Waste Solidification Demonstration Program: Experimental Techniques for Characterization of High Level Radioactive Solidified Waste*, BNWL-1425. 1970. (Prepared for Atomic Energy Commission.)

Bell, F. G. *Engineering Geology and Construction*. London: Spon Press, 2004.

Bergeron, Kenneth D. *Tritium on Ice: The Dangerous New Alliance of Nuclear Weapons and Nuclear Power*. Cambridge, MA: MIT Press, 2002.

Berlin, Robert E., and Catherine C. Stanton. *Radioactive Waste Management*. New York: John Wiley and Sons, 1989.

Black, Brian. *Nature and the Environment in Twentieth Century American Life*. Westport, CT: Greenwood Press, 2006.

Blake, E. Michael. "Where Nuclear Reactors Can (and Can't) Be Built." *Nuclear News*, 23 (November 2006).

Bodvarsson, G. S., W. Boyle, R. Patterson, and D. Williams. "Overview of Scientific Investigations at Yucca Mountain: The Potential Repository for High-Level Nuclear Waste." 38 *Journal of Contaminant Hydrology* 3 (1999).

Boffey, Phillip M. "Radioactive Waste Site Search Gets into Deep Water." 190 *Science* 361 (1975).

Boholm, Åsa, and Ragnar Löfstedt, eds. *Facility Siting: Risk, Power, and Identity in Land Use Planning*. Sterling, VA: Earthscan, 2004.

Bonstead, Angela. "EPA's Mixed Approach to Mixed Waste." 8 *Environmental Lawyer* 521 (2002).

Boyle, Elizabeth H. "Political Frames and Legal Activity: The Case of Nuclear Power in Four Countries." 32 *Law and Society Review* 141 (1998).

Brenner, Michael J. *Nuclear Power and Non-Proliferation: The Remaking of U.S. Policy*. New York: Cambridge University Press, 1974.

Breyer, Stephen G., Richard B. Stewart, Cass R. Sunstein, and Adrian Vermeule. *Administrative Law and Regulatory Policy: Problems, Text, and Cases*. 6th ed. Austin, TX: Aspen Publishing, 2006.

Broad, William J. "A Mountain of Trouble." *New York Times Magazine*, 37 (November 18, 1990).

Brody, Julia G., and Judy K. Fleishman. "Sources of Public Concern about Nuclear Waste Disposal in Texas Agricultural Communities." In *Public Reactions to Nuclear Waste: Citizens' Views of Repository Siting*, ed. Riley E. Dunlap, Michael E. Kraft, and Eugene A. Rosa, 115. Durham, NC: Duke University Press, 1993.

Buccino, Sharon. "NEPA under Assault: Congressional and Administrative Proposals Would Weaken Environmental Review and Public Participation." 12 *New York University Environmental Law Journal* 50 (2003).

Bunn, Matthew. *Assessing the Benefits, Costs, and Risks of Near-Term Reprocessing and Alternatives*. Conference paper. Presented at Forty-Seventh Annual Meeting of the Institute for Nuclear Materials Management, Nashville, TN, July 16, 2006.

Bunn, Matthew, Steve Fetter, John Holdren, and Bob van der Zwaan, Belfer Center for Science and International Affairs, Kennedy School of Government. *The Economics of Reprocessing vs. Direct Disposal of Spent Nuclear Fuel*. Cambridge, MA: Belfer Center, 2003. belfercenter.ksg.harvard.edu/publication/2089/economics_of_reprocessing_vs_direct_disposal_of_spent_nuclear_fuel.html?breadcrumb=%2Fpublication%2F806%2Feconomics_of_reprocessing_versus_direct_disposal_of_spent_nuclear_fuel.

Calabresi, Steven, and Kevin Rhodes. "The Structural Constitution: Unitary Executive, Plural Judiciary." 102 *Harvard Law Review* 1153 (1992).

Callan, L. Joseph, Executive Director of Operations, Nuclear Regulatory Commission. *Classification of Hanford Low-Activity Waste Fraction as Incidental*, SECY-97-083. 1997.

Calmet, Dominique P. "Ocean Disposal of Radioactive Waste: Status Report." 31 *IAEA Bulletin* 48 (1989).

Campbell, John L. "The State and the Nuclear Waste Crisis: An Institutional Analysis of Policy Constraints." 34 *Social Problems* 18 (1987).

Campbell, Katherine, Andrew Wolfsberg, June Fabryka-Martin, and Donald Sweetkind. "Chlorine-36 Data at Yucca Mountain: Statistical Tests of Conceptual Models for Unsaturated-zone Flow." 62–63 *Journal of Contaminant Hydrology* 43 (2003).

Campbell, Patricia L., and Jim Haried. "Financing and Licensing Nuclear Energy Parks." 24 *Natural Resources and Environment* 42 (Fall 2009).

Canepa, Joseph F. "Implementation of State-Federal Agreements: Observations and Suggestions from New Mexico." In *Waste Management '85: Waste Isolation in the U.S.—Technical Programs and Public Education*, ed. Roy G. Post, vol. 3. Tucson: Arizona Board of Regents, 1985.

Carter, Luther J. "The Path to Yucca Mountain and Beyond." In *Uncertainty Underground: Yucca Mountain and the Nation's High-Level Nuclear Waste*, ed. Allison M. Macfarlane and Rodney C. Ewing, 381. Cambridge, MA: MIT Press, 2006.

———. "Congressional Committees Ponder Whether to Give States a Right of Veto over Radioactive Waste Repositories." 200 *Science* 1136 (1978).

———. *Nuclear Imperatives and Public Trust: Dealing with Radioactive Waste*. Washington, DC: Resources for the Future, 1987.

———. "Nuclear Wastes: The Science of Geologic Disposal Seen as Weak." 200 *Science* 1135 (1978).

———. "WIPP Goes Ahead, Amid Controversy." 222 *Science* 1104 (1983).

Chaturvedi, Lokesh. *The Karst and Related Issues at the Waste Isolation Pilot Plant*. 2009. www.wipp.energy.gov/library/Karst_Chaturvedi_062309.pdf.

Chuang, Jane. "Who Should Win the Garbage Wars? Lessons from the Low-Level Radioactive Waste Policy Act." 72 *Fordham Law Review* 2403 (2004).

Cochran, Thomas B., National Resources Defense Council. "How Safe is Yucca Mountain?" Conference paper. Remarks at "Symposium on Uncertainty in Long-Term Planning: Nuclear Waste Management, A Case Study." 2008. docs.nrdc.org/nuclear/files/nuc_08010701A.pdf.

———. Statement on "Environmental, Safety, and Economic Implications of Nuclear Power," 06-IEP-[1N]. *Energy Report: Nuclear Power*, 2007 Workshops (June 28, 2007).

Cochran, Thomas B., and Geoffrey H. Fettus. "NRDC's Perspective on the Nuclear Waste Dilemma." 8 *Environmental Law Reporter* 10791 (2010).

Cohen, Bernard L. "The Situation at West Valley." 104 *Public Utilities Fortnightly* 26 (1979).

Cohen, Linda R., Mathew D. McCubbins, and Francis McCall Rosenbluth. "The Politics of Nuclear Power in Japan and the United States." In *Structure and Policy in Japan and the United States*, ed. Peter F. Cowhey and Mathew D. McCubbins, 177. Cambridge, UK: Cambridge University Press, 1995.

Colglazier, E. W. "Evidential, Ethical, and Policy Disputes: Admissible Evidence in Radioactive Waste Management." In *Acceptable Evidence: Science and Values in Risk Management*, ed. Deborah G. Mayo and Rachelle D. Hollander, 137. New York: Oxford University Press, 1991.

Colglazier, E. W., and R. B. Langum. "Policy Conflicts in the Process for Siting Nuclear Waste Repositories." 13 *Annual Review of Energy* 317 (1988).

Cook, Constance E. *Nuclear Power and Legal Advocacy: The Environmentalists and the Courts*. Lexington, MA: Lexington Books, 1980.

Cooper, Benjamin, Senate Committee on Energy and Natural Resources. "Monitored Retrievable Storage." In Roy G. Post, ed., *Waste Management '85: Waste Isolation in the US—Technical Programs and Public Education*, vol. 3. 1985.

Cotton, Thomas A. "Nuclear Waste Story: Setting the Stage." In *Uncertainty Underground: Yucca Mountain and the Nation's High-Level Nuclear Waste*, ed. Allison M. Macfarlane and Rodney C. Ewing, 29. Cambridge, MA: MIT Press, 2006.

Craft, Elena S., Aquel W. Abu-Qare, Meghan M. Flaherty, Melissa C. Garofolo, Heather L. Rincavage, and Mohamed B. Abou-Donia. "Depleted and Natural Uranium: Chemistry and Toxicological Effects." 7 *Journal of Toxicology and Environmental Health* 297 (2004).

Cragin, Susan. *Nuclear Nebraska: The Remarkable Story of the Little County That Couldn't Be Bought*. New York: AMACOM, 2007.

Craig, Paul P. "High-Level Nuclear Waste: The Status of Yucca Mountain." 24 *Annual Review of Energy and the Environment* 461 (1999).

Crawford, Mark. "DOE, States Reheat Nuclear Waste Debate." 230 *Science* 150 (1985).

Croff, Allen G. "Risk-Informed Radioactive Waste Classification and Reclassification." 91 *Health Physics* 449 (2006).

Cynkar, Robert J. "Constitutional Conflicts on Public Land: Dumping on Federalism." 75 *University of Colorado Law Review* 1261 (2004).

Dalrymple, Will. "GNEP Turns to the World." *Nuclear Engineering International* 9 (May 15, 2009).

Davidon, Ann Morrissett. "The U.S. Anti-Nuclear Movement." *Bulletin of the Atomic Scientists* 45 (Dec. 1979).

Dernbach, John, ed. *Stumbling toward Sustainability*. Washington, DC: Environmental Law Institute, 2002.

Donovan, E. Jay. "Kentucky Bills Would End Moratorium on Nuclear Power Plant Construction." Heartland Institute (May 1, 2008). www.heartland.org/policybot/results/23071/Kentucky_Bills_Would_End_Moratorium_on_Nuclear_Power_Plant_Construction.html.

Downey, Gary L. "Politics and Technology in Repository Siting: Military Versus Commercial Wastes at WIPP, 1972–1985." 7 *Technology in Society* 47 (1985).

Dunlap, Riley E., Michael E. Kraft, and Eugene A. Rosa, eds. *Public Reactions to Nuclear Waste: Citizens' Views of Repository Siting*. Durham, NC: Duke University Press, 1993.

Easterling, Douglas. "Fair Rules for Siting a High-Level Nuclear Waste Repository." 11 *Journal of Policy Analysis and Management* 442 (1992).

Easterling, Douglas, and Howard Kunreuther. *The Dilemma of Siting a High-Level Nuclear Waste Repository*. Boston: Kluwer Academic Publishers, 1995.

Edelstein, Michael R. "Hanford: The Closed City and Its Downwind Victims." In *Cultures of Contamination: Legacies of Pollution in Russia and the U.S.*, ed. Michael R. Edelstein, Maria Tysiachniouk, and Lyudmila V. Smirnova, 253. Oxford: Elsevier, 2007.

Electric Power Research Institute. *Spent Nuclear Fuel Transportation: An Overview*. Palo Alto, CA: 2004.

Elam, Mark, and Göran Sundqvist. *Stakeholder Involvement in Swedish Nuclear Waste Management*, SKI Report 2007:2. 2007.

———. "The Swedish KBS Project: A Last Word in Nuclear Fuel Safety Prepares to Conquer the World." 12 *Journal of Risk Research* 969 (2009).

Emmons, William R., and Gregory E. Sierra. "Incentives Askew?" *Regulation Magazine* 22 (Winter 2004).

"Energy Research Agency Replaces AEC." *Science News*, 248 (October 19, 1974).

English, Mary R. *Siting Low-Level Radioactive Waste Disposal Facilities: The Public Policy Dilemma*. Westport, CT: Quorum Books, 1992.

Erickson, Jon D., Duane Chapman, and Ronald E. Johnny. "Monitored Retrievable Storage of Spent Nuclear Fuel in Indian Country: Liability, Sovereignty, and Socioeconomics." 19 *American Indian Law Review* 73 (1994).

Farsetta, Diane. "Nuclear Industry Targets State Laws" (March 27, 2009). www.commondreams.org/view/2009/03/27.

Feiveson, Harold A. "A Skeptic's View of Nuclear Energy." 4 *Daedalus* 60 (2009).

Flynn, James, and Paul Slovic. "Yucca Mountain: A Crisis for Policy—Prospects for America's High-Level Nuclear Waste Program." 20 *Annual Review of Energy and the Environment* 83 (1995).

Flynn, James, Paul Slovic, and C. K. Mertz. "The Nevada Initiative: A Risk Communication Fiasco." 13 *Risk Analysis* 497 (1993).

Froomkin, Michael. "Reinventing the Government Corporation." 3 *University of Illinois Law Review* 543 (1995).

Garrick, B. John. "Contemporary Issues in Risk-Informed Decision Making on the Disposition of Radioactive Waste." 91 *Health Physics* 430 (2006).

Garrick, B. John, and Stan Kaplan. "A Decision Theory Perspective on the Disposal of High-Level Radioactive Waste." 19 *Risk Analysis* 903 (1999).

Gerrard, Michael B. "Fear and Loathing in the Siting of Hazardous and Radioactive Waste Facilities: A Comprehensive Approach to a Misperceived Crisis." 68 *Tulane Law Review* 1047 (1994).

Gershey, Edward L., Robert C. Klein, Esmeralda Party, and Amy Wilkerson. *Low-Level Radioactive Waste: From Cradle to Grave*. New York: Van Nostrand Reinhold, 1990.

Gervers, John H. "The NIMBY Syndrome: Is It Inevitable?" 29 *Environment* 18 (1987).

Gillette, Robert. "Plutonium (I): Questions of Health in a New Industry." 185 *Science* 1031 (1974).

Giugni, Marco. *Social Protest and Policy Change: Ecology, Antinuclear, and Peace Movements in Comparative Perspective*. Lanham, MD: Rowman and Littlefield, 2004.

Gowda, M. V. Rajeev, and Doug Easterling. "Nuclear Waste and Native America: The MRS Siting Exercise." 9 *Risk: Health, Safety, and Environment* 229 (1998).

Greenberg, Michael R. "NIMBY, CLAMP, and the Location of New Nuclear-Related Facilities: U.S. National and 11 Site-Specific Surveys." 29 *Risk Analysis* 1242 (2009).

Greenberg, Michael R., Karen W. Lowrie, Joanna Burger, Charles Powers, Michael Gochfeld, and Henry Mayer. "Nuclear Waste and Public Worries: Public Perceptions of the United States' Major Nuclear Weapons Legacy Sites." 14 *Human Ecology Review* 1 (2007).

———. "Preferences for Alternative Risk Management Policies at the United States Major Nuclear Weapons Legacy Sites." 50 *Journal of Environmental Planning and Management* 187 (2007).

———. "The Ultimate LULU? Public Reaction to New Nuclear Activities at Major Weapons Sites." 73 *Journal of the American Planning Association* 346 (2007).

Greenberg, Michael R., Karen W. Lowrie, Henry J. Mayer, and Bernadette M. West. *The Reporter's Handbook on Nuclear Materials, Energy, and Waste Management*. Nashville: Vanderbilt University Press, 2009.

Greenwood, Ted. "Nuclear Waste Management in the United States." In *The Politics of Nuclear Waste*, ed. E. William Colglazier Jr., 20. New York: Pergamon Press, 1982.

Gronlund, Lisbeth, David Lochbaum, and Edwin Lyman, Union of Concerned Scientists. *Nuclear Power in a Warming World: Assessing the Risks, Addressing the Challenges*. Cambridge, MA: Union of Concerned Scientists, 2007.

Gross, John Karl. "Nuclear Native America: Nuclear

Waste and Liability on the Skull Valley Goshute Reservation," 7 *Boston University Journal of Science and Technology Law* 140 (2001).

Gunter, Paul, Beyond Nuclear. *Leak First, Fix Later: Uncontrolled and Unmonitored Radioactive Releases from Nuclear Power Plants*. 2010. www.beyondnuclear.org/storage/documents/LeakFirst_FixLater_BeyondNuclear_April182010_FINAL.pdf.

Haggerty, Bernard P. "'TRU' Cooperative Regulatory Federalism: Radioactive Waste Transportation Safety in the West." 22 *Journal of Land, Resources, and Environmental Law* 41 (2002).

Halgraves, Robert, and Ralph Mohr. "Liquid Fluoride Thorium Reactors." 98 *American Scientist* 304 (2010).

Hancock, Don. "Symposium: The Nuclear West—Which Road to the Future?" 24 *Journal of Land, Resources, and Environmental Law* 29 (2004).

Hardin, Jason. "Tipping the Scales: Why Congress and the President Should Create a Federal Interim Storage Facility for High-Level Radioactive Waste." 19 *Journal of Land, Resources, and Environmental Law* 293 (1999).

Hawkins, A. R. "Regulation at Hanford: A Case Study." In *Challenges in Radiation Protection and Nuclear Safety Regulation of the Nuclear Industry*, ed. Malgorzata K. Sneve and Mikhail F. Kiselev. New York: Springer, 2008.

Hertzberg, Hendrick. "Some Nukes." *New Yorker*, 19 (March 22, 2010).

Hess, Gerald F. "Hanford: Cleaning Up the Most Contaminated Place in the United States." 38 *Arizona Law Review* 165 (1996).

Hoffman, E. A., and W. M. Stacey. "Nuclear and Fuel Cycle Analysis for a Fusion Transmutation of Waste Reactor." 63–64 *Fusion Engineering and Design* 87 (2002).

Hollister, Charles D., and Steven Nadis. "Burial of Radioactive Waste under the Seabed." *Scientific American* 60 (January 1998).

Hylko, James M. "How to Solve the Used Nuclear Fuel Storage Problem." 152(8) *Power* 58 (August 2008).

"In Comments to Blue Ribbon Panel, N.M. Officials Open to More Waste." 22 *Weapons Complex Monitor*, no. 4–5, 11 (Jan. 28, 2011).

Institute for Twenty-First Century Energy, U.S. Chamber of Commerce. *Revisiting America's Nuclear Waste Policy*. 2009. www.energyxxi.org/reports/Nuclear_Waste_Policy.pdf.

———. *Classification of Radioactive Waste: A Safety Guide*, Safety Series No. 111-G-1.1. 1994.

———. *Geological Disposal of Radioactive Waste: Safety Requirements*, No. WS-R-4. 2006.

International Atomic Energy Agency. *Geological Disposal of Radioactive Waste: Technological Implications for Retrievability*, No. NW-T-1.19. 2009.

———. "Multinational Fuel Bank Proposal Reaches Key Milestone." March 6, 2009. www.iaea.org/NewsCenter/News/2009/fbankmilestone.html.

———. *Predisposal Management of High Level Radioactive Waste*, WS-G-2.6. 2003.

———. *Radioactive Waste Management Profiles No. 9: A Compilation of Data from the Net Enabled Waste Management Database (NEWMDB)*. 2008.

———. *Thorium Fuel Cycle: Potential Benefits and Challenges*, IAEA-TECDOC-1450. 2005.

Jacob, Gerald. *Site Unseen: The Politics of Siting a Nuclear Waste Repository*. Pittsburgh: University of Pittsburgh Press, 1990.

Jaffe, Louis L. "The Effective Limits of the Administrative Process: A Reevaluation." 67 *Harvard Law Review* 1105 (1954).

———. *Judicial Control of Administrative Action*. Boston: Little, Brown, 1965.

Johnson, Jeff. "Yucca Mountain." 80 *Chemical and Engineering News* 20 (July 8, 2002).

Jones, Richard M. "FY 2010 Energy and Water Development Appropriations Bill: Nuclear Waste." *FYI: The AIP Bulletin of Science Policy News* (October 6, 2009). www.aip.org/fyi/2009/119.html.

Joskow, Paul L., and John E. Parsons. "The Economic Future of Nuclear Power." 138 *Daedalus* 45 (2009).

"The Kansas Geologists and the AEC." *Science News*, 161 (March 6, 1971).

Kasperson, Roger E., Gerald Berk, David Pijawka, Alan B. Sharaf, and James Wood. "Public Opposition to Nuclear Energy: Retrospect and Prospect." 5 *Science, Technology, and Human Values* 11 (1980).

Katz, James E. "The Uses of Scientific Evidence in Congressional Policymaking: The Clinch River Breeder Reactor." 9 *Science, Technology, and Human Values* 51 (1984).

Keanneally, Roger M., and John H. Kessler. "Behavior of Spent Fuel and Safety-Related Components in Dry Cask Storage Systems." Conference paper. International Association for Structural Mechanics in Reactor Technology Conference—SMiRT-16. 2001. www.iasmirt.org/SMiRT16/W1309.pdf.

Keating, William Thomas. "Politics, Energy, and the Environment: The Role of Technology Assessment." 19 *American Behavioral Scientist* 37 (1975).

Keeney, Ralph L. "An Analysis of the Portfolio of Sites to Characterize for Selecting a Nuclear Repository." 7 *Risk Analysis* 195 (1987).

Keheley, Tom, Areva. *AREVA Use of CFD in Fuel Assembly Design and Licensing*. Presentation for High-End Computing for Nuclear Fission Science and Engineering, Salt Lake City, UT. February 23, 2006. www.inl.gov/cams/workshops/highendcomputing/d/areva_use_of_cfd_in_fuel_assembly_design_and_licensing.pdf.

Kemp, David D. *Exploring Environmental Issues: An Integrated Approach*. New York: Routledge, Taylor and Francis Group (2004).

Kerr, Richard A. "For Radioactive Waste from Weapons, a Home at Last." 283 *Science* 1626 (1999).

Kessler, John H., and Robin K. McGuire. "Total System Performance Assessment for Waste Disposal Using a Logic Tree Approach." 19 *Risk Analysis* 915 (1999).

Kiefer, Wayne E. "Low-Level Radioactive Waste Issues in Michigan: 1980–2000." 33 *Michigan Academician* 343 (2002).

Kojo, Matti. "The Strategy of Site Selection for the Spent Nuclear Fuel Repository in Finland." In *The Renewal* of *Nuclear Power in Finland*, ed. Matti Kojo and Tapio Litmanen, 161. New York: Palgrave Macmillan, 2009.

Kojo, Matti, Mika Kari, and Tapio Litmanen. "The Socio-Economic and Communication Challenges of Spent Nuclear Fuel Management in Finland." 52 *Progress in Nuclear Energy* 168 (2010).

Kraft, Michael E., and Bruce B. Clary. "Citizen Participation and the NIMBY Syndrome: Public Response to Radioactive Waste Disposal." 44 *Western Political Quarterly* 299 (1991).

Kravets, David. "With No Long-Term Solution, Nuclear Pallbearers Bury Waste in America's Backyard." *Wired* (March 16, 2009). www.wired.com/politics/security/news/2009/03/nuclear.

Kunreuther, Howard. "Voluntary Procedures for Siting Noxious Facilities: Lotteries, Auctions, and Benefit Sharing." In *Hazardous Waste Siting and Democratic Choice*, ed. Don Munton, 338. Washington, DC: Georgetown University Press, 1996.

Lambright, W. Henry. "Changing Course: Admiral James Watkins and the DOE Nuclear Weapons Complex." In *Security in a Changing World: Case Studies in U.S. National Security Management*, ed. Volker C. Franke, 55. Westport, CT: Praeger Paperback, 2002.

Lazarus, Richard J. *The Making of Environmental Law*. Chicago: University of Chicago Press, 2004.

Lefevre, Stephen R. "Trials of Termination: President Carter and the Breeder Reactor Program." 15 *Presidential Studies Quarterly* 330 (1985).

Lehtonen, Markku. "Opening Up or Closing Down Radioactive Waste Management Policy? Debates on Reversibility and Retrievability in Finland, France, and the United Kingdom." 1 *Risk, Hazards and Crisis in Public Policy* 139 (2010).

Leonard, Rebecca. "All Mixed Up about Mixed Waste." 32 *Environmental Lawyer* 219 (2002).

Light, Alfred R. "The Hidden Dimension of the National Energy Plan: Executive Policy Direction in Nuclear Waste Management." 9 *Publius* 169 (1979).

Lindemyer, Jeff. "The Global Nuclear Energy Partnership: Proliferation Concerns and Implications." 16 *Nonproliferation Review* 79 (2009).

Linsley, Gordon. "International Standards Related to the Classification and Deregulation of Radioactive Waste." 91 *Health Physics* 470 (2006).

Lochbaum, David, Union of Concerned Scientists. *U.S. Nuclear Plants in the 21st Century: The Risk of a Lifetime*. 2004.

Lyman, Edwin S., Union of Concerned Scientists. *The Global Nuclear Energy Partnership: Will It Advance Nonproliferation or Undermine It?* 2006.

Macfarlane, Allison. "Underlying Yucca Mountain: The Interplay of Geology and Policy in Nuclear Waste Disposal." 33 *Social Studies of Science* 783 (2003).

Marshall, Elliot. "Clinch River Dies." 222 *Science* 590 (1983).

Mason, J. B., K. Wolf, K. Ryan, S. Roesener, M. Cowen, D. Schmoker, P. Bacala, and B. Landman, THOR Treatment Technologies, LLC. *Steam Reforming Application for Treatment of DOE Sodium Bearing Tank Wastes at Idaho National Laboratory for Idaho Cleanup Project*. 2006. www.inl.gov/technicalpublications/Documents/3303727.pdf.

Mauro, Frank J., and Glenn Yago. "Government Targeting in Economic Development: The New York Experience." 2 *Publius* 63 (Spring 1989).

Mayer, Henry J., Charles W. Powers, Michael Gochfeld, James H. Clarke, Joanna Burger, and Lisa S. Bliss. *Progress and Challenges: The Cost of Cleaning Up the Nation's Nuclear Weapons Legacy Wastes*. Draft manuscript (copy on file with author). 2010.

Mayer, Henry J., Charles W. Powers, Michael Gochfeld, James H. Clarke, Joanna Burger, David S. Kosson, and Lisa S. Bliss. *Challenges of Nuclear Weapons Waste* (CRESP Report). 2010.

Mayo, Deborah G., and Rachelle D. Hollander, eds. *Acceptable Evidence: Science and Values in Risk Management*. New York: Oxford University Press, 1991.

Mazur, Allan, and Beverlie Conant. "Controversy over a Local Nuclear Waste Repository." 8 *Social Studies of Science* 235 (1978).

McCutcheon, Chuck. *Nuclear Reactions: The Politics of Opening a Radioactive Waste Disposal Site*. Albuquerque: University of New Mexico Press, 2002.

Metz, William. "Reprocessing: How Necessary Is It for the Near Term?" 196 *Science* 43 (1977).

Millan, Stan, and Andrew J. Harrison Jr. "A Primer on Hazardous Materials Transportation Law of the 1990s: The Awakening." 22 *Environmental Law Reporter* 10,583 (1992).

Mintz, Joel, A. *Enforcement at the EPA: High Stakes and Hard Choices*. Austin: University of Texas Press, 1995.

Montange, Charles H. "Federal Nuclear Waste Disposal Policy." 27 *Natural Resources Journal* 309 (1987).

———. "The Initial Environmental Assessments for the Nuclear Waste Repository under Section 112 of the Nuclear Waste Policy Act." 4 *UCLA Journal of Environmental Law and Policy* 187 (1985).

Mora, Carl J., Sandia National Laboratories. *Sandia and the Waste Isolation Pilot Plant, 1974–1999*. Presented at Southwest Oral History Association Meeting, Long Beach, CA. April 28, 2000.

Mostaghel, Deborah M. "The Low-Level Radioactive Waste Policy Amendments Act: An Overview." 43 *DePaul Law Review* 379 (1994).

Munton, Don, ed. *Hazardous Waste Siting and Democratic Choice*. Washington, DC: Georgetown University Press, 1996.

Murchison, Kenneth M. "Does NEPA Matter? An Analysis of the Historical Development and Contemporary Significance of the National Environmental Policy Act." 18 *University of Richmond Law Review* 557 (1983).

Murphy, Arthur W., and D. Bruce La Pierre. "Nuclear 'Moratorium' Legislation in the States and the Supremacy Clause: A Case of Express Preemption." 76 *Columbia Law Review* 392 (1976).

Natural Resources Defense Council. *DOE's Nuclear Energy Research Programs Threaten National Security*. 2003. www.nrdc.org/nuclear/bush/freprocessing.asp.

———. *Position Paper: Commercial Nuclear Power*. 2005.

Neill, Helen R., and Robert H. Neill. "Shallow Buried Transuranic Waste: A Comparison of Remediation Alternatives at Los Alamos National Laboratory." 49 *Natural Resources Journal* 151 (2009).

Newberry, William F. "The Rise and Fall and Rise and Fall of American Public Policy on Disposal of Low-Level Radioactive Waste." 3 *South Carolina Environmental Law Journal* 43 (1993).

No! The Coalition against High-Level Nuclear Waste. *White Paper Regarding Opposition to the High-Level Nuclear Waste Storage Facility Proposed by Private Fuel Storage on the Skull Valley Band of Goshute Indian Reservation, Skull Valley, Utah.* 2000.

Noyes, Robert. *Nuclear Waste Cleanup Technology and Opportunities.* Park Ridge, NJ: Noyes Publications, 1995

Nuclear Energy Institute. *Nuclear Waste Disposal for the Future: The Potential of Reprocessing and Recycling.* 2006. www.nei.org/resourcesandstats/documentlibrary/nuclearwastedisposal/whitepaper/reprocessingandrecycling.

———. *Regulatory Framework for Recycling Nuclear Fuel, Enclosure 1.* 2008.

Nuclear Energy Study Group, American Physical Society. *Consolidated Interim Storage of Commercial Spent Nuclear Fuel: A Technical and Programmatic Assessment.* 2007.

Nuclear Waste Management Organization. *Choosing a Way Forward: The Future Management of Canada's Used Nuclear Fuel.* 2005.

———. *Implementing Adaptive Phased Management: 2010–2014.* 2010.

———. *Moving Forward Together: Designing the Process for Selecting a Site.* 2009.

Nukewatch. "Fact Sheet: The Yucca Mountain Nuclear Waste Burial Proposal—Suitability Disqualified." 2008. www.nukewatch.com/Wisconsinsdebate/yuccamountainfactsheet.pdf.

Organisation for Economic Co-operation and Development, Nuclear Energy Agency. *Confidence in the Long-Term Safety of Deep Geological Repositories Radioactive Waste Management: Its Development and Communication.* 1999.

———. *Disposal of Radioactive Waste: Can Long-Term Safety Be Evaluated?* 1991.

———. *Geological Disposal of Radioactive Waste: Review of Developments in the Last Decade.* Paris: OECD Publishing, 1999.

———. *Nuclear Legislation in OECD Countries: Finland—Regulatory and Institutional Framework for Nuclear Activities.* 2010. www.nea.fr/law/legislation/.

———. *Partnering for Long-Term Management of Radioactive Waste—Evolution and Current Practice in Thirteen Countries.* 2010.

———. *Radioactive Waste Repositories and Host Regions: Envisaging the Future Together.* 2010.

———. *Stakeholder Participation in Radiological Decision Making: Processes and Implications—Case Studies for the Third Villigen Workshop.* Villigen, Switzerland. 2003.

Organisation for Economic Co-operation and Development Nuclear Energy Agency, International Atomic Energy Agency. *Uranium 2007: Resources, Production, and Demand.* Paris: OECD Publishing, 2008.

"Passing of the AEC." *Science News,* 55 (January 25, 1975).

Pelham, Ann. "Government Groping with Problem of Atomic Reactor Waste Disposal." 35 *Congressional Quarterly Weekly Report* 2555 (1977).

Pollack, Joshua. "Time for a Test-Ban Bargain." *Bulletin of the Atomic Scientists* (November 30, 2009). www.thebulletin.org/web-edition/columnists/joshua-pollack/time-test-ban-bargain.

Powers, Charles W., Joanna Burger, David Kosson, Michael Gochfeld, and David Barnes, eds., Consortium for Risk Evaluation with Stakeholder Participation. *Amchitka Independent Science Assessment: Biological and Geophysical Aspects of Potential Radionuclide Exposure in the Amchitka Marine Environment.* 2005.

Project on Managing the Atom, John F. Kennedy School of Government. *The Economics of Reprocessing versus Direct Disposal of Spent Nuclear Fuel.* 2003.

Rabe, Barry G., William Gunderson, Hilary Frazier, and John M. Gilroy. "NIMBY and Maybe: Conflict and Cooperation in the Siting of Low-Level Radioactive Waste Disposal Facilities in the United States and Canada." 24 *Environmental Law* 73 (1994).

Rabe, Barry G., William Gunderson, and Peter T. Harbage. "Alternatives to NIMBY Gridlock: Voluntary Approaches to Radioactive Waste Facility Siting in Canada and the United States." In *Hazardous Waste Siting and Democratic Choice,* ed. Don Munton, 84. Washington, DC: Georgetown University Press, 1996.

Rechard, Rob P. "Historical Relationship between Performance Assessment for Radioactive Waste Disposal and Other Types of Risk Assessment." 19 *Risk Analysis* 763 (1999).

Richardson, Gary. "Perspective of a Former Idaho Trout Farmer." 10 *Science for Democratic Action* (Nov. 2001). www.ieer.org/sdafiles/vol_10/10-1/index.html.

Rissmiller, Kent. "Equality of Status, Inequality of Result: State Power and High-Level Radioactive Waste." 23 *Publius* 103 (1993).

Roe, David. *Dynamos and Virgins.* New York: Random House, 1984.

Rogers, Kenneth A. "Fire in the Hole: A Review of National Spent Nuclear Fuel Disposal Policy." 51 *Progress in Nuclear Energy* 281 (2008).

Rosa, Eugene A., Riley E. Dunlap, and Michael E. Kraft. "Prospects for Public Acceptance of a High-Level Nuclear Waste Repository in the United States: Summary and Implications." In *Public Reactions to Nuclear Waste: Citizens' Views of Repository Siting,* ed. Riley E. Dunlap, Michael E. Kraft, and Eugene A. Rosa, 291. Durham, NC: Duke University Press, 1993.

Rosa, Eugene A., and James F. Short. "The Importance of Context in Siting Controversies." In *Facility Siting: Risk, Power, and Identity in Land Use Planning,* ed.

Åsa Boholm, and Ragnar Löfstedt. Sterling, VA: Earthscan, 2004.

Rosenstein, Matthew A., and William R. Roy, eds. *"Plan D" for Spent Nuclear Fuel*. Champaign: Program in Arms Control, Disarmament, and International Security, University of Illinois at Urbana-Champaign, 2009.

Ross, David P. "Yucca Mountain and Reversing the Irreversible: The Need for Monitored Retrievable Storage in a Permanent Repository." 25 *Vermont Law Review* 815 (2001).

Rucker, Dale F., Matthew K. Silva, and Lokesh Chaturvedi. Environmental Evaluation Group. *Performance Assessment Issues to Be Resolved at the Waste Isolation Pilot Plant*. Presented at Eighth Bi-Annual Spectrum Conference, Chattanooga, TN. September 27, 2000.

Runyon, Cheryl L. *Low-Level Radioactive Waste: A Legislator's Guide*. Denver: National Conference of State Legislatures, 1994.

Sachs, Noah. "The Mescalero Apache Indians and Monitored Retrievable Storage of Spent Nuclear Fuel: A Study in Environmental Ethics." 36 *Natural Resources Journal* 881 (1996).

Sadler, Susan J. "Standards for Containers, Tanks, Incinerators, Boilers, and Furnaces." In *The RCRA Practice Manual*, ed. Theodore L. Garrett, 217. Chicago: American Bar Association, 2004.

Schaefer, James. *State Opposition to Federal Nuclear Waste Repository Siting: A Case Study of Wisconsin, 1976–1988*. Green Bay: Center for Public Affairs, University of Wisconsin, 1988.

Shapiro, Saul B. "Citizen Trust and Government Cover-up: Refining the Doctrine of Fraudulent Concealment." 95 *Yale Law Journal* 1477 (1986).

Shoesmith, David W. "Waste Package Corrosion." In *Uncertainty Underground: Yucca Mountain and the Nation's High-Level Nuclear Waste*, ed. Allison M. Macfarlane and Rodney C. Ewing, 301. Cambridge, MA: MIT Press, 2006.

Shropshire, David, Kent Williams, J. D. Smith, and Brent Boore, Idaho National Laboratory. *Advanced Fuel Cycle Economic Sensitivity Analysis*, INL/EXT-06-11947. 2006.

Sigmon, E. Brent. "Achieving a Negotiated Compensation Agreement in Siting: The MRS Case." 6 *Journal of Policy Analysis and Management* 170 (1987).

Silva, Matthew K., Dale F. Rucker, and Lokesh Chaturvedi. "Resolution of the Long-Term Performance Issues at the Waste Isolation Pilot Plant." 19 *Risk Analysis* 1003 (1999).

Skibine, Alex Tallchief. "High Level Nuclear Waste on Indian Reservations: Pushing the Tribal Sovereignty Envelope to the Edge?" 21 *Journal of Land, Resources, and Environmental Law* 287 (2001).

Socolow, Robert H., and Alexander Glaser. "Balancing Risks: Nuclear Energy and Climate Change." 138 *Daedalus* 31 (2009).

Socolow, Robert H., and Stephen W. Pacala. "Stabilization Wedges: Solving the Climate Change Problem for the Next Fifty Years with Current Technologies." 305 *Science* 968 (2004).

Solomon, Barry. "High-Level Radioactive Waste Management in the USA." 12 *Journal of Risk Research* 1009 (2009).

Sovacool, Benjamin K. "Valuing the Greenhouse Gas Emissions from Nuclear Power: A Critical Survey." 36 *Energy Policy* 2950 (2008).

Stacy, Susan M. *Proving the Principle: A History of the Idaho National Engineering and Environmental Laboratory, 1949–1999*. Washington, DC: U.S. Government Printing Office, 2000.

Stahl, David. "Drip Shield and Backfill." In *Uncertainty Underground: Yucca Mountain and the Nation's High-Level Nuclear Waste*, ed. Allison M. Macfarlane and Rodney C. Ewing, 301. Cambridge, MA: MIT Press, 2006.

Stewart, Richard B. "Environmental Regulation Under Uncertainty." 10 *Research in Law and Economics* 71 (2002).

———. "Solving the U.S. Nuclear Waste Dilemma." 8 *Environmental Law Reporter* 10783 (2010).

Stucker, Jan Collins. "Nuclear White Elephant: Attempted Revival of a Lost Cause." *New Republic*, 15 (January 20, 1982).

Sun, Marjorie. "Radwaste Dump WIPPs Up a Controversy." 215 *Science* 1483 (1982).

Swainston, Harry W. "The Characterization of Yucca Mountain: The Status of the Controversy." 2 *Federal Facilities Environmental Journal* 151 (1991).

Swan, R. J., and M. E. Lakes, Fluor, Inc. *Challenges with Retrieving Transuranic Waste from the Hanford Burial Grounds*, HNF-32573-FP. 2007. (Prepared for Assistant Secretary for Environmental Management, Department of Energy.)

Thies, Daniel. "Recent Development: The Decline of the Court of Federal Claims in *Nebraska Public Power District v. United States*." 33 *Harvard Journal of Law and Public Policy* 1203 (2010).

Thompson, Stuart C. "The Hazardous Materials Transportation Act: Chemicals at Uncertain Crossroads." 15 *Transportation Law Journal* 411 (1987).

Titus, A. Costandina. "Bullfrog County: A Nevada Response to Federal Nuclear-Waste Disposal Policy." 20 *Publius* 123 (1990).

———. "Governmental Responsibility for Victims of Atomic Testing: A Chronicle of the Politics of Compensation." 8 *Journal of Health Politics, Policy, and Law* 277 (1983).

Tomero, Leonor. "The Future of GNEP: Domestic Stakeholders." *Bulletin of the Atomic Scientists* (August 8, 2008). www.thebulletin.org/web-edition/reports/the-future-of-gnep/the-future-of-gnep-domestic-stakeholders.

Tugwell, Franklin. *The Energy Crisis and the American Political Economy*. Stanford, CA: Stanford University Press, 1988.

Union of Concerned Scientists. *Reprocessing and Nuclear Wastes Fact Sheet: Reprocessing Would Increase Total*

Volume of Radioactive Waste. 2009. www.ucsusa.org/assets/documents/nwgs/reprocessing-and-nuclear.pdf.

Urban, Christopher J. "EPA's Hazardous Waste Identification Rule for Process Waste Gone Haywire, Again." 9 *Villanova Environmental Law Journal* 99 (1998).

Urban Environmental Research, LLC. *Lessons Learned from New Mexico's Experience with the Development of a Nuclear Waste Repository*. 2001. (Prepared for Nuclear Waste Division, Department of Comprehensive Planning, Clark County, New Mexico.)

Van Ness Feldman, P.C. *Federal Commitment Regarding Used Fuel and High-Level Wastes*. 2010. (Prepared for Blue Ribbon Commission on America's Nuclear Future.)

Vari, A., Patricia Reagan-Cirincione, and J. L. Mumpower. *LLRW Disposal Facility Siting: Successes and Failures in Six Countries*. Berlin: Springer, 1994.

Von Hippel, Frank, International Panel on Fissile Materials. *Managing Spent Fuel in the United States: The Illogic of Reprocessing*. Princeton: International Panel on Fissile Materials, 2007.

Walker, J. Samuel. *Containing the Atom: Nuclear Regulation in a Changing Environment, 1963–1971*. Berkeley: University of California Press, 1992.

———. *Permissible Dose: A History of Radiation Protection in the Twentieth Century*. Berkeley: University of California Press, 2000.

———. *The Road to Yucca Mountain: The Development of Radioactive Waste Policy in the United States*. Berkeley: University of California Press, 2009.

———. *Three Mile Island: A Nuclear Crisis in Historical Perspective*. Berkeley: University of California Press, 2004.

Wall, Annemarie. "Going Nowhere in the Nuke of Time: Breach of the Yucca Contract, Nuclear Waste Policy Act Fallout and Shelter in Private Interim Storage." 12 *Albany Law Environmental Outlook Journal* 138 (2007).

Werner, James D. "Radioactive Waste." In *Stumbling toward Sustainability*, ed. John C. Dernbach, 479. Washington, DC: Environmental Law Institute, 2002.

Wernicke, Brian, James L. Davis, Richard A. Bennett, Pedro Elósegui, Mark J. Abolins, Robert J. Brady, Martha A. House, Nathan A. Niemi, and J. Kent Snow. "Anomalous Strain Accumulation in the Yucca Mountain Area, Nevada." 279 *Science* 2096 (1998).

Winograd, Isaac J. "Radioactive Waste Disposal in Thick Unsaturated Zones." 212 *Science* 1457 (1981).

Wisconsin Legislative Reference Bureau. *Rethinking the Moratorium on Nuclear Energy*, Brief 06-7. 2006. legis.wisconsin.gov/lrb/pubs/wb/06wb7.pdf.

Wohlstetter, Albert. "Spreading the Bomb without Quite Breaking the Rules." 25 *Foreign Policy* 88 (Winter 1976).

World Nuclear Association. *Mixed Oxide (MOX) Fuel*. 2009. www.world-nuclear.org/info/inf29.html.

———. *Processing of Used Nuclear Fuel*. 2010. www.world-nuclear.org/info/inf69.html.

———. *Radioactive Waste Management: Nuclear Waste Disposal*. 2009. www.world-nuclear.org/info/inf04.html.

Yablokov, Alexey V., Vassily B. Nesterenko, and Alexey V. Nesterenko. *Chernobyl: Consequences of the Catastrophe for People and the Environment*. New York: New York Academy of Sciences, 2009.

Yates, Richard F. "Preemption under the Atomic Energy Act of 1954: Permissible State Regulation of Nuclear Facilities' Location, Transportation of Radioactive Materials, and Radioactive Waste Disposal," 11 *Tulsa Law Journal* 397 (1976).

Yellin, Joel. "The Nuclear Regulatory Commission's Reactor Safety Study." 7 *Bell Journal of Economics* 317 (1976).

Zeller, E. J., D. F. Saunders, and E. E. Angino. "Putting Radioactive Waste on Ice: A Proposal for an International Radionuclide Depository in Antarctica." *Bulletin of the Atomic Scientists* 4 (January 1973).

Zinberg, Dorothy. "The Public and Nuclear Waste Management." *Bulletin of the Atomic Scientists* 34 (January 1979).

Federal and State Agency Documents

Advisory Committee on Nuclear Waste, Nuclear Regulatory Commission. *History and Framework of Commercial Low-Level Radioactive Waste Management in the United States*, NUREG-1853. 2007.

Advisory Panel on Alternative Means of Financing and Managing Radioactive Waste Facilities, Department of Energy. *Managing Nuclear Waste—A Better Idea: A Report to the U.S. Secretary of Energy*. 1984.

Allen, Lawrence E., Matthew K. Silva, and James K. Channell, Environmental Evaluation Group. *Identification of Issues Relevant to the First Recertification of WIPP*, EEG-83. 2002.

Anderson, Roger D. *Report to Sandia Laboratories on Deep Dissolution of Salt, Northern Delaware Basin, New Mexico*. 1978.

Andrews, Anthony, Congressional Research Service. *Spent Nuclear Fuel Storage Locations and Inventory*, RS22001. 2004.

Army Environmental Policy Institute. *Health and Environmental Consequences of Depleted Uranium Use in the U.S. Army: Technical Report*. 1995.

Board on Energy and Environmental Systems, National Research Council of the National Academies. *Review of DOE's Nuclear Energy Research and Development Program*. 2008.

Board on Radioactive Waste Management, National Research Council of the National Academies. *One Step at a Time: The Staged Development of Geologic Repositories for High-Level Radioactive Waste*. 2006.

———. *Safety and Security of Commercial Spent Nuclear Fuel Storage: Public Report*. 2006.

Bredehoeft, J. D., A. W. England, D. B. Stewart, N. J. Trask, and I. J. Winograd, U.S. Geological Survey. *Geologic Disposal of High-Level Radioactive Wastes—Earth Sciences Perspectives*, CIRC-799. 1978.

Buck, Alice L., Department of Energy. *A History of the Atomic Energy Commission*, DOE/ES-0003/1. 1983.

Carlsbad Area Office, Department of Energy. *Pioneering Nuclear Waste Disposal*, DOE/CAO-00-3124. 2000.

Chalmers, James, Doug Easterling, James Flynn, Catherine Fowler, John Gervers, Robert Halstead, Roger Kasperson et al., Nevada Agency for Nuclear Projects, Office of the Governor. *State of Nevada Socioeconomic Studies of Yucca Mountain 1986–1992: An Annotated Guide and Research Summary*, NWPO-SE-056-93. 1993.

Channell, James K., John C. Rodgers, and Robert H. Neill, Environmental Evaluation Group. *Adequacy of TRUPACT-I Design for Transporting Contact-Handled Transuranic Wastes to WIPP*, EEG-33. 1986.

Channell, James K., and Ben A. Walker, Environmental Evaluation Group. *Evaluation of Risks and Waste Characterization Requirements for the Transuranic Waste Emplaced in WIPP during 1999*, EEG-75, DOE/AL58309-75. 2000.

Chaturvedi, Lokesh, and James K. Channell. *The Rustler Formation as a Transport Medium for Contaminated Groundwater*, EEG-32. 1985.

Collins, T. E., and G. Hubbard, Nuclear Regulatory Commission. *Technical Study of Spent Fuel Pool Accident Risk at Decommissioning Nuclear Power Plants*, NUREG-1738. 2001.

Commission on Geosciences, Environment, and Resources, National Research Council of the National Academies. *Disposition of High-Level Radioactive Waste through Geological Isolation: Development, Current Status, and Technical and Policy Challenges.* Discussion paper prepared for a workshop at the Arnold and Mabel Beckman Center of the National Academies, Irvine, California, November 4–5, 1999.

———. *Nuclear Wastes: Technologies for Separation and Transmutation.* 1996.

———. *Radioactive Waste Repository Licensing: Synopsis of a Symposium Sponsored by the Board on Radioactive Waste Management.* 1992.

———. *Rethinking High-Level Radioactive Waste Disposal: A Position Statement of the Board on Radioactive Waste Management.* 1990.

Commission on Natural Resources, National Research Council of the National Academies. *Radioactive Waste Management at the Savannah River Plant: A Technical Review.* 1981.

Committee on Alternatives for Controlling the Release of Solid Materials from Nuclear Regulatory Commission–Licensed Facilities, National Research Council of the National Academies. *The Disposition Dilemma: Controlling the Release of Solid Materials from Nuclear Regulatory Commission–Licensed Facilities.* 2002.

Committee on Disposition of High-Level Radioactive Waste through Geological Isolation, National Research Council of the National Academies. *Disposition of High-Level Waste and Spent Nuclear Fuel: The Continuing Societal and Technical Challenges.* 2001.

Committee on End Points for Spent Nuclear Fuel and High-Level Radioactive Waste in Russia and the United States, National Research Council of the National Academies. *End Points for Spent Nuclear Fuel and High-Level Radioactive Waste in Russia and the United States.* 2003.

Committee on Geologic Aspects of Radioactive Waste Disposal, National Research Council of the National Academies. *Report to the Division of Reactor Development and Technology, United States Atomic Energy Commission.* 1966.

Committee on Improving Practices for Regulating and Managing Low-Activity Radioactive Waste, National Research Council of the National Academies. *Improving the Regulation and Management of Low-Activity Radioactive Wastes.* 2006.

Committee on the Management of Certain Radioactive Waste Streams Stored in Tanks at Three Department of Energy Sites, National Research Council of the National Academies. *Tank Waste Retrieval, Processing, and On-Site Disposal at Three Department of Energy Sites: Final Report.* 2006.

Committee on Review of DOE's Nuclear Energy Research and Development Program, National Research Council of the National Academies. *Review of DOE's Nuclear Energy Research and Development Program*, BEES-J-05-01-A. 2008.

Committee on Risk-Based Approaches for Disposition of Transuranic and High-Level Radioactive Waste, National Research Council of the National Academies. *Risk and Decisions about Disposition of Transuranic and High-Level Radioactive Waste.* 2005.

Committee on Science and Technology for Countering Terrorism, National Research Council of the National Academies. *Making the Nation Safer: The Role of Science and Technology in Countering Terrorism.* 2002.

Committee on Separations Technology and Transmutation Systems, National Research Council of the National Academies. *Nuclear Wastes: Technologies for Separation and Transmutation.* 1996.

Committee on Technical Bases for Yucca Mountain Standards, National Research Council of the National Academies. *Technical Bases for Yucca Mountain Standards.* 1995.

Committee on Transportation of Radioactive Waste, National Research Council of the National Academies. *Going the Distance? The Safe Transport of Nuclear Fuel and High-Level Radioactive Waste.* 2006.

Committee on Waste Disposal, National Research Council of the National Academies. *The Disposal of Radioactive Waste on Land.* 1957.

Committee on the Waste Isolation Pilot Plant, National Research Council of the National Academies. *Improving Operations and Long-Term Safety of the WIPP: Final Report.* 2001.

———. *The Waste Isolation Pilot Plant: A Potential Solution for the Disposal of Transuranic Waste.* 1996.

Department of Energy. *A Monitored Retrievable Storage Facility: Technical Background Information*, DOE/RW-0311P. 1991.

———. *Draft Environmental Impact Statement: Waste Isolation Pilot Plant*, DOE/EIS-0026-D. 1979.

———. *Draft Title 40 CFR 191 Compliance Certification*

Application for the Waste Isolation Pilot Plant, DRAFT-DOE/CAO-2056. 1995.

———. *Final Environmental Impact Statement for a Geologic Repository for the Disposal of Spent Nuclear Fuel and High-Level Radioactive Waste at Yucca Mountain, Nye County, Nevada*, DOE-EIS-0250. 2002.

———. *Final Environmental Impact Statement for a Rail Alignment for the Construction and Operation of a Railroad in Nevada to a Geologic Repository at Yucca Mountain, Nye County, Nevada*, DOE/EIS-0369. 2008.

———. *Final Environmental Impact Statement: Management of Commercially Generated Radioactive Waste*, vol. 1, DOE/EIS-0046F. 1980.

———. *Final Environmental Impact Statement: Waste Isolation Pilot Plant*, DOE-EIS-0026. 1980.

———. *Final Supplement Environmental Impact Statement: Waste Isolation Pilot Plant*, DOE-EIS-0026-FS. 1990.

———. *Final Supplemental Environmental Impact Statement for a Geologic Repository for the Disposal of Spent Nuclear Fuel and High-Level Radioactive Waste at Yucca Mountain, Nye County, Nevada*, DOE/EIS-0250F-S1. 2008.

———. *Generation IV Nuclear Energy Systems Initiative*. 2006. www.nuclear.energy.gov/pdfFiles/GENIV.pdf.

———. *Implementation Guide for Use with DOE M 435.1-1, DOE G 435.1-1*. 1999.

———. *National Transportation Program: Transporting Radioactive Materials—Answers to Your Questions*. 1999.

———. *National TRU Waste Management Plan*, Revision 3, DOE/NTP-96-1204. 2002.

———. *Plutonium Recovery from Spent Fuel Reprocessing by Nuclear Fuel Services at West Valley, New York, from 1966 to 1972*. 1996. www.osti.gov/opennet/document/purecov/nfsrepo.html.

———. *Radioactive Waste Management Manual*, DOE M 435.1-1. 1999.

———. *Reassessment of the Civilian Radioactive Waste Management Program: Report to the Congress by the Secretary of Energy*, DOE/RW-0247. 1989.

———. *Recommendation by the Secretary of Energy Regarding the Suitability of the Yucca Mountain Site for a Repository under the Nuclear Waste Policy Act of 1982*. 2002. www.energy.gov/media/Secretary_s_Recommendation_Report.pdf.

———. *Recommendations for the Management of Greater-Than-Class-C Low-Level Radioactive Waste*, DOE/NE-077. 1987.

———. *Record of Decision: Program of Research and Development for Management and Disposal of Commercially Generated Radioactive Wastes*, DOE/EIS-0046-D. 1981.

———. *The Remote-Handled Transuranic Waste Program*. 2007. www.wipp.energy.gov/fctshts/RH_TRU.pdf.

———. *Report of Task Force for Review of Nuclear Waste Management*, DOE ER-0004-D (Draft). 1978.

———. *Report to the Secretary of Energy on the Conclusions and Recommendations of the Advisory Panel on Alternative Means of Financing and Managing (AMFM) Radioactive Waste Management Facilities*. 1985.

———. *Revised Draft Environmental Impact Statement for Decommissioning and/or Long-Term Stewardship at the West Valley Demonstration Project and Western New York Nuclear Service Center*, DOE/EIS-0226-D (Revised). 2008.

———. *Rocky Flats History: Beyond the Buildings*. 2003. www.lm.doe.gov/WorkArea/linkit.aspx?LinkIdentifier=id&ItemID=3026.

———. *Savannah River Site High-Level Waste Tank Closure Final Environmental Impact Statement*, DOE/EIS-0303. 1994.

———. *Spent Nuclear Fuel Recycling Program Plan: Report to Congress*. 2006.

———. *Statement for Use of Dedicated Trains for Waste Shipments to Yucca Mountain*. 2005.

———. *Supplement Analysis for Transportation of Transuranic Waste from the Mound Plant to Savannah River Site for Storage, Characterization, and Repackaging*, DOE/EIS-0200-SA02. 2001.

———. *Tank Waste System Integrated Project Team Feasibility Report*, SRNL-TR-2009-00313. 2009.

———. *Title 40 CFR Part 191 Subparts B and C Compliance Recertification Application for the Waste Isolation Pilot Plant*, DOE/WIPP 09-3424. 2009.

———. *Viability Assessment of a Repository at Yucca Mountain*, DOE/RW-0508. 1998.

———. *Waste Isolation Pilot Plant Disposal Phase Final Supplemental Environmental Impact Statement*, DOE/EIS-0026-FS2. 1997.

———. *Waste Isolation Pilot Plant Hazardous Waste Facility Permit Renewal Application*. 2009. www.wipp.energy.gov/library/rcrapermit/final_renewal_application_9_09.htm.

———. *Waste to Be Consolidated at Idaho Site before Shipment to WIPP*. 2008. www.wipp.energy.gov/fctshts/Consolidation.pdf.

———. *Yucca Mountain Repository License Application for Construction Authorization*. 2008.

Department of Transportation. *Guidelines for Selecting Preferred Highway Routes for Highway Route Controlled Quantity Shipments of Radioactive Materials*, DOT/RSPA/HMS/92-02. 1992.

Energy Information Agency, Department of Energy. *Federal Financial Interventions and Subsidies in Energy Markets 2007*, SR/CNEAF/2008-01. 2008.

Environmental Evaluation Group. *Radiological Health Review of the Final Environmental Impact Statement, Waste Isolation Pilot Plant*, DOE/EIS-0026. 1981.

———. *Review of the Draft Supplement Environmental Impact Statement, DOE Waste Isolation Pilot Plant* DOE/EIS-0026-DS, EEG-41. 1989.

———. *Review of the WIPP Draft Application to Show Compliance with EPA Transuranic Waste Disposal Standards*, DOE/AL 58309-61. 1996.

Environmental Protection Agency. *Environmental Impact Statement Comments: Management of Commercial High-Level and Transuranium-Contaminated Radioactive Waste*, D-AEC-A00107-00. 1974.

———. *RCRA Orientation Manual*, EPA 530-R-07-010. 2008.

———. *Technical Support Document for Section 194.14/15—Evaluation of Karst at the WIPP Site.* 2006.

Environmental Protection Agency, Department of Energy, and State of Idaho. *Federal Facility Agreement and Consent Order*, 1088-06-29-120. 1991.

Esh, David W., Anna H. Bradford, Kristina L. Banovac, and B. Jennifer Davis, Nuclear Regulatory Commission. *Risks and Uncertainties Associated with High-Level Waste Tank Closure.* 2002.

Flynn, James H., C. K. Mertz, and Paul Slovic, Nevada Agency for Nuclear Projects, Office of the Governor. *Yucca Mountain Socioeconomic Project—The 1991 Nevada State Telephone Survey: Key Findings*, NWPO-SE-036-91. 1991.

———. *Yucca Mountain Socioeconomic Project—The Autumn, 1993 Nevada State Telephone Survey: Key Findings*, NWPO-SE-058-94. 1994.

———. *Yucca Mountain Socioeconomic Project—The Spring, 1993 Nevada State Telephone Survey: Key Findings*, NWPO-SE-057-93. 1993.

———. *Yucca Mountain Socioeconomic Project—The 1994 Nevada State Telephone Survey: Key Findings*, NWPO-SE-062-94. 1994.

Fultz, Keith O., Senior Associate Director, General Accounting Office. *Status of the Department of Energy's Waste Isolation Pilot Plant*, GAO/T-RCED 88-63. 1988.

General Accounting Office. *Institutional Relations under the Nuclear Waste Policy Act of 1982*, RCED-87-14. 1987.

———. *Issues Related to the Closing of the Nuclear Fuel Services, Incorporated, Reprocessing Plant at West Valley, New York Valley Plant*, EMD-77-27. 1977.

———. *The Liquid Metal Fast Breeder Reactor: Options for Deciding Future Pace and Direction*, GAO/EMD-82-79. 1982.

———. *Nuclear Waste: Comprehensive Review of the Disposal Program Is Needed*, GAO/RCED-94-299. 1994.

———. *Nuclear Waste: Quarterly Report on DOE's Nuclear Waste Program as of December 31, 1986*, GAO/RCED-87-95FS. 1987.

———. *Nuclear Waste: Quarterly Report on DOE's Nuclear Waste Program as of June 30, 1986*, GAO/RCED-86-206FS. 1986.

———. *Nuclear Waste: Repository Work Should Not Proceed until Quality Assurance Is Adequate*, GAO/RCED-88-159. 1988.

———. *Nuclear Waste: Status of DOE's Implementation of the Nuclear Waste Policy Act*, RCED-87-17. 1987.

———. *Radioactive Waste: Status of Commercial Low-Level Waste Facilities*, GAO/RCED-95-67. 1995.

———. *Spent Nuclear Fuel: Options Exist to Further Enhance Security*, GAO-03-426. 2003.

———. *Status of the Department of Energy's Implementation of the Nuclear Waste Policy Act of 1982 as of September 30, 1984*, GAO/RCED-85-42. 1984.

———. *U.S. Fast Breeder Reactor Program Needs Direction*, GAO/EMD-80-81. 1980.

———. *Waste Cleanup: Status and Implications of DOE's Compliance Agreements*, GAO-02-567. 2002.

Gilbertson, Mark A., Deputy Assistant Secretary for Engineering and Technology, Department of Energy. "Energy Parks Initiative: Leveraging Assets to Increase the Taxpayer's Return on Investment." Presentation at Public Information Workshop: Energy Parks Concept at SRS, August 18, 2009. sro .srs.gov/energypark.htm.

Global Nuclear Energy Partnership. Steering Group Action Plan. 2007.

Government Accountability Office. *Department of Energy: Fundamental Reassessment Needed to Address Major Mission, Structure, and Accountability Problems.* 2001.

———. *Department of Energy: Further Actions Are Needed to Strengthen Contract Management for Major Projects.* 2005.

———. *Global Nuclear Energy Partnership: DOE Should Reassess Its Approach to Designing and Building Spent Nuclear Fuel Recycling Facilities*, GAO-08-483. 2008.

———. *Low-Level Radioactive Waste: Disposal Availability Adequate in the Short Term, but Oversight Needed to Identify Any Future Shortfalls*, GAO-04-604. 2004.

———. *Low-Level Radioactive Waste: Status of Disposal Availability in the United States and Other Countries*, GAO-08-813T. 2008

———. *Low-Level Radioactive Waste Management: Approaches Used by Foreign Countries May Provide Useful Lessons for Managing U.S. Radioactive Waste*, GAO-07-221. 2007.

———. *Nuclear Security: Actions Taken by NRC to Strengthen Its Licensing Process for Sealed Radioactive Sources Are Not Effective*, GAO-07-1038T. 2007.

———. *Nuclear Waste: Actions Needed to Address Persistent Concerns with Efforts to Close Underground Radioactive Waste Tanks at DOE's Savannah River Site*, GAO-10-816. 2009.

———. *Nuclear Waste: DOE's Environmental Management Initiatives Report Is Incomplete*, GAO-09-697R. 2009.

———. *Nuclear Waste: Uncertainties and Questions about Costs and Risks Persist with DOE's Tank Waste Cleanup Strategy at Hanford*, GAO-09-913. 2009.

———. *Nuclear Waste Management: Key Attributes, Challenges, and Costs for the Yucca Mountain Repository and Two Potential Alternatives*, GAO-10-48. 2009.

———. *Plans for Addressing Most Buried Transuranic Wastes Are Not Final, and Preliminary Cost Estimates Will Likely Increase*, GAO-07-761. 2007.

———. *Recovery Act: Most DOE Cleanup Projects Appear to Be Meeting Cost and Schedule Targets, but Assessing Impact of Spending Remains a Challenge*, GAO-10-784. 2010.

———. *Securing U.S. Nuclear Materials: Poor Planning Has Complicated DOE's Plutonium Consolidation Efforts*, GAO-06-164T. 2005.

Hanford Health Information Network, Washington State Department of Health. *The Release of Radioactive Material from Hanford: 1944–1972.* Spokane, WA: Gonzaga University, 2000.

Hubbard, Marion King, National Research Council of the National Academies. *Energy Resources: A Report to the Committee on Natural Resources of the National Academy of Sciences—National Research Council of the National Academies.* 1962.

Institute for Policy Studies. *Radioactive Wastes and the Global Nuclear Energy Partnership.* 2007.

Interagency Review Group on Nuclear Waste Management. *Report to the President,* TID-29442. 1979.

Keister, Marsha, and Kathryn McBridge, Idaho National Laboratory. *Spent Nuclear Fuel Transportation: An Examination of Potential Lessons Learned from Prior Shipping Campaigns,* INL/EXT-06-11223. 2006.

Keller, R. F., Comptroller General, General Accounting Office. *DOE Policies in Connection with Construction and Operation of the Waste Isolation Pilot Plant at Carlsbad, New Mexico,* B-192999. 1979.

Knecht, Dieter A., J. H. Valentine, A. J. Luptak, M. D. Staiger, H. H. Loo, and T. L. Wichmann, Idaho National Laboratories. *Options for Determining Equivalent MTHM for DOE High-Level Waste,* INEEL/EXT-99-00317. 1999.

Kocher, David C., and Allen G. Croff, Oak Ridge National Laboratory. *A Proposed Classification System for High-Level and Other Radioactive Wastes,* ORNL/TM-10289. 1987.

Lomenick, T. F., Oak Ridge National Laboratory. *The Siting Record: An Account of the Programs of Federal Agencies and Events That Have Led to the Selection of a Potential Site for a Geologic Repository for High-Level Radioactive Waste,* ORNL/TM-12940. 1996.

Lowenthal, Micah D., Lawrence Livermore National Laboratory. *Radioactive-Waste Classification in the United States: History and Current Predicaments,* UCRL-CR-128127. 1997.

Maggiore, Peter, Secretary, New Mexico Environment Department. *Final Order of the Secretary of the New Mexico Environment Department: In the Matter of the Final Permit Issued to the United States Department of Energy and Westinghouse Electric Company Waste Isolation Division for a Hazardous Waste Permit for the Waste Isolation Pilot Plant,* HRM 98-04 (P). 1999.

Marcinowski, Frank, Office of Environmental Management, Department of Energy. *Overview of DOE's Spent Nuclear Fuel and High Level Waste.* 2010. brc.gov/pdfFiles/Environmental_Management_BRC_03252010.pdf.

"Memorandum between the Western Governors and the U.S. Department of Energy: Regional Protocol for the Safe and Uneventful Transportation of Transuranic (TRU) Waste" (February 25, 2003).

"Memorandum between the Western Governors and the U.S. Department of Energy: Regional Protocol for the Safe and Uneventful Transportation of Transuranic (TRU) Waste" (June 15, 2009).

Mercer, J. W., and B. R. Orr, U.S. Geological Survey. *Review and Analysis of Hydrogeologic Conditions near the Site of a Potential Nuclear-Waste Repository, Eddy and Lea Counties, New Mexico,* USGS-OFR-77-123. 1977.

Molecke, Martin A., Sandia National Laboratories. *A Comparison of Brines Relevant to Nuclear Waste Experimentation,* SAND83-0516. 1983.

Monitored Retrievable Storage Review Commission. *Nuclear Waste: Is There a Need for Federal Interim Storage?* 1989.

Mushkatel, Alvin H., K. David Pijawka, Patricia Jones, and Nivi Ibitayo, Nevada Agency for Nuclear Projects, Office of the Governor. *Governmental Trust and Risk Perceptions Related to the High-Level Nuclear Waste Repository: Analysis of Survey Results and Focus Groups,* NWPO-SE-052-92. 1992.

National Cancer Institute, National Institutes of Health. *Estimated Exposures and Thyroid Doses Received by the American People from Iodine-131 in Fallout following Nevada Atmospheric Nuclear Bomb Tests.* 1999.

National Council on Radiation Protection and Measurements. *Risk-Based Classification of Radioactive and Hazardous Chemical Wastes,* Report No. 139. 2002.

National Governors Association. *Low-Level Waste: A Program for Action.* 1980.

National Nuclear Security Administration, Department of Energy. *Draft Nonproliferation Impact Assessment for the Global Nuclear Energy Partnership: Programmatic Alternatives.* 2008.

Neill, Robert H., James K. Channell, Lokesh Chaturvedi, Marshall S. Little, Kenneth Rehfeldt, and Peter Spiegler, Environmental Evaluation Group. *Evaluation of the Suitability of the WIPP Site,* EEG-23. 1983.

Neill, Robert H., and Lokesh Chaturvedi, Environmental Evaluation Group. *Status of the WIPP Project.* 1991.

Neill, Robert H., Lokesh Chaturvedi, William W. L. Lee, Thomas M. Clemo, Matthew K. Silva, Jim W. Kenney, William T. Bartlett et al., Environmental Evaluation Group. *Review of the WIPP Draft Application to Show Compliance with EPA Transuranic Waste Disposal Standards,* EEG-61, DOE/AL/58309-61. 1996.

Nelson, Roger, Carlsbad Field Office, Department of Energy. "WIPP-DOE's Transuranic Waste Strategy and Future Packaging and Shipping Initiatives." Presentation at the Third Annual RadWaste Summit, September 10, 2008. goneri.nuc.berkeley.edu/tokyo/2009-01-15_Nelson1.pdf.

———. "WIPP Site Selection and Early Site Studies." Presentation at the International Atomic Energy Agency Training Course: General Training on Methodologies for Geologic Disposal in North America, November 6, 2008. www.nuc.berkeley.edu/files/WIPPHistory.ppt.

Nevada Agency for Nuclear Projects, Office of the Governor. *A Mountain of Trouble: A Nation at Risk—Report on Impacts of the Proposed Yucca Mountain High-Level Nuclear Waste Program,* vol. 1. 2002.

———. *Executive Summary of the State of Nevada*

Socioeconomic Studies: Biannual Report, 1993–1995, NWPO-SE-063-95. 1995.

———. *Yucca Mountain Socioeconomic Project Preliminary Findings: 1989 Nevada State Telephone Survey*, NWPO-SE-025-89. 1989.

Nevada Commission on Nuclear Projects. *Report and Recommendations of the Nevada Commission on Nuclear Projects*, Nevada Legislative Report 42-09. 2009.

Nevada Site Office, Department of Energy. *Nuclear Timeline*, DOE/NV-1243. 2008.

New Mexico Environment Department. *Waste Isolation Pilot Plant: Hazardous Waste Facility Permit*, NM 4890139088-TSDF. 2006.

New York Department of Environmental Conservation. *West Valley: History and Future*. 2008.

Nuclear Energy Institute. *U.S. State-by-State Commercial Nuclear Used Fuel and Payments to the Nuclear Waste Fund*. 2010. www.nei.org/resourcesandstats/documentlibrary/nuclearwastedisposal/graphicsand charts/usstatebystateusedfuelandpaymentstonwf/.

Nuclear Regulatory Commission. *A Pilot Probabilistic Risk Assessment of a Dry Cask Storage System at a Nuclear Power Plant*, NUREG-1864. 2007.

———. *Blending of Low-Level Radioactive Waste*, SECY-10-0043. 2010.

———. *Environmental Survey of the Reprocessing and Waste Management Portions of the LWR Fuel Cycle: A Task Force Report*, NUREG-0116. 1976.

———. *Final Update of the Commission's Waste Confidence Decision*, SECY-09-0090. 2010.

———. *NRC Staff Guidance for Activities Related to U.S. Department of Energy Waste Determinations*, NUREG-1854. 2007.

———. *Radioactive Waste: Production, Storage, Disposal*, NUREG/BR-0216. 2002.

———. *Regulatory Structure for Spent Fuel Reprocessing*, SECY-08-0134. 2008.

———. *Safety of Spent Nuclear Fuel*, NUREG/BR-0292. 2003.

———. *Update on Reprocessing Regulatory Framework: Summary of Gap Analysis*, SECY-09-0082. 2009.

Nuclear Waste Repository Project Office, Nye County, Nevada. *Final Report: Rail Transportation Economic Impact Evaluation and Planning Study for the Caliente and Mina Corridors*. 2007.

Nuclear Waste Technical Review Board. *1997 Findings and Recommendations: Report to the U.S. Congress and the Secretary of Energy*. 1998.

———. *Fifth Report to the U.S. Congress and the U.S. Secretary of Energy*. 1992.

———. *Moving beyond the Yucca Mountain Viability Assessment: A Report to the U.S. Congress and the Secretary of Energy*. 1999.

———. *Report to the U.S. Congress and the Secretary of Energy: January to December 1999*. 2000.

———. *Sixth Report to the U.S. Congress and the U.S. Secretary of Energy*. 1992.

———. *Survey of National Programs for Managing High-Level Radioactive Waste and Spent Nuclear Fuel: A Report to Congress and the Secretary of Energy*. 2009.

Office of the Chief Financial Officer, Department of Energy. *FY 2007 Congressional Budget Request: Budget Highlights*, DOE-CF-009. 2006.

———. *FY 2008 Congressional Budget Request: Budget Highlights*, DOE-CF-021. 2007.

———. *FY 2011 Congressional Budget Request: Budget Highlights*, DOE/CF-0046. 2010.

Office of Civilian Radioactive Waste Management, Department of Energy. *Annual Report to Congress*, DOE/RW-0216. 1989.

———. *Integrated Monitored Retrievable Storage/Repository Comparative Study*, RHO-BW-CR-154P. 1985.

———. *Mission Plan for the Civilian Radioactive Waste Management Program*, DOE/RW-0005. 1985.

———. *A Multiattribute Utility Analysis of Sites Nominated for Characterization for the First Radioactive-Waste Repository: A Decision-Aiding Methodology*, DOE/RW-0074. 1986.

———. *National Transportation Plan*, DOE/RW-0603. 2009.

———. *Recommendation by the Secretary of Energy of Candidate Sites for Site Characterization*, DOE/S-0048. 1986.

———. *Recommendation by the Secretary of Energy of Candidate Sites for Site Characterization for the First Radioactive-Waste Repository*, DOE/S-0048. 1986.

———. *Report to Congress on the Demonstration of the Interim Storage of Spent Nuclear Fuel from Decommissioned Nuclear Power Reactor Sites*, DOE/RW-0596. 2008.

———. *Report to the President and the Congress by the Secretary of Energy on the Need for a Second Repository*, DOE/RW-0595. 2008.

———. *Screening and Identification of Sites for a Proposed Monitored Retrievable Storage Facility*, DOE/RW-0023. 1985.

———. *Yucca Mountain Repository License Application: Safety Analysis Report*, DOE/RW-0573. 2008.

———. *Yucca Mountain Science and Engineering Report: Technical Information Supporting Site Recommendation Consideration—Revision 1*, DOE/RW-0539-1. 2002.

Office of Environmental Management, Department of Energy. *Linking Legacies: Connecting the Cold War Nuclear Weapons Production Processes to Their Environmental Consequences*, DOE/EM-0319. 1997.

———. *Radioactive Material Transportation Practices Manual for Use with DOE O 460.2A, DOE M 460.2-1A*. June 2008.

———. "Recovery Act Funds $24 Million in Technology Projects." *Environmental Management Recovery Act News Flash*, 1 (June 30, 2010).

———. *Reduction of EM Footprint and Establishment of Energy Parks*. 2008. www.energyca.org/PDF/FootprintReduction.pdf.

———. *Report to Congress: Status of Environmental Management Initiatives to Accelerate the Reduction of Environmental Risks and Challenges Posed by the Legacy of the Cold War*. 2009.

———. *Updated Appendices to the Status of Environmental Management Initiatives to Accelerate* the Reduction of Environmental Risks and Challenges Posed by the Legacy of the Cold War. 2010.

Office of Fuel Cycle Management, Department of Energy. *Global Nuclear Energy Partnership Strategic Plan*, GNEP-167312. 2007.

Office of the General Counsel, Nuclear Regulatory Commission. *Nuclear Regulatory Legislation: 110th Congress; 2nd Session*, vol. 1, no. 8, NUREG-0980. 2009.

Office of Management and Budget. *Budget of the United States Government, Fiscal Year 2011: Department of Energy Budget Overview.* 2010. www.whitehouse.gov/omb/budget/fy2011/assets/energy.pdf.

Office of Nuclear Energy, Department of Energy. *Draft Global Nuclear Energy Partnership Programmatic Environmental Impact Statement,* DOE/EIS-0396. 2008.

Office of Radiation and Indoor Air, Environmental Protection Agency. *Radiation Protection at EPA: The First Thirty Years*, EPA 402-B-00-001. 2000.

Office of River Protection, Department of Energy. *High Aluminum HLW Glasses for Hanford's WTP*, ORP-42448-A. 2009.

Office of State Programs, Nuclear Regulatory Commission. *Means for Improving State Participation in the Siting, Licensing, and Development of Federal Nuclear Waste Facilities,* NUREG-0539. 1979.

Office of Technology Assessment. *Comparative Analysis of the ERDA Plan and Program*, OTA-E-28. 1976.

———. *Long-Lived Legacy: Managing High-Level and Transuranic Waste at the DOE Nuclear Weapons Complex*, OTA-BP-O-83. 1991.

———. *Managing Commercial High-Level Radioactive Waste.* 1982.

———. *Managing the Nation's Commercial High-Level Radioactive Waste,* OTA-O-171. 1985.

———. *Partnerships under Pressure: Managing Commercial Low-Level Radioactive Waste,* OTA-O-426. 1989.

Oregon Department of Energy. *Radioactive Material Transport in Oregon 2009: Report to State and Local Government.* 2010. www.oregon.gov/ENERGY/NUCSAF/docs/2009TransportReport.pdf.

Panel on Social and Economic Aspects of Radioactive Waste Management, National Research Council of the National Academies. *Social and Economic Aspects of Radioactive Waste Disposal: Considerations for Institutional Management.* 1984.

Picket, John B., and Stephen W. Norford, Westinghouse Savannah River Co. *First Commercial U.S. Mixed Waste Vitrification Facility: Permits, Readiness Reviews, and Delisting of Final Wasteform*, WSRC-MS-98-00314. 1998.

Pierce, William G., and Ernest I. Rich, U.S. Geological Survey. *Summary of Rock Salt Deposits in the United States as Possible Disposal Sites for Radioactive Wastes*, USGS-TEI-725. 1958.

Radioactive Waste Management Committee, Nuclear Energy Agency. *Reversibility and Retrievability for the Deep Disposal of High-Level Radioactive Waste and Spent Fuel: Intermediate Findings and Discussion Document of the NEA R&R Project.* 2010.

Rechard, Rob P., WIPP Performance Assessment Department, Department of Energy. *Milestones for Disposal of Radioactive Waste at the Waste Isolation Pilot Plant (WIPP) in the United States.* 1998.

Rice, Eric E., and Claude C. Priest. "An Overview of Nuclear Waste Disposal in Space." In Hofmann, Peter L., and John J. Breslin, eds., Department of Energy. *The Technology of High-Level Nuclear Waste Disposal: Advances in the Science and Engineering of the Management of High-Level Nuclear Waste*, vol. 1, 370, DOE/TIC-4621. 1981.

Robertson, John B., U.S. Geological Survey. "Geologic Problems at Low-Level Radioactive Waste-Disposal Sites." In *Groundwater Contamination*, Panel on Groundwater Contamination, National Research Council of the National Academies, 104. Washington, DC: National Academy Press, 1984.

Sandia National Laboratories. *Greater-Than-Class C Radioactive Waste and DOE Greater-Than-Class C-Like Waste Inventory Estimates.* 2007.

Schneider, Mycle, and Yves Marignac, International Panel on Fissile Materials. *Spent Nuclear Fuel Reprocessing in France.* 2008.

"Second Modification to the July 1, 1981 'Agreement for Consultation and Cooperation' on WIPP by the State of New Mexico and the U.S. Department of Energy." 1987.

South Carolina Department of Health and Environmental Control. *Commercial Low-Level Radioactive Waste Disposal in South Carolina*. CR-000907. 2007.

Staats, Elmer B., Comptroller General, Government Accountability Office. *Nuclear Energy's Dilemma: Disposing of Hazardous Radioactive Waste Safely*, EMD-77-41. 1977.

Steering Group, Global Nuclear Energy Partnership. *Action Plan.* 2007. www.ne.doe.gov/pdfFiles/GNEP_action_plan.pdf.

Swedish National Council for Nuclear Waste Report. *Nuclear Waste State of the Art Report 2010: Challenges for the Final Repository Programme*, SOU 2010:6. 2010.

Swedish Nuclear Fuel and Waste Management Company (SKB). *Äspö Hard Rock Laboratory Annual Report 2006*, TR 07-10. 2007.

Task Force on an Alternative Program Strategy, Department of Energy. *A Proposed Alternative Strategy for the Department of Energy's Civilian Radioactive Waste Management Program.* 1993.

Task Force on Nuclear Waste, Western Governors' Association. *Report to Congress: Transport of Transuranic Wastes to the Waste Isolation Pilot Plant—State Concerns and Proposed Solutions.* 1989.

Task Force on Radioactive Waste Management, Department of Energy. *Earning Public Trust and Confidence: Requisites for Managing Radioactive Wastes*, SEAB 95000302. 1993.

Texas Commission on Environmental Quality. *Radioactive Material License No. R04100.* 2009.

Total System Performance Assessment Peer Review

Panel, Department of Energy. *Final Report Total System Performance Assessment Peer Review Panel.* 1999.

Utah Department of Environmental Quality. *High Level Nuclear Waste Storage in Utah.* 2006. www.deq.utah.gov/Issues/no_high_level_waste/documents/pdocs/06_PFS_Factsheet.pdf.

———. *State of Utah Fact Sheet on PFS's Right-of-Way Applications to Use Public Lands for the Transport, Storage, and Transfer of High Level Nuclear Waste.* 2006. www.deq.utah.gov/Issues/no_high_level_waste/documents/bdocs/06_04_BLM_Issues.pdf.

Western Governors' Association. *WIPP Transportation Safety Program Implementation Guide.* 2003

Williams, Alice C., Department of Energy. "West Valley Spent Nuclear Fuel Shipment." Presentation at INMM Spent Fuel Management Seminar XXI, January 21, 2004. www.nwtrb.gov/meetings/2004/jan/williams.pdf.

WIPP Transportation Technical Advisory Group, Western Governors' Association. *Waste Isolation Pilot Plant Rail Transportation Safety Program Implementation Guide* (Draft). May 2004.

Congressional Documents

Andrews, Anthony, Congressional Research Service. *Nuclear Fuel Reprocessing: U.S. Policy Development.* 2008.

———. *Spent Nuclear Fuel Storage Locations and Inventory.* 2004.

Committee on Appropriations, House of Representatives. *2004 Departments of Transportation and Treasury and Independent Agencies Appropriations Bill,* H.R. Rep. No. 108-243. 2003.

———. *Consolidated Appropriations Act, 2008: Committee Print.* 2008.

Committee on Appropriations, Senate. *1990 Energy and Water Development Appropriation Bill,* S. Rep. No. 101-83. 1989.

———. *2007 Energy and Water Appropriations Bill,* S. Rep. No. 109-274. 2006.

Committee on Armed Services, House of Representatives. *Department of Energy National Security and Military Applications of Nuclear Authorization Act of 1980,* H.R. Rep. No. 96-162. 1979.

———. *Nuclear Waste Policy Act of 1982,* H.R. Rep. No. 97-491, pt. 2. 1982.

———. *Waste Isolation Pilot Plant Land Withdrawal Act of 1991,* H.R. Rep. No. 102-241, pt. 2. 1991.

Committee on Armed Services, Senate. *National Defense Authorization Act for Fiscal Year 1989,* S. Rep. No. 100-326. 1988.

Committee on Energy and Commerce, House of Representatives. *Hazardous Materials Transportation Act Uniform Safety Amendments Act of 1990,* H.R. Rep. No. 101-444, pt. 1. 1990.

———. *Low-Level Radioactive Waste Policy Amendments Act of 1985,* H.R. Rep. No. 99-314. 1985.

———. *Texas Low-Level Radioactive Waste Disposal Compact Consent Act,* H.R. Rep. No. 104-148. 1995.

Committee on Energy and Resources, House of Representatives. *Waste Isolation Pilot Plant Land Withdrawal Act,* H.R. Rep. No. 102-241, pt. 3. 1991.

Committee on Interior and Insular Affairs, House of Representatives. *Comprehensive National Energy Policy,* H.R. Rep. No. 102-474, pt. 8. 1991.

———. *Nuclear Waste Policy Act of 1982,* H.R. Rep. No. 97-491, pt. 1. 1982.

Committee on Science and Technology, House of Representatives. *West Valley Demonstration Project Act,* H.R. Rep. No. 96-1100. 1980.

Conference Committee, House of Representatives. *U.S. Department of Energy National Security and Military Applications of Nuclear Energy Authorization Act of 1980: Conference Report (to Accompany S. 673),* H.R. Rep. No. 96-702. 1979.

———. *Energy Policy Act of 1991,* Conference Report (to Accompany H.R. 776), H.R. Rep. No. 102-1018. 1991.

———. *Ronald W. Reagan National Defense Authorization Act for Fiscal Year 2005,* Conference Report, H.R. Rep. No. 108-76. 2004.

Congressional Budget Office. *Cleaning Up the Department of Energy's Nuclear Weapons Complex.* 1994.

Holt, Mark, Congressional Research Service. *Civilian Nuclear Spent Fuel: Temporary Storage Options,* 96-212 ENR. 1998.

———. *Civilian Nuclear Waste Disposal.* 2008

———. *Civilian Nuclear Waste Disposal,* RL-33461. 2009.

———. *Civilian Nuclear Waste Disposal.* 2010.

———. *Nuclear Waste Disposal: Alternatives to Yucca Mountain,* CRS-R40202. 2009.

———. *Nuclear Weapons Production Complex: Environmental Compliance and Waste Management,* IB90074. 1997. digital.library.unt.edu/ark:/67531/metacrs433/m1/.

Joint Committee on Atomic Energy, House of Representatives. *Amendments to the Atomic Energy Act of 1954,* H. Rep. No. 86-1125. 1959.

Parker, Larry, and Mark Holt, Congressional Research Service. *Nuclear Power: Outlook for New U.S. Reactors,* RL33442. 2007.

Index

Page numbers in bold refer to tables and illustrations.

9/11 Commission Act (2007), 128, 132

Abraham, Spencer, 69
accelerator-produced radioactive materials (ARM), 18, 21, 94, 95
Adams, Martha, 228
Administrative Procedure Act (APA, 1946), 52–53, 181
Advanced Energy Initiative, 246. *See also* Global Nuclear Energy Partnership (GNEP)
Advanced Mixed Waste Treatment Project (Idaho), 134
Agence nationale pour la gestion des déchets radioactifs (ANDRA), 292
Agreement States, 19, 57, 62, 145, 147. *See also specific states*
AIFs (assured isolation facilities), 158–59
Aiken County, South Carolina, 228, 229
Alabama Power Co. v. DOE, 332n556
Alaska, 31
Alexander, Lamar, 200
Allied General Nuclear Services, 46. *See also* Barnwell Plant (South Carolina)
aluminum, 111
American Nuclear Energy Council, 215
American Physical Society, 234, 240, 241, 342n5
American Recovery and Reinvestment Act (2009), 184
americium-241 (Am-241), 93, 94
Anaya, Toney, 173, 174, 354n46
ANDRA (Agence nationale pour la gestion des déchets radioactifs), 292
Andrews, Texas
　LLW disposal at, 154–55, 157–58, 159, 258
　Texas Compact and, 63, 154, 160, 279, 284
Andrus, Cecil, 61, 139, 177
APA (Administrative Procedure Act, 1946), 52–53, 181
Apodaca, Jerry, 164
Apostolakis, George, 229
Areva, 91, 92, 385n158
Argonne National Laboratory, 74, 139

Arizona, 151
assured isolation facilities (AIFs), 158–59
Atlantic Compact, 151, 152, 154
Atomic and Space Development Authority (New York), 45, 243
Atomic Energy Act of 1946 (AEA), 8, 9, 17, 20
Atomic Energy Act of 1954 (AEA)
　civilian nuclear power and, 18–19, 22
　exempt waste and, 120
　HLW definition and regulation, 102
　mixed wastes and, 112
　role of states, 18–19, 145
　on special nuclear material, 95
　transportation and, 124
Atomic Energy Commission (AEC)
　Agreement States and, 18–19
　breeder reactors and, 47
　EPA and, 10, 33–34, 298–99, 301
　ERA and, 34–35, 298
　HLW definition, 100, 103
　HLW disposal and, 25–27
　litigation and, 9, 49–53
　LLW disposal and, 21, 27, 42, 62, 145, 146, 160–61
　nuclear materials and, 20
　nuclear power and, 17–18, 25
　nuclear weapons testing and, 31
　private facilities and, 21
　repository-siting process and, 28–29
　reprocessing and, 44–45, 242, 243, 245–46, 255
　transportation and, 124–25, 129
　TRU disposal and, 27–28, 42, 99
　on waste disposal, 8, 22
　weapons production complex, 10, 17
　WIR and, 338–39n117
　See also Project Salt Vault (Lyons, Kansas)
Atoms for Peace initiative, 18, 133
Australia, 7, 82, 246, 268

Babcock and Wilcox Company, 316n12
Baltimore Gas & Elec. v. NRDC, 53

413

Barnwell Plant (South Carolina)
    Carter administration and, 164, 244
    construction, 46
    EnergySolutions and, 159
    LLW disposal at, 146, 147, 149, 158, 160, 258
    restriction to in-compact waste, 63, 154, 155, 157
    South Carolina and, 151
    termination, 46–47, 245, 271
Barrows, Larry, 355–56n90
Bartlett, John W., 206, 219
Beatty, Nevada, 146, 147, 149, 154
becquerel (Bq), 86
Belgium, 294, 333n584
BIA (Bureau of Indian Affairs), 80, 135, 237–38
Bingaman, Jeff, 60–61, 171–72
BLM (Bureau of Land Management), 80, 135, 174, 213, 237–38
Blue Ribbon Commission on America's Nuclear Future
    Hanford and, 314
    members, **273**
    mission, 232, 272–74
    Nye County and, 227–28
    on reprocessing, 268
    on SNF storage options, 235, 239, 253, 267
    task, 3, 16, 74, 258
    WIPP and, 185
Bodman, Samuel W., 246
Bonneville Power Administration (BPA), 393n149
BRC (below regulatory concern) wastes, 94, 97, 120–22, 159, 304, 334n2
breeder reactors, 18, 47–48, 49, **90**, 386n178. *See also* Clinch River Breeder Reactor (CRBR, Tennessee)
Brookhaven National Laboratory, 74
Bullfrog County, Nevada, 212, 216
Bulloch v. United States, 31
Burciaga, Guerrero, 172
Bureau of Indian Affairs (BIA), 80, 135, 237–38
Bureau of Land Management (BLM), 80, 135, 174, 213, 237–38
Burns, Stephen, 272
Bush, George W., administration
    on federal liability, 78
    GNEP and, 2, 80, 92, 242, 246
    MOX fuel and, 92
    NFMDA and, 113
    PFS storage facility and, 238
    support of nuclear power, 75
    Yucca Mountain repository and, 69, 222
byproduct material, 20, 336n38

C&C (Consultation and Cooperation) Agreements
    Nevada and, 198, 210, 288–89
    New Mexico and, 60–61, 171–73, 174–75, 178, 194, 288
    NWPA on, 194, 203–4, 206, 209
    Washington and, 204, 366n97
California
    LLW disposal and, 151, 153, 284
    opposition to new nuclear plants, 55, 322n271
    Pacific Gas & Electric Co. decision and, 19, 316n22
    on transportation, 138
    Yucca license withdrawal and, 228
Calvert Cliffs' Coordinating Committee v. AEC, 50
Canada
    repository-siting process and, 292
    uranium and, 7, 82, 246, 250, 268
    See also Nuclear Waste Management Organization (NWMO, Canada)
cancer
    LCFs, 138, 233, 334n609
    nuclear weapons testing and, 31
Cannon, Howard, 195
Carlsbad, New Mexico
    role in WIPP development, 8, 29, 162, 184–85, 256, 283, 287, 391–92n93
    on SNF and HLW at WIPP, 240, 263
    See also WIPP (Waste Isolation Pilot Plant, Carlsbad, New Mexico)
Carruthers, Garrey, 61, 175, 177
Carter administration
    CRBR and, 48
    on new nuclear power plants, 60
    on nuclear waste management, 15, 165, 188
    reprocessing and, 2, 9, 27, 46–47, **96**, 163–64, 242, 244–45, 255
    WIPP and, 60, 167
    See also Interagency Review Group on Nuclear Waste Management (IRG)
Catawba Nuclear Station (South Carolina), 92
Central Compact, 152–53, 158–59, 350n69
cesium-137 (Cs-137)
    characteristics, 86, 89
    disposal of, 23, 47, 81, 82, 89, 92, 110–11, 115, 247–48, 264
    IAEA framework and, 119
    production of, 97
    UREX+ and, 247–48, 249
Channell, James, 181
Chernobyl disaster (Ukraine), 32, 38, 176, 198
Chevron v. NRDC, 101
China
    reprocessing and, 245, 268, 333n584
    support of nuclear power, 330n523
CH-TRU (contact-handled transuranic waste)
    definition of, 99
    EM and, 43
    at WIPP, 134, 139, 182, 260, 284
Chu, Steven
    Blue Ribbon Commission on America's Nuclear Future and, 16, 74

on dry-cask storage, 80, 261, 265
NWF and, 78
on reprocessing, 81, 268
Yucca Mountain repository and, 70, 227–28, 379n448
Church, Frank, 28
Citizens for Alternatives to Radioactive Dumping, 182
Clark County, Nevada, 212, 228
Clinch River Breeder Reactor (CRBR, Tennessee), 47–48, 200, 236
Clinton, Hillary, 109
Clinton administration, 179, 302–3
Clive facility (Utah)
   capacity, 157
   EnergySolutions and, 159
   LLW disposal at, 46, 63, 121, 154, 155–56, 158, 160, 258, 259
   transportation to, 133
Coast Guard, 343n13
Cochran, Thomas, 68
Colorado
   HLW and SNF in, 264–65, 332n565
   HLW reclassification initiatives and, 109
   nuclear weapons testing in, 31
   on transportation, 138
Commercial Vehicle Safety Alliance (CVSA), 132
compacts, 63, 147–56, **149**, **152**, 160–61, 256, 259, 279
Comprehensive Environmental Response, Compensation, and Liability Act (CERCLA), 35–36, 39–40, 111
Connecticut
   nuclear power plants moratoriums and, 55–56
   opposition to new nuclear plants, 324n319
   on SNF storage, 265
   See also Atlantic Compact
Consolidated Appropriations Act (2008), 239
Consortium for Risk Evaluation with Stakeholder Participation (CRESP), 44, 286, 315n2
Consultation and Cooperation Agreements. *See* C&C (Consultation and Cooperation) Agreements
Council of State Governments, Eastern Regional Conference, 131
Council of State Governments, Midwest, 131
CRBR (Clinch River Breeder Reactor, Tennessee), 47–48, 200, 236
CRESP (Consortium for Risk Evaluation with Stakeholder Participation), 44, 286, 315n2
curie (Ci), 86

Davis-Besse Nuclear Power Station (Ohio), 302
Decision Research, 215
deep borehole disposal, 57
deep geological repositories
   in Finland and Sweden, 290–92
   for HLW, 23, 24, 25–27

IRG on, 9, 188, 190
NAS on, 3, 8, 254–55
risks and, 87
See also repository-siting process; *specific sites*
Defense Nuclear Facilities Safety Board, 40
Department of Commerce (DOC), 206
Department of Defense (DOD), 37, 129, 336n40
Department of Energy (DOE)
   cleanup agreements and progress, 2, 32, 35–44, **39**, **41**, 116, 232, 311–14, **313**
   consolidated storage and, 239, 240
   ERDA and, 34, 298
   FEIS and, 58–59
   GTCC waste disposal and, 4, 98, 156–57
   HLW definition, regulation, and reclassification initiatives, 101, 102–3, 105–11
   LLW disposal and, 94, 95, 145, 156, 158, 259
   management and self-regulation, 117, 293, 299, 301
   mixed wastes and, 111–13, 299
   MRS facilities and, 235–36
   packaging and, 138–39
   Radioactive Waste Management Manual, 102, 106–8, 109–10
   Reagan administration and, 206
   retrievability and, 264
   SNF and, 100
   transportation and, 123–24, 129, 132–34, 136–42, 143
   TRU definition and regulation, 99
   utilities and, 77–78
   See also C&C (Consultation and Cooperation) Agreements; Global Nuclear Energy Partnership (GNEP); Office of Environmental Management (EM); *specific facilities*
Department of Homeland Security (DHS), 128
Department of Justice (DOJ), 78, 172
Department of the Interior (DOI)
   PFS storage facility and, 135, 238, 266
   Reagan administration and, 206
   WIPP and, 29, 174, 176, 178
Department of Transportation (DOT)
   IRG on, 58
   nuclear waste transportation and, 123–29, 130–32, 133–34, 137, 141–42
   WGA and, 177
depleted uranium (DU), 86, 91, 94–95
Deutch, John M., 56, 165. *See also* Interagency Review Group on Nuclear Waste Management (IRG)
DHS (Department of Homeland Security), 128
Diablo Canyon plant (California), 32
DOC (Department of Commerce), 206
DOD (Department of Defense), 37, 129, 336n40
DOE. *See* Department of Energy (DOE)
DOI. *See* Department of the Interior (DOI)
DOJ (Department of Justice), 78, 172

Domenici, Pete, 170, 354n58
DOT. *See* Department of Transportation (DOT)
Dresden Generating Station (Illinois), 316n12
dry-cask storage, 54, 79–80, 217, 232–35, 240, 261, 265, 275
Duke Energy, 92
Durenberger, David, 37

Edwards, James B., 47
EEG. *See* Environmental Evaluation Group (EEG)
Egypt, 83, 251
Eisenhower administration, 18, 133
EM. *See* Office of Environmental Management (EM)
Energy and Water Appropriations Act (1997), 220, 221
Energy and Water Development Appropriations Act (1990), 213
Energy Policy Act of 1992 (EnPA), 67–68, 120, 217, 224
Energy Policy Act of 2005 (EnPA), 18, 75, 94, **96**
Energy Reorganization Act of 1974 (ERA)
 creation of NRC and ERDA, 10, 34–35, 100, 298, 301
 on HLW disposal, 105
 on LLW disposal, 145
 transportation and, 124
Energy Research and Development Administration (ERDA)
 creation of, 10, 34, 298
 repository-siting process and, 29, 60, 164, 196
 WIPP and, 28, 60, 162–63, 165
EnergySolutions, 155–56, 159. *See also* Barnwell Plant (South Carolina); Clive facility (Utah)
EnPA (Energy Policy Act, 1992), 67–68, 120, 217, 224
EnPA (Energy Policy Act, 2005), 18, 75, 94, **96**
Envirocare (now EnergySolutions). *See* EnergySolutions
Environmental Defense Fund, 322n271
Environmental Evaluation Group (EEG)
 on compliance certification application, 180
 funding, 60, 173, 182, 287, 290
 mission, 168–69, 256
 on SEIS for WIPP, 177
 on TRUPACT, 139
environmental movement
 legislation and, 32–33
 rise of, 30–31
 Rocky Flats fire and, 27–28
 SNF storage and, 189
 waste regulation and, 116
 See also Environmental Protection Agency (EPA)
Environmental Policy Institute, 33
Environmental Protection Agency (EPA)
 AEC and, 10, 33–34, 298–99, 301
 on BRC, 120, 304
 FFCA and, 38

 Hanford cleanup program and, 311–14, **313**
 on LAMW, 159
 mixed wastes and, 21, 111–13
 Nuclear Energy Institute v. EPA, 69–70, 223–26, 300–301, 302
 radiation standards, 34, 62, 65, **71**, 298–99, 300–301, 302–3
 RCRA/CERCLA and, 35, 40
 repository-siting process and, 191, 206–7
 Rocky Flats and, 38, 61, 177
 SARA and, 36
 TRU definition and regulation, 99
 WIPP and, 61–62, 68, 179–82, 260, 299
 Yucca-specific standards and, 67–70, **71**, **72**, 101, 217, 224, 285, 299
EPA. *See* Environmental Protection Agency (EPA)
ERA. *See* Energy Reorganization Act of 1974 (ERA)
ERDA. *See* Energy Research and Development Administration (ERDA)
ERDA Authorization Act (1978), 46, 164
Erkins, Bob, 28
Esmeralda County, Nevada, 212, 216, 227, 228
European Union (EU), 25, 251
EW (exempt waste), 118, 120

fast breeder reactors, 47–48, 49
Fast Flux Test Facility, 47
fast neutron reactors (FNRs), 92, 247, 248, 335n13
FBI (Federal Bureau of Investigation), 38, 61, 177–78
Federal Facilities Compliance Act of 1992 (FFCA), 36–39, 111, 112, 299–300
Federal Land Policy and Management Act of 1976 (FLPMA), 163, 172, 174, 175–76, 178
Federal Railroad Safety Act of 1970 (FRSA), 125, 127
FEIS (Final Environmental Impact Statement), 58–59
Fernald Feed Materials Production Center (Ohio), 39, 44
FFCA (Federal Facilities Compliance Act of 1992), 36–39, 111, 112, 299–300
Finland
 repositories and, 283, 290–91, 292, 294
 reprocessing and, 333n584
 retrievability and, 269
fissionable material, 20. *See also* special nuclear material
FLPMA (Federal Land Policy and Management Act of 1976), 163, 172, 174, 175–76, 178
FOIA (Freedom of Information Act of 1966), 33, 163, 198, 225, 286
Foley, Tom, 67, 195, 208
Ford administration, 2, 46, 60, 243, 255
Fort St. Vrain Generating Station, 74
France
 repositories and, 292

reprocessing and, 80, 83, 91, 245
retrievability and, 269
uranium leasing and, 246
waste classification and regulation, 118
waste management and, 294
Freedom of Information Act of 1966 (FOIA), 33, 163, 198, 225, 286
Friends of the Earth, 49
FRSA (Federal Railroad Safety Act of 1970), 125, 127
Fukushima I Nuclear Power Plant (Japan), 12–13, 44, 76, 79, 330n523, 332n568
Fultz, Keith, 176

GAO (Government Accountability Office, previously General Accounting Office)
    on dry-cask storage, 234
    on LLW disposal, 158
    on Los Alamos, 184
    on NRC, 301
    on NWPA, 201, 366n96
    on on-site storage, 234
    on pool storage, 79
    on repository-siting process, 197, 205, 327n432
    on transportation, 133, 241
    on WIPP, 60, 164, 176
Gardner, Booth, 370n200
General Electric, 45, 316n12. *See also* Midwest Fuel Recovery Plant (Morris, Illinois)
Georgia, 199
Germany, 294, 333n584
Gervers, John, 371n229
Getty Oil Company, 45. *See also* West Valley Reprocessing Plant (New York)
Gilinsky, Victor, 222
Global Nuclear Energy Partnership (GNEP)
    advanced fuel technology and, 80–81
    George W. Bush administration and, 2, 80, 92, 242, 246
    criticism of, 2, 81–83, 242, 248–52
    on extended surface storage, 264
    local communities and, 240, 283
    Obama administration and, 2, 81, 252, 268
    UREX+ system, 92, 247–49
    wastes and, 246–47
GNEP. *See* Global Nuclear Energy Partnership (GNEP)
Going the Distance?, 124, 130
Gore, Al, 200
Gorton, Slade, 365n68
Goshutes. *See* Skull Valley Band of the Goshute Nation storage facility (PFS)
Graham, Lindsey, 109
GTCC (greater-than-class-C) waste
    definition of, 20–21, 94, 95, 97, 99, 145
    disposal of, 4, 7, 95, 97, 98, 105, 156, 303–4
    IAEA framework and, 119
    from reprocessing, 82, 92, 249

sources of, 156–57
WIPP and, 263
Guinn, Kenny, 223

Hamilton, Ray, 172
Hanford Education Action League, 33
Hanford Site (Washington)
    cleanup, 39, 42, 43–44, 252, 311–14, **313**, 321n85
    FOIA and, 33, 198
    HLW disposal at, 27, 28, 74, 134, 261
    HLW reclassification initiatives and, 103, 105, 110–11, 342n201
    leaks at, 21–22, 26
    LLW disposal at, 27, 43, 156, 158
    repository-siting process and, 196, 197–98
    SNF disposal at, 43, 74
    transportation and, 139
    TRU disposal at, 27, 183
    weapons production at, 17
Harris, Stuart, 314
Hartsville Nuclear Plant (Tennessee), 200
HASC (House Armed Services Committee), 101, 166, 170, 301
Hazardous and Solid Waste Amendments of 1984 (HSWA), 112
Hazardous Materials Transportation Act of 1975 (HMTA), 125–26, 129
Hazardous Materials Transportation Uniform Safety Act of 1990 (HMTUSA), 127–28, 129, 131, 132
heavy-water reactors, 17, **90**
Hess, Harry, 25–26
HLW (high-level waste)
    definition, classification, and regulation, 20, 93, 100–102
    at DOE facilities, 40, 42, 74
    IAEA framework and, 118, 119
    reclassification initiatives, 102–11
    from reprocessing, 20, 22–23, 26–27, 47, 91, 249, 254
    transportation, 123–24, 127, 134, 135, 140–42
    from weapons production, 20, 22–23
HLW (high-level waste) storage and disposal
    consolidated interim storage, 274–75
    current situation and options, 6, 79, 231–32, 253, 260–61, 264–65, 275–76
    at DOE facilities, 21, 42–43, 134, 261, 264–65, 278
    FEIS on, 58–59
    IRG on, 56–57
    NWPA on, 64–66, 190
    options and methods, 22–25, 93, 101, 303
    Reagan administration on, 195
    repository-siting process, 25–27
    Sweden and, 291–92
    at West Valley, 134, 265
HMTA (Hazardous Materials Transportation Act of 1975), 125–26, 129

HMTUSA (Hazardous Materials Transportation Uniform Safety Act of 1990), 127–28, 129, 131, 132
HSWA (Hazardous and Solid Waste Amendments of 1984), 112

IAEA (International Atomic Energy Agency)
    on BRC, 334n2
    on fuel bank, 83, 251
    on geological repositories, 25
    waste classification framework, 117–22, 303
ICC (Interstate Commerce Commission), 125, 126, 129
ice sheet disposal, 24, 59
Idaho
    HLW and SNF in, 43, 264–65, 332n565
    HLW reclassification initiatives and, 109
    transportation of TRU and, 134, 139
    TRU from Rocky Flats and, 28, 59, 61, 177, 256
    TRU waste at INL and, 183
IFNEC (International Framework for Nuclear Energy Cooperation), 252. *See also* Global Nuclear Energy Partnership (GNEP)
Illinois, 55–56, 153–54. *See also* Midwest Fuel Recovery Plant (Morris, Illinois); Sheffield, Illinois
India, 46, 244, 330n523
informed public assent, 281–82
INL (Idaho National Laboratory)
    breeder reactor at, 18
    cleanup, 39, 42, 43–44, 321n85
    HLW at, 74, 134, 261
    HLW reclassification initiatives and, 103, 105, 109–11
    LLW disposal at, 27, 43, 156, 352n38
    SNF disposal at, 74, 133, 244
    transportation and, 133, 134, 139
    TRU disposal at, 27, 28, 139, 183
    TRU from Rocky Flats and, 27–28, 61, 256
    on uranium prices, 250
    weapons production at, 17
Interagency Review Group on Nuclear Waste Management (IRG)
    on geological repositories, 9, 188, 190
    on HLW and SNF disposal, 56–57, **59**, 295
    legislation and, 255–56
    on LLW disposal, 7, 57–58, **59**, 147–48, 160
    mission, 3, 56, 276–77
    on repositories and siting process, 57, 275
    on transportation, 58, **59**
    on TRU disposal, 57
    on WIPP, 56, 166, 263, 356n96
intermediate scale facilities (ISFs), 57, 354n55
intermodal transport facilities (ITFs), 135
International Framework for Nuclear Energy Cooperation (IFNEC), 252. *See also* Global Nuclear Energy Partnership (GNEP)

Interstate Commerce Commission (ICC), 125, 126, 129
Inyo County, California, 212
Iran, 83, 251, 268
Isaacs, Tom, xv, 278, 315n4, 391n84
ISFs (intermediate scale facilities), 57, 354n55
Italy, 83, 251
ITFs (intermodal transport facilities), 135

Jackson, Henry M., 166–67, 193, 326n415
Jaczko, Gregory, 159, 229
Jaffe, Louis, 302
Japan
    Areva and, 385n158
    Fukushima nuclear accident, 12–13, 44, 76, 79, 330n523, 332n568
    licensing process in, 51
    reprocessing and, 233, 234, 235, 239, 245, 268, **270**, 275
    uranium leasing and, 246
    waste classification and regulation, 118
Johnson, Gary, 181
Johnston, J. Bennett, 67, 68, 207–8, 213
Joint Timbisha Shoshone Tribal Group, 228
Joskow, Paul L., 76

Kansas, nuclear power plants moratoriums and, 55–56. *See also* Project Salt Vault (Lyons, Kansas)
karst, 169
Kentucky, 55–56, 316n22
King, Bruce, 163, 167
King, Gary, 185
Kuwait, 251

Lakes Environmental Association, 199
LAMW (low-activity mixed waste), 159
landfill disposal
    BRC wastes and, 94, 97, 120–21, 159, 304
    in Europe, 118, 159
    LLW and, 62, 86, 92, 93, 97, 145, 299, 313
    TRU and, 99
    waste from weapons production, 17
latent cancer fatalities (LCF), 138, 233, 334n609
Leboeuf, Lamb, Green & Macrae, LLP v. Abraham, 215
Leckband, Susan, 314
light-water reactors (LWRs), 75, 88, **90**, 92
LILW (low- to intermediate-level waste), 118–19
liquid metal fast breeder reactors (LMFBRs), 47, 48
List, Robert, 216
LLRW (low-level radioactive waste), 145. *See also* LLW (low-level waste)
LLRWPA. *See* Low-Level Radioactive Waste Policy Act of 1980 (LLRWPA)
LLRWPAA. *See* Low-Level Radioactive Waste Policy Amendments Act of 1985 (LLRWPAA)

LLW (low-level waste)
    Cs-137 and Sr-90 as, 89, 92
    definition, classification, and regulation, 20–21, 93–94, 95, **98**, 115–16, 145, 304
    IAEA framework and, 119
    production of, 20, 93, 145–46, 254
    from reprocessing, 82, 92, 249
    transportation, 123–24, 133, 135, 145
LLW (low-level waste) storage and disposal
    at commercial disposal facilities, 21, 146–47
    compacts and, 147–56, **152**
    current situation and options, 154–61, 258–60
    DOE and, 95
    at DOE facilities, 40, 42–43, 62–63
    IRG on, 7, 57, 58
    NRC and, 95
    NRC on, 97–98
    options and methods, 94
locally undesirable land uses (LULUs), 263
London Convention (Convention on the Prevention of Marine Pollution by Dumping of Wastes and Other Matter, 1972), 24
Los Alamos Scientific Laboratory (New Mexico)
    cleanup, 321n85
    LLW disposal at, 27, 43, 156
    transportation of TRU to WIPP, 137, 139, 179
    TRU disposal at, 27, 183, 184
    weapons production at, 17
    on Yucca Mountain repository, 219, 220
Lott, Trent, 366n105
Louisiana
    designation of Yucca and, 207
    repository-siting process and, 29, 67, 196, 204
    See also Johnston, J. Bennett
Loux, Robert, 222, 224
low-activity mixed waste (LAMW), 159
low-level radioactive waste (LLRW), 145. *See also* LLW (low-level waste)
Low-Level Radioactive Waste Policy Act of 1980 (LLRWPA)
    on BRC, 120
    compacts and, 63
    as failure, 4, 160–61
    legislative history, 58, 147
    on LLW disposal, 147–48, 150–51
    NGA and, 256
    on transportation, 126–27
    on TRU disposal, 99
Low-Level Radioactive Waste Policy Amendments Act of 1985 (LLRWPAA)
    DOE and, 158
    on LAMW, 159
    LLW definition, 95, 99
    on LLW disposal, 95, 149–54, 155, 259
    on role of states, 63, 95
Luken, Charles, 37
Luth, William, 165

Lynchburg Research Center (Virginia), 316n12
Lyons, Kansas. *See* Project Salt Vault (Lyons, Kansas)

Magwood, William, 229
Maine, 55–56, 78, 199, 265
Manhattan Project, 31, 254
Marine Protection, Research, and Sanctuaries Act (1972), 100
Marshall Islands, 31
Martinez, Susana, 185
Maryland, 50
Massachusetts, 55–56, 154, 265
Masto, G. Catherine, 226
Maxey Flats, Kentucky, 146, 147
McCone, John A., 25–26
McCormack, Mike, 189
McGovern, George, 356n96
Megatons for Megawatts program, 92, 333n601
Mescalero Apaches, 236–37, 284
Metzenbaum, Howard, 37, 320n170
Meyers, Robert, 225–26
Michigan
    dry-cask regulations and, 217
    LLW disposal and, 152, 154
    repository-siting process and, 29, 196
    utilities and, 78
Midwest Compact, 152
Midwest Fuel Recovery Plant (Morris, Illinois)
    reprocessing at, 45, 244
    storage at, 133, 235, 261, 330n512
Minnesota, 19, 55–56, 199, 316n22. *See also* Prairie Island Nuclear Power Plant (Red Wing, Minnesota)
Mississippi
    nuclear weapons testing in, 31
    repository-siting process and, 29, 195, 196, 197, 204, 366n105
mixed wastes, 11, 21, 36, 111–13, 299–300. *See also* LAMW (low-activity mixed waste)
Montana, 55–56
Montgomery, Sonny, 203
Morris, Illinois. *See* Midwest Fuel Recovery Plant (Morris, Illinois)
Mound Laboratories (Miamisburg, Ohio), 44, 134
MOX (mixed oxide) fuel
    cost, 82–83, 91, 245, 250
    plutonium and, 95
    PUREX process and, 91–92, 242
    vs. UREX+ mixture, 247
    See also PUREX (plutonium uranium extraction)
MRS (monitored retrievable storage) facilities
    federal facilities, 235–37, 238
    NWPA and NWPAA on, 66, 67, 80, 193, 194, 200–201, 266, 299
    potential sites, 200–201

Nader, Ralph, 49
Nader v. NRC, 49
Nader v. Ray, 49
NARM (naturally occurring and accelerator-produced radioactive material), 21
NAS (National Academy of Sciences)
  Advisory Committee on Geologic Aspects of Radioactive Waste Disposal, 26
  Advisory Committee on Nuclear Waste Report, 22–23, 25
  Board on Radioactive Waste Management, 25, 79, 198, 302
  Commission on Geosciences, Environment, and Resources, 24–25, 249
  Committee on Improving Practices for Regulating and Managing Low-Activity Radioactive Waste, 118, 120–21, 159–60, 304
  Committee on Risk-Based Approaches for Disposition of Transuranic and High-Level Radioactive Waste, 102–3
  Committee on Science and Technology for Countering Terrorism, 233–34, 265
  Committee on the Management of Certain Radioactive Waste Streams Stored in Tanks at Three Department of Energy Sites, 109–11
  Committee on the WIPP, 169, 179
  Committee on Transportation of Radioactive Waste, 124, 130, 134, 135–36, 140–43
  on deep geological repositories, 3, 8, 254–55
  on GNEP, 2, 81–82
  Panel on Social and Economic Aspects of Radioactive Waste Management, 202
  on Yucca Mountain repository, 68–69
National Defense Authorization Act (1989), 168
National Defense Authorization Act (1997), 113
National Defense Authorization Act (2005), 108–10, 117
National Defense Authorization Act (2009), 333n599
National Environmental Policy Act of 1969 (NEPA), 30, 32–33, 50, 52, 177
National Governors Association (NGA), 58, 148, **149**, 150–51, 160–61, 189, 256, 259
National Security and Military Applications of Nuclear Energy Authorization Act of 1980. *See* WIPP Authorization Act (1980)
National Waste Terminal Storage Program, 196
Native American tribes
  MRS facilities and, 236
  repository-siting process and, 191, 193–95, 203
  on SNF and HLW disposal, 189
  transportation and, 131
naturally occurring and accelerator-produced radioactive material (NARM), 18, 21, 94, 95, **96**
naturally occurring radioactive material (NORM), 18, 21, 94, 95, **96**
Natural Resources Defense Council (NRDC)
  CRBR and, 48
  litigation and, 50, 53, 54, 101
  on SNF storage options, 235
NCRPM (National Council of Radiation Protection and Measurement), 121
NEA (Nuclear Energy Agency), 25, 27
Nebraska, 152–54, 284
Nelson, Ben, 153, 154
NEPA. *See* National Environmental Policy Act of 1969 (NEPA)
neptunium, 99
Nevada
  C&C agreements and, 198, 210, 288–89
  EnPA and, 68, 217
  LLW disposal and, 150
  MRS facilities and, 236
  notice-of-disapproval and, 67, 69, 203
  opposition to Yucca Mountain, 4, 9, 10, 11, 15–16, 187, 209–14, 222–25, 228–30, 257, 283, 287–88
  public opinion in, 214–16, **216**
  repository-siting process and, 29, 187, 195, 196–97, 198, 201, 204, 208–9
  transportation to Yucca and, 132, 141, 142
  See also Beatty, Nevada; NTS (Nevada Test Site); Yucca Mountain nuclear waste repository (Nevada)
Nevada Commission on Nuclear Projects (NCNP), 197, 208
Nevada v. Burford, 213
Nevada v. Dept. of Energy, 212, 225, 345n98
Nevada v. Herrington, 196–97, 211
Nevada v. NRC, 225
Nevada v. Watkins, 212–14, 280
New Hampshire, 199, 265. *See also* Seabrook Station (New Hampshire)
New Jersey, 55–56, 147. *See also* Atlantic Compact
New Mexico
  C&C agreements and, 60–61, 171–73, 174–75, 178, 194, 288
  on HLW and SNF disposal at WIPP, 79, 164–65, 231–32, 240, 241, 261, 263, 284
  land withdrawal for WIPP and, 175–78
  legislative initiatives and litigation, 60–61, 62, 256
  MRS facilities and, 236–37
  nuclear weapons testing in, 31
  repository-siting process and, 163
  role in WIPP, 10, 11, 29, 60–61, 165–66, 167, 168–69, 285, 287, 300
  transportation to WIPP and, 129, 136–37, 138–39, 143
  See also Carlsbad, New Mexico; Environmental Evaluation Group (EEG); Los Alamos Scientific Laboratory (New Mexico); WIPP (Waste Isolation Pilot Plant, Carlsbad, New Mexico)

New York
    HLW reclassification initiatives and, 109
    LLW disposal and, 150, 154
    NRC and, 53
    repository-siting process and, 29
    reprocessing and, 243
    See also West Valley Reprocessing Plant (New York)

New York Times
    on DOE's facilities, 38, 215
    on repositories, 217
    on Rocky Flats fire, 28
    on Yucca Mountain repository, 220–21

New York v. United States, 150, 160, 259
NFMDA (Nuclear Fuel Management and Disposal Act of 2006), 113
NFS (Nuclear Fuel Services), 45, 243–44. *See also* West Valley Reprocessing Plant (New York)
Nixon administration, 246
    creation of EPA, 33
North Carolina, 152, 154, 199, 284, 350n47. *See also* Atlantic Compact; Southeast Compact
North Dakota, 151
Northern States Power Company v. Minnesota, 19, 316n22
North Korea, 268
Northwest Compact, 154, 155–56
Norway, 251
notice-of-disapproval mechanism
    Nevada and, 67, 69, 203
    NWPA and NWPAA on, 194–95
    NWPA on, 191
NPL National Priorities List (CERCLA), 36, 183–84
NRC. *See* Nuclear Regulatory Commission (NRC)
NRDC. *See* Natural Resources Defense Council (NRDC)
NRDC v. Abraham, 107–8, 109, 116
NTS (Nevada Test Site)
    LLW disposal at, 43, 46, 156, 158
    Nevada and, 187, 283
    nuclear weapons testing at, 31, 33, 219
Nuclear Energy Agency (NEA), 25, 27
Nuclear Energy Institute, 233, 252
Nuclear Energy Institute v. EPA, 69–70, 223–26, 300–301, 302
Nuclear Fuel Management and Disposal Act of 2006 (NFMDA), 113
Nuclear Fuel Waste Act of 2002 (Canada), 292
nuclear power
    cost, 76
    opposition to, 8, 15, 30, 31–33, 48–56, 76, 255, 265–66
    plants, **51**
    private sector and, 21
    support of, 75–77
nuclear power reactors, 75–76. *See also specific types of reactors*
nuclear proliferation, 82, 83, **90**, 91, 247, 250–51

Nuclear Regulatory Commission (NRC)
    Agreement States and, 19, 57, 145, 147
    on BRC, 94, 120, 304
    creation of, 10, 34, 298
    criticism of, 301–2
    disposal of decommissioned reactors and, 93
    on dry-cask storage, 217, 232–33, 234
    on DU, 94–95
    EPA standards and, 299, 300–301
    on extended surface storage, 264
    on GTCC waste disposal, 98, 105, 157
    HLW definition and regulation, 100, 101–2, 104
    HLW reclassification initiatives and, 103–6
    licensing process and litigation, 49–54
    licensing standards, **72**, 75
    LLW classification and regulation, 95, 145, 147, 158–59
    LLW disposal and, 57, 62, 94, 97–98, 147, 150, 158, 161, 299
    mixed wastes and, 21, 112
    NARM and, 18, 94, **96**
    on pool storage, 79
    on repositories, 65, 219–20
    repository-siting process and, 191–92
    on reprocessing, 252
    retrievability and, 264
    on SNF storage and disposal, 60, 261, 265, 272, 363n23
    transportation and, 101, 102, 123–24, 128, 129–30, 133–34, 137
    on TRU disposal, 99
    on TRUPACT, 139
    Waste Confidence Decision (1984), 54, 195, 216, 390n48
    WIPP and, 62
    York Committee for a Safe Environment v. NRC, 49–50
    Yucca Mountain license application and withdrawal, 15–16, 69–70, 73, 187, 225–26, 227–29, 231, 257
    Yucca Mountain repository and, 218, 224, 299
Nuclear Threat Initiative, 83, 251
nuclear waste
    classification, 7, 20–21, 84–85, 93–95, **96**, 113–22, 303–5
    financing and, 296–98
    new entity for management, 293–96
    regulation, 7–8, 298–303, **300**
    risks, 4–5, 85, 115–16
    See also specific wastes
Nuclear Waste Fund (NWF), 77, 78, 192, 194, 209, 271, 297
Nuclear Waste Management Organization (NWMO, Canada), 278, 292, 294, 297
Nuclear Waste Policy Act Amendments of 1987 (NWPAA)
    designation of Yucca, 9, 15, 66–67, 187, 207–9, 213, 219–22, 222–25, 257, 279, 281

Nuclear Waste Policy Act Amendments of 1987, *continued*
   as failure, 230
   on funding, 211–12
   on MRS facilities, 67, 80, 194, 200–201, 236–37, 238
   NRC and, 299
   on role of states, 5
Nuclear Waste Policy Act of 1982 (NWPA)
   on C&C agreements, 194, 203–4, 206, 209
   on consolidated storage, 66
   on costs, 65
   as failure, 3–4, 230, 257
   on funding, 211
   HLW definition and regulation, 64–66, 101, 102, 103, 256
   legislative history and background, 101, 188–89, 276–77
   on MRS facilities, 66, 80, 193, 200–201, 266, 299
   on repositories, 3–4, 9, 15, 74, 262, 264, 299, 300–301
   on repository-siting process, 15, 64–65, 186, 189–93, 209
   on retrievability, 268
   on role of states and tribes in siting, 5, 193–95, 209
   SNF definition and regulation, 64–66, 77, 99, 100, 235–36, 237, 238, 239, 256
   on transportation, 65, 123, 129
   WIPP and, 167, 193
   on Yucca Mountain repository, 16, 70, 73, 157, 226, 261
Nuclear Waste Project Office (NWPO, Nevada), 210
Nuclear Waste Technical Review Board (NWTRB), 208–9, 218, 221, 386n184
nuclear weapons
   DU and, 94
   once-through thermal fuel cycle and, 91
   opposition to, 31
   testing, 31
   wastes and, 254
NWF (Nuclear Waste Fund), 77, 78, 192, 194, 209, 271, 297
NWMO (Nuclear Waste Management Organization, Canada), 278, 292, 294, 297
NWTRB (Nuclear Waste Technical Review Board), 208–9, 218, 221, 386n184
Nye County, Nevada, 216, 227–28

Oak Ridge National Laboratory (ORNL)
   cleanup, 39, 43, 321n85
   LLW disposal at, 27, 43, 156, 352n38
   MRS and, 200
   TRU disposal at, 27, 183
   weapons production at, 17
   See also Project Salt Vault (Lyons, Kansas)
Obama, Barack, Yucca Mountain repository and, 4, 9, 15, 70, 226–27, 257
Obama administration
   climate and energy legislation and, 76
   global uranium fuel bank, 83, 252
   GNEP and, 2, 81, 252, 268
   reprocessing and, 242
   support of nuclear power, 2, 16, 73, 75, 257, 267–68
   WIPP and, 184
   Yucca Mountain repository and, 2, 4, 9, 15–16, 70, 187, 227–30, 257, 260–61
   See also Blue Ribbon Commission on America's Nuclear Future
ocean dumping, 21, 23, 100
Office of Civilian Radioactive Waste Management (OCRWM), 65, 66, 74, 206
Office of Environmental Management (EM), 38, 40–44, **41**, 74, 116, 312–14
Office of Management and Budget (OMB), 29, 34, 299
Office of Nuclear Waste Negotiator (ONWN), 80, 236–37, 240, 279, 283–84
Office of Technology Assessment (OTA), 201–3, 295, 319n114, 326n424
Ohio, 159. *See also* Fernald Feed Materials Production Center (Ohio); Mound Laboratories (Miamisburg, Ohio)
O'Leary, Hazel, 179
O'Leary, John, 168–69
once-through thermal fuel cycle
   electricity generation and, 88–91
   vs. reprocessing, 82, 245
   wastes generated by, 88–91, 249
   weapons production and, 91
ONKALO (Finland), 291
ONWN (Office of Nuclear Waste Negotiator), 80, 236–37, 240, 279, 283–84
Oregon, 55–56, 105–6
ORNL (Oak Ridge National Laboratory). *See* Oak Ridge National Laboratory (ORNL)
Ostendorff, William, 229
OTA (Office of Technology Assessment), 201–3, 295, 319n114, 326n424

Pacific Gas & Electric Co. v. State Energy Resources Conservation and Development Commission, 19, 55, 316n22
Paducah Gaseous Diffusion Plant (Kentucky), 43
Paradox Basin, Utah, 367n114
Parker, Frank, xv, 318n101, 327n434, 372n254
Parsons, John E., 76
Peach Bottom Atomic Power Station (Pennsylvania), 49–50
Penner, Rudolph, 48
Pennsylvania, 55–56. *See also* Peach Bottom Atomic Power Station (Pennsylvania); Shippingport Atomic Power Station (Pennsylvania)

Permian Basin, 25, 28
PG&E, 322n271
Philadelphia v. New Jersey, 147
plutonium, 43–44, 88–89, 91–92, 95, 99
plutonium-239 (Pu-239), 86, 91, 335n10
plutonium-240 (Pu-240), 89, **90**
Plutonium Management and Disposition Agreement (2000), 245
pool storage, 7, 43, 54, 79, 89–91, 233–34, 261, 265
Posiva Oy, 290–91
Power Reactor Development Co. v. International Union of Electric, Radio and Machine Workers, 49
Prairie Island Nuclear Power Plant (Red Wing, Minnesota), 54, 265
Price, Melvin, 60, 62, 166, 167
Price-Anderson Act (1957), 173, 204, 357n131
Private Fuel Storage (PFS). *See* Skull Valley Band of the Goshute Nation storage facility (PFS)
Private Ownership of Special Nuclear Materials Act (1964), 316n7
Project Plowshares, 283, 392n99
Project Salt Vault (Lyons, Kansas), 26, 28–29, 165, 256
PUREX (plutonium uranium extraction)
    description, 91–92, 242
    in Europe, 80
    vs. UREX+, 247
    See also MOX (mixed oxide) fuel

Radioactive Waste Consultation Act (1979), 166, 168
Radioactive Waste Management Committee (NEA), 27
Radioactive Waste Management Manual (DOE M 435.1), 102, 106–8, 109–10
radioactivity, 85–86, **87**
RCRA (Resource Conservation and Recovery Act), 21, 35–36, 37–38, 39–40, 111–13, 159, 180–81, 300
Reagan administration
    CRBR and, 48
    on defense waste disposal, 64, 195, 299, 364n54
    DOE and, 206
    reprocessing and, 245
    WIPP and, 172
Red Wing, Minnesota. *See* Prairie Island Nuclear Power Plant (Red Wing, Minnesota)
Reicher, Dan, 179
Reid, Harry, 9, 70, 187, 215, 222–23, 229, 257, 328n450
rem (roentgen equivalent in humans), 86, 344n66
repository-siting process
    AEC/ERDA and, 28–29, 60, 164, 196
    crisis and failure of, 201–7
    current situation and options, 262–63, 274, 277, 296
    EPA and, 191, 206–7
    for HLW, 25–27, 57
    implementation of, 65–66, 195–200
    IRG on, 57, 275
    litigation and, 196–97
    MRS facilities and, 200–201
    NRC and, 191–92
    NWPA on, 15, 64–65, 186–87, 189–93, 209
    public participation and trust in, 194, 205–7, 280–86
    role of states and tribes in, 29, 191, 193–95, 203–7, 209, 278–79, 280–81, 286–90
    strategies, 278–80
    for TRU, 28–30, 57, 162–63
    See also specific states and facilities
reprocessing
    AEC and, 44–45, 242, 243, 245–46, 255
    George W. Bush administration on, 242
    Carter administration and, 2, 9, 27, 46–47, **96**, 163–64, 242, 244–45, 255
    cost and risks, 7, 27
    EM and, 43
    Ford administration and, 2, 46, 60, 244, 255
    HLW from, 21, 22–23, 26–27, 47
    nuclear power and, 22
    vs. once-through thermal fuel cycle, 245
    Reagan administration and, 245
    SNF heat and, 247–48
    support and criticism of, 267–68, 269
    transmutation and, 25
    TRU from, 47, 99
    weapons production and, 91
    See also Global Nuclear Energy Partnership (GNEP); PUREX (plutonium uranium extraction); UREX+ reprocessing system; *specific plants*
Resource Conservation and Recovery Act (RCRA), 21, 35–36, 37–38, 39–40, 111–13, 159, 180–81, 300
retrievability
    ice sheet disposal and, 24
    NWPA on, 190
    repositories and, 264, 268–71, **270**
    TRU and, 28
    See also MRS (monitored retrievable storage) facilities
Rhode Island, on SNF storage, 265
RH-TRU (remote-handled transuranic waste)
    definition of, 99
    EM and, 43
    from UREX+, 249
    at WIPP, 134, 139, 182, 260, 284
Richardson, Bill, 181
Richland, Washington, 146, 147, 149, 154, 155, 160, 258
Rockefeller, Nelson, 45, 243
Rockwell International, 177. *See also* Rocky Flats nuclear weapons facility

Rocky Flats nuclear weapons facility
    1969 fire, 8, 27–28, 59, 61, 162, 256
    cleanup, 39, 44
    EPA and FBI raid at, 38, 61, 177–78
    Idaho and, 28, 59, 61, 177, 256
    SSNM from, 43
    transportation of TRU to WIPP, 139
Rocky Mountain Compact, 154, 156
Romer, Roy, 61, 177
Roser, Herman, 392n116
Rusche, Ben, 198
Russia, 80, 82, 92, 245, 246

Sacramento Municipal Utilities District v. United States, 78
Safe Drinking Water Act of 1974 (SDWA), 62, 68, 180
saltstone, 111
Samuelson, Don, 28
Sandia National Laboratory
    defense LLW disposal at, 27
    defense TRU disposal at, 27
    SNF at, 74
    on TRUPACT, 138–39
    WIPP and, 33, 163, 164, 169
SARA (Superfund Amendments and Reauthorization Act of 1986), 36, 38, 111, 146
Schaefer, Bob, 320n170
Schlesinger, James, 60, 164–65, 166
SDWA (Safe Drinking Water Act of 1974), 62, 68, 180
seabed burial, 23, 24, 25, 57, 59
Seaborg, Glenn T., 22, 28
Seabrook Station (New Hampshire), 32
Sell, Clay, 246
Sheffield, Illinois, LLW disposal at, 146, 147
Shippingport Atomic Power Station (Pennsylvania), 18, 316n12
Shoesmith, David, 376n396
sievert (Sv), 86
SKB (Sweden), 269, 291–92
Skull Valley Band of the Goshute Nation storage facility (PFS)
    capacity, 80
    construction, 237–38
    DOI and, 135, 238, 266
    ONWN and, 284
    study grant and, 236
    support and opposition to, 283
    transportation to, 135
    Utah and, 80, 237, 238
SNF (spent nuclear fuel)
    accidents and, 134–35
    characteristics, **90**
    definition, classification, and regulation, 20, 93, 99–100
    heat generation and, 87, 247–48
    IAEA framework and, 119
    production of, 20, 88–89, 91, 99–100
    transportation and, 123–24, 127, 133–34, 135, 140–42, 241
SNF (spent nuclear fuel) storage and disposal
    Carter on, 188
    consolidated interim storage, 6, 235–41, 266–67, 274–75
    costs and funding, 239–40, 271–72, 297–98
    current situation and options, 6–7, 73–74, 79, 80, 231–32, 253, 260–61, 264–67, 275–76, 277–78
    at DOE facilities, 21, 40, 43, 74, 80, 217, 261, 264–65, 278
    dry-cask storage, 54, 79–80, 217, 232–35, 240, 261, 265, 275
    federal liability, 77–78
    FEIS on, 58–59
    Finland and, 290–91
    IRG on, 56–57
    NWPA on, 64–66, 190
    opposition to, 234–35, 240–41
    options and methods, 89–91, 93
    pool storage, 7, 43, 54, 79, 89–91, 233–34, 261, 265
    at reactors, 6, 54, 79, 232–35, 261, 265–66, 278
    repository delay and, 216–17
    retrievability, 190, 268–71
    risks, 233–34
    space disposal, 24
    WIPP and, 163–66, 263
South Africa, 83, 251, 330n523
South Carolina
    HLW reclassification initiatives and, 109
    LLW disposal and, 147, 150, 151, 154, 350n47
    opposition to nuclear power and, 55
    Yucca license withdrawal and, 73, 228, 229, 257
    See also Atlantic Compact
South Dakota, 151
Southeast Compact, 152, 154, 350n47
Southern Company, 73
Southern States Energy Board (SSEB), 131
Southwest Compact, 151
Southwest Research Information Center, 33, 163, 198, 354n45
space disposal, 24, 25, 57, 59
Spain, 118
special nuclear material, 18, 20, 95
SRS (Savannah River Site)
    cleanup, 39, 42, 43, 44, 321n85
    HLW reclassification initiatives and, 103, 105, 109–11, 342n201
    HLW storage and disposal at, 27, 28, 74, 134, 261
    LLW disposal at, 27, 43, 156
    reprocessing and, 43, 92, 245
    SNF storage at, 74
    SSNM and, 43–44
    transportation and, 134, 139

TRU storage at, 183
weapons production at, 17
on Yucca Mountain repository, 220
SSEB (Southern States Energy Board), 131
SSNM (surplus special nuclear material), 43–44
Stahl, David, 376n396
states
   BRC wastes and, 94
   consolidated storage and, 236, 238, 267
   inducements for, 265, 267, 284
   LLW disposal and, 95, 147–56
   nuclear power plants moratoriums and, 10, 55–56, 75
   nuclear waste regulation and, 18–19
   RCRA and, 36
   repository-siting process and, 5, 10, 29, 191, 193–95, 203–7, 209, 278–79, 280–81, 286–90
   on SNF and HLW disposal, 189
   transportation and, 123, 125–31, 133–34, 141, 238
   See also Agreement States; C&C (Consultation and Cooperation) Agreements; *specific states*
steam reformation, 111
storage and disposal of nuclear waste
   incidents, 21–22
   methods, 21, 86–87
   See also specific methods and wastes
strontium-90 (Sr-90)
   characteristics, 89
   disposal of, 23, 47, 81, 82, 89, 92, 110–11, 115, 247–48, 264
   IAEA framework and, 119
   UREX+ and, 247–48, 249
Sununu, John, 199
super-compaction, 158
Superfund. *See* Comprehensive Environmental Response, Compensation, and Liability Act (CERCLA)
Superfund Amendments and Reauthorization Act of 1986 (SARA), 36, 38, 111, 146
Surface Transportation Board (STB), 132, 142
surplus special nuclear material (SSNM), 43–44
Svinicki, Kristine, 229
Sweden
   HLW repository in, 291–92, 294
   repository-siting process and, 283
   reprocessing and, 333n584
   retrievability and, 269
   waste classification and regulation, 118
Swift, Al, 208
Switzerland, 333n584
Synroc, 334n7
Sys. Fuel Inc. v. United States, 331n550
Szymanski, Jerry S., 214

Tennessee, 67, 200, 236. *See also* Clinch River Breeder Reactor (CRBR, Tennessee)
Tennessee Valley Authority (TVA), 200, 393n150
Tennessee v. Herrington, 204, 205
terrorism
   reprocessing and, 91, 251
   storage and, 5, 7, 79, 233–34, 265
   transportation and, 7, 124, 128, 142–43, 241
Texas
   LLW disposal and, 154–55, 157, 258, 350n64
   repository-siting process and, 29, 67, 153, 195, 196, 198, 201, 204, 208
   See also Andrews, Texas
Texas Compact
   Andrews facility and, 63, 154, 160, 279, 284
   members, 63, 150, 151, 259
Texas v. Department of Energy, 196
thorium fuel cycle, 17, 336n43
Three Mile Island, 9, 32, 55, 133, 255
transmutation, 24–25, 57
transportation
   accidents and incidents, 134–35, 138, 139, 140, 182
   interim storage and, 241
   IRG on, 58
   NRC and, 101, 102
   NWPA on, 65
   packaging and, 128, 129–30, 137, 138–39
   regulation, 123–33, 145
   risks, 142–43
   states and tribes and, 123, 125–31, 133–34, 141, 238
   terrorism and, 7, 124, 128, 142–43, 241
TRU (transuranic wastes)
   definition, classification, and regulation, 20, 93, 98–99, 324n327
   from electricity generation, 88–89
   IAEA framework and, 119
   from reprocessing, 47, 93, 99, 247, 249, 252
   from Rocky Flats, 27–28, 59, 61, 177, 256
   transportation, 123, 134, 135, 241
   from weapons production, 20, 27, 91, 99, 254
   See also CH-TRU (contact-handled transuranic waste); RH-TRU (remote-handled transuranic waste)
TRU (transuranic wastes) storage and disposal
   at DOE facilities, 21, 40, 42, 43, 181–82, 183–84, **183**, 321n211
   FEIS on, 58–59
   IRG on, 57
   options and methods, 93, 99, 256
   at WIPP, 16, 43, 57, 99, 134, 260
TRUPACT (transuranic packaging container), 137, 138–39
TVA (Tennessee Valley Authority), 200, 393n150

Udall, Tom, 180, 181
Union of Concerned Scientists (UCS), 73, 79, 233–34, 235, 265

Union of Concerned Scientists v. AEC, 49
United Arab Emirates, 251
United Kingdom
    reprocessing and, 91, 245, 333n584
    uranium leasing and, 246
    waste classification and regulation, 118, 294
United States v. Kentucky, 316n22
uranium
    Canada and, 7, 82, 246, 250, 268
    landfill disposal and, 92
    once-through thermal fuel cycle and, 88
    prices, 82–83, 250, 309
    See also depleted uranium (DU)
uranium-233 (U-233), 95
uranium-235 (U-235), 88, 89, 94, 95
uranium-236 (U-236), 89
uranium-238 (U-238), 86, 88, 94
uranium fuel banks, 83, 251, 252
uranium mill tailings (UMT), 21, 93, 94, 155
UREX+ reprocessing system, 92, 247–49
U.S. Ecology, 153, 158–59. *See also* Richland, Washington
Utah
    LLW disposal and, 155–56, 159
    PFS storage facility and, 80, 237, 238
    repository-siting process and, 196, 197
    transportation and, 238
    See also Skull Valley Band of the Goshute Nation storage facility (PFS)
utilities
    opposition to nuclear power and, 50–51
    waste management and, 64–66, 77–78, 235, 239–40, 266, 271–72, 293–94, 297

Valhi, 159. *See also* Andrews, Texas
Vermont, 76, 155, 265. *See also* Texas Compact
Vermont Yankee Nuclear Power Corp. v. NRDC, 50, 53
Vermont Yankee Nuclear Power Plant (Vernon, Vermont), 52–53, 54, 76, 272
Virginia, 199
Vucanovich, Barbara, 68

Washington
    C&C agreements and, 204, 366n97
    Hanford cleanup program and, 311–14, **313**
    HLW reclassification initiatives and, 105–6
    LLW disposal and, 150
    nuclear waste storage and, 353n22
    repository-siting process and, 29, 67, 195, 196, 204, 208
    Richland facility and, 154
    Yucca license withdrawal and, 73, 229, 257
    See also Hanford Site (Washington); Richland, Washington
waste incidental to reprocessing (WIR), 106–8, 110–11
Waste Treatment and Immobilization Plant (WTP), 312

Watkins, James, 38, 167, 177–78
Western Governors' Association (WGA), 131, 139–40, 143, 177
Westinghouse Electric Company, 316n12
West Valley Demonstration Project Act (WVDPA, 1980), 45, 100–101, 244, 262
West Valley Nuclear Services Company, 45
West Valley Reprocessing Plant (New York)
    cleanup, 42, 43, 45–46, 244
    closure, 9, 45, 255, 271
    HLW at, 26, **96**, 134, 243–44, 262, 265
    HLW reclassification initiatives and, 109
    LLW disposal at, 146–47
    reprocessing at, 45, **96**, 242–43
    SNF at, 26, 243–44
    TRU at, 26, 43, 93
West Virginia, 55–56
White Pine County, Nevada, 228, 229
WIPP (Waste Isolation Pilot Plant, Carlsbad, New Mexico)
    capacity, 181–82
    FOIA and, 33, 163
    funding, 184
    GTCC waste and, 157, 263
    HLW disposal and, 174, 193, 231–32, 261, 263, 284
    IRG on, 56, 166
    litigation and, 182–84
    mixed wastes and, 113
    repository-siting process and, 29–30, 162–63, 165
    restriction to defense TRU waste, 163–67, 256–57, 265
    retrievability and, 264
    shipments to, 43, 62, 134, 139, 182
    SNF disposal and, 59–60, 79, 193, 231–32, 240, 241, 263, 284
    as success, 4, 5, 184–85, 281, 282
    timetable, 201
    transportation to, 129, 135, 136–40, 143, 173–74, 177, 179, 182, 241
    TRU disposal at, 16, 43, 57, 99, 134, 260
    See also Environmental Evaluation Group (EEG)
WIPP Authorization Act (1980)
    on C&C agreements, 171
    on disposal of defense waste, 60, 62, 166–67, 193, 195
WIPP Guide, 140
WIPPLWA (Waste Isolation Pilot Plant Land Withdrawal Act of 1992)
    EPA certification and, 61–62, 68, 179–82, 260, 299
    restriction to defense TRU waste, 60, 167, 175
    on transportation, 129, 137, 139, 179–80
WIR (waste incidental to reprocessing), 106–8, 110–11
Wisconsin, 55–56, 197, 199
Wis. Elec. Power Co. v. United States, 331n550

Wright, Jim, 67, 195, 208
WVDPA (West Valley Demonstration Project Act, 1980), 45, 100–101, 244, 262

Xcel Energy, 265. *See also* Prairie Island Nuclear Power Plant (Red Wing, Minnesota)

Yakima Indian Nation, 105–6
York Committee for a Safe Environment v. NRC, 49–50
Yucca Mountain nuclear waste repository (Nevada)
    capacity, 74, 193, 263–64
    characteristics, 218–19
    designation by Congress, 9, 15, 66–67, 187, 207–9, 213, 222–25, 257, 279, 281
    environmental standards, 67–70, **71**, **72**, 101, 217, 224, 285, 299
    as failure, 5
    funding, 271
    GTCC waste and, 157
    license application and withdrawal, 15–16, 69–70, 73, 187, 225–26, 227–29, 231, 257
    local support, 227, 282
    Nevada opposition to, 4, 9, 10, 11, 15–16, 187, 209–14, 222–25, 228–30, 257, 283, 287–88
    NFMDA and, 113
    NRC and, 101
    Obama presidential campaign and administration, 2, 4, 9, 15–16, 70, 187, 226–30, 257, 260–61
    regulation, methodologies, and design, 219–22
    repository-siting process and, 65, 207
    reprocessing and, 247–48
    retrievability and, 264
    SNF and HLW storage and disposal at, 391n87
    timetable, 2–3, 231, 262
    transportation to, 132, 135, 140–42

zirconium 93 (Zr-93), 89